Natural Resources
Ecology, Economics, and Policy

Natural Resources
Ecology, Economics, and Policy

JERRY L. HOLECHEK
RICHARD A. COLE
JAMES T. FISHER
RAUL VALDEZ
New Mexico State University

Prentice Hall, Upper Saddle River, NJ 07458

Library of Congress Cataloging-in-Publication Data

Natural resources : ecology, economics, and policy / Jerry L. Holechek ... [et al.].
      p. cm.
    Includes bibliographical references and index.
    ISBN 0-13-896077-1
      1. Natural resources. 2. Natural resources--Government policy. 3. Ecology. I.
Holechek, Jerry L.

HC85 .N37 2000
333.7--dc21   99-055590

Publisher: *Charles E. Stewart, Jr.*
Associate Editor: *Kate Linsner*
Managing Editor: *Mary Carnis*
Production Liaison: *Eileen O'Sullivan*
Production Editor: *Lori Harvey, Carlisle Publishers Services*
Director of Manufacturing & Production: *Bruce Johnson*
Manufacturing Manager: *Ed O'Dougherty*
Marketing Manager: *Ben Leonard*
Cover Design: *Wanda España*
Formatting/Page make-up: *Carlisle Communications, Ltd.*
Printer/Binder: *Courier Westford*

Printed in the United States of America

10   9   8   7   6   5   4   3   2   1

ISBN 0-13-896077-1

Prentice-Hall International (UK) Limited, *London*
Prentice-Hall of Australia Pty. Limited, *Sydney*
Prentice-Hall Canada Inc., *Toronto*
Prentice-Hall Hispanoamericana, S.A., *Mexico*
Prentice-Hall of India Private Limited, *New Delhi*
Prentice-Hall of Japan, Inc., *Tokyo*
Pearson Education Asia Pte. Ltd., *Singapore*
Editora Prentice-Hall do Brasil, Ltda., *Rio de Janeiro*

Contents

Preface   **xix**

Chapter 1   **Natural Resources: An Overview**   **1**

Natural Resources Defined   1
    Abiotic and biotic resources   2
    Ecosystems defined   2
    Resource consumption, use, and renewal   2
Natural Resource Management   5
    Management objectives   5
    Private and public resource management   5
    An integrated approach to management   6
The Need for Natural Resource Management   7
    The human population increase   7
Land Area and Land Uses   8
    Natural resource issues   9
An Optimistic View of the Future   12
Other Sources of Information   13
Organization of the Book   13
    Section 1: Management foundations   13
    Section 2: Air, water, and land resources   13
    Section 3: The land-based renewable resources   14
    Section 4: The wild living resources   14
    Section 5: The mineral and energy resources   14
    Section 6: Integration of natural resources management   14
    Literature Cited   15

SECTION 1   **MANAGEMENT FOUNDATIONS**   **17**

Chapter 2   **The Historical Perspective**   **19**

Colonization (up to 1776)   19
    Ecology and landscape   19
    Economics   23
    Policy   25
Westward Expansion (1776–1860)   26
    Ecology and landscape   26
    Economics   28
    Policy   29
The Gilded Age (1861–1899)   32
    Ecology and landscape   32

Economics   33
Policy   37
The Progressive Period (1900–1945)   40
Ecology and landscape   40
Economics   41
Policy   44
Neoprogressive Period (1945–Present)   48
Ecology and landscape   48
Economics   50
Policy   52
Literature Cited   57

Chapter 3   Basic Ecology   60
Introduction   60
Populations   61
Population identity   61
Distribution and dispersal   61
Factors affecting population distribution   62
Population habitat and physical niche   63
Population Structure and Dynamics   63
Population structure   63
Population dynamics   63
Population Energetics, Biomass, and Production   65
Energetics   65
Population biomass   66
Production, Productivity, and P/B ratio   67
Population Strategies   67
Communities   68
Definition   68
Community Structure and Functions   69
Community habitat   69
Biodiversity   69
Trophic functions   71
Competitive interactions within communities   74
Interaction among biotic communities   76
Ecosystems   77
Ecosystems defined   77
Ecosystem structure and function   78
Energy flow   79
Material flow, storage, and cycling   82
Succession   87
Ecosystem manipulators: The human component   94
Literature Cited   96

Chapter 4   Conservation Economics   99
The Problem   99
Hard Choices   99
Timber versus spotted owls (Pacific NW)   100
Cheap energy versus functioning ecosystems   100
Clean air versus cheap steel   100
Endangered wolves versus livestock and wildlife   101
Endangered fish versus food   101
Hydropower versus recreation   101
Electric power versus forests   101
Mining versus recreation on public lands   102

Riparian ecosystems versus livestock 102
Hydro-dam relicensing 102
Choices Displace Opportunities 102
Economics is about choices 102
Importance of good decisions 104
What the market mechanism does 104
How the market solves the three problems 104
Who controls the market? 106
The invisible hand of the market 106
Can profit incentives protect natural resources? 106
How markets settle conservation questions 107
When are markets best at promoting conservation? 107
Limits of the market 108
The Future 108
Harnessing the Power of the Market 109
Concepts for incentive-based pollution control 109
Pollution taxes 109
Marketable waste emission permit systems 110
Applications of Incentive-Based Regulations 112
Pollution taxes 112
Marketable permit systems 112
Governments and Conservation 114
How government settle conservation questions 114
Limits of government in settling conservation questions 114
Rent seeking 114
Taxpayer ignorance 114
Empire building 114
Abuses of economic analysis 115
What government can do to protect natural resources 115
Why measure benefits and costs? 116
How to Measure Benefits and Costs 117
Basic principles 117
Implementation principles 117
Implementation practices, incremental benefits and costs 118
Measurement problems 118
Scope and Limits of BCA 118
The bad news 118
The good news 119
A summary of BCA 119
Conclusion 120
Literature Cited 120

Chapter 5   Planning, Policy, and Administration   122

Elements of Organizational Function 122
Identifying and solving problems 122
Forces in organizational integration 122
General approaches to management 124
Management Systems 124
Organizations as management systems 124
Management system boundaries 126
The Internal Management Environment 128
Resources, use, and management 128
Opportunity and use 129
Managing for result, not process 129
External environment 129
Measuring management performance and success 130

Public, Private, and Advocacy Systems   131
   Public   131
   Business   131
   Advocacy   132
Planning Process   133
   Inventorying planning environments   133
   Reviewing mandates   134
   Reviewing mission   134
   Environmental scanning   134
   Stakeholders   135
   SWOT analysis   135
   Forecasting   135
Focusing Management Intent   135
   Mission   135
   Planning vision   136
   Goals   136
   Issues   136
   Strategic planning   136
   Setting objectives   137
   Long-range planning   137
   Operations planning   137
   Organizational performance evaluation   138
Organizational Planning Boundaries   138
   Project planning   138
   Program planning   139
   Comprehensive organizational planning   139
Regional (Geographic) Planning   140
   Politically bounded planning   140
   Naturally bounded planning   140
   Integrated resource management planning   140
Policy   141
   What is policy?   141
   Roles of federal, state, and local governments   142
   Budget policy   142
   Public and private policy   142
   U.S. Constitution and policy development   143
Making Policy   143
   Legislation   143
   Common elements of law   144
Organizational Administration   145
   Integrating organizational activities   145
   Information flow and authority   145
   Information management   147
   Networking   147
Summary   147
Literature Cited   148

SECTION 2   AIR, WATER, AND LAND RESOURCES   149

Chapter 6   Atmospheric Resources and Climate   151

Introduction   151
Climatic Factors and Elements   151
Climatic Factors   152
   Atmospheric composition and pressure   152
   Solar radiation   152

Earth latitude, shape, and rotation 153
Distribution of continents 153
Topography 154
Marine currents 155
Interactions between climatic factors 156
Vegetation as a climatic factor 156
Climatic Elements 156
Wind 156
Temperature 157
Precipitation 158
Precipitation systems 161
Humidity 162
Types of Climate 162
Equable climate 162
Desert climate 163
Polar climate 163
Mediterranean climate 163
Continental climate 163
Tropical wet and dry climate 163
Climatic Types in the United States 163
The Pacific climate 163
The Great Basin climate 164
The Southwestern climate 165
The Plains climate 166
The Eastern climate 166
The Florida climate 166
Climatic Instability and Natural Resource Management 166
World climatic history 167
Climatic lessons from the past 168
Climatic Change and Human Activities 169
The greenhouse effect and radiation balance 169
Global warming and fossil fuels 171
The implications of global warming 172
Possible strategies for managing global climate change ... 173
Acid precipitation 173
Ozone depletion 175
Desertification 175
Literature Cited 176

Chapter 7   Water Resources   **178**

Introduction 178
The Properties of Water 180
Physical attributes 180
Water Forms and Distribution 181
The planet's reserves 181
Glacial ice 181
Lakes 182
Rivers 182
Oceans 182
Groundwater 182
Water Resources and Management 183
The hydrologic cycle 183
The watershed 183
What is watershed management? 184
Watershed processes 185

Managing land use practices   187
Roles of wetland and riparian areas   189
Special watershed management techniques   190
Multipurpose Water Resource Management   192
The water resource management agencies   192
Integrated management objectives   192
Water management engineering   194
Water Quality Management   197
Types of water quality problems   197
Surface water impairment   197
Disease-causing organisms   198
Nutrients   199
Silts and suspended solids   199
Biochemical oxygen demand (BOD)   200
Salinity and other dissolved solids   200
Toxic materials   200
Acid mine drainage   200
Thermal discharges   201
Approaches to Water Quality Management   201
Classification of water pollution sources   201
The watershed approach   201
Water treatment   201
Home water treatment   203
Water Uses in the United States   203
Types of water use   203
Off-stream uses   205
In-stream uses   207
Regional trends in water use and consumption   207
Water Use Problems and Conflicts   207
Overpumping   208
Water allocation and wildlife habitat   208
Salinization   210
Who owns the water?   211
Meeting Water Demand in the Twenty-First Century   213
Water conservation   213
Reclamation of sewage water   213
Development of groundwater   213
Desalinization   213
Developing salt-resistant crops   213
Developing drought-resistant crops   213
Rainmaking   213
Harvesting icebergs   214
Long distance water transport   214
Improved integration of water use   215
Water in the nation's future   215
Literature Cited   215

Chapter 8   Soil: The Basic Land Resource   219

Importance of Soil   219
The soil profile   219
Soil formation   220
Soil characteristics   222
Soil classification   224
Soil erosion   229
Soil erosion in the United States   231

Controlling Soil Erosion   234
    Conservation tillage   234
    Maintaining soil fertility   240
The USDA-Natural Resources Conservation Service   241
Soil Policy   242
Literature Cited   245

Chapter 9   Ecosystems of the United States   246

Introduction   246
Terrestrial Ecosystems   246
    Overview   246
    Biomes of the United States   253
Deserts   260
    Hot desert   260
    Cold desert   263
Woodlands   265
    Pinon-juniper woodland   265
    Mountain browse   266
    Oak woodland   267
    Western coniferous forest   268
    Southern pine forest   269
    Eastern deciduous forest   270
Tundra   271
    Alpine tundra   271
Aquatic Ecosystems   272
    Wetland ecosystems   272
    Stream and river ecosystems   279
    Estuarine ecosystems   282
    Marine ecosystems   284
Literature Cited   288

SECTION 3   THE LAND-BASED RENEWABLE RESOURCES   291

Chapter 10   Forests and Forestry   293

Introduction   293
Tree Structure and Function   294
Forest Products   296
Forest Distribution   297
    Boreal forest (Taiga)   298
    Temperate deciduous forests   299
    Temperate coniferous forests   300
    Temperate mixed forests   300
    Temperate broad-leaved evergreen forests   300
    Tropical evergreen forests   300
    Tropical deciduous forests   302
Forests of the United States   302
    Major forest types   302
Forest Land Area in the U.S.   306
The Status of the U.S. Timber Resource   306
    Commercial forest land   307
    Timber productivity across the regions   308
    Timber removals   308
    Who owns the nation's forests?   308
Forest Management   309
    Forest management defined   309

Criteria used to classify stands   310
Age and size distribution   311
Species composition   312
Classification based on density   312
Classification based on site quality   313
The concept of shade tolerance   314
Stand Management   315
Silvicultural systems   315
Cutting and reproduction methods   315
Intermediate Treatments   319
Emergency Cuttings or Thinnings   319
Timber Harvesting   320
Tree Planting   320
Forest Protection   321
Forest fire   321
Insects   321
Disease   324
Exotic pests   324
Abiotic factors   324
Forest Ecosystem Management   324
Ecosystem management defined   324
The clear-cutting controversy   326
Ecosystem management and fire   327
Gap dynamics   328
Forest succession and ecosystems management   330
Global Forest Problems   330
Recycling Wastepaper   331
Reducing Paper Use   332
Concluding Remarks   332
Literature Cited   332

Chapter 11   Rangeland and Range Management   335

Rangelands Defined   335
Rangeland Management Defined   336
Basic concepts   336
Types of Rangeland   337
Historical Perspective   339
Rangeland Ecology   340
Grazing effects on range plants   340
Rangeland Condition and Trend   341
Range Animal Ecology   342
Comparative digestive systems   342
Forage selection by different ungulates   343
Comparative nutritive value of grasses, forbs, and shrubs   343
Use of nutritional knowledge in management   345
Animal suitability for different rangelands   346
Rangeland Management   346
Importance of correct stocking rate   346
Improving livestock distribution   348
Grazing systems   349
Rangeland Livestock Production   356
Importance of the West   357
Livestock management during drought   357
Poisonous plant problems   358
Controlling Rangeland Vegetation   358

Government Policy 359
 Importance of federal lands 360
 The Vale program 361
 Policy changes 362
Range Management and the Future 364
Literature Cited 366

Chapter 12 Farmland and Food Production 368

Introduction 368
A Brief History 369
Major Types of Agriculture 369
The Green Revolution 370
Agriculture Problems in the United States 372
 Increasing regulation 374
 Urban sprawl 374
 Rising production costs 375
 Declining exports 376
 Atmospheric pollution 377
 Restricted water supplies 377
 Excess capacity 377
Agriculture in Developing Countries 377
Pesticide Controversies 379
Sustainable Agriculture 382
Farmland Policy 382
 Farm programs in other countries 387
 Trends in current policy 388
Eliminating Government Involvement in Agriculture:
 The New Zealand Case 389
 Conclusion 390
Literature Cited 390

Chapter 13 Outdoor Recreation 392

Introduction 392
What is Recreation? 393
The Importance of Outdoor Recreation 395
Attributes of Outdoor Recreation 395
Resource Conflicts and Resolution 396
Historical Perspectives 397
 Federal management 397
 Early attitudes 397
 Recent trends 398
Federal Recreational Management 399
 National Forest Service 399
 National Park Service 400
 Federal water resources agencies 403
Other Recreational Management 403
 State and local government 403
 Private recreation opportunities and tourism 404
Recreational Challenges on Public Lands 404
 Importance of public land recreation 404
 Subdividing private grazing lands 404
 Agriculture on the urban interface 406
 Scenic beauty and range management 406
 Public opinion and management of federal rangelands 408
 Recreation and ranching 408

Managing recreation costs on public lands   408
Conflict resolution in multiple-use decisions   410
Outdoor Recreational Management   410
Interfacing people with resources   410
Recreational planning   410
Recreational economics   411
Integrative management planning   412
Recreational resource managers   412
Visitor management   413
Natural resource management   414
Information service management   415
Future Demands for Outdoor Recreation   415
Literature Cited   417

Chapter 14   Urban Land-Use Management   420

The Urban Landscape   420
Resources and services expectations   420
The Urban Ecosystem   421
Urban form and function   424
Integrating urban form with natural functions   425
Development of Urban Infrastructure   428
Public and private partnership   428
Transportation   430
Utilities   431
Housing subsidies   432
The superhighways   433
Urban Growth and Decline   434
Urban growth   434
Urban decline   435
The role of the automobile   435
Urban Sprawl and Downtown Renewal   436
Urban Land-Use Planning   438
Historic land-use development   438
Contemporary Urban Land-Use Planning   441
Planning professionals   441
Nongovernment stakeholders   442
Government role in planning   442
Managing the Urban Ecosystem   444
Integrating urban ecosystem services   444
Land   445
Water   449
Air   454
Regional Planning Challenges   458
Literature Cited   458

SECTION 4   THE WILD LIVING RESOURCES   461

Chapter 15   Wildlife Conservation and Management   463

Introduction   463
Wildlife Values and Conflicts   463
The controversial resource   463
Wildlife authority   464
What Is Wildlife?   465
The professional concept of wildlife   465
The public concept of wildlife   465

Management Philosophy: Wildlife Conservation   466
  Values of wildlife   467
Historical and Legislative Perspectives of Wildlife Conservation   469
  Foods and other goods   469
  Recreation   472
  Animal damage control   474
  Management perspectives in the twentieth century   476
  Threatened and endangered species   480
Responsibilities of a Wildlife Manager   481
  Changing emphasis   481
  Population assessment and management   482
  Habitat assessment and management   483
  Resource demand assessment and management   483
  Resource user satisfaction   484
  Education and research   484
Contemporary Concepts in Wildlife Management   485
  Managing wildlife supply and demand   485
  The role of public input   485
  Categorizing wildlife for management   486
  Habitat management   486
  Habitat management strategies   492
  Managing wildlife populations   494
Commercialization and Wildlife Management   497
Challenges and Trends in Wildlife Management   500
Literature Cited   502

Chapter 16   Fishery Conservation and Management   504

The Fishery Resource   504
Fishing for Food and Other Goods   506
  Cultural importance   506
  Early fishery allocation   507
  North American fisheries   508
  Technological revolution   508
  Reaching food fishery limits   511
  The aquacultural potential   513
Recreational Fishing   514
  Cultural significance   514
  Growth of sportfishing in the U.S.   514
  The Limits to sportfishing   516
Fisheries Biodiversity Issues   517
  A growing concern   517
  Biodiversity values   518
Fishery Science and Management   519
  Early management emphases   519
  Modern fishery management   519
  New management principles   523
The Fishery Professional   525
  Changing expectations   525
  Bioassessment and management   526
  Habitat assessment and management   527
  Resource demand assessment and management   528
  Resource user satisfaction   528
  Management administration   529
  Original research   529
  Education   529

Future Fishery Issues   530
Literature Cited   530

Chapter 17   Biodiversity and Endangered Species Management   533

The Dilemma   533
Biodiversity   534
What is biodiversity?   534
The need for international cooperation   535
The escalating loss of biodiversity   535
Biodiversity services and value   536
Endangered Species   538
What are endangered species?   538
Causes of extinction   539
Endangered Species Policy and Management   543
Legislation and treaties   543
Habitat conservation plans   544
Management needed to maintain biodiversity   545
Endangered Ecosystems   546
Literature Cited   547

SECTION 5   THE MINERAL AND ENERGY RESOURCES   549

Chapter 18   Mineral Resources   551

Introduction   551
Geological Foundations   551
The realm of minerals and rocks   551
Plate tectonics   553
Plate boundaries   554
The rock cycle   555
Mineral deposits   556
Mineral distribution and abundance   557
Strategic and critical minerals   558
Mining and Mineral Extraction   558
What is an ore?   558
Surface mining   559
Subsurface mining   560
Processing   561
Important Metallic Minerals   562
Iron   562
Aluminum   563
Copper   563
Lead   563
Zinc   563
Gold and silver   563
Nonmetallic resources   564
Recycling   564
Recycling metals   564
Iron and steel   564
Aluminum   566
Copper   566
Lead   567
Environmental Concerns with Mining Activities   567
Geological exploration   567
Mining extraction   567

Mineral processing 569
Future mineral availability 569
Literature Cited 570

Chapter 19   Nonrenewable Energy Resources **572**

Introduction 572
Energy Use 572
Historical perspective 572
Energy use today 573
Electrical energy 575
Energy in the home, business, and industry 576
Energy used for transportation 577
Energy Production 578
Coal 578
Oil and gas 584
Nuclear energy 592
Nonrenewable Energy and the Environment 599
Nuclear plant accidents 599
Terrorism 602
Environmental degradation 602
The Immediate Future 603
Literature Cited 603

**Chapter 20   Renewable Energy: The Sustainable Path to
a Secure Energy Future** **605**

Introduction 605
Renewable Energy Sources 607
Solar energy 607
Wind energy 611
Hydropower 615
Biomass 617
Geothermal energy 621
Additional renewable energy sources 624
Concluding Remarks 626
Literature Cited 628

**SECTION 6   INTEGRATION OF NATURAL RESOURCES
MANAGEMENT** **631**

Chapter 21   Natural Resources and International Development **633**

Problems with Third World Development 633
National unity 634
Market-oriented economy 635
Democratic form of government 635
Sound education system 636
Protection of property rights 636
Opportunity for social and economic mobility 637
Level of economic growth exceeds level of population growth 637
Growth Strategies 639
Development Options 639
Natural Resources versus Entrepreneurship 641
International Policy 641
Literature Cited 643

Chapter 22   Economics and Economic Systems   **644**

Economic Terminology   644
Basic Economic Principles   644
    Market theories   645
    The Marx alternative to markets   646
    Mixed economies   647
    The most successful economy in the world   648
Problems with Market Economies: The Business Cycle   650
    Stages of the business cycle   651
    Depressions in the United States   653
    Keynesian economic approach   655
    The economy of the 1980s and 1990s   657
Creative Destruction and Human Progress   658
Problems with Centrally Planned Economies   660
    Environmental problems in the former Soviet Union   661
    The collapse of communism   661
Problems with Mixed Economies   662
Importance of International Trade and Competition   664
    International trade   664
    International competition   665
Economies in the Twenty-First Century   666
Literature Cited   666

Chapter 23   Sustainable Development, Technology, and the Future   **668**

Overview   668
Sustainable Development   670
    Defining sustainable development   670
    National sustainable development goals   670
National Strategies for Sustainable Development   671
    Sustainable development and conservation   672
    Information deficiencies and sustainable development   672
    Sustainable development in U.S. river corridors   673
    Land control and sustainable development   674
Ecosystem Management   675
    Defining ecosystem management   675
    Integrating resources management into ecosystems management   676
    Ecosystem health and adaptive management   677
Technology   678
    The importance of computers   678
    Systems analysis   678
    Models   679
    Artificial intelligence   680
    Virtual reality   680
    Economic analysis   681
The Future   681
Literature Cited   685

Glossary   **689**

Index   **721**

Preface

The purpose of this book is to introduce students to the science of natural resource management, coupling the latest concepts and technology with proven traditional approaches. We hope our intended audience includes managers on public and private lands, forestors, wildlife biologists, marine biologists, earth scientists, farmers, ranchers, hydrologists, urban planners, environmental scientists, conservation biologists, economists, politicians, and the growing segment of the public interested in natural resource management. We have tried to provide a comprehensive text for those concerned with natural resource management, not only in the United States but in other parts of the world as well.

Our approach has involved coupling fundamental topics such as plant ecology, soil science, climatology, economics, and policy with the most recent research. Some traditional concepts and viewpoints on natural resource management have been substantially altered as a result of new findings. This is particularly true in areas such as forestry, range management, plant ecology, energy conservation, urban planning, and wildlife management.

The management of natural resources has become more integrated through its 100 year history. Our textbook on natural resource management attempts to integrate ecological, economic, and policy factors into a functional, applied framework. We believe sound natural resource management requires an understanding of the interactions between natural and social processes. While the natural processes are predominantly ecological, the relevant social processes are primarily economic and political. We consider economics, the measurement of relative value, a critical part of natural resource management decision making. Therefore we have made economics an important component of our book, in contrast to many of the other books on natural resource management. Political policy effects the will of society in regard to choices among how natural resources will be conserved, produced, managed, and allocated to society.

We recognize that many natural resources cannot be readily valued monetarily, such as scenic beauty, open space, natural wonders, and rare plant and animal species, but contribute greatly to human quality of life. It is the intent of our book to help managers make more rational choices between material and spiritual well-being. At the same time, the future welfare of society must be considered, as well as the present.

Because a diversity of benefits is usually the object of most natural resource management, today's resource managers need a broad understanding of various natural resource categories, as well as specialization in certain specific areas. Forestors, wildlife biologists, geologists, range managers, environmental scientists, etc. need a basic understanding of all the other disciplines. Most comprehensive natural resource management is done through team work and requires strong communication and interpersonal skills. Therefore we have made planning an important component of our book.

While many books on the environment and natural resources have been somewhat pessimistic about the future on planet earth, our book presents a view of guarded optimism. Although the human population of the world continues to increase, major breakthroughs in technology show great promise for alleviating resource scarcity, pollution, and degradation. Improved political economic systems are resulting in more equitable distribution of resources to human populations in many parts of the world. At the same time, we readily acknowledge that elimination of rain forests, reductions in biodiversity, global warming, loss of open space, loss of the ozone layer, and the human population increase are all important natural resource challenges. However, we believe that through science, technology, universal education, and application of proven socio-economic principles humanity can overcome these daunting problems.

We freely acknowledge that this text will not convey everything there is to know about natural resource management. However, we believe it will provide the reader with a powerful understanding of how basic principles of ecology and economics can be integrated to successfully solve natural resource problems.

We have drawn heavily from many other great books on various aspects of natural resource management to develop our subject. We express deep gratitude to Raymond Dasman, G. Tyler Miller, Jr., Bradley Schiller, Oliver S. Owen, Daniel D. Chiras, Eric G. Bolen, William L. Robinson, Bernard J. Nebel, Richard T. Wright, and Ronald D. Knutson, whose textbooks provided the basis for coverage of many of our subjects.

We received both encouragement and helpful criticism from many of our colleagues. Special thanks are given to William H. Fleming, James Teer, and James D. Yoakum for their valuable suggestions on our manuscript.

JERRY L. HOLECHEK
RICHARD A. COLE
JAMES T. FISHER
RAUL VALDEZ

CHAPTER 1

Natural Resources: An Overview

We live in an era of increasing social and ecological awareness. Profound questions exist about whether or not healthy societies can be sustained without destruction of the natural resource bases they depend upon. When considering the answers to these questions, understanding of the interactions between natural and social processes is required. While the natural processes are predominantly ecological, the relevant social processes are primarily economic and political. Thus the content of this book goes beyond the typical inventory of natural resources encountered in many introductory texts to a larger view of natural resource ecology, economics, and policy. This treatment introduces students to the problems and some of the potential solutions they can expect to encounter as professionals or as informed citizens in the field of natural resource management.

NATURAL RESOURCES DEFINED

In this book we apply an inclusive definition to natural resources. They are divided into basic categories: air, water, soil, energy, minerals, farmland, forest land, rangeland, parks and wildlands, urban lands, wildlife, fisheries, and biodiversity. Under our definition, natural resources are the raw materials used to satisfy human needs. Humanity has grown to over 5 billion because of the mobilization of natural resources required to meet basic needs for food, water, habitation, infrastructure, and other material well-being. Natural resources include nonliving and living things that interact and affect human welfare. Natural resources are differentiated from goods manufactured using human ingenuity and labor. Trees, fish, and food crops are natural resources in contrast to furniture and prepared meals, which incorporate substantial human inputs. Solar radiation, wind, coal, and water are natural resources, but the electricity produced from them is not. Sand and lime are natural resources, but the concrete made from them is not. Similarly, the plastics, fibers, glasses, and various refined metals that compose many of our manufactured goods are not natural resources even though they require natural resources for their production.

As in all things, the boundaries between natural and artificial (an artifact of human modification) are often difficult to define. Whereas wild crops are definitely natural resources, cultivated crops are less so. Yet cultivated crops make an important contribution to local landscapes, support wildlife, and influence other natural processes. Similarly, while land is the most fundamental of natural resources, the buildings, highways, dams, railroads, and other engineered structures set on the land are not. Also among artificial structures are lawns, gardens, parks, vacant lots, waterways, and urbanized landscapes that grade into more wild environments. We include land used for farm and urban development among our natural resources because they form important resource and environmental links between humanity and the natural world.

Abiotic and Biotic Resources

Abiotic resources include minerals, air, water, solar radiation, and atomic radiation. Some abiotic resources (nonliving), such as oil and coal, were formed from once-living matter exposed to anaerobic environments under pressure for millions of years. Biotic resources are living organisms such as fish, wildlife, range grasses, and trees. Soil resources include a mix of biotic and abiotic properties that have unique attributes such as arability, fertility, and drainage.

Ecosystems Defined

We define ecosystem as a physical area with unifying ecological characteristics on which humans have placed boundaries for management purposes. An ecosystem is the interactive complex of biotic (plants, animals, microbes) and abiotic (e.g., soils, topography, climate) components occupying the defined area. For water resources, the watershed is used as the common management unit. For range and forest resources, vegetation in various topographical settings forms convenient boundaries for management. Particular plant and animal communities are associated with various ecosystems (Figure 1.1). Whatever boundaries are placed on ecosystems, the interactions among the biotic and abiotic components sustain a wholeness that is greater than the sum of the component parts. Figure 1.2 shows the components of a grassland ecosystem and the products it produces that are usable to humans.

An important concept in natural resource management is that ecosystems differ greatly in their resistance to overuse. For example, heavily grazed rangelands in the tallgrass prairie of Kansas can return to pristine or climax conditions within 5–10 years through natural processes if soil erosion has not occurred. However, in the Chihuahuan Desert of New Mexico, 20 years or more may be needed to overcome the effects of poorly controlled livestock grazing. Generally the resistance of terrestrial ecosystems to abuse increases as precipitation and temperature increase. Water and heat regulate biochemical activity and ecological rates of adjustment and recovery.

Resource Consumption, Use, and Renewal

Resources may be consumed during use or remain to be used again. When coal is burned and grass is eaten by livestock they are consumed and cannot be used again in the original form. A wilderness park may be visited, endangered species preserved, or a fish caught and returned without consumption. They could be used over and over with human benefit from each use.

This flow of materials often leads to recycling and renewal of resources through ecological processes. Burning of wood and coal, for example, generates carbon dioxide, which influences heat balance in the atmosphere and provides a resource for more photosynthesis and more wood production. The carbon dioxide may cycle into phytoplankton, forming a basis for fishery renewal. Fish and trees are renewable resources, as long as reproductive populations are maintained. Coal is a nonrenewable resource—at least within time frames relevant to resource managers. Similarly, genetic diversity is a renewable resource as long as the ecosystem functions needed to sustain various life forms are retained. While unmined iron and copper ore are not renewable, their manufactured products can be recycled, reducing

FIGURE 1.1 Different levels of organization in the living systems of the world (adapted from Miller 1990 by John N. Smith).

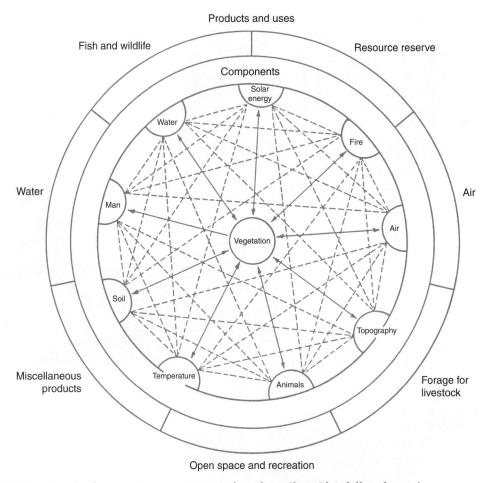

FIGURE 1.2   Grassland ecosystem components and products (from Blaisdell et al. 1970).

the demand on ore reserves. The renewable resources include soils, range, farm products, forests, wildlife, fisheries, water, air, genetic information, and wildlands. These resources can be managed to meet basic human needs on a sustainable basis. However, their overuse by humans can also cause depletion or degradation beyond repair. Overgrazing or improper farming practices can destroy within a few years an inch of fertile topsoil that took 1,000 years to form.

Nonrenewable resources occur in fixed quantity, and so from a practical standpoint cannot be regenerated by technology. These resources include most energy sources (coal, oil, natural gas, uranium), nonmetallic minerals (magnesium, phosphates, potash), and metallic minerals (gold, silver, copper). During the 1970s there was a concern that depletion of fossil fuels and some minerals such as copper would constrain economic development and improvement in the human condition. These concerns were somewhat alleviated during the 1980s and 1990s with substitution, recycling, miniaturization, improved detection, improved extraction, and higher mechanical efficiency. An example of substitution involves modern telecommunications equipment. Fiber optics that use silica as the basic structural component have largely replaced copper wiring. In the next 10–30 years, electrical cars that use batteries rechargeable from the sun's energy will probably replace cars that use fossil fuels. Within the last 20 years, the per capita consumption of fossil fuels was greatly reduced by improved vehicle fuel efficiency. Between 1973 and the late 1990s average fuel efficiency of new American cars nearly doubled, increasing from 13 miles per gallon to over 20 miles per gallon. These are just a few of the many examples of how technology has reduced problems of nonrenewable resource scarcity. Historically, the power of technology to solve natural resource problems has always been underestimated.

NATURAL RESOURCE MANAGEMENT

Management Objectives

Natural resource management is the manipulation of ecosystem components for human benefit. Management philosophy was once dominated by resource exploitation: take all that you can in a cost-effective manner and move on. Exploitation philosophy is shortsighted and tends to ignore future human needs. It still remains a predominant philosophy in some parts of the world. Conservation philosophy, which now predominates in the developed world, emphasizes the efficient use of natural resources while protecting their capability for renewal.

As a rule, the object of natural resource management is to obtain the optimum combination of goods and services for society on a sustained basis. Goods include all material resources. Services include energy resources and other non-material functions. For example, intact wetland ecosystems can function to provide a number of pollution, water supply, and flood control services while also sustaining genetic diversity. Natural resource management as we define it has two basic aspects: (1) sustaining or restoring the integrity of the energy, air, water, microbial, plant, and animal complex (the ecosystem complex), and (2) supplying products such as wood, food, water, minerals, energy, and wildlife consistent with demand and sustained production far into the future. In theory and practice renewable resources can be used while the integrity of the system that sustains their renewal is preserved.

Natural resource management is fundamentally concerned with ecosystem interactions—that is the interactions of energy, air, water, mineral, microbial, plant, and animal components. A key concept in natural resource management is that any human activity influencing one component will also affect the other components. Through research, many ecosystem interactions have been found to be predictable enough for reliable management results. Other processes remain to be discovered and clarified for improved management.

Private and Public Resource Management

Natural resource ecology, economics, and policy are closely linked with concepts of public and private resource ownership. While air, water, and wildlife have mobility, land is stationary. That simple reality allows land resources to be easily claimed and developed for either private or public gain, whereas water, air, and wildlife are more likely to be part of a more public domain. It has been only recently in the history of civilization that common individuals had the right to own land and other property independently of their ruler or their tribal chief.

Most anthropological evidence indicates tribal organization dominated human group behavior in prehistoric societies before the Agricultural Age began about 10,000 years ago. Before recorded history and written titles existed, occupancy and use of land were determined by tribal decision. Tribes decided "policy" in councils of various form, which represented various family or clan interests. Although ownership and trade of personal clothes, utensils, and art commonly occurred, there was no well-defined concept of land or resource ownership nor system of written title and laws protecting ownership of personal items. Settlement was usually temporary and possessions were few enough to allow easy movement elsewhere. There was little need for more than oral claim to personal property.

Once settlement became permanent and possessions accumulated, a more effective means evolved for protecting possessions and recording trade. As populations grew around reliable sources of cultivated food, division of labor and hierarchal organization evolved. During most of recorded history, resources were claimed by the "State," usually by kings, emperors, high priests, or other rulers at the top of the hierarchy who equated themselves with the "State." They developed the rules by which society lived and usually accorded themselves ultimate power backed by a warrior class. Although each member of the public used property, it typically was under the totalitarian control of the privileged few.

Land is the primary setting for most changes in the use of natural resources. History shows that informed people have been most willing to invest in sustaining the value of their own property, but they have tended to take for granted or to avoid responsibility for the perpetuation of environmental resources shared in common—the air, water, soil, and wildlife. Even if a farmer wanted to recover soil eroded from his land by wind or water, how could he ever reclaim it? On the other hand, why would he want it known that the eroded soil was his? He might be forced to compensate for the pollution his activities caused. Perhaps the pond he built filled with sediment because of some neighbor's "thoughtless" land management. Did he contribute to similar property degradation on other neighboring lands? This conundrum exists because ownership fails to completely constrain natural movements of air, water, and other material.

Our contemporary sense of environmental quality remains focused on air, water, and wildlife qualities more than in the quality of privately owned land. Air and water movements are the primary means by which ecological impacts, such as soil displacement and contamination, occur. Because these processes are not constrained by boundaries, environmental management is usually a shared public responsibility that is poorly served by independent private interests. The environment is composed of natural resources and its management is part of natural resource management. Land development and use is both the basic source of wealth—to individuals and to nations—and also the origin of most air, water, and other resource degradation. Because of it's ecological, economic, and policy implications, the history of land use control is central in the history of natural resource management.

An Integrated Approach to Management

The management of natural resources has become more integrated throughout history. This textbook attempts to integrate historical, ecological, economic, and policy factors into a functional framework. We consider economics—the measurement of relative value—a critical part of natural resource management decision making. Economics provides a means for comparing values assigned by people in order to best allocate scarce resources among competing uses for the greatest human benefit. Without a basic knowledge of economic principles, any selection of natural resource management alternatives will likely be far from ideal. However, not all valuations can be readily converted to a monetary standard. The monetary value for protecting genetic resources, for example, is not accurately estimable at this time. Although protective laws remain popular among most U.S. citizens, an inability to directly compare values of alternative uses of natural resources is the basis for substantial controversy in decision making. An important goal is the inclusion of all resources under a single monetary standard.

Most federal lands in the United States are managed to optimize public welfare. The harmonious use of public land for more than one purpose—such as water, timber, wildlife, and livestock production—is defined as multiple use. Multiple use philosophy also is becoming more common for resources managed privately because it usually leads to the greatest sustained benefit to the owners as well as to the greatest benefit for "the customers." Because a diversity of benefits is usually an object of most natural resource management systems, contemporary resource managers need a broad understanding of the various natural resource categories, as well as specialization in certain areas. Thus foresters, fisheries biologists, geologists, range managers, and environmental scientists need a basic understanding of all the other disciplines. Most comprehensive resource management is done through teamwork and requires strong communication and interpersonal skills.

THE NEED FOR NATURAL RESOURCE MANAGEMENT

The Human Population Increase

Many scientists consider the most pressing problem confronting humankind to be the tremendous increase in the human population expected in the next 50 years. About 10,000 years ago, the world's human population was about 1 million people (Chrispeels and Sadava 1977). It had reached 250 million about the time of Christ, and 1 billion in 1850. Doubling time to 2 billion took only 80 years and 4 billion was reached in only 46 more years. The world human population in 1998 was 6 billion (USDC 1999) when the annual growth rate was 1.4%. Although the human population growth rate began declining in the 1980s, the world population could reach 11 billion or more before stabilizing.

The decreased population growth rate has occurred nearly exclusively in developed countries with high per capita incomes. Birthrates remain high in the developing South American, Central American, and African countries (Table 1.1). Religious beliefs, family labor needs, and lack of education about birth control largely explain the high birthrates in developing countries compared to those of developed countries. In many developed countries, including the United States, food production has kept far ahead of the national population growth rate. Since the early 1980s, food surpluses have been one of the major problems confronting agriculture in the United States (Schiller 2000). However, food shortages remain an important problem in many developing countries because they do not have the resources and trade goods to exchange for food produced in the United States and elsewhere. This imbalance in wealth and world trade is expected to continue for many decades into the future—in part because of high population growth in the poorest countries.

Trends in human population growth and economic development will have a considerable influence on how various countries will use natural resources in coming years. Although the emphasis may shift among natural resource products, rapid expansion in the human population will undoubtedly make natural resource management more important than ever before, and with different emphases than in the past. Recent international conferences about atmospheric change, decreasing biodiversity,

TABLE 1.1   Growth and Density of the Human Population in Various Parts of the World in 1998

	Percent Net Growth per Year	Doubling Time (Years)	Population per Square Mile	Population per Square Kilometer
World	1.4	51	114	45
Africa	2.7	27	66	26
Kenya	2.4	30	128	50
Sudan	2.9	25	34	13
China	1.0	72	336	131
India	1.7	42	829	325
Europe	0.3	240	270	106
North America	1.2	60	40	16
Canada	1.2	60	8	3
Mexico	1.9	38	129	51
U.S.A.	1.0	72	75	29
Oceania	1.5	48	9	4
Australia	1.1	65	6	2
New Zealand	1.1	65	34	13
South America	1.7	42	48	19
Argentina	1.1	65	33	13
Brazil	1.2	60	50	20
Fromer U.S.S.R.	0.3	—	22	9

Source: USDC 1999 and FAO 1998.

declining ocean fisheries and other resource issues would have been unheard of at the beginning of this century. International trade already is being influenced by the relative value different countries place on these resources.

Although part of the solution to emerging resource problems may rest with decreased population growth rate, much of it also resides with more efficient and effective resource management. History shows that the most effective way to encourage reduced population growth is to improve the quality of life through education, democratic process, and technology. Technology is a double-sided sword, however, which has contributed to many existing problems while providing solutions for many others. Although resource depletion and degradation are challenging problems, recent progress shows they are not entirely overwhelming. Improved natural resource management is one strategy that can contribute to problem solution.

LAND AREA AND LAND USES

About 11% of the land in the world is farmed; 24% is in permanent pasture; 31% is forest or woodland; and deserts, glaciated areas, high mountain peaks and urbanized/industrialized land comprise the remaining 34% (Table 1.2). The total land covered by buildings, parking lots, highways, and other infrastructure is between 3 and 4%.

The total land area in the United States is about 1.91 billion acres (2.26 billion with Alaska). About 20% or 382 million acres is usable crop production. Cities, homes, factories, airports, highways, and other human structures account for 97 million acres or 5% of the total. Around 65% remains as forest and rangeland for wood, meat, fiber, mineral, and energy production (Figure 1.3).

In the U.S. a shear lack of land will probably not compromise human welfare over the next 50–100 years. However, improper planning of land use over the past 50 years has unnecessarily increased congestion, pollution, and infrastructure costs and caused a severe decline in the open space that most Americans value. In our opinion much better land use planning and landscape design will be critical to maintaining and improving human quality of life in the twenty-first century.

TABLE 1.2   Percentages of Farmland, Grassland, and Forest Land for Selected Countries and Regions in 1997

	Farmland	Permanent Pastureland (Grassland)	Forest and Woodland
World	11	25	29
Africa	6	30	23
Kenya	4	7	4
Sudan	5	24	20
China	10	42	13
India	52	4	20
Europe	28	17	32
North America	13	17	32
Canada	5	3	35
Mexico	13	38	21
United States	20	26	31
Oceania	5	56	18
Australia	6	60	14
New Zealand	2	53	26
South America	6	28	46
Argentina	10	51	21
Brazil	7	22	58
Former U.S.S.R.	10	17	41

Source: Based on FAO (1998) data.

Natural Resource Issues

Many issues have developed out of imbalances in resource supply and demand. Resource sustainability requires understanding of the underlying mechanisms determining continued supply and the effect of human demand on those mechanisms. For renewable resources, much of that understanding comes out of ecological and economic science. Ecology is the science that provides understanding of resource renewal rates. Economics is the science that examines relationship between resource supply and demand. Policy is the means by which societies organize guidance in balancing supply and demand based on ecological and economic information. At the policy level, there is an emerging sense that resource supply and demand questions need to be addressed in an ecosystems context.

Natural resource management has a long history with its modern roots established during the last century when a period of unchecked resource exploitation in the industrialized world resulted in widespread polluted air, contaminated water, declining wildlife and fisheries, depleted forests, degraded rangelands, and blighted urban areas. Natural resource management by the turn of the century centered on conservation of water, forest, fishery, wildlife, and outdoor recreational resources. Conservation philosophy was vindicated by the "dust bowl" era when the social costs of farmland mismanagement in terms of human health and property destruction became undeniable. The role of mass communication in shaping public opinion was beginning to have its impact through movie house documentaries and radio.

Following World War II (WWII), the development and rapid acceptance of television amplified the spread of information, including growing concern about environmental contamination by organic wastes, oil, toxic metals, synthetic chemical compounds, and atomic radiation. A period of unprecedented economic growth and optimism followed WWII as the world recovered from mass destruction. A "green revolution" in agriculture promised to solve growing food shortages. But a more pessimistic view emerged in the 1960s when environmental scientists began to present their growing concerns in popular books and through the mass media (television). In this period they were able to greatly influence the public toward greater environmental regulation. Other scientists observed that population growth was spurred to highs by the green revolution and that

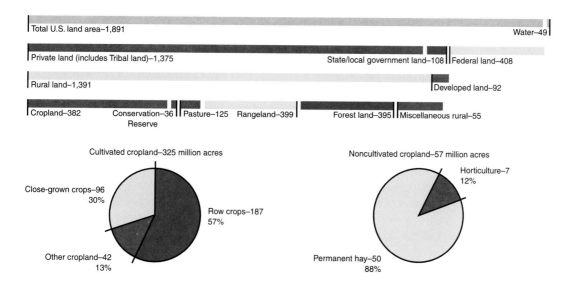

Millions of acres in the 48 contiguous states, Hawaii, Puerto Rico, and the U.S. Virgin Islands.

FIGURE 1.3   Total land area and land in the contiguous United States (from *A Geography of Hope.* 1997. USDA-Natural Resources Conservation Service).

fertilizers and pesticides contributed to a growing environmental threat. Numerous pieces of regulatory legislation highlighted by the National Environmental Policy Act (1969) and the Endangered Species Act (1973) were passed by the United States Congress to halt environmental decline. Food, mineral, and energy costs rose while the world population continued to increase. The book *The Limits to Growth* from the Club of Rome (1972) used a computer model to show how human population increase and natural resource depletion could lead to widespread hunger and great environmental decline if trends remained the same. People responded with continued support for enforcing new environmental regulations, which soon had positive impacts on water and air quality.

During the 1980s, the economist Julian Simon made the case (*The Ultimate Resource, The Resourceful Earth*) that technology could potentially solve most problems related to resource scarcity if socioeconomic systems rewarded innovation and efficiency. The fact that the 1980s and 1990s have been characterized by declining real (inflation adjusted) prices for food, minerals, and energy resources and lower pollution levels in developed countries has given credibility to many of Simon's arguments, at least in the short run. However, environmental degradation and declining human conditions in parts of Africa, Asia, Latin America, and eastern Europe have been used as counter-arguments against Simon's more positive outlook based on continued technological advances.

As the world approaches the twenty-first century, there is general agreement that developed countries have made considerable progress in solving problems that relate to natural resource scarcity and pollution of water and land resources. However, depletion of ozone in the outer atmosphere, the "greenhouse effect" (atmospheric warming from increased CO_2 levels), and rapid depletion of the world's genetic diversity remain issues of controversy and serious concern. Whereas point sources of pollution appear under control, there are many problems related to water resources, such as ground water contamination and depletion, which remain unresolved.

In developing countries, the destruction of the rain forest, loss of endangered species, loss of open space, and desertification have become major natural resource issues. In some of these countries, slowing the rate of population growth and increasing the rate of economic development have shown the potential to improve the human condition and slow or halt environmental decline. However, flawed socioeconomic systems and exploding human populations remain as major impediments to progress in many developing countries.

In the remainder of this section we briefly highlight the key issues for the various categories of natural resources.

Soils.   Current primary problems are erosion and soil contamination from pesticides and herbicides. However, new tillage and grazing practices can greatly reduce erosion. Biological engineering and changes in cropping systems are reducing the use of herbicides and pesticides. The most serious soil problems are in developing countries and center around erosion and salinization.

Water.   This resource in the future may be the biggest constraint on human population growth and economic development. Considerable progress has been made in cleaning up our rivers and lakes but ground water contamination is still a serious problem. Ground water depletion is becoming a limiting factor to economic growth in many parts of the world.

Air.   Generally there have been improvements in most U.S. cities, and the problem of acid rain appears to have been exaggerated. Global warming and ozone depletion are serious global problems that have not been resolved.

Biological Diversity.   The loss of biological diversity is accelerating. Much of this problem centers around destruction of the rain forest in tropical developing countries because of rapid human population growth and flawed socio-economic systems. One potential solution is to couple economic compensation for conservation to sustainable rain forest management programs. While rain forests are especially problematic, biodiversity loss is a concern in many other ecosystems.

Minerals. Generally the scarcity of minerals does not appear to be an immediate or long-term problem. The former Soviet Union has vast reserves of most metallic minerals that are now becoming available in the world market place as result of change from a communist to a market economy. Substitution is reducing the need for many scarce metals such as copper and silver. Recycling of metals such as iron and aluminum has many advantages beyond extending supplies. These include energy conservation, pollution reduction, and reduction in disposal problems. There will be an increased emphasis on the repair of environmental damages after mineral extraction in the U.S. and other developed countries.

Energy. New technology and globalization of the world economy (free trade) have reduced the concern over energy scarcity that existed in the 1970s. Solar energy, nuclear energy, and wind driven energy can become important replacements for fossil fuels. Since the early 1970s, increases have occurred in the efficiency of fossil fuel use in automobiles, industrial use, and home heating. The major challenges now center around minimization of air pollution and finding better methods for disposal of hazardous (nuclear) waste. However, the scarcity problem could reemerge sometime in the twenty-first century.

Farmland. The concern over lack of food in the 1970s was replaced by a surplus of food in world markets in the 1980s. However, starvation remains a problem in many developing countries because of high human population growth rates and inefficient agricultural systems. In developed countries, the big challenges are preserving prime farmland and minimizing the use of herbicides and pesticides. In developing countries, flawed socioeconomic systems and rapid population growth are the root cause of soil erosion and food scarcity problems.

Rangeland. The historic problem on rangelands has been overuse of forage resources. Generally, economic benefits of sustainable grazing practices have not been well understood in either developed or developing countries. Government subsidies and public ownership of grazing lands have often undermined sustainable grazing practices. Improved knowledge of grazing management outcomes and better transfer of knowledge to range users has much potential to improve this natural resource both in the U.S. and other countries. We believe preservation of open space will be the biggest rangeland challenge in the twenty-first century.

Forest Land. In the U.S., forest land problems center around maintaining mature (old growth) woodlands and minimizing environmental damage from road construction and other logging operations. In developing countries, deforestation is a serious problem, particularly in tropical areas. Linking economic aid to changes in natural resource management in developing countries could have many benefits. Rain forest preservation is thought to be crucial in slowing global warming from the greenhouse effect.

Urban Land Use. Many of the resource management problems associated with cities have improved. Most notably the Clean Air Act and Clean Water Act have reduced point source emissions. However many problems remain, especially from nonpoint source emissions, urban sprawl, and city center decay. These are complex issues that contribute greatly to climate change, loss of biodiversity in the developed world, and increased human stress. Many fundamental problems remain in the underdeveloped nations.

Recreation.   Tourism has become the world-class means of recreation. Ecotourism is an especially attractive joint solution to protecting ecosystem integrity while sustaining economic development. However, much remains to be done to create and maintain market-based ecotourism. Many unique wilderness areas are threatened with extinction. In the U.S., which is rich with recreational land, the biggest challenge is to provide affordable access to uncrowded recreational land close to where people live.

Wildlife.   In the U.S., many wildlife success stories have occurred over the past 20 years. Changes in government policies in the mid-1980s have reversed the decline in farmland wildlife. The net loss of wetlands has been nearly halted. Improved wetland management is causing a comeback in waterfowl and other wetland animals. Urban wildlife programs have been successful in creating favorable conditions for many species of birds and mammals in towns and cities, but careful animal damage control is an increasing challenge. Although some species remain endangered, such as the northern spotted owl and desert tortoise, market incentives show potential to alleviate some problems in the U.S. In developing countries, the future of many wildlife species is less than optimistic. However, even here market forces could solve many wildlife problems when allowed to operate.

Fisheries.   Overexploitation continues to be the major problem confronting ocean fishery resources. Aquaculture has been successfully used for production of salmon, trout, shrimps, oysters, and so on, but it is not practical for large ocean fish such as tuna. Better international agreements are needed to ensure sustainability of large ocean fishes and mammals (whales). Bans on ocean dumping of waste and better control of oil spills would help many species of marine fishes and mammals.

Natural Parks and Wildlands.   Throughout the 1900s large areas have been set aside in the U.S. as wilderness and natural parks. The debate will continue on how much wilderness should be set aside. These areas preserve places with great scenic beauty and/or high natural biological diversity. Generally these areas do not contribute much to the national timber, energy, and mineral resources. Private groups have shown increasing willingness to buy up unique areas both in the developed and in the developing countries. We expect that the private sector will play an even greater role in natural resource preservation in the twenty-first century. However, for the large tracts of federal lands in the U.S. with no unique esthetic or biological value, there could be an opposite trend toward privatization.

AN OPTIMISTIC VIEW OF THE FUTURE

While many studies of the world natural resources in the decades since World War II have emphasized the limits to world growth, such as Club of Rome (1972), our position is more optimistic. Recent global trends in technology, education, acceptance of free market economies, economic productivity, and global concern for sustainability show promise for alleviating resource scarcity and degradation while providing improved quality of life. Over the past 20 years, many measures of environmental health on our planet show improvement, particularly in developed countries. Improvements in energy efficiency, reductions in air pollution, reductions in water pollution, increases in farm yields, reduction in pesticide and herbicide use, increases in farmland and urban wildlife, and a reversal in wetlands loss are some positive natural resource changes in the U.S. The elimination of rain forests, desertification, reductions in biodiversity, watershed pollution, the loss of the ozone layer, global warming, and human population explosion are all important natural resource problems that remain to be solved. However, through science, technology, universal education, and the application of proven socioeconomic principles, we believe humanity can overcome these daunting problems.

OTHER SOURCES OF INFORMATION

We acknowledge this text will not convey everything there is to know about natural resource management. However, we believe it will provide the reader with a powerful understanding of how basic principles of ecology and economics can be integrated to solve natural resource problems. At the same time we encourage students and practicing natural resource managers to read the many other excellent books that exist on this subject. Some of the books that have influenced us and have been repeatedly cited in this work include Dasmann (1976), Miller (1990), Schiller (1994, 2000), Owen and Chiras (1995), Owen et al. (1998), and Nebel and Wright (1996, 1998).

ORGANIZATION OF THE BOOK

This book is organized into six sections. Each section forms a basis for more effective learning in the sections that follow. Summaries of the foundational knowledge areas underlying natural resource management are followed by sections that describe resource management in more detail. A final section expands the national focus of the book to a more global perspective and a more integrative emphasis. A glossary of terms is provided.

Section 1: Management Foundations

The first section introduces basic concepts in ecology, economics, policy, and planning process as they pertain to natural resource management. We emphasize those concepts that underlay the development of the chapters that follow. The lead chapter in the section is a brief history of natural resource ecology, economics, and policy in the U.S. Understanding of history leading up to the present state is a prerequisite for understanding contemporary natural resource management. The second chapter, on ecology, summarizes important concepts in population, community, and ecosystems ecology, with an emphasis on principles that underlay management of renewable living resources. The next chapter focuses on the role of economics in aiding management decisions and policy development directed at the goal of sustained benefit from natural resources. The management operations addressed in later chapters are preceded by management planning and policy actions described in the last chapter of this section. As the chapters in this section of the book will show, most contemporary natural resource managers need a broad educational background in both natural and social sciences in addition to specialization in certain aspects.

Section 2: Air, Water, and Land Resources

The second section introduces the basic resources associated with the atmosphere, hydrosphere, lithosphere, and biosphere. The chapter on climate discusses basic climatic processes and the various climatic types of the world. It includes a brief section on climatic history. Climatic problems that confront natural resource managers in the twenty-first century, such as the greenhouse effect, are addressed. Water is the basic resource addressed in Chapter 2, which describes its characteristics and shows how it is important for maintaining life processes, transporting natural materials, and navigating. It illustrates the hydrologic cycle, which does much to connect the atmosphere, hydrosphere, and lithosphere into geophysical processes that sustain the biosphere. The last two chapters treat the lithosphere and the biosphere. The lithosphere provides both raw minerals and soils. Soil not only is the source of physical support and nutrition for most terrestrial and much aquatic production, it also provides the foundation for most terrestrial engineering, such as roads and buildings. Because of the diverse services provided by soils, urban development often competes with farming, forestry, and range use. The process of soil development, soil erosion, and management of soil resources are treated here. The last chapter in this section describes predominant ecosystem types defined mostly by the combination of topography (especially the distributions of surface water and dry land) and vegetation. Ecosystem attributes determine to great extent the types of resource development, use, and management concerns. At this juncture, the reader may continue on to Section 3 or skip to Section 5 before returning to Sections 3 and 4.

Section 3: The Land-Based Renewable Resources

The third section summarizes the development of resources we most frequently associate with living processes in the biosphere, including the urban habitats of humanity. A chapter on forest resources leads this section, followed by another on range resources. These two chapters go a long way toward covering the management of resources derived from noncultivated vegetation. An important chapter on farmland summarizes the state of cultivated plants and other farm resources with an emphasis on farm policy. Another chapter on rural parks, wilderness, and other reservations in which the vegetation is managed for purposes of human recreation completes most of the rural land survey of resources and their management. A chapter on urban landscapes addresses the use of land for human industry, residence, infrastructure and recreation and discusses the extent to which this use competes with other uses. Certain urban land uses fall undeniably into renewable resource categories, such as parks, parkways, private lawns, and roadside vegetation. In addition, the land itself is a renewable source of space in which old structures may be removed and replaced with either natural or artificial structure. Development of all of the land-based resources has environmental effects addressed in each of the chapters.

Section 4: The Wild Living Resources

Chapters 15, 16, and 17 discuss those wild resources that are held in trust by the public. The chapter topics include wildlife, fisheries, and biodiversity. Terrestrial wildlife and biodiversity are most obviously dependent on the way the land-based resources are managed, but fishery and aquatic biodiversity also are impacted through the watershed processes that link land and water. The wildlife chapter discusses commercial and recreational resources as well as animal damage control. The fishery chapter summarizes the state of food and recreational fisheries, and management directed at declining aquatic biodiversity. The biodiversity chapter summarizes the dilemma of declining biodiversity and management directed at preserving genetic diversity and ecosystem services associated with diversity.

Section 5: The Mineral and Energy Resources

The first chapter in this section treats the minerals of the lithosphere, which supplies many of the metals and other elements that have become so important in contemporary technology and architecture. Also supplied are the sands, gravels, clays, and other earth materials so important for sustaining the human habitat of concrete, tile, ceramics, and glass. The lithosphere also provides the inorganic nutrients used in fertilizers, such as phosphorus and potassium. The second chapter in this section is about nonrenewable energy resources and describes the state of the fossil and atomic fuels, which are the sources of energy most used in industrial and municipal processes. The final chapter in this section treats the renewable forms of energy including solar energy, hydropower, and wind energy. This introduction also defines the role of solar energy as the primary driving force in the world's ecosystems.

Section 6: Integration of Natural Resources Management

Trends in technology development; global, social, and natural processes; and the incorporation of more of a systems process in management all contribute to a growing emphasis on integrated natural resource management. This section contains three chapters that extend concepts of natural resource management in the U.S. to a global perspective. A common theme emerges for global natural resource management: the need for more integrated management. The first chapter discusses those attributes that appear to be needed to establish sustainable societies based on wise development of natural resources. The next chapter in this section reviews the performance of economic systems in the global perspective, emphasizing the trends in the world today. The last chapter summarizes the most recent trends in integrated resource management in support of sustainable development and the global issues that are likely to dominate natural resource management during the next few decades.

LITERATURE CITED

Blaisdell, J. P., V. L. Duvall, R. W. Harris, R. D. Lloyd, and E. H. Reid. 1970. Range research to meet new challenges and goals. *Journal of Range Management* 22:227–234.

Chrispeels, M. J. and D. Sadava. 1977. *Plants, food, and people.* San Francisco, CA: W. H. Freeman.

Club of Rome. 1972. *The limits to growth.* New York, NY: Universe Books.

Dasmann, R. F. 1976. *Environmental science.* 4th edition. New York, NY: John Wiley & Sons.

Food and Agriculture Organization (FAO). 1998. *Production yearbook.* United Nations FAO Statistics Series, No. 51. Rome, Italy.

Miller, G. T. 1990. *Resource conservation and management.* Belmont, CA: Wadsworth.

Nebel, B. J. and R. T. Wright. 1996. *Environmental science: The way the world works.* 4th edition. Upper Saddle River, NJ: Prentice-Hall.

Nebel, B. J. and R. T. Wright. 1998. *Environmental science: The way the world works.* 5th edition. Upper Saddle River, NJ: Prentice-Hall.

Owen, O. S. and D. D. Chiras. 1995. *Natural resource conservation.* 6th edition. Upper Saddle River, NJ: Prentice-Hall.

Owen, O. S., D. D. Chiras, and J.P. Reganold. 1998. *Natural resource conservation.* 7th edition. Upper Saddle River, NJ: Prentice-Hall.

Schiller, B. R. 1994. *The economy today.* 6th edition. New York, NY: McGraw-Hill.

Schiller, B. R. 2000. *The economy today.* 8th edition. New York, NY: McGraw-Hill.

Simon, J. L. 1981. *The ultimate resource.* Princeton, NJ: Princeton University Press.

Simon, J. L. and A. Kahn, eds. 1984. *The resourceful earth.* New York, NY: Basil Blackwell.

United States Department of Commerce (USDC). 1999. *Statistical Abstract of the United States*, 117th edition. U.S. Department of Commerce, Bureau of Census.

SECTION 1

Management Foundations

Prerequisite knowledge for a full introduction to natural resource management must include basic principles of ecology, economics, and organizational management. Chapter 2 starts this section with a historical overview of natural resource development in the U.S., pointing out the importance of ecology, economics, and policy in natural resource management. Chapter 3 is an introduction to ecology, which is the study of interactions among living organisms and their environment. Ecology is an integrative science guiding sustainable natural resource development. It addresses the natural and managed regulation of processes determining rates of living resource production and the impacts of resource development on the many services ecological systems provide to humanity. Principles of natural resource conservation economics are introduced in Chapter 4 to show how economic study helps reveal how best to distribute natural resources for more total benefit to people in present and future generations. Organizational planning process, described in Chapter 5, includes a variety of approaches used to identify the most effective management policies in advance of making policy decisions. The most effective natural resource planning integrates ecological and economic analyses to determine those management policies and strategies that protect those ecological processes sustaining the greatest benefit over present and future generations. This section establishes a foundation for the specialty chapters that follow.

CHAPTER 2

The Historical Perspective

"One of the few good descriptions of *Homo sapiens* is that he is, above all, a change making animal. The evidence of that lies in what he has done—his history" (Roberts 1993). History also shows us that the human condition is a product of previous change wrought by predecessors and a source of insight about future legacies. The history of natural resource development and management is a study of change in human interactions with their natural resources. One of the most valuable benefits of understanding history is that it can prevent repeating past mistakes. In this chapter, we review the history of natural resource management in the United States from the colonial era to the present. We also consider challenges confronting natural resource managers in the twenty-first century.

COLONIZATION (UP TO 1776)

Ecology and Landscape

Over 10,000 years ago, the Northern Hemisphere was in an ecologically dynamic time. Ice, fire, flood, drought, and other natural events caused by widespread climate change widely altered landscapes. Continental glaciers were in retreat and the land bridge from Asia to Alaska had submerged. This isolated the earliest human colonists of the Americas until European discovery thousands of years later. In this same period agriculture began in Asia (Roberts 1993). Livestock husbandry originated over 11,000 years ago in Asia Minor. Inventions of axes, adzes, and other long-handled implements were critical in advancing the rate of clearing and cultivation. Agriculture appears to have advanced most rapidly in arid ecosystems along fertile river bottoms with reliable sources of water for irrigation.

Agriculture did not become established in North America until about 2,500 years ago (Butzer 1990, Roberts 1993). Irrigated agriculture developed in the arid Southwest along the easily worked floodplains of desert rivers. Woodland agriculture also developed throughout much of the warmer eastern deciduous forests at lower elevations where soils were relatively easily worked (Figure 2.1). There is less indication of extensive agriculture in the open prairies, where the thick grass roots resisted

FIGURE 2.1 This early engraving graphically portrayed an Algonquian Indian village as it appeared to European colonists. Farming was widely practiced by eastern Native Americans (courtesy of National Park Service).

cultivation and bison and other game were plentiful alternatives. Native Americans depended on hunting, fishing, and foraging for calorie-rich fruits and seeds. For many tribes, mobility was critical for survival. Their ultimate adaptation to ecological and social stress was to move elsewhere. This mobility was accomplished with minimum intertribal conflict at population densities less than 1% of current levels. However, when tribes met intertribal conflicts often occurred over resources.

During the time of European colonization in the late sixteenth century, North America held an abundance of natural resources unmatched in western Europe, where many resources were depleted. The eastern coastal plain south of New England and river floodplains were ideal for farming once cleared of forest. The deciduous hardwoods made effective fuel, durable furniture, lumber, and tools while straighter softwoods made unexcelled ship masts, fences, and housing material. Under the forested Midwest lay rich glacial tills, which as farmland would be exceeded only by tallgrass prairie. Farther west the grasslands grew more arid, shorter, and more bunched, but served to nourish huge numbers of bison and other grazing animals.

The fisheries of the North Atlantic and North Pacific were among the richest anywhere (Waterman 1975). Early European explorers described "countless numbers" of large animals roaming the North American continent (Kline 1997, Bolen and Robinson 1995), which were recorded in art as well as journals. Bison, elk, turkey, deer, waterfowl, wild pigeons, and other game animals were widely distributed and commonly encountered. An estimated 60 million bison, 10 million turkey, and 2 billion passenger pigeons inhabited the continent. Stoddard et al. (1975) calculated the equivalent of over 67 million animal units (cow and calf equivalent) of wild grazing animals on the central Great Plains grasslands, which is the same amount of livestock

FIGURE 2.2 J.M. Stanley was an early artist-natural historian who painted this scene of a herd of bison near Lake Jessie in eastern North Dakota (courtesy of U.S. Dept. of Interior).

grazing pressure existing there today. The great abundance of fur-bearing animals provided incentives for the earliest exploration by European colonists. Beaver were especially widely distributed and eventually became the most popular of furs. By the seventeenth century the monarchies of Spain, France, and England had established American colonies in anticipation of wealth from resource extraction.

The conditions discovered by European colonists had not been stable, but responded to dynamic ecological processes. Weather was variable, especially in the heartland where droughts were frequent and seasons were extreme. Lightning-caused fire was a widespread natural phenomenon, which sustained a dynamic and diverse pattern of vegetation types and associated animal communities. Animal populations sometimes fluctuated dramatically because of disease, severe winters, and drought. In the larger river valleys, floods periodically reshaped river habitats and sustained a dynamic mix of river channels, wetlands, riparian forests, and small lakes and ponds. Fire was frequent in prairies and drier forests where it created a mosaic of fire-adapted grasslands, savannahs, and forests. Accounts by early European explorers indicate that bison (Figure 2.2) at least temporally overgrazed local areas (England and De Vos 1969). Prairie grouse and waterfowl populations may have been suppressed by lack of vegetation cover (Kirsch and Kruse 1972). Unlike most present-day livestock, however, bison moved freely over the range leaving vast areas ungrazed and able to recover until they returned years later (England and De Vos 1969).

European colonists found a vast forest with many openings and parklike savannahs caused by fires and aboriginal settlements. The mix of forest and upland openings supported an abundance of wild game including deer, turkey, grouse, doves, pigeons, squirrels, and rabbits (Clepper 1966). Extensive wetlands, lakes, and streams provided habitat for large flocks of waterfowl, various fish species, and fur-bearing beaver, muskrat, raccoon, otter, and mink. Because these clearings attracted the densest game populations and were easiest to farm, they were the first areas settled by European colonists on the East Coast.

The first cattle, sheep, and horses were brought into the arid Southwest from Mexico by Coronado in 1540. By the seventeenth century the Spanish had established numerous settlements near perennial sources of water. Escaped cattle and horses from Spanish settlements multiplied rapidly in the mild climate and abundant forage. Feral horses dramatically changed the lives of many tribes, who came to depend on them for their survival. Spanish colonial life remained agrarian long after the industrial age was established on the East Coast in the eighteenth century.

Unlike Native Americans, the first impulse of European colonists was to remain on the land to defend and "improve" their claim by further clearing, planting more crops, and grazing more livestock. The European concept of private claim, based in written law, encouraged permanent settlement. One reason certain game animals were rapidly depleted following settlement was their concentration near forest openings where early farming was most feasible. Wild game grew more scarce around farm margins as farms proliferated and farmers continued to hunt to augment farm produce. As the Europeans advanced across the continent, the progressive reduction of fish and game resources contributed to a wave of conflict with Native Americans.

Permanent establishment provided reproductive advantages that allowed the population of European colonists to expand while those of Native Americans declined. Introduction of Eurasian diseases decimated Native Americans. Early European immigrants produced a variety of native crops such as corn, squash, and tobacco. Whereas Native Americans tilled little and moved frequently to new areas, European tilling exposed soil to erosion and continuous cultivation that eventually depleted fertility. Typically, settlers moved to virgin land only after the soil productivity declined beyond that required for family subsistence. In the northeast, many "hill farms" were located on infertile and stony ground. Cotton and tobacco grown by the southern colonists were especially exploitive of soil resources. Decreased agricultural productivity was a major motivator for western expansion.

Significant Events in Colonial America

~10,000 B.C.	Last of American colonization over land bridge from Asia.
~1,000 A.D.	Norse in Newfoundland form the first European settlement in North America. It fails to persist.
~1440	Portuguese probably discover fishing grounds at Grand Banks off Newfoundland.
1497	Cabot is the first to seek Northwest Passage to Asia in a 350-year quest that opens northern fur trade and fisheries.
1540	Seeking gold, Spanish introduce cattle, goats, and horses to the Southwest.
1609	The first successful planting and harvest of tobacco by European colonists followed the advice of American Indians. This established the colonies' most important cash crop.
1620	Following John Smith's discovery of fisheries, the Plymouth Company formed to take advantage of New England fisheries. Mayflower Compact established a basis for participatory local government.
1623	The first saltworks, sawmill, and cattle were established in New England.
1647	The first public education was required of towns by the Massachusetts colony.
1664	Transfer of lands from the Crown to the Duke of York initiated land speculation, sale, and transfer to private ownership.
1705	British initiated laws that restricted colonial trade only to England and levied taxes on goods including rice, molasses, pitch, tar, rosin, turpentine, and hemp and, later, copper, furs, tobacco, sugar, textiles, coffee, indigo, and wine.
1709	The first copper mine was established in Connecticut followed by the first ironworks in 1714 in Virginia.
1715	Nantucket had become a whaling center for offshore sperm whales and established a resource for a large candle industry started in 1750.
1724	The first rice irrigation greatly expands production in South Carolina.

1730	New York establishes pottery and stoneware industry followed in 1739 by glass factories in New Jersey and Pennsylvania.
1750	England passes the Iron Act, which limits colonial steel industry to protect England's industry from competition.
1760	One third of all English ships were being built in the colonies with cheap timber and other maritime supplies.
1764	England passes the Sugar Act, which is one of a series of tax laws on resources and goods leading to the American Revolution.

Economics

Independence and Wealth.   The discovery of bountiful natural resources by Europeans encouraged colonization and land claim in the New World despite great odds. Abundant and favorable climate and soils at the time of colonization formed the foundation for the great prosperity the nation experiences today. Much of present American character and culture is based on the development of a vast and diverse supply of natural resources (Clepper 1966). The early establishment of market economies, democratic governance, and individual right to titled land ownership established a strong philosophical foundation for the new nation once independence was declared.

East coast European colonists were favorably positioned to adopt a strong market economy and democratic local government even before independence from Britain was declared in 1776. Private property was made widely available by the British Crown to entice stable settlement. Local and international trade soon became the basis for thriving business and the establishment of industry along the East Coast. The early colonists developed valued trade with Great Britain for North American resources in short supply, such as ship timber and furs.

The Origins of the Market Economy.   The foundations for complex market economies extend back to the earliest cities, where labor diversified into specialized services and commerce flourished. The specialization of services required ways to exchange goods and services in kind. The first forms of money used for exchange originated over 4,500 years ago while the first metal coins appeared about 2,700 years ago (Roberts 1993). The first true cities of diverse industrial, religious, commercial, and residential structure arose in the Fertile Crescent of western Asia more than 5,000 years ago. An apparent precursor to urban life was development of agriculture at locations with dependable water and conducive climates. Settlements grew into cities around fortified centers of administration and commerce, which became the first political "states." Professional militaries were needed to protect cities and to mount offensives elsewhere. Metal refining advanced with discovery of ore smelting in Eurasia and mines became strategically important for equipping armies. Writing and records keeping appeared soon after the first cities. Specialized artisan services evolved from simple wood, clay, metal working, and textile weaving.

Highly organized bulk trade came about after permanent settlement was secured around stable agriculture. Trade generated wealth and wealth generated taxes for the rulers, who encouraged more trade. Overland bulk trade was difficult because of the rudimentary animal drawn carts that were used. Waterborne trade was much more efficient and had much to do with the development of the Phoenician, Greek, and Roman Empires along the shores of the Mediterranean.

Although marketplaces were common in ancient Mesopotamia, Greece, Rome, and other civilizations, the important attributes of a capitalist free-market system had not fully evolved in any of them. Heilbroner and Thurow (1994) identify several attributes of a capitalistic society, which the ancient empires failed to exhibit. None of them left evidence that *all* people had a legally protected right to private property.

The privileged classes alone were allotted that right. Also, there was no central market system independent of the whims of nobility or priesthood and no organized means for property exchange, money lending, or employment. The economies were traditional barter or command economies, dictated by the privileged few. Most peasants had to work when it was demanded of them.

European Feudalism.   The redistribution of titled land from privileged class to commoner appears to have first occurred in Europe after the feudal system disintegrated following the Renaissance (fourteenth to sixteenth centuries). From 700 A.D. to 1200 A.D., much of Europe was split into small feudal states, organized around warlords who sustained a stable balance of power among themselves, monarchies, and the Christian Church. This feudal system exploited the serfs, farmers, and artisans who had to pay land-use tribute to the lords, churches, and kings (Heilbroner and Thurow 1994). Serfs shared access to the lords' estates in areas known as "commons." Serfs often were prohibited from hunting, fishing, or cutting wood on much of the lords' estates, forcing overuse of the commons. This practice left a strong cultural distaste for the private ownership of fish and wildlife. It also was to later influence the way in which many local communities in the North American colonies managed their common-use areas.

Market economies began a slow emergence in Europe following the Christian Crusades to Palestine (Roberts 1993). Western Europeans were introduced to Middle Eastern goods such as silks, spices, and perfumes. Some of the crusaders remained at ports on the eastern Mediterranean where they bought and sold goods for shipping to Venice and ports farther west. A new merchant class in Europe began to accumulate wealth independent from the land. In the sixteenth century the bubonic plague and trade from the New World and Asia destabilized the feudal economy and began replacing it with a market economy (Heilbroner and Thurow 1994). The plague caused a shortage of labor, which favored more liberal treatment of serfs. Eventually, reduced productivity and lower revenues from serfs drove the feudal landlords into debt. Merchants often took land title from the impoverished landlords in lieu of money or barter.

European Market Economy and Industrial Revolution.   The merchants split the land into smaller parcels and sold it for profit to a growing number of middle class artisans and merchants. They also invested part of their profit in more innovative and larger operations, hiring displaced serfs with valuable skills. Literacy increased rapidly after invention of the printing press in 1440. As the size of businesses grew, competition grew among merchants, forcing more investment in size, innovation, and efficiency. Once personal income was tied to innovation, invention accelerated into an Industrial Revolution. The market economy in England had become established long before Adam Smith first described its functioning in his classic book, *The Wealth of Nations,* published in 1776.

The Colonial Economy.   The economy of the New World was elemental at first; based mostly in raw natural resources such as game meat, furs and hides, agriculture, fisheries, and timber. In this era, people were few and natural resource supplies seemed unlimited. Much of the economy was based on direct barter of goods produced from local natural resources. They used game animals, birds, and fish for food and clothing (fur and leather). Although the forests of eastern North America had to be cleared for farming, they provided an abundance of wood for homes, furniture, utensils, charcoal, and boats (Clepper 1966). In addition to wood, the forests provided nuts, maple syrup, fruit, soap ingredients, tar, turpentine, rosin, and other staples. These pursuits demanded only hand tools, foot travel, and work animals.

Economies within the colonies were local and often quite isolated when situated any distance from water transport. Roads were little more than wide trails, except for a few freight routes between major cities (Oliver 1956). Trade in the north de-

FIGURE 2.3 Gristmills for water-powered grinding of grain into flour, similar to this one in nineteenth-century Georgia, were common in late colonial America before independence (courtesy of National Park Service).

veloped around whaling and fisheries, saltworks for fish preservation, ship building (various forest products), rudimentary textile mills, granaries, and glass and metal industries. Development of mineral resources was slow, with some mining of copper and iron in New England. In the South, slavery became the basis for a labor intensive agriculture emphasizing exportable tobacco, rice, indigo (a black clothing dye), cotton, and sugar.

The main sources of power in the colonial New World were wind power for boats, animal power for tillage, and water power for millwork (Figure 2.3). Fire for kilns, furnaces, stoves, and fireplaces was fueled by ubiquitous wood, including charcoal. Pursuit of wealth often translated into ownership of lands suitable for logging, farming, and settlements. Men like Benjamin Franklin grew wealthy from speculative land purchase and profitable resale (Freidenberg 1992).

Numerous issues precipitated the American Revolution in 1776. Most important was a desire by the colonists to be the masters of their own resources at home and in international market places. The British government was bureaucratic and slow to respond to citizen needs in the colonies. It resisted demands for sale of inland private lands to replace "worn out" lands. Colonists resented "taxation without representation" on goods entering and leaving the colonies (Jenkins 1997). The English aristocracy viewed the colonies as a huge source of natural resources and trade with outpost civilizations too fragile for independent governance.

Policy

Democracy, Land Ownership, and Resource Regulation.   The earliest European settlers of the New World confronted a vast wilderness that provided them with both opportunities and challenges. Successful settlement, survival, and development of sustained civilization were driving concerns. A puritanical skepticism of secular pleasure and aesthetics was widespread. Except to avoid local depletion, resource protection was not a prominent concern (Kline 1997). Regulation of resource use was a controversial issue because preservation in Europe had favored the privileged few who had inherited their wealth. This wariness extended to lands and waters used in common, such as local fisheries, hunting grounds, sources of fuelwood, and irrigation water. Elected town leadership and democratic town meetings became the basis for deciding how local natural resources would be used.

In response to local resource depletion, early in the colonial period local townships established ordinances restricting fuelwood, fish, and game harvest. Other restrictions were ordered by the English Crown to reserve the best of certain resources. As early as 1691, the charter establishing the colonial province of Massachusetts contained a provision for reserving certain "royal trees" for the crown (Clepper 1966). These trees, the largest white pines, were valuable for shipbuilding. Such laws made sense to the colonies because the ships built from the pines carried trade between North America and Europe.

Other attempts to conserve the future timber supply in colonial America involved municipal penalties for unauthorized tree cutting, enactment of laws against setting forest fires, and stipulation that for every five acres cleared, one acre must be left in forest. Wild fish and game were assigned to common ownership early in colonial history, in contrast to European consignment to private control. This public behavior came in response to the widely despised aristocratic privilege in Europe. Most ordinances remained a local discretion. Except for a few edicts from England, there was no regional regulation of resource use until long after national Independence in 1783.

Town governments were responsible for trails, roads, navigation channels, public education, law enforcement, court proceedings, common defense, and common response to disasters from fire, storm, flood, and disease. Common-use areas, or "commons," were established by consistent public use of navigable waters, footpaths, highways, grazing areas, fuelwood, recreational areas, and meeting areas. These local government actions were to become the foundation for the public natural resource policy now serving the U.S.

WESTWARD EXPANSION (1776–1860)

Ecology and Landscape

The Nation in 1776 was mostly agrarian with industry centered in coastal river cities and towns where water power and boat transport were available. Farming required a large amount of cultivated land. The colonial farmers had to be nearly self-sufficient (Clepper 1966). They raised a wide variety of crops that might include two or more grains, vegetables, fruits, and nuts. Large numbers of livestock were kept for pork, wool, poultry, eggs, and leather. Several horses and oxen were needed for transportation and to pull farm implements. Lands too poor for cultivation usually were maintained as pasture. By 1800 only small islands of virgin timber remained east of the Appalachians. These virgin lands typically were on mountaintops and in swamps. By 1825 nearly all of the Great Smokey Mountains had been cleared for grazing or farming. Cultivation was even more intensive in the South under cotton and tobacco culture. Many farmers gave up because the land they cleared could not sustain a family. They left for better lands farther west or industrial employment along the coast. These abandoned lands were typically in rugged terrain or areas prone to drought, flooding, fire, or erosion. Abandoned lands began ecological succession into second-growth forests, some of which are now over two centuries old.

Significant Events During Western Expansion

1768	A series of treaties with American Indian tribes opens lands west of the Appalachians for settlement and land speculation.
1769	Steam engine invented.
1776	Adam Smith publishes *The Wealth of Nations* and initiates the theory of market economics.
1784	Russians establish permanent settlement in Alaska and Spain closes lower Mississippi to American navigation.
1785	Land ordinance legislation establishes township basis for public domain allocation.
1787	The U.S. Constitution is ratified and the first steamboats are operated on the Delaware and Potomac Rivers.
1796	Federal land disposal policy is established emphasizing privatization of public domain at low prices.
1803	The Louisiana Purchase opens the Mississippi to commerce and is followed by the Lewis and Clark expedition to the West Coast.
1806	The national road from Maryland to Illinois is initiated and completed 22 years later.
1824	Engineers of the War Department are authorized to maintain major waterways for navigation establishing barge canals.
1825	The Erie Canal is finished, linking the Atlantic with the Great Lakes. Steam locomotive is invented.
1830	The railroads are rapidly extending westward from eastern states.
1833	Economic depression encourages westward movement.
1836	The first wagon trains cross what will become the Oregon and Santa Fe Trails.
1844	Telegraph invented.
1845	Iron bridges were invented and greatly advance highways and railroads.
1846	Sewing machine invented.
1848	After the Mexican War, acquisitions from Spain in 1848 and 1853 complete what will become the 48 contiguous states of the U.S.
1849	The California gold rush is first of many western mining "rushes."
1850	Herbert organizes the New York Game Protection Association.
1859	First major oil extraction and development of the Bessemer process for steel production.

Overland travel was arduous and long, requiring several days for what now takes a few hours on interstate highways. Waterways were the most reliable and cheapest travel routes. People concentrated along them as they moved across the Appalachians and down the Ohio River valley.

In 1800 wilderness dominated landscapes west of the Appalachian Mountains. Various accounts in the 1700s and early 1800s marveled at the abundant wildlife, fishery, grassland, and woodland resources in the present states of Ohio, Kentucky, Tennessee, Indiana, Illinois, and Missouri (Clepper 1966). Small outposts of European civilization occurred in the Ohio and Mississippi river valleys and in what is now Florida, Louisiana, New Mexico, and California.

FIGURE 2.4   During the first few decades of the nineteenth century, many miles of canal were constructed for mule-towed freight barges. As railroads developed, interest in developing more canals waned (courtesy of Railroads Memorial Museum, Altoona, PA).

Economics

The invention of the steam engine by James Watt in 1769 was a critical aspect of the industrial age. It drastically improved water and land transportation (steamboats, trains) starting in the 1820s (Oliver 1956). Labor saving innovations resulted in many job losses in eastern industrial centers from 1830 through 1850 (Hibbard 1924). Deplorable working conditions caused many laborers to quit and move back to the land once they learned that fertile Midwestern farmland could be acquired from the federal government at a cheap price. The discovery of gold in California in 1849 was the first of many important mineral discoveries that encouraged people with poor labor and farm prospects to go westward.

The fur trade was a tremendous incentive for early western expansion into the vast forests of the upper Midwest and the mountainous far West. By the Civil War (1861), beaver and many other furbearers were growing scarce.

Barge canals (Figure 2.4) were dug to link the Great Lakes, Mississippi tributaries, and the Atlantic Ocean. They made Buffalo and Pittsburgh pivotally important trade centers, followed by Chicago, Cleveland, Detroit, and Saint Louis. Most canals were dug between 1817 and 1840, when further development was eclipsed by railroad development.

Following the expedition of Lewis and Clark in 1803–1805, a chronological pattern of settlement occurred during westward expansion (Clepper 1966). In advance of transportation improvements, fur trappers entered the wilderness either before it was acquired from other governments or soon thereafter. Trading posts were established, linked by improved trails and then roads. Once transportation links were established, the U.S. government sent soldiers to protect commerce. Next came the farmers, who created an agricultural bonanza first in the forest openings and then in the flat prairie soils of the central Great Plains. Next physicians, merchants, blacksmiths, and other service providers moved to the posts and forts, forming nuclei for small communities.

Flat boats, poled upstream and floated downstream, opened limited commerce into the wilderness along the main rivers. By the late 1820s, paddle-wheel steamboats churned the Mississippi, Ohio, and Missouri Rivers (Figure 2.5) (Oliver 1956). After 1830, railroads were rapidly established through much of the East up to the edge of the Great Plains. Chicago and Saint Louis quickly grew to large cities based

FIGURE 2.5 By the Civil War many steamboats used inland waters; especially the Mississippi and Ohio Rivers (courtesy of U.S. National Park Service).

on resource processing and commerce. Highways remained poor except for a few good toll roads in the East. Deep rivers remained major barriers to cross-country travel. During the 1840s and 1850s, many wagon train expeditions were organized to follow the Oregon and Santa Fe Trails westward to new settlements (Figure 2.6).

During the 1820 to 1860 period, numerous new inventions improved manufacturing and home life. The process of steel production was greatly improved, which was critical to expansion of the railroads. Invention of the suspension bridge in 1848 greatly enhanced railroad and highway construction and maintenance. Sewing machines, photography, vulcanized rubber, improved looms, and canned goods were invented. The petroleum industry got its start in 1859 at Titusville, Pennsylvania, replacing whale oil for lamps and other uses (Figure 2.7).

Policy

Challenges For The New Nation. After the War of Independence, national security, credibility, and solvency were federal preoccupations. Claims to the continent were split among several nations, often at war. There was concern that British and Spanish territories might attract trade from Midwestern settlers if the states were not quickly linked to Midwest territories. Public land was plentiful as the nation grew in size (Figure 2.8). The primary motivation for the Louisiana Purchase from France in 1803 was control of the Mississippi and its resources (Jenkins 1997). Subsequent exploration by Lewis and Clark in 1804–1805 revealed immense resources as an added bonus. Additional land purchases and treaties with Spain, Great Britain, and Russia continued until the nation's geography was completed with the purchase of Alaska

FIGURE 2.6   A bone-weary family rests on the prairie along a wagon trail (courtesy of U.S. National Park Service).

FIGURE 2.7   An oil field developed in Pennsylvania during the Civil War (courtesy of Drake Well Museum).

in 1867. Most of the land acquired by the United States was owned and administered by the federal government until there was enough settlement to grant statehood. Because of the high demand for private land, sale of public domain was an attractive alternative to taxes for raising federal revenues (Hibbard 1926).

The land acquisition and disposal policy of the early federal government was critical in determining how natural resources were developed and managed. Federal land policy was guided by a strategy that (1) improved national security along U.S. borders, (2) gained revenue to pay back loans, and (3) responded to public demands for more private land. Congress passed land disposal legislation in 1796 and 1800 in accordance with that strategy. The resulting policy, charging a low price for public

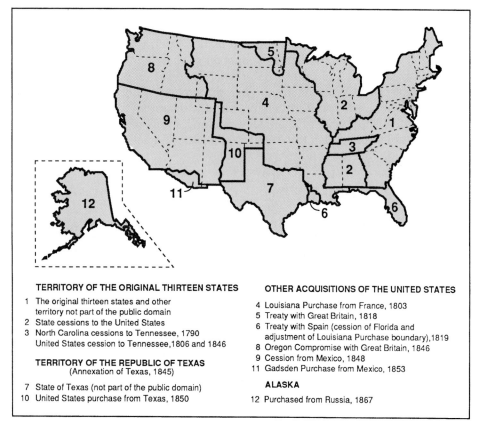

FIGURE 2.8 The most basic of natural resources—land—was added to the public domain of the U.S. over a period of 77 years as the nation expanded westward (courtesy of U.S. Dept. of Interior-Bureau of Land Management).

domain land, was a compromise between eastern establishment demand for federal revenue and western settler demand for free land (Hibbard 1924). Because of limited federal funds, management basically involved military exploration and general survey of new lands as they were acquired.

The Early Environmental Movement. There was growing awareness about soil degradation in Europe by the eighteenth century, but few farmers knew anything about conservation practices. George Washington was well aware of the problem (Oliver 1956). When Thomas Jefferson returned to his Virginia farm in 1794 after several years in Europe, he was appalled by the soil erosion. He immediately applied a program of crop rotation, use of legumes, use of manure fertilizer, and contour plowing to rebuild his fields. Jefferson was an amateur scientist and a strong advocate for agricultural education in colleges. However, his ideas on agricultural colleges and educational extension did not become a reality until after the Civil War.

As the Industrial Revolution progressed in Europe and the eastern U.S., a new attitude toward wild nature began to emerge among the artists and intellectuals of cultured society (Kline 1997). It was closely aligned with the ideals behind democratic government. In this intellectual Age of Enlightenment, the scientific, aesthetic, and spiritual appeal of wild nature grew as it became tamed and less threatening. Led by George Catlin, Ralph Waldo Emerson, Henry Thoreau, and Thomas Cole, romantic art and transcendentalist essays became very popular among East Coast intellectuals. It contributed to the growth of early tourism, sport hunting and fishing, and suburban development in areas bordering East Coast cities. A growing network of railroads, horse-drawn trolleys, and steamboats provided the means for tourism. Catlin and Thoreau made the early public cases for national nature preserves (Reiger 1975), which were to first materialize in the establishment of Yellowstone National Park.

Early complaints about diminished hunting and fishing sport were voiced well before the Civil War (Reiger 1975). Henry William Herbert, a spokesman for sportsmen in New York, complained about unlimited market hunting of wildlife. He expressed the view that game animals had greater value for sport than meat. Just before the Civil War, he organized the first of numerous private conservation organizations, the New York State Game Protective Association. Their primary goals were to curtail market hunting, enforce protective fish and game laws, eliminate barriers to spawning fish movement, and to replenish depleted fisheries by stocking. Many of them became fish culturists who would later establish the American Fisheries Society.

THE GILDED AGE (1861–1899)

Ecology and Landscape

Ninety years of population expansion and settlement following the War of Independence had greatly modified the U.S. landscape by the onset of the Civil War in 1861. A growing network of highways, railroads, canals, and natural waterways connected much of the continent outside the mountainous West, the high plains, and "North woods." People were most concentrated in the river valleys and coastal plains because they provided rich farmland and water for power, transportation, and waste removal. Settlement remained sparse over most of the interior. However, trappers and hunters had reached far into the wilderness and taken a high toll of the wildlife resources near waterways, roads, trails, and settlements.

Much of the deciduous forest lands of the western Appalachian and Great Lakes states had been cleared and converted to farmland. Fuelwood supplies were becoming scarce enough to encourage transition to coal use. By 1870 coal mining had begun to influence local landscapes in Pennsylvania. The pine forests of New England were nearly logged out, but large tracts remained in upstate Maine, New York, Pennsylvania, and the upper Great Lakes region. Many of the cotton and tobacco lands of the Southeast had been abandoned because of nutrient depletion and erosion (Figure 2.9). The forests and arid desert of the mountainous West remained mostly wilderness. Only small areas of timber around coastal cities had been cut on the West Coast, but logging accelerated following the discovery of gold in California. Damage from

FIGURE 2.9 Erosion became a chronic and widespread problem. After the Civil War much of the land in the South was farmed by tenants who had little incentive for soil conservation (courtesy of U.S. National Park Service).

natural flooding aggravated by watershed deterioration began to increase as human populations concentrated in flood-prone areas.

White-tailed deer and turkey had become scarce in settled areas and several species of anadromous fish were becoming rare. Beaver had been trapped out of most of its range south of Canada (Bolen and Robinson 1995). With the demise of beaver, millions of beaver ponds disappeared, reducing the number of wetlands and altering the hydrology of watersheds.

Significant Events in the Gilded Age

1861	Onset of the Civil War and massive mobilization and destruction of U.S. resources.
1862	The Morrill Act establishes Land Grant Colleges that lead in natural resources education.
1862	The Homestead Act gives 160 acres with stipulations it must be farmed.
1862	The Transcontinental Railroad Act is passed. It establishes mass transit across the U.S.
1864	George P. Marsh publishes first major soil and water conservation work *Man in Nature.*
1865	The Civil War comes to a close.
1867	Alaska is purchased from Russia virtually completing U.S. domain.
1868	Completion of the Transcontinental Railroad.
1870	The American Fish Culturists Association becomes the first conservation-oriented professional organization and predecessor of the American Fisheries Society.
1871	The U.S. Bureau of Fisheries is formed establishing a federal conservation mission.
1872	Yellowstone National Park is established by an Act of Congress.
1872	The General Mining Law of 1872 encourages public domain mine development.
1876	Telephone invented.
1879	The internal combustion engine and the electric light invented.
1883	The Civil Service Act establishes technical competency requirement for government employees.
1886	The National Audubon Society founded under leadership of sportsman-editor, George Bird Grinnell, two years before the Sierra Club is formed by John Muir.
1886	New York is the first city electrified from a central power plant.
1890	The Sherman Antitrust Act becomes basis for later trust-busting under Theodore Roosevelt.
1891	The Forest Reserves Act authorizes National Forests.
1898	Smith publishes first scientific account of widespread overgrazing on western rangelands.

Economics

Success in the Civil War clearly depended on technological advantage. Railroads, iron-clad ships, telegraph, photography, various weapons, preserved foods, and other inventions were further developed to win the war. The Civil War laid the foundation for the most inventive period in history (1870s). Rail and highway transportation difficulties initiated a larger role for the federal government in regulating commerce after the war.

FIGURE 2.10 During the Gilded Age, railroad magnates held great wealth and political influence as illustrated here by the cartoonist Thomas Nast in *Harper's Weekly* (July 10, 1886) (courtesy of Library of Congress).

The Gilded Age of the 1870s grew out of technological innovation and aggressive resource development. Many of the technologies so important to us today were first developed including electric dynamos, internal combustion engines, refrigeration, automated looms, automated food canning, and telephone communication. Kerosene lighting was soon replaced by electric lighting generated at central power stations and distributed by electric transmission lines. This occurred first in New York City in 1882, but in a few years areas as remote as Montana, Idaho, and Utah had electricity (Oliver 1956).

Big business propelled the nation to world economic leadership by 1900, but at an alarming cost to natural and human resources. Powerful business leaders gained notoriety as "the robber barons" because of their ruthless tactics used to gain business advantage. Government regulation designed to promote fair competition was virtually nonexistent. The press was emerging as an important force in American politics. Cartoonists, such as Thomas Nast, depicted the hold the railroad industry had over the U.S. Senate (Figure 2.10). The most successful businessmen ruthlessly eliminated their competition, particularly in the steel, oil, railroad, and finance industries. They formed monopolizing "trusts" that avoided state regulations and often controlled resource extraction, transportation, and refinement (Jenkins 1997). They bribed and otherwise influenced many government officials. Although economic development substantially increased total national wealth, it was distributed dispro-portionately in favor of the few business empires.

As the frontier "closed" in the 1890s, the prevailing national sense of unlimited resources was reexamined. In this period, westward migrants were generally disappointed by the marginal productivity of the semiarid lands they homesteaded. Therefore, the closing of the frontier in the 1890s restricted one major recourse (homesteading) to unfavorable industrial conditions. Left with poor homestead alternatives, the closing of the frontier helped stimulate political movements for improved employment conditions in industry.

Both in Europe and the U.S., employees were almost completely at the mercy of employers (Jenkins 1997). Children often started work at age 10. Pay was meager. Ten-hour days (or more) and six days a week, without break or vacation, was the typical regimen. There were no work benefits, such as unemployment insurance or compensation for injury on the job, and working conditions often were hazardous. In this environment it was not surprising that a labor movement would grow and the downside of unregulated capitalism would be questioned. Karl Marx wrote about the numerous social problems caused by unregulated capitalism (see Chapter 21). He advocated much more government regulation in favor of the "working class" (Heilbroner and Thurow 1994). Growing "threats" of social revolution were influential in determining the "progressive" course of the U.S. federal government during the early twentieth century. It also set the stage for the rise of communism in Eurasia and the associated international stress that has dominated much of twentieth-century politics.

By the end of the nineteenth century, the entire country was served by a network of linked highways, waterways, and railways designed to move mineral and agricultural resources from privately claimed lands to markets all over the country and to international ports. Railroads throughout the nation's interior were used extensively to transport minerals, grain, livestock, timber, game, and manufactured products. Steam power had almost totally replaced sail on the Great Lakes, major rivers, coasts, and oceans. Urban electrification was escalating. The first important conversion of hydropower to electricity was completed in 1895 at Niagara Falls (Oliver 1956).

Logging in the Great Lakes and northwestern states accelerated as railroads made transport of building material from sawmill to destinations across the country much cheaper. Wildfires devastated many areas because combustible timber wastes were routinely left to catch fire. Destructive flooding increased downstream from exploited watersheds. The Johnstown flood in western Pennsylvania was a particularly deadly rallying event for the growing forest conservation movement.

Following the Civil War until the late 1880s, livestock use of western rangeland escalated in response to eastern market demands for meat and improved rail transport. The livestock industry rapidly developed in the 1870s as a result of East Coast and European investment (Holechek et al. 1998). The severe winter of 1885–1886 killed many cattle in the Northern Great Plains. Severe drought during 1891–1892 resulted in heavy losses from starvation in the Southwest. These events depressed investment interest, devastated livestock producers, and substantially degraded rangeland productivity (Holechek et al. 1998).

Demand for game meat was high and market hunting flourished despite growing criticism (Clepper 1966). The railroads made it possible to rapidly ship large quantities of game animals, hides, and even bison bones to industrial centers. Market hunters almost completely eliminated bison between 1870 and 1876. Prairie chickens, ducks, geese, cranes, plovers, and pigeons were shipped by the trainload from the prairies to processing centers in Chicago and Saint Louis (Figures 2.11 and 2.12). Despite new protective state laws, poaching continued to contribute to reduced densities of whitetail deer, turkey, waterfowl, plumed birds, passenger pigeons, trout, and other fish in eastern forests, wetlands, and streams. The exploitation continued mostly because of insufficient education and law enforcement. By 1900, even the farmland game species of the southern and Midwestern states were beginning to decline from overhunting. Elk, bighorn sheep, and pronghorn antelope were scarce or

FIGURE 2.11 Sale of game animals at the Fulton Market, New York City, 1978 (courtesy of Johnson 1964 and North Dakota Game and Fish Department).

FIGURE 2.12 Mixed bag of small game taken by two hunters near Stump Lake, North Dakota in the 1890s (courtesy of Johnson 1964 and North Dakota Game and Fish Department). By the late nineteenth century, most states had enacted protective game laws or were considering them to prevent overhunting.

eliminated throughout much of their range. In many locations, household consumption contributed more to decline than market hunting.

Policy

Public Land Disposal.   Territorial lands and unclaimed federal lands remaining within state boundaries were opened to free settlement with the passage of the Homestead Act of 1862. This allowed a settler to obtain 160 acres after farming the land for five years or for $1.25 an acre after six months of residence. It resulted in rapid transfer of public lands in the Great Plains to private ownership.

Many early settlers tried to farm dry lands better suited to grazing. The land grants were too small to provide the minimum grazing land needed for a family ranch. Inappropriate land grants led to many conflicts between farmers and ranchers, which peaked in the 1890s. There was much fraudulent accumulation of grazing land and frequent violence. By 1900 only public domain lands of low productivity remained unsettled. Much land in the western states had been degraded by overgrazing and unsuccessful cultivation.

The federal government adopted the agricultural education philosophy of Thomas Jefferson with passage of the Morrill Act in 1862. It established a system of Land Grant Colleges on public domain lands provided to each state or territory. The colleges were to have a technical emphasis in agriculture and engineering. These institutions were to play a successful role in training extension specialists, research scientists, and future farmers and ranchers.

Influential Early Environmentalists.   As exploitation escalated, a small group of far-sighted individuals emerged to eventually change American attitudes and government policies toward renewable resources. They were a mix of artists, literary figures, magazine editors, wealthy sportsmen, scientists, and concerned politicians who built on the philosophy of earlier romantics and concerned sportsmen. Their interests in improved management of natural environments were based on utilitarian, recreational, spiritual, and esthetic values (Clepper 1966). George Perkins Marsh (Figure 2.13) published a book that called attention to degradation of soil, vegetation, and watersheds in many locations of the world (Marsh 1864). He established many of the basic principles of soil and watershed conservation (Clepper 1966). Marsh also described how managed cutting, followed by replanting, would save soil, water, and timber resources. The predecessor of the American Fisheries Society, the American Fish Culturists Association, was formed in 1870 with a strong conservation platform. In 1871, the U.S. Bureau of Fisheries was formed to assess food fisheries status and to develop conservation approaches. They emphasized fry stocking, building fish passages at dams, and outlawing market fishing in certain waters (Scott 1875).

Artists were very influential in shaping public attitudes (Kline 1997). Albert Bierstadt, Frederick Church, Thomas Moran, and numerous others captured the wilderness beauty in their art (Novak 1995). Exhibits of these paintings attracted thousands of influential viewers and sold for high prices. Early photographers, such as William Jackson, also began to chronicle wilderness aesthetics in black and white (Figure 2.14). These artists saw great beauty in wild nature and communicated directly to the influential educated class (Novak 1995).

The writers who influenced change in renewable natural resource management were of two philosophical schools: those who espoused the conservation of resources and those who advocated preservation of nature. Outdoor-sport writers, led by George Bird Grinnell, educated the public on the need for sound conservation policies (Reiger 1975). Grinnell also was largely instrumental in establishing the National Audubon Society in 1886. In the West, John Muir led an effort to protect wilderness in National Parks from logging and dam building. Muir's followers promoted a zoned approach to resource use with liberal preservation of wilderness in parks and other reserves. Muir was instrumental in the establishment of Yosemite Park in 1890 (Figure 2.15) and he was a founder of the Sierra Club (Kline 1997).

GEORGE P. MARCH
1801-1882
Interpreter of Man and Nature

JOHN MUIR
1828-1914
Preservationist of Natural Resources

PRESIDENT THEODORE ROOSEVELT
1858-1919
"The Conservation President"

ALDO LEOPOLD
1886-1948
Scientific Wildlife Management

FIGURE 2.13   Important men in the early natural resource conservation movement (courtesy of U.S. Department of Agriculture and University of Wisconsin).

FIGURE 2.14 Photograph of Yellowstone Falls taken by William Jackson, a famous early photographer of western landmarks who influenced public land policy (courtesy of Library of Congress).

FIGURE 2.15 Early photograph of Yosemite Valley, which gained much public support for the National Park Service. The efforts of John Muir, who founded the Sierra Club and initiated a public land preservation movement, played a key role in the establishment of National Parks and National Park Service (courtesy of U.S. National Park Service).

Forest Preservation.   By the early 1870s rapid depletion of timber began to raise serious concerns (Reiger 1975, Williams 1989, Kline 1997). In 1873 the American Association for the Advancement of Science formally asked the U.S. Congress to pass laws for forest protection. In 1875, the American Forestry Association was formed. Congress created the Federal Division of Forestry in 1881 in the Department of Interior, but it was later transferred to the Department of Agriculture. In 1882, Massachusetts authorized acquisition of forests near major cities to protect water supplies. Similar actions soon followed in other states along the Atlantic seaboard. Watershed protection grew in the wake of increasingly disastrous floods in logged areas. New York set aside the Adirondack and Catskill Reserves in 1888. Pennsylvania began to set aside lands in the heavily logged upper Susquehanna watershed. The first serious commitment of the federal government to forest resource conservation was shown with passage of the Forest Reserves Act of 1891. This act allowed the president to set aside specific areas of public land, which were to become the National Forests. The act was strengthened in 1897 when specific provisions relating to watershed protection were attached to it.

Mineral and Water Resources.   The General Mining Law of 1872 first established a federal policy supporting development of mineral resources on public lands. It made suitable public lands available for mineral exploration and purchase ($2.50 to $5.00 per acre), if mined. It permitted development without any royalty or other compensation to the U.S. The law continues to the present time without significant modification. It is controversial because it subsidizes development that is often environmentally damaging.

Increasing demand for inland and coastal navigational improvements resulted in establishment of the River and Harbor Act of 1875. This act, and subsequent updates, authorized the Corps of Engineers to initiate and maintain many projects aiding navigation, including channel dredging and various kinds of engineered structures. Following exploration of Southwestern water resources by John Wesley Powell, the U.S. Geologic Survey was established in the Department of Interior in 1879. Its primary responsibilities were to map public lands, classify land for appropriate use, and monitor water resources. This established a basis for federal water supply management. In 1899, the Corps of Engineers was authorized to regulate modification of navigable waters, including dumping of wastes into them.

Parks and Recreation.   As America grew more affluent, wilderness and natural wonders acquired significant recreational value. Tourism boomed on the wheels of a rapidly growing railroad network. Untouched wilderness remained only in the most rugged, remote, and harsh terrain. Some of the first successful government preservation efforts were directed toward the more awe inspiring areas, including the upper Yellowstone in Wyoming, the Hot Springs of Arkansas, Niagara Falls in New York, Mackinac Island in Michigan, the Adirondack Preserve in New York, and Yosemite Valley in California. The national parks were first established with the Yellowstone Act of 1872, which set aside 2.25 million acres (Yellowstone National Park) to be kept in natural condition as "a pleasuring ground" for the people of the United States.

THE PROGRESSIVE PERIOD (1900–1945) _____

Ecology and Landscape

Tremendous ecological transformation occurred in the U.S. during the nineteenth century. By the end of the century, two-thirds of the continent was settled and substantially developed wherever water and other necessities were consistently available. A network of water and rail based transportation linked agriculture, logging, and mining across the nation to manufacturing centers on the nation's large rivers and coasts. Native Americans were allocated reservation lands amounting to a small fraction of what they once roamed.

By 1900, most of the natural landscape of the eastern and central U.S. had been altered and often degraded by logging and farming. The vast forests and prairies of the eastern and central U.S. had been placed under cultivation (Williams 1989). Large areas of northern pine forests had burned following wasteful logging practices. Overgrazing was rampant in the high plains and western mountain ranges. Inappropriate farming in the semiarid plains led to massive soil erosion and farm community failure during the "dust bowl" drought of the early 1930s. The Northwestern forest had been extensively cut at lower elevations and much of it converted to farms. Much of the cotton and tobacco farmland of the South was badly eroded and had been abandoned because of degraded fertility and gully erosion.

The remaining undeveloped natural areas in the U.S. were either too rugged, too dry, or too wet to develop. Much fertile arid land and wetland near river floodplains remained undeveloped because of the inability of the States to muster the resources to provide irrigation or drainage. Flooding was becoming more of a problem as clearing and agriculture altered the watershed capacity for slowing runoff rates and as more people settled the floodplains. Levees had been built on some of the major river systems to improve navigation and to help manage flooding.

The dramatic change in vegetation, human numbers, and hunting and fishing pressure was associated with widespread decline of certain wildlife and fish species. Elk, pronghorn, and turkey had been decimated and bison were nearly extirpated. Passenger pigeons were about to go extinct. Whereas natural forest and grassland species declined, species adapted to farmland conditions flourished, including bobwhite quail, cottontail rabbits, redwing blackbirds, fox squirrels, and various rodents.

Exotic species were becoming common by accident and design. Urban areas and farmland were especially altered by planting of exotic crops, gardens, and trees. Many European plant species became naturalized. Some were to become pest species, such as dandelions, purple loosestrife, and water hyacinths. Salt cedar, Russian olive, Russian thistle (tumbleweed), and cheat grass invaded the arid West. Also from Europe, rock doves (common pigeons), house sparrows, starlings, house mice, and Norway rats became ubiquitous members of urban and farm wildlife. Foresters, first educated in Europe, had begun to import "more manageable" European species for development of large plantations in the Northeast. Fisheries professionals stocked European brown trout and carp using hatchery methods from Europe. The extensive development of canals in the nineteenth century allowed exchanges of previously isolated aquatic faunas. Construction of the Welland Canal around Niagara Falls contributed to overfishing and invasion of the upper Great Lakes by sea lamprey and alewives.

Economics

Technology. The progressive movement led the nation into a new era based on high-minded social goals, scientific innovation, and more regulation by federal and state governments. Human welfare continued to improve during this period and wealth was more equally distributed throughout society as labor organized and as the federal government increasingly regulated big business. With help from the Civil Service Act of 1883, which required appropriate technical expertise for government employees, the federal government became more actively involved in natural resource development and management, especially of forest, range, and water resources. The completion of the Panama Canal in 1909 was an unprecedented engineering accomplishment of immense benefit. It became a model for many other large government water resources projects and public works.

The Progressive Period

1901	Theodore Roosevelt becomes president of the U.S. and leads the "progressive movement" toward the "greatest good for the greatest number" through strong government.
1902	The Reclamation Act establishes federal role in developing water resources in arid West resulting in some of the largest reservoirs ever built.
1903	The Wright brothers invent motorized air travel.
1905	The first wildlife refuge authorized and first federal water pollution law regulates refuse dumping in navigable waters.
1905	The U.S. Forest Service formed under direction of Gifford Pinchot, who coins the word conservation and promotes "wise use" and sustainable development of resources.
1914	The First industrial refrigerators greatly improve food transport and storage and first permanent federal income tax provides unprecedented policy leverage.
1916	Migratory Bird Treaty Act establishes coordinated management of waterfowl and other birds with Canada.
1916	The Federal Highway Act of 1916 ushered in national highway planning and transcontinental "motoring."
1924	The first wilderness is set aside by the Forest Service and Aldo Leopold and Robert Marshall work toward founding of the Wilderness Society in 1935.
1929	Television is invented and the stock market crashes for complex reasons associated with liberal lending policies and overproduction of cars and appliances.
1933	Bad farming practice and the dust bowl drought hammer agricultural production and aggravate the nation's worst economic depression. The U.S. Soil Conservation Service is first authorized as the Soil Erosion Service.
1934	The Taylor Grazing Act creates the first government control over grazing on public domain lands.
1935	The term ecosystem is coined by Tansley, emphasizing links between biotic communities and abiotic environment.
1936	The Omnibus Flood Control Act establishes a nationwide federal leadership role in flood control and onset of dam construction and improved watershed management.
1940	The U.S. Fish and Wildlife Service is created in the Department of the Interior.

Technology advanced at a tremendous rate. Many improvements were made in metallurgy, mechanics, building materials, and farm practices leading to larger, stronger, more efficient, and safer ways to manage, extract, move, refine, and otherwise develop natural resources. Invention of the radio greatly expanded interstate communication. Advances based on the internal combustion engine, motorized flight, and electric generators lead to public pressure on government to assure that appropriate infrastructure was developed to facilitate interstate commerce.

Automobile manufacturers, led by Henry Ford, developed mass production. In 1908, the Model T Ford became the first car affordable to the middle class. Automobile ownership became as much a part of American aspiration as owning a home. Highways improved markedly with the paving of the U.S. highway system. The first hard-surface road crossed the continent in 1913. People began to "motor" across the

FIGURE 2.16   The decades around 1900 were among the most inventive in history. Thomas Edison was influential in establishing central electrical supply and incandescent lighting (courtesy of U.S. National Park Service).

country and fresh farm goods became widely available as a trucking industry burgeoned. Many new products were now ordered cost-effectively through national catalogs and shipped via rail and truck to local retailers or directly to the home.

Thanks to Thomas Edison, one of America's greatest inventors, electrification progressed rapidly at the turn of the century (Figure 2.16). Electrification powered the first household radios, electric stoves, washing machines, and industrial refrigerators just before World War I (1914–1918). Air conditioning was introduced to movie theaters in the late 1920s. Pop-up toasters, televisions, household refrigerators, stereo systems, garbage disposals, electric razors, blenders, tape recorders, and fluorescent lighting soon followed. The cities benefited from electrification first and the poorest rural regions were last. During the depression years (1930s), the federal government started the Tennessee Valley Authority and the Bonneville Power Authority to provide electricity by hydropower in depressed rural areas.

Numerous medical advances led to much improved understanding of diseases and how to treat them (the development of penicillin and insulin). Draining of wetlands reduced mosquito-borne diseases such as malaria. Public health organizations became more influential and improved waste disposal practices reduced typhoid, cholera, and other deadly diseases.

Depression and Conservation.   The Progressive period was a time of great economic change and stress. The key event was the 1930s' depression, which started with the stock market crash in 1929. The depression was brought on by a complex assortment of factors, but its root cause was overinvestment in production capacity and excessive use of credit (debt) (see Chapter 22). The depression coincided with severe drought in the Great Plains, which brought about massive farm failures. Economic growth came to a standstill in the early 1930s. Depression dominated national policy until declaration of war in 1941.

After the stock market crash in 1929, the economy spiraled downward as personal income shrank, trade declined, and international markets sank into economic depression. Based in part on theories of John Maynard Keynes, the federal government led by Franklin Roosevelt used government tax and spend policies for the joint solution of

FIGURE 2.17   Construction of a lock and dam structure on the upper Mississippi River in 1935, served both to stimulate work during the Great Depression and to improve water resources management (courtesy of U.S. Dept. of Interior).

environmental and economic problems (Heilbroner and Thurow 1994). Disastrous floods and soil erosion focused legislative attention on improved water and soil management. President Roosevelt promoted projects for irrigation, flood control, and hydropower development through reservoir construction. These projects provided much needed work (Figure 2.17). Roosevelt's administration believed careful construction could conserve living resources, such as migratory fishes, by proper engineering. He encouraged establishment of the Civilian Conservation Service and other federally sponsored work programs with joint social and environmental improvement goals.

Although the effectiveness of Roosevelt's "New Deal" prior to World War II has been questioned, the federal government would never step back to the passive role it had taken before 1900. World War II reinforced the role of federal government in the U.S. economy when it mobilized and guided the largest conversion of natural resources into war technology experienced up to that time. An economic triangle of big labor, big business, and big government was firmly established and interacted to determine the future course of natural resource development.

Policy

Progressive Emphases.   The main goal of the federal government during the Progressive movement before World War I was to encourage sustained use of natural resources and to assure that the benefits would be distributed equitably among the people. The conservation ideal was natural resource management for the greatest good of the greatest number over the long run. Because public lands were much easier to influence than private lands, the federal government shifted away from a policy of public land disposal to public land management.

The conservation movement of the Progressive Era was highly utilitarian, which fit with prevailing public sentiment. Progressive policy emphasized resource development that generated jobs, affordable housing, food, safety, and health. Outdoor recreation was considered among the important uses of public lands and waters. This was due to the growing economic value of tourism and general recognition that outdoor recreation was good for public health.

The movement promoting natural resources management had split into several camps by the time Theodore Roosevelt became president in 1901. One camp, led by

John Muir, favored preservation. He advocated zoned public land use, in which a large proportion of public lands would be preserved in a pristine state for low-intensity recreational use, epitomized in the National Parks. In contrast, Gifford Pinchot favored multiple resource uses on public lands, and believed all resources should be developed to their full potential. Pinchot's philosophy gained the greatest following within the progressive movement. However, Muir's philosophy was very influential in establishing a large national park system and, later, a network of wilderness areas on other public lands.

In the progressive view of conservation promoted by Pinchot, nature was the servant of human needs and resources were "wasted" if not used. Therefore, the concept of conservation included management of natural threats to the development of land and water resource use. An important weakness in this utilitarian conservation philosophy was poor understanding of the value of ecosystem biodiversity. This led to the endangerment of numerous species.

While the federal government took control of natural resource management on public lands, it also acquired a powerful tool for influencing private behavior—the income tax. Although temporarily enacted during the Civil War, the federal income tax did not become permanently established until 1914. Through tax law, Congress could redistribute wealth and encourage desired behaviors in the private sector. Income taxes provided a dependable source of revenue for financing a federal government of expanding size.

Water and Mineral Resources.   Transfer of public domain to State control under the Swamp Acts and Desert Land Acts had resulted in little development of those lands as required. The Reclamation Act of 1902 authorized federal construction of large dams and water delivery systems to irrigate arid, but otherwise fertile, desert floodplains. The Bureau of Reclamation originated under this act and continues to operate and maintain many irrigation projects in the western U.S.

Maintenance and development of improved navigational resources remained an important emphasis of the Corps of Engineers. Amendments to the Refuse Act of 1899 in 1905 authorized the Corps of Engineers to enforce prohibition of garbage and other solid waste dumping into navigable waters. This was the first significant federal antipollution legislation. The Rivers and Harbors Act of 1917 provided the Corps with the authority to conduct studies on flood control and related problems. It authorized flood control projects along the Mississippi and Sacramento-San Joaquin Rivers. The federal government assumed primary responsibility for flood control along those same rivers via the Flood Control Act of 1928.

Flood control authority was extended to all areas of the nation by the Omnibus Flood Control Act of 1936. It gave the secretary of agriculture authority to study watershed process and develop programs to retard runoff and soil erosion through the newly created "Soil Conservation Service." The Corps was authorized to build many large dams and levees for flood control and navigation purposes. This law was the first to require gross benefits of government projects to exceed costs.

Wetland drainage and filling accelerated early in the century in the name of progressive land conservation. Wetland mismanagement was symptomatic of the general ecological ignorance and expediency that pervaded government decision process. Except for waterfowl and certain fisheries, wetlands were thought of as wastelands because they impeded land resource development. There was little appreciation of the human benefits they provided, such as habitat for waterfowl, groundwater maintenance, and water quality maintenance. Their filling rate accelerated when mosquitoes were discovered to cause malaria and yellow fever. Many wetlands were channelized and filled with dredge material as the Corps of Engineers pursued its mission of improved navigation in coastal and riverine environments. Numerous wetlands were drained under government programs that served agricultural land users. Federal wetland management policy grew more controversial as more was learned about the beneficial roles of wetlands in supporting fish and wildlife habitat.

The Mineral Leasing Act of 1920 initiated the first significant change in Federal mineral policy since the General Mining Law of 1872. It placed fossil fuels and fertilizer minerals under a leasing system that provided monetary compensation to the U.S. It required the land to be returned to U.S. control after mining was completed. This law provided a basis for the later expansion of government and enforced reclamation of mined lands.

Forest and Park Resources.   By the early 1900s the value of forests for wood, watershed protection, and outdoor recreation was well recognized. Part of the solution to flooding was linked to reforestation and other watershed improvement. Federal forest management was strengthened in 1905 by formation of the U.S. Forest Service headed by Gifford Pinchot. He was an articulate politician who had a considerable influence on Theodore Roosevelt and the conservation movement.

Most of the large national parks were established during the Progressive movement. They were formally brought together in one coordinated management system with formation of the National Park Service in 1916. The National Park Service was also authorized to manage the national monuments, which had been established under the Antiquities Act of 1906. It permitted preservation of lands with unique scientific or historic value through Presidential proclamation. President Theodore Roosevelt, an ardent outdoor recreationist and conservationist, aggressively set aside lands for national forests, parks, and monuments.

Personnel within the Forest Service were among the first advocates of wilderness protection. Aldo Leopold, a Forest Service employee, was instrumental in establishing the first Forest Service wilderness in 1924. He and another Forest Service employee, Robert Marshall, were founding members of the Wilderness Society in 1935. Marshall was a forester, explorer, and writer who played a critical role in strengthening the Sierra Club and National Parks Association as political forces in the management of public lands.

Range and Soil Resources.   Government intervention into grazing problems on the western range began in 1898 when the Department of Interior granted grazing permits to limit the number of livestock on federal lands. Even so, much of the public rangelands continued to degrade (Smith 1898, Stoddard et al. 1975). After 1905, the U.S. Forest Service established a process of forage allotment on national forests. Because of high demand for beef during World War I, severe overgrazing took place between 1915 and 1920.

During the early 1930s' drought, the United States reaped the consequences of 60 years of range exploitation. Vast areas of the Great Plains had been plowed to take advantage of temporarily favorable climatic conditions and high grain prices. After the drought, millions of acres of plowed land returned to grazing use. The drought also precipitated passage of The Taylor Grazing Act of 1934. This act placed administration of remaining public domain land under the Grazing Service, which later became the Bureau of Land Management. It resulted in allocation of grazing privileges on the public domain lands that had not been transferred to private holdings or to agency management authority. Prior to the Taylor Grazing Act, there was no control over how public domain lands unassigned to the Forest Service and National Park Service were to be used. If one person did not exploit the forage in an area, his or her neighbor probably would. This provides an excellent example of what Hardin (1968) later called "the tragedy of the commons" (Figure 2.18).

Until the 1930s, government policies continued to be oriented toward converting the remaining public domain land into private ownership under pressure from western states. The failure to recognize that large areas of the West could not sustain farming resulted in great land damage, social upheaval, and economic loss. The Enlarged Homestead Act of 1909 was passed to encourage settlement of the westernmost states. It increased homestead size to 320 acres, with the requirement that one-fourth be cultivated. This amount of land was insufficient for ranching and encouraged plowing of

FIGURE 2.18   "Tragedy of the commons" (from Holechek et al. 1998) (drawing by John N. Smith).

rangeland that would not sustain cultivation. The Stockraising Homestead Act of 1916 granted 640 acres for raising 50 cows. In most areas, this act caused rangeland destruction because 640 acres would not provide sufficient forage for 50 cows.

In 1928 Dr. Hugh Hammond Bennet wrote the first U.S. Department of Agriculture Bulletin on soil erosion entitled "Soil Erosion, a National Menace." The severe drought on the Great Plains confirmed Bennet's warnings about impending disaster from soil erosion. Dust was blown thousands of miles, damaging buildings, human health, and economic activity. In 1933 Congress created the Soil Erosion Service in the Department of Interior, which in 1935 became the Soil Conservation Service in the Department of Agriculture.

Fish and Wildlife.   In 1900, unregulated market hunting was brought to closure after decades of massive slaughter of shore and wading birds to provide feathers for women's hats. Public outrage resulted in passage of the Lacey Act of 1900, which made interstate shipment of game killed in violation of state law also a federal offense (Trefethen 1966). In the same year, the Protection of Migratory Game and Insectivorous Birds Act was passed. It was followed in 1916 by the Migratory Bird Treaty Act. The two laws placed all migratory birds under federal protection, ended spring waterfowl hunting, and ended hunting for scarce shorebirds.

The first of many national wildlife refuges was established in 1905. Some species responded quite rapidly to the increased protection. Deer populations exploded in many areas after 1910. By the 1920s, major die-offs of deer occurred in various states where they had increased to such an extent that they exceeded their food supply (Trefethen 1966). Such incidents and growing public interest demonstrated a need for improved wildlife management. In the 1920s, Aldo Leopold published initial studies and theories on game management. His classic textbook *Game Management* was published in 1933. These and other works formed the basis for the new profession of wildlife management and led to the establishment of The Wildlife Society in 1939.

During the 1930s, several wildlife laws were enacted to aid farmers through tax breaks and conservation incentives, which improved waterfowl and upland game habitats. The Predatory Mammal Control Program of 1931 promoted development of methods of "eradication, suppression, or bringing under control" various mammals viewed as pests, including mountain lions, wolves, and other species now protected.

The Duck Stamp Act of 1934 required all waterfowl hunters under age 16 to purchase a $1.00 stamp, with the proceeds to be used to fund creation and restoration of wetland habitat. The Wildlife Restoration Projects Act of 1937 returned to the states revenues collected from federal excise tax on sporting arms and ammunition. To qualify for revenues the states agreed to use the funds for wildlife management and to apply state hunting license fees to fish and wildlife management. Prior to this law, it was common for state legislatures to use license fees for unrelated programs (Clepper 1966). In 1940, the U.S. Fish and Wildlife Service was established in the Department of Interior by consolidating the Bureau of Biological Survey and the Bureau of Fisheries. This basic organization of government wildlife management still exists today.

Federal fisheries legislation was less active and targeted specific fishery resources. Growing competition for fishery resources on the open seas led to establishment of the International Halibut Commission in 1909 and the North Pacific Salmon Commission in 1937. These international agreements between Canada and the U.S. established a pattern for future ocean fishery management. However, they failed to halt the decline of many valued fish species.

World War II Impacts on Natural Resource Management. The Progressive period was brought to closure by the ending of World War II in 1945. During the war, changes in natural resource policy were few. The war split the twentieth century in two. Before the war, the U.S. looked mostly inward to solve problems at home. Even though the U.S. was becoming an important part of the global economy, national interests were primarily self-centered. The war forced Americans to deal with global problems. Rapid advances in communication further prevented a return to isolationist self-concern. World War II was the culmination of the industrial age and the harbinger for the information age. Inventions before and during the war were rapidly transformed into the diverse computing, communications, and knowledge-based industries of today.

NEOPROGRESSIVE PERIOD (1945–PRESENT)

Ecology and Landscape

The Western Hemisphere was spared the destruction of World War II. By the end of the war, the amount of land in public and private ownership had stabilized with most of the public land west of the Great Plains (Tables 2.1 and 2.2). Much marginal farmland was converted back to grassland or woodland with remaining farmland being more intensively cultivated. After the exodus of dryland farmers from the dust bowl in the high plains, range use became more compatible with sparse and uncertain rainfall. Groundwater resources were developed to irrigate many areas in the 11 western states. Groundwater depletion became a serious problem in several regions. To accommodate larger equipment and more efficient land use, farm fields also were enlarged and were

TABLE 2.1   Present Custodianship of Federal Lands in the United States

	Acres (Millions)	Percent of Total
Total land in the United States	2,271	100
Not owned by federal government	1,621	71
Owned by federal government	650	29
Bureau of Land Management	268	12
Forest Service	191	6
Fish and Wildlife Service	86	4
National Park Service	73	3
Department of the Army	12	1
Bureau of Reclamation	4	<1
Department of the Air Force	<1	<1
Corps of Engineers	9	<1
Bureau of Indian Affairs	3	<1
Department of the Navy	3	<1
Other Agencies	4	<1

Source: Holechek et al. 1998 (from USDI reports).

TABLE 2.2   Percent of Land in the 18 Western States Presently in Federal Ownership[a]

State	Percentage	State	Percentage
Alaska	68	New Mexico	31
Arizona	47	North Dakota	4
California	44	Oklahoma	2
Colorado	36	Oregon	52
Idaho	62	South Dakota	6
Kansas	<1	Texas	1
Montana	28	Utah	64
Nebraska	1	Washington	24
Nevada	83	Wyoming	48

[a]28% of the United States in federal ownership
Source: Holechek et al. 1998 (from USDI reports).

more consistently planted to the same crops. Wild areas left between farmed lands were gradually eliminated, leaving the farmed landscape more monotonous.

Before WWII, the U.S. was fairly distinctly split into localized urban-industrial landscapes and widespread rural landscapes. Cities were compact and built around business and market centers. Even in large cities like New York, a person could drive to woodland, farms, and small towns in a short time. After the war, a new phenomenon of suburban growth began as people desired larger backyards, closer proximity to schools, and escape from urban blight and crime. Migration to the suburbs was facilitated by the development of interstate highways starting in 1956. During the 1970s, a small countercurrent of young and idealistic professionals began to move back into the inner city to buy cheap housing for renovation. Conditions have slowly improved in many blighted urban areas while suburban sprawl continues unabated.

Over the course of the last five decades, the nation's landscape has been drastically altered by urbanization. Much of the farmland adjacent to larger cities has been converted to urban and suburban landscape. Natural ecosystems have become increasingly fragmented by roads and developed areas. In general, people in the U.S. have continued to move westward and southward. This trend has been driven by the relatively cheap land, cheap labor, mild winters, and appealing environment. The widespread development of water resources also has been important.

Water and air quality declined after WWII until the 1970s when environmental legislation slowed that trend and to some extent reversed it. However, air quality remains in violation of standards in many cities. Diffuse sources of water pollution are still a problem in many parts of the nation. Numerous mines, waste dumps, and abandoned industrial sites continue to leak contaminants. As the industrial age gave way to the information age, many industrial areas were abandoned leaving large expanses of barren land and crumbling structures that became known as "brownfields."

In recent years, there has been much interest in dedicating land to biodiversity preservation. Nongovernment organizations, led by the Nature Conservancy, have been instrumental in purchasing lands with exceptional natural value to add to public lands or to retain under nongovernment control. The leading federal land management agencies include the U.S. Forest Service, Bureau of Land Management, Fish and Wildlife Service, and National Park Service (Table 2.1). The U.S. Fish and Wildlife Service has initiated a nationwide inventory and analysis of land use to identify areas where biodiversity protection is needed.

Landscape changes in the undeveloped world have accelerated tremendously since WWII as human populations and development have expanded. Trends there mirror changes in the U.S. during the nineteenth century. Large areas of tropical forests have been cut and converted to other land uses, often with disastrous results. Desertification is occurring in semiarid nations as a consequence of overgrazing and overharvesting of fuel woods. Soil erosion has accelerated in many locations, decreasing the life of reservoirs built for flood control, electrification, and irrigation. Numerous cities have become huge, congested, and badly polluted as people have left the degraded land for employment. Air and water quality deteriorated tremendously in parts of industrialized Eastern Europe under communism.

National and international goals increasingly consider ecosystem protection and restoration as important elements for future human welfare. However, the continuing trend toward fragmentation of land ownership into many small parcels makes large-scale ecosystem restoration a challenging problem. Although most people in the U.S. favor ecosystem restoration, few will readily sell their personal land, even when fairly compensated. Therefore, where restoration is sometimes most needed, the costs are tremendous and often prohibitive.

Economics

After World War II, U.S. policy goals broadened to enhance worldwide welfare through democracy and free enterprise in an international market economy. The federal government played a more active role in almost every aspect of American life and in international politics. During the two decades following WWII, the U.S. had unprecedented economic growth because of its advantages in world trade. This period also was a time of rapid population growth. The baby boom following the war was to have disproportionate influence on later policy and economics. People were better educated than ever before, in part because of government investment in education, but also because of the escalating availability of information. In planning for their childrens' happiness and success, parents emphasized education, fun, beauty, health, safety, security, and material welfare.

The Neoprogressive Period

1941	Onset of U.S. involvement in World War II and massive commitment of natural and human resources.
1945	The atomic bomb is exploded and WWII ends with U.S. infrastructure and natural resources basically intact and a Cold War continuing for the next half century.
1946	The U.S. baby boom begins along with a two-decade period of exceptional economic growth.
1948	The first nationwide TV broadcast and Vogt (1948) publishes a seminal study of world population problems.
1949	Xerox develops first photocopier and Aldo Leopold publishes a land-use ethic in *A Sand County Almanac*.

1951	The first commercial computer takes up a large room.
1952	The first hydrogen bomb is exploded in the Pacific.
1953	Eugene Odum publishes *The Fundamentals of Ecology,* the first systems-oriented ecology textbook.
1955	The transistor is invented.
1956	The Interstate Highway Act is justified for nuclear attack evacuation, wartime mobility, and economic development.
1957	*Sputnik,* the first space satellite is launched by Russia, pressing the race for dominance in space.
1960	The Multiple-Use Act authorizes the U.S. Forest Service to manage for sustained yield and multiple uses of resources.
1962	Rachel Carson publishes *Silent Spring,* an indictment of pesticide abuse that galvanizes the environmental movement.
1964	The Wilderness Act authorizes a public wilderness system.
1965	The Water Resources Planning Act establishes benefits-based planning in a watershed context.
1968	Paul Erlich publishes *The Population Bomb.*
1969	The National Environmental Policy Act ushers in the "green" decade for laws pertaining to clean air, clean water, endangered species, fisheries conservation, and hazardous material cleanup.
1977	The first personal computer is invented.
1993	Web browser software opens up "information superhighway."
1999	The first dam is breached in Maine under federal order to restore natural flow for anadramous fisheries.

Television ushered in the information age and much improved awareness of national and world events. Photocopying, invented in 1949, resulted in a large leap in extending information. The first commercial computer was built in 1951. UNIVAC, as it was called, filled a large room but had less computing power than today's laptop portable. The Internet, developed first for security and research purposes, became the "information superhighway" starting in 1993. Modern computers ushered in the era of database management. They made it practical to use complex mathematical models for resource management. Connected with computers, many resources are now inventoried and monitored using geographic information systems (GIS).

Soon after the war, scientists began lobbying for peaceful applications of atomic power for electric-power generation. Atomic power plants first showed up in the landscape in the 1950s (Figure 2.19). Atomic power, along with wind, hydro, and fossil fuel power, collectively provided cheap electricity for running an increasing array of laborsaving devices. Atomic power was promoted because it produced no air pollution, assuming accidental releases of radioactive emissions did not occur.

By the early 1960s people were better off in the U.S. than any previous time. Construction of the federally supported interstate highway system in the late 1950s made access to the nation's public lands and historic monuments much easier. Visitation rates skyrocketed at national parks, national forests, federally built reservoirs, and lands managed by the bureau of land management (Table 2.3). Participation in sightseeing and outdoor recreation increased rapidly and the tourist industry mushroomed.

FIGURE 2.19 Atomic power plants began development soon after WWII and were touted for their low environmental impact. However, accidental radioactive releases, costly safeguards, and difficulty in developing socially acceptable radioactive waste storage have altered the once rosy prospects for fission-produced power (courtesy of U.S. Dept. of Interior).

TABLE 2.3   Estimated Total Visitor Days (1,000) and Hunter Days (1,000) of Recreational Use on Public Lands Under the Jurisdiction of the Bureau of Land Management in 1965 and 1980

	Visitor Days		Hunter Days	
	1965	1980	1965	1980
Arizona	808	7,430	120	150
California	3,162	16,969	290	1,237
Colorado	1,382	4,572	249	601
Idaho	700	9,421	130	402
New Mexico	659	6,978	114	306
Oregon	3,699	4,266	361	994
Utah	558	2,811	240	335
Wyoming	119	4,097	55	361
	11,087	56,544	1,559	4,386

Source: USDI reports.

Policy

Environmental Policy.   Agricultural and ecological research in the U.S. accelerated after World War II to help address world food shortages and potential environmental threats from atomic radiation and other contaminants. The awesome power of the atomic and hydrogen bombs both inspired and frightened the scientific community. It lobbied for improved understanding of atomic potential and risks. Ecological studies sponsored by the Atomic Energy Commission quickly revealed how contaminated materials were mobilized and moved through ecosystems, often to undesirable locations in human and wildlife food webs (Golley 1993). Agricultural research over the next two decades generated the green revolution based on a combination of improved plant breeding, tilling, pesticide application, and fertilization techniques. Use of pesticides and fertilizers escalated without much concern over side effects except in research laboratories. However, health scientists soon verified previously suspected threats of chemical toxicity.

FIGURE 2.20 Air pollution was a serious health problem in many industrial cities of the U.S. before and following World War II. Pittsburgh was among the worst examples (courtesy of U.S. Dept. of Interior).

Soon after the war William Vogt (1948) published *Road to Survival,* a seminal work on human ecology and population pressures. Aldo Leopold's philosophy of a land ethic was published in his book, *A Sand County Almanac* (Leopold 1949). He was among the first to call public attention to the need for sustainable land use. Although Leopold greatly influenced other thinkers, he had little direct impact on natural resource and environmental policy at that time. The press, however, was beginning to identify oil, radiation, and chemical pollution problems. Air pollution was of increasing concern especially in the industrial cities (Figure 2.20). Scientific information about environmental contamination gradually accumulated. In 1962, Rachel Carson published *Silent Spring*, which was a pivotal work on pesticide contamination. Secretary of Interior Steward Udall attracted attention with his 1963 book *The Quiet Crisis.* It called for comprehensive environmental planning and environmental regulation by government. In 1968, Paul Erlich published *The Population Bomb,* which was widely read. This book focused on how human population growth was jeopardizing ecological support systems.

By the 1960s, numerous federal agencies were responsible for various monitoring, research, and regulatory programs pertaining to the environment. The National Environmental Policy Act (NEPA) of 1969 established the present approach to environmental regulation. It was followed by authorization of the Environmental Protection Agency (EPA) in 1970, which pulled together the environmental functions of numerous agencies under a single administration. NEPA requires government and private agencies to draft environmental impact statements for proposed actions affecting federal lands. It mandated that the federal government provide an acceptable environment for future generations.

Earth Day was celebrated for the first time on April 22, 1970. This was a grassroots event organized outside of the mainstream environmental organizations of the time. Its success showed the extent of environmental concern in the baby boom generation.

Amendments to the Clean Air Act of 1970 established standards for a much larger number of air contaminants than those first identified in the Clean Air Act of 1963. It required industry to install scrubbers to remove noxious emissions and automobiles to be fitted with catalytic converters. The Federal Water Pollution Control Act of 1972 mandated restoration of water quality to standards that supported fishing, swimming, and other recreational uses. It created a public works program for funding the development of domestic sewage treatment plants. Amendments to the Clean Water Act in 1977 granted control of wetland filling to the U.S. Army Corps of Engineers. Its intent was to decrease the rate of wetland loss through protection, mitigation, restoration, and creation. These laws applied to private lands as well as public lands. The Ocean Dumping Act of 1972 and the Safe Water Drinking Act of 1974 added to the improvement of water resources.

Better management of toxic materials was addressed by the Toxic Substance Control Act of 1976 and the Resource Conservation and Recovery Act of 1976. These acts established safe handling procedures for wastes and encouraged recycling. The Environmental Response, Compensation, and Liability Act of 1980 established a trust fund (the "superfund") to recover waste storage areas associated with mines, industry, and military installations.

Water and Mineral Resources. The Water Resources Research Act of 1964 established a cooperative state and federal water research program. The Water Resources Planning Act of 1965 facilitated comprehensive river basin planning. It resulted in development of methodology for evaluating the benefits and costs of federally funded water projects. It was also the first time in water resources management that environmental protection was considered. Both economic and environmental benefits were to be considered. The Water Resources Development Act of 1986 elevated ecosystem protection and restoration to higher consideration in the Army Corps of Engineers mission. The Soil Conservation Service broadened its mission and became the Natural Resources Conservation Service in 1994.

After years of strip mining that left parts of Appalachia with barren lands, acid streams, and acid lakes, Congress passed the Surface Mining Control and Reclamation Act in 1977. It requires reclamation of surface mines to productive use after mining is completed (Figure 2.21).

Forest and Range Management. Livestock grazing on federal rangelands was reduced 25% between 1960 and 1992 (Table 2.4). Range improvement on private lands accelerated during the 1980s and 1990s because of improved information and education programs by state and federal agencies and improved education of modern ranchers. Future range management on public lands is likely to place greater emphasis on water, wildlife, recreation, and biodiversity and less emphasis on livestock production. Because of the natural role of fire in range ecosystems, it is likely to receive greater use in manipulating rangeland vegetation.

Since the Taylor Grazing Act of 1934, there have been several movements by private interest groups to transfer remaining federal lands into private ownership. The most recent attempt, known as the "sagebrush" rebellion, occurred in the late 1970s and early 1980s. These movements have been unsuccessful because of heavy public opposition. The majority of citizens in the United States believe public lands are part of their natural heritage and should be retained in federal ownership. The Federal Land Policy and Management Act of 1976 made government retention of existing public land official policy.

The early 1960s were characterized by considerable change in the philosophy of public land management. The Multiple Use Act of 1960 mandated that Forest Service lands be managed for grazing, wildlife, timber, recreation, and mineral development, rather than for a single use. Previously, range research and management had been geared toward producing forage for livestock. This change in mandate had considerable influence on range and forest management practices. Multiple use was defined as management for the combination of uses that "best meet the needs of the American people."

The Forest Management Act of 1976 mandated forest management plans emphasizing multiple use. It specified that plans should provide for diversity of plant and animal communities. Wilderness was recognized as an important use. The act set

FIGURE 2.21 Much of northern Appalachia was scarred by abandoned coal mines, a practice that was stopped by the Reclamation Act of 1977. The photo on the top is a typical abandoned mine and on the bottom is a reclaimed coal mine (courtesy of Office of Surface Mining and Reclamation and Enforcement).

TABLE 2.4 Trend (1960–1993) in Actual Animal Unit Months (thousands) of Livestock Grazing on Bureau of Land Management and Forest Service Lands

Type of Livestock	Bureau of Land Management				
	1960	1970	1980	1990	1993
Cattle	10,277	13,368	8,984	9,327	8,486
Horses & burros	198	400	72	72	62
Sheep & goats	4,790	2,741	1,351	1,445	1,211
Total AUMs	15,265	16,209	10,407	10,844	9,759
	Forest Service				
Cattle	8,084	—	7,594	—	6,784
Sheep	1,083	—	893	—	805
Horses & burros	210	—	176	—	130
Total AUMs	9,377	—	8,663	—	7,719

Sources: U.S. Department of Interior annual reports and U.S. Forest Service annual reports (from Holechek et al. 1998).

many limits on timber harvest, addressing concerns about water quality, soil erosion, and fish and wildlife habitat.

Parks and Recreation. The Wilderness Act of 1964 strengthened the existing Forest Service policy of protecting lands with exceptional wilderness attributes. It extended consideration to a larger range of federal lands. In 1965, the Water Conservation Fund was authorized, which used offshore oil well revenues to purchase new public recreation areas. The Wild and Scenic Rivers Act of 1968 was passed to protect some of the free-flowing rivers from dam building and other modification. In the same year, the National Trails System Act was passed to maintain the scenic quality of major trails within the National Forests.

Fish and Wildlife. The Fish and Wildlife Coordination Act in 1946, as amended in 1958, provided that fish and wildlife conservation receive consideration in water resource development. The Fish Restoration and Management Projects Act of 1950 established a federal tax on fishing equipment. The U.S. Fish and Wildlife Service redistributes these funds to state agencies for sportfishery development. It was expanded in 1985 in response to a tremendous increase in sportfishing.

The Fishery Conservation and Management Act of 1976 recognized growing problems with depleted ocean stocks of commercial fish. It established a 200-mile offshore authority for managing coastal fisheries more effectively. However, massive closures of fisheries in the early 1990s demonstrate that this law has failed to prevent overharvest.

The Endangered Species Act of 1973 and the Marine Mammals Protection Act of 1976 were designed to comprehensively protect all wild organisms from human-caused extinction. The Endangered Species Act authorized the U.S. Fish and Wildlife Service to identify threatened and endangered species, regulate their take, and recover their populations to unthreatened levels. Its power is most concentrated in government authority to protect habitat needed for species recovery regardless of whether the land is publicly or privately owned. The act was later amended to allow development of habitat conservation plans by landowners as a way to provide some flexibility. The Endangered Species Act also established in federal law an ecosystem perspective for management.

Present Status. While attempts to weaken environmental laws typically have met public disfavor, the public has resisted more costly environmental initiatives during the past two decades. Few federal environmental laws have originated since 1980, although many laws have been amended, usually to strengthen or make them more workable.

Amendments and agreements have added a dimension of ecosystem restoration and management to the activities of federal government, such as in the Water Resources Development Act of 1986 and 1996 and the Food and Agriculture Conservation and Trade Act of 1990. These actions are oriented toward protecting biodiversity from further deterioration. Wetlands have received exceptional restoration attention. Agencies like the Natural Resources Conservation Service (Soil Conservation Service before 1994) and the Army Corps of Engineers now restore wetlands that they once degraded or eliminated (Figure 2.22). Much of the future of natural resource management is likely to be associated with restoration ecosystem management and the emerging concept of sustainable development.

Since the 1960s the roles of nongovernment agencies have increased tremendously in influencing government environmental policy (Coggins 1993). They have acted through the courts to protect existing law and to encourage agencies to follow through as authorized. The National Wildlife Federation, Environmental Defense Fund, Sierra Club, World Wildlife Fund, and Wilderness Society have been especially influential in protecting their special interests through lobbying and litigation. This has precipitated policy recommendations to review environmental law enacted

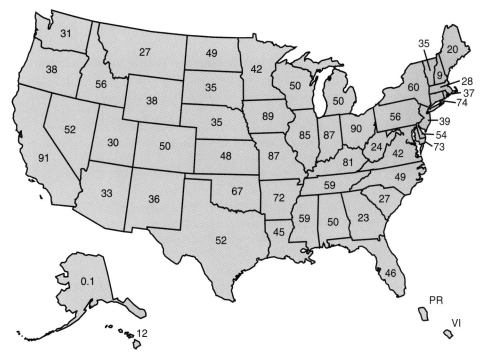

FIGURE 2.22 Percentage of Wetlands Acreage Lost, 1780s–1980s (after Dahl and Johnson 1991). About half of the U.S. wetlands have been lost from filling and draining. Policy now enforced by the U.S. Army Corps of Engineers and the Environmental Protection Agency seeks to prevent any net wetland loss.

during what has been called the Environmental Overreach Period (Martz 1993). The main objective is to move management of resources away from polarizing litigation toward cooperative management approaches.

Coggins (1993) believes U.S. natural resource policy has too many land zoning categories, agencies, allocation mechanism, statutes, and regulations and "not enough consistency, clarity, uniformity, efficiency, or comprehensibility." In summary, according to Coggins (1993): "The present legal structure is not the product of conscious design; rather, it is simply the residue of more or less ad hoc developments over two centuries."

The twentieth century is ending with growing awareness of natural resources problems and opportunities. Evidence increasingly indicates that humanity is causing costly changes in the biosphere, although the implications are not always clear. The major challenge to natural resource managers in the twenty-first century will be how to sustain economic development in a world where many resources are in rapid decline. Improving education, science, technology, and social ethics provide many opportunities for improved management of natural resources, including some restoration of what formerly existed.

LITERATURE CITED

Bolen, E. G. and W. L. Robinson. 1995. *Wildlife ecology and management.* 3rd edition. Englewood Cliffs, NJ: Prentice-Hall.

Boorstin, Daniel J. 1987. *Hidden history.* New York, NY: Harper and Row.

Boyer, P. S. 1995. *The enduring vision.* Lexington, MA: D.C. Heath and Company.

Bramwell, A. 1989. *Ecology in the 20th century.* New Haven, CT: Yale University Press.

Butzer, K. W. 1990. The Indian legacy in the American landscape (pp.27–50). In *The making of the American landscape,* edited by M. P. Conzen. New York, NY: Routledge.

Carson, R. 1962. *Silent spring.* Boston, MA: Houghton Mifflin.

Clepper, H. 1966. *Origins of American conservation.* New York, NY: The Ronald Press.

Coggins, George C. 1993. Trends in public land law (a title the inaccuracy of which should become manifest) (pp. 49–65). In *Natural resources policy and law: Trends and directions,* edited by Lawrence J. MacDonald and Sarah F. Bates. Washington, DC: Island Press.

Conlin, J. R. 1993. *The American past.* Fort Worth, TX: Harcourt Brace.

Conzen, M. P., ed. 1990. *The making of the American landscape.* New York, NY: Routledge.

Cronan, W. 1991. *Nature's metropolis: Chicago and the great West.* New York, NY: W. W. Norton.

Crosby, A. W. 1986. *Ecological imperialism: The biological expansion of Europe, 900–1900.* New York, NY: Cambridge University Press.

Dahl, T. E. and C. E. Johnson. 1991. *Status and trends of wetlands in the conterminous United States, mid-1970s to mid-1980s.* Washington, DC: U.S. Department of the Interior, Fish and Wildlife Service.

Davis, S. M. and J. C. Ogden, eds. 1994. *Everglades: The ecosystem and its restoration.* Delray Beach, FL: St. Lucie Press.

England, R. E. and A. De Vos. 1969. Influence of animals on pristine conditions of the Canadian grasslands. *Journal of Range Management* 22:87–94.

Erlich, P. R. 1968. *The population bomb.* New York, NY: Ballantine Books.

Feldman, David L. 1991. *Water resources management in search of an environmental ethic.* Baltimore, MD: Johns Hopkins University Press.

Freidenberg, D. M. 1992. *Life, liberty and the pursuit of land.* New York, NY: Prometheus Books.

Golley, F. B. 1993. *A history of the ecosystem concept in ecology.* New Haven, CT: Yale University Press.

Graham, F. 1971. *Man's dominion: The story of conservation in America.* New York, NY: M. Evans and Company.

Hardin, G. 1968. *Tragedy of the commons.* Science 162:1243–1248.

Heilbroner, Robert and Lester Thurow. 1994. *Economics explained.* 3rd edition. New York, NY: Simon and Schuster.

Hibbard, H. B. 1924. *A history of the public land policies.* New York, NY: Macmillan. (Recopyrighted and published by the University of Wisconsin Press, Madison, in 1965)

Holechek, J. L., R. Peiper, and C. Herbel. 1998. *Range management principles and practices,* 3rd edition. Upper Saddle River, NJ: Prentice-Hall.

Jenkins, P. 1997. *A history of the United States.* New York, NY: St. Martin's Press.

Johnson, M. D. 1964. *Feathers from the prairie.* Bismark, ND: North Dakota Game and Fish Department.

Kirsch, L. J. and A. D. Kruse. 1972. *Prairie fires and wildlife.* Bismark, ND: North Dakota Game and Fish Department.

Kline, B. J. 1997. *First along the river: A brief history of the U.S. environmental movement.* San Francisco, CA: Acada Books.

Knight, Richard L. and Sarah F. Bates. 1995. *A new century for natural resources management.* Washington, DC: Island Press.

Leopold, A. 1933. *Game management.* New York, NY: Scribner.

Leopold, A. 1949. *A Sand County almanac and sketches here and there.* New York, NY: Oxford University Press.

Marsh, G. P. 1864. *Man and nature; or, physical geography as modified by human nature.* New York, NY: Charles Scribner.

Martin, P. S. and R. G. Klein, eds. 1984. *Quarternary extinctions: A prehistoric revolution.* Tuscon, AZ: University of Arizona Press.

Martz, Clyde, O. 1993. Natural resources lawman historical perspective (pp. 22–48). In *Natural resources policy and law: Trends and directions,* edited by L. J. MacDonnell and S. F. Bates. Washington, DC: Island Press.

Nester, W. R. 1997. *The war for America's natural resources.* New York, NY: St. Martin's Press.

Novak, B. 1995. *Nature and culture: American landscape and painting 1825–1875.* New York, NY: Oxford University Press.

Oliver, J. W. 1956. *History of American technology.* New York, NY: The Ronald Press.

Reiger, J. F. 1975. *American sportsmen and the origins of conservation.* New York, NY: Winchester Press.

Roberts, J. M. 1993. *A short history of the world.* New York, NY: Oxford University Press.

Scheffer, V. B. 1991. *The shaping of environmentalism in America.* Seattle, WA: University of Washington Press.

Schubert, F. N. 1980. *Vanguard of expansion: Army engineers in the trans-Mississippi West, 1819–1879.* Office of the Chief of Engineers. Washington, DC: Superintendent of Documents, U.S. Government Printing Office.

Scott, G. 1875. *Fishing in American waters.* New York, NY: Harper.

Smith, J. G. 1898. Grazing problems in the Southwest and how to meet them. U.S. Dep. Agric. Div. Agrost. Bull. 16:1–47.

Stoddard, L. A., A. D. Smith, and T. W. Box. 1975. *Range management.* 3rd edition. New York, NY: McGraw-Hill.

Thompson, P. E. 1970. The first fifty years—the exciting ones (pp. 1–11). In *A century of fisheries in North America,* edited by N. G. Benson. Bethesda, MD: The American Fisheries Society.

Trefethen, J. B. 1966. Wildlife regulation and restoration. In *Origins of American conservation,* edited by H. Clepper. New York, NY: The Ronald Press.

Trefethen, J. B. 1975. *An American crusade for wildlife.* New York, NY: Winchester Press.

Turner, F. J. 1963. *The significance of the frontier in American history,* edited by H. P. Simonson. Frederick Unger Publications.

Udall, S. L. 1963. *The quiet crisis.* New York, NY: Holt, Rinehart and Winston.

Vogt, W. 1948. *Road to survival.* New York, NY: William Sloan Associates, Inc.

Waterman, C. F. 1975. *Fishing in America.* New York, NY: Holt, Rinehart and Winston.

White, L., Jr. 1967. The historical roots of our ecological crisis. *Science* 125:1205.

Williams, M. 1989. *Americans and their forests: A historical geography.* New York, NY: Cambridge University Press.

CHAPTER 3

Basic Ecology

INTRODUCTION

Natural resource management depends on ecological science, or ecology as it is commonly called. Ecology is the knowledge and study of interrelationships among organisms and their environment (Colinvaux 1993, Ricklefs 1996). Ecology became identified as a scientific discipline with a definite body of accumulated knowledge at the beginning of the twentieth century. The foundations were laid in the 1700s with the development of biogeography and a classification system for living things. Darwin in 1859 developed his theory of evolution—a discourse on how interactions between organisms and their environment could result in adaptive change. An early pioneer in applied ecology, George Perkins Marsh, wrote an astute assessment of watershed processes and their relationship to vegetation conditions and soil erosion (Marsh 1865). The first major integrative work in aquatic ecology was published by Forbes in 1888, who conceived of "the lake as a microcosm." He established a basis for what would eventually become ecosystem science.

Frederick Clements stood out among early terrestrial biologists because his 1907 book, *Plant Physiology and Ecology,* was the first to formally organize ecology into a discrete subdiscipline (see Clements 1907). Allee et al. (1949) summarized the rapidly evolving field of animal ecology. Andrewartha and Birch (1954) added substantially to the science of population ecology. Odum (1959) organized the best synthesis and conceptual treatment of ecosystems science available at that time. Numerous other general ecology books have been completed since then, but Colinvaux (1993) and Ricklefs (1996) are recent standouts.

Ecology is usually approached from three standpoints: the population (*autecology*), the community (*synecology*), and the ecosystem. Each examines organismic interaction with environment at a different scale. While early plant ecologists emphasized the community approach, animal ecologists emphasized the population approach. Ecosystems ecology emerged as an approach for unifying plant, animal, and environmental ecology. This chapter starts with an introduction to population and community ecology, and then culminates with a summary of important ecosystem processes and concepts.

POPULATIONS

Population Identity

Organisms are identified by how well their morphological, physiological, behavioral, and genetic attributes match other organisms. Those individuals that commonly exchange genetic material under natural conditions are assigned to the same species. Even within species, however, subspecies are often recognized. A population is defined as a group of individuals with similar traits and potential for reproductive interaction. They typically share the same locality and resources (Figure 3.1).

Distribution and Dispersal

Populations are distributed in the three dimensions of space in ways that reflect their life requirements. Arboreal species, adapted to life in trees, differ in distribution from fossorial species adapted to burrowing life. Similarly, coastal oceanic species rarely overlap in vertical distribution with the species adapted to the darkness and pressure of the oceanic abyss. At a still finer scale, fossorial species may be limited only to that volume of earth that has the proper moisture and particle sizes suitable for burrowing.

Species that are broadly adapted to the environment tend to be more widely distributed than those that are narrowly adapted. However, the sharpness of environmental change also is important in determining distributions. For the most specialized species, even subtle differences in the environment may limit distribution. Changes caused by fire, flooding, grazing, and so on create or reduce connections between isolated islands of suitable environment. Many species depend on a variety of environments for activities such as feeding, resting, and rearing of young.

Population dispersal is defined as the movement away from place of birth. It is the means by which populations successfully reach isolated areas suitable for colonization. Populations within species often are isolated from other populations for variable periods of time depending on the environmental dynamics in the area. Population abundance also changes as distributions expand and contract with change in the environment. However, the extent of change depends on the population *density,* or numbers per unit area or volume. Density of large organisms is most frequently estimated as the total number of individuals per unit area (e.g., number/hectare). For small aquatic and soil organisms, numbers in a volume of water or soil is typically measured. Populations may spread out and contract without change in abundance, causing their density to change. Therefore, the total number of organisms in a population

FIGURE 3.1 A population of bison on mixed prairie grassland in South Dakota (courtesy of Holechek et al. 1998).

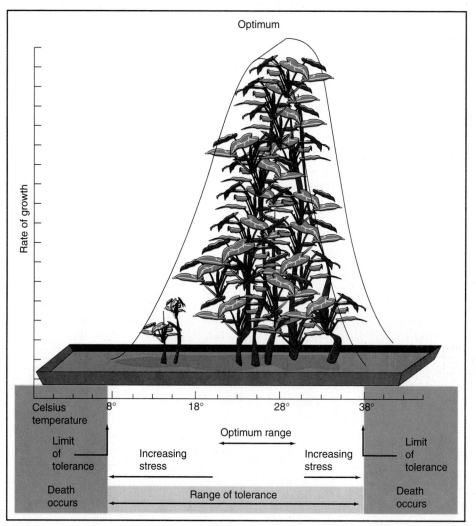

FIGURE 3.2   For every factor influencing growth, reproduction, and survival, there is an optimum condition. Above and below the optimum, stress increases until survival becomes impossible at the limits of tolerance (from Nebel and Wright 1996).

is the product of their average density within the occupied range times the area of the range. Density can vary greatly within the range. A herding population, for example, can have high density within the herd and low density throughout the range.

Factors Affecting Population Distribution

Understanding of population distributions requires knowledge of environmental regulating factors. Two basic concepts underlie most explanations of observed distributions. During the nineteenth century, Justus Liebig discovered that crops planted in a field conducive to their growth did not grow if a single nutrient was missing or in short supply. From these observations he concluded that a group of similar organisms is limited by the required resource that is in shortest supply. More recently it has been discovered that factors sometime interact in combination to establish a limiting minimum. The concept was later expanded by Shelford to include maximum as well as minimum environmental tolerances of the population (Colinvaux 1993).

From these concepts it was deduced that a population is able to exist and reproduce successfully only within a definite range of environmental conditions (Figure 3.2). The sum of the limiting conditions is referred to as the *environmental resistance*. Odum (1959) summarized ways in which organism tolerances to environmental conditions were most likely to be expressed. An organism may have a wide range of tolerance for one factor and a narrow range for another. Organisms with a wide tolerance for all factors are likely to be widely distributed. When conditions are not optimal for a species with respect to one ecological factor, the limit of tolerance may be reached for all factors. Tolerance is usually least broad during the reproductive period. Common variables limiting population distributions are temperature, water, light, substrate condition (e.g., soil, rock, sediment), nutrients, toxins, salinity, fire, and oxygen (in aquatic environments).

Population Habitat and Physical Niche

In population ecology, habitat and niche are two widely used words that can be conceptually confusing. Habitat usually is thought of as the site or place occupied by a population. It is characterized by obvious terrain and vegetation features (forest, desert scrub, marsh, or stream habitat). Most terrestrial and wetland habitats are identified more specifically by predominant vegetation (oak forest, cypress swamp) whereas aquatic habitat is more often identified by physical and chemical form (e.g., freshwater lake, coastal ocean).

The species niche is defined as all the environmental factors that a species needs to survive, stay healthy, and reproduce in a ecosystem. Numerous models of habitat suitability for various populations have been developed for use by the U.S. Fish and Wildlife Service and other agencies. They guide habitat protection, creation, and restoration based on the physical niche of the desired species. These models typically include limiting variables. Habitat determines the potential distribution of a population, but often not all suitable habitat is occupied. This is because of past extirpation of the population or because the habitat is new and never previously inhabited.

Habitat is more dynamic than the concept of place or site might imply. Environmental variables in the habitat often change daily (e.g., light, temperature), seasonally (moisture, temperature), or less predictably (fire, extreme flooding). Many populations are adapted to variation, which often triggers life-cycle events, such as reproduction, dormancy, migration, and feeding. Variability of habitat is required for life-cycle completion of many species. The reproduction of many aquatic species is triggered by light, temperature, or flooding. Numerous terrestrial plants reproduce only after exposure to fire. Reduction of environmental variation frequently leads to lower species diversity.

POPULATION STRUCTURE AND DYNAMICS _____

Population Structure

Population structure is defined as the distribution of the members among age categories and sexes. Unstable populations are indicated by bulges in the age distribution (Figure 3.3). A bulge among older ages indicates a population that has failed to reproduce enough to replace loss and is decreasing in number. A bulge in the youngest ages indicates a population poised for population increase (Figure 3.3, Mexico). Each species has a characteristic sex ratio for a stable population and variation from that ratio also indicates an unstable population. A higher ratio of females to males may signal a higher potential reproduction rate. Fecundity is gaged by the number of embryos per adult or per female in the population. A complete set of data describing the numbers of population members in age and sex categories can tell managers much about the history and future of the population.

Population Dynamics

Population structure depends on birth or germination rate (increase in number), death rate (decrease in number), immigration (invasion) rate (increase in number), and emigration (dispersal) rate (decrease in number). Birth and germination rates, or natality, are rates at which new members are added to the population. The death rate,

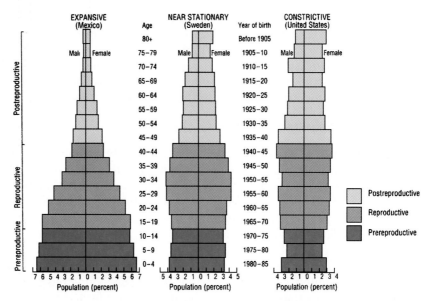

FIGURE 3.3   Age structures of human populations in Mexico, Sweden, and the United States (from Owen and Chiras 1995).

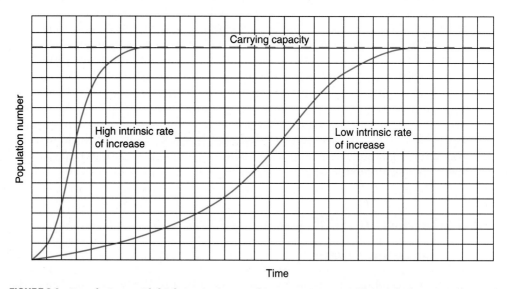

FIGURE 3.4   Populations with high intrinsic rate of increase more quickly reach the environmental carrying capacity when first introduced to an area than species with low intrinsic rate of increase.

or mortality, is the rate that members are subtracted from the living population. When natality exceeds mortality, the population increases. In a stable population natality about equals mortality.

Populations have an *intrinsic rate of increase* determined by individual growth rates as well as numbers (Figure 3.4). Members of populations grow by adding body mass (typically measured by weight and less commonly by volume). The intrinsic rate of population increase depends on fecundity (number of offspring produced) per adult and growth rate to maturity. For populations with a high potential number of progeny per adult, rapid growth rate, and early maturation, the intrinsic rate of population increase is high. The realized population increase in numbers also will be high when the death rate is low. Stable populations with a high intrinsic rate of increase must have a compensating high mortality rate. Populations with high intrinsic rate of increase have high *biotic potential*.

a)

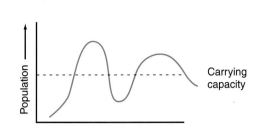

b)

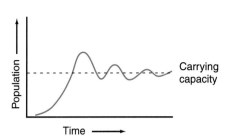

FIGURE 3.5   Two of the many possible types of population fluctuations around carrying capacity. Depending on species and circumstances, populations can vary widely in density and in time (a) or vary within narrow limits and quite regularly (b).

A very general relationship exists between the maximum size obtained by individuals in the population and the intrinsic rate of population increase. Organisms of large adult size typically have slower growth, lower fecundity, and a longer maturation time than smaller organisms. They also tend to have lower death rates. Healthy female elephants, for example, take years to grow to maturity and may live as adults for decades, but they typically have one baby every three or four years. Healthy female mice, in contrast, take only weeks to grow to maturity, typically survive only for months, have several young in each litter, and have a litter every few weeks. In general, smaller organisms live shorter lives and have a higher population turnover rate than larger organisms.

Populations in optimum habitat typically increase rapidly from low initial number. This type of situation often exists in changeable environments that suddenly become optimal after a period of difficult conditions, such as extended drought. It also exists when a population first gets established in a new and optimum habitat. After an initial lag, a newly established population typically reaches a maximum rate of increase, then the rate of increase slows, and finally reaches a somewhat stable number. At that point the population has reached the carrying capacity of the habitat (Figures 3.4 and 3.5). As the population becomes more dense with respect to the resources available, competition for resources creates a "feed back" that slows the population increase to zero. Depending on conditions, the population number may remain stable at the carrying capacity or collapse and fluctuate (Figure 3.5).

Carrying capacity often varies because environmental factors such as climate vary and because the population can depress its own resources (Figure 3.6). Terrestrial plants, for example, often deplete the soil of nutrients as they accumulate in their tissues. Plant-eating animals often become so abundant they depress the plant regeneration. Ultimately the animal population declines from lack of food, the plants regenerate, and the cycle repeats. Climatic changes can alter soil water availability, which can limit plant growth. Because of these dynamics, population numbers are rarely constant. Typically they are in continuous adjustment to changing habitat conditions.

POPULATION ENERGETICS, BIOMASS, AND PRODUCTION

Energetics

Energy typically is expressed in calories or joules (J) received per unit area per unit time (e.g., Cal/m^2/day). Green plants use light and soil nutrients to create living matter, or *biomass* (also sometimes referred to as *standing crop*). Otherwise the energy used for metabolism comes from the chemical bonds in organic compounds.

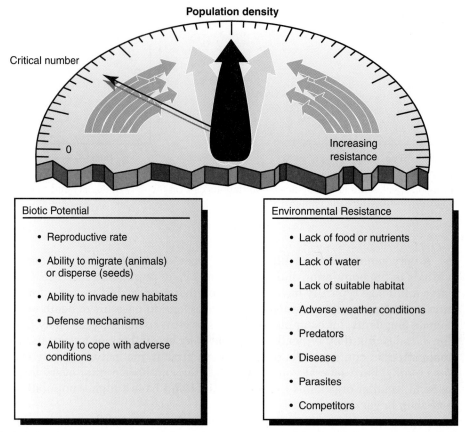

FIGURE 3.6   A stable population in nature is the result of a balance between factors tending to increase population and factors tending to decrease population (environmental resistance) (from Nebel and Wright 1996).

Not all energy used by organisms within a population, through feeding or otherwise, is actually *assimilated* or retained by the organism (Figure 3.7). For animals, some unassimilated matter is egested as feces. Efficiencies of assimilation vary among organisms depending on the quality of the energy source available to them. For plants, assimilation of light by chlorophyll depends on the wavelengths of light available to them. This is determined by how much of the light has been absorbed by the atmosphere, overhead tree canopy, or overlying water. For animals, assimilation efficiency depends mostly on the organic composition of the foods they eat. Once nutritional energy is assimilated, some is used in body maintenance and respiration, and the rest goes into net production. In turn, net production is partitioned among reproduction, growth, and various secretions necessary for life.

Population Biomass

Biomass is defined as the amount of matter contained in the population of organisms at a given time. It is commonly expressed in total weight including water, or dry weight, or as an element such as carbon or nitrogen (Table 3.1). To compare populations, biomass is expressed on a weight per unit area basis, such as grams of carbon per square meter ($g\ C/m^2$). The biomass of a population can be estimated by multiplying population numerical density times mean individual weight or by summing the weights of all individuals. Biomass often is expressed as a mean for some determined period, most typically one year. Because water content is highly variable among organisms, weights of different organisms are usually compared on a dry matter basis. To obtain dry matter, samples are dried to eliminate the water content.

Biomass often is expressed as carbon weight because carbon is the "backbone" or primary element in organic matter. It can be readily measured, and is linked closely to calo-

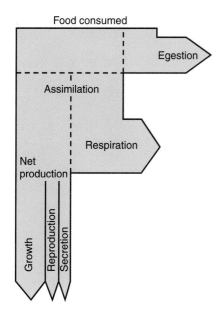

FIGURE 3.7   Energy partitioning among metabolic pathways of animals in a population. Of the food energy consumed, a fraction passes through to be egested and a fraction is assimilated into each member of the population. The fraction assimilated is partitioned between respiratory metabolism and population net production. The net production is the portion of the population that serves as a food base for predators.

TABLE 3.1   Common Measures of Biomass and Productivity Expressed for a Hypothetical Organism, Showing General Relationships Among Dry Weight, Calories, and Carbon

| Measure | Population Parameter | | |
	Mean Biomass	Productivity	P/B
dry weight	1000 g/m^2	5,500 g/m^2/year	5.5
carbon	400 g C/m^2	2,200 g C/m^2/year	5.5
calories	4,000 Cal/m^2	22.0 KCal/m^2/year	5.5

ries of energy contained in the biomass. Carbon is usually measured by burning weighed samples and measuring the carbon dioxide gas emitted using gas spectrophotometry. Calories are determined from the heat released upon combustion in a bomb calorimeter. The choice of biomass expression depends on the objectives of study or management.

Production, Productivity, and P/B Ratio

Organic production is the total generation of organic matter over some time period. Productivity is the rate of biomass generation. Thus, a production of 2,000 kg over 100 days has an average productivity of 20 kg/day. (Table 3.1). P/B ratio is the productivity divided by the mean biomass of a population for a given period of time. The P/B indicates the rate of population tissue turnover. P/B is related to mean size (biomass/individual). Small species typically have higher P/B ratios than large species.

POPULATION STRATEGIES

Populations have adapted to their habitats through two main strategies. One strategy is to become generally adapted to wide variation in the environment and give up some efficiency in the use of specific food and cover. Species that adopt this strategy are called generalists. The other strategy is to become highly adapted to the efficient use of specific resources in the ecosystem (Figure 3.8). These species are called specialists. Specialists are more likely to do poorly in unstable ecosystems. Widely distributed species are more likely to be generalists than specialists. Many species exhibit characteristics of both extremes.

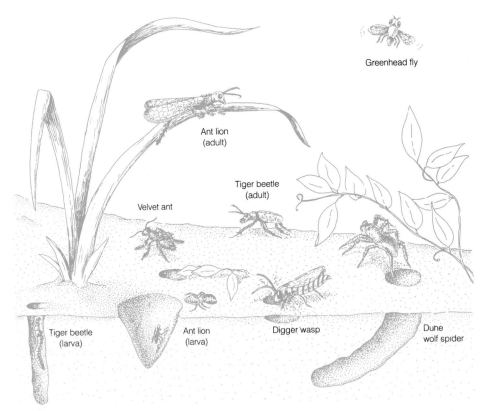

FIGURE 3.8   Specialization in insect use of sand dune habitat (from Lerman 1986). Specialists often are adapted to specific strata in an ecosystem. Sand dune specialists, for example, are narrowly adapted to subterranean, ground surface, dune vegetation, and arial conditions.

Resilience refers to the capability of a population to "bounce back" quickly from stress. Many resilient populations are also quite tolerant of environmental variation and can disperse and colonize new areas rapidly. Generalists typically have greater resilience than specialists. Specialists tend to be better adapted than generalists to habitats with stable carrying capacity.

Populations with similar feeding behavior are sometimes lumped together, and are referred to as *guilds* after the professional unions of middle-age Europe. The concept of guilds has been most applied to birds, fish, and aquatic invertebrates. Organisms in the same guild are not necessarily related taxonomically. Because they are related through function, guild members typically compete for the same resources. However, they have often evolved subtle but effective means for sharing these resources.

Certain *keystone species* have outsized impacts and influence on the diversity of populations that can be supported in an ecosystem. Animals that shape their physical environment to exceptional degree can greatly influence the physical niches available for species evolution. Prominent species in this category include the dominant plants (aspen, cattail) and animals, such as muskrat (burrows and eats cattails), beaver (creates ponds and eats aspen), and of course humans (deconstruct, construct, reconstruct, and otherwise manage resources). It is often unclear how much effect a species exerts on other populations until after the species is greatly reduced or eliminated.

COMMUNITIES

Definition

A biotic community is an association of interacting populations occupying a given locality for a particular time period (Figure 3.9). Particular biotic communities are associated with prairies, woodlands, marshes, lakes, croplands, oceans, and so on. Com-

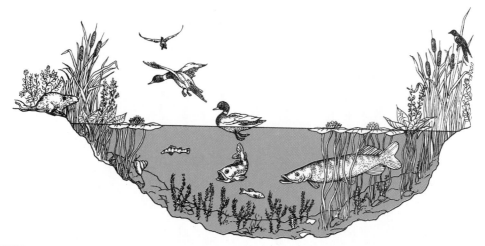

FIGURE 3.9 A community often is depicted by the dominant plants and animals, as indicated here for a marsh-pond community; however, most communities are composed of many more less obvious species (from Owen and Chiras 1995).

munities usually include plants, animals, fungi, bacteria, and other major groups of organisms. Some communities, however, are reduced to fewer groups because of harsh environments, such as hot springs and caves.

COMMUNITY STRUCTURE AND FUNCTIONS

Community Habitat

Community habitat refers to the physical environment (climate, terrain, soil, water, and so on) that interacts with the community to provide needed resources. Many ecologists have come to accept the plant-oriented definition for community habitat provided by Daubenmire (1968). Under his definition, habitat is "an area characterized by a specific set of climatic, edaphic (substrate resources), topographic, and biotic factors." It implies potential for occupancy by a specific type of vegetation (the climax), if undisturbed by humans, fire, or weather. Because plants are so important in the distribution of most animals, the definition of community habitat used for plants can also serve for the animals in the community.

Communities interact to form larger geographical groupings, or *biomes*, indicated by similar vegetation type. A shortgrass prairie biome, for example, includes numerous communities due to the edaphic, topographic, and microclimatic variation. Soil moisture is an important prairie variable, for example. Plants adapted to very moist soils occur at one extreme and many upland plants at the other extreme. Yet the biome vegetation is characteristically short and herbaceous throughout. The shortgrass prairie biome is characteristic habitat for adapted animal species, such as pronghorn antelope, prairie dogs, and bison in the shortgrass prairie (see Chapter 9).

A stand is a homogenous unit of vegetation that differs from surrounding vegetation occurring within a biome (Figure 3.10). Stands are dominated by one or a few species of plants. Stands result from differences in the environmental requirements of the various plant species and from environmental diversity. Tree plantations and crop land are examples of stands created by humans. Particular animal species are associated with specific stands of vegetation. Beaver, for example, are most abundant where stands of preferred food species, such as willows and aspen, are common.

Biodiversity

Biodiversity refers to the variety of species, or *species richness,* within the community. In more elaborate measures of diversity the relative *evenness of abundance* also

FIGURE 3.10 A stand of longleaf pine trees in North Carolina (courtesy of U.S. Forest Service).

contributes to the index of diversity. A community dominated by one or two species is less diverse than a community where the species are all relatively uncommon and more equal in abundance. Often species richness and evenness of abundance are correlated. In areas known for their high numbers of species, such as rainforests, no single species stands out in abundance. In such situations there is a low probability that any single species will dominate community functions. Regardless of what may happen to one species, community functions will be sustained. Species redundancy is central to the idea that diverse communities are functionally more stable.

The least diverse community, in theory, is composed of one species (monocultures). Communities are rarely completely monotonous, although many species in what appear to be single-stand communities (e.g., cattail marsh, aspen grove) are obscure and rare. A community may be relatively rich in species and yet exhibit low diversity. Even in cultivated fields of corn or wheat, the diversity of invertebrates, microbes and vertebrates can be high, although the crop comprises most of the biomass.

Communities dominated by monotypic stands encourage species that will consume or parasitize them. Muskrats, for example, will consume cattails and tip the balance more in favor of other plants in the marsh. Humans have learned to exploit the most usable species in ecosystems. Cropland monocultures are the extreme in this direction. One price paid for large continuous crop monocultures is costly competition with pest species and depletion of soil nutrients (see Chapters 8 and 12).

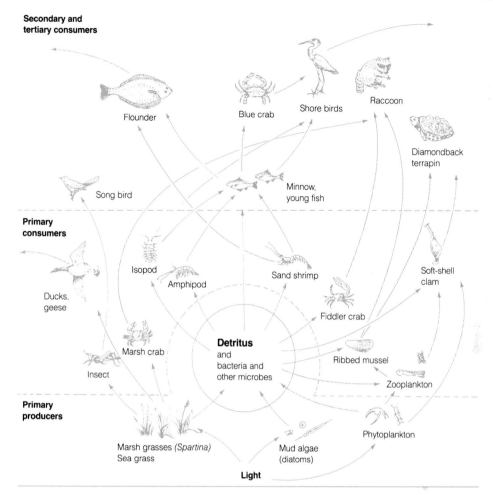

FIGURE 3.11 The various components of an estuarine community (from Lerman 1986). A community is composed of all of the primary producer, consumer, and decomposer organisms interacting in a given locality and at a given time.

The number of different communities present in a region and their arrangement also adds an important dimension to diversity. In some regions, species composition may change little throughout the entire area. In other areas of similar size, there may be many discrete communities.

Trophic Functions

Community functions are roles played by the different components of the community. Community biotic components include producers, consumers, and decomposers (Figure 3.11). The most basic community functions have to do with energy and material transfers among producer, consumer, and decomposer components. This will be discussed along with some management implications.

Producers. *Primary producers* are plants, certain bacteria, and algae containing the pigment chlorophyll, which can convert solar energy to chemical energy through *photosynthesis.* The primary producers support, directly or indirectly, all other groups of organisms. *Primary productivity* has been defined as "the rate of organic matter storage by photosynthetic and chemosynthetic activity of producer organisms in the form of organic substances which can be used as food materials" (Odum 1959).

Photosynthesis is the process by which primary producers convert energy from the sun, carbon dioxide from the atmosphere, and water and minerals from the soil into nutritional resources for maintenance and growth (Figure 3.12). Many smaller

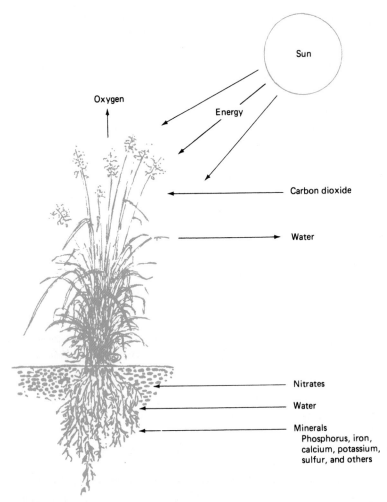

FIGURE 3.12 Materials used by grass plant for photosynthesis (drawing by John N. Smith) (from Holechek et al. 1998).

primary producers have no specialized tissues and take up nutrients across their simple-cell surfaces. Oceanic and lake phytoplankton are small primary producers in this category responsible for most of the earth's primary production. Large terrestrial plants have leaves, roots, and vascular transport tissues (e.g., wood), that perform specialized life functions. Photosynthesis occurs primarily in the leaves (and sometimes the stems) where the pigment chlorophyll captures light that drives the formation of sugars, starches, fats, proteins, and other food compounds. Carbon dioxide from the air is taken up through leaf openings, called stomata. Plant roots obtain nitrogen, phosphorus, potassium, iron, and other nutrients from soil and aquatic sediments.

Many factors determine the rate of photosynthesis (primary productivity). These include the area of leaf or cell surface, intensity and quality of light, amount of carbon dioxide in the air or water, physiological efficiency of the plant, soil or water nutrient concentrations, water supply, and temperature. The amount of food manufactured is proportional to the amount of chlorophyll exposed to light. Light is often the limiting factor for primary production in ocean and lake depths, and under dense canopies of vegetation. Even where light is plentiful, photosynthesis is below potential because some other resource, such as nitrogen or phosphorus, is limiting. Under optimal conditions plant communities can use about 2% of the light that reaches them. However, under most conditions solar energy conversion efficiency is much lower. In many terrestrial communities water is the most frequent limiting factor. In most terrestrial and aquatic communities, inorganic nutrients, such as nitrogen and/or phosphorus, are likely to be most limiting.

Carbon dioxide in the air rarely limits terrestrial primary production. Carbon dioxide is more likely to be limiting in aquatic environments where it is less available as a gas dissolved in water than in air. More often, however, other nutrients limit photosynthesis. Two of the most common limiting nutrients are nitrogen and phosphorus, which often are the basis for commercial fertilizers. In the open ocean, iron is most likely to limit production. A variety of other inorganic nutrients can be limiting in certain circumstances and for particular groups of primary producers. For example, the availability of silica can determine the relative abundance of diatoms (a form of algae with a silica shell) in fresh waters.

Water is a necessary chemical constituent of photosynthesis. It provides hydrogen and, in terrestrial plants, it serves to keep the stomata open and the plant turgid. When the stomata close from water stress, carbon dioxide is denied entry. Water is also important in transporting dissolved minerals from the soil into and through the plant. For truly aquatic primary producers, water provides the environment for nutrient supply and light transmission as well as physical support. Water depth above and below the ground surface is key to the distribution of many aquatic, wetland, riparian (in river floodplains), and upland primary producers. Also, variation in the seasonal supply of water is important. This is because plant need for water varies with life stage and environmental conditions. Certain types of wetlands, for example, only exist where there is regular seasonal flooding and drying.

Physiological rates differ among species and are controlled by temperature. When low temperature forms ice, water is no longer available for uptake. Freezing within cells and tissues can, upon expansion into ice, disrupt cell walls and badly damage or kill plants. Concentrations of salts in plants can protect them against freezing.

Numerous other factors are influential in determining the rates of primary production for the entire community and for different species within the community. Even the casual observer will notice changes in plants as they walk a path from higher and dryer environments to moister lowlands, wetlands, and open water. Gradients of light, moisture, nutrients, toxic materials, acidity, salinity, sediment particle size, and so on combine to determine the relative productivity and survival of different species. The primary producer components change along these gradients, influencing the consumers and decomposers that depend upon them for food, cover, and other needs.

There can be many subgroupings within the primary producer component of any particular community. In a forest community, for example, the tallest trees function in the upper most stratum where they dominate the light. Understory trees and shrubs typically use light with greater efficiency than those in the overstory. In the soil, root systems of different plants occupy varying depths, which permits more efficient use of stored water and nutrients. Where communities are more structurally uniform, as in the phytoplankton community of oceans, the spatial differentiation into functional groups is less obvious than in forests or grasslands.

Decomposers. Decomposers break down and absorb dead organic matter (detritus). The decomposer organisms include mostly bacteria and fungi, which possess enzyme systems necessary to break down resistant organic materials. In the process, organic matter is reduced to its inorganic components, which then become available once again for primary producer uptake. Without decomposers, elements would eventually be completely tied up in undecomposed organic material, and community function would stop.

Certain bacterial decomposers are among the most widely distributed and adapted life-forms. They are found in any environment where organic production occurs. Different species of decomposers vary in their tolerance to environmental conditions. There are significant differences among the bacteria and fungi, which are the most important of the decomposers. These diverse groups complement one another in their different approaches to decomposition. Bacteria group into three types with respect to their oxygen requirements: aerobic, facultative aerobic-anaerobic, and anaerobic. Whereas aerobic bacteria cannot tolerate an environment without oxygen, anaerobic bacteria are killed by oxygen. Bacteria of some kind occur in nearly all environments. Unlike fungi, many bacteria can function in anaerobic as well as aerobic environments, although the rate of decomposition falls sharply as oxygen concentration approaches zero.

Fungi have an advantage over bacteria when the organic matter is in larger masses, such as woody plant matter and soil humus, and is resistant to weathering into smaller particles. Fungi have threadlike hyphae, which can penetrate deep into organic tissue. In contrast, bacteria use the strategy of small spherical and rod shapes to develop a high contact surface area for the decomposition process.

Until raw sewage treatment became nearly universal in the U.S., many urban rivers were anoxic (without oxygen) because of overloading with raw human wastes (Perry and Vanderklein 1996). Oxygen, an important rate regulator of organic decay, was depleted by the waste overload. Deep soils and sediments of wetland and aquatic ecosystems typically remain continuously anaerobic and accumulate organic matter. Through geologic time this organic matter may eventually become oil and natural gas. The decomposition rate also slows in acidic environments. Calcium appears to be an especially important limiting nutrient under acidic conditions. Where decomposition is slow, bogs, bog waters, and other "brown-water" lakes, streams, and swamps develop and create unique ecosystems with many life-forms (Horne and Goldman 1994). The brown stain in the water is caused by an accumulation of large organic molecules that decompose very slowly in the calcium-poor environment.

Consumers.   Consumers are mostly animals, which ingest other living things and convert the energy into biomass (Figure 3.13). They are categorized into functional subcategories (primary, secondary, tertiary) based on the number of feeding transformations (or *trophic levels*) that have occurred (Figure 3.13). *Herbivores* feed directly on primary producers. They include cattle, grain-feeding birds, various insects, and molluscs. *Carnivores* are those animals that eat other animals including herbivores and other carnivores. Cats, snakes, adult frogs, many fish, and numerous invertebrates are strictly carnivorous. Many animals are *omnivores,* which feed on both plants and animals. Humans, chickens, turtles, bears, tadpoles, carp, and crayfish are all examples of omnivores.

Detritivores are consumers that ingest dead organic matter and decomposers (Figure 3.14). Some of the most obvious detritivores are ants, termites, and nematodes, which are critical, but often overlooked, components of terrestrial ecosystems (Paris 1969). Aquatic detrital food chains are particularly important in woodland streams, ponds, and bogs with rich supplies of organic detritus. Many aquatic insects, molluscs, worms, and crustaceans are detritivores. Consumers and decomposers compete for organic foods while complementing each other in converting organic matter back to inorganic elements.

Competitive Interactions Within Communities

Competition has been defined in many ways by different ecologists (Risser 1969, Ricklefs 1996). We define competition as two or more organisms trying to gain one or more scarce, required resources from the same ecosystem.

Generally, the factors for which competition occurs among plants are water, nutrients, light, oxygen, and carbon dioxide (Risser 1969). In some cases, plant competition may occur for pollinating agents or for suitable germination sites. For animals, competition typically occurs for food and habitat. Competition may occur among species (interspecific) or among individuals of the same species (intraspecific). The results of competition may manifest at the individual or at the population level. Individuals may show the effects of competition by reduced growth rate, reduced vigor, and death. At the population level, competition may be demonstrated by reduced density or by changing patterns of distribution (Risser 1969). Organisms often avoid competition by using the same resources at different spatial locations or seasons during the year (Figure 3.15).

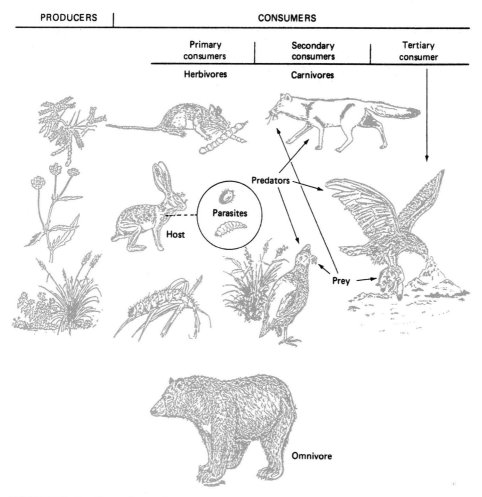

FIGURE 3.13 Feeding relationships among plants and animals (redrawn from Nebel 1981 by John N. Smith).

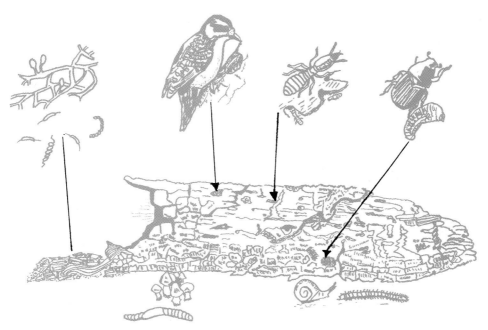

FIGURE 3.14 Various detritivores make use of the organic matter in a fallen tree (adapted from Miller 1990 and Nebel and Wright 1996).

FIGURE 3.15   American robins (left) and American woodcocks (right) both feed largely on earthworms, but find their prey in different ways and in different locations, thereby avoiding direct competition for the same food resource (from Bolen and Robinson 1995, design by R. Miller).

Interactions Among Biotic Communities

Communities are interactive in a variety of ways. Through wind, flight, and other movement organisms disperse from one community to another. Many mobile organisms are only temporary community members. Migratory animals are obvious examples. Many larger species of vertebrates, such as ducks and geese, require a different habitat for reproduction and early life stages than for other life stages. They move seasonally among communities as their need for resources changes. They can become very influential forces, as evidenced by the seasonal impact migratory grazing herds can have on grassland vegetation. Similarly, huge schools of planktivorous fish can have a local short-term impact on plankton populations as they move through lakes and oceans. Wild geese and duck populations have adapted to and affect biotic communities thousands of miles apart in order for them to use different summer and winter habitat. These migratory species link communities that are otherwise separated by large distances.

Communities are bridged by less mobile species at their *ecotones,* or areas where two communities come together (Figure 3.16). Ecotones are of particular importance to terrestrial vertebrates, which typically require more than one community to meet their daily needs. The three main types of ecotones include simple, gradual, and mosaic ecotones. The simple ecotone involves an abrupt change from one community to another (Figure 3.16). The gradual ecotone refers to an extended area where species from two or more communities gradually blend together. The mosaic ecotone involves a blend of two or more communities (Figure 3.16).

Simple ecotones are most common in desert and forested areas. They are usually caused by a sharp change of either soils or topography. Wetlands often form simple ecotones between terrestrial and aquatic communities. Gradual ecotones are characteristic of prairie areas where soils and topography change slowly across the landscape. Ecotones are especially gradual in the open oceans where temperature variation is the predominant environmental gradient. Mosaic ecotones are common in forest and grasslands where storm, fire, and groundwater close to the surface cause patchy distributions of vegetation. Whittaker (1975) discovered that many communities do not exhibit truly discrete identities at all, especially where they occur along continuous environmental gradients, such as mountain slopes.

When a community is partially or entirely eliminated by catastrophe, or newly created, the colonization of the area can occur either though ecological or evolutionary process. Many ocean islands formed from volcanic activity have unique flora and fauna as a consequence of extended isolation coupled with long periods of existence. Hawaii and the Galápagos Islands are excellent examples. Recently formed islands are colonized by organisms that are present in the vicinity or somehow (transport by humans) make it to island shores. Much has been learned for example from Krakatoa, an island near Sumatra, which was entirely reformed by a volcano in the late eighteenth century. It has since been colonized by many species found in the vicinity.

FIGURE 3.16 A mosaic ecotone in central New Mexico where a woodland and grassland community come together. Another simple ecotone occurs along the fenced area in the foreground due to differences in methods of cattle grazing.

Some basic principles of *island biogeography* are important in *landscape ecology* and *conservation biology.* Organisms dispersing from continents reach large islands faster than they reach small islands the same distance offshore. Organisms colonize islands closer to shore faster than ones farther offshore. These principles pertain even when islands are something other than land in water, such as lakes surrounded by land or patches of different vegetation in a "sea," or *matrix,* of another vegetation type. There are many examples of vegetational *patchiness,* such as swamps surrounded by upland forests, marshes and pothole lakes surrounded by upland prairies, and mountain forests surrounded by desert lowlands.

Living things are often used to manage communities. Manipulation of animal grazers, such as domestic livestock and grass carp, can control the form and composition of terrestrial and aquatic plant communities (Figure 3.16). Introductions of plants and animals alters the value of living resources, sometimes positively and often negatively. Invasive plants and animals often are keystone species that can displace native species and greatly alter community structure and function. Because invasive species are usually widespread and displace rarer species of limited distribution, they can reduce genetic diversity.

ECOSYSTEMS

Ecosystems Defined

Definitions of ecosystem usually emphasize interactions among the living and nonliving components occupying some defined space (Colinvaux 1993, Ricklefs 1996) (Figure 3.17). Forbes (1888) expressed this concept in terms of the lake "microcosm"—a miniature world among many on earth. Ecosystems can be very small, such as the intestinal community living in a flea, or as large as the biosphere that forms an envelope around the entire earth. Most practitioners of ecosystem management think of each ecosystem as a "functional unit consisting of the organisms (including man) and environmental variables of a specific area" (Van Dyne 1966). Ecosystem functions are time

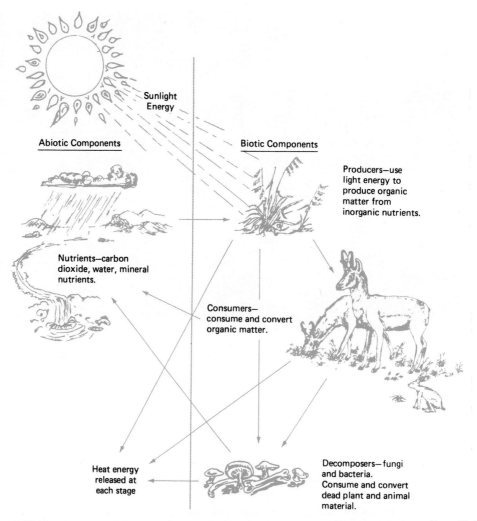

FIGURE 3.17   General diagram showing interactions among ecosystem components (from Holechek et al. 1998). An ecosystem is defined by the interaction of abiotic components in the habitat with biotic components in the community (drawing by John N. Smith).

dependent processes, such as organic matter production, mineral cycling, and succession following disturbance. Structure is the form of the ecosystems at any point in time.

ECOSYSTEM STRUCTURE AND FUNCTION

We define ecosystem structure as the spatial arrangement of the living and nonliving environment in a particular area. Structure results from the relative abundance, spatial arrangement, and form of the ecosystem's many component parts. The abiotic structural components consist mainly of matter in the rock, soil, water, and air. The structure of soil, for example, determines how much water, oxygen, and nutrients can reach plant roots and soil organisms. The hydrogen ion content of soil and water determines its acidity and alkalinity. It is important in determining the distribution and abundance of many species and the availability of some nutrients. Availability of inorganic nutrients such as nitrogen, potassium, and sulfur determines distribution, production, and abundance of primary producers.

Ecosystem functions include energy flows; material storage, flow, and cycling; genetic controls on life processes (genetic diversity); and successional processes. We will discuss each of these functions.

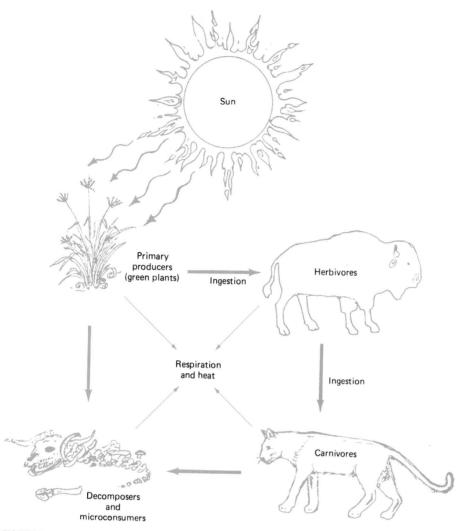

FIGURE 3.18 Generalized diagram showing energy flow through a grassland ecosystem (drawing by John N. Smith) (from Holechek et al. 1998).

Energy Flow

The sun is the basic source of energy for all life on planet earth. Solar energy is received by grasses, forbs, and shrubs and is transferred by the process of photosynthesis into stored chemical energy in plant tissues. When herbivores eat plant tissue, they gain energy stored in plant tissues through the process of digestion (Figure 3.18). Carnivores, in turn, eat other animals and derive their energy requirements from their food. However, energy is dissipated at each step in the food chain through respiration.

Energy flow throughout the ecosystem operates under the first law of thermodynamics: Energy can be neither created nor destroyed, only changed in form. The organisms at each step in the food chain are not completely efficient at harvesting all available food resources, so energy retention diminishes considerably at each trophic level (Figure 3.19). Once energy dissipates to heat and is radiated back to space, it can never be recovered and reused. Thus energy flow is a one-way path and must be continually refueled by energy from the sun.

Trophic-Dynamics. Trophic-dynamics remains one of the earliest and most promising ways to understand the diverse interactions within ecosystems. Lindeman (1942) was the first to describe the concept of feeding (trophic) levels in communities based on the number of times food energy had been transformed from one life-form to another.

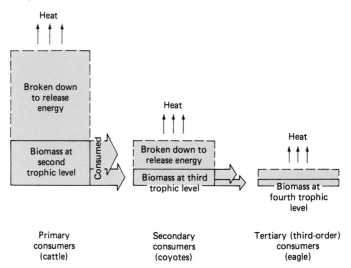

FIGURE 3.19   The loss of energy as it moves through trophic levels (from Nebel 1981).

Primary producers make up the first level, herbivores the second, first-level carnivores the third, and so on until the top carnivore level is reached (Figure 3.18).

Most ecosystems have 4 to 6 trophic levels. For example, lake phytoplankton (primary producers) support zooplankton (herbivores), which are eaten by small minnows (first-level carnivores), which feed yellow perch (second-level carnivores), which are consumed by northern pike (third-level carnivores), which are eaten by people (top-level carnivores), which most often decompose without being consumed. A species rarely occupies just one level. Many species are omnivores that feed from different levels. Other species go through life stage changes in which one life stage feeds in one level and another life stage feeds in another level. Many frogs, which are carnivorous as adults, have herbivorous or omnivorous larvae (tadpoles). Many top-level carnivores progressively feed at higher levels as they grow to a larger size.

Although it is far from a constant rule, the size of consumers tends to increase with position among trophic levels. The primary exception is in grassland and savannah communities, which have exceptionally large herbivores (i.e., bison, elephants, rhinoceros). However, the largest biomass of terrestrial herbivores, even in grasslands, is found in the ants, worms, and other invertebrate species.

Trophic Efficiency.   The conversion of production in one consumer level to another consumer level is inefficient. A few studies have been conducted to estimate magnitudes of energy flow in ecosystems. These studies show that less than 2% of the solar radiation received yearly by primary producers is converted to production through photosynthesis, and less than 1% is the rule. Only a relatively small portion of the above-ground primary productivity (about 10%) is harvested by herbivores, including livestock (Lewis 1971, Pieper 1983).

An average of about 10% of the energy of each previous trophic level is retained by the next trophic level. By the time it reaches the top carnivore level, very little is left. For example, if 1,000 kg/m^2 is produced by plants, only 10 g/m^2 is retained at the first carnivore level. From a practical standpoint, this means that humans retain only 1.0% of the energy if they eat only pigs or chickens fed grain compared to 10.0% conversion if they eat the wheat or corn directly (Figure 3.20). The point here is that the world can support far more humans if they eat only plant material compared to if they eat a large amount of meat.

Species tend to become progressively more specialized as they become larger. However, the biomass present is not necessarily indicative of the capability of a species to recover from stress. In world fisheries, rapid decimation by harvest has occurred for what appeared to be very abundant populations. Because they have low

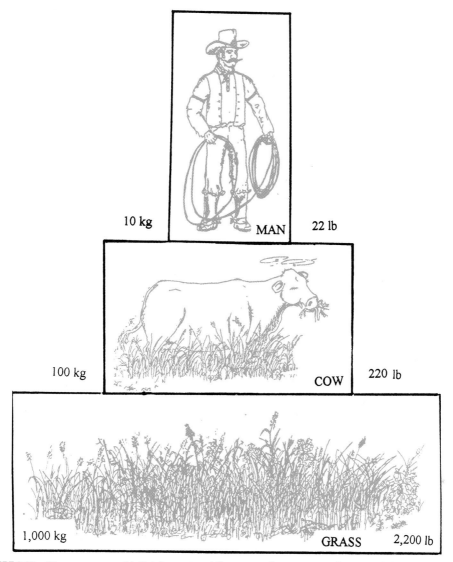

10 kg   MAN   22 lb

100 kg   COW   220 lb

1,000 kg   GRASS   2,200 lb

FIGURE 3.20 Biomass pyramid. It takes 1,000 kilograms of grass to produce 100 kilograms of beef needed to produce 10 kilograms of human biomass (adapted from Owen and Chiras 1995 by John N. Smith).

intrinsic rates of increases, the relatively high biomass in large fish species is a misleading indicator of production and sustainable yield. The list of the endangered species of the world has a remarkable number of large carnivorous species. Large compared to small species, require much more area of food producing habitat to support viable populations. Habitat fragmentation tends to affect the viability of larger species more than smaller species.

Consumers have developed many adaptations to take advantage of foods of all sizes and types. Some of the most productive consumers are detritivores—among the most impressive being earthworms in moist organic soils. One of the most important ways of reducing particle size is by consuming it and egesting fecal matter. Feces provide a rich substrate for decomposers. Thus, many detritivores specialize in consuming fecal matter (copraphagy).

Humans, of course, occupy dominant positions in food webs. Some ecologists believe that as the world becomes more densely populated people will be forced to eat more plant material and less meat. This belief is based on the idea that food demand will outpace technological developments. However, crop production so far has increased faster than the human population. Past advances have depended heavily on high amounts of fossil fuel, fertilizers, and pesticides (see Chapter 12). This approach is contributing to environmental damage, such as global warming and eutrophication of rivers and marshes. Thus, associated with any technological advance are trade-offs that need to be considered as societies determine their priorities. Natural resource managers increasingly will be involved in quantifying the benefits and costs associated with new technologies and providing that information to the public for policy decisions.

Resource Partitioning.   Within trophic levels, populations compete for food while avoiding predation. They also complement each other through unique specializations that give them advantages under certain conditions. Because most ecosystems are dynamic and environments vary, a large number of niches are available. As new species evolve, other species evolve to respond to them as a food resource, competitor or predator. In this manner, the energy resources are partitioned among many specialist species, each with higher ecological efficiencies than their more generalized predecessors. The energy resources of the habitat are used more efficiently as diversification occurs, and the number of trophic levels increases.

Material Flow, Storage, and Cycling

The second basic functional process of ecosystems is material flow, storage, and cycling. This function includes physical, chemical, and biological processes, which are discussed below.

Material Flow.   Materials flow both actively and passively into, through, and out of ecosystems (Figure 3.21). Water, wind, and living organisms are important transport agents. Water and wind act to reshape the earth through erosion. They redistribute the eroded material into deposits which form valley alluvium, outwash plains, floodplains, and sediments. Water also dissolves minerals and transports them through the soil, wetlands, streams, oceans, and in the water vapor of the atmosphere. These suspended and dissolved materials create the nonliving foundation for supporting

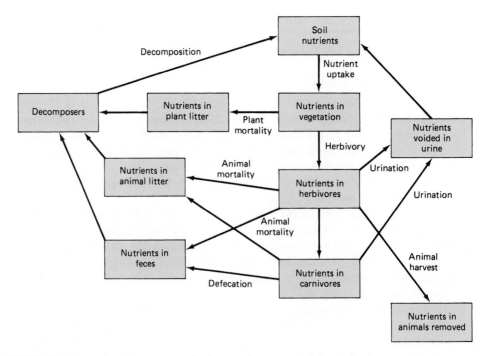

FIGURE 3.21   Generalized nutrient cycle diagram for a terrestrial grassland ecosystem (from Pieper 1977).

biotic communities but also include inhibiting contaminants. Individual elements are absorbed by primary producers and are synthesized into living tissues. Herbivores consume plant tissue and the elements contained there. Some plant material is not consumed by herbivores and is instead broken down by decomposers and microconsumers. The chemical elements are then returned to the soil. For herbaceous plants these turnover rates can be rather rapid, but, for woody plants, elements may be tied up in plant material for many years. The feces and urine of all consumers are eventually returned to the soil, where they may be taken up by the plants.

Salmon, for example, are important transport pathways for returning nutrients to the infertile streams where their young are born. By dying after reproduction their decayed carcasses enrich the system that supports the fry and fingerlings.

Unlike energy, which degrades and becomes unusable as it flows through ecosystems, chemical elements are reusable and cycle through the various compartments. Many cycles are completed in hours or days. Other cycles take millions of years to complete. Whereas many material cycles include only solid forms, others involve both solid and gaseous forms. Few materials become liquid or remain that way for long. Water is an important exception. Among the more important material cycles are those for carbon, oxygen, nitrogen, phosphorus, and iron, which we will briefly discuss.

Carbon and Oxygen. The primary element in organic matter is carbon. The original source of atmospheric carbon dioxide is from volcanic emissions from the earth's interior. However, some carbon dioxide comes from carbonate rocks like limestone and dolomite. Carbon in fossil organic matter, such as coal and oil, can be returned to the atmosphere through natural combustion or burning by humans (Figure 3.22). Much of the earth's carbon is fixed in plant communities, especially in the wood of forests. Carbon dioxide builds up in the atmosphere when forests are cut and oxidized (burned). In tropical rain forests cutting and burning is now occurring at a more rapid rate than photosynthesis is locking up carbon in new organic matter. The accumulation of carbon dioxide in the atmosphere from forest cutting and fossil fuel burning is believed to be causing atmospheric warming, or the green house effect (Freedman 1995) (see chapter 6).

Nitrogen and Sulfur. Nitrogen and sulfur are two nutrients of exceptional importance in ecosystem dynamics (Figure 3.23). Nitrogen is the most abundant gas in the atmosphere, but it is very scarce in the lithosphere (rock) and hydrosphere. Only certain bacterial forms can fix atmospheric nitrogen into organic nitrogen. Some of them, such as Cyanobacteria (commonly called blue-green algae), carry out photosynthesis in moist soils and in water. Other bacteria live in a symbiotic relationship with root nodes of leguminous plants. Nitrogen occurs in nitrate and ammonia forms.

Atmospheric nitrogen is also oxidized to nitrate-nitrogen by lightning. Nitrate-nitrogen is taken up by rooted plants, phytoplankton, and other primary producers. Ammonia nitrogen is a by-product of protein consumption (in waste products) and decomposition. Both ammonia and nitrate-nitrogen are very soluble in water, rapidly leached from soils, and are the main forms of nitrogen transported by streams and rivers. All nutrients, including nitrogen, cycle repeatedly through primary production and decomposition. In river channel and floodplain sediments this process is called *nutrient spiraling*. As it spirals through streams, rivers, wetlands, lakes, and estuaries, much of the nitrogen carried downstream returns to the atmosphere. Where decomposing matter gets trapped in anaerobic environments, such as deep sediments and wet soils, ammonia is reduced to gaseous nitrogen. It leaks back into the atmosphere in the form of small bubbles.

Nitrogen can be toxic in high concentrations. The fixation of atmospheric nitrogen by certain planktonic bacteria (Cyanobacteria) causes accumulation of toxic by-products. Nitrogen emitted with combustion products of petroleum contributes to the formation of acids in water vapor, causing acid rain.

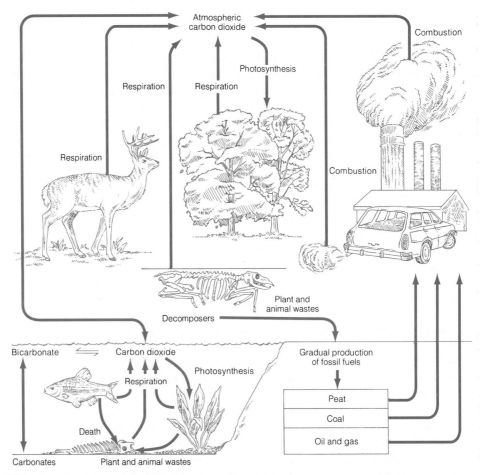

FIGURE 3.22   The carbon cycle (from Moran et al. 1980).

Sulfur is unlikely to be a limiting nutrient, but can reach toxic levels in certain aquatic environments. Wetlands are especially important in storing sulfur where it is reduced to hydrogen sulfide gas, which is toxic in high concentrations. The gas is oxidized rapidly when slowly released to the atmosphere in wetland gassing. There it dissolves in rain and is washed back to the land where it contributes to primary production. Human natural resource management has increased sulfur mobilization and redistributed it from the lithosphere to the atmosphere. Sulfur is artificially increased in the atmosphere by man's combustion of coal and oil. Under these conditions it can contribute to acid rain (see chapter 6). Exposing sulfur-bearing minerals to the air during surface coal mining results in acid mine drainage, a major water pollution problem.

Phosphorus and Iron.   Many nutrients are metals and related elements that have no gaseous phase. Among them, phosphorus and iron are most often limiting nutrients. Their compounds remain locked within the lithosphere and hydrosphere. They are part of the *sedimentary cycle,* which is the long geological process of continental erosion, sedimentation into depressions (usually the ocean), and uplift in new continents. Phosphorus is especially likely to limit primary production at high elevations in areas of slowly weathering rock such as granites and schists, where both iron and nitrogen are more available (Figure 3.24). However, iron and phosphorus typically have closely associated geochemical transformations. In an oxidized state, phosphorus tends to form complexes with iron. Both elements precipitate and fall to the bottom of deep lakes and oceans where turbulence is low.

Because of erosion, rivers and streams transport phosphorus and iron toward lower elevations through nutrient spiraling. Stream turbulence keeps oxidized iron

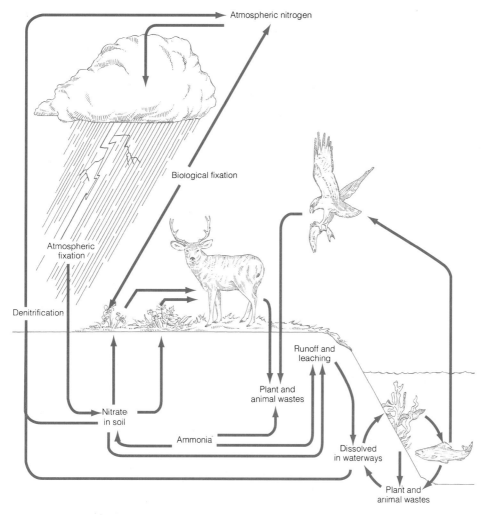

FIGURE 3.23   The nitrogen cycle (from Moran et al. 1980).

and phosphorus in suspension where they are available for biotic uptake. Once the rivers reach estuaries and oceans, much of the phosphorus and iron fall to the bottom either through chemical change or through biological uptake and settling of dead organisms. Iron is typically more trapped than phosphorus. It is usually the more limiting nutrient in the open ocean.

Where phosphorus and iron reach lakes, they gradually precipitate and fall to the bottom wherever the water is well aerated. Precipitation may be biological, through uptake by plankton which die and settle out, or through chemical and physical processes. Iron oxides and phosphates form complexes that gradually build mass and settle to the bottom. As long as water remains oxygenated, the precipitate will remain on the bottom unless it is disturbed by the wind. In some lakes, wind disturbance is an important means for recycling phosphorus to the surface where it can be taken up once again. Many lakes are deep enough to resist full mixing to the bottom for part or all of the year. They form a stratum of more dense water over the bottom—the hypolimnion—which is easily depleted of oxygen by decomposition when enough dead matter settles.

Relatively nutrient rich, *eutrophic,* lakes often produce enough organic matter to overwhelm all of the oxygen resources in the bottom stratum. Once this bottom layer becomes anoxic, changes in the chemical environment result in oxygen being stripped from phosphates and iron oxides and the two elements going back into solution. If the stratification is permanent, the two nutrients remain well below the surface. If the stratification is temporary, as it is in most U.S. lakes, the nutrients are returned to the

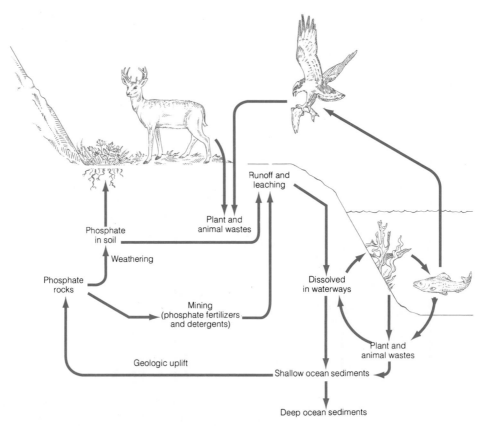

FIGURE 3.24   The phosphorus cycle (from Moran et al. 1980).

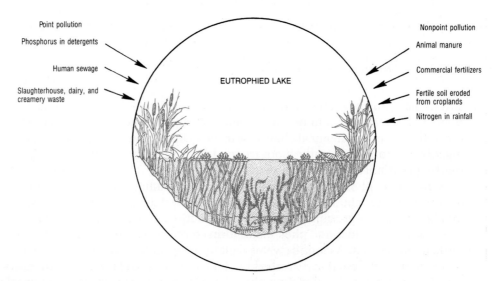

FIGURE 3.25   Anthropogenic sources of nutrients causing harmful eutrophication (from Owen et al. 1998).

surface upon remixing where they are available for more primary production. Most large lakes are efficient traps for nitrogen and phosphorus. *Cultural eutrophication* is a form of pollution that results from excess release of nutrients to natural waters (Figure 3.25). Intense production causes imbalances between consumers and decomposers, and results in severe oxygen depletion. This reduces the value of aquatic resources. Incomplete sewage treatment and overfertilized farmland are two main sources of phosphorus pollution (Perry and Vanderklein 1996).

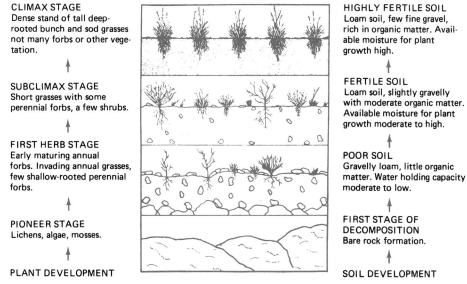

CLIMAX STAGE
Dense stand of tall deep-
rooted bunch and sod grasses
not many forbs or other vegetation.

↑

SUBCLIMAX STAGE
Short grasses with some
perennial forbs, a few shrubs.

↑

FIRST HERB STAGE
Early maturing annual
forbs. Invading annual grasses,
few shallow-rooted perennial
forbs.

↑

PIONEER STAGE
Lichens, algae, mosses.

↑

PLANT DEVELOPMENT

HIGHLY FERTILE SOIL
Loam soil, few fine gravel,
rich in organic matter. Available moisture for plant
growth high.

↑

FERTILE SOIL
Loam soil, slightly gravelly
with moderate organic matter.
Available moisture for plant
growth moderate to high.

↑

POOR SOIL
Gravelly loam, little organic
matter. Water holding capacity
moderate to low.

↑

FIRST STAGE OF
DECOMPOSITION
Bare rock formation.

↑

SOIL DEVELOPMENT

FIGURE 3.26   Primary succession on grassland rangelands (from Gay 1965).

Succession

Succession is defined as the replacement of one biotic community by another. Clements (1916) provided the first detailed description of plant succession. In recent years, some of Clement's theories have been challenged, but his ideas still remain the foundation for modern plant ecology. In the classical sense, plant succession in both terrestrial and aquatic ecosystems involves the replacement of one plant community by another until the final community is reached. This final, somewhat stable community is often called the *climax*. Although a given habitat type may support several successional plant communities, it will support only one climax plant community.

A climax community is considered to exist on an area if significant changes in species composition do not occur within 500 years (Barbour et al. 1987). Climax plant communities are not static, but the changes that do occur involve small random shifts in numbers and/or relative biomass that fluctuate around a long-term mean. Directional, cumulative, nonrandom vegetative changes that occur within a period of 500 years are referred to as *seral stages* (Barbour et al. 1987). The term *sere* is used to describe an entire progression of seral stages from the first community that occupies bare ground to the final climax vegetation.

Ideas concerning successional changes and stability have been quite controversial, and the classical view of succession has been subject to numerous interpretations. Under the classical view of succession, plants are the overwhelming biological force driving successional rate and direction. The bacteria, fungi, microbial algae, and animals play dependent and secondary roles. In recent years, the importance of animal pollinators, grazers, soil makers, and decomposers has become more appreciated.

Terrestrial Primary Succession.   Succession may start from any point along the sere. Primary succession starts from bare rock and open water (primary areas). On bare ground suitable microhabitats are first colonized by lichens, algae, moss, bacteria, fungi, and animals (Figure 3.26). A rudimentary soil forms as rock is weathered, and water and organic matter are added from the partially decomposed organisms. Seeds from nearby plants are carried by wind, water, or animals to the newly formed soil where some germinate and establish the first vascular plants. Gradually more advanced soil animals and decomposers colonize the area, accelerating organic conversion, nutrient cycling, soil aeration, and water retention.

The earliest plant colonizers are often annual plants (generalists) that can survive under rather harsh conditions, but die at the end of the growing season. With further weathering and soil formation perennial plants become established. Herbaceous plants typically precede woody plants. Each assemblage of plants influences soil and microclimate, making it more suitable for plants that have higher requirements for water, nutrients, and other resources.

Early colonist species alter the environment to an extent that it no longer is suitable for them and they are replaced by better adapted (more specialized) plants. Total biomass of all life-forms, total energy storage in organic matter, structural diversity, and rate of mineral cycling increase as succession proceeds (Lewis 1969). The processes of terrestrial primary succession are summarized as follows:

1. The development of soil from parent materials
2. Increasing longevity with successional advance
3. Replacement of species with broad ecological requirements by those occupying narrow niches complementary with other species
4. Greater accumulation of living tissue and litter per unit area
5. Modification of microenvironmental extremes
6. Change in size of plants from small to large
7. Increase in the number of pathways of energy flow
8. More nutrients tied up in living and dead organic matter
9. Greater resistance to fluctuation in the controlling factors

Periodicity and *plant stratification* tend to be highly developed as the climax community is approached. Periodicity involves different species making peak nutritional demands at different times during the year. Stratification is the development of layers of plants with different heights and rooting depths. These changes allow for a larger number of species, more pathways for energy flow, greater stability of total energy flow, and greater total productivity. As the habitat becomes more diverse, species become more specialized in habitat use and there is more efficient use of light, water, space, and nutrition. The microhabitat becomes less extreme as temperature, moisture, air movement, and other environmental factors become less variable and the community becomes more resistant to whatever fluctuations occur.

Whereas productivity generally increases as succession advances, it commonly declines as the climax is approached. Under climax conditions, nutrients are tied up in wood and other organic matter. Soil moisture is often quickly reduced by uptake from the larger plants. This, in turn, reduces the length of the growing season. Dense growth can reduce the efficiency of light use per unit of leaf area. In grasslands, excessive mulch accumulation can chemically and physically retard productivity. These observations suggest that terrestrial successional stages just preceding the climax support the highest production of goods such as timber and red meat. Genetic diversity is likely to be greatest in the late seral stages, but total regional diversity will be maximized where a mix of different seral stages exist.

Aquatic Primary Succession.   Aquatic succession depends on the initial depth and nutrient concentration of the lake (Figure 3.27). In deep infertile waters, succession is extremely slow if it occurs at all. Studies of such lakes indicate that the total community production is respired away over an annual cycle and inconsequential amounts of organic matter are stored. The lake will never fill to a point where rooted plants can get established. Even more extreme conditions exist in the open ocean. However, production in fertile lakes usually exceeds respiration and dead organic matter accumulates on the bottom. This can eventually allow rooted vascular plants to become established. Once this occurs, the rate of filling may accelerate and the lake will become a marsh or swamp with emergent plants. Because emergent plants

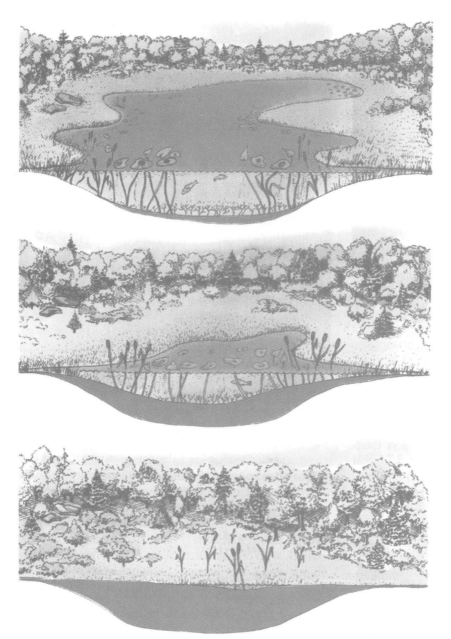

FIGURE 3.27 Primary succession in ponds and lakes involves a process in which they are gradually filled and invaded by surrounding land vegetation (from Nebel and Wright 1996).

transpire water, they can contribute to lowering the water level. Most commonly, soil erosion from the watershed surrounding a lake contributes to filling. It accelerates the establishment of conditions favoring rooted plant communities and causes hydric succession toward a terrestrial climax. Thus, a close link exists between the successional stages in the terrestrial watershed and succession in ponds and lakes. If erosion accelerates in the watershed, succession in the receiving waters will accelerate. In another form of terrestrial encroachment, vegetation extends into the pond or lake from the sides where it dies and accumulates. Eventually the pond is filled with peaty decaying organic matter, and becomes a terrestrial environment.

Secondary Succession. Secondary succession is defined as vegetational change that occurs after some type of disturbance such as fire, grazing, windstorms, flooding, or logging. Secondary succession generally occurs much faster than primary succession, and generally in a more predictable fashion. Although primary succession is of

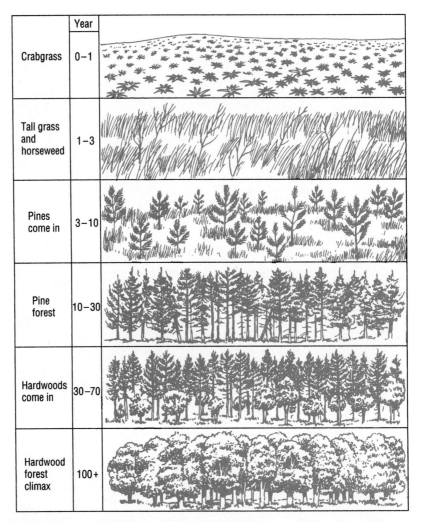

	Year	
Crabgrass	0–1	
Tall grass and horseweed	1–3	
Pines come in	3–10	
Pine forest	10–30	
Hardwoods come in	30–70	
Hardwood forest climax	100+	

FIGURE 3.28   Secondary ecological succession of plant communities on abandoned cotton fields in Georgia over a 100-year period (from Owen and Chiras 1995).

educational interest, it typically plays a much smaller role in natural resource management than secondary succession. The variability in secondary succession is reduced as the climax is approached (Huschle and Hironaka 1980). Figure 3.28 shows the stages of secondary succession that occur in the abandoned cotton fields in the Piedmont region of Georgia.

Figure 3.29 depicts secondary successional stages following fire in piñon-juniper woodlands in three different areas. Succession is somewhat predictable because the changes are very similar. A skeleton of dead trees remains after a fire, but these areas are soon occupied by annuals, and then by some perennial grasses and forbs. The herbaceous vegetation is invaded by some shrubs and eventually the climax piñon-juniper woodland. Two of the diagrams show what happens when fire occurs during one of the seral stages. The intensity of the fire, characteristics of the fuel, time of year, and so on, all influence the damage the fire does. Rates of terrestrial succession are driven mostly by moisture and temperature. In humid ecosystems, such as the tallgrass prairie in the eastern Great Plains, recovery to climax following disturbance can occur within five to ten years if severe erosion and soil nutrient depletion has not occurred. In the more dry Chihuahuan Desert, recovery from disturbance may require 20 or more years.

Climax Theory.   The final stage of succession has been viewed in different ways by different authors. Some have viewed the climax as "stable" while others have viewed it in "dynamic equilibrium" with the environment (Figure 3.30). Clements (1916)

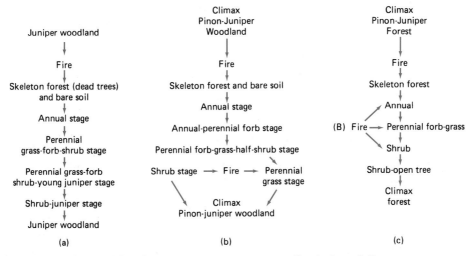

FIGURE 3.29 Successional pathways in piñon-juniper woodlands from different areas: (a) Central Utah (Barney and Frischknecht 1974), (b) Arizona (Arnold et al. 1964), and (c) Colorado (Erdman 1970).

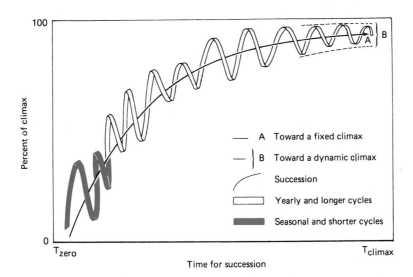

FIGURE 3.30 A model of succession asymptotically approaching a dynamic climax (B), which includes climatic variations associated with daily, seasonal, and yearly phenomena. The absolute climax might be defined as the midpoint of these variations at A (from Heady 1973).

viewed the climax as ultimately controlled primarily by climate—mostly precipitation and temperature intensities, amounts and temporal patterns. This he referred to as the "climate climax." He considered large areas of landscape as having the same climax to be governed by the same climate. Development of climax vegetation was considered a very slow process on the same time scale as geologic changes. He also recognized that periodic disturbance such as fire could sustain conditions deviating from the climate climax, which he referred to as *dis*climax stages.

Retrogression away from climax can occur from disturbances outside the community, such as fire, storms, grazing, disease, cultivation, and so on. However, some retrogressions to earlier stages can be caused by the community itself. In the tallgrass prairie of the central Great Plains, excessive mulch accumulation has caused the productivity of the climax grasses to be reduced and the composition to shift to earlier stages (Holechek et al. 1998). High levels of mulch tie up excessive amounts of nutrients, physically inhibits plant growth, delays plant growth in the spring by depressing soil temperatures, and provides habitat for pathogens and insects that can be harmful to the climax plants. There is evidence of similar community controlled retrogression in the Chihuahuan Desert and in various forest ecosystems.

Although fire and grazing are often considered forces that cause retrogression, there is also much evidence that they play a crucial role in maintaining the climax vegetation in many forest and grassland ecosystems. The timing, intensity, and frequency of both fire and grazing are critical factors in determining whether they result in retrogression or progression (Holechek et al. 1998).

Other authors considered other types of climax in addition to climatic climax, such as edaphic climax, where soil characteristics may have an influential role in the type of vegetation that develops. This was sometimes called the "polyclimax" viewpoint. Others considered vegetational development more a matter of chance, with discrete communities difficult to discern. This "individualistic" concept later gave way to the idea of the "continuum," which stated that the distribution of each species was independent of that of other species and that these distributions overlapped with each other (Curtis 1959). Whittaker (1967) found that species were often distributed along environmental gradients and that individual communities were only delineated somewhat arbitrarily along the gradient. In other words, ecotones were very gradual and extended in many locations. Therefore, discrete communities were hard to define. It is also obvious, however, that some ecotones are quite sharp between adjacent plant communities and discrete communities are easily recognized by the dominant plants in the community. The difficulty of identifying discrete communities increases as more rare species are included in the community.

Fire and Succession.   Fire has been a dominant force in the evolution of most of the world's grasslands and forests, as well as some desert shrublands. Grasses and forbs are favored by periodic fires (every 3–10 years), but shrubs are suppressed (Figure 3.31). Huge areas of continuous, flammable grassland vegetation occurred in the Central Great Plains of the U.S. before the advent of settlement in the mid-1800s (Barbour et al. 1987). Early accounts describe how summer thunderstorms moving across the prairie would rapidly ignite a series of fires in their path. Until the advent of fences, roads, farming, and livestock grazing these fires could spread long distances until they burned out due to lack of fuel or a weather change. Native Americans burned the grasslands of the Great Plains to improve conditions for game animals. The nutritious lush regrowth that occurs shortly after a fire is a strong attractant to bison, elk, deer, pronghorn, and a wide variety of other life.

Most grasslands appear dependent on fire for recycling nutrients from vegetation to soil to new biomass (Mutch 1970). Ponderosa pine, the primary timber tree in the western U.S., has heavy, fire resistant bark. Periodic fires cleared out the small trees and shrubs, maintaining most ponderosa pine areas as an open parkland. In the southeastern U.S., longleaf pine, another important timber tree, depends on fire for its maintenance. Fire is necessary to reduce understory competition, prepare a mineral seedbed, reduce fungal disease, and open the forest canopy (Barbour et al. 1987).

Grazing and Succession.   Many grassland ecosystems in the world evolved under heavy grazing pressure from large animals. During this process, plants developed defense mechanisms and animals have adapted to varying degrees to these defenses. Some of the mechanisms plants use to protect themselves from herbivory include low stature, delayed elevation of growing points, high amounts of seed production, synthesis of poisonous compounds, synthesis of compounds that reduce palatability, and growth of appendages such as thorns and spines that protect the plant (Holechek et al. 1998). Generally all these mechanisms require a reallocation of energy from growth to defense. Therefore they tend to make the grazing-resistant plants less competitive with plants not having these defense mechanisms when grazing does not occur, or occurs at light to moderate levels (Figure 3.32). Holechek et al. (1998) reviewed evidence showing that light to moderate grazing tends to cause progression toward the climax and is important along with fire in maintaining a high seral or cli-

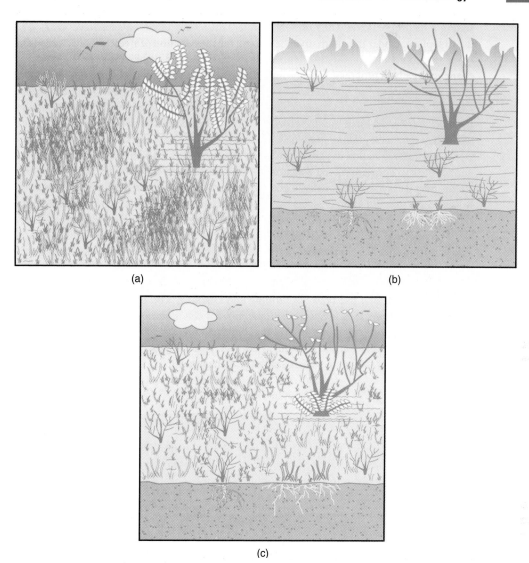

(a)   (b)

(c)

FIGURE 3.31 Before fire control and livestock grazing, desert grasslands of the southwestern U.S. were maintained by fires that periodically reduced mesquite and other shrubs. The buildup in vegetation (a) creates a fuel load that will carry fire (b). After the fire (c), the perennial grasses gain a competitive advantage over the mesquite because they have dense fibrous root systems that allow them to regenerate more quickly than the shrubs (from Holechek et al. 1998).

max conditions in mesic grasslands and many semiarid communities. Progressive effects of controlled grazing are less evident in more arid grassland communities.

Even under protection from large herbivores, vegetation is dynamic and fluctuates in response to other controlling factors, especially climate. Figure 3.33 shows fluctuations in black grama basal cover over time on the Jornada Experimental Range in southern New Mexico for nearly a 40-year period, under protection from grazing, and three intensities of grazing. In this study, heavy grazing represented utilization of more than 55% for black grama; intermediate grazing, 40% to 55% utilization; and conservative grazing, less than 40% utilization. Even without grazing, black grama cover was extremely low during the mid-1920s. Black grama cover was highest when the plants were grazed conservatively.

Management Applications. Knowledge of plant succession processes is essential to forest, rangeland, wildlife, and national park managers. Many private natural areas are now being purchased throughout the world by conservancy groups, such as The Nature Conservancy, who will need a strong knowledge of plant succession in order

FIGURE 3.32   Rangelands in Colorado showing the long-term effect of heavy and moderate cattle grazing on vegetation composition. The heavily grazed area in the foreground is dominated by big sagebrush while the moderately grazed area in the background is dominated by western wheatgrass. Big sagebrush has volatile oils that are toxic to cattle.

to obtain their goals. Knowing which seral stages and what arrangement of seral stages are needed and how to obtain them is a critical part of forest, range, and wildlife management. In a few cases, a passive approach of merely letting the forces of nature cause progression to a higher seral stage or perpetuate the climax may be satisfactory. However, in most situations, active management—such as use of grazing, burning, logging, and/or cultivation—will be required to obtain the desired mix of plant communities within a reasonable time frame.

Ecosystem Manipulators: The Human Component

Humans are a keystone species in a class of their own. They are the only species that consciously plans and manipulates ecosystem components for their own benefit. Unfortunately, we humans are not perfect and have lost many benefits because of past mistakes. Human activities affect all trophic levels, and have had both positive and negative influences on natural processes and functions. In the past, intentional manipulations of ecosystems have been largely directed towards increasing those populations that contributed to amount of wood, meat, plant foods, fiber, minerals, energy resources, and recreational resources with little regard for the diversity of other organisms and their interactions with the environment (Figure 3.34).

More recently, managers have recognized that all organisms are intertwined in the pyramid of life. Therefore, any management practice affecting one species will, in turn, affect all others. This has lead to the present policy of multiple-use management on public lands in the United States.

Multiple use involves managing ecosystems for livestock, lumber, wildlife, minerals, water, fisheries, recreation, biodiversity, and other ecosystem services on a sustained basis. Obviously, no acre of land can be used for maximum production of all goods and services. However, most land units will provide two or more products and regionally most if not all can be produced. The product or products emphasized will depend primarily on the characteristics of the ecosystem involved and demands by society.

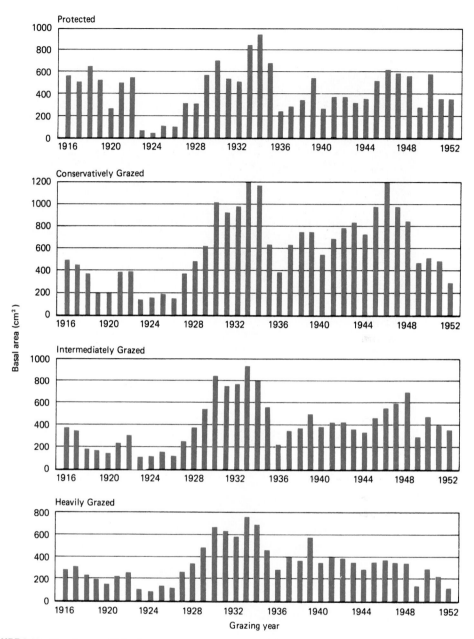

FIGURE 3.33 Basal area of black grama on areas protected from grazing and at three intensities of grazing on the Jornada Experimental Range, southern New Mexico, 1916–1953 (from Paulsen and Ares 1962).

Lately, the need for more effective management of all natural resources has lead to a holistic approach called ecosystem management (e.g., Vogt et al. 1997). Although there will continue to be a need for specialization, there will also be a need for awareness of how each specialty fits into an ecosystem context. Because ecosystem structure and function are so complex, natural resource researchers in the past have taken a mostly "reductionist" approach to ecosystem management. This involves using a one-factor approach, such as studying the influence of grazing on individual plants, the influence of fertilization on vegetation, and the influence of fishing on fish abundance. However, any management practice influences many components rather than just the intended one (Crisp 1964).

In the future, the demands placed on the earth's ecosystems for all products will accelerate because of a rapidly expanding human population on a shrinking land

FIGURE 3.34 In the 1970s agricultural systems in the United States typically involved planting large areas of the same crop year after year (top photo). In the 1990s there has been a strong trend toward more emphasis on diversified farming, soil and water conservation, and wildlife habitat improvement (bottom photo) [photos courtesy of U.S. Dept. of Agriculture (*top*) and U.S. Dept. of Interior, Bureau of Reclamation (*bottom*)].

base. This means management must be much more intensive than in the past. Each ecosystem responds differently to management. However, the same basic ecological principles apply to all ecosystems.

LITERATURE CITED

Allan, J. D. 1995. *Stream ecology.* New York, NY: Chapman & Hall.

Allee, W. C., A. S. Emerson, O. Park, T. Park, and K. P. Schmidt. 1949. *Principles of animal ecology.* Philadelphia, PA: Saunders.

Andrewartha, H. G. and C. L. Birch. 1954. *The distribution and abundance of animals.* Chicago, IL: University of Chicago Press.

Arnold, J. F., P. A. Jameson, and E. H. Reid. 1964. *The piñon-juniper type of Arizona: Effects of grazing, fire, and tree control.* U.S. Dept. Agric. Rept. Res. Rep. 84.

Barbour, M. G., J. H. Burk, and W. D. Pitts. 1987. *Terrestrial plant ecology.* 2nd edition. Menlo Park, CA: Benjamin-Cummings.

Barney, M. and N. C. Frischknecht. 1974. Vegetation, changes following fire in the piñon-juniper type of west-central Utah. *Journal of Range Management* 27:91–96.

Bebee, A. and A. Brennan. 1997. *First ecology.* New York, NY: Chapman & Hall.

Bolen, E. G. and W. L. Robinson. 1995. *Wildlife ecology and management.* 3rd edition. Englewood Cliffs, NJ: Prentice-Hall.

Carpenter, S. R. and J. F. Kitchell. 1993. *The trophic cascade in lakes.* New York, NY: Cambridge University Press.

Clements, F. E. 1907. *Plant physiology and ecology.* New York, NY: Henry Holt & Company.

Clements, F. E. 1916. *Plant succession.* Carnegie Inst. Wahs., Publ. 242.

Cole, G. A. 1994. *Textbook of limnology.* 4th edition. Prospect Heights, IL: Waveland Press.

Colinvaux, P. A. 1993. *Ecology 2.* New York, NY: John Wiley & Sons.

Crisp, D. J., ed. 1964. *Grazing in terrestrial and marine environments.* British Ecological Society Symposium Number 4. Blackwell Sci. Public. Ltd., Oxford.

Curtis, J. T. 1959. *The vegetation of Wisconsin.* Madison, WI: Univ. of Wisconsin Press.

Daubenmire, R. 1968. *Plant communities: A textbook of plant synecology.* New York, NY: Harper & Row.

Day, J. W., C. A. S. Hall, W. M. Kemp, and A. Yanez-Arancibia. 1988. *Estuarine ecology.* New York, NY: John Wiley & Sons.

Erdman, J. A. 1970. *Piñon-juniper succession after natural fires on residual soils of Mesa Vedge, Colorado.* Brigham Young Univ. Sci. Bull. Biol. Ser. 11:1–24.

Forman, R. T. T. and M. Godron. 1986. *Landscape ecology.* New York, NY: John Wiley & Sons.

Freedman, B. 1995. *Environmental ecology: The ecological effects of pollution, disturbance, and other stresses.* 2nd edition. New York, NY: Academic Press.

Gay, C. 1965. *Range management: How and why.* N. Mex. State. Univ. Coop. Ext. Circ. 376.

Heady, H. F. 1973. Structure and function of climax. In *Arid shrublands,* Proc. 3rd Workshop, U.S./Australia Rangelands Panel. Denver, CO: Society for Range Management.

Holechek, J. L., R. D. Pieper, and C. H. Herbel. 1998. *Range management principles and practices,* 3rd edition. Upper Saddle River, NJ: Prentice Hall.

Horne, A. J. and C. R. Goldman. 1994. *Limnology.* 2nd edition. New York, NY: McGraw-Hill.

Huschle, G. and M. Hironaka. 1980. Classification and ordination of plant communities. *Journal of Range Management* 33:179–182.

Jenny, H. 1980. *The soil resource origin and behavior.* New York, NY: Springer-Verlag.

Johnston, C. A. 1998. *Geographic information systems in ecology.* Malden, MA: Blackwell Science.

Lerman, M. 1986. *Marine biology: Environment, diversity, and ecology.* Menlo Park, CA: The Benjamin/Cummings.

Lewis, J. K. 1969. Range management viewed in the ecosystem framework. In *The ecosystem concept in natural resource management,* edited by G. M. VanDyne. Academic Press, Inc., New York.

Lewis, J. L. 1971. The grassland biome: A synthesis of structure and function. In *Preliminary analysis of structure and function in grasslands,* edited by N. R. French. Range Science Series No. 10, Colorado State University, Ft. Collins.

Lindeman, R. L. 1942. The trophic-dynamic aspect of ecology. *Ecology* 23:399–418.

MacArthur, R. H. and E. O. Wilson. 1967. *The theory of island biogeography.* Princeton, NJ: Princeton University Press.

Marsh, G. P. 1865. *Man and nature; or, physical geography as modified by human nature.* New York, NY: Charles Scribner.

Miller, E. C. 1938. *Plant physiology.* 2nd edition. New York, NY: McGraw-Hill.

Moran, J. M., M. D. Morgan, and J. H. Wiersma. 1980. *Introduction to environmental science.* San Francisco, CA: W. H. Freeman and Company.

Mutch, R. E. 1970. Wildland fires and ecosystems—a hypothesis. *Ecology* 51:1046–1051.

Nebel, B. J. 1981. *Environmental science: The way the world works.* Englewood Cliffs, NJ: Prentice Hall.

Nebel, B. J. and R. T. Wright. 1996. *Environmental science: The way the world works.* 5th edition. Upper Saddle River, NJ: Prentice Hall.

Nebel, B. J. and R. T. Wright. 1998. *Environmental science: The way the world works.* 6th edition. Upper Saddle River, NJ: Prentice Hall.

Odum, E. P. 1959. *Fundamentals of ecology.* 3rd edition. Philadelphia, PA: W. B. Saunders.

Owen, O. S. and D. D. Chiras. 1995. *Natural resource conservation.* 8th edition. Englewood Cliffs, NJ: Prentice Hall.

Owen, O. S. D. D. Chiras, and J. D. Reganold. 1998. *Natural resource conservation.* 7th edition. Upper Saddle River, NJ: Prentice Hall.

Paris, O. H. 1969. The function of soil fauna in grassland ecosystems. In *The grassland ecosystems: A pro synthesis,* edited by R. L. Dix and R. G. Beidleman. Range Science Series No. 2, Colorado State University, Ft. Collins.

Paulsen, H. A. and F. N. Ares. 1962. *Grazing values and management of black grama and tobosa grasslands and associated shrub ranges of the Southwest.* U.S. Dept. Agric. Tech. Bull. 1270.

Perry, James and Elizabeth Vanderklein. 1996. *Water quality: Management of a natural resource.* Cambridge, MA: Blackwell Science.

Pieper, R. D. 1977. *Effects of herbivores on nutrient cycling and distribution.* Proc. 2nd U.S./Australia Rangeland Panel. Australia Rangeland Society, Western Australia.

Pieper, R. D. 1983. *Consumption rates of desert grassland herbivores.* Proc. 14th International Grassland Congress, edited by J. A. Smith and V. W. Hays. Boulder, CO: Westview Press.

Primack, R. B. 1993. *Essentials of conservation biology.* Sunderland, MA: Sinauer Associates.

Ricklefs, R. E. 1996. *The economy of nature: A textbook in basic ecology.* 4th edition. New York, NY: W. H. Freeman.

Risser, P. G. 1969. Competitive relationships among herbaceous grassland plants. *Botanical Review* 35:251–284.

Rosenzweig, M. L. 1995. *Species diversity in space and time.* New York, NY: Cambridge University Press.

Van Dyne, G. M. 1966. *Ecosystems, systems ecology, and systems ecologists.* ORNL-3957, Oak Ridge, TN: Oak Ridge National Laboratory.

Vogt, K. A., J. C. Gordon, J. P. Wargo, D. J. Vogt, H. Asbjornsen, P. A. Palmiotto, H. J. Clark, J. L. O'Hara, W. S. Keeton, T. Petel-Weynand, and E. Witten. 1997. *Ecosystems: Balancing science with management.* New York, NY: Springer.

Wetzel, R. G. 1983. *Limnology.* 2nd edition. New York, NY: Saunders College Publishing.

Whittaker, R. H. 1967. Gradient analysis of vegetation. *Biological Review* 42:207–264.

Whittaker, R. H. 1975. *Communities and ecosystems.* 2nd edition. New York, NY: Macmillan.

Wilson, E. O., ed. 1988. *Biodiversity.* Washington, DC: National Academy Press.

CHAPTER 4

Conservation Economics

THE PROBLEM

Natural resources including land, water, wildlife, and minerals, can be preserved for future generations or developed into commercial products for current generations. These resources typically have many competing uses in any one time period, and there are many different periods in which they can be used. For example, ancient redwoods in northern California can be cut down to produce lumber for houses, they can be left standing for later timber harvest, or they can be designated as wilderness. Conservation economics deals with the problem of how to best distribute natural resources to provide more total benefit to people in current and future generations.

HARD CHOICES

Natural resource economics is about making hard choices when no single choice is best for all people in all generations. Since the beginning of recorded time, governments have tried to design institutions such as laws, policies, and regulations to help make better choices in meeting society's objectives. Government natural resource agencies in the U.S. are expected to improve the general welfare by choosing wisely among natural resource conservation proposals.

Adam Smith (1776), in his book, *The Wealth of Nations,* was the first to demonstrate that *markets* are efficient mechanisms for allocating resources to maximize human benefits to current and future generations. This theme has been an important one in economic science ever since. For example, Howe et al. (1986) recently proposed that water markets be used for conserving water in the western U.S. Under his approach people who need water could buy it from somebody who was willing to sell.

In this chapter, we apply basic economic principles to natural resource conservation. Underlying the chapter are two basic assumptions: First, environmental and natural resource management policies should produce the greatest possible total human benefits for present and future generations; and second, these policies should produce a fair distribution of those benefits. We will begin by considering hard management choices.

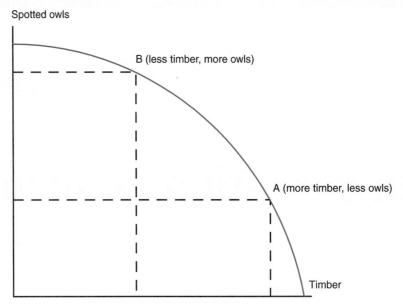

FIGURE 4.1   The trade-off between logs and spotted owls.

Timber versus Spotted Owls (Pacific NW)

In the Pacific Northwest there are only so many acres of land with ancient forests, called old growth trees. These trees are exceptionally valuable because they provide timber and high paying jobs for loggers. However, the same trees also provide a critical habitat component for northern spotted owls. This species is so rare that in 1990 it was listed as an endangered species. Because the total supply of old growth forest land is scarce, it is essential that we make the right decisions in allocating it to these and other uses. Underlying the issue of logs versus owls is the more fundamental one of measuring the relative value associated with each use (Figure 4.1).

We are now at point A in Figure 4.1, producing lots of logs and few spotted owls. Montgomery and Brown (1992) have found that ensuring owl survival requires that high quality forest not be logged; i.e., that we move to point B, producing more owls and less timber. Choosing correctly in this situation depends on whether or not the benefits gained from moving to point B from point A outweigh the costs. Nobody knows the answer for certain, but a good part of it will be determined by attempts to measure those benefits and costs. Many hard choices like this are faced by natural resource managers.

Cheap Energy versus Functioning Ecosystems

How much more are you willing to pay in the price of gasoline to ensure that ecosystems are being protected? The U.S. Congress has considered passing laws to require oil companies to build double-hulled tankers so that accidents do not cause an ecological catastrophe. But that policy would raise the price of gasoline, because the double-hulled tankers are ultimately paid for by people who buy gasoline.

Clean Air versus Cheap Steel

People in many of America's biggest cities live with air pollution every day. This pollution can be cleaned up, but the costs are high. The Clean Air Act of 1970 and later amendments are strong laws restricting air emissions in many parts of our economy. Achieving air pollution control objectives set up by Congress in the 1990 amendments to the Clean Air Act will require an estimated $25 to $50 billion each year to be spent by both the public and private sectors. Before the 1990 amendments, there were fewer than 20 regulated pollutants. The Clean Air Act currently regulates more than 380 pollutants at a considerable cost to industry. The steel industry is particularly hard hit by the new laws. They will require the steel industry to spend additional billions of dollars over the next 10 years. This will add substantially to the steel industry's operating costs and to the cost of steel products we all buy, like new cars. So Americans must face the question: Do we want cheaper steel or cleaner air?

Endangered Wolves versus Livestock and Wildlife

How much are you willing to pay in higher beef and lamb prices and decreased abundance of deer and elk to restore wolves to native habitat? Recent reintroductions of wolves in Yellowstone Park and in Arizona may result in lower sustainable abundances of large wildlife. We all may have to pay a price in fewer deer, elk, and other animals watched and hunted to restore wolves. Wolves also eat cattle and sheep. So reintroduced wolves could reduce the U.S. beef and lamb supplies in our grocery stores, which will increase the price of beef and lamb.

Endangered Fish versus Food

The silvery minnow of central New Mexico was recently declared an endangered species. The little fish lives in a small stretch of the Rio Grande between Albuquerque and Socorro, New Mexico. The cost of preserving the minnow is much higher than the income lost by the few fish bait dealers who might catch and sell them. Before the minnow was declared endangered, most of the river commonly went dry for part of the summer. This occurred because the major use of the riverbed was to move irrigation water from Cochiti Reservoir above Albuquerque to downstream farmers who needed the water to grow crops in the desert. Continuing to let the river go dry in the summer may increase the scarcity of the minnow and possibly cause extinction unless people find some other way to manage the river. The hard choice facing the people of New Mexico and the U.S. is how much food production should be given up to save the minnow. Another hard choice is who should compensate farmers in water or money if they are not allowed to take water from the river to save the minnow.

Hydropower versus Recreation

Glen Canyon Dam, which backs up the Colorado River on the Arizona-Utah border, was completed in the early 1960s by the U.S. Bureau of Reclamation, mostly to produce cheap electricity. The best time to let water go over the dam's turbines for power production is on hot summer afternoons, when power demand is high because people need it to cool down their sweltering Phoenix and Las Vegas homes. However, the best time to let high flows go past the dam for environmental benefits is in March, when the value of hydropower is nearly zero. Water flow in March reproduces natural spring runoff needed by riparian vegetation downstream.

Here is the hard choice. By sending water past the dam in March for environmental benefits, Phoenix and Las Vegas lose some high-valued summer electricity that must be made up with expensive coal burning. However, by holding the high flows back until summer, all the March environmental benefits are lost. As federal taxpayers, we and the Bureau of Reclamation must try to make the optimum choice.

Electric Power versus Forests

Much of the western United States is endowed with abundant and high quality deposits of low sulfur coal. An important use of that coal is burning it to produce electricity. An unfortunate by-product of burning that coal is sulfur dioxide, which travels in the air thousands of miles to the north and east. When it falls back to earth as acid rain, the precipitation combines with the chemical to produce sulfur dioxide, which is thought to damage forests and lakes in Canada and the northeastern U.S. How much are we prepared to pay in the value of lost forest products to ensure a continued cheap supply of coal-fired electric power?

Mining versus Recreation on Public Lands

People need coal and mineral products for their daily lives. However, mining activities of various kinds can conflict with outdoor recreation. One widely publicized conflict emerged in the late 1970s in southern Utah between coal development and nearby Bryce Canyon National Park. One of the proposed coal fields would damage outdoor recreation by impairing scenic values at the southern end of the park. On the basis of an economic study that estimated recreational values displaced by the proposed mine (Haspel and Johnson 1982) by then Secretary of the Interior, Cecil Andrus, the coal mining project was shut down.

Riparian Ecosystems versus Livestock

Most of the southwestern United States is arid, which means that water for agriculture, cities, and wildlife is scarce and unreliable. Livestock in the deserts of Arizona and New Mexico congregate near the streams because that's where water and grass are most available. Livestock that are allowed to graze these areas affect wildlife and their habitat, such as the endangered southwestern willow flycatcher. Ensuring the survival of the flycatcher may require taking cattle off a large percentage of these streamside areas. So we are faced with the question of how many cattle we are willing to give up and how much more we are willing to pay for beef to guarantee survival of the flycatcher and other species that require dense riparian (streamside) habitats.

Hydro-Dam Relicensing

The Federal Energy Regulatory Commission (FERC) licenses all nonfederal power plants in the U.S. Many of these licenses will soon come up for renewal. Changes in the amount and timing of water used for power generation can affect other uses of water downstream, such as recreation, irrigation, and endangered species. The relicensing process gives environmental groups the opportunity to seek downstream flows more compatible with needs of outdoor recreation and endangered species. New licenses may contain terms and conditions relating to wildlife protection and enhancement. Typically, minimum flows will be required downstream from dams. In developing these terms and conditions, FERC must, by law, give equal consideration to power and nonpower values. So as a nation we are faced with the hard choice: Do we want cheaper electricity or more suitable flows for downstream recreation, fish, and endangered species?

CHOICES DISPLACE OPPORTUNITIES

Few decisions are easy to make because most entail costs as well as gains. Socrates as well as other historical figures teach us that people live happiest lives when they continually reevaluate impacts of their choices, learn from their mistakes, and make better future choices. Socrates' famous quote is "the unexamined life is not worth living." Nowhere is this more true than in natural resource management. The common thread among all of the hard choices presented above is that every public or private choice involving natural resources creates certain opportunities while displacing others. The opportunity gained is the benefit. The opportunity displaced is the cost. Economics helps us think about which choices we should make to receive the maximum benefits with the least cost.

Economics Is About Choices

All civil societies invest a large amount of time, energy, and money in the *design of institutions* for assuring that good decisions are continually made, carried out, and judged. By institutions, we mean practices or organizations for governing peoples' behavior. The two main institutions used for solving a society's economic problems are *markets* and *government controls.* The way society goes about solving its economic problems tells us a lot about the way they approach their natural resource conservation problems. All societies face three basic economic problems that are about making choices (Figure 4.2).

First, societies need to determine **what** commodities will be produced and in what quantities. For example, will we use our public and private rangeland and forest land to produce only large quantities of timber and livestock with no outdoor recreation, water, and habitat (living area) for endangered species? Or will we use those lands to produce smaller quantities of livestock and timber, but more outdoor recreation, water, and endangered species habitat? Will we allocate our rivers to produce electric power and water for industry with nothing for recreation and endangered species, or will we use those rivers to produce smaller amounts of a greater number of products? Finally, should we let our rivers run wild to the ocean and take the inevitable floods and droughts that result, or will we take taxpayer resources (money, land, and water) and allocate them to building dams to reduce flood and drought damages?

Next, each society needs to determine **how** will it produce all these goods. By this we mean who, what resources, and what technologies will be used. For example, will businesses use environmentally friendly technologies for producing paper that keeps pulp from being dumped in blue-ribbon trout streams, but makes a ream of paper cost $15, or will businesses use environmentally damaging technology that reduces the price of a ream of paper to $3, but dump paper wastes into streams and rivers? Will houses be produced at low cost with lumber harvested cheaply from big trees that displace spotted owls, or will they be produced with more expensive bricks, stone, and adobe while preserving spotted owl habitat?

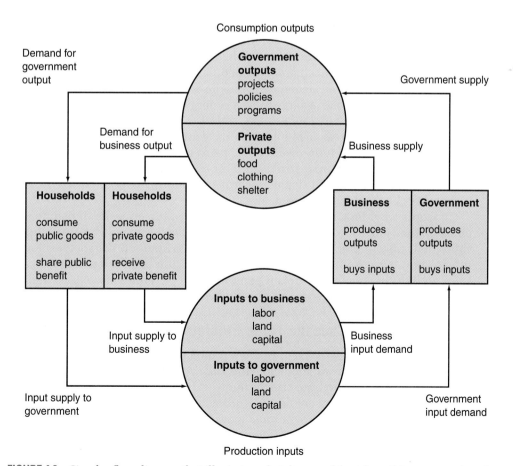

FIGURE 4.2 Circular flow diagram that illustrates what, how, and for whom things are produced in an economic system.

Last, all societies must decide for **whom** the goods shall be produced. In other words, who will enjoy the benefits of the nation's products? To put it differently, this asks who gets the income needed to buy those produced commodities. Will forest product workers receive high wages and live in expensive houses in unpolluted environments, or will sawmill owners pay low wages to the workers, reserving high profits and a clean environment for themselves? Will laws be passed that do not protect the environment, thus reduce wages to environmental science students and increase incomes earned by lumberjacks and ranchers, or will we pass laws that require environmentally friendly technologies that increase salaries of biologists and reduce incomes for farmers and miners?

Importance of Good Decisions

Where natural resources are unusually scarce or where their demand is high, it is essential we make the right decisions because the cost of making mistakes is so high. For example, in desert climates like Albuquerque, Phoenix, or Las Vegas, water is scarce. Therefore, it is essential to design and apply sound plans for finding, storing, saving, and using water. Similarly, the cost of a failed effort to save a species from extinction is high—the species is forever lost. Finally, in nations where protein is scarce, such as much of Africa, or in situations of war or famine, policies that trade off cattle ranching with other activities, such as tourism, must be carefully thought out because the wrong policy may cause many people to starve.

What the Market Mechanism Does

Mechanics. The market is more than a place where goods are traded like the New York Stock Exchange or the local livestock or automobile auction. It is a process by which buyers and sellers of any good or service interact to determine its availability (quantity) and value (price). The driving force of the market is the notion that everything has its price: each commodity, each service, and each natural resource.

How does the market accommodate peoples' desire for more of something, which economists call increased demand? If people want to buy more of something, such as thick, juicy, beef steaks, a surge of new orders will be placed for them (Figure 4.3). As buyers scramble to buy more steaks, sellers raise the price because they can do it without losing customers. The higher price sends a signal for all to see. This signal lures ranchers to raise more beef cattle, packers to pack more steaks, and grocery stores to buy and stock more steaks on their scarce shelf space. The market process, through the price and profit signal, rewards people who respond to buyers' desire for more steaks by producing more steaks.

If people want less, the market process does just the opposite. If people want less chicken, they place fewer orders for chicken. Chicken sellers, who are anxious to get rid of their unsold chickens rather than have them spoil on the shelves, lower their price. They then produce and sell less chicken. Instead they use those scarce resources, like chicken feed, chicken pens, chicken packinghouse space, and chicken shelf space for something else (e.g., beef production).

How the Market Solves the Three Problems

What Things Are Produced. What things are produced is determined by consumers' dollar votes. If consumers want more housing, they vote with their dollars by buying more housing and less of something else. In the world of politics our votes lead to action—they signal who will represent us and, indirectly, what actions we want our representatives to take. Dollar votes are in some ways similar to political votes. The size of dollar votes attracts added resources into producing more of what people desire.

How do those dollar votes attract resources to produce what people want? Business firms are enticed into producing goods in high demand by the higher potential to earn profit. Profit is what remains after a business subtracts its total cost from its

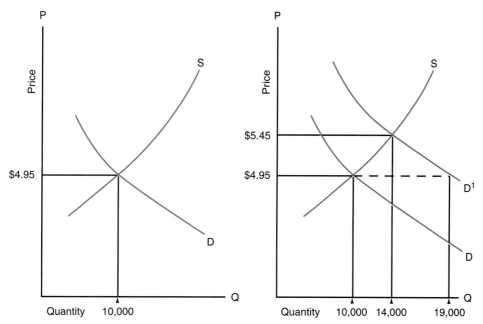

FIGURE 4.3 Market equilibrium for steaks before and after an increase in demand.

total revenue. The opposite of profit is loss. Profit is the main reason businesses exist. It is the main measure of their success and lack of it is the main measure of their failure.

How Are Things Produced. By this question, we ask what methods of production will be used to produce things. The way things are produced is determined by the competition among different producers who must all minimize costs to stay in business. All businesses compete for a limited number of customers and their dollar votes. Any business that fails to minimize its production costs will need to charge higher prices to its customers than a competing business that successfully minimizes its costs. Therefore, businesses that fail to minimize costs lose business to competitors who do minimize costs.

Producers maximize profits by minimizing their costs. To keep costs at a minimum, they minimize the use of expensive inputs and maximize use of cheap inputs. Agriculture provides a good example of how this works in different parts of the world. In the U.S., land is relatively plentiful and cheap and labor is scarce, so land prices are relatively low and wages are high. For American farmers, the low price of land and high labor price signal the use of laborsaving technology for cotton, wheat, and livestock production. Where labor is abundant and land is scarce, as in Asia, farmers find they can minimize their costs by producing cotton, wheat, and livestock with labor-intensive technology.

For Whom Are Things Produced. This question means "who gets how much of what is produced?" In a market economy, goods are produced for people with income to buy them. The distribution of income in a market economy is determined by who owns productive resources (land, natural resources, labor, and capital) and how well the production meets the wants of consumers.

Each person's income in a market economy equals the price of the product or service he provides times the quantity he chooses to sell to the market. The distribution of income among people depends on the quantities of all goods or services they sell, and the prices of each unit. Ranchers who sell the most beef with the least cost will accumulate the most wealth. This is why ranchers oppose hikes in grazing fees and reductions in numbers of livestock that they can graze on public lands. It is also why cattle ranchers want you to eat as much beef as you can buy rather than chicken, pork, or fish.

Who Controls the Market?

No one person controls the market. The three bosses of the marketplace are consumers, technology, and profits. Consumers are directed by their natural or learned preferences. They vote with dollars for those preferences. TV marketers try to change your preferences, so you want to buy what they have to sell. Jeans producers will try to convince you that if you buy their jeans, you'll be sophisticated, better looking, and well-loved. They will be delighted if you alter your preferences and spend your dollar votes on their jeans.

Technology sets limits on what can be produced and at what cost. About 150 years ago, lonely, tired, cold, wet California gold miners wanted to talk to their families in the eastern United States. The lucky ones were willing to pay lots of gold dust for those intimate conversations. But until the telephone was invented and the lines stretched, they could not do it, no matter how high the demand. Demand must be backed by the technology to carry it out.

Profits reward private businesses who make and deliver the goods in highest demand to consumers at the least cost. Profits lure firms into areas where consumers want to buy more goods and to exit enterprises where consumers want less. They also reward businesses who use the cheapest methods of production.

The Invisible Hand of the Market

Adam Smith wrote a classic book in 1776 titled *The Wealth of Nations.* In it he emphasized the importance of self-interest as a human motivator. This fact has led some people to call economics the dismal science. Smith tells us that every individual uses his resources to produce the greatest value for himself. Smith's idea was that in selfishly pursuing only our own personal interests, we all are led as if by an invisible hand to accomplish the best good for everyone.

Translated, this means that producers who attempt to maximize their own profits, consumers who buy where prices are the lowest, and workers who go where wages are highest do a great service to society. Further, they do it more efficiently than government bureaucracies or social planners who would try to do the same thing. Profit has been criticized because it is associated with greed and self-interest. Still, profit incentives are in some ways like a well-oiled machine that can perform some amazing tasks for us.

For society, profit encourages businesses to produce what people want to buy, motivates them to produce using the least costly methods, and rewards them if they conserve on scarce resources. The profit signal tells businesses how to get rich and tells them when they are rich. In short, the profit motive, combined with free market competition for customers, ensures that products are made at minimum costs and with maximum choices for us as consumers.

Can Profit Incentives Protect Natural Resources?

Thousands of farmers, oil companies, foresters, ranchers, wildlife managers, and others ask daily whether profit can protect natural resources. The answer depends on whether or not there are property rights (i.e., rights to own, use, buy, and trade natural resources) (see Anderson and Leal 1991). Where there are well-defined property rights to natural resources, the profit motive helps to protect private forests, private grazing lands, and private water. The following examples help explain why.

High timber prices encourage forest owners to plant more trees so they have something to harvest for the future. In places where wildlife can be privately owned (e.g., Texas), high tourist and hunter expenditures encourage landowners to invest in conservation of desirable wildlife.

Where water rights are secure and are marketable, high water prices encourage water right owners to use their scarce water carefully and/or to save it for possible sale to somebody else. Establishing an emergency drought water bank helped California get through a severe drought in the early 1990s with minimum hardship. California established high prices for farmers' water, if the farmers agreed to deposit it into a bank for resale to big cities. The California Water Banking Law included legal assurance that depositing water into the bank would not change farmers' permanent water right. This legislation was cheap, easy, and effective.

How Markets Settle Conservation Questions

By and large in the U.S., we accept free markets for allocating natural resources between the present and the future. There is little likelihood that this practice will be abandoned (Herfindahl 1961). We do not, for example, impose a tax on copper production in order to save copper for use by future generations. We do it to raise revenue to run the government. The question of the total quantity of minerals extracted over time is left to the mining company, the smelter, and the refiner. Likewise, the production and distribution between present and future generations of food buyers is largely in the hands of private farmers, grain distributors, food processors, and retail grocery stores.

The market settles conservation questions by discovering, renewing, producing, and selling natural resources over time to maximize their owners' current and future incomes. A private forest is a good example. The opportunity for current and future profit tells the private forester when to plant new trees, when to harvest old trees, and how long to wait for little trees to grow into big ones. Will these profit-maximizing decisions lead to depletion of their forests? If the private forest owner has secure property rights, the answer is a resounding "no." Private forest owners have a strong financial incentive to take care of their trees. Cutting down all their trees and selling them today without replanting means they will have nothing to sell in future years.

The profit system can also help provide for future national timber needs. If a private forest owner expects population to be growing and people to keep buying houses with wood frames, they can plan on high future demands for timber and, as a result, high future prices. So they have every incentive to plant more trees now, wait for more tree growth, and harvest fewer trees now and save more for the future when they can get a higher price. Their personal self-interest acts like an invisible hand in providing for future generations' timber needs.

When Are Markets Best at Promoting Conservation?

Markets are best at settling conservation questions in those situations when all potential resource users can effectively voice their demands in a monetary form. Take forests for example. If they were auctioned so that private bidders could effectively voice their demands for all forest products, including spotted owls, timber, livestock, water, and recreation, then the market system would make it profitable for businesses to produce the right mix of timber, spotted owls, and other forest products.

The following is a good general principle: When all resource users are able to voice their demands monetarily, prices of all resources are set by current and anticipated future demands for and supplies of those resources. When that happens, anybody who decides to use a particular resource at a particular time is faced with a price equal to the value of other uses displaced by that use. Then natural resources will be priced at their real cost, and private decisions will be made to use a resource now only when the public benefit of the use exceeds all public costs (Ward 1998). In that very special case, the market institution can be trusted to conserve natural resources.

Limits of the Market

There is one major limit of relying on the market system to settle conservation questions. Markets only account for private income produced by privately owned resources; they sometimes fail to account for all human benefits produced by natural resources. Many publicly owned resources, such as air and wildlife, are unpriced. The market system has little incentive to conserve them because it is unprofitable to do so. So the market system fails to conserve and therefore depletes or wastes any natural resource that is priced at less than its real cost.

Underpriced Resources.   One good example of the market system failing to conserve an underpriced resource is reflected in the price of coal used to produce electric power in the western U.S. In an unregulated market, burning coal to produce electric power causes acid rain, which damages forest and lake resources in the northeastern U.S. and in Canada (see chapter 6). Damages to these resources are not included in the cost of the coal-fired electricity charged to customers in Phoenix and Las Vegas. So electricity buyers in those cities are not facing the full cost of their power. Where there are underpriced resources, unregulated markets fail to signal people with the real cost of their actions. This discourages conservation.

Spillover Benefits.   Spillover benefits are derived by society as a direct by-product of some economic trade between a business and its customer. An example of a spillover benefit is the pulp mill owner who considers spending money on controlling pollution from his pulp mill to reduce discharges into a blue-ribbon trout stream. Even if the benefits to society exceed the costs to the mill owner, he will probably decide against the investment because he cannot make any money selling the cleaner water to fishermen. Therefore, the mill fails to invest in pollution controls and the trout stream remains polluted.

Spillover Costs.   Spillover costs are a third example of where market incentives discourage conservation. Spillover costs are costs sustained by society as a direct by-product of some economic trade between a business and its customer. A good example is provided by the logging company that clear-cuts a patch of forest and is not charged for silting up the downstream river. The effect of the muddy stream is of no concern to the logger if he does not have to pay for the downstream damages. But these costs are of prime importance to downstream water users because they have to live with the siltier water or clean it up. The presence of spillover costs dull the potential power of market incentives to promote conservation.

In the case of either spillover benefits or spillover costs, these externalities do not enter into the market-pricing calculations of the parties undertaking the activity. So letting the market system settle conservation questions blocks conservation whenever natural resources are priced lower than their real cost. Put more simply, underpriced resources are inadequately conserved by the market system.

THE FUTURE

In future years, three problems will magnify the limitations of the unregulated market to conserve natural resources in the U.S. One is the increase in population, especially in the dry western U.S., with its dramatic effect on the supply of water and open space. The other factor is the increasing per capita demand for many forms of outdoor recreation, such as camping, hiking, and fishing. The third is the increased value people assign to endangered species.

HARNESSING THE POWER OF THE MARKET

Despite their limits, market forces have considerable power that can be harnessed to promote natural resource conservation (Anderson and Leal 1991). The idea is simple—business responds to profit opportunity. So making conservation profitable to business is a sound approach to natural resource management.

Concepts for Incentive-Based Pollution Control

By the 1960s, pollution problems in the U.S. were severe enough to warrant interest in finding ways to deal with them. One way of dealing with the pollution problem was environmental legislation, which reached a peak in the "green" decade of the 1970s. Politicians generally preferred to address environmental problems by passing laws that set limits on acceptable pollution. Their preferred method was "command and control" rules and regulations, such as design standards, which require firms to use prescribed technologies. Another example is performance standards, which specify the maximum amount of pollution that individual firms can dump into the environment.

Economists proposed that market incentives were a better way to control pollution than government regulations for two reasons. First, market forces could be harnessed to make pollution expensive to polluters. The effect of this is to both reduce pollution and make it profitable for businesses to search for production technologies that reduce pollution. Second, market forces would be used to lower the costs of pollution control by eliminating regulations and leaving decisions about the details of reducing pollution up to individual firms. The two most popular ways for harnessing the power of the market are *pollution taxes* and *marketable permit systems*.

Pollution Taxes

A pollution tax is levied on firms that discharge waste such as raw sewage into environments such as streams, lakes, the atmosphere, or the soil. The tax is set equal to the damages the firm does. The power of the market is harnessed by setting the tax level according to the damage that the firm's pollution causes to society when dumped into the environment. A firm that emits a pollutant with low damages gets a low tax, but pollution sources that impose high costs to society are taxed more heavily. This is basically a resource use charge—just like paying to use the facilities in a national park.

An important advantage of pollution tax compared to rules and regulations is that the tax raises the costs of making the polluting product. Therefore, market forces make it profitable for the firm to invest in pollution control measures, as long as the waste dumping charge is more than the cost of controlling pollution. By investing in measures to reduce pollution, the firm reduces its costs of production because the pollution tax charge falls by more than the cost of installing pollution control. Another advantage of pollution taxes is that they encourage firms to find the least cost measure to reduce pollution. If there are several technologies for controlling emissions of raw sewage into a river, all of which reduce the pollution tax by X dollars, the firm has a profit incentive to find the cheapest pollution control technology. By contrast, regulations on pollution emissions specifying what kind of pollution control technologies must be used are considerably more expensive because they don't give firms the freedom to search for ways to minimize the cost of compliance (Field 1994). With the market incentive, firms are encouraged to minimize their total pollution taxes and the cost of technologies required to reduce the taxes.

Pollution taxes also have advantages over fines commensurate with damages done. This is because fines typically involve court battles and presume some measure of guilt. Pollution taxes are treated like any other cost of production, and the firm will take steps to invest in pollution control that minimize the tax.

Some of the better-known pollution taxes in the U.S. include the excise taxes on hazardous chemicals, enacted in 1980 to fund the Environmental Protection Agency's superfund. A related reason for the superfund concept was that pollution problems often could not be assigned to a specific polluter. Similarly, after the 1989 *Exxon Valdez* oil spill, Congress imposed an added tax on petroleum products to pay for the Oil Spill Liability Trust Fund. One could argue that this added tax reflected the expected cost of ecological damages from future oil spills. More recently Congress has considered legislation to impose a carbon tax on fossil fuels like gasoline which, when burned, release carbon dioxide into the atmosphere, possibly increasing the greenhouse effect and contributing to global warming.

Pollution taxes are not perfect. To make them an effective way for dealing with pollution, policymakers need to know the cost of the damages caused by the pollutant before setting the level of the tax. Otherwise it would be easy to set the tax too low or too high. In fact, for several reasons it is difficult to know the cost of damages caused by the pollutant. First, it is hard to determine which activities (e.g., oil, chemical production) cause how much pollution. It's also hard to determine how much of the pollutant stays in the environment for how long, and how harmful the pollutant is to people who are exposed to it. It is also most difficult to determine the monetary cost of damage to health (Turner, Pearce, and Bateman 1993, Chapter 12). Should those costs be based on lost lifetime income, lost days of work, or people's willingness to pay to avoid the pollution? This makes it difficult to determine what pollution taxes should be. Another problem with pollution taxes is that if one country imposes them on its own economy, they put their own firms at a disadvantage compared to foreign competitors.

Marketable Waste Emission Permit Systems

The idea of marketable waste emission permits is simple: The government decides on how much waste the atmosphere or water body can assimilate safely (i.e., is nonpolluting) then issues waste emission permits (quotas) for no more than that amount (Figure 4.4).

This method of preventing pollution harnesses market forces by allowing the emission permits to be tradeable—that is, they can be bought and sold in the open market. Emission permits are especially attractive for many reasons (Pearce and Turner 1990): They are cheaper than command and control regulations, they allow for firms entering or leaving the area, they can be bought by environmental groups, they adapt to inflation or other changes in values, they adapt to geography, and they have low administrative costs.

Lower Cost.   Marketable permits let those who emit wastes (emittors) trade permits for cash. So emittors with low costs for abatement will find it cheaper to abate their emissions rather than buying permits. However, emittors whose technology is locked in and who find abatement very expensive will find it cheaper to buy the marketable permits from somebody. By giving emittors a chance to trade cash for permits on the open market, the total cost of preventing pollution is reduced compared to the regulatory approach of setting standards.

Entering or Leaving the Area.   Suppose we are dealing with a high growth area like southern California, and new emittors move in. The effect of the new emittors coming to town is to increase the demand for permits. If the total capacity of the environment to assimilate wastes is unchanged, the authorities may wish to keep the total number of permits fixed to allow 100,000 tons per month (Figure 4.4). If new emittors want to come to town and do business, they will invest in the cheapest combination of buying new permits and waste emission abatement. Similarly, if existing emittors decide to leave the area, their waste emission permits can be marketed to

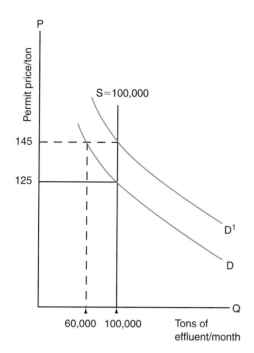

FIGURE 4.4 Supply and demand for marketable effluent permits, before and after new firms move into an area.

others. If, for example, the government discovered that the pollution standard needed tightening up to 60,000 tons per month (Figure 4.4) they could buy some of the permits themselves, thereby taking permits out of the market. So the permit system makes it easy for the government and for individual waste emittors to adapt to changes in scientific knowledge or economic need.

Environmentally Friendly.   An open market in waste emission permits allows anybody to buy them, including groups like environmentalists, sportsmen, or departments of game and fish. If an environmental group such as the Sierra Club wanted to lower waste emission in the area, they could enter the market and buy some of the permits, keeping them out of the market for as long as they wished. If the group was a private fishing club and the permit was to limit waste emission that impaired a water body's capacity to support certain fish, they could allocate some of their annual operating budget to buy permits, thus saving the need to stock large numbers of fish in poor quality waters.

Adapts to Economic Changes.   In periods of high inflation or rapid changes in technology in the polluting industries, the initial allocation of permits may change in value or need to be changed to adapt to new technology. Because the price of the permits automatically adjusts to changes in supply or demand, these outside changes can be handled efficiently and quickly.

Geographical and Spatial Flexibility.   The region in question may have complex dispersions of wastes, which may vary considerably with weather patterns, time of year, and the kind of firms that buy the permits. In principle, these conditions can all be handled under a permit system. Different quantities of permits could be issued for different seasons of the year, different weather patterns, different types of wastes, and more, and each kind of permit could operate in a fairly independent market. As long as the markets were open, flexible prices would clear the market and the costs of abating waste emission could be minimized.

Low Administrative Costs.   Under a pollution tax system, the authority must estimate the cost of pollution damages and the costs of various abatement technologies.

However, the authority can easily make mistakes that have high economic and political costs. Under a permit system, the authority does not need to estimate the cost of complying, but they need to know how many permits to initially issue. The waste emitting industries themselves are the only ones who need to know the details about the cost of abatement. They can buy the permits at the going price if abatement is expensive, and invest in abatement if the permits are more expensive.

APPLICATIONS OF INCENTIVE-BASED REGULATIONS

Pollution Taxes

Solid Waste.   The main U.S. experience with real-world pollution taxes has been in the arena of ordinary garbage disposal. In Highbridge, New Jersey, the city altered its flat fee for garbage hauling of $280 per year with a price that increased with the amount of trash picked up by haulers (Hanley 1988). Each 30-gallon can set out at a curbside each week had to have a sticker before it would be hauled away. Each household could buy 52 stickers for $140, about $2.75 per sticker. The effect of the experiment in raising the price of garbage hauling was to reduce the amount of trash put outside people's houses by 25%. Because the cost of disposing large recyclable items like used furniture could use four or more stickers, more of it was recycled than before.

Water Pollution.   Europe has more experience with pollution taxes than the U.S. According to Hahn (1989), Germany, France, and Holland have all set charges for various water pollutants to compensate for their social costs. The effect of the taxes has been twofold—to encourage pollution reduction and to help pay for the cost of cleaning up pollution. France (1960s) and Germany (1970s) have both enacted emission charges on water pollution. These charges seem to be accepted by the French people as a way of doing business to improve water quality. In Germany, the charges seem to be improving water quality (Hahn 1989). In Holland, the effect of water pollution taxes (administered on both volume and concentration of pollutants) has been to cause a slow but steady increase in water quality (Bressers 1983).

Environmental Adders for Electric Power.   Electrical power generation makes use of unpriced environmental resources. However, it is difficult to attach monetary values to these damages. So some have suggested adding taxes to the price of electric power, called "adders." However, these taxes are used only for selected electric power capacity expansion projects and are not actually charged to the consumer. The advantage of using adders is to promote cleaner electric power sources through design and selection. Utility regulators in more than half the states are either using adders or are considering using them. However, because the monetary value selected for the adder can be arbitrary and because the consumer never actually pays for it, the debate on the use of adders is far from over and is likely to continue (U.S. Department of Energy 1995).

Marketable Permit Systems

The U.S. has had some experience with marketable permit systems since the mid-1970s. Amendments to the 1990 Clean Air Act have many features designed to clean up air pollution as efficiently and cheaply as possible. It allows businesses to make choices on the best way to reach pollution cleanup goals. Because businesses must make a profit, they have an incentive to clean up their pollution at the lowest cost possible. These new flexible programs are called market or market-based approaches. For example, the acid rain cleanup program offers businesses choices as to how they reach their waste reduction goals and includes waste allowances that can be traded, bought, and sold. Consider the following examples of marketable permit systems.

Lead Banking. Lead causes developmental problems in children and heart disease and strokes in adults. Lead in gasoline is dispersed into the atmosphere when burned by cars. In 1985, the Environmental Protection Agency (EPA) set up a program that required lower lead standards for leaded gasoline. To reduce the very high cost imposed on industry of complying with these standards, EPA set up a lead banking program (Hahn 1989). Under the banking program, gasoline refiners who were able to reduce the lead content in their gasoline by more than the legally required standard for a given period of time were allowed to bank the difference in the form of "lead credits." These credits could then be used or sold in any following time period up to 1987, at which time all refiners had to meet the new standard. This proved to be a successful program. More than half of refiners participated in the banking program. Hahn found that the cost savings from the program may have exceeded $228 million. The costs were saved by allowing old refineries, for whom complying was expensive, to trade banked credits with more modern refineries who could comply at a lower cost. Requiring all refineries to meet the standards, without allowance for the differential cost of compliance, would have been considerably more expensive.

Emissions Trading. The EPA has been experimenting with emissions trading since 1976 for five air pollutants. These include volatile organic compounds (VOCs), sulfur dioxide, carbon monoxide, nitrogen oxides, and particulates. The EPA allows credits to be earned whenever a polluting source reduces emissions by more than the legally required amount.

Sulfur Dioxide Allowance Trading. The 1990 Clean Air Amendments set up an ambitious system of tradable permits for emissions of sulfur dioxide (SO_2), a chemical that can cause acid rain. Acid rain is formed when SO_2 and nitrogen oxide, released when fuel is burned, are transformed into sulfuric and nitric acids. These acids ultimately return to the earth dissolved in rain. Scientific studies over the last 20 years have shown that acid rain causes increased levels of acid in lakes and streams. Scientists suspect, but do not know for sure, that these acids may also damage forests. The acids can also erode buildings, bridges, and statues.

In establishing the Clean Air Amendments, Congress wanted to reduce SO_2 emissions by 10 million tons from 1980 levels. The EPA harnessed the power of the market by allowing sulfur dioxide allowances to be traded. Namely, the holders of allowances for SO_2 emissions—mostly electric utilities east of the Mississippi River—were allowed to transfer their permits among one another. Therefore, utilities who could reduce their emissions at the least cost had an incentive to do so. They were also able to sell their unused allowances to utilities who would save money by purchasing the allowance rather than reducing emissions. Allowances can also be banked for later use. The result of this banking plan was that the incremental cost of reducing SO_2 emissions was made equal across all sources. Making these incremental costs equal across all pollution sources has the effect of achieving total SO_2 abatement at minimum total cost. Utilities can also offer SO_2 emission allowances for sale at an annual government-sponsored auction.

This SO_2 allowance trading program has turned out to be quite successful. Desired reductions in emissions have been accomplished and even exceeded. More important, total SO_2 abatement costs have been much less than what they would have been without the trading provisions. Stavins (1998) estimates that tradeable permits for SO_2 have resulted in a cost savings of nearly $1 billion per year over the regulatory alternatives that Congress had considered in earlier years.

GOVERNMENTS AND CONSERVATION

How Government Settles Conservation Questions

Where government gets involved in conservation, the instruments for resolving natural resource conflicts enter the political arena. Participants in this arena include elected officials, interested voters, government agency employees, lobbyists, the press, environmentalists, agriculture, mining, and more. Obviously *some* sort of resolution comes out of this process, but it is not always clear if the results promote better conservation and human welfare.

Limits of Government In Settling Conservation Questions

Despite the good intentions behind attempts to set up government conservation policies and plans, four kinds of "government failures" stand in the way of conservation. These failures include rent seeking, taxpayer ignorance, empire building, and abuses of economic analysis.

Rent Seeking

Rent seeking occurs when special interests use their private resources in lobbying and other activities aimed at getting government to pass laws that economically benefit the group. Successful rent seeking (i.e., getting the group's favorite laws passed) increases net benefits for the special interest group, typically at a cost to the larger society. An example with a long history is the iron triangle, a group who sponsored and completed western water projects. This triangle consisted of western politicians, irrigated agriculture in connection with banks and real estate, and the federal water management agencies.

Taxpayer Ignorance

Taxpayer ignorance occurs because we find it more bothersome than beneficial to stay informed about issues that do not affect us directly. For example, New York taxpayers have little interest in investing the time and money to become informed on proposed legislation that would develop a new irrigation reservoir on the Colorado River. This is because the reservoir costs the New York taxpayer only pennies per person. It simply is not worth their time to scrutinize public actions that have little direct bearing on them. Many federal water projects were built that had little chance of promoting the national interest simply because the local beneficiaries wanted the legislation that paid for the project and the rest of the nation simply did not care enough.

Empire Building

Empire building involves a government agency purposely increasing its own power and influence for its own sake. Empire building, especially when combined with a bureaucratic desire for job security, can impede programs aimed at conservation. Few government agencies or employees can resist the temptation to increase their own influence or power. They often do too much of it and for too long.

Government gets away with building empires because there is no market check to stop them. Private business loses money if people are not willing to pay for the products produced by their expansion. The owner of a bicycle shop will not open new stores unless is enough customer demand for bikes to support them. However, if the government builds too many dams or drains too many swamps, there are no negative profits signals that require them to stop building and draining. The only support needed for government projects is a legislative majority, which can be bought by the small minority by financing political campaigns of key legislators.

Once initiated, government programs and agencies can take on a life of their own. The same agencies that made mistakes based on scientific ignorance were "rewarded" with a chance to correct those mistakes. After World War II, for example, the U.S. Congress gave the Army Corps of Engineers and Natural Resources Conservation Service (formerly the Soil Conservation Service) large budgets to drain and fill wetlands. Beginning in the 1980s, these agencies received budgets from Congress to create or restore wetlands. This problem typically starts in social impatience with careful research and planning, which often is looked upon by special interests as costly "foot dragging."

Abuses of Economic Analysis

The final government failure comes from agencies that ignore the economic consequences of their decisions. Sometimes a government agency chooses to ignore important costs or benefits because it is in the business of promoting one type of resource use (Herfindahl 1961). In fact, stimulating use of the resource may be implied in the agency's mission statement. Beginning in the early twentieth century, hundreds of dams were built in the western United States by the Bureau of Reclamation and Corps of Engineers, whose missions are, respectively, irrigation development and flood protection.

Certain costs or benefits were ignored in major decisions because they could not be quantified in monetary terms or could be described only in vague nonquantitative terms. Several dams built in the 1950s and 1960s were subjected to poor benefit-cost analyses that assigned high values to flat-water recreation behind a dam, while presuming that whitewater recreation and wildlife habitat displaced by the dam had zero value, simply because those values were hard to measure.

Government programs promote conservation only if benefits and costs are properly measured and used. Only then is the public protected against rent seeking by special interests, taxpayer ignorance, and empire building. Nevertheless, economic analysis is expensive and often difficult to understand. It is seen by some as unreliable, and is unpopular with politicians. In summary, government regulations and public resource management have the potential to promote inferior conservation plans compared to the imperfect market system, whenever proposed public actions are not based on objective economic analysis.

What Government Can Do to Protect Natural Resources

Benefit-Cost Analysis (BCA) is an economic tool for comparing in monetary terms the desirable and undesirable impacts of proposed natural resource policies where desirability is measured in human welfare. It is a method for ranking the economic performance of natural resource projects, policies, and programs in which impacts are measured in nontechnical terms and estimated by scientific methods. It is a way to compare, in common units, all the gains and losses to people resulting from some action.

BCA organizes information in a way that promotes the conduct of rational policy analysis. Rational policy analysis considers all the relevant alternatives, and identifies and evaluates all the consequences that would follow from the adoption of each alternative. It then selects that alternative which would be preferable in terms of maximizing society's overall welfare.

In the Flood Control Act of 1936, Congress required that the benefits of water development projects, "to whomsoever they may accrue," should exceed all the costs. Since that time, BCA has been used widely to decide whether any federal water project should be started. In more recent years BCA has been used to help make three kinds of public natural resource decisions: (1) a simple ranking of the comparative benefits of several possible actions, (2) the optimal size or scale of a project produced by a decision, and (3) the optimal timing or sequencing of several elements of a decision.

BCA uses a simple decision rule. If, for some proposed plan, the sum of its human benefits exceeds the sum of the human costs by a larger amount than any other action with the same aim, the proposed action should be adopted. Otherwise it should not. This assumes that all people are treated equally, so that an additional dollar of benefit accruing to a rich person is valued the same as a dollar of cost paid by a poor person. An advantage of BCA is that the dollar, as a unit of measure for ratings, is easily understood by everyone.

For the concept of revenue to the private firm, BCA substitutes the notion of benefit to society. For the cost of the private firm, BCA substitutes the concept of opportunity cost—or the value of benefits displaced when resources are taken from other economic activities to support the proposed plan in question. For the profit of the business firm, BCA substitutes the concept of net benefit, which means the difference between benefit and cost.

BCA values all natural resource services in the common denominator—money. Consider the huge range of services produced by natural resource decisions: electric power, water quantity, water quality, critical habitat for endangered species, timber, forage for cattle and wildlife, recreation, fish, minerals, and others. By expressing all these outputs in a common denominator, it allows all natural resource policy decisions to be compared on a level playing field. Although most natural resource benefits can be valued in monetary terms, some values remain resistant to monetization.

Why Measure Benefits and Costs?

Why spend all the effort and money necessary to conduct a benefit-cost analysis? Why not use plain honest-to-goodness profit and loss accounting used by any business?

One answer is that what counts as a benefit or a loss to one part of the nation—to one or more persons or groups—does not necessarily count as a benefit or loss to the whole nation. Therefore, with BCA, we are concerned with the economic welfare of the nation as a whole rather than some small part of it. BCA for natural resource decisions asks the same question that private companies ask of their accountants. However, instead of asking whether the owners of a private enterprise will be made better off by the company's proposed decision, BCA asks whether society as a whole will be made better off by undertaking the action.

BCA helps make better management decisions by using a time-tested economic framework for organizing economic data. In natural resource management, there are plenty of ways governments can make bad decisions. Management by tradition is a widely applied method for making poor decisions. Resource agencies pursue activities, such as stocking X pounds of trout at a Y water or leasing Z board feet of timber on some national forest, simply because that is what they have done for years. Without information on benefits and costs, resource agencies often have no choice but to manage by tradition.

Another way to make bad decisions is through fear of change. In his small but mighty book entitled *The Prince,* the fifteenth-century Italian writer Nicolo Machiavelli warned us there is nothing more difficult to take in hand, more perilous to conduct, or more uncertain of its success than to take the lead in the introduction of a new order of things. This is because the innovator faces as enemies all who have done well under the old conditions, and lukewarm defenders in those who may do well under the new. While change is resisted by virtually everybody, BCA gives a quantitative way to decide which changes are worthwhile by government and which ones should be left to the private sector or screened out altogether.

Poor conservation decisions are promoted by the necessity of government resource agencies to appease special interests. These special interests abound. In the western U.S., for example, Congress sets prices for federal timber, water, energy, and livestock forage below market levels because powerful special interests benefit economically, not because these low prices are good economic decisions for the nation as a whole.

Finally, unsound decisions happen when public resource managers' personal biases are allowed to influence decisions. America is "a nation of laws and not men." In the realm of natural resource policy this means we do not want some local Forest Service official deciding the mix of grazing and recreation on federal lands based on personal biases or political connections. The decision should be made through the consistent and objective application of time-tested economic principles. These principles should be set up at the national level and applied to resources and resource issues at the local level.

HOW TO MEASURE BENEFITS AND COSTS

Basic Principles

Willingness to Pay.   In principle, *the benefit of any natural resource conservation decision is measured by what people are willing to pay for it. Costs are benefits displaced by the decision.* These concepts apply to both goods and services that are sold in markets and those that are not.

Consider the case of market goods. Suppose the U.S. secretary of interior proposes limiting logging in the Southwest to produce a more critical habitat for the Mexican spotted owl. The benefits lost (costs) from such a policy are timber. Timber is sold in markets. So the total cost is the sum of prices times quantities of timber displaced by the decision.

As another example, consider the example of a nonmarket good, in which Congress considers passing a law that requires shipbuilders to build double-hulled oil tankers in order to reduce future damages from oil spills. The benefit is the value of fish not killed, beaches not fouled by spilled oil, marine wildlife not killed, and ecosystems not damaged. While the concept applies to both market and nonmarket goods, for the latter, prices typically must be estimated by some indirect method. This might involve contingent valuation surveys, in which people are asked directly what the resource is worth to them.

Scarcity, Demand, and Benefits.   Both market and nonmarket goods share an important characteristic. When they are scarce and in high demand, they are in great value, and the benefits of policies that supply more of them are quite high. For example, a policy that brings water to growing cities in the desert (which is scarce) to produce high quality drinking water (which is in high demand) is likely to produce great benefits. Similarly, the policy that brings water to a humid, cool region like Minnesota (where water is abundant) to produce habitat for an obscure species (which is in low demand) is likely to produce negative human benefits.

How Benefit and Cost Information Should Be Used.   Benefits and costs are measured and reported to facilitate sound conservation decisions. By sound, we mean the decisions are worth more to a nation than they cost. There are many wrong ways to use information about benefits and costs, but only one right way.

Implementation Principles

An Almost Right Way.   Requiring the ratio of total benefits to total costs to exceed 1.0 is almost right. According to that rule, resources should be committed to support a conservation decision if the ratio of the present value of benefits to the present value of costs exceeds 1.0. This method is almost right but not quite. It is not quite right because it is too conservative. Only having the benefit-cost ratio equaling 1.0 lets one implement decisions for which benefits are no greater than costs. A private business cannot stay in business if its revenues do not exceed its costs by a safe margin. For public natural resource decisions, benefits should exceed costs by the greatest amount possible.

The term "benefit-cost ratio" is widely heard in policy circles such as in Washington, D.C. It is easily understood by Congress and the press because it is expressed as dollars of benefit per dollar of cost. When Congress and other decision makers enact laws or make appropriations for their projects, they want to know if the benefit-cost ratio exceeds 1.0. Once the ratio exceeds 1.0, these decision makers may consider the economics to be acceptable and then use its political attractiveness to make the final decision.

The Right Way.   The *maximum net present value* (MNPV) gage is the one right way to use information on benefits and costs. According to this rule, natural resource conservation should be carried out to maximize the present value of net benefits. If carried out, the MNPV rule will produce the decision that maximizes the economic performance of the available resources. With the MNPV decision rule, only plans with net present value greater than zero are accepted as economically performing. If a management decision has a negative net present value, society gives up more than it gets over the life of the action, or society is paying more for the plan than it is worth.

Implementation Practices, Incremental Benefits and Costs

The incremental (marginal) benefits of a proposed decision should always be compared to its incremental cost. How does marginal benefit translate into the natural resource management decision process? Depending on restrictions imposed by law or custom, managers and policymakers with a goal of economic efficiency emphasize those products generating the greatest additional public benefit for the added cost of management.

If additional logging resulted in less additional benefit than the added cost of implementing the logging, efficient actions would make other choices, such as outdoor recreation, livestock grazing, flood reduction, or improved functioning of ecosystems to support clean water, clean air, and biodiversity. As long as the gain in benefits exceeds the added costs compared to logging, other actions should be selected over more logging. Managers can increase total economic benefits by choosing policies for which the marginal benefits exceed marginal costs.

Measurement Problems

Unfortunately, all of the benefits and costs generated by management are not equally easy to measure. Some benefits cannot be accurately estimated given the present state of economic knowledge. For example, the benefits from restoring ecosystems and recovering most endangered species cannot yet be accurately estimated. Also, the real cost of managing for timber, grazing, recreation, and flood reduction may not be estimated because the negative side effects on other ecosystem services are not fully known. That does not mean ecosystems and endangered species are worth zero or worth whatever they cost to produce. The continuation of the Endangered Species Act, along with a majority of public support, suggests its value to society is at least as great as its costs. Thus, in the political arena, the willingness to pay for the benefits of nonpriced goods and services is negotiated through the political process and legislation. The legislative decision about how much to regulate use or to allocate to non-monetized benefits is harder than it might be with some reliably monetized benefit.

SCOPE AND LIMITS OF BCA _____

The Bad News

BCA is difficult for most people to understand. Most politicians and other decision makers do not really understand it, nor does the average voter. It is technical and requires a good deal of economic theory. So people who read a BCA report either must accept the numbers on faith or ignore them.

Measurement problems plague BCA. Dollar values of things like endangered species habitats are hard to measure because one cannot buy them at the downtown grocery store. So the accuracy of monetary values of natural resource programs is questionable when the natural resource is not traded in commercial markets. Many of those nonmarket benefits require a contingent valuation analysis. Here people are asked what they would be willing to pay for the service even though such payment is rarely required and people have no previous payment experience. For that reason many decision makers do not believe some contingent valuation numbers. Therefore, BCA often ignores most important parts of government projects, such as saving wildlife habitat or wilderness land, because these resources are not traded in markets. So, unless a lot is spent on estimating benefits of the decision, the most important values are not necessarily included in the BCA.

Ethical problems also plague BCA. It may be morally wrong to use BCA to evaluate policies such as the Clean Air Act, which restricts technologies used by coal companies. How can the added length of miners' lives be valued in cold money terms? When a politician or other decision maker sees dollar values placed on such a proposed piece of legislation, it may confuse rather than enlighten. Being fair to future generations is another ethical issue poorly handled by BCA. Valuing future benefits requires using a discount rate to deflate future benefits so they can be expressed in a common denominator with the current generations' benefits. Any discount rate greater than zero may be unfair because it assigns a lower value to future generations than to current generations.

Some say that BCA is unfair because it ignores the distribution of benefits and costs. It only measures total benefits and total costs of an action. A proposed policy that does well on a BCA may be consistent with an economic arrangement that makes the rich richer and the poor poorer.

Finally, government agencies who do their own BCAs are like foxes guarding the hen house. These agencies have every incentive to cheat by inflating the benefits of their programs and ignoring costs. By inflating their own program's benefits, they may get a bigger budget in the future. Federal water construction agencies, such as the Army Corps of Engineers and Bureau of Reclamation, have been accused of this practice in past years.

The Good News

BCA is an objective way to judge proposed policies that connects economic principles to policy actions. It is based on an established body of economic theory that has been scrutinized, debated, and improved by many economists over the past 100 years. It attempts to measure the values of all the people who benefit from and who pay for a government policy decision, and not just a select few. Therefore, it is democratic and not elitist. By valuing impacts in dollars, it measures impacts in units that are understandable to both decision makers and the average adult. Other policy evaluation measures use some sort of indices, which typically are arbitrary. There is a vast body of literature on applications of BCA to various natural resource conservation issues.

Regardless of how BCA information is abused, the information allows the public, government employees, and elected officials to separate truly inferior alternatives from ones that have some merit. It can also be used to help isolate economically weak features of proposed plans that should be dropped, so that the stronger ones can be carried out.

Summary of BCA

The use of BCA permits government to use economic information to design better natural resource conservation proposals, defined as those for which the economic benefits exceed the costs. It can play an important role in legislative and regulatory policy debates on how natural resources should be managed. Although formal BCA should not be viewed as either required or sufficient by itself for designing good public policy, it can provide a powerful framework for consistently organizing diverse information. In this way it can greatly improve the process and therefore, the outcome of natural resource policy analysis (Arrow et al. 1996). Moreover, even if the data available for a BCA are poor, the steps required to carry it out force analysts to raise the right questions. It is more important to measure the right thing in a crude sort of way (doing the right thing) than to be measuring the wrong thing with impressive refinement (doing the thing right).

CONCLUSION

Economics gives us insight about and better understanding of human behavior. Through that understanding, we can design more efficient and equitable private and government programs for natural resource management. By influencing that behavior, particularly with economic incentives, we can better meet the needs of current and future generations of people. Economics also informs us about how we can better allocate scarce resources among people competing for them. If you think a career in natural resource conservation may be for you, we believe the investment you make in economics courses will produce a high return (high BCA). For further reading on natural resource economics, we recommend Anderson and Leal (1991), Field (1994), Tietenberg (1996), and Schiller (2000).

LITERATURE CITED

Anderson, J. L. and D. L. Leal. 1991. *Free market environmentalism.* Boulder, CO: Westview Press.

Arrow, Kenneth J., Maureen L. Cropper, George C. Eads, Robert W. Hahn, Lester B. Lave, Roger G. Noll, Paul R. Portney, Milton Russell, Richard Schmalensee, V. Kerry Smith, and Robert N. Stavins. 1996. Is there a role for benefit-cost analysis in environmental, health, and safety regulation? *Science* 272 (April 12):221–222.

Bressers, J. 1983. The effectiveness of Dutch water quality policy. *Universiteit Twente,* The Netherlands.

Field, Barry C. 1994. *Environmental economics: An introduction.* New York, NY: McGraw-Hill.

Forbes, S. A. 1888. *The lake as a microcosm.* Bulletin Peoria Scientific Association 1888:77–87.

Goodstein, Eban S. 1998. *Economics and the environment.* 2nd edition. Upper Saddle River, NJ: Prentice-Hall.

Hahn, R. W. 1989. Economic prescriptions for environmental problems: How the patient followed the doctor's orders. *Journal of Economic Perspectives* 3(2):95–114.

Hall, D. C., 1998. *Public choice and water rate design.* Department of Economics, California State University, Long Beach, prepared for the workshop on the political economy of water pricing implementation. *The World Bank,* November 3–5.

Hanley, R. 1988. Pay by bag trash disposal really pays, town learns. *New York Times,* November 24, p. B1.

Haspel, A. E. and F. R. Johnson. 1982. Multiple destination trip bias in recreation benefit estimation. *Land economics.* 58:361–372.

Herfindahl, Orris. 1961. What is conservation? (pp. 1–12). In *Three studies in mineral economics, resources for the future.* Washington, DC: U.S. Government Printing Office.

Howe, C. W., D. R. Schurmeire, and W. D. Shaw. 1986. Innovative approaches to water allocation: The potential for water markets. *Water Resources Research* 22(4): 439–445.

Kahn, James R. 1997. *The economic approach to environmental and natural resources,* 2nd edition. Orlando, FL: The Dryden Press.

Krutilla, John. 1967. Conservation reconsidered. *American Economic Review* (September):777–786.

Machiavelli, Nicolo. 1505. *The prince.* Florence, Italy.

Montgomery, C. and G. M. Brown. 1992. Economics of species preservation: The spotted owl case. *Contemporary Policy Issues* 10:1–12.

Pearce, David W. and R. Kerry Turner. 1990. *Economics of natural resources and the environment.* Baltimore, MD: The Johns Hopkins University Press.

Pinchot, Gifford. 1947. *Breaking new ground.* New York, NY: Harcourt, Brace and Co.

Prato, Tony. 1998. *Natural resource and environmental economics.* Ames, IA: Iowa State University Press.

Samuelson, Paul and William D. Nordhaus. 1985. *Economics.* 12th edition. New York, NY: McGraw-Hill.

Schiller, B. S. 2000. *The economy today.* 8th edition. New York, NY: Random House.

Smith, A. 1776. *The wealth of nations.* London, England.

Stavins, R. N. 1998. What can we learn from the grand policy experiment? Lessons from SO_2 allowance trading. *Journal of Economic Perspectives* 12(3):69–88.

Tietenberg, Tom. 1996. *Environmental and natural resource economics.* 4th edition. New York, NY: HarperCollins.

Turner, R. Kerry, David Pearce, and Ian Bateman. 1993. *Environmental economics: An elementary introduction.* Baltimore, MD: The Johns Hopkins University Press.

U.S. Department of Energy. 1995. Electricity generation and environmental externalities: Case studies. Energy Information Administration, Office of Coal, Nuclear, Electric and Alternate Fuels, Coal and Electric Analysis Branch, DOE/EIA-0598, Distribution Category UC-950, September.

Van Kooten, G. Cornelis. 1993. *Land resource economics and sustainable development: Economic policies and the common good.* Vancouver, BC: University of British Columbia Press.

Wantrup, S. V. C. 1968. *Conservation: Economics and policies.* Berkeley, CA: University of California Agricultural Experiment Station.

Ward, Frank A. 1998. Economics of water conservation. *New Mexico Journal of Science* 38:127–139.

CHAPTER 5

Planning, Policy, and Administration

ELEMENTS OF ORGANIZATIONAL FUNCTION

Identifying and Solving Problems

Planning, policy, and administration are elements in an orderly process designed to identify and solve problems. Although we think of them as organizational elements, they also apply to the individual. Chances are you have already encountered a problem today, which you somehow managed through planning and action. In our daily routines we encounter many small problems, which we either solve quickly, set aside, or ignore. Sleeping too late, for example, requires problem identification (I could be late!), searching for solutions (What choices do I have?), choosing a solution (skipping breakfast or making up an excuse), applying the solution (going hungry), and evaluating the result (Was I on time?). These are basic elements in all planning and implementation of solutions to problems (Figure 5.1).

In identifying the problem and taking stock, we often forecast conditions and consequences (What if the traffic is bad?). In determining a course of action, we usually weigh alternatives available to us and pick the best one for our circumstances (skip shower, breakfast, or dog walking). A thorough evaluation may lead to rethinking how we wake ourselves up. This could lead to a new personal policy—a procedural rule—for a new wake-up routine (set two alarm clocks). Permanent adjustments of our personal affairs will probably require careful administration; i.e., facilitating more effective activities.

Forces in Organizational Integration

Planning, policy, and administration are complementary elements in problem identification and solution and are essential forces for effective organization function (Figure 5.2). Natural resource management organizations have to identify problems and the means for solving them before acting out actions and evaluating the results. The care taken in this process can greatly improve management effectiveness. Management can be defined simply as the effective implementation of plans. Planning is the process of identifying problems or issues and their possible solutions, then mapping out actions and evaluating results. Policy is guidance developed by an or-

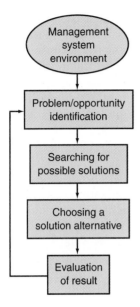

FIGURE 5.1 Problem identification, resolution, and evaluation.

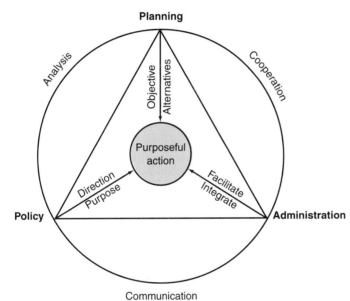

FIGURE 5.2 Components of effective management functioning by organizations.

ganization to carry out plans. Administration facilitates coordination of actions into an organizational integrity that serves a purpose held in common by all members of the organization. While the degree of formality varies among organizations, planning, policy, and administration are essential functions of any successful management organization.

Planning, policy, and administration determine organizational effectiveness in accomplishing purpose from the natural, human, and technical resources on hand. All members of organizations have roles and stakes in the outcome from organizational effort. While styles of management are diverse, all organizations need to coordinate the planning and actions of individual members into an effective organizational integrity if they are to efficiently solve their problems and take advantage of important opportunities.

Contemporary natural resources management concepts have roots in science and its applications. Management starts with identification of problems, establishes scientific alternatives for problem solution, and picks the most cost-effective solution for implementation. The organized planning process strongly parallels the scientific method, which emphasizes testing of alternatives (hypotheses), evaluating results, drawing conclusions and integrating them into scientific knowledge. Adaptive management is an approach to natural resource management that promotes continuous scientific evaluation of management effectiveness and continual adaptation of management to conditions as new knowledge is gained (Walters 1986).

General Approaches to Management

The dynamics of environments inside and outside of an organization has much to do with determining the general approach taken to management. In a changing environment, like that of most natural resource organizations, planning is an indispensable and continuous process. In slowly changing environments, relatively little investment is needed for planning. The planning process can be incremental and aimed at stabilizing the organizations impact (Braybrooke and Lindblom 1963). In such environments, the goal of the agency is to maintain the status quo. However, in a time of change, organizations can end up "out of sync" with forces in the management environment and lose effectiveness and influence.

Organizations seeking adaptation to change frequently use the so-called rational planning approach. It formulates goals and objectives, designs and analyzes alternative plans, implements the most effective plan, and evaluates for degree of success (Kaiser et al. 1995). It identifies gradually emerging issues that may be opportunities or problems and incorporates them into the process of goal and objective formulation. However, in more dynamic environments, an approach that focuses on rapidly emerging issues and their strategic management has advantages over one that focuses on goals (Bryson 1995).

The need for strong communication skills grows as the complexity of organizational tasks grows and involves more people. In such situations, the planning process often becomes a more public and formal organizational effort to keep management from dissolving into disarray or chaos. Effective organizational planning integrates personal planning into a coherent institutional process directed toward accomplishing organizational purpose. Excellent communication skills are a prerequisite to linking personal planning with others in an organizational planning process.

MANAGEMENT SYSTEMS

Organizations as Management Systems

Despite the "action" emphasis at professional entry level, career advancement typically depends on understanding the management system. In contrast to the static structural chart of organizations often used to demonstrate the organizational "system," we emphasize the importance of functions in organizations. What organizations do is more important than how they appear on a chart. The master function is management, including the "hands-off" management of preservation. Planning, operations, policy, and administration are the main functions subsidiary to management. Each member of an organization makes his or her contribution to the organization, usually in functional "bundles" called programs and projects (or something similar in concept). These functional bundles form management subsystems within the organizational system. Each subsystem requires a manager (commonly project and program managers) to oversee planning, policy, and administrative links to the larger system.

Natural resource management systems include government, advocacy, professional, and business organizations, all of which interact in more-or-less predictable

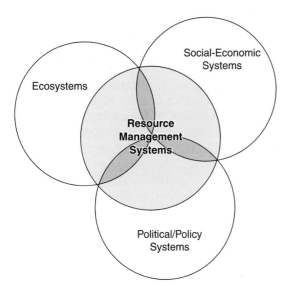

FIGURE 5.3 The basic components of natural resource management systems.

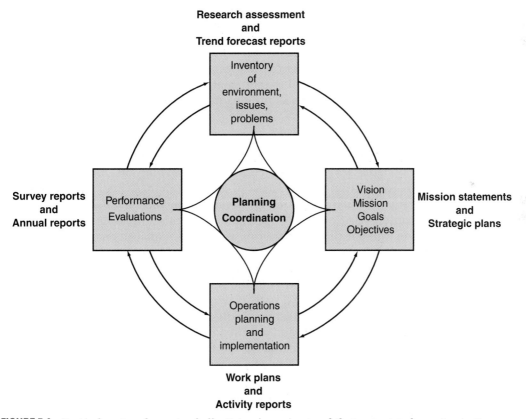

FIGURE 5.4 Basic planning elements of effective management and their associated communication products. While the process of planning and implementation progresses forward through inventory, objective setting, operations, and evaluation, there is learning feedback at each step in the process. Each step generates communication designed to facilitate the management process.

ways to bring about positive benefits from natural resources (Figure 5.3). The hands-on manipulation of resources is only one function of the management system, one which is guided by planning, policy, and administrative functions.

While many specific approaches are taken to management, the same four basic elements are common to all effective management processes (Figure 5.4): (1) inventorying the planning environment; (2) focusing management intents through mission, goals, and objectives; (3) allocating resources to appropriate operations; and (4) evaluating outcomes (Lorange and Vancil 1977, Crowe 1983, Koteen 1989). Planning requires an inventory of existing conditions in the management environment, including the resources at hand to do the job and those stakeholders who will be affected by it. Inventory often includes forecasting, based on trend analysis or other information, and developing different plans for a series of possible future conditions. Planning focuses on transforming a vision of future desirable conditions through statements of mission and goals into specific objectives and the strategies and tactics used for their accomplishment. Operations are designed to accomplish the objectives that will realize the vision. Evaluation is needed to judge progress made toward realization of the desired future condition. An evaluation provides new information for modifying the inventory and the planning process that follows. In this way, the planning process cycles refined and improved information for organization adaption.

Each organization forms a planning and management system, which interacts with other management systems in environments held in common and around purposes of mutual interest (Lorange and Vancil 1977). Management systems also overlap and interact with ecosystems and economic systems to redirect their processes toward planned outputs and outcomes (Figure 5.4). Without the generally predictable behavior of ecosystems, socioeconomic systems, and management systems, chaos would reign and civilization would not exist as we know it. Appreciation for underlying systems process is fundamental to effective management and more meaningful contribution of each member of the organization. Many management systems do not interface or integrate as well as they might and one of the growing challenges in natural resource management is improving management integration.

Management System Boundaries

Unique natural resource management systems exist for diverse organizations including private businesses, advocacy groups, government, and any other type or combination of organizations. The process of planning in these management systems starts with the legal mandates limiting the organization and the mission determined by the organization (Bryson 1995). Mandate and mission establish boundaries around a unique internal planning environment—the management domain in each management system (Figure 5.5). The mission, either explicitly or implicitly, respects natural limits to organization accomplishment (avoids the impossible), and determines organization purpose within the scope of authority and natural limits. Basic authorities are granted by law, such as purposes defined for agencies, the right to private property, and all of the responsibilities attending citizenship.

Management mandate and mission usually identify the specific resource uses and geographic area to be developed, regulated, or otherwise influenced by the organization. The U.S. Forest Service, for example, is bounded by its legal authority to manage all national forest natural resources according to legally authorized policies. Within the geographical limits of its authority, the Forest Service has a multipurpose management authority over grazing, timber, recreation, biodiversity, watershed, and other resources, which it administers. In contrast, the U.S. Environmental Protection Agency (EPA) is primarily authorized to prevent environmental degradation through enforcement of regulatory policy applied throughout all of the U.S. Whereas it is a single-purpose organization, limited to environmental protection, the authority of the EPA is geographically unlimited within the U.S. The U.S. Fish and Wildlife Service has a more complex authority, including management jurisdiction over an extensive refuge system, regulatory enforcement of numerous federal fish and wildlife laws, and administration of funding to states gained through federal taxation on fishing and hunting equipment and supplies. State, county, and municipal agencies have no jurisdiction beyond their political boundaries, but otherwise are similar to federal agencies in the way their management systems are bounded.

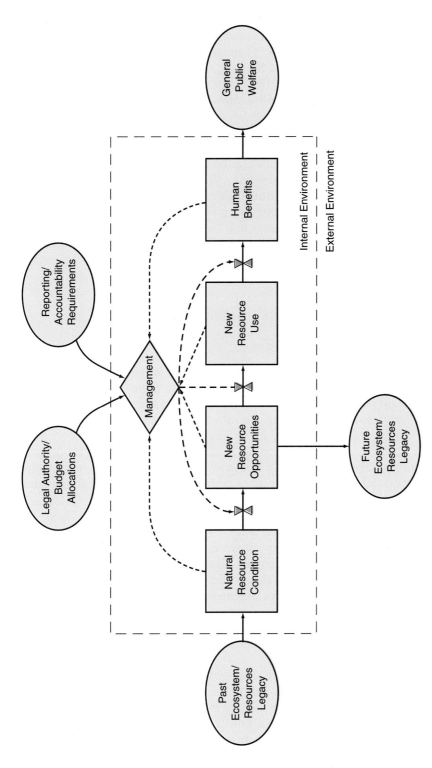

FIGURE 5.5 The management system. Management "drives" a process of assessing resource condition and developing new resource opportunities based on expected resource use and human benefits. Factors in the external environment provide ecological and social inputs that enable management system function. System outputs effect the future condition of the ecosystem and resources as well as general public welfare.

Organizations outside government also have management system boundaries defined by property right and other law. A private corporation holding land is authorized through recognition of private property rights to manage the ecosystem associated with that property within legal constraints imposed by government. In the U.S., for example, a private land holder cannot, without permit, manage or harm the wildlife, which is a public resource. Through the Bill of Rights, U.S. advocacy organizations, such as the Wilderness Society, can influence law and policy through a legally sanctioned process, usually involving public hearings, media advertisement in the public interest, and publication. Some nongovernment organizations buy and reserve lands for a specific resource use. The Nature Conservancy has been especially prominent in setting aside reservations for their mission of biodiversity protection.

Natural resource management often is based on geographical units, regardless of whether an organization is authorized control over land through ownership or through the public trust. Most natural resource management systems classify resources through some kind of geographically based reference system. Management units often are split up on a regional basis and management plans often define the boundaries of each separate area to be affected by management, and to what degree.

THE INTERNAL MANAGEMENT ENVIRONMENT

Resources, Use, and Management

Whether private or public, managers operate with the intent of serving the interests of people who use the resources, either to profit from them or to benefit from them according to legal mandate. Private businesses may manage natural resources to sell to customers or to use themselves in manufacture of refined products that are then sold. Many owners of timberlands, for example, open them to sale and extraction while other owners extract and develop lumber or more finished wood products, such as furniture. Private businesses compete in the market for customers and in so doing are driven to benefit the customers while they too benefit through continued employment and profit (see chapter 4).

Government agencies are mandated through their authorizing laws to serve resource-user "customers." However, it is often more difficult to determine exactly who the customer is. When the Forest Service, for example, sells timber, the logging company stands out as a customer among all of the taxpayers who are also affected by the income generated from the sale and from possible environmental costs associated with the sale. Like many large businesses, agencies often produce many different products for different groups of resource consumers. However, because of their typical broad mandates, government agencies have no choice but to serve all customers uniformly well while private businesses can specialize or diversify as they choose within broad regulatory limits.

Agencies are placed in the difficult position of optimally serving all users for maximum public benefit. Whereas competition is the mechanism relied on to assure customer benefit from private management systems, the operation of a public resource agency is much more of a public affair open to scrutiny and criticism. The diverse demands made on public agencies, including accountability through a complex process of public review and comment, often adds layers of cost to resource management that private companies can avoid. Without competition to drive better customer service, the customers have little choice but to add expensive public checks and balances to the cost of operation. Therefore, private management is preferable wherever the products are clearly priced and competition can flourish.

Opportunity and Use

Opportunities may be provided in a diversity of ways other than increasing the production level and sustainability of resources (Figure 5.5). Providing more access often is a critical element in providing opportunity. For example, roads, boat ramps, improved navigation channels, trails, railroads, and airports all increase access. Provision of facilitative services can be important, such as providing power, fuel, food services, housing alternatives, and a diverse array of other wants and needs. Management effectiveness in conveying benefits to customers often depends on how well a mix of opportunities are provided.

Resource opportunities may or may not be developed for use. Whereas some opportunities are intended for immediate consumptive or nonconsumptive use, other opportunities are intended for possible future use (use options) or simply to provide knowledge of continued existence. These nonuse opportunities are provided by forbidding any present destructive use, whether it be consumptive or not. Endangered species protection is one of the better examples of option value protection, which often is compatible with low intensity recreational, scientific, or educational use.

Managing For Result, Not Process

Simply providing opportunity does not guarantee use where it is desired. When a private company misses the mark, it soon suffers as competitors take their business away. However, agencies that manage public resources are not so easily focused on the customer because they get paid in any case. Too many public resource management organizations have emphasized provision of opportunities over customer benefits, using the premise "if you build it, they will come." As a consequence numerous underused roads, campgrounds, harbors, boat ramps, reservoirs, and the like were built in place of more useful provision of opportunities elsewhere. Resources developed for a use that never materializes benefit no one. In fact, the net effect can conceivably be negative because management money is diverted from development of beneficial opportunities elsewhere.

External Environment

No management system is independent of other systems. Each management system depends on inputs from outside the system and, in turn, each management system's outputs contribute to the inputs of other systems beyond their boundaries. For natural resource management systems, most of the more influential inputs and outputs are ecological and economic.

The ecosystem condition "inherited" by an agency or firm is a fundamental determinant of the resource condition, economic demands placed on resource use and cost of management. No where is this more evident than in water resource management. An agency situated high in the watershed has a much different ecosystem perspective than management agencies located lower in the watershed in large rivers or estuaries. Upper watershed agencies are more concerned about output impacts than about changes in inputs from external sources. Similar kinds of connecting flows among the jurisdictions of management systems occur along plant and animal dispersal routes, animal migration routes, coastal and oceanic currents, and prevailing winds.

Thus, some of the more vexing resource problems in recent years pertain to large animal migration and dispersal from public to private lands, downwind acid rain fallout across state and international borders, international incongruities in managing migratory fish and birds, and progressive water quality deterioration from upper to lower watershed ecosystems (Figure 5.6). Simply documenting the characteristics of ecological flows between management systems is basic to solving some of the related management problems. Keeping track of management impacts on ecological outputs is a fundamental need. Surprisingly, the need for such documentation is

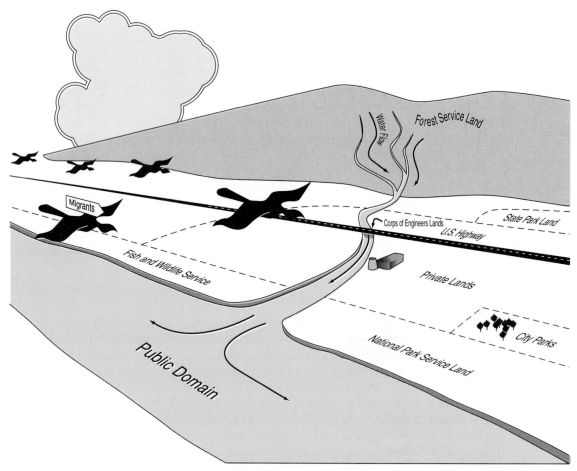

FIGURE 5.6   Political authorities often cut across natural processes, such as weather, water flow, and animal migration. When these ecosystem processes are ignored within political planning boundaries, problems usually result. Integrated planning is needed to contend with such problems.

often overlooked because of the tendency of most management systems to "externalize" the costs of the problems they caused because the impacts appeared elsewhere—"out of sight, out of mind." Much air and water pollution was caused as a consequence of such closed-systems thinking. The import-export connections between management systems often are the focus of mediating regulation administered by an overlying management system. The Environmental Protection Agency, U.S. Fish and Wildlife Service, and National Marine Fisheries Service are good examples of nationwide regulatory management systems superimposed over the public and private land-based management systems.

Measuring Management Performance and Success

Effective management monitors opportunity provision and use, and focuses on public and customer satisfaction to evaluate management success. This information is used by management to reorganize goals and strategies for accomplishing organizational purpose. The information also may be reported to the concerned public and oversight groups. Private firms can gauge success by the income and profit they sustain while monitoring goods and services production for improved efficiency. Much of this type of monitoring is required by law for income tax purposes, but some of this data is collected for cost management.

Public agency providers can do the same. However, there is not the same force of law and personal gain involved to motivate monitoring. Whereas budget expenditure records are required by law, there is no equivalent of business profit records for gauging the success of public organizations in providing services. Whereas government agencies are good at monitoring the opportunities provided and their costs, they are much more inconsistent in monitoring public use and satisfaction. Osborne and Gaebler (1992) wrote an influential book entitled *Reinventing Government,* which emphasizes outcomes assessment in the form of public "customer" satisfaction as the way to gauge government effectiveness. The Government Performance and Results Act of 1993 requires all government organizations to report the outcomes generated by federal expenditure. It is too soon to judge the success of that legislation, as agencies wrestle with what performance assessment means in terms of opportunity outputs and public-satisfaction outcomes. However, long-term government trends indicate a gradual but steady shift of agencies away from near-sighted provision of opportunity to public-service satisfaction.

PUBLIC, PRIVATE, AND ADVOCACY SYSTEMS

Public

Numerous federal, state, and local government agencies make up an interactive complex of natural resource management systems (Table 5.1). Some agencies regulate the activities of other agencies, for example, as the EPA does under the authority of the National Environmental Policy Act of 1969. Most of these federal agencies form management subsystems within cabinet departments in the executive branch of federal government. Departments of Agriculture, Energy, and Interior are most consistently organized around natural resource management, but with some exceptions. Other departments play small roles in natural resource management, but few play no role at all. The numerous federal management systems can interact through complementary and overlapping jurisdictions, oversight, coordinative legislation, executive orders, various directives, and networking through professional organizations, workshops, publications, Internet services, and other communication. Because of the diverse purposes authorized by departments and agencies, the subsystems do not integrate as well as they might into a single federal natural-resource organization.

There has been continuous discussion about possible rearrangement of agencies and departments. The last major change in federal government pertaining to natural resource management was to integrate most of the diverse environmental protection functions into the Environmental Protection Agency, which has yet to gain cabinet status after years of consideration. Insufficient integration of natural resource planning process at state and local agencies also has impeded coordination among local, state, and federal management systems, interfering with the development of effective planning process.

Business

Private business management systems are similar to government systems in their basic attributes, the main difference being the importance of competitive forces in their function. The nature of market-driven business systems is described in more detail in Chapter 4. Unlike government agencies, profit-motivated businesses have investors who expect to realize substantial profit from company activity. Private businesses are quite diverse in size and complexity of goods and services they sell in the marketplace. Most businesses are small and usually specialized in providing a narrow range of goods and services. Common types of small businesses oriented around natural resources management are ranches, farms, engineering and construction firms, landscape and gardening services, travel services (restaurants, shops, gas stations/garages, hotels/motels), bus and other transportation companies, logging companies and sawmills, guiding and tour services, equipment and retail services, outdoor recreational services (ski slopes, marinas, golf courses, recreational vehicle rentals), mineral extraction companies, and local power and water supply companies.

Larger companies may be "vertically assembled" based on ownership, extraction, transportation, delivery, and refinement of resources. Large mineral-resource industries often have an extensive network of landholdings, extraction equipment, refineries, and product delivery services. Most land-based businesses focus on development of specific resources, but some have diversified. There has been increasing interest in developing recreational services on privately held rangeland and forest land, for example.

Advocacy

Advocacy groups are extremely diverse and represent a wide variety of public and private land-use concerns. Advocacy groups are nongovernment organizations that

TABLE 5.1   Major Federal Agencies with Management Authority Over Natural Resources

U.S. Agency	Department	Natural Resource Authorities
Army Corps of Engineers (Civil Works)	Defense	Water resources management for navigation maintenance, flood damage reduction, recreation, environmental improvement, hydroelectric, and other water use; also enforces wetland protection law and administers public lands
Bureau of Land Management	Interior	BLM public lands management for range, forest, watershed, recreation, mineral, fish and wildlife, and other authorized use
Bureau of Reclamation	Interior	Water resources management for irrigation supply, recreation, environmental improvement, hydroelectric, and other water use; also administers public lands associated with water resource projects
Department of Defense	Defense	Authority over resources on military installations
Environmental Protection Agency	Stands alone	Enforces U.S. environmental protection and restoration law and administers public water treatment facilities programs
Federal Highway Administration	Transportation	Oversees development and maintenance of U.S. highway system and associated lands
Federal Power Authority	Energy	Oversees development and coordination of U.S. energy production
Fish and Wildlife Service	Interior	FWS land management of the fish and wildlife refuge system, enforcement of U.S. fish and wildlife law including Endangered Species Act, and monitors fish and wildlife resources
Forest Service	Agriculture	USFS public lands management for forest, range, watershed, recreation, mineral, fish and wildlife, and other authorized use
Geological Survey	Interior	Federal natural resources research and mapping agency including hydrological, geological, and biological processes
National Marine Fishery Service	Commerce	Monitors marine resources and administers international marine fisheries and marine mammal laws including Endangered Species Act
National Park Service	Interior	NPS public lands management including national parks, monuments, recreational shore, wild rivers, and other reservations of natural and cultural heritage
Natural Resources Conservation Service	Agriculture	Agricultural resources services associated with water supply and drainage, soil maintenance, erosion control, and farmland, fish, and wildlife habitat management
Office of Surface Mining, Reclamation and Enforcement	Interior	Oversees and enforces federal legislation pertaining to environmental protection from surface mining and reclamation of lands for productive use following mining
Tennessee Valley Authority	Energy	Water resources management for hydroelectric, flood damage reduction, recreation, fish and wildlife, and other uses in the Tennessee River Basin

are typically nonprofit organizations. They are motivated typically to shape public or private policies through various strategies including marketing their positions through communications media, business boycotts, demonstrations, court actions, strategic property purchases, and, rarely, unlawful sabotage. Over the past several decades, they have become the primary avenue for representation of stakeholder interests in outcomes pertaining to public resources management. Environmental advocacy groups have been especially effective in promoting their special interests.

The goals of advocacy groups are organized with special interest objectives in mind, but sophisticated organizations realize they can be more successful in a situation where none of the stakeholders are perceived to be total losers. Public management systems are increasingly involving advocacy groups early in the planning process to help work through difficult issues, especially those pertaining to private land use and development.

PLANNING PROCESS

Inventorying Planning Environments

Inventory is the process of gathering and analyzing information. It sets the stage for development of vision and its translation into reality (Figure 5.7). Inventory activities go by numerous titles including survey, monitoring, research, intelligence, environmental scanning, cost accounting, forecasting, trend analysis, systems analysis, issue and problem identification, SWOT analysis, stakeholder analysis, and information resources and management. Problems and opportunities are identified early through continuous and astute inventory methods. For natural resource organizations, inventory centers on the state of natural resources, their development and use, and the profits or benefits derived from them. The most basic intelligence pertains to changes in the "balance" between resource supply and resource demand, changes in the influential planning environment, such as laws and other mandates,

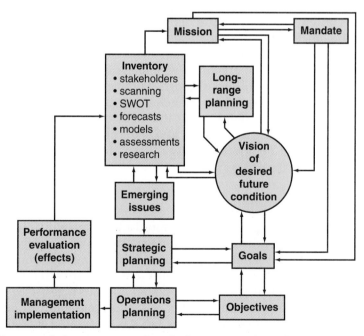

FIGURE 5.7   The general planning process by which mission and mandates are accomplished. Planning starts with a vision of a desired future condition and identification of goals and objectives that will realize the vision through operations planning and implementation. An inventory of existing conditions, including emerging issues, is completed to develop effective strategy.

organizational condition, and stakeholder needs. Perhaps most importantly, inventory helps to clarify visions of desired future resource conditions. Inventory examines the state of things both inside and outside the management system.

Reviewing Mandates

Planning inventory often starts with a review of the laws or the mandate authorizing actions or regulating action (Bryson 1995). For public organizations, laws provide authorities to work toward certain ends. But authorization laws often are quite broad, leaving latitude for the organization to create a sense of purpose or mission that extends beyond existing law. Planning often inventories the legal mandate, organizational authorization, and policies before proceeding into new initiatives.

Private organizations also are bound by laws that limit their operations to socially acceptable practices. These laws often originate from government action designed to promote constitutional ideals, which strive to protect individual rights while assuring improvement in the general public welfare. Most organizational legal conflicts arise over private right versus the general public welfare. Whereas legal mandates both authorize and regulate the operations of public-service organizations, they only regulate the operations of private organizations.

Some extralegal public expectations from government can approach the force of legal mandate. Public opinion may at times be strong enough to approach the force of law. Many state fish and wildlife agencies, for example, are legally mandated to serve the needs of all state citizens. Although management of fish and wildlife resources benefit many nonconsumptive users, a select group of hunters and fishers typically pays most of the management costs in many states. As a consequence, the desires of hunters and fishers can act as an extralegal "mandate" whenever they differ collectively from those of other citizens.

Reviewing Mission

The organizational mission is the fundamental communication of organizational purpose among members and all others affected by its performance. The mission usually identifies who is to be served by the organization and how and where that service is to be provided. An effective mission statement conveys an accurate sense of organizational attitude toward its employees and the world it intends to influence. Inspirational mission statements are motivational and impressive, contributing to morale of the organization membership and the respect of stakeholders, if the organization follows through. For agencies, a good mission gives focus to authorities and frames values in such a way that goals may flow naturally from it. Because the mission statement is the main communication of intent between the organization and the internal and external management environments, it always needs to be reviewed as new information is received through inventory (Figure 5.7).

Environmental Scanning

Environmental scanning is a sweeping approach to reviewing planning environment trends relevant to organization management. Trends and emerging issues that have significance as management opportunities or problems are brought to the planning process for identification of the most effective alternative. This process usually requires perusal of trade magazines, newsletters, various digests, legal briefs, news media, press releases, Internet home pages, and other communications from within and among diverse stakeholder groups. The focus is on finding emerging issues—potential problems, solutions to problems, or other opportunities in their earliest stages. These are tracked and incorporated into the planning process as they more fully emerge or are discarded as they fade in importance.

Stakeholders

Stakeholders are individuals and groups affected by actions of the organization. Stakeholders are best assessed and brought into the planning process as early as possible (Bryson 1995). Certain stakeholder groups often are the object of organizational performance as service recipients, promoters, or detractors.

SWOT Analysis

Effective planning often requires SWOT analysis of the strengths (S) and weaknesses (W) of the organization and the opportunities (O) and threats (T) originating outside the organization (Bryson 1995). Forecasts of change are an important part of effective planning. The effects of resource demand on continued supply, for example, depend on trends in both resource supply and demand. Forecasting analysis allows prediction of future supply based on estimated demand. From this information, stakeholder benefits often can be tracked. Inventory also includes monitoring of all organizational costs incurred by project activity, allowing benefits to be compared to costs.

Forecasting

Forecasting is the process of predicting future trends and/or future conditions. The future usually is dependent on too many events outside management influence for perfect forecasting. The most useful approach is to plan for several feasible alternative futures. Planning for alternative futures introduces an element of flexibility and proactive process in anticipation of change.

Mathematical models often are used to integrate information into analytical tools for forecasting the effects of different management policies. Typically the models represent relationships among variables determining resource supply and demand. Development of precise predictive models can greatly reduce inventory costs, but often with substantial investment in the developmental research. Predictive models exist, for example, to characterize water resources, fish and wildlife habitat, timber production, range production, viability of rare species, and nonrenewable resource depletion. Many organizations use qualitative approaches for analyzing future change. Some models are comprehensive enough to represent whole systems and are used for systems analysis. Sometimes physical models, including natural ecosystems, are used as a reference for the planning process.

FOCUSING MANAGEMENT INTENT _____

Mission

The process of focusing management intent into implementation of a plan typically starts with the statement of organizational mission and values. When well stated, mission and value statements guide the formation of planning vision, goals, and objectives for organizational accomplishment. Basic elements of all U.S. government missions follow from the mandates of the U.S. Constitution, which emphasizes improvement of the general welfare, including the interests of future generations. For those agencies responsible for managing renewable resources, the mission usually identifies resource conservation or sustainability as a guiding philosophy. In response to political influences, most agency missions historically have focused their service on the most active special interests identified in their authorities: timber and recreation in the Forest Service, agricultural irrigation in the Bureau of Reclamation, navigation and flood-damage reduction in the Army Corps of Engineers, livestock in the Bureau of Land Management, environmentalists in the Environmental Protection Agency, hunters and conservationists in the Fish and Wildlife Service, and many others.

Planning Vision

At the center of systematic planning (Figure 5.7) is the focusing of a vision of desired future condition based on achievable tasks (Bryson 1995). The establishment of an acceptable vision of desired future condition, or outcome, may be the most critical part of effective natural resource planning process (Bryson 1995). If the vision is vague or unrealistic, much time and money may be spent ineffectively. Without visions of what needs to be accomplished, an organization can drift through the management system environment without having much effect.

A planning challenge is building consensus among diverse stakeholders in the development of a *shared vision* of desired future condition of resources. The approach to building consensus is to provide an environment that favors the open exchange of information among the resource experts knowledgeable about trends and stakeholder's needs. A variety of approaches have been developed to facilitate an orderly discussion of issues. A critical part of the discussion hinges on the knowledge of how benefits will be distributed among stakeholders as a consequence of the policies selected. Visions can be developed at different organization levels ranging from small everyday projects to the vision behind organizational mission.

Goals

Goals are more general descriptions of elements in the vision to be accomplished and the general means to be used in accomplishment. They do not specify the results precisely. Where organizational visions of successful management are reasonably clear, goals are relatively easily developed in pursuit of vision realization. The goals of public resource agencies usually emphasize generations of public benefit while sustaining resources. Private enterprise emphasizes a larger market share, a greater profit, and better employee income and benefits. Goals also identify the approaches that are to be used in pursuing the mission, including education, regulation, research, and management. Goals typically are not scheduled for a specific completion date and sometimes remain unchanged for many years. Goals are most easily developed in stable planning environments where visions of desirable future conditions are clear (Bryson 1995).

Issues

Even in the most stable planning environments, organizational threats and opportunities emerge. Issues usually can be disassembled into problems that need solution if the organization is to most efficiently pursue its purpose. A special area of the goal-driven planning process deals with issue management (Bryson 1995); threats are mitigated and opportunities are exploited.

In organizations with unstable planning environments and vague visions of organizational success, most planning is dominated by disruptive threats. Many resource organizations have become less stable in recent decades as conflicting demands of diverse stakeholder groups have grown in intensity. In those situations, visions of success often become muddled and issues drive the planning process. Strategies and tactics are developed to mitigate or exploit each issue. As greater control is developed over issues, they may become the focus of new goals and new visions of success (Bryson 1995). But until then, the issues replace the goals as targets for planning accomplishment.

Strategic Planning

Strategy is a proactive, as opposed to reactive, and adaptive, as opposed to standard, approach to solving problems and taking advantage of opportunities (Bryson 1995). Strategic planning typically addresses time horizons beyond the present budget planning period. It usually extends less than three years forward. These shifts may derive from past natural resource practices, new knowledge, changes in law, changes in customer and public preferences, new technologies, natural catastrophes, and new trends identified during inventory of planning environments. Strategic planning is most effective where long-range planning has provided sound information about major resource trends.

The U.S. Forest Service, for example, uses 50-year planning horizons based on timber cutting cycles. Even 50 years is far short of the centuries needed to reestablish old-growth forests. Similarly, recent legislation has promoted ecosystem restoration, which also requires far-off visions of desired future conditions and strategies for their realization. Strategic planning needs to be flexible, adaptable to changes in the planning environment, and continuous. Strategic planning is focused more on relative program emphasis and new program needs than it is on project planning. Numerous government agencies have recently reexamined management goals to accommodate the need to sustain biodiversity on public lands and to incorporate the diverse interests of public stakeholders at an early planning stage.

Setting Objectives

Objectives are developed to focus tasks on measurable achievement. In contrast with the open-ended time frames typical of goals and issues, objectives have deadlines for specific accomplishments. The specific procedures used to pursue objectives compose the operational tactics—or *operations*. Typically, a suite of specific objectives is used based on indicators of the desired future condition. For example, a restored wetland may be the planning objective. A completely functional wetland may be anticipated if the objectives detail the important attributes of the landscape, the water supply, nutrient and sediment supply, and the species that are envisioned in the structure and function of that wetland.

Long-Range Planning

Long-range planning involves analyzing trends and then setting far-off goals, and/or avoiding strategies that are ultimately incompatible with desired long-range outcomes. Since most natural resource management is intent on improving human quality of life, long-range targets typically focus on sustaining natural resource options for future use. For natural resource management, knowledge of resource abundance, renewal, extraction rates, and their environmental regulation are critical elements in estimating long-range status. Long-range analysis and planning addresses generic issues about trends in resource supply and demand, availability of substitute resources, and the need for regulating demand to sustain resource supply. It also examines the recovery times needed to sustain renewable resources at levels compatible with social demand.

Operations Planning

Operations planning links the organization budget to plan implementation. In operations planning, project objectives are clearly defined and budgets assigned to each area of management directed at project completion. The typical target of the most detailed operations planning is the budget year—which may be based on the calendar year or some other initial and ending date. When done well, a number of alternative approaches to objective accomplishment are compared and the most cost-effective alternative is used to accomplish each project.

Operations planning focuses on objectives formulated for completion according to schedule. An important part of operations planning is the assignment of personnel time to scheduled tasks. Within the Forest Service, budgeted projects for recreational service might include numerous outputs like road improvement, installation of new campgrounds, development of nature trails, improvement of user law compliance, development of brochures, and research reports. Operations planning usually follows a predictable annual cycle in both public and private management systems and is linked to strategic planning.

Organizational Performance Evaluation

Without evaluating accomplishment, there is no way to confirm management success. Thorough evaluation usually requires a well-organized monitoring process, which minimally monitors management costs and benefits. Performance evaluation may also monitor some of the intermediate management outputs designed to provide customer opportunities. The main reason for this monitoring is to determine inefficiencies in providing benefits and how to improve upon them. Performance can be measured in terms of process outputs and in terms of customer satisfaction or benefits. Job performance is usually based on completion of assigned tasks, but ultimate organizational performance is gaged on benefits provided. It is possible for an individual to have a high performance rating for output production while benefits fall. The problem was in choosing the job to be done in the first place, which often is not the prerogative of the person who does the job.

Private businesses have no question about the utility of performance evaluation. Without it, they would have no knowledge of how much income exceeded costs and the proper base for taxes and distributing profits. Government agencies have had more problems evaluating performance in benefit terms, tending to count opportunities provided (e.g., roads built, wildlife counted) or process rates (publications produced, law violations cited) rather than benefits generated. Although many agencies are required by their authorities to generate more benefit than cost, techniques for estimating benefits were not available for certain activities until very recently (e.g., recreational use). Other activities, such as much environmental protection and improvement, cannot be monetized to estimate benefit in a way that can be compared directly to cost. Instead, the relative significance of the activity needs to be gaged as a proxy for benefit. The Government Performance and Results Act of 1993 institutionalized more consistency in federal government with respect to performance assessment.

ORGANIZATIONAL PLANNING BOUNDARIES

Practicality dictates that planning be manageable in scope. Boundaries may be set on planning process in a number of diverse ways. Planning is frequently delineated by organizational activities including project, program, and comprehensive organizational activities. Planning also is frequently organized geographically or regionally. Integrative management planning process facilitates organizational interaction.

Most organizations divide their labor according to program and project boundaries. Program and project planning often are done separately, but involve interactive processes. Management system boundaries often are based on the dimensions and relationships of work activities organized to assemble materials, energy, and information into a new or improved output or "product." The output may be a good or a service of some kind, or a combination of goods and services.

Project Planning

Project planning typically focuses on organizing natural, human, and fiscal resources into the desired kind and amount of output in which some form of accomplishment can be measured. The most obvious types of projects are physical constructs of some kind—for example, dams, levees, boat ramps, toilets, campgrounds, roads, city parks, trails, and habitats. Projects also may include new methods for identifying and extracting natural resources, ways to refine raw materials, and development of improved information services in the form of books, videos, and advertising in various media. Conceptually at least, there are no activities in organizations that could not be assigned to a project.

Projects form a basic unit for planning and operations in natural resource management. They typically set up objectives for accomplishment with deadlines. Project boundaries often cross administrative lines, and require careful coordination usually through a project manager.

Many organizations separate "routine operations" from projects. Because routine operations typically are not expected to have clear outputs and outcomes, they usually are not considered projects. However, some management organizations attempt to integrate most, if not all, activities into projects. Even operations as general as administration contribute to different projects. That contribution must be integrated into project cost if the real cost of the project is to be estimated. Certain administrative initiatives may be categorized into projects of their own. Most, if not all, operations can be linked to accomplishable objectives. This approach provides targets for improvement or, at very least, sustained performance. Otherwise the routine operations can steadily degrade in performance, diverting important resources away from innovative accomplishment.

Many government projects include a mix of agencies and private contractors in project development. A federal water resource project, for example, typically involves local and federal agency "partnering" and contracting with private services, all of whom need to be coordinated through project planning process. Most projects also have a variety of stakeholders who usually need to be involved in planning process, especially early in the development of a shared vision.

Program Planning

Programs typically are organized around customer service and associated benefits. A state conservation organization, for example, may have sportfishery, commercial fishery, wildlife, forest, range, and mineral-resource programs, each with numerous projects. Programs often are linked to specific sources of funding, but may be funded out of a larger pool. In either case, program managers need to argue for funding based on the benefits that will be produced from the program. Commonly, programs form around the special interests of certain stakeholder categories, which makes for easier analysis of program effectiveness than when they cut across linkages between resources, opportunities, uses, and benefits. However, many agencies are organized that way, with planning in one program and implementation in other programs.

Programs envelop groups of projects with similar attributes and purposes into a larger organizational decision process. A program-project matrix may be used to organize project activities into program areas. Each program includes projects pertaining to outcomes desired of each program and program managers typically seek those projects that contribute most to program performance in generating outcomes as planned. When left up to the program management, an important function is to decide how to most cost-effectively distribute budgets among projects so as to generate the greatest public benefit or profit. Program management also provides centralized support services to all projects, thereby cutting project costs.

Comprehensive Organizational Planning

Comprehensive management planning coordinates across groups of programs within an organization (Crowe 1983). Comprehensive planning addresses decisions about future program emphasis, the development of new programs, and the retirement of others. Comprehensive planning seeks integration among all programs under the organizational umbrella. It may seek to develop new and/or terminate old programs. Much of the demand for government programs is expressed in new laws. Over recent decades, for example, natural resource agencies have responded to new laws by directing more management resources to programs that emphasize protection of resource options (e.g. endangered species) and ecological services (e.g., flood-damage reduction, erosion control, and fish and wildlife habitat).

In most planning processes, long range, strategic, and operations planning grade together. Strategic planning uses outputs from long-range planning to help identify appropriate goals, program emphases and strategies. Operations planning uses outputs from strategic planning to formulate project objectives and tasks for their accomplishments. The most effective planning involves continuously adding output information to the planning inventory and adapting plans to new conditions. This is called adaptive management.

REGIONAL (GEOGRAPHIC) PLANNING

A common way to place boundaries on planning and management systems is through geographic or regional boundaries. The boundaries are most commonly defined by political boundaries, but are increasingly being defined by natural boundaries.

Politically Bounded Planning

Political boundaries traditionally have been emphasized in the identification of planning regions, such as city, county, state, and national boundaries; the boundaries associated with different government agency authorities; and private land boundaries. Urban planning often ends at the city limits, but increasingly it is extended to a metropolitan area and beyond (Kaiser et al. 1995, Cullingworth 1997). Regional planning usually is chosen when the object of planning is land and water resource use that requires linkages across political boundaries. Before the 1960s, most regional planning was economically motivated, and often focused on one or two forms of development, as was the Tennessee River basin. It was developed for electric power generation by the federal government based on economic improvement of an especially depressed area. Transportation planning also is regionally based. Regional planning for utilities and transportation provision were among the first ways business and governments integrated activities through the planning process; therefore, they have also been very influential in determining the way in which the nation has developed during the twentieth century. However, more recent regional planning has incorporated environmental quality considerations in response to a rapidly degrading environment and corrective laws passed mostly in the 1960s and 1970s.

Naturally Bounded Planning

Regional planning is increasingly adopting principles of landscape ecology into the planning process (Steiner 1991, Dramstad et al. 1996). When this is done, it emphasizes natural connections among ecosystems and more natural bounding within and among planning regions. Management systems also may be naturally bounded at least in part, such as by watersheds, coastal zones, geological formations, soil types, and vegetation types. Coastal zone planning has been encouraged by the Coastal Zone Management Act of 1972 and is done by most coastal states (Kaiser et al. 1995). Watershed planning became useful for water resources management when the Water Resources Planning Act of 1965 established river basins to delineate national water resources planning. Within the past several years, the EPA resurrected a watershed approach, which is gaining greater attention among many natural resources agencies at this time. Many resource management agencies apply natural boundaries, such as watersheds and vegetation types, within their authorized political boundaries.

Integrated Resource Management Planning

Integrated resource management and planning seeks the development of a more holistic approach to planning across organizational planning boundaries. Many of the natural resource agencies in the past interacted relatively little except through policy mandates. They have been encouraged toward more integration by laws that require greater cooperation and coordination. At the heart of integrated resource management is the development of a shared vision of desirable future condition. Policy development in the form of new laws and their interpretation into rules and guidelines is a critical part of this integrated management process. The problem has taken on an international perspective, most recently with concern over ocean fisheries and global warming. The new concept of ecosystem management (see Chapter 23) demands a more integrated effort than has been required previously. This will be especially true if a shared vision of future landscape conditions is to be realized for large watersheds, oceans, and other large ecosystems.

Several decades ago a form of urban regional planning called comprehensive planning was first initiated to integrate across local political jurisdictions. The process has had mixed success, depending greatly on the extent that states have developed planning guidelines (Kaiser et al. 1995). Attempts at integrated planning have been derailed when the participating organizations came to the table with their own mostly inflexible plans. By then they had invested too much in the planning process to modify it readily. Integrated planning often is a long process involving any number of stakeholders who can stymie a shared vision. Increasing emphasis on protection of personal property rights has been a major contributor to slowdown in integrated land-use planning. Although difficult and often prolonged, integrated resource management planning is the most promising approach to many pervasive problems, such as urban sprawl and habitat fragmentation.

POLICY

What is Policy?

Policy is the means by which organizations guide behavior within and outside their membership. For many organizations, this begins with the writing of a charter, constitution, or other document defining why the organization exists and generally how it will act in pursuit of its purposes. Much policy is written to guide the interpretation of law into specific guidance, rules, standards, regulations, and other actions. In the U.S., the Constitution is the foundational law guiding all public policy. Because planning guides pursuit of purpose, effective policy depends on effective planning. Because policy guides organizational behavior, including planning, effective planning depends on effective policy. The two are closely aligned.

Both private and public policies contribute to natural resource development and management in the U.S. and many other countries. Private policy is determined through a variety of organizational management styles, which must also respond to relevant public policies. Public policy is developed and administered by government through a system of laws, regulations, rules, and standards. An important function of the U.S. Supreme Court is to assure that all public and private policy is consistent with the U.S. Constitution. All state constitutions and laws must be consistent with the U.S. Constitution as well as any county, municipal, or other government law.

Most laws place authority for their enforcement in a specified branch of government. These laws often authorize an agency to develop regulations and management standards. Most laws pertaining to natural resource management authorize discretion to agencies in developing and enforcing regulations, but usually require public involvement in the process. For natural resource management, this process usually requires agency knowledge of resource availability, renewal rates, if any, demand for consumptive and nonconsumptive use, and the environmental and social implications of use. Some authorities are linked closely to public lands assigned to agencies for management responsibility, such as the U.S. Forest Service. Other authorities are less directly tied to land ownership, such as water resource management by the U.S. Army Corps of Engineers, Bureau of Reclamation, and the Natural Resource Conservation Service, but usually involve land-use agreements if not public acquisition. The regulatory authorities of some agencies, such as the Environmental Protection Agency and the U.S. Fish and Wildlife Service, extend to all public and private lands and waters.

Roles of Federal, State, and Local Governments

The need for more policy consistency and effectiveness was one reason why federal policy eventually came to dominate natural resource management. Natural resource public policy was locally developed and applied before federal government became active in developing resource policy at the end of the last century. A mix of local, regional, national, and international policy now exists, creating a complicated mix of jurisdictions and regulations. This mix requires coordination in regional management and complicates management across jurisdictions. Federal policy development often has benefited from earlier state legislation that attempted to deal with natural resource issues. Many state and local natural resource laws were enacted before the federal government addressed these issues. These regional laws often served as models for the development of national law.

During the twentieth century, the federal government assumed leadership over state and local governments in developing comprehensive natural resource policies. Many of the issues associated with natural resources required a large view of national trends, such as universal resource depletion and environmental degradation. The federal government has taken the lead because local governments had neither the broad perspective nor the resources to coordinate legal solutions to universal issues. Also, most public land ownership resides with the federal government. State and local governments control relatively little public land.

Budget Policy

Although legal authority guides agency function, budgetary authority drives it by paying salaries and other operation costs. All public policy is contingent upon appropriate allocation of budget among government functions. Many agency authorities are limited by the size of the budget allocated to them. Budget development and justification is a crucial aspect of all organizational activity. Budget justification begins with how effectively salary is used to attain authorized outcomes. Job description and performance assessment often is tied back to the anticipated performance of the agency.

Budget policy varies among different private and public organizations, often complicating the coordination of activities. Although federal government budgets from October 1 through September 30, the budget year for other organizations commonly start in January or July. Planning for budget allocation is diverse, even among agencies within governments. These complicating differences reflect past independence in the planning and management process, many of which have contributed to existing resource management problems.

Public and Private Policy

Public policy takes precedence over private policy. Public policy can effect human behavior through either punishment or reward. Prohibitive laws allow no personal choice and administer punishment when violated. In contrast, tax policy often provides incentives for desired behavior. Tax policies that mimic the private market place provide positive incentives in place of the negative incentives from threats of fine and imprisonment. Policy can be relaxed, or laissez faire, leaving much personal choice, or quite restrictive, leaving little personal choice. With respect to natural resource management, public policy has become increasingly restrictive as resources in general have grown scarcer and user actions have had increasing impact on each other.

This is no clearer than with air quality laws. Clean air, a resource "owned" in common and supporting human health and happiness, has grown much scarcer because the atmosphere is used as a sink for industrial byproducts. The growth of population and industrial development has made clean air more scarce. The determination of resource value and scarcity depended mostly on government-sponsored research. Until recently there has been a heavy reliance on regulating emission amounts under threat of fines. More recently, as knowledge has grown, there is growing interest in selling permits to allow set amounts of waste emission at permit prices that would encourage reduction of waste emission (see Chapter 4). The permits could be sold to others. This approach allows more choice and relies more on market incentives.

U.S. Constitution and Policy Development

The U.S. Constitution guides all policy development in the U.S. Its interpretation is the source of some of the more vexing social problems pertaining to resource management. Constitutionally compatible laws only provide equal opportunity to gain economic and other resources. They do not assure equal economic resource allocation among citizens. In this way, the Constitution supports a free enterprise system based on fair competition.

The Constitution was established by the U.S. government to improve the general welfare of citizens. This imperative establishes guidance for government decisions that affect all aspects of welfare and has influenced the development of economic and other measures of benefit derived from government action. Although the intent of welfare improvement seems clear, there is much room for interpretation. General welfare can be promoted overall while harming individuals. Thus an important issue underlying many natural resource management decisions is the determination of how much benefit is generated from management with respect to the costs.

The Constitution also asserts protection of liberty for both present and future generations of citizens. A broad interpretation of this assertion precludes avoidable deprivation of the rights of future citizens to the use of public resources. Unnecessary destruction of resources reduces the freedom to choose among options for pursuing opportunity. Thus the government in the U.S. exists to improve the general welfare of its citizens while preserving the freedom of future generations. These constitutional mandates establish the goals of public natural resource management in the U.S.

The Bill of Rights and many of the amendments that follow clarify personal privileges of citizenship. Through these amendments, the Constitution respects individual rights to pursue opportunities as they see fit as long as they do not threaten the rights of others. The government is obligated to consider the desires of all public sectors.

Whereas the Constitution grants the right to private property, it also grants the right of the government to purchase upon demand (condemnation authority) and at fair price, any private land deemed necessary for public functions. This provision protects options for developing beneficial public infrastructure such as transportation systems and other common-use areas. More recently this has been the basis for demanding compensation for decreased land values associated with enforcement of the Endangered Species Act. This controversial interpretation will probably remain a policy issue for some time to come. An amendment to the act allows a habitat conservation plan to be developed by the land owner as a way to mitigate the restrictive effect on the land owner, but problems remain to be settled.

MAKING POLICY

Legislation

The legal system in the U.S. and other democracies often works imperfectly, but it has been an effective way to communicate and coordinate principals for group behavior. Although many people equate law with constraint of personal behavior, it is a means by which society resolves conflicts among individuals and groups and fosters opportunity. Law making is an incremental adaptive process. Most long established natural resource law has been amended numerous times and has had various sections repealed.

Most laws are developed through interactions among legislators, legislative aids, special interest groups, and agencies authorized to carry out the policies. The press plays an important role by communicating the issues. Special interest groups usually assign volunteers or pay lobbyists to represent their interests. The word lobbyist derives from a time when special interest representatives cornered legislators in the lobbies of state and federal capitols to influence legislation. Although one-on-one interaction remains, especially in state and local government, more often lobbyists bring their issues to legislative aids, where legislative concepts are first discussed, outlined, and drafted. Special interest groups also affect legislative action through financial support of campaign activities of legislators. Government agencies also influence legislation based on past experience in use of authority (Rourke 1984).

The special interest approach to legislation has advantages and disadvantages. The main advantage is that special interests are able to organize behind spokespersons who efficiently influence legislative change. As long as all interests are well represented by special interests, this often is the most efficient way to systematically sort through the pros and cons of each issue before legislative action. A pure democracy is vulnerable to apathy and ignorance about issues. For many issues, the majority of people are disinterested and not very well informed. Disadvantages arise when interests are low profile but widespread, without articulate representation, and their financial backing is overwhelmed by wealthier interests. They also exist when the agencies authorized to regulate excesses of special interests instead become advocate of the special interests.

In response to what has been seen as too much special interest manipulation of the policy process, an increasing amount of state and local policy is developed directly by public referenda. Setting up a vote by referendum typically requires a public petition of significant size and approval of the legislature. This strategy may be effective when the issue has general support and limited opposition from special interests. However, the ultimate social effectiveness of legislation by referendum remains to be decided. The federal government does not as yet present referendums to the public for vote.

Common Elements of Law

Natural resource law is voluminous and diverse. Many laws include a brief justification for the law. Numerous laws are reminders of the need to improve the general welfare and assure that public benefit exceeds management cost. Programs authorized by law often have a spending limit attached to them. While most laws typically authorize a specific agency to take leadership, many comprehensive laws direct coordination among agencies and the public.

Most authorization law directs development of specific codes; that is, the standards and regulations to be determined and obeyed. Such things as pollutant classification and concentration, wetland qualities, hunting and fishing regulations, engineering specifications, grazing rates, and timber harvest rates are recommended by technical employees in the field for approval at the highest levels. In contemporary proceedings regarding rules and regulations, the public frequently is invited to comment in advance. Technical recommendations are not always accepted by the public. The communication skills needed to extend information in a meaningful way to the public are in high demand.

Heads of agencies most typically are appointed by the president, governor, or other chief executive officer, but some state agency officials are elected. Often an oversight commission or board is appointed or elected to assure the public is well served. Although public representation is the idea behind directorships, commissions, boards, and other authorized governing bodies, public input is rarely polled directly using contemporary techniques that minimize bias. The system continues to function primarily through special interest persuasion.

ORGANIZATIONAL ADMINISTRATION

Integrating Organizational Activities

Organizational planning and natural resource management are integrative processes both impeded and facilitated by organizational structure. Organizational units are formed from the integration of tasks performed by individuals into a work product that is more than the sum of the individual tasks. The effectiveness of planning and management is influenced by organizational structure. Reorganization often is an unavoidable adjustment to new conditions and responsibilities or to more effectively meet established responsibilities. However, reorganization also is disruptive of established links among units resulting in trade-offs. Each reorganization must be weighed against the anticipated improvement.

In large organizations, units of increasing size envelop smaller units. Each unit functions to produce refined outputs from less refined inputs passed on to it by other units. Outputs may be material products like lumber produced from inputs of logs, sawmill facilities, payroll, and human resources. They may be information inputs, such as output expectations generated from plans, additional training needed to meet plan expectations, and activity schedules of complementary organizational units. Many units in large organizations exist for the purpose of refining raw information into more refined or applicable outputs through data acquisition, analysis, and conceptual synthesis. Each functional unit is in turn integrated into larger units with similar functions to form the entire organization. Large organizations, like the executive branch of the U.S. government, have many organizational branches and many subunit supervisory levels. The uppermost level is occupied by the organizational leader, a Chief Executive Officer (CEO). At the next lower level, organizational units branch off into functional groupings.

Natural resource management is administered and planned by numerous private enterprises at national, regional, and local governments. The coordination of management activities to accomplish organizational goals is enhanced by an integrative and comprehensive organizational planning process. Plans are meaningless without adept administration, which leads the organization in developing and executing organizational plans. Administration facilitates the planning process and plan realization through coordination and communication within and among administrative divisions and units. The organization of these units usually derives from the diverse responsibilities assigned to people throughout the organization and from the geographic location of the activities.

The executive branch of the U.S. government, for example, is divided into departments, each of which has a central focus. Most of these departments are involved with natural resource issues in one way or another. However, natural resources are the primary focus of the Departments of Agriculture and Interior. The Department of Interior, for example, has a wide variety of natural resource responsibilities mixed with certain social responsibilities. It has broad authority over fish, wildlife, certain water development, rangeland, and mineral development, especially with respect to public lands and waters. In the Department of the Interior, the Fish and Wildlife Service is administered through programs and geographic regions—each area of which has a similar organizational structure. Administrative regions are determined primarily by the boundaries of states composing a region.

Information Flow and Authority

Each position within an organization is assigned a certain amount of authority over its own function and the function of supervisees in a lower level. Authority is delegated downward from upper administrative levels to lower levels in supervisory lines (Figure 5.8). Authority delegation determines the decision discretion for each position and is usually described for the position. Supervisory authority is indicated by the subordinate positions assigned in the job description or in modifications made

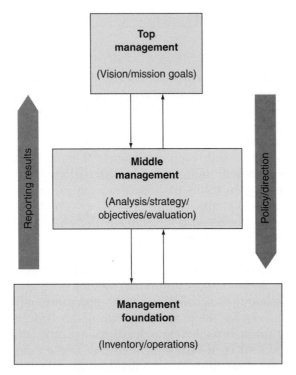

FIGURE 5.8 Information flow in a management organization. In general, policy directives and guidance flow from upper levels of management to lower levels. Information in the form of various written and oral reports with recommendations move upward from the foundational "field"-level of management.

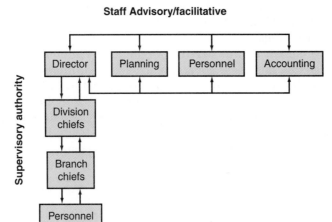

FIGURE 5.9 General lines of authority and information flow within a natural resource management organization. Levels below the Director also may have support staff with specialized skills.

from positions of higher authority (Figure 5.9). The chief administrative officer of an organization has the most comprehensive position in the planning and operations process. Planning, the decision process, and operations become more focused as one progresses from the top of the organizational pyramid to the bottom. The implications of decisions and operations become increasingly broad with progress up each rung of the administrative ladder.

Staff positions stand in support of line authority, but are not in the direct chain of command. Planners, advisors, and other special service positions often reside outside the chain of command in a "side-bar" support role. They respond to the requests of line authority with analyses and recommendations. They may have their own supervisees outside the line of authority, but do not make command decisions that reach outside the staff unit.

Administrative structures exist to bring more cohesiveness to people operating toward the completion of a single or small set of outputs. An appropriate organization facilitates an efficient and an effective decision process leading to more desirable work outputs. The form or structure of a management organization has a lot to do with how efficiently organizations function.

Information Management

All natural resource organizations are faced with a tremendous information resource from which they need to distill relevant information for an effective decision process. Regardless of other functions, each subsystem must use, manage, and add to information needed for organizational function. The structure of a management organization facilitates an efficient communication process when form follows function. In efficient organizations, each administrative unit is a subsystem designed for information input, processing, and output. Input information travels "bottom-up" to the most comprehensive level of the decision process. Only information appropriate for organizational function is accepted as input at each level in the administrative ladder. The "noise" is filtered out. Reporting needs to be readable, comprehensive, to the point, and concise.

Linkages among administrative units serve as conduits for information flow upward with appropriate filtration at each level. In this way, an effective organization passes an optimum quality and quantity of information to the uppermost administrative levels. Obviously, to work well, each administrative unit must be appropriately informed about organizational priorities and about the resources and conditions that exist inside and outside the organization. Priorities and policy guidance flow top-down from the command level of the CEO.

Organizational flexibility must be counterbalanced with enough stability to sustain effective pursuit of purpose. Bureaucracy is a form of organizational arteriosclerosis that occurs when structure becomes too rigid to adapt to functional requirements. Like blood moving through stiff arteries and damaged kidneys, information flow is slowed and inappropriately filtered. In such organizations, priorities and policies change little despite dissatisfaction with organizational outputs. Bureaucracy is relatively rare in private profit-making organizations because they compete poorly and go out of business. Government agencies can persist much longer in an inflexible inefficient state—until they no longer can be tolerated and have budgets cut or are entirely eliminated or reorganized.

Networking

In the past, organizational structure controlled information flow much more than it does now. Horizontal information flow moved within the same administrative level, most often to people occupying the same common work space. Vertical information flow occurred between supervisors and supervisees within an administrative line of authority. Opportunities for information to flow obliquely and to jump levels was possible, but more limited. With the development of contemporary electronic networking, information has the potential for moving vertically, horizontally, and obliquely among workers within organizational structures. Information flow also can readily jump intermediate levels in either direction. It can also jump from organization to organization. This technically mediated freedom of information flow is a boon to integrated resource management. However, there is also more opportunity for misinformation to be passed. Whereas organizational vigilance over information flow was once more centralized, technological networking forces more individual responsibility on all personnel.

SUMMARY

Planning, policy, and administration control the functions of management systems operations. They determine the effectiveness and efficiency with which all organizations pursue their purposes. Whether business, government, or advocacy groups, organizations operate based on principles held in common. Most organizations function to benefit humanity in some way. They use a variety of strategies to determine the most appropriate distribution of benefits among members of society. Evaluation of result is necessary for determining the degree of success and for improving the efficiency and effectiveness of the management process.

In an era of increasing complexity, many organizations divide functions into program and project subsystems. They integrate activities across divisions within organizations and across organizations. Both private and public organizations have concentrated more on their internal environment in the past than will be necessary in the future. Advances in communications and computing technology have enhanced possibilities for more effective integration. Among the biggest challenges are determining and executing that level of organizational integration needed for maximum sustained benefit to society.

LITERATURE CITED

Braybrooke, D. and C. Lindblom. 1963. *A strategy of decision.* New York: Macmillan.

Bryson, J. M. 1995. *Strategic planning for public and nonprofit organizations.* San Francisco: Jossey-Bass.

Crowe, H. A. 1983. *Comprehensive planning for fish and wildlife.* Wyoming Department of Game and Fish.

Cullingworth, B. 1997. *Planning in the U.S.A.: Policies, issues, and processes.* New York, NY: Routledge.

Dramstad, W. E., J. D. Olson, and R. T. Forman. 1996. *Landscape ecology principles in landscape architecture and land-use planning.* Washington, DC: Island Press.

Garvin, A. 1996. *The American city: What works, what doesn't.* New York, NY: McGraw-Hill.

Kaiser, E. J., D. R. Godshalk, and F. Stuart Chapin, Jr. 1995. *Urban land use planning.* 4th edition. Urbana, IL: University of Illinois Press.

Koteen, J. 1989. *Strategic management in public and non-profit organizations.* New York, NY: Praeger.

Lorange, P. and R. F. Vancil. 1977. *Strategic planning systems.* Englewood Cliffs, NJ: Prentice-Hall.

MacDonald, L. J. and S. F. Bates. 1993. *Natural resources policy and law: Trends and directions.* Washington, DC: Island Press.

Moreau, D. H. 1998. A giant step backward. *Water resources update,* Council on Water Resources. Issue No. 110.

Osborne, D. and T. Gaebler. 1992. *Reinventing government.* Reading, Mass: Addison-Wesley.

Rourke, F. E. 1984. *Bureaucracy, politics and public policy,* 3rd edition. Boston: Little, Brown and Company.

Steiner, F. 1991. *The living landscape: An ecological approach to landscape planning.* New York: McGraw-Hill.

Walters, C. 1986. *Adaptive management of natural resources.* New York, NY: McGraw-Hill.

Air, Water, and Land Resources

Air, water, and land interact via earth systems to provide the necessities for terrestrial life, including the basic needs of humanity. In Chapter 6, the atmosphere is shown to be a reservoir for weather and the life-essential gasses, such as oxygen, carbon dioxide, and water vapor. Atmospheric water is linked to the land through the hydrosphere—the realm of water presented in Chapter 7. Water is the universal solvent in which the many materials essential to and affecting life processes are suspended, dissolved, and transported. Soils, discussed in Chapter 8, act with air, water, and the diverse living ecosystems of the earth, described in Chapter 9, to support forest, range, farm, park, urban, wildlife, fishery, and other renewable resources. Soil is the product of and the foundation for ecosystem process. Erosion at rates in equilibrium with undisturbed ecosystem process contributes to the flow of nutrients and sediments supporting life in wetland and aquatic environments. Accelerated erosion caused by careless use of the soil resource diminishes both the soil's fertility and the useful productivity of aquatic ecosystems. Human conversion of the carbon held in vegetation and fossil fuels to carbon dioxide in the atmosphere may be altering the heat content of the biosphere and changing the balance of liquid water, water vapor, and ice in the biosphere. The redistribution of materials from natural resource development and its consequences will be a central concern of natural resource management in future decades.

CHAPTER 6

Atmospheric Resources and Climate

INTRODUCTION

The atmosphere holds many resources of direct value to humanity and influences the sustainability of several others. The atmosphere stores oxygen and nitrogen, essential elements for all life. It plays a critical role in the hydrologic cycle, which sustains fresh water supplies. It can assimilate a certain amount of gaseous waste products. The atmosphere maintains temperatures conducive to life processes. It is the source of climate, which interacts with water and earth to determine the attributes and functioning of the world's ecosystems. Climate refers to the seasonal and annual patterns of atmospheric conditions over relatively large regions of the earth's surface. Weather is the state of the atmosphere at a particular time and place with regard to temperature, humidity, cloudiness, precipitation, and wind. Climatic variation is an important source of uncertainty in management outcomes for many natural resources.

A sound understanding of atmospheric processes and climate is essential for management of living resources. In this chapter, we will discuss basic principles pertaining to the atmosphere, then use those principles to explain how they influence the attributes of living resources in different geographical areas of the United States and elsewhere. Human impacts on climatic stability will also be discussed. Our coverage of this subject is derived in part from Trewartha (1961), Rumney (1968), Ricklefs (1976), Roberts and Lansford (1976), Godish (1997), Holechek et al. (1998), and McKnight (1999).

CLIMATIC FACTORS AND ELEMENTS

Climate for a given area is primarily a function of the seasonal and annual patterns of atmospheric elements; particularly wind, temperature, and precipitation. The most important factors determining climate are atmospheric composition and pressure, solar radiation, earth shape, latitude, continental distribution, topography, marine currents, and earth's orbital revolution around the sun. To a lesser extent, the vegetation in an area influences the climate. Humans have impacted local climates by cutting forests, building dams, and building factories that release large amounts of carbon dioxide (CO_2) and other compounds into the atmosphere. Volcanoes have been among the primary natural factors that have modified local and world climatic conditions for short time periods.

CLIMATIC FACTORS

Atmospheric Composition and Pressure

About 99% of the earth's atmosphere is made up of nitrogen and oxygen (Figure 6.1). Among the remaining 1% are water vapor, carbon dioxide, ozone, methane, and various other gasses. It is also composed of colloidal aerosols, dust, and ions in aqueous solution. The atmosphere is most dense near the earth's surface. Density decreases at higher altitudes. Nitrogen, oxygen, and water vapor are concentrated close to the earth's surface and decrease in density with movement upward from earth. Natural ozone forms a layer in the outer atmosphere. Because gasses contract at cooler temperatures and expand at warmer temperatures, heating and cooling changes density and causes mixing as cool air sinks and warm air rises. This also causes areas of low and high barometric pressure and changing pressures at any one spot.

Solar Radiation

Solar radiation is composed of electromagnetic waves of energy characterized by wave length. Only a small fraction of solar radiation is visible light with wavelengths between about 450 and 700 nanometers. The human eye recognizes different wavelengths of visible light as different colors. Wavelengths shorter than the visible range fall into the ultraviolet range, and longer wavelengths are infrared radiation. In the visible range, blue light is shorter than red light with greens and yellows in between. The sun radiates wavelengths in unequal amounts and more red is radiated than blue. However, blue light has about twice the energy of red light.

These attributes of solar radiation become important when they interact with gasses, aerosols, and particles in the atmosphere. Dust particles tend to reflect light

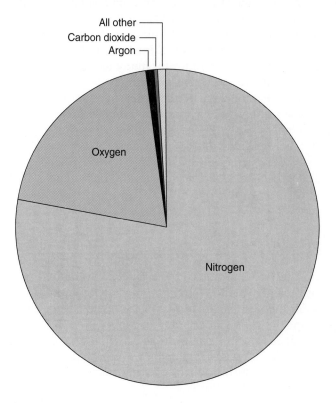

FIGURE 6.1   Proportional volume of the gaseous components of the atmosphere (from McKnight 1990, 1999).

uniformly back to outer space, where it is lost from the earth forever. Aerosols of sulfur and other compounds also reflect most visible wavelengths. Therefore, dust and aerosols cause atmospheric cooling. Other compounds reflect and absorb light more selectively. Ozone also intercepts, absorbs, and reflects high energy ultraviolet light and acts as a shield to this potentially harmful radiation. Infrared radiation is absorbed by water vapor, carbon dioxide, methane, and a variety of anthropogenic gasses called greenhouse gasses. The greenhouse gasses maintain a higher temperature at the surface of the earth than would occur without them.

Earth Latitude, Shape, and Rotation

Radiation reaches the earth with greatest intensity at the equator. With distance from the equator, measured by latitude, the slope of the earth's spherical surface increases the angle of incident radiation. As a consequence, the intensity of radiation per unit area of landscape decreases from the equator to the poles. Because radiation is less intense at higher latitudes, less is absorbed and transformed to heat.

In general, the earth's mean annual climate is warmest at the equator and becomes progressively cooler toward the poles. Light becomes increasingly periodic with movement from the equator to the poles. The sun's highest position in the Tropics is directly overhead while near the poles it is closer to the horizon or entirely behind it. Major geographical patterns in temperature, rainfall, and wind are caused by the uneven distribution of the sun's energy over the earth's surface as a result of the earth's spherical shape.

The ability of air to hold water increases as it warms. Each 10°C rise in temperature nearly doubles the rate of evaporation from a water surface. This explains why warm tropical air around the equator can hold much more water than arctic air. Warming air expands and rises until cooled to the temperature of surrounding air. The process of cooling with gain in altitude causes condensation and precipitation. Tropical air is very humid compared to polar areas because water is cycled rapidly through the atmosphere over the tropics. The high temperatures cause evaporation from the oceans and evapotranspiration (a combination of plant transpiration and evaporation) from the continents.

As the tropical air rises and cools, water condenses and precipitates almost daily. The rotational spin of the earth around its axis moves faster than the atmosphere and causes easterly tradewinds away from the equator. As the rising air moves toward the middle latitudes it cools, condenses into clouds, releases precipitation, and dries out. Therefore, the middle latitudes of the continents are occupied mostly by arid lands.

Distribution of Continents

The distribution of continents affects climate because water evaporates more rapidly from exposed surfaces of water than from land (Figure 6.2). Therefore, inland areas generally have much lower precipitation than coastal areas. The greater area of water surface also accounts for the greater rainfall in the Southern Hemisphere compared to

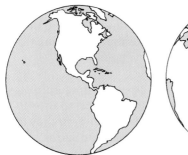

FIGURE 6.2   The distribution of land and water (from McKnight 1990, 1999).

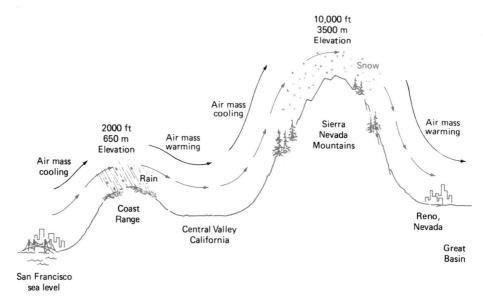

FIGURE 6.3   As air masses move inland from the Pacific Ocean, they rapidly lose their moisture due to cooling as a result of elevation increases from the Coast Range and Sierra Nevada Mountains. This is called the *orographic effect* (from Holechek et al. 1998) (drawing by John N. Smith).

the Northern Hemisphere. Oceans and lakes comprise 81% of the Southern Hemisphere compared to 61% of the Northern Hemisphere. In addition, the ability of the continents to absorb and store heat is less than that of the oceans. As a result, areas inland from the ocean vary much more in seasonal temperature than areas near the ocean. In addition, ocean heat storage causes a lag effect in coastal temperatures, causing warmer autumns and cooler springs than occur inland.

Topography

Altitude.   Topography influences climate because as air gains altitude it cools, contracts, and concentrates water vapor. This results in condensation and precipitation. As air descends the leeward slopes of the mountains it warms, expands, and dries out. Areas on the leeward side of mountains are usually quite dry. The Great Basin area of the western United States is quite arid because it lies between the Rocky Mountains to the east and the Cascades to the west (Figure 6.3).

Topography interacts with temperature and precipitation to determine the climate for any area. Much of the world is characterized by large topographic variation. This is particularly true in North America. The wide difference in topography between the eastern and western United States largely explains the difference in climate and vegetation. Hopkins (1938) related altitude, latitude, and longitude to vegetation in a law known as *Hopkins's bioclimatic law*. The interpretation of Hopkins's law reveals that a 305 m increase in altitude will result in essentially the same phenological (relationship between climate and plant growth cycle) changes that would be encountered traveling 107 km north with no increase in elevation. This law applies to many parts of the western United States, where extremes in plant development during the spring may occur within a few kilometers because of elevation differences.

Aspect.   *Aspect* refers to the directional orientation of slopes. Temperature on slopes in North America increases as aspect changes from the north to east to west to south. The aspect of the slopes has considerable influence on the vegetation they support. Where warmer south- and west-facing slopes support grasses and herbs, the cooler north- and east-facing slopes often support trees (Figure 6.4).

FIGURE 6.4   Grasslands occur on south-facing slopes and forest lands occur on north-facing slopes in northeastern Oregon.

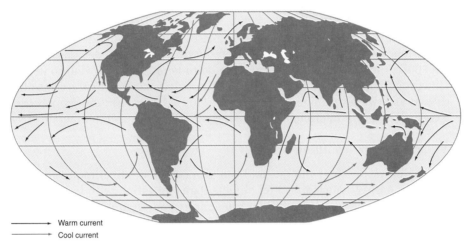

— Warm current
— Cool current

FIGURE 6.5   The major ocean currents. Water movement generally proceeds clockwise in the Northern Hemisphere and counterclockwise in the Southern Hemisphere (from McKnight 1990, 1999).

Degree of Slope.   Degree of slope is of considerable importance in natural resources because it affects both vegetation productivity and human capability to harvest these resources. In land surveys, slope is commonly expressed in percent. As slope increases, vegetation productivity declines per unit of precipitation because less water enters the soil and more runs off as overland flow.

Marine Currents

The primary role of marine currents in climate determination is through heat transfer. Because of the same rotational forces that effect wind movements, marine currents move clockwise north of the equator and counterclockwise south of the equator (Figure 6.5). Cold water moves along the western coast of the continents toward the Tropics while warm water from the Tropics moves along the eastern coasts of the continents away from the Tropics. Therefore, the Pacific coast of southern California and Mexico is drier than Gulf and Atlantic coasts of the same latitude in the eastern U.S.

Interactions Between Climatic Factors

The climatic factors and elements interact to help determine the various world ecosystems. It is generally known that rainfall during warm months promotes more plant growth than during cold months. In the Northern Mixed Prairie country of western North Dakota, 16 inches of precipitation is more effective than 25 inches in the southern Mixed Prairie of western Oklahoma, although precipitation patterns are similar during the summer. This is because evaporation and transpiration (evaporation of moisture from leaves) is much less during the summer in North Dakota. Evapotranspiration (evaporation and transpiration) losses increase with decreases in humidity and increases in temperature. Weaver and Clements (1938) developed a system of classification of vegetation potential based on the ratio of precipitation to evaporation (P/E). Under their system, an area characterized by a P/E ratio of less than 0.2 would be a desert, 0.2–0.4 a shrubland, 0.4–0.6 a grassland dominated by short or midgrasses, 0.6–0.8 a grassland dominated by tall grasses, 0.8–0.10 a deciduous forest, and greater than 1.0 would be a coniferous forest.

In regions with uniform climate, topography can cause considerable environmental variation. Slope and aspect determine the moisture content and temperature of the soil in mountainous areas. Soils on steep slopes are well drained. This often creates moisture stress for plants that would not occur on flat areas. South-facing slopes are more xeric (dry) than north-facing slopes because they are exposed to direct heating from the sun. For the same reason, west-facing slopes are more xeric than east-facing slopes. In the Tropics, the influence of solar aspect disappears because the sun passes directly overhead.

Mean air temperatures drop with increases in altitude and latitude. Every one kilometer increase in elevation is characterized by about a 6°C decrease in temperature. An 800-kilometer (500-mile) increase in latitude results in a 6°C drop in temperature. However, high elevation tropical areas are not characterized by the same climatic conditions as low elevation temperate areas. This is because in the Tropics there is little seasonal variation in climate while large variations occur in temperate areas.

Vegetation as a Climatic Factor

Vegetation responds to climate and can also contribute to climate. Forest vegetation typically has a greater impact on climate than vegetation with less biomass and vertical structure. Vegetation modifies topography and, in doing so, alters wind patterns and changes the light transmission to the ground. Forest interiors are darker, more humid, more stable in temperature, and less windy than adjacent deforested areas. Precipitation intensity is also moderated. Transpiration from vegetation dries the soil and adds humidity to air above the forest. In addition, vegetation sequesters (holds) carbon in biomass, peat, and other stored organic matter, thereby influencing the pool of carbon dioxide in the atmosphere. Photosynthesis removes carbon dioxide from the atmosphere and generates oxygen as a by-product.

CLIMATIC ELEMENTS

Wind

Wind, like precipitation and temperature, is caused by the solar radiation interacting with the earth's rotation and gravity. Rising air in the Tropics moves both north and south in the upper layers of the atmosphere. The rotation of the earth on its axis results in a deflection of the rising air to the right in the Northern Hemisphere and to the left in the Southern Hemisphere (Figure 6.6). This deflective force is called the Coriolis force. The region of rising air at the equator is referred to as the doldrums because there is almost no directional air movement. The early sailing ships were stranded for long time periods in this zone.

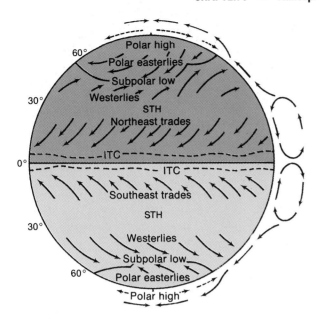

FIGURE 6.6 The general circulation of the atmosphere disregarding the major land masses (from McKnight 1990, 1999).

Wind can substantially reduce precipitation effectiveness by increasing soil evaporation losses and increasing plant transpiration. Wind has little influence on soil moisture below 20 to 30 cm (Veihmeyer 1938). The highest wind velocities occur in flat terrain with few trees. The Great Plains in the central United States are characterized by high wind velocities, particularly in most of Washington, Oregon, and California, where topographic barriers are at a maximum. Hot summer winds during dry years greatly magnify the effect of drought in the Great Plains by increasing both soil moisture losses and plant transpiration.

Temperature

Temperatures in different seasons vary substantially in different parts of the world (Figure 6.7). In tropical areas, monthly and yearly temperatures show little variation. Seasonal variation increases with latitude and distance inland from oceans and large lakes. Extreme cold reduces the availability of moisture to vegetation by freezing the soil. Temperature also affects enzyme and other biochemical reaction rates.

For example, in the mountain areas of the northwestern United States, temperature can have as much influence as precipitation on annual variation in productivity of vegetation. Nearly all native plants in this region are cool-season plants. They grow to limited extent during intermittent warm periods in the winter, but most of their growth occurs in the spring. Years with above-average spring temperatures are usually characterized by above-average plant growth even though total precipitation may be below average. However, in years when below-freezing temperatures occur late in spring, plants often grow more slowly regardless of precipitation amount. This often occurs because much of the moisture in the soil is evaporated by wind before temperatures are warm enough for high rates of growth. When high wind velocities occur, considerable snow may be evaporated directly from the snowpack on high-elevation areas. This is referred to as sublimation. In most years, summers are too dry for plant growth in the Northwest. In the northern portions of the Great Plains, particularly Montana and North Dakota, spring temperatures vary considerably from year to year and can have a substantial influence on productivity of cool-season prairie grasses.

Above-average temperatures usually occur concurrently with drought in the Great Plains. This increases evaporation of the limited soil moisture and greatly magnifies the effect of drought.

Frost-Free Period. On high mountainous areas (over 2000 m elevation) in the western United States, the frost-free period at many locations is less than 100 days. This

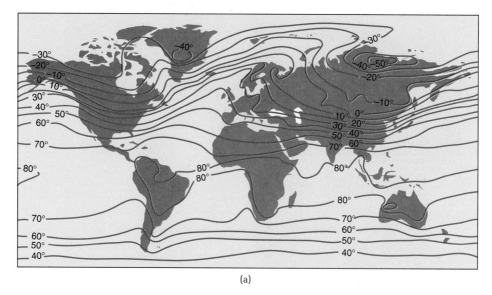

(a)

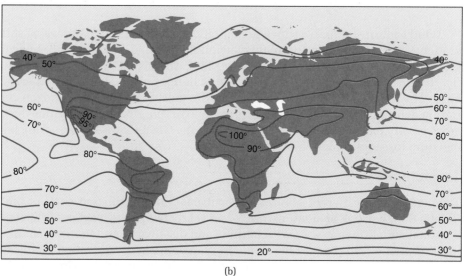

(b)

FIGURE 6.7   Mean January (a) and July (b) temperatures (degrees Fahrenheit) for the world (from McKnight 1990, 1999).

gives plants only a very short period to complete their growth cycle. The low temperatures on these lands are a much bigger constraint on vegetation productivity than is precipitation.

Precipitation

Precipitation is the most important single factor determining the type of vegetation and its productivity in arid to semiarid terrestrial areas. Vegetation productivity increases rapidly as precipitation increases up to about 500 mm per year. Above 500 mm of precipitation per year, soil characteristics can assume much greater importance than precipitation in determining vegetation productivity. The amount of water held in the soil is significantly important. Wetlands are among the most productive ecosystems. Critical characteristics of precipitation that affect ecosystems are the total amount, the distribution, the form, and the annual variability. The average annual precipitation for various parts of the world is shown in Figure 6.8.

Precipitation Variability.   The growth-limiting effects of relatively low annual precipitation in the western U.S. is made more severe by relatively high year-to-year

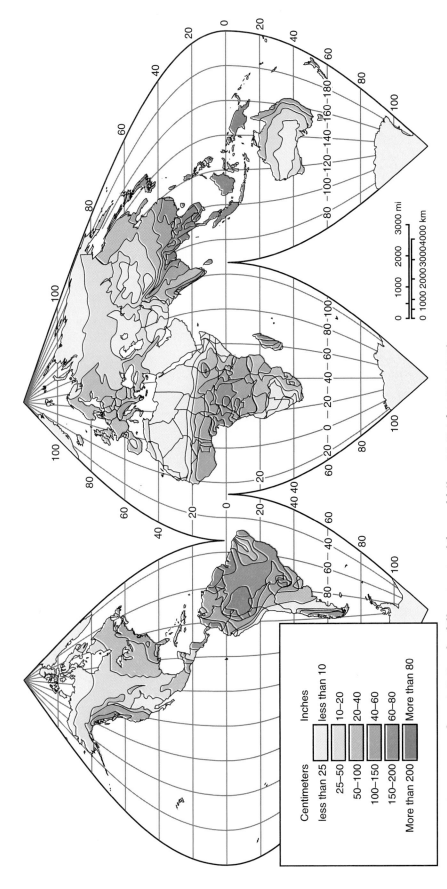

FIGURE 6.8 Average annual precipitation for different parts of the world (from McKnight 1990, 1999).

Centimeters	Inches
less than 25	less than 10
25–50	10–20
50–100	20–40
100–150	40–60
150–200	60–80
More than 200	More than 80

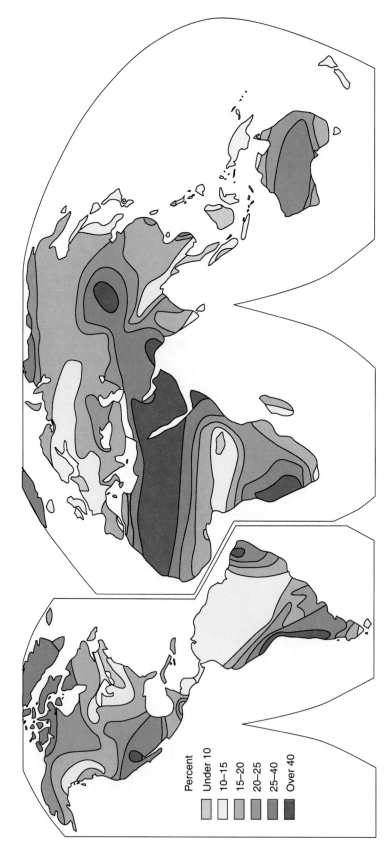

FIGURE 6.9 Precipitation variability in the world indicated by the yearly percentage departure from the average (from McKnight 1990, 1999).

Percent
Under 10
10–15
15–20
20–25
25–40
Over 40

variability (McKnight 1990, Figure 6.9). The variability in precipitation increases rapidly as the annual total drops below 450 mm per year (Conrad and Pollak 1950). Even slight reductions from normal precipitation can cause severe reductions in plant yields in areas with less than 300 mm of precipitation, while much greater reductions in precipitation may have no influence on plant yields in areas with over 800 mm of precipitation (Klages 1942). Annual variability in timing of precipitation can be more important than variability in the total amount that occurs.

Drought. *Drought* is defined as prolonged dry weather, which is indicated when annual precipitation is less than 75% of average annual amount (Holechek et al. 1998). Using this criterion, over the past 54 years (1944–1998) drought in the United States has occurred in 43% of the years in the Southwest, 13% of the years in the Northwest, 21% of the years in the northern Great Plains, and 27% of the years in the southern Great Plains. The most severe and widespread drought in the Great Plains region of the United States in recorded history occurred between 1933 and 1935. In the southwestern United States the most severe drought occurred between 1951 and 1956. These droughts all had tremendous influences on native vegetation.

Precipitation Systems

Four types of storm systems typify local precipitation conditions. These include orographic, frontal, convective, and convergent storms (McKnight 1990). Knowledge of these storm systems is useful for understanding the climatic causes of different ecosystems.

Frontal Storm Systems. Frontal storms are caused by one air mass displacing another. When this happens, the warmer air mass is forced upwards (McKnight 1990, Figure 6.10). The line of contact between the two air masses is termed a front. Frontal types may be distinguished by the way they move relative to the air masses involved. If warm air is replacing cold air, it is termed a warm front. If cold air is replacing warm air, it is a cold front. A front that is not moving is termed a stationary front. The leading edge of contact between two air masses is the front itself. Warm fronts tend to result in mild,

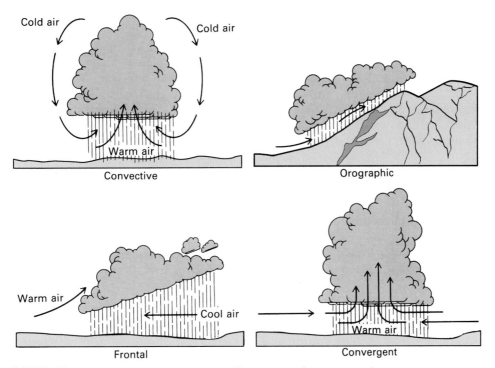

FIGURE 6.10 The four basic types of storms (from McKnight 1990, 1999).

prolonged storms because the warm air moves gradually into the upper cold air. In contrast, cold fronts usually result in intense, short-duration storms because the dense cold air moves very rapidly into the above layer of warm air. Frontal storms are the primary source of precipitation in the prairie country of the central United States.

Orographic Storms.   Orographic storms are caused by the lifting of a moving air mass over a topographic barrier. This type of storm causes much of the precipitation in the Appalachian, Rocky, Cascade, Sierra Nevada, and Coastal mountains in the United States. This type of storm results from adiabatic cooling, which is the cooling of an air mass with increased elevation. The adiabatic cooling rate of dry air is about 10°C with each kilometer of increase in elevation. The adiabatic cooling rate is less for wet air because it can hold more heat. With increasing temperature, the adiabatic cooling rate decreases while the opposite is true of decreasing temperature. Orographic storms can cause considerable climatic variation between mountainous areas in close proximity.

Convective Storms.   Convective storms result from a strong differential heating of the earth's surface, which causes an upward lift of associated air. This type of storm is common during the summer in the southwestern United States, and is usually a warm-weather phenomena. Reflection is a primary factor causing affected air to become warmer than its surroundings. The affected air rises very quickly and rapid condensation releases heat preventing the air from cooling to surrounding air. The condensation continues until the raindrops are heavy enough to fall through the updraft. This results in a localized high-intensity storm sometimes called a "cloudburst."

Convergent Storms.   Convergent storms are less common than the other three storm types. They occur when air converges and results in a general uplift because of crowding (McKnight 1990). Convergent storms are frequently associated with convection and are most common in low latitudes.

Humidity

Humidity refers to the amount of moisture in the air. It is usually expressed as *relative humidity,* which is the percentage of the maximum quantity of moisture that the air can hold at the prevailing temperature. Cold air can hold less moisture than warm air. Rates of water evaporation and plant transpiration increase as the relative humidity decreases. Therefore, areas with high humidity give greater plant growth per unit of precipitation than do areas where humidity is low. The lowest summer humidities in the United States occur at low elevations in Nevada, Arizona, New Mexico, and California. Relative humidities increase with movement eastward across the Great Plains toward the Atlantic Ocean. The highest humidities occur in the southeastern states, particularly in Florida.

TYPES OF CLIMATE

Several approaches have been used in classification of climate. The system described by Dansereau (1957) has been accepted by many ecologists in recent years. This system involves categorization of climate into six broad types based on annual regimens that produce definite types of vegetation.

Equable Climate

This climatic type is characterized by lack of seasonality, or in other words, little annual fluctuation in temperature and precipitation. The most clear-cut cases of the equable climate are in central Africa (Sudan, Chad, Niger) and northern South America (Brazil).

Desert Climate

A low ratio of precipitation to evaporation (< 0.2) characterizes desert climate. Temperatures may vary seasonally, but precipitation is low year around. Variability in precipitation from year to year is usually quite high. Vegetation is typically small shrubs, succulents, and ephemerals. Examples of this climatic type occur in the Mojave Desert in California, the Great Basin in Nevada and Utah, the Sonoran Desert of southern Arizona, and the Chihuahuan Desert of southern New Mexico.

Polar Climate

Areas with a polar climate have a cool summer with a very cold winter. Here average temperatures of the warmest months are below 10°C. Vegetation is characterized by low-growing grasses, sedges, and lichens. High mountain alpine areas and the tundra in northern North America and Asia have this kind of climate.

Mediterranean Climate

This type of climate is manifested in a cool, wet winter with a moderately warm, dry summer. This climate is, in general, mild. It is found at midlatitudes on west coasts. This climate favors evergreen woody shrubs with broad leaves. Central and southwestern California have this kind of climate, as does Spain and Morocco.

Continental Climate

This kind of climate has considerable difference in temperature between the warmest and coldest months, unlike the types previously discussed. Several temperature and precipitation variations occur within this climate type, which result in different kinds of vegetation. The interior of North America and Eurasia have continental climates.

Tropical Wet and Dry Climate

The tropical wet and dry climate characterize southeastern Asia, central Africa, India, northern Australia, and parts of South America. This pattern is described by Wallen and Gwynne (1978) and Heady and Heady (1982). It is characterized by high temperatures throughout the year, with sharply defined wet and dry periods. Considerable variability exists in precipitation between years in most areas receiving this pattern. Temperatures below freezing occur only on high mountain areas. Tropical wet and dry climates with 130 to 250 mm of precipitation typically support desert shrubs, grasses, and cactus. Where annual rainfall is between 250 and 500 mm an open savanna of grasses and trees occurs. A dry forest with large trees and scattered shrubs occurs in the 500- to 1140-mm precipitation zone. Tropical forest dominates most landscapes that have over 1140 mm of precipitation.

CLIMATIC TYPES IN THE UNITED STATES

The major patterns characterizing the climate of the United States were placed in six categories by Humphrey (1962). These include the Pacific pattern, the Great Basin pattern, the Southwestern pattern, the Plains pattern, the Eastern pattern, and the Florida pattern (Figure 6.11). These will be discussed individually following Humphrey (1962) and Holechek et al. (1998).

The Pacific Climate

The pacific climate characterizes the Pacific coastal areas of California, Oregon, Washington, and British Columbia. It extends inland to the Cascade and Sierra Nevada Mountains. Summers are quite dry and winters are wet (Figure 6.12). The percentage of precipitation occurring during the winter increases from north to

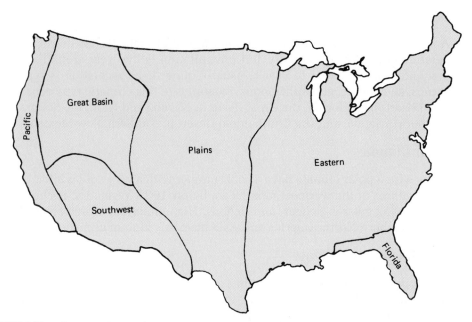

FIGURE 6.11 Climate types in the United States (from Holechek et al. 1998) (based on the classification of Humphrey 1962).

south, although total precipitation decreases. Precipitation also decreases rapidly with movement inland. The percentage of the total precipitation that falls during summer ranges from about 10% in northern Washington to 1% in most of California. The temperatures associated with this pattern are mild near the coast but become progressively colder with movement inland.

The vegetation of the southern part of this climate type, in California and southwestern Oregon, is primarily chaparral and grassland. In Washington, British Columbia, and northwestern Oregon, dense coniferous forest occurs as a result of the high total rainfall (900 to 2500 mm). Because this entire area is characterized by little summer rainfall, the herbaceous species are almost entirely cool-season. Grasslands occur in low-elevation areas with 250 to 460 mm of annual precipitation. The warmer temperatures associated with these areas permit growth of the cool-season grasses in the winter and spring when moisture is available. Other parts of the world with this pattern of climate and similar vegetation include southern Europe around the Mediterranean Sea and the coastal areas of South Africa, Australia, and southwestern South America.

The Great Basin Climate

The Great Basin climate occurs largely in the states of Washington, Oregon, Idaho, Nevada, Utah, and portions of eastern California. In California, Oregon, and Washington, this pattern occurs east of the Sierra-Cascade Mountains. The Great Basin region has much less precipitation than the Pacific region, but the pattern of precipitation is similar. Like the Pacific climate, summers are the driest period. However, more precipitation falls in the spring, and summers are less droughty than in the Pacific region. Total precipitation ranges from 500 mm in northeastern Oregon and eastern Washington to 100 mm in Nevada. Low precipitation is largely in reflection of the rain-shadow effect of the Sierra-Cascade and Rocky Mountains. Winter temperatures over much of this region are cold, with the exception of palous prairie in Oregon and Washington.

Shrubs of the genera *Artemisia, Atriplex,* and *Juniperus* dominate most of this region because they can use moisture stored deep in the soil profile during the summer dry period. Grasses in the region are generally all of cool-season type because of the

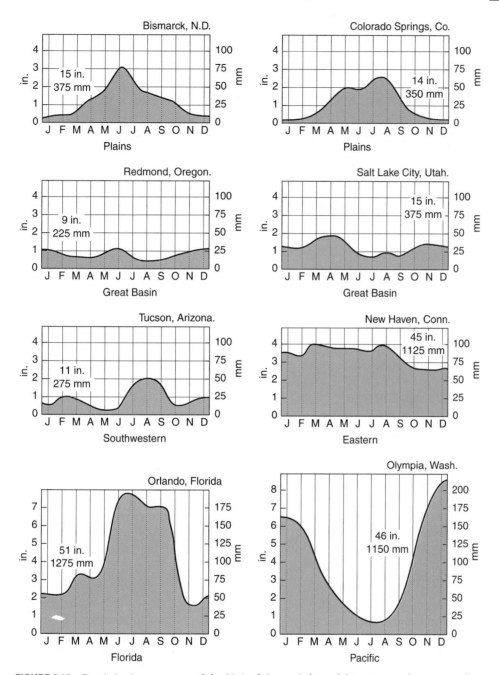

FIGURE 6.12   Precipitation patterns of the United States (adapted from Trewartha 1961 and Holechek et al. 1998).

lack of summer moisture. Portions of the Middle East, Soviet Union, South Africa, Australia, and western South America have this pattern of climate and similar associated vegetation.

The Southwestern Climate

The southwestern climate occurs in New Mexico, Arizona, southern Utah and Nevada, and southwestern Texas. This region has a bimodal precipitation pattern with winter precipitation, spring drought, summer precipitation, and fall drought. This bimodal pattern has maximum development in Arizona. Temperatures are warm throughout this region during the entire year. Total precipitation ranges from 150 to 460 mm. Pacific air moving in from the west causes the winter rains, which are primarily frontal storms. Summer rains are primarily convectional storms that result from air moving in from the Gulf of Mexico. These storms are local, usually of short duration, and very intense. Their effectiveness in generating forage growth is much lower than that of the more frequent, light-intensity storms in the central Great Plains.

Shrubs in the genera *Prosopis, Larrea, Opuntia,* and *Acacia* dominate many areas because they are well adapted to warm temperatures and long dry periods. Nearly all of the perennial grasses in this region are warm-season plants and grow as the result of summer rainfall. Winter precipitation favors cool-season annual forbs.

The Plains Climate

The plains precipitation pattern characterizes the central United States. Total precipitation ranges from 250 to 900 mm. Maximum precipitation delivery occurs in the late spring and summer, with a moderate to light amount in the fall. The least amount of precipitation occurs during the winter. It is significant that most of the precipitation in the central plains area comes from frequent, light-intensity rains during the spring-summer growing season.

This climatic pattern in large part explains why the central United States is a grassland. Because precipitation occurs during both the spring and summer growing season, this region is dominated by a mixture of warm- and cool-season grasses. Other parts of the world with this type of climate and similar vegetation include the lee side of the Andes in South America and continental, mid-latitude Europe and Asia.

The Eastern Climate

The eastern climate covers all of the eastern United States with the exception of the Florida peninsula. It is distinguished by uniformity in amount of precipitation during each month. Over most of the East the precipitation ranges from 900 to 1400 mm, with greater amounts in the southern portion and the mountains. With movement northward, an increasing percentage of the precipitation comes as snow.

A large amount of precipitation with even distribution in a temperate climate provides favorable conditions for deciduous trees. This explains why much of the natural vegetation is deciduous forest. Like the eastern U.S., northern Europe, including the British Isles, has this pattern of climate and similar vegetation.

The Florida Climate

The Florida climate is restricted to the Florida peninsula. This area receives very heavy rainfall (250 mm or more per month) during the summer period from June to October, but the period November to May is quite dry. Annual total precipitation for this area ranges from 1270 to 1500 mm. Because of the heavy rainfall and warm climate, this climatic type is the most productive of all regions for forage. However, the rainfall also leaches the soil, and several minerals are deficient in the forage.

CLIMATIC INSTABILITY AND NATURAL RESOURCE MANAGEMENT

One of the questions asked by many natural resource managers, particularly those involved with farmlands and rangelands, is whether technology in the twenty-first century will make humans invulnerable to climatic instability. This question may have even greater significance now than in the past because the world's population of 5.8 billion people is four times that in 1900. It will likely double within the next 50 years to 12 billion. Any failure on the part of technology or unfavorable change in world climatic conditions could spell doom for literally billions of people because of severe reduction in food production.

World Climatic History

World and local climatic conditions can vary among decades, centuries, and mille-niums. An interesting review of world climatic conditions over the last one billion years is provided by Roberts and Lansford (1979). They provide evidence the world has gone through many cycles of cold and warm periods (Figure 6.13). Over the last million years, the world has been warming, but it is still in an ice age compared to the last 50–100 million years.

The Last 25,000 Years. Between 22,000 and 14,000 years ago, the world experienced extensive continental glaciation. About 12,000 years ago these ice sheets began to re-treat. By 7,000 to 8,000 years ago, the ice sheets in the Northern Hemisphere had re-treated to their present extent. It appears that the period from 7,000 to 5,000 years ago was warmer than now with exceptionally cold intervals occurring about 2,800 and 3,500 years ago (Roberts and Lansford 1979).

The Last 1,000 Years. According to Lamb (1972) the earth was in a dry, warm period about 1,000 years ago. During this time the Viking voyages to Iceland and Greenland occurred. Apparently vineyards flourished in England indicating higher summer temperatures than at present.

A 200-year cooling trend began around 1200 A.D. and was characterized by re-markably severe European winters. Vineyards in both continental Europe and En-gland fell on hard times (Roberts and Lansford 1979). The Viking colonies in Iceland

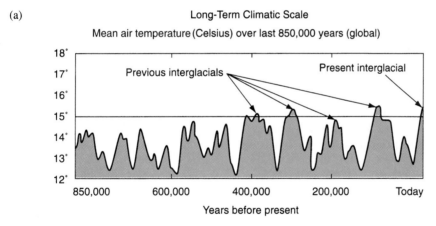

(a)

Long-Term Climatic Scale

Mean air temperature (Celsius) over last 850,000 years (global)

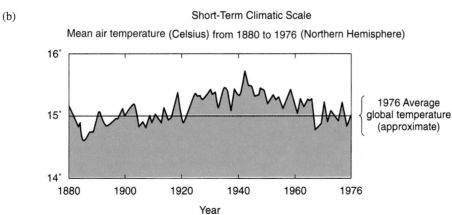

(b)

Short-Term Climatic Scale

Mean air temperature (Celsius) from 1880 to 1976 (Northern Hemisphere)

FIGURE 6.13 These two graphs of global temperature fluctuations show that the earth's climate is variable on both long (a) and comparatively short (b) time scales (National Center for Atmospheric Research, from National Academy of Sciences and other sources) (from Roberts and Lansford 1979).

and Greenland declined because of an increase in ocean ice and colder, more stormy inland weather.

In the 1400 to 1550 period, the climate became more favorable, but then a 300-year cold spell set in known as the "Little Ice Age." During this period vineyards disappeared from England and farming became impossible in the mountainous portions of western Europe.

In much of Europe, the year 1816 was one of the very worst of the "Little Ice Age." It was characterized by a wet winter, a long cold spring, and a rainy summer. Widespread famine occurred across England and Europe because of failure of the wheat crop. This year was known as "The year without a summer in the U.S." Heavy snow in June and frost in July and August occurred in the northeastern U.S. (Roberts and Lansford 1979). In northern New England, the corn crop was nearly wiped out. Widespread hunger was prevalent during the cold winter that followed.

The Last 100 Years.   Since the 1850s, a warming trend has occurred over the Northern Hemisphere. However, there was a 20-year period between the 1940s and 1960s when temperatures decreased slightly, but the upward trend resumed in the 1970s and appears to be still underway.

La Nina–El Nino Cycle.   At roughly 22-year intervals a climatic phenomena known as the "La Nina–El Nino Cycle" affects the central Great Plains and southwestern U.S. Basically it involves 11 wet years (El Nino) and 11 dry years (La Nina). It is responsible for the eight extended droughts in the region that occurred in the 1830s, 1850s, 1870s, 1890s, 1910s, 1930s, 1950s, 1970s, and may now be in effect. The explanation centers around spots on the sun that change their polarity every 11 years. This in turn causes a cooling in the South American Pacific Ocean, which results in high pressure systems that deflect air flows from the Gulf of Mexico.

During drought periods in the Great Plains when La Nina is in effect, other parts of the world such as China, India, and Russia appear to also experience unfavorable climatic conditions. This was particularly true in the early 1970s when world food production fell 2% (1972) after increasing every year since 1945. In the mid-1990s unfavorable climatic conditions due to La Nina effects in the U.S., China, Russia, India, and Australia again depressed world grain production and caused run-ups in world grain prices.

Climatic Lessons from the Past

Knowledge of the climatic history of various parts of the world can be invaluable in planning for future adversity. Although it cannot be known for sure that past patterns will be repeated, it is certain that the climate in a particular area will fluctuate and ultimately change if enough time passes.

Failure to heed climatic lessons from the past caused the "dust bowl" of the 1930s in the Great Plains of the U.S. During the late 1910s and 1920s, millions of acres of native grasslands in the Great Plains were plowed for large wheat growing operations. Memories of the homesteaders who had been wiped out in past droughts, such as in 1893–1894 and 1910–1915, were forgotten. When the worst of all Great Plains droughts began in 1930, everyone was completely unprepared. Thousands of tons of topsoil blew away and, at the peak in 1934, the dust darkened the sky in places as far away as New York City. Starting in 1935, a wide array of government programs went into effect to repair the damage. Much farmland was returned to pasture, trees were planted to provide windbreaks, and farming techniques were developed to maintain crop residue for soil protection. However, by the mid-1950s, when drought struck again, many areas had been converted from grassland to cropland. After a 10- to 12-year period of application of conservation practices following the drought of the 1950s, there was another plow-out in response to world grain shortages and high prices in the early 1970s.

In the mid-1980s, low grain prices and the threat of another dust bowl led to a new wave of federal government programs aimed at retiring erodible lands to permanent grasslands and incentives for conservation tillage (see Chapters 8 and 12). It is encouraging that the mid 1990s run-up in grain prices did not trigger major expansion of crop culture in the Great Plains. However, only time will tell how well the lessons from the climatic history of the Great Plains have really been learned.

CLIMATIC CHANGE AND HUMAN ACTIVITIES

During the past 20 years, there has been much concern that the earth's climate will become warmer because of human activities. Acid precipitation and ozone depletion are other climate problems associated with human activities. Another concern is that certain areas are becoming more arid (desertification) because of destruction of the vegetation by grazing, woodcutting, and cultivation. We will discuss these concerns and their validity.

The Greenhouse Effect and Radiation Balance

The "greenhouse effect" refers to an increase in atmospheric temperature caused by increasing amounts of carbon dioxide and certain other gasses that absorb and trap heat radiation, which normally escapes from the earth (Godish 1997). Various activities by humans, such as burning fossil fuels and cutting forests, are thought to be increasing carbon dioxide levels in the atmosphere and, thus, contributing to the "greenhouse effect."

The earth's atmosphere reflects, absorbs, and emits radiation. Through radiation balance, it maintains surface temperature on the earth about 60°F higher than would otherwise occur. Without this environmental service, life on earth could not exist as we know it. Past fluctuations of atmospheric temperatures have occurred up to at least 9°F since world continental glaciation about 15,000 years ago (Houghton et al. 1996, OSTP 1997). The causes for past climate changes are unclear. However, dramatic changes can occur within centuries.

Climate change results from the interaction of incoming solar radiation and outgoing infrared radiation with changing concentrations of gasses and other material in the atmosphere. Unaltered albedo is one source of climate change. Albedo is the reflectivity of sunlight from the earth's surface. Ice, snow, bare soils, clouds, aerosols, and atmospheric dust have high albedos. As the climate changes, albedo changes further. Warming creates more water vapor, which forms more clouds and melts more snow and ice.

External forcing factors that drive the climate system include, most importantly, the sun's energy output and the earth's rotation. However, such things as volcanic events and human activities are also important. Human activities—such as logging, farming, grazing, and urbanization—have altered the face of the earth, changing its albedo. Human activities generate certain gasses, aerosols, and particles that can influence climate by changing both the albedo and the greenhouse effect.

The radiation and energy balance of the earth is determined mostly by atmospheric composition and earth features. Nitrogen and oxygen, which make up 99% of atmospheric gas, reflect and absorb incoming light, but are transparent to infrared energy leaving the earth (Figure 6.14). In contrast, greenhouse gasses absorb departing infrared energy and reradiate it in all directions, returning some of it back to the earth. Heat builds up as the proportion of infrared radiation returned to the earth increases. The primary greenhouse gasses causing the greenhouse effect are water vapor, carbon dioxide, methane, nitrous oxide, and ozone. They are products of both natural and human activities.

Rice paddies, industry, fossil fuel use, biomass burning, landfills, and agriculture all contribute greenhouse gasses. Carbon dioxide is released in largest quantities from

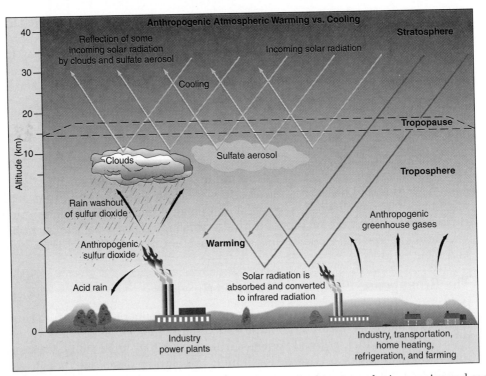

FIGURE 6.14   Human (anthropogenic) factors that are involved in atmospheric warming and cooling. Increases in some gasses, such as carbon dioxide, promote global warming while others, such as sulfur dioxide, cause cooling (from Nebel and Wright 1996).

combustion of fossil fuels and burning of forests. Methane is released from anaerobic locations, such as in landfills, rice paddies, natural wetlands, and feedlot wastes. Nitrous oxide and ozone are products of automobile and other internal combustion. Large sinks of carbon occur in the oceans, where carbon is mixed into and stored in the abyss with residence times of centuries or more. It is also stored in marsh sediments, peat bogs, and forests in huge quantities. If these stores of carbon were released, they would greatly increase infrared emission back to the earth and accelerate global warming.

Although water vapor also is a product of human activity, the overwhelming source is evapotranspiration from the earth's land and water masses. Water vapor increases following some type of change that results in greater temperature, such as increased concentration of carbon dioxide and methane in the atmosphere. Higher temperatures increase evaporation and raise the average humidity. Because water vapor is a highly effective greenhouse gas, its increased formation in the atmosphere further increases the earth's surface temperature.

Evidence from glacial ice cores indicates that the earth's climate has fluctuated drastically as the various factors that drive it have changed. Human activity is implicated in documented carbon dioxide increase and associated change in atmospheric temperature.

Not all scientists are in agreement that the "greenhouse effect" is due entirely to human activities. Some believe it is more of a recovery process from the "little ice age" (1550–1850) brought about by natural forces. As recently as the 1980s some scientists were sounding alarm about a dangerous trend towards global cooling. There is so much uncertainty about the greenhouse effect that few conclusions can be drawn. However, we are encouraged that global emissions of major greenhouse gases have begun to decline (World Resources Institute 1994). This appears to have resulted from improved manufacturing processes, reduced carbon emissions from automobiles, and a lower rate of tropical deforestation. However, even with the present decline, greenhouse gases will continue to accumulate in the atmosphere maintaining the warming potential and risk of global climatic change.

Global Average Temperature

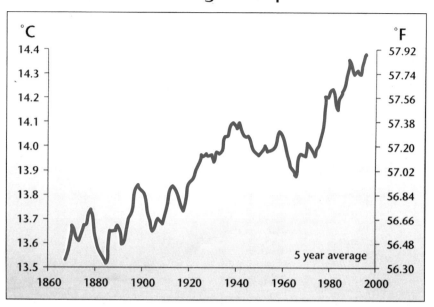

FIGURE 6.15 The global average temperature has risen by about 1°F over the last century (from OSTP 1997).

Global Warming and Fossil Fuels

There is enough evidence to convince many atmospheric scientists that global warming is underway in part because of fossil fuel use (Houghton et al. 1996, OSTP 1997, Godish 1997). This evidence is derived from both empirical observations and predictions from mathematical models based on understanding of underlying processes. However, model predictions of change rates, eventual climatic outcomes, and consequences of different proposed management scenarios remain somewhat uncertain. Global warming has many implications for natural resource management. Concern over global warming is likely to remain a dominant issue in natural resource management for decades to come.

Mean surface temperature of the earth has increased about 1°F during the past century (Houghton et al. 1996, OSTP 1997). The trend line suggests future increase at the same or greater rate (Figure 6.15). Projected changes in mean temperature over the next century range from 2° to 6.5°F. This variation in estimated temperature increase is based on differences in the assumptions used in different mathematical models for temperature prediction. The main cause for this increased temperature is believed to be increasing concentrations of the greenhouse gasses previously discussed.

One of the most important greenhouse gasses, carbon dioxide, has been increasing since about 1860 (Figure 6.16). The data in Figure 6.16 are from direct measurements at Mauna Loa Observatory in Hawaii since 1960, and on carbon dioxide concentrations trapped in glacial ice of known age before 1960. Close correlations between temperature and atmospheric carbon dioxide concentration have been derived from study of bubbles in glacial ice up to 160,000 years old. Because carbon dioxide absorbs infrared radiation leaving the earth and emits some of it back to the earth, it increases temperatures at the surface of the earth (Figure 6.16). Other greenhouse grasses that could be contributing to temperature increase include water vapor, methane, and nitrous oxide.

Past increases in atmospheric carbon dioxide started about the time (1850s) that industry began using significant quantities of coal for production of steel and other

Carbon Dioxide Concentrations

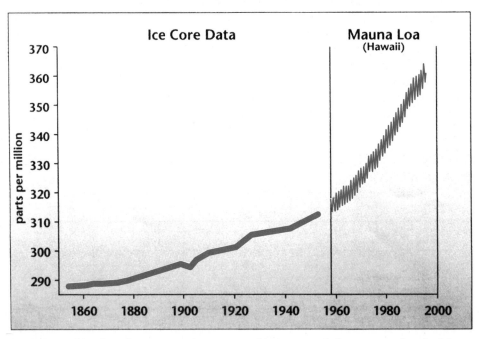

FIGURE 6.16 Carbon dioxide concentration measured in ice cores before 1960 and at the Mauna Loa Observatory since 1958 (from OSTP 1997).

material goods. Annual world emissions of carbon from combustion of oil, gasoline, coal, natural gas, and other fossil fuels now exceed six billion tons, of which the U.S. contributes 22%. An additional one to two billion tons of carbon are emitted from forest cutting and burning mostly outside the U.S. in the developing world. Future emissions are projected to nearly double over the next 35 years. Because much of the future increase will occur in the developing world, the U.S. percentage is projected to fall to about 15% by 2035.

The Implications of Global Warming

The resource management implications of global warming are staggering. There are many threats and some opportunities. One expected change is a rise in ocean water level of 15 to 90 cm with a most probable rise of 50 cm by the year 2100. The rise in water level will occur because water expands as it warms and more glacial ice will melt. Obviously, wherever the ocean inundates present human habitation, it will degrade private property value. Similarly, public property set aside for coastal wildlife refuges may no longer provide the necessary habitat. Estuarine and salt-marsh ecosystems would shift inland at rates depending on the coastal slope and existing engineering, such as sea walls. In coastal areas of low slope, presently dry property is projected to become shallow wetland.

A global temperature rise will increase evaporation from land and water surfaces. One net result is more water in the atmosphere and a greater greenhouse effect from the increased water vapor. This phenomenon is one reason why in some models the rate of warming increases at an increasing rate. Because land heats more than water per unit of radiation absorbed, heat is expected to become more concentrated over the central continents. The northern part of the Northern Hemisphere is expected to have the greatest warming because a high proportion of the continental land mass is located there. Inland freshwater is expected to decrease while coastal freshwater runoff could increase substantially in certain regions. While some estuaries will become fresher, others will become more saline.

Continental heating is predicted to accentuate the rate of hydrologic cycles and cause greater seasonal extremes of temperature and water runoff. The incidence of intense storms already has increased about 20% over the past century and is expected to continue to increase as mean temperature rises. This will result in greater wind and flood damage if people continue to concentrate in coastal zones and river flood plains.

Because terrestrial plant and animal distributions are primarily determined by temperature and moisture, global climate change is expected to result in the shifting of whole ecosystems. This means that many lands now managed for certain forest, range, farm, recreation, wildlife, and biodiversity values will have to be managed for a new set of resource values. Places that formerly had optimum temperature and moisture for certain crops could become much drier and other locations may become much wetter. The rich wheat and corn belts of the central plains states could shift northward into Canada over the next century. This would increase agricultural productivity in Canada while decreasing it in the U.S. A similar shift northward could occur for the prairie pothole ponds important for waterfowl production. The regional changes based on prediction models vary substantially at this time. There is a need for improved models and management strategies for different regions under different scenarios of global warming.

Possible Strategies for Managing Global Climate Change

Various strategies for dealing with global climate change are now under discussion. The three main strategies are to slow emissions of greenhouse gasses, manage ecosystems to favor greater sequestration of carbon in carbon "sinks," and manage ecosystems to allow species to adapt to changes as they occur. Even if emissions could be totally stopped at this time, the lags in heat absorption by the oceans probably will result in a continued rise in temperature for many decades. Adaptation appears to be necessary at least to the extent that thermal lags cannot be prevented.

Greenhouse gas emissions can be slowed through a combination of technology improvements and conservation. The use of noncarbon alternatives to energy production, such as solar and wind energy, is part of the first strategy. Preservation of carbon "sinks" in forests and peaty wetlands, as well as resource management to restore and create new carbon sinks, are important aspects of the second strategy. Adaptation includes improved understanding of climate change, improved forecasts of impacts, and changes in resource management that mitigate damage and take advantage of possible benefits. One benefit is that a rise in carbon dioxide concentration will increase productivity of many plants. This might have beneficial impacts on some woodland, cropland, and rangeland ecosystems if other environmental factors remain favorable.

Acid Precipitation

Atmospheric precipitation has a natural weak acidity caused by the interaction of water and carbon dioxide which forms carbonic acid. Acidity is determined by the concentration of hydrogen ions in solution. About three decades ago, hydrogen ion concentrations greater than expected from natural sources were first reported for North American precipitation (Godish 1997). Acid rain was first discovered in Europe and later linked to industrial emissions into the atmosphere. Further study revealed that snow could be more acidic than expected from natural causes. Many lakes were found to be unnaturally acidic. Elevated hydrogen ion concentrations were found in clouds, dew, and atmospheric crust. By 1980, research scientists had mapped hydrogen ion concentrations in precipitation. They found the greatest concentrations over the eastern Great Lakes, upper Midwest, and mid-Atlantic states (Figure 6.17).

The major source of acid rain is combustion of fossil fuels used to generate electricity, power automobiles, and fire metal smelters and industrial boilers. The major source of acidity, sulfuric acid, is associated with sulfate aerosols. The sources

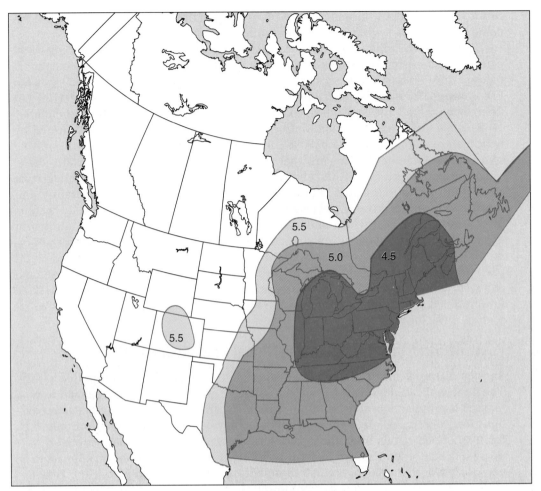

FIGURE 6.17   Regions receiving acid rain deposition. The numbers refer to the pH of the acid deposition (from Nebel and Wright 1996).

of sulfate aerosols are mostly from high-sulfur coal. Nitric acid, a minor contributor, is produced mostly from oxides in automobile exhausts. The oxides of sulfur and nitrogen can be transported many hundreds of miles before they form acid precipitation. About half of the acidic precipitation from the U.S. ends up in Canada and some Canadian emissions end up in the U.S.

Most of the ecosystems sensitive to acid rain exist in Canada. However, significant sensitive areas also exist in the Northeastern and North Central U.S., and in isolated high mountain areas. As lakes become more acidic, their productivity and biodiversity decline. Losses of fish populations have been associated with elevated acidity at numerous sites. So far, forest and crop production do not appear to be significantly affected by acid rain, although gradual depletions of nutrients may in time reduce productivity of exposed terrestrial ecosystems. Poor air quality due in part to acidity contributes to respiratory problems in humans.

Sulfur emissions are controlled by regulations of the U.S. Environmental Protection Agency and government regulatory agencies in other nations. Sulfates are removed from coal both before and after combustion using a number of processes (see Godish 1997 for details). Less effective means are available for removing nitrogen oxides.

One ramification of removing sulfate aerosols to prevent acid precipitation is that their associated albedo is reduced. Early model projections of greenhouse warming failed to consider this artificial source of atmospheric cooling and overestimated the warming. However, recent information suggests there could be more removal of sulfate aerosols by emissions treatment and, therefore, less reduction of global warming than in the recent past.

Ozone Depletion

In the late 1960s, there existed the possibility that ozone, O_3, at the outer edge of the stratosphere could be depleted by high-altitude flight and nuclear weapons testing (Godish 1997). This was disturbing because ozone shields the earth's surface from potentially harmful ultraviolet radiation. Because of these theories, U.S. policy steered away from both activities. In the 1970s, scientists proposed a theory that chlorofluorocarbons used in aerosols, refrigerants, and foam propellants could deplete ozone once chlorine, a strong oxidant, was released and the molecules decomposed. Later, they theorized that atmospheric accumulation of N_2O—a product of fertilizer breakdown—could react with and deplete the ozone.

In the late 1980s, a scientist working with data accumulated in Antarctica discovered what appeared to be depletion of ozone over the South Pole. Subsequent study revealed concentrations of chlorine in the area of the "ozone hole." Since discovery of the ozone hole over Antarctica, its size has increased. Recent measurements indicate that ozone depletion may be occurring over the entire globe. Ozone depletion is of great concern because of the role stratospheric ozone plays in radiation balance. It is especially effective in absorbing ultraviolet radiation that can cause cell destruction and related diseases—including skin cancer. Studies in Canada indicate that ultraviolet light has increased at the earth's surface, as would be expected with significant ozone depletion. Ultraviolet light regulates plant growth and has been implicated with depression of phytoplankton photosynthesis in oceans near Antarctica. These observations have caused concern over the sustainability of primary productivity in both land and water.

Desertification

Desertification is the formation of desert-like conditions, largely through human actions, in areas that do not have desert climates. Biological productivity declines while the prevailing climatic conditions are thought to remain constant. Human activities implicated as causes of desertification include uncontrolled livestock grazing, burning, woodcutting, and temporary cultivation and abandonment of semiarid to arid lands (Figure 6.18). Africa has been the focal point of concern over desertification during the past 20 years because of continued droughts in the Sahel region. These droughts have caused tremendous losses of livestock and human hardship.

FIGURE 6.18 Overgrazing by cattle and drought have contributed to desertification in the arid Sahel of Africa.

Percent

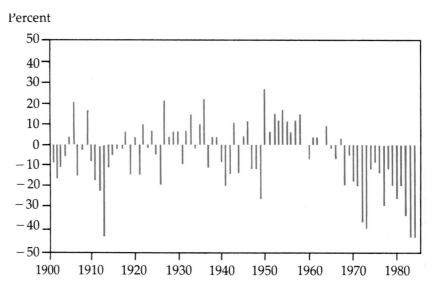

FIGURE 6.19   Rainfall fluctuations in the Sahel and Sudan. Expressed as a percent departure from the long-term mean, 1901–1984 (from Brown and Wolf 1985).

Comprehensive reviews on the subject of desertification are provided by the United Nations (1977), Glantz (1977), Postel (1989), and Mouat and Hutchinson (1995).

Past climatological data show that drought has been a recurring phenomenon in the Sahel (Wallen and Gwynne 1978, Winstanley 1983) (Figure 6.19). However, the effects of drought on the vegetation have been magnified in recent years because of rapidly increasing human and livestock populations. There is also some evidence that recent droughts have been more severe than those in the past (Winstanley 1983).

As human populations increase in semiarid to arid areas, desertification may become an important problem in other parts of the world as well as Africa (Mouat and Hutchinson 1995). Application of soil and range management practices have considerable potential to reduce or reverse the desertification problem in many areas (see Chapters 8 and 11).

LITERATURE CITED

Brown, L. R. and W. C. Wolf. 1985. *Reversing Africa's decline.* World Watch Paper No. 65. Washington, DC: World Watch Institute.

Conrad, V. and L. W. Pollak. 1950. *Methods in climatology.* Cambridge: Harvard University Press.

Dansereau, P. 1957. *Biography: An ecological perspective.* New York: The Donald Press Co., 394 pp.

Finch, V. C. and G. T. Trewartha. 1942. *Elements of geography: Physical and cultural.* New York: McGraw-Hill Book Co., 823 pp.

Glantz, M.H., ed. 1977. *Desertification: Environmental degradation in and around arid lands.* Boulder, CO: Westview Press.

Godish, T. 1997. *Air quality.* 3rd edition. New York: Lewis Publishers.

Heady, H. F. and E. B. Heady. 1982. *Range and wildlife management in the Tropics.* New York: Longman.

Holechek, J. L., R. D. Pieper, and C. H. Herbel. 1998. *Range management principles and practices.* 3rd edition. Upper Saddle River, NJ: Prentice-Hall.

Hopkins, A. D. 1938. *Bioclimatics: A science of life and climate relations.* U.S. Dept. Agr. Misc. Pub. 280. 188 pp.

Houghton, J. T., L. G. Meira Filho, B. A. Callander, N. Harris, A. Kattenberg, and K. Maskell. 1996. *Climate change 1995: The science of climate change.* Published for the Intergovernmental Panel on Climate Change. New York: Cambridge University Press.

Humphrey, R. R. 1962. *Range ecology.* New York: Ronald Press, 234 pp.

Klages, K. H. W. 1942. *Ecological crop geography.* New York: Macmillan.

Lamb, H. H. 1972. *Climate: Present, past, and future.* London: Methuen Publishing Co.

McKnight, T. L. 1990. *Physical geography: A landscape appreciated.* 3rd edition. Upper Saddle River, NJ: Prentice-Hall.

McKnight, T. L. 1999. *Physical geography: A landscape appreciation.* 6th edition. Upper Saddle River, NJ: Prentice-Hall.

Miller, G. T. 1990. *Resource conservation and management.* Belmont, CA: Wadsworth Publishing Company.

Mouat, D. A. and C. F. Hutchinson, eds. 1995. *Desertification in developed countries.* London: Kluwer Academic Publishers.

Nebel, B. J. and R. T. Wright. 1996. *Environmental science: The way the world works.* 5th edition. Upper Saddle River, NJ: Prentice-Hall.

Nebel, B. J. and R. T. Wright. 1998. *Environmental science: The way the world works.* 6th edition. Upper Saddle River, NJ: Prentice-Hall.

Office of Science and Technology Policy. 1997. *Climate change: State of knowledge.* Executive Office of the President. Washington, DC.

Postel, S. 1989. *Halting land degradation. State of the World 1989.* World Watch Institute Report. New York: W. W. Norton.

Ricklefs, R. E. 1976. *The economy of nature.* Portland, OR: Chiron Press.

Roberts, W. O. and H. Lansford. 1979. *The climate mandate.* San Francisco, CA: W. H. Freeman.

Rumney, G. R. 1968. *Climatology and the world's climates.* New York: Macmillan.

Trewartha, G. 1961. *The earth's problem climates.* Madison: University of Wisconsin Press.

United Nations. 1977. *Desertification: Its causes and consequences.* Oxford: Pergamon Press.

U.S. Department of Agriculture (USDA). 1987. *The Jornada Experimental Range, Las Cruces, New Mexico.* U.S. Dep. Agric. Misc. Publ. 689–479.

Veihmeyer, F. J. 1938. Evaporation from soils and transportation. *American Geophysical Union Transaction.* 1938:612–619.

Wallen, C. C. and M. D. Gwynne. 1978. Drought—A challenge to rangeland management. Proceedings International Rangeland Congress 1:21–32.

Weaver, J. E. and F. E. Clements. 1938. *Plant ecology.* 2nd edition. New York, NY: McGraw-Hill.

Winstanley, D. 1983. Desertification: A climatological perspective (pp. 185–213). In *Origin and evolution of deserts,* edited by S. G. Wells and D. R. Naragan. Albuquerque, NM: University of New Mexico Press.

World Resources Institute. 1994. *World Resources 1994–1995.* New York: Oxford University Press.

CHAPTER 7

Water Resources

INTRODUCTION

In the fall of 1997, the world watched with excitement as images of the surface of Mars were captured by robotic equipment. Such images inspired a sense of awe but also gave us deep appreciation for the earth's living landscapes. More specifically, such images make us more appreciative of water, a life-sustaining resource that has been conspicuously absent in space probe images of other planets. In contrast, the photographs of Earth taken by astronauts show a strikingly blue planet with water covering more than two-thirds of its surface. Water is a critical component of all life. It can be beautiful, powerful, and destructive (Figure 7.1).

Like many other renewable natural resources, supplies of water can be both scarce and abundant at the same time and place, depending on the required quality. An example is Los Angeles, where seawater is abundant but usable "fresh" water is limited. While water in the oceans is inexhaustible, without processing to remove salts it is unsuitable for most human uses. Lack of potable water is becoming a major obstacle to further urbanization of the Southwestern United States. It is also the primary constraint on the development of several African, Latin American, and Asian countries. In many parts of the world, demand for water of usable quality is far greater than available supplies. Problems of loss in biodiversity are often associated with man's control of water (drainage of marshes, building dams, contamination of estuaries).

Modern man uses water for agriculture, transportation, industrial processes, waste disposal, power generation, and recreation as well as for consumption. The tremendous advance of civilization over the past 200 years has depended heavily on the capability to control floods, develop irrigation systems, and use water as a source of power (steam engines). Without question, the destiny of humankind in the twenty-first century will depend heavily on being able to solve problems relating to water.

This chapter provides an overview of water uses, water resources management (including watershed management), water quality management, and water needs in the United States. We also discuss the hydrologic cycle and how it interacts with atmospheric and watershed processes. The reader is referred to World Resources Institute (1994) for more detailed analysis of world water problems. For more detailed

(a)

(b)

(c)

FIGURE 7.1 Water is vital to life on planet Earth. It can be (a) beautiful, (b) powerful, and (c) destructive (courtesy of the United States Department of Agriculture).

discussions of watershed management, water quality problems, and other issues related to water resources, we refer the reader to Gleick (1993), Heathcoate (1988), Perry and Vanderklein (1996), Nebel and Wright (1998), and Owen et al. (1998).

THE PROPERTIES OF WATER

Physical Attributes

Water supports life on our planet because of its unique physical and chemical properties. Several of its properties contribute to the massive transfers of energy that take place in the earth's atmosphere and oceans. In turn, these energy transfers lead to a more even heat balance away from the earth's equator. They maintain the continuous circulation of moisture that renews the freshwater supply throughout the globe. To perform its critical roles, water must remain in a fluid state over wide extremes of temperature. It can do this because it has a rather high boiling point (100°C, 212°F) and a low freezing point (0°C, 32°F). Because of its solvent properties, water contributes to biogeochemical cycling as it carries dissolved minerals from land to sea (Smil 1997). On a much smaller scale, minerals cycle through individual ecosystems in the same manner.

The ability of water to buffer the temperature of the earth's surface can be attributed to its high heat of vaporization and capacity for storing heat. Because liquid water can store a large amount of heat, large bodies of water do not warm or cool rapidly. Without the vast volume of ocean water, earth temperatures would vary more.

Water shows remarkable changes as it moves through its liquid, solid, and vapor states. Liquid water is the only common substance that expands rather than contracts when it freezes. As a result, ice is less dense than liquid water and floats above water when oceans, lakes and rivers freeze. Consequently, water bodies do not freeze solid in cold climates and aquatic life is more abundant than would otherwise be the case. Water releases heat when it changes from liquid to ice and from steam to liquid. The extremely high surface tension of water is created by the forces that attract water molecules to one another. This property allows liquid water to adhere to soil and move through plants. Water's low viscosity makes it an outstanding medium for distributing nutrients through soils and plants as well as for distributing heat in ocean eddies and currents.

Finally, water contributes the hydrogen necessary for photosynthesis to occur in organisms ranging from simple algae to giant redwoods. Without photosynthesis, the earth's atmospheric oxygen and its living occupants would not exist. Water is also essential in maintaining the circulatory systems and temperature equilibrium in higher plants and animals.

Many of water's unique thermodynamic and solvent properties derive from the way the hydrogen and oxygen atoms join to form water molecules that become connected. Hydrogen and oxygen atoms bond by sharing electrons (i.e., they form a covalent bond). Rather than being on opposite ends of a linear molecule, the hydrogen atoms attach to oxygen at an angle forming a "wishbone." This allows negatively charged electrons to spend more time on the oxygen end (e.g., apex of the wishbone) of the molecule. Because the hydrogen ends are positively charged, a water molecule's electrical charge distribution is asymmetrical. It is this property that allows water to separate polar solute molecules. More specifically, water's polarity allows it to dissolve and carry many nutrients through the tissues of living organisms and to flush waste products from those tissues.

Another attribute of special significance is water's cohesiveness. This means that water molecules tend to join together and behave as a much larger molecule. This effect is created by hydrogen bonds which join the hydrogen of one molecule to the oxygen of another. Hydrogen bonds are much weaker than covalent bonds. However, a large number of hydrogen bonds acting in unison have a strong binding effect.

WATER FORMS AND DISTRIBUTION

The Planet's Reserves

Water is distributed throughout the planet, but most (97.4%) occurs as ocean water (Figure 7.2). The sea covers 71% of the total area to an average depth of 4 kilometers (2.5 miles). Its massive size is maintained by return flows from rivers and melting ice. Seawater is highly saline and can not be used as a source for drinking water or for crop irrigation. Sodium chloride and magnesium sulfate are the most abundant dissolved salts found in ocean water. These salts, as well as a host of other substances, are carried to the sea by rivers. Salts are released from the land masses through erosion and weathering processes. Across the continents there is considerable variation in the amount of runoff each river contributes to the ocean.

Glacial Ice

About 10% of the earth's land surface is covered by glacial ice. Most (85%) glacial ice is found in Antarctica and Greenland (Table 7.1). Much of the remaining portion is scattered across temperate mountain valleys. As glacial ice has advanced and retreated over geologic time, the proportion of freshwater found on the planet in a frozen state has varied. At present, glacial ice comprises a rather small portion of the planet's total water (1.7%) (fresh and saline). However, it accounts for 68.7% of its freshwater. If all this ice suddenly melted, the oceans would rise about 60 meters (200 feet). Some of the world's largest cities—such as London, Tokyo, New York, Seattle, and Los Angeles—would be inundated.

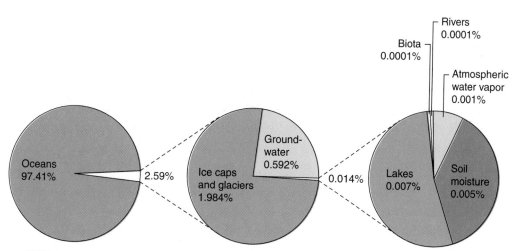

FIGURE 7.2 Distribution of the world's water (courtesy of United States Department of Agriculture). Source: David H. Speidel and Allen F. Agnew, "The World Water Budget," in *Perspectives on Water Uses and Abuses,* David H. Speidel, Lon C. Ruedisilli, and Allen F. Agnew, eds. (Oxford University Press, New York, 1988).

TABLE 7.1 **The Distribution of the World's Ice by Geographic Region**

Glacier Region	Area of Ice-Covered Surface (10^3 km^2)
North polar regions	2,000
Temperate countries of the Northern Hemisphere	190
Tropical countries	0.1
Temperate countries of the Southern Hemisphere	26
Antarctica	14,000
Total of the earth	16,000[a]

[a]Total given in the original table does not add to total of entries because of rounding.
Source: Gleick 1993.

Lakes

A variety of geologic events such as earthquakes (e.g., Lake Nyasa in East Africa), volcanic activity (e.g., Crater Lake, Oregon) and glaciation (e.g., the Great Lakes) have led to the creation of lakes (Figure 7.1). Freshwater lakes are found mostly in regions of ancient and recent glaciations and in regions of large tectonic fractures in the earth's crust. Lakes do not persist indefinitely over geologic time. Increased aridity, caused by climatic shift, has caused numerous massive lakes to become spacious salt beds. Utah's Great Salt Lake is what remains of ancient Lake Bonneville. Lakes also disappear as they slowly fill with sediments and become colonized by plants adapted to a hydric environment.

In terms of total water volume, lakes appear insignificant alongside oceans. Nevertheless, various lakes throughout the world cover large areas (e.g., The Saint Lawrence Great Lakes, Aral Sea, Lake Victoria). Nearly half (48%) of the world's large lakes (i.e., greater than 500 km^2) are found in North America (Herdendorf 1990). In the U.S. alone there are 41 million acres of lakes and reservoirs. While the number of reservoirs built for recreation outnumber all other purposes, most large reservoirs have been built primarily for water supply and flood control. Most large reservoirs are multipurpose structures.

Rivers

Rivers account for much less of Earth's total water than lakes. However, they are primary arteries of water flows across continents. Rivers such as the Amazon are massive in their width and depth and greatly influence the biology of some continents. Though less awesome in size, rivers traversing the world's arid regions provide water that can radically change ecosystems and settlements. Variation of flow is natural for most large rivers, which typically form floodplains that are periodically inundated. Rivers unconstrained by engineered controls (e.g., dams, levees) continuously cut new channels through their floodplains and create a mix of backwater lakes and wetlands, which support diverse natural communities. In the U.S., some rivers, such as the Mississippi, remind us that their path is self-determined in spite of the best efforts at flood control. In the U.S., rivers and streams have a total length of about 3.5 million miles. However, only about one-third of these flow continuously.

Oceans

The oceans occupy the largest area and volume of water among the different types on Earth. A salinity averaging about 35 parts per thousand (3.5%) makes ocean water useless for drinking and agriculture unless it is desalinated. Although ocean water desalinization is technically feasible, high cost makes it a last resort. The oceans are important for their fisheries, other biological resources, minerals, recreation, and transportation uses. The deep oceans remain among the least explored areas on the earth.

Groundwater

A significant portion of the world's water is underground where it must be accessed through excavation and deep wells. The quality and quantity of ground water is highly variable among the different regions of continents and the world. Much groundwater is too saline for drinking water or irrigation. Some of the best quality water resources, however, occur underground where it must be developed for use.

WATER RESOURCES MANAGEMENT

The Hydrologic Cycle

The process by which energy from the sun causes water from the land and oceans to vaporize into the atmosphere, condense as the result of cooling, return to the earth as precipitation, and flow to points where it vaporizes once again is termed the "hydrologic cycle" (Figure 7.3). Other important sources are from transpiration (moisture released from plant parts, mainly leaves) and evaporation from streams, lakes, and land surfaces. Man can influence the hydrologic cycle by modifying vegetation and soil. These modifications, which can have a major impact on both water yield and quality, are discussed later in the chapter. That part of the hydrologic cycle that occurs in the landscape is often referred to as watershed process.

The Watershed

A watershed is the total area of land that contributes runoff water above a given point on a waterway. The areas drained by rivers and streams are called drainage basins, or watersheds. Watersheds are hierarchical; that is, small watersheds contribute to large watersheds in a progressive accumulation of watershed area as

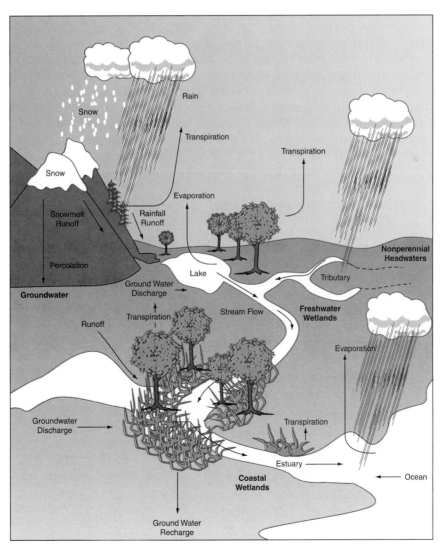

FIGURE 7.3 The hydrologic cycle (courtesy of United States Department of Agriculture).

drainages join. Hydrologists sometimes refer to first-, second-, third-order and larger stream channels as they describe watersheds. A second-order stream channel forms when two first-order channels join, and so on. Depending on management concern, one could choose a massive watershed of high order (e.g., tenth order), such as the area drained by the Mississippi (an area encompassing numerous states), or a watershed served by a minor tributary of low order.

Precipitation that falls to the watershed surface and enters the ground is called infiltration. The portion of the moisture that fails to infiltrate is surface runoff. After water infiltrates and water moves downward in the soil, it eventually reaches an impervious layer of rock or dense clay where it accumulates until it completely fills all the spaces above the impervious layer. This accumulated water is called groundwater, and its upper surface is the water table (see Figure 7.3). Where surface slope intersects the groundwater table, groundwater resurfaces as springs. Stream channels often cut into aquifers and gain flow from groundwater. Because water flowing through the ground is filtered of particulates, often enriched with dissolved nutrients, and has relatively constant flow rate and temperature, springs can have a large impact on water quality. Any activity that reduces infiltration and increases surface runoff diminishes the usually beneficial effects of springwater inputs. Where streams are perched above the groundwater table, they lose surface flow through bottom infiltration.

Groundwater flows through layers of porous material called aquifers. Sand and gravel layers provide good aquifers because of their permeability. In contrast, aquifers comprised of mostly clay or crystalline rocks are poor aquifers. Except for fissures that allow spring flow to the surface, artesian aquifers are overlain by some type of impervious matter such as clay or shale. Artesian aquifers often form fountain-like springs caused by the pressure from the weight of overlying mass. Many thousands of years ago—during periods of wetter climate—some parts of the world, such as Israel and the southern Great Plains, accumulated massive amounts of water deep beneath the earth's surface. These are called fossil aquifers, and water withdrawal from them is commonly referred to as "water mining." Groundwater also typically accompanies rivers with well-developed floodplains. There it often sustains riparian vegetation, adjacent to the river, and isolated ponds and wetlands during periods of low river flow.

What Is Watershed Management?

Watershed management involves the manipulation of soil and vegetation on a particular area (a watershed) to reduce erosion, improve water quality, and/or increase water supplies. Basic approaches to watershed management involve control of water infiltration, runoff, erosion, and pollution. Specific watershed management practices include conversion of vegetation type, construction of small reservoirs, water harvesting, water spreading, and control of livestock grazing, farming, logging, recreation, and urban land use practices. Allen and Sharpe (1960) summarized the objectives of the water resources manager as follows:

A. Reduce floods and high water flows
   1. Maintain high watershed infiltration
   2. Encourage deep water percolation
   3. Reduce snow-melting rates
   4. Reduce surges by mechanical barriers
B. Control water quality
   1. Maintain or increase soil cover (e.g., vegetation, leaf litter)
   2. Improve watershed infiltration
   3. Protect stream banks and beds against erosion
   4. Reduce water temperatures by maintaining or increasing shade over streams (enhances fish habitat)

C. Increase water yields

   1. Reduce evaporation and transpiration losses

   2. Reduce losses to precipitation interception by vegetation

   3. Increase snow accumulation

   4. Decrease evaporation from streams and reservoirs

D. Alter timing of water yield

   1. Reduce snow melt in winter and prolong it in spring

   2. Spread water into subsurface storage areas (meadows) for later return to streams

   3. Raise lake outlets or create reservoirs for increased temporary storage

E. Control and prevent erosion

   1. Reduce exposure of bare soils

      a. Prevent unnaturally destructive fires

      b. Regulate grazing

      c. Discourage cultivation of eroded slopes

      d. Plant vegetation

   2. Retard runoff

      a. Construct check dams and contour trenches

      b. Restore beaver

      c. Cover eroded areas with brush, litter, or other debris

More recently, watershed management includes the objective of restoring watersheds to a previous natural condition whenever that strategy is most cost effective in providing the greatest beneficial use of watershed services. Wetland restoration, for example, has been emphasized in part because of wetland effects on flood reduction and water quality improvement.

Watershed Processes

Infiltration and Percolation. Water moves into the soil by *infiltration* and through the soil by *percolation.* After a raindrop reaches the soil surface, it can infiltrate the soil, evaporate, or become a part of overland flow. The primary factors influencing infiltration rate are intensity of precipitation, amount and kind of vegetation cover, and soil surface properties (texture, structure, and organic matter) (Figure 7.4). Fine-textured soils (clays) generally have lower infiltration and percolation rates than do coarse-textured soils (sands). However, some fine-textured soils have infiltration rates comparable to those of coarse texture. This results from the cementing of fine particles into aggregates that act like larger particles.

When infiltration is high, much of the precipitation is stored in the soil for plant use, and part of it may percolate to groundwater where it can be recovered from springs or wells. Conversely, high rates of surface runoff contribute to soil loss and flooding. The most manageable factor influencing infiltration is vegetation cover. When raindrops fall on unprotected soil, they dislodge soil particles and erode away the soil surface. As plant cover declines, infiltration decreases (Table 7.2). This reduces soil moisture available for forage production and contributes to desertification of arid areas.

When percolation is high, as in sandy soils, little water is stored for long in the soil. Such soils dry quickly following rains and the vegetation they support is especially vulnerable to drought. Optimum water infiltration and storage typically occur on gentle slopes in soils with a mix of particle sizes and organic matter.

Water repellency has been found under a variety of climates and vegetation types, including many in the U.S. (DeBano 1969, DeBano and Rice 1973). It affects the hydrologic cycle by reducing infiltration rates (DeBano 1969). Water repellent soils are

VEGETATION TYPE

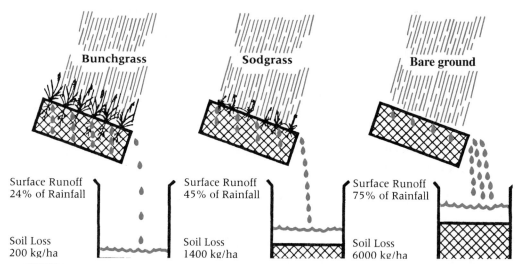

FIGURE 7.4   Influence of vegetation type on sediment loss, surface runoff, and rainfall infiltration from 10 cm of rain in 30 minutes (adapted from W. H. Blackburn et al. 1986 by Knight 1993).

TABLE 7.2   Standing Air-dry Herbage, Mulch, and Rate of Water Intake on Heavily, Moderately, and Lightly Grazed Watersheds in South Dakota

Grazing Intensity	Total Herbage (Lb/Acre)	Mulch (Lb/Acre)	Water Intake Rate (In/Hr)
Heavy	900	456	1.05
Moderate	1,345	399	1.69
Light	1,869	1,100	2.95

Source: Data from Rauzi and Hanson 1966.

present with and without burning, but they are often associated with a burn. On burned watersheds the water repellent layer occurs just below the surface of the soil. The thickness of the layer depends on the intensity of the fire and the type and amount of mulch. Plants with high levels of volatile oils are most often associated with water repellency (De Bano 1989, Dennis 1989).

Runoff and Erosion.   Surface runoff is initiated when the amount of precipitation exceeds the infiltration and storage capacity of the soil. The primary factor influencing runoff is the amount of vegetation available to retard water movement over the soil surface (Figure 7.5). Runoff declines as soil cover increases.

Erosion and sediment (suspended mineral materials in water) deposition are major problems caused by excessive runoff. Maximum sediment yields occur at about 250 mm (10 in.) of annual precipitation. A decrease occurs below 250 mm precipitation because of insufficient runoff for sediment transport. Above 250 mm precipitation, erosion is less because increased vegetation reduces precipitation impact and roots the soil in place. Excessive sediment is economically significant because it is detrimentally deposited on land and plants, reduces depression storage capacities, causes increased flood hazard, and pollutes stored water supplies (Higgins et al. 1989, Walling and Webb 1996). On the other hand, some "normal" natural erosion is necessary to sustain river floodplains, deltas, and estuaries.

Geologic erosion is normal erosion for a natural environment undisturbed by humans. Human disturbances such as overgrazing, logging, farming, and road construction, can cause accelerated erosion, which proceeds at a higher rate than normal

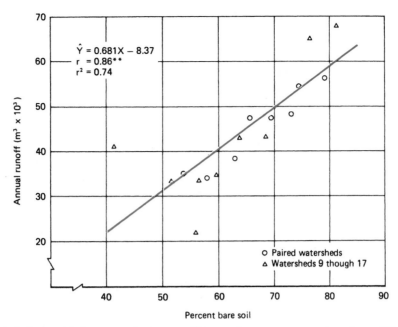

FIGURE 7.5 Relationship between percentage of bare soil versus average annual runoff from 17 watersheds in western Colorado (from Branson and Owen 1970, cited in Branson et al. 1981). The r^2 of 0.74 indicates that 74% of the variation in annual runoff can be explained by the fraction of bare soil in these 17 watersheds.

geologic erosion. Differentiating the two types of erosion is a challenging problem to the watershed manager. Rangelands often have high rates of geologic erosion because of steep slopes, aridity, thin soils, and a sparse vegetation cover.

Accelerated erosion occurs when human activities destroy the vegetation cover that retards soil loss from the forces of water and wind (Figure 7.6). The best protection against erosion is to establish and maintain a good vegetative cover. The inverse relationship between accelerated erosion and plant cover is well established (Osborn 1956, Marston 1958, Thurow et al. 1986). Accelerated erosion is the most severe consequence of watershed mismanagement because replenishment of lost soil is a slow process. Several hundred years are required to form an inch of soil; therefore, losses of soil result in nearly permanent reductions of rangeland, farmland, woodland, and watershed productivity. During the initial phases, changes in watershed management can often bring accelerated erosion under control. However, in the more advanced stages, costly measures are often necessary, such as mechanical structures and revegetation.

Managing Land Use Practices

Farming. Erosion and runoff control practices on farmlands include conservation tillage, terracing, strip cropping, contour farming, gully reclamation, and windbreaks. These practices are discussed in considerable detail in Chapter 8. In addition, runoff of fertilizers and pesticide residues are major sources of water pollution. We also refer the reader to Brady and Weil (1996) for in-depth discussions on farmland watershed management.

Logging. Properly designed forest cutting and log removal can increase downstream water yields while minimizing erosion. Selective cutting and shelter wood cutting, in contrast to clear cutting (removal of all trees on the logged area), leave some of the trees to protect the site from soil erosion (see Chapter 10). Clear cutting forests in small strips or patches, as opposed to large blocks, can minimize soil erosion problems. Because forest logging roads are major sources of sediment, road closure and forest restoration are important means for erosion management.

FIGURE 7.6   Severe gullying and loss of soil from unsound management practices in Mississippi. This area was a cotton field 35 years prior to when the picture was taken (courtesy of U.S. Department of Agriculture, Natural Resources Conservation Service).

FIGURE 7.7   The overgrazed range on the left shows severe water erosion compared to the moderately grazed range on the right. The soil surface is nearly a foot lower and dominated by rocks.

Livestock Grazing.   The key to maintaining healthy hydrological conditions on rangelands is through grazing practices that develop and maintain a good plant cover (Holechek et al. 1998). Perennial grasses, because of their high basal area and excellent soil binding properties, play the critical role in watershed stability. Moderate stocking rates in conjunction with practices that promote even livestock distribution over the range are the best approaches to maintaining a good perennial grass cover (Figure 7.7). The success of any grazing program geared toward watershed maintenance and enhancement is best measured by the residue of living and dead vegeta-

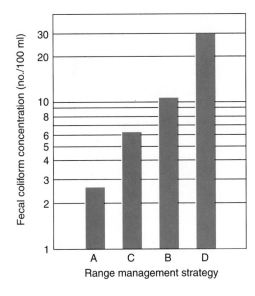

FIGURE 7.8 Fecal coliform bacteria loadings for four range management strategies for the period 1979–1984 in central Oregon (A = no grazing, B = grazing without management to improve livestock distribution, C = grazing with management to improve livestock distribution, D = grazing with cultural practices to increase forage and improve livestock distribution) (from Tiedemann et al. 1987).

tion (mulch) it maintains on the site throughout the year. In Chapter 11 we discuss the positive impacts of moderate stocking rates on economic returns and provide guidelines for determining minimum vegetation standing crop residues for different rangeland types in the United States.

Fecal wastes from range livestock grazing can cause local pollution problems (Figure 7.8). More serious problems can occur at livestock feedlots. Fecal coliform bacteria counts in water have been used as an indicator of infectious bacterial contamination (Wadleigh 1968). However, the coliform bacteria themselves are not pathologically harmful. Poorly managed livestock operations have caused increased bacterial pollution in rangeland streams (see review by Holechek et al. 1998). The severity depends on livestock numbers, timing of grazing, frequency of grazing, and access to the stream. In Wyoming, livestock grazing, recreational activities, and wildlife all contributed to increased coliform bacterial loads in stream waters (Hussey et al. 1986). Practices that improve livestock distribution and attract livestock away from streamside areas are recommended based on Tiedemann et al. (1987) and Binkley and Brown (1993). The reader is referred to Holechek et al. (1998) for more detailed information on livestock grazing management in watersheds.

Outdoor Recreation.   Areas of intense recreational use are sources of accelerated erosion, especially on steep slopes and fine soils. For parks and other areas used intensively for sports, turf management not only provides a safe substrate but also holds the soil in place. Controlling access to hikers and recreational vehicles also is a key strategy.

Urban Land Use.   Urban development can create great disturbance of the watershed and accelerate erosion from building and road construction. The high fraction of impenetrable watershed surface in urban environments can aggravate downstream flooding and alter groundwater recharge. Intense use of fertilizers, pesticides, and other elements in cities causes major water quality problems from "nonpoint" sources.

Roles of Wetland and Riparian Areas

Wetlands.   Wetlands have been recognized in protective legislation (e.g., the Clean Water Act) for the many services they perform pertaining to water resources. Among them, wetlands provide for water storage and natural treatment that filters out sediment and traps excessive nutrients. Natural inland wetlands tend to be exceptionally effective recharge zones for clean groundwater. They also reduce the erosive energy of river and coastal flooding.

FIGURE 7.9 Livestock tend to concentrate in riparian areas, causing damage to streambanks and fecal contamination of water. The fenced area on the left has been excluded from grazing. Grazing practices that disperse rather than concentrate livestock are recommended (courtesy of William Platts).

Riparian Zones.   Riparian management is extremely critical for fish and wildlife populations and other uses (Thomas 1986). Riparian zones usually receive a disproportionate amount of the grazing, urban development, farming, recreation use, and forest cutting impact. Fish and wildlife populations have been severely affected by uncontrolled grazing, road construction, farming, and forest practices in riparian zones (Figure 7.9). A summary of 20 studies by Platts (1982) was consistent (with one exception) in showing that poorly controlled grazing degraded the streamside environment and the local fishery (Figure 7.10). Reviews of how livestock grazing can be manipulated to reduce adverse impacts on riparian areas are provided by Heady and Child (1994) and Holechek et al. (1998).

Riparian zones often are the most productive sites in a region because floodplains frequently have rich soils with plentiful moisture. They have a greater diversity of plant and wildlife species than adjoining ecosystems. Healthy riparian systems purify water as it moves through the vegetation by retaining sediment and by retaining water in aquifers beneath the floodplain. Riparian zones often are a diverse mix of wetland and upland vegetation, all of which is linked closely with the floodplain groundwater. Maintaining proper amounts of herbaceous vegetation is a critical part of increasing sediment deposition and enhancing channel restoration in small stream systems (Clary et al. 1996). Many wildlife species are dependent upon the diverse habitat niches in riparian areas. They also serve as a focal point for many forms of recreation (camping, fishing, picnicking, bird watching, and so on).

Special Watershed Management Techniques

Vegetation-type Conversion.   Several studies reviewed by Branson et al. (1981) show that conversion of shrubland or woodland to herbaceous vegetation can greatly increase water yields (Figure 7.11). This is because grasses and forbs generally transpire much less water than do woody plants. Water yield for rangelands has been improved with chemical and mechanical brush control (see reviews by Vallentine 1989 and Holechek et al. 1998). Methodologies for vegetation-type conversion are discussed by Vallentine (1989), Heady and Child (1994), and Holechek et al. (1998).

Water Harvesting.   Water harvesting is the process of collecting and storing precipitation for beneficial uses from land that has been treated to increase runoff. The process has been used in Israel to grow olives, apricots, and other crops. In the United States

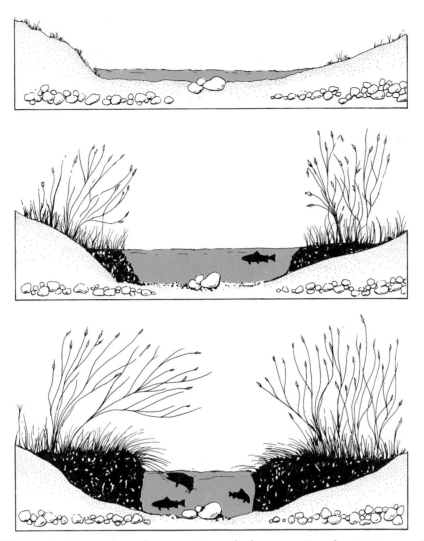

FIGURE 7.10 Typical stream channel cross section under heavy grazing, after 2 to 3 years without grazing, and after 5 to 10 years without grazing (from Bowers et al. 1979).

FIGURE 7.11 Piñon-juniper woodland in New Mexico where small areas have been converted to grassland to increase forage for livestock and wildlife and increase water yields.

and Australia, water harvesting is used to provide drinking water for livestock and wildlife and to create wildlife habitat on arid ranges. Water harvesting structures in the United States are called rain traps, catchment basins, paved drainage basins, trick tanks, and guzzlers.

Water Spreading.   Water spreading is a technique that involves diversion of water from drainages onto the surrounding landscape through a system of dikes, dams, or ditches (Branson et al. 1981). The diversion of runoff water with earthen banks to areas favorable for cultivation dates back to the early inhabitants of the Middle East and South America. Water spreading on rangelands has three main functions: (1) increasing forage production by spreading of floodwater, (2) reducing erosion in drainageways, and (3) reducing downstream flooding and sedimentation. Stream channels (dry most of the time, but flowing for short periods) generally provide the water supply for water spreading schemes. Water ponding has been used in Australia to reclaim bare areas.

MULTIPURPOSE WATER RESOURCE MANAGEMENT

The Water Resource Management Agencies

Five agencies are responsible for much of the federal management and research of water resources in the U.S. The Army Corps of Engineers is the oldest of the resource management agencies, starting early in the nineteenth century. In addition to its original authority for improving navigation of rivers, harbors, and coastal waters, the Corps has assumed the lead in flood damage reduction. In recent years, it has assumed an environmental improvement mission element equal in emphasis to navigation and flood damage reduction. The Bureau of Reclamation is mainly responsible for developing and managing surface water irrigation systems, mostly in the arid West. The Natural Resources Conservation Service is authorized to help manage waters on private agricultural lands, including building small reservoirs and drainage systems. They emphasize erosion control and environmental protection and improvement. The Tennessee Valley Authority was authorized to operate a system of reservoirs and connecting waters in the Tennessee River. They emphasize hydropower and navigation development among numerous other purposes. In addition to the federal management agencies, the U.S. Geological Survey is the main research arm of the federal resource agencies. It operates a branch that concentrates on monitoring water supply and has numerous stations located throughout the U.S. It recently consolidated biological research into a new Biological Resources Division. These federal agencies coordinate with state water resource organizations, often in cooperatively funded projects. Other water resources management is conducted by local governments and private utilities, mostly for power production, but also for irrigation and municipal water supply.

Integrated Management Objectives

The federal water resource agencies develop multiobjective projects for efficient provision of water resource benefit. That means that wherever feasible, the projects provide for all significant needs. Therefore, large reservoirs often are operated for an assortment of water storage (e.g., for irrigation and municipal supply), flood reduction, hydroelectric production, navigation, recreation, and environmental protection. It is not uncommon for two or more agencies to operate in the same river basins. After years of sometimes working at cross-purposes, they are now expected to integrate activities within a basin context. Careful monitoring by the Geological Survey and development of water-routing models (mostly by the Corps of Engineers) have been particularly helpful in making integration possible. We will briefly discuss the objectives of multipurpose water resource management.

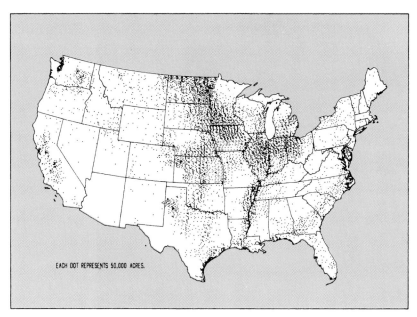

FIGURE 7.12 Areas in the United States with flooding problems (from USDA-Natural Resources Conservation Service).

Flood Damage Reduction. Recurrent flooding of rivers has been both beneficial and detrimental to the development of civilization. Ancient societies flourished along the Nile River in Egypt and the Tigris and Euphrates Rivers in Iraq because the periodic floods sustained rich floodplain soils for agriculture. On the other hand, major floods throughout history have caused massive property destruction and loss of human life. This problem has been compounded with the advance of civilization and its associated activities. Cultivation of erodible lands, deforestation, overgrazing, mining, and urbanization all contribute to flooding problems because they remove vegetation and soil, thereby reducing the land's ability to infiltrate and retain water. In the United States, about 96 billion dollars have been spent on flood control during the twentieth century. Areas where major flooding problems occur are shown in Figure 7.12. Various watershed practices to control floods center around maintaining vegetation cover on the watershed, dam construction, levee construction, dredging, and stream channelization.

Irrigation and Water Supply. The largest reservoirs in the U.S. were built for irrigation purposes in the western U.S. Irrigation systems include dams for water storage and water diversion, channels and piping to deliver the water to use areas, and control gates and valves for allocating water to each user. Irrigation management also includes vegetation management to reduce transpiration loss and to increase water yield.

Navigation. A system of U.S. waterways is now maintained in most of the large rivers and coastal zones of the Gulf of Mexico and the Atlantic. In addition, harbors are maintained along the ocean coasts and in the Laurentian Great Lakes. Navigation is one of the cheapest and environmentally cleanest forms of bulk transport. Many locks have been built and are operated by the U.S. Army Corps of Engineers to facilitate navigation past rapids, shoals and dams. In addition, thousands of miles of river, harbor, and coastal channels are dredged to sustain navigation.

Recreation. Recreation also is a purpose of water resources management. It includes boating, swimming, fishing, and water-based wildlife observation and hunting. Marinas, docks, boat ramps, and numerous other facilities support recreational activity.

Environmental Protection and Improvement. Since enactment of the National Environmental Policy Act of 1969, all federal agencies have been responsible for protecting

FIGURE 7.13 The Hoover Dam on the Colorado River was one of the first large multipurpose dams built during the 1930s "Depression Era" (courtesy of U.S. Bureau of Reclamation).

the environment. They are required to adopt the least environmentally damaging cost alternative within reason. In recent decades, the water resources agencies also have adopted missions for improving the environment through restoration of impaired ecosystems. A difficult challenge, however, is balancing environmental protection and improvement with more tangible benefits that have quantifiable monetary values.

Water Management Engineering

Reservoir Construction.   Reservoirs are a valuable means of capturing water from rain and snowmelt that would otherwise go unused for irrigation, municipal supplies, hydropower, recreation, or navigation (Figure 7.13). They also permit control of river flooding and thereby allow people to farm and live on fertile flood plains downstream. Recreation, hydroelectric power, and irrigation water are other important products from reservoirs.

Dams and reservoirs often are controversial because they have significant costs as well as benefits. Construction is usually expensive; however, as long as the benefits exceed costs, the expense is justified. Most reservoirs will ultimately fill up with sediment after a 25- to 300-year time span (Figure 7.14). As long as that loss is included in the analysis the benefits still may justify the dam. Vast areas of scenic beauty, fish and wildlife habitat, and farmland have been altered by dams. On the other hand, beauty and productive habitat have been created by other dams. The reservoirs behind the world's 13,000 dams over 50 feet in height have displaced millions of people (Miller 1990). In China, the massive Three Gorges Dam project will require resettlement of 3.3 million people; however, the power and flood damage reduction are expected to benefit many millions more.

Dams have greatly contributed to the 90% decline in salmon populations in the northwestern United States while providing inexpensive navigation and electricity.

(a)

(b)

FIGURE 7.14 The Mono Reservoir in the Santa Barbara was built to stop silt flowing into the Gibraltar Reservoir. By 1938, it had completely silted in (photo a). In 1949, vegetation had invaded the silted area (photo b) and woodland was forming (courtesy of the U.S. Forest Service).

Some outdoor sports enthusiasts believe dams replace desirable forms of water recreation (white-water canoeing, kyaking, rafting, stream fishing) with less desirable, more "artificial" forms (motorboating, sailboating, lake fishing). However, many other people use and benefit from reservoir recreation nationwide; especially boaters and anglers. Unsafe dams in populated areas are a hazard to human safety. Dam failure in Rapid City, South Dakota killed 237 people while another in Buffalo Creek, West Virginia killed 125 (Miller 1990); however, thousands of lives have been protected by flood-control dams. The U.S. Army Corps of Engineers, which is the foremost U.S. agency involved with flood damage reduction, now espouses the use of nonstructural (i.e., without dams and levees) solutions where it is in the nation's interest. Dams built and operated under federal auspices undergo a careful benefit-cost analysis to ensure wise public investments. These analyses include environmental costs and benefits.

Levee Construction.   Levees are dikes usually of earth or concrete erected along the banks of a river to prevent flooding. Most major rivers in the U.S. have some levee development along them. They have been heavily used along the Mississippi River with mixed results. They were originally built on the Mississippi River to facilitate barge traffic by forcing flood waters to cut deeper central channels. They have also permitted extensive development (agriculture, industrial, residential) of floodplains that would not otherwise have occurred. Levees are also criticized because they sometimes increase flood severity by concentrating river flow downstream, they can destroy floodplain wetlands that store flood waters and provide habitat for wildlife, they often impair the aesthetic value of the river, they can encourage unwise floodplain urban development, and they are expensive to maintain (Owen and Chiras 1995). As the 1993 Mississippi River flood and the 1996 Red River flood showed, levees also fail.

In recent years, government agencies (U.S. Army Corp of Engineers, the Bureau of Reclamation, Natural Resources Conservation Service) have increasingly adopted the view that restricting development in floodplains and relocating people away from floodplains is more practical than floodplain protection, at least in areas of relatively low population density. In other areas, such as around large cities, even larger and more effective levees may be needed in the future. Climate change predictions are especially relevant in this regard because of the connections between flooding, navigation needs, and other water resource management.

Dredging.   Dredging involves widening and/or deepening river channels and coastal harbors to facilitate navigation and reduce flooding. Dredging also maintains the intracoastal canal, which parallels much of the Gulf and Atlantic coasts. This practice prevents navigation channels from clogging with silt, but often impairs aesthetic values and can be damaging to wildlife habitat. It can contribute to degradation and destruction of wetlands bordering the river by reducing water inflow into these areas. The Corps of Engineers oversees most dredging activities for navigation improvement and the filling of waters feeding into navigable waters. They have developed programs for developing beneficial use of dredged material, including creation of wetlands and islands for fish and wildlife habitat.

Stream Drainage Channelization.   Stream channelization involves the straightening and deepening of streams to reduce flooding and make more land available to agriculture

FIGURE 7.15 Long stretches of the Rio Grande in southern New Mexico have been channelized to increase farmland, and to deliver water more quickly to irrigation uses. Channelization also has adversely impacted wildlife and fish populations and impaired the river's aesthetic quality.

or other development (Figure 7.15). The United States Department of Agriculture Natural Resources Conservation Service has directed the stream channelization program. Basically, channelization converts meandering, vegetated streams into straight ditches. The NRCS has often constructed small reservoirs on channelized streams, which can benefit wildlife. However, channelization has many drawbacks that include loss of trees and shrubs associated with the streambank, increase in water temperatures because stream shading from trees and shrubs is reduced, loss in wildlife diversity, and lower nutrient levels in the stream (Owen and Chiras 1995). In many cases, downstream flooding increases because the dissipating effect of streamside vegetation and meandering is reduced or eliminated. There has been considerable opposition to stream channelization by the Natural Resources Conservation Service in the 1990s. This has caused the government to scale back this program as the evidence accumulates indicating that recent benefits have failed to exceed costs.

WATER QUALITY MANAGEMENT

Types of Water Quality Problems

Surface water quality degradation is widely publicized, but has not become a major limitation to water availability or use nationwide. A relative abundance of good quality surface water still exists, although water quality problems have developed in some areas. Under the Clean Water Act and other legislation, The Environmental Protection Agency monitors the conditions of the nation's lakes, streams, and estuaries with the aim of identifying and regulating pollutants at their sources. About one-third of our lakes, streams, and rivers show at least some impairment.

The five leading causes of surface water impairment include bacteria, siltation, nutrients, oxygen-depleting substances, and metals. Because rivers, lakes, and estuaries differ greatly in their physical attributes, specific categories of pollutants affect them differently (Table 7.3). Health-threatening bacteria, for example, are the leading cause of impairment in river waters but nutrients are the prime cause of lake and estuary impairment.

The EPA has also identified the five leading sources of surface water pollutants (Table 7.4). These include agriculture, municipal sewage-treatment plants, aquatic habitat modification, urban runoff/storm sewers, and resource extraction. They impact lakes, rivers, and estuaries with different intensities. Agriculture, because of its role in producing nutrient and sediment problems, is the leading cause of lake and river impairment. Estuaries suffer most from urban runoff and storm sewers that carry petroleum contaminants. Estuaries are also highly vulnerable to eutrophication caused by treated sewage waste. While secondary sewage treatment removes organic pollutants, much of the inorganic nutrient remains.

Surface Water Impairment

Water pollution problems are most severe along coastal areas, much of the Midwest, and in the eastern U.S. (Figure 7.16). In the U.S., there are countless forms of pollutants. We

TABLE 7.3 **Five Leading Causes of Water Quality Impairment**

Rank	Rivers	Lakes	Estuaries
1	Bacteria	Nutrients	Nutrients
2	Siltation	Siltation	Bacteria
3	Nutrients	Oxygen-depleting substances	Oxygen-depleting substances
4	Oxygen-depleting substances	Metals	Habitat alterations
5	Metals	Suspended solids	Oil and grease

Source: EPA 1995.

TABLE 7.4   Five Leading Sources of Water Quality Impairment Related to Human Activities

Rank	Rivers	Lakes	Estuaries
1	Agriculture	Agriculture	Urban runoff/storm sewers
2	Municipal sewage treatment plants	Municipal sewage treatment plants	Municipal sewage treatment plants
3	Hydrologic/habitat modification	Urban runoff/storm sewers	Agriculture
4	Urban runoff/storm sewers	Unspecified nonpoint sources	Industrial point sources
5	Resource extraction	Hydrologic/habitat modification	Petroleum activities

Source: EPA 1995.

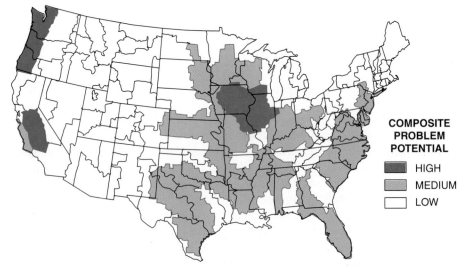

COMPOSITE
PROBLEM
POTENTIAL

■ HIGH
▨ MEDIUM
□ LOW

FIGURE 7.16 Composite potential surface water pollution associated with agriculture in the United States. Isolated locations may have greater composite problem potential than indicated at the scale used on the map (from USDA Natural Resources Conservation Service).

have grouped them into six major categories, which include disease-causing organisms, nutrients, silts and suspended solids, biochemical oxygen demand, and toxics. Each category will be discussed.

Disease-Causing Organisms

The pollutants of prime concern in this category are bacteria and other pathogens such as viruses and protozoa. Waterborne pathogens cause human illnesses that range from typhoid and dysentery to minor respiratory and skin diseases. Pathogens can enter waters through a number of routes, including inadequately treated sewage, stormwater drains, septic systems, runoff from livestock pens, and sewage dumped overboard from recreational boats. Municipal waste can carry numerous pathogenic microorganisms (Nash 1993). Widespread decreases in fecal bacteria have occurred as municipal waste treatment has improved. In 1993 and 1994, for example, only 30 reported disease outbreaks were associated with drinking water, 23 were associated with public drinking water supplies, and seven were with private wells. This low incidence reflects the effectiveness of sewage treatment overseen by the EPA and local government agencies.

Under rare conditions, certain protozoan pathogens such as *Cryptosporidium* can pass through water treatment filtration and disinfection processes in sufficient numbers to cause health problems. *Cryptosporidium* causes cryptosporidiosis, a gastrointestinal disease. An outbreak of this pathogen occurred in Milwaukee, Wisconsin in 1993. More than 400,000 persons contracted the disease, over 4,000 were hospitalized, and more than 50 died.

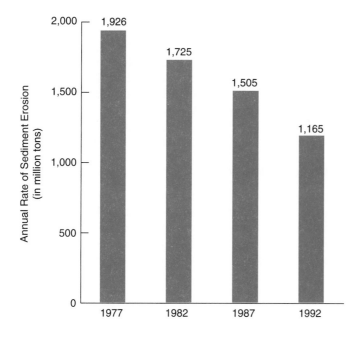

FIGURE 7.17 Sediment erosion from cropland has decreased over the period from 1977 through 1992 (EPA 1994, Environmental Indicators of Water Quality in the United States).

Nutrients

Nutrients can reach such high concentrations in water that its ability to support certain native species is altered. Nutrients can overstimulate the growth of algae and aquatic plants in a process known as cultural eutrophication. In advanced cases, cultural eutrophication interferes with the recreational use of lakes, increases the cost of water treatment before it can be used, and reduces diversity in native fish, plant, and animal populations. It is one of the leading problems facing our nation's lakes and reservoirs. Phosphorus and nitrogen are among the common nutrients that cause eutrophication (see Chapter 3). They are carried along into streams and lakes by runoff and wastewater. The origin of nitrate in surface waters can also be traced to atmospheric deposition which involves nitrogen oxide emissions from power plants and motor vehicles. Nutrient enrichment is caused by a wide variety of land use practices, including treated and untreated sewage wastewater, feedlots, concentrations of range livestock, farm and yard fertilizers, overabundance of certain wildlife (e.g., geese), and any disturbance that accelerates watershed erosion.

Nitrate-contaminated drinking water creates a grave risk for humans. Nitrates get into groundwater from many anthropogenic causes. Often they are associated with septic tanks that release them into groundwater serving private wells. Contamination can also be caused by animal wastes from farm feedlots as well as heavy applications of nitrogen fertilizer applications to sandy soils. Infants are the group at greatest risk. In some susceptible babies under six months old, nitrates can be converted to nitrites by stomach bacteria. These nitrites can then bind with blood hemoglobin and prevent oxygen from circulating to the rest of the body. The disease can produce "blue baby symptoms" and may prove fatal. Nitrates are not usually harmful to adults or older children.

Silts and Suspended Solids

As fine silt and suspended solids accumulate in aquatic habitats, fish have difficulty respiring and their eggs are smothered. Turbidity reduces plant productivity to the extent that photosynthesis is impaired by reduced sunlight. Recreational opportunities can also be lost. In a water quality context, sediment usually refers to soil particles that enter the water column from eroding land. Sediment consists of particles of all sizes, including fine clay particles, silt, sand, and gravel. Suspended sediments can often be traced to agricultural sources. However, progress has been made toward reducing sediment loading from cropland (Figure 7.17).

Biochemical Oxygen Demand (BOD)

The decomposition of organic materials reduces the availability of dissolved oxygen that is crucial to respiration of fish and aquatic invertebrates. Industrial sources currently contribute about one-third of this biochemical oxygen demand (BOD) load nationwide. Most industrial BOD load reduction occurred in the mid 1970s shortly after the Federal Water Pollution Control Act of 1972 was passed. Municipal reductions occurred later, in the early 1980s. Within a decade after the 1972 legislation was passed, municipal loads of BOD decreased 46% and industrial BOD loads decreased 71% nationally.

Salinity and Other Dissolved Solids

Salts typically make up most of the total dissolved solids in water. Salinity impairs the use of water for drinking and crop irrigation, and adversely affect aquatic ecosystems. In the West, surface waters have been contaminated with salts by irrigated agriculture. In coastal areas, saltwater intrusion from ocean sources also causes local contamination of groundwaters where too much is pumped for use.

Toxic Materials

Toxic materials can have adverse effects when present in extremely low concentrations. They can cause death, mutation, or reproductive failure in fish and wildlife, and may pose carcinogenic or other health threats to humans. They may be persistent or dissipate quickly. More than 60,000 commercial chemical substances are currently in use in the U.S. They potentially threaten our water if handled irresponsibly. About 3.5 billion pounds of formulated pesticide products are used each year. Pesticides, including insecticides and herbicides, are applied extensively to crop, pasture, and forest land throughout the nation. In urban areas, they are used on lawns and gardens, and to exterminate pests in buildings and homes. The greatest release of pesticides has been on farms.

Mining operations contribute directly and indirectly to heavy metal toxicity. However, the mining industry has cleaned up abandoned mines and controlled discharges from active ones. The silver, lead, and copper mines found in the western U.S. can directly contribute metal-laden runoff through tailings piles and mine seepage. Many of the worst problems come from mines operated and abandoned many decades ago. Pollution from abandoned mines is addressed in the Federal Surface Mining Control and Reclamation Act of 1977. Discharges from active mines are regulated by discharge permits issued by EPA and states. Many states also use best management practices (BMPs) to reduce the pollution threats posed by mines. Best management practices applied to mining, forestry, and other activities are designed to prevent rather than correct water quality problems.

Aerial sources of toxic metals also originate from fossil-fuel ignition and wind erosion. In addition to natural sources, coal-burning power plants appear to be a source of mercury contamination. Aerial sources come by way of both dry (dust) and wet precipitation. One of the worst cases of air pollution occurs in the Owens River Valley in southern California. There, the dried out bed of an extinct lake is a source for tremendous storms of contaminated dust, which are transported long distances. Because the lake is extinct as a consequence of water transfer from the Owens River to Los Angeles, there is a movement to return at least some of the water for lake restoration. However, there is also resistance because Los Angeles now depends on the water.

Polychlorinated biphenyls (PCBs) are used to cool electrical transformers. They were discharged into water systems prior to regulatory controls in the 1970s. PCB contamination of stream sediments has led to moratoria in 15 states on the consumption of fish caught in streams below points of known PCB discharges.

Acid Mine Drainage

Acid mine drainage problems occur when sulfur-bearing minerals are exposed to water and atmospheric oxygen, which forms sulfuric acid. Acidic water draining or seep-ing from mines dissolve metals from geologic formations and carry them into waterways. On entering more alkaline streams, iron compounds dissolved in the acidic water precipitate and impede production of bottom-dwelling aquatic organisms.

Thermal Discharges

Power plants and industry release cooling water with elevated temperature. The discharged waters often average 10°C or greater. When ambient temperatures are warm, this heating can block river migrations of fish or trap fish in areas where they are exposed to pollutants or other harm. In winter, fish often are attracted to the warmed water. Although this can result in excellent fisheries, it also leaves fish vulnerable to thermal shock if the power plants have to shut down operation. For these reasons among many, very large cooling towers often are used to cool water before it is returned. In other instances, large cooling ponds are built and dedicated to cooling purposes and compatible uses.

APPROACHES TO WATER QUALITY MANAGEMENT

Classification of Water Pollution Sources

Anthropogenic pollution can be categorized as emanating from either municipal, industrial, or agricultural sources (Figure 7.18). Pollutants can be further classified as originating from either point or nonpoint sources (Table 7.5). Point sources discharge pollutants directly into surface waters from a conveyance. Point sources include municipal sewage treatment plants, industrial facilities, and combined sewer overflows. Nonpoint sources deliver pollutants to surface waters from diffuse origins. Nonpoint sources include urban runoff, agricultural runoff, and atmospheric deposition of contaminants in air pollution. Habitat alterations, such as construction, dredging, and streambank channelization, can also degrade water quality.

In the mid 1970s, point-source pollution was believed to be more serious than nonpoint pollution. As the large point-source discharges were brought into compliance, it became more obvious that nonpoint sources were also an important cause of water quality problems. Therefore, more recent efforts to improve the quality of the nation's water have given greater attention to nonpoint pollution.

The Watershed Approach

The Environmental Protection Agency has begun applying a watershed approach to water quality problems. Nationwide, about 22% of all watersheds need capital investments to restore water quality. About one-half need improved management to attain long-term water resource goals. Watershed plans integrate programs for control of point and nonpoint sources. They provide decision-makers with opportunity to protect and restore habitat for aquatic life (e.g. wetlands). The watershed approach to pollution control fosters more comprehensive environmental solutions, more cost-effective programs, and improved public involvement.

Water Treatment

Water suppliers use a variety of treatment processes to remove contaminants from drinking water. Water typically passes through a series of treatment processes to remove common contaminants. Specific treatments may be needed at some locations but not others. The most commonly used processes include filtration, flocculation and sedimentation, and disinfection (Figure 7.19). Some treatment programs also include ion exchange and adsorption. Flocculation refers to water treatment processes that combine small particles into larger particles which settle out of the water as sediment. Filtration removes remaining particles from the water supply. Ion exchange

Everybody is Somebody's Neighbor

Development in formerly rural, agricultural areas is placing increased pressures on watersheds. The growth in developed land and specifically urban and suburban land has natural resource implications far beyond loss of productive agricultural land. With development comes paved surfaces, automobile traffic, and residential chemical use, among others.

SEPTIC SEEP

Like a full sponge, aging septic drain fields that treat sewage by slow filtration, and overloaded sludge-holding tanks can leak bacteria, nitrate, and liquid poisons into groundwater. Homeowners with septic systems can reduce this risk by pumping tanks regularly and avoid introducing solvents or other potential pollutants into the septic system.

WHEN THE SKY FALLS

Even though a watershed may be free of smokestacks, winds may still bring acidic substances from surrounding cities and industries as well as nitrogen from automobile exhaust and phosphorous from windblown soil.

ON THE FARM

As suburban sprawl intensifies, farm numbers are dwindling in many formerly rural watersheds. Remaining farms can help to protect the watershed by improving pesticide and nutrient management, fencing livestock away from streams, and making use of natural predators in pest control to reduce pesticide use.

LEAFY BUFFERS

Lacking a cushion of wetlands, streams can still be partly shielded from runoff and sediments. Setbacks from lakes and creeks and planting of waterside shrubs and other vegetation can help to trap sediment, slow flow, and provide shade and wildlife habitat.

Art by C. Bruce Morser
Source: Adapted from National Geographic, February 1996.

URBAN OOZE

As fields are paved for roads and parking lots, rainfall moves faster off the land. This torrent picks up debris and pollutants and can cause flooding, scour riverbanks, and prevent the slow filtration of water needed to recharge groundwater.

CONSTRUCTION

Soil erosion from development can be controlled with filter fences and water diversions, or trapped in sediment basins. Protective buffers can be planted or existing waterside vegetation maintained to further reduce sediment loss to nearby streams and rivers.

NATURAL FILTERS

Key to a healthy watershed, low-lying wetlands trap runoff and filter its sediments through natural vegetation. Protecting and restoring wetlands offer opportunities to increase the extent of these natural filtration systems.

FORESTS

Logging can cause serious sediment problems for streams. Soil erosion from clear cut slopes and access roads can contribute large amounts of sediment to nearby streams and rivers. Greenways along streams and cutting practices that leave tree roots in the soil can help to trap sediment.

SEDIMENT TRAPS

Large development sites can install sediment traps that catch stormwater and control runoff. Ponds may be two-tiered: one with an impervious lining to settle out sediment and potential pollutants and another that promotes slow infiltration of rainwater into the aquifer. Sediment ponds may also provide habitat for certain waterfowl species.

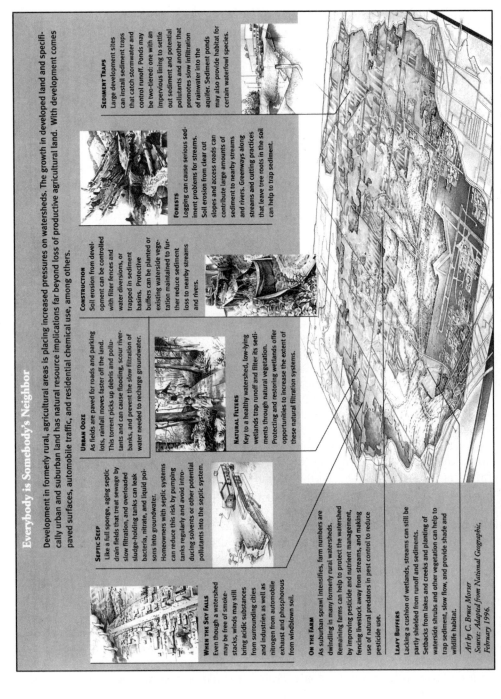

FIGURE 7.18 Examples of watershed sources of pollutants and conditions that prevent or reduce pollution loads (from *A Geography of Hope*, 1997; United States Department of Agriculture, Natural Resources Conservation Service).

TABLE 7.5   Examples of Point and Nonpoint Pollution Sources

Point Sources

Industrial	Pulp and paper mills, chemical manufacturers, steel plants, metal process and product manufacturers, textile manufacturers, food processing plants
Municipal	Publicly owned sewage treatment plants, which can receive discharges from industrial facilities or businesses as well as homes and public services
Combined sewers	Single facilities that treat both storm water and sanitary sewage, which may become overloaded during storm events and discharge untreated wastes into surface waters

Nonpoint Sources

Storm sewers/urban runoff	Runoff from impervious surfaces including streets, parking lots, buildings, lawns, and other paved areas.
Agricultural	Crop production, pastures, rangeland, feedlots, other animal holding areas
Silvicultural	Forest management, tree harvesting, logging road construction
Construction	Land development, road construction
Resource extraction	Mining, petroleum drilling, runoff from mine tailing sites
Land disposal	Leachate or discharge from septic tanks, landfills, and hazardous waste sites
Hydrologic modification	Channelization, dredging, dam construction, streambank modification

Source: EPA 1995.

processes are used to remove inorganic constituents if they cannot be removed adequately by filtration or sedimentation.

Chlorine, chloramines, or chlorine dioxide are commonly used for disinfection. Residual concentrations can be maintained to guard against biological contamination in the water distribution system. Ozone is a powerful disinfectant, but it is not effective in controlling biological contaminants in distribution pipes.

Home Water Treatment

Some homeowners choose to add further protection to their drinking water. Numerous appliances are commercially available for this purpose. Home filtration units use activated carbon filters which adsorb organic contaminants and constituents that cause taste and odor problems. They also remove chlorination by-products, some cleaning solvents, and pesticides. They do not remove metals such as lead and copper. Ion exchange units can be used to remove minerals, particularly calcium and magnesium, and are sold for water softening. Some ion exchange softening units remove radium and barium from water. Ion exchange systems that employ activated alumina are used to remove fluoride and arsenate from water. Reverse osmosis treatment units generally remove a more diverse list of contaminants than other systems. They can remove nitrates, sodium, other dissolved inorganics, and organic compounds. Distillation units boil water and condense the resulting steam to create distilled water.

WATER USES IN THE UNITED STATES

Types of Water Use

Since 1900, the U.S. population has increased 200% and now numbers about 260 million. Over the same period, however, per capita water use increased about 650%. This reflects tremendous growth in all human uses for water. Water is used either in place in the stream (*in-stream use*) or it is withdrawn for *off-stream use*. In-stream water uses do not require the diversion or withdrawal of water from its source. In-stream uses include navigation, hydropower generation, fish and wildlife propagation, recreational activities, and the maintenance of estuary salinity. The quantity of freshwater withdrawn in the U.S. for all uses increased steadily from 1950 to 1980 (Figure 7.20).

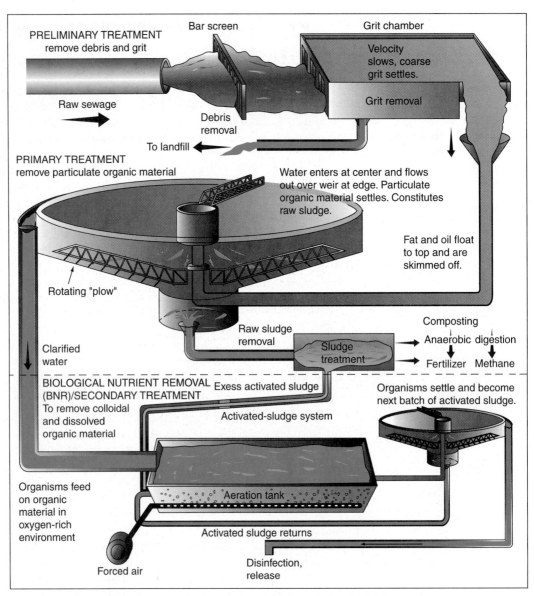

PRELIMINARY TREATMENT
remove debris and grit

Bar screen

Grit chamber

Velocity slows, coarse grit settles.

Grit removal

Raw sewage

Debris removal

To landfill

PRIMARY TREATMENT
remove particulate organic material

Water enters at center and flows out over weir at edge. Particulate organic material settles. Constitutes raw sludge.

Fat and oil float to top and are skimmed off.

Rotating "plow"

Raw sludge removal

Composting

Anaerobic digestion

Sludge treatment

Fertilizer   Methane

Clarified water

BIOLOGICAL NUTRIENT REMOVAL
(BNR)/SECONDARY TREATMENT
To remove colloidal and dissolved organic material

Exess activated sludge

Organisms settle and become next batch of activated sludge.

Activated-sludge system

Organisms feed on organic material in oxygen-rich environment

Aeration tank

Activated sludge returns

Forced air

Disinfection, release

FIGURE 7.19   Example of wastewater treatment (from Nebel and Wright 1998).

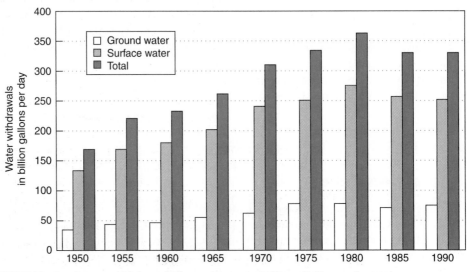

FIGURE 7.20   Freshwater withdrawals from 1950 to 1990 (from Solley et al. 1993).

Each year, considerably more water was obtained from surface than from ground sources. From 1980 to 1990, the total amount withdrawn from surface and groundwater sources actually decreased slightly.

Consumptive use of water precludes further use downstream. Virtually all water use involves some conversion to water vapor through transpiration, perspiration, and evaporation. Other water use diverts water from the original river flow to another watershed and thus is no longer available to downstream users. From the standpoint of downstream users, the conversions and transfer out of the stream is tantamount to water "consumption." It is no longer available to them for use. Many types of water use return most of the water withdrawn from streams, lakes, or groundwater. In cities, most water is returned through sewers and storm drainage systems after it is poured or flushed down drains and runs off roofs, parking lots, and streets. The returned water may be altered in quality, but can be improved through treatment. In crop production, some water is consumed as it evaporates from the soil surface or is transpired. Water also is consumed by manufacturing and food processing industries. Water consumption typically is greatest in arid environments where evapotranspiration exceeds precipitation.

Average per-capita use in 1990 was 1,340 gal/day of freshwater. Declines in withdrawals from 1980 to 1990 reflect numerous factors. Higher energy prices, more efficient application techniques, extent of irrigation development, regional droughts, increased competition for water, domestic water conservation, and water recycling all played a part in reversing the steady increase in water use that occurred in the preceding decades. About 17% of all water used in the U.S. is saline.

Off-Stream Uses

The United States Geological Survey (USGS) periodically reports water withdrawals by category. Especially useful are reports prepared by Solley et al. (1988), and USGS (1993). For simplicity, we will confine the discussion to thermoelectric, agricultural, industrial, and domestic categories.

Thermoelectric.   Water in this category is used mainly for condenser cooling in thermoelectric power plants, which rely on steam production. Steam electric generation requires water boiling by burning fossil fuels (e.g., coal, gas, oil) or by nuclear fission. The steam cycle is maintained by condensing the steam, once it has been used, and boiling it again for steam regeneration. Thermoelectric cooling can use saline as well as freshwater. In 1990, about 131 billion gallons of freshwater were used each day in thermoelectric power production (Figure 7.21). This is roughly equivalent to the flow of ten Mississippi rivers. More than 30% of the water used for thermoelectric cooling is saline.

Agriculture.   Irrigation is the primary use of the nation's freshwater (Figure 7.21). It accounts for 84% of the freshwater consumed each day. Irrigation was used in the U.S. by

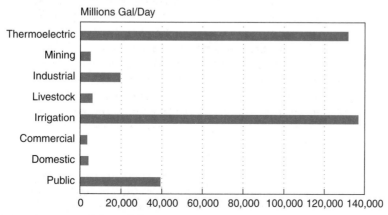

FIGURE 7.21   Freshwater withdrawals by use category (from Solley et al. 1993).

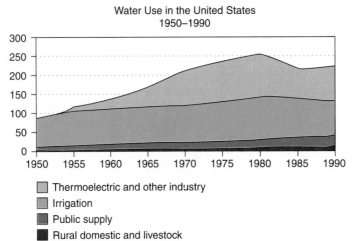

FIGURE 7.22   Trends in water use 1950–1990 by use category (from Solley et al. 1993).

Native Americans before European colonization. Irrigation water has been used to increase crop yield and to increase the number of plantings each year. Without irrigation, growers would have to confine crop production to seasons of high precipitation. Today, irrigation is used on 15% of the nation's total cropland and it contributes 38% to the total value of all crops. For individual crops, irrigated acreage as a share of total acreage is most significant for rice (100%), orchards (81%), vegetables (64%), and cotton (36%).

Irrigation water can be supplied by farms themselves or by irrigation companies or districts. Water is applied to crops by two general methods. Flood irrigation includes applications made by flooding or by using furrows. Nonflood irrigation is done with center pivot, traveling gun, trickle, and drip systems. Irrigation water statistics include water used to irrigate public and private golf courses as well as water applied to farm and horticultural crops. They exclude irrigation of lawns and gardens, which is a substantial use of water in urban settings.

Total irrigation withdrawals increased progressively from 1965 to 1980 when use peaked (Figure 7.22). From 1980 to 1985 irrigation actually decreased 9% and in 1990 it had not returned to 1980 levels. Most of the decrease in total water use is due to decreased irrigation use as a consequence of more efficient techniques and retirement of some lands from irrigation agriculture.

The water used by livestock is included in the irrigation category but represents an insignificant amount. Livestock use includes water withdrawn for feed lots, dairies, and other farm needs.

Industry and Mining.   Industries use water primarily for cooling and washing, and for creating all sorts of products. Water is used, for example, to produce computer chips, paper, steel and chemicals, and prepared foods. Mining uses water for extraction, sorting, and processing. In 1990, industries and mining accounted for 4% of the fresh water withdrawn. One-third of the water used by mining is saline as is 15% of the water used for industrial purposes.

Domestic and Commercial.   Domestic water use includes water used by households for drinking, food preparation, bathing, washing clothes and dishes, flushing toilets, and watering lawns and gardens. Commercial water use includes water for motels, hotels, restaurants, office buildings, other commercial facilities, and public institutions (e.g., government buildings and lands, schools, museums, military bases, auditoriums, stadiums). In 1990, domestic and commercial uses accounted for 7% of all water consumed.

Most water for domestic and commercial uses is provided by public-supply facilities, which divert, treat, and deliver the water. In 1990, domestic deliveries averaged 105 gallons per day for each person served, the same as during 1985. The per-capita use has remained about the same for the last decade owing to conservation efforts including the installation of additional meters and water-conserving plumbing fixtures, and regulated irrigation of lawns and gardens.

In-Stream Uses

Important in-stream uses include hydropower production, recreational uses, and maintenance of ecosystem functions. While off-stream uses typically are more consumptive and environmentally problematic, not all in-stream uses are without environmental effects.

Hydropower.   In 1990, hydroelectric plants produced about 11% of the nation's total power output. From 1985 to 1990, in-stream use increased about 8%. Nature largely determines how much energy can be produced from hydropower (dams) each year. When surface flow decreases during periods of drought, so does energy production. The Pacific Northwest region uses by far the most water for hydroelectric power production. Hydropower production plants can have major impacts on natural ecosystems. Much of the damage done to migrating salmonid populations in the Northwest is caused by hydropower plants, which block upstream passage and kill a proportion of the young fish as they move downstream through the turbines. However, hydropower neither causes climate change nor thermal pollution. A major challenge is developing improved compatibility, if possible, with fish passage.

Navigation.   Historically, the most important in-stream use of water was navigation and it continues to be a critical use to this day. Thousands of miles of navigable waterways are maintained in the U.S. as well as other parts of the world. Without water transportation, the cost of many bulk goods would be significantly greater.

Recreation.   In-stream flows support fish and wildlife, boating, rafting, and other forms of water-based recreation. The values of such use can be substantially greater than the prevailing off-stream use values. However, in-stream uses for recreation have been historically belittled or ignored in many western states where water is owned and used mostly in off-stream settings. In recent years, most states have adopted laws to provide for in-stream flow in support of recreational use.

Ecosystem Support.   In addition to recreation, maintenance of in-stream flow is justified to maintain natural biodiversity and ecosystem function, especially in streams supporting threatened or endangered species. Most states have laws that provide in some way for minimum in-stream flow needs.

Regional Trends In Water Use and Consumption

Groundwater depletion is an important problem in the southwestern United States (Figure 7.23). Domestic uses of surface freshwater are more concentrated in the eastern states, which are more heavily populated. Industrial uses are greater in the East and are especially concentrated in the northeastern states. Irrigation is the predominant use in the West. Here supplemental irrigation is needed to assure reliable crop production. Among all states, California leads the U.S. in water consumption, with 20% of total water consumed. This high use results from a combination of large population demand and large irrigation demand.

WATER USE PROBLEMS AND CONFLICTS _____

Water is a hot issue today owing to controversies and problems surrounding its use. Basically, the problem is too little water relative to various demands. More now than ever, the public and user groups are keenly interested in water issues. Space will not permit discussion of all of the important water issues confronting the nation or even

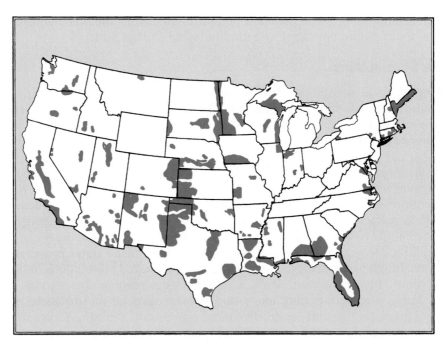

FIGURE 7.23   Regions where groundwater depletion is an important problem (United States Department of Agriculture).

a specific region (see Postel 1996, EPA 1995, Perry and Vanderklein 1996). However, we will briefly examine some of the more prominent problems and issues.

Overpumping

One important problem in several regions is that water is being withdrawn faster than nature is restoring it. In such cases, it can be said that water is being mined. This is happening in geographic locations as remote as Lubbock, Texas, in the High Plains (Supalla et al. 1982), Phoenix, Arizona, and Florida. As water is pumped to meet the growing demands of irrigated agriculture and escalating urban growth, land actually subsides often damaging buildings, roads, and other structures. East of Phoenix, water tables have dropped more than 120 meters. If groundwater withdrawals continue at current rates, water tables in Albuquerque, New Mexico will drop an additional 20 meters by 2020 (Postel 1996).

Substantial groundwater withdrawals for irrigation began in the southern High Plains in the early 1940s and spread to the middle High Plains by the 1950s. Heavy pumping reached the northern High Plains in the 1960s. The Ogallala formation, a massive aquifer in several plains states (Figure 7.24), is being rapidly depleted by overpumping. As it declines, large areas of productive farmland are reverting to dryland agriculture or rangeland. Along coastlines, over pumping essentially sucks saltwater into groundwater systems by a process called "saltwater intrusion." This has occurred, for example, along the coast of California and along the southeastern coast (Florida).

Water Allocation and Wildlife Habitat

A major conflict is how to satisfy recreation, aesthetic, and biodiversity protection needs for water while continuing to meet the growing needs of agricultural irrigation and fast growing western cities. In this regard, California's Central Valley faces tremendous challenges. On the one hand, California produces nearly 50% of the nation's fruits, nuts, and vegetables. Its agriculture provides a $20-billion-a-year industry that continues to employ people as manufacturing jobs are lost. On the other hand, the Central Valley is the most heavily pumped contiguous area in the U.S. California must somehow divide its water resources equitably. Rivers and wetlands have been deprived over previous

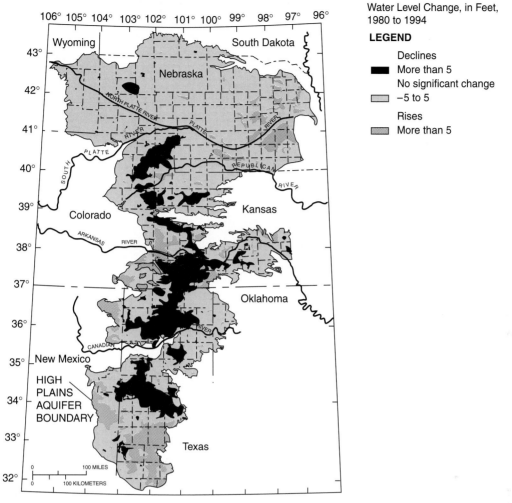

FIGURE 7.24 Water level change for the Ogallala aquifer (from United States Department of Agriculture, Natural Resources Conservation Service).

decades in favor of agriculture, industry, and household users who are rapidly growing in numbers. California has one of the fastest growing human populations in the U.S.

Over the past decade, society has become more aware of the damage done to river and wetland ecosystems. The Florida Everglades serves as an excellent example of what could be lost if efforts are not successful in restoring water flows to natural areas. The Everglades is the only spot in the conterminous United States where truly tropical and subtropical ecosystems can be observed. As land development and agriculture have rerouted flows from northern areas, the Everglades have shrunk in size and biotic diversity. Pollutants carried into the Everglades pose an added threat to its diverse ecosystems. Because of the Everglade's sensitivity, Congress has supported development of a plan to restore the Everglades under a partnership of state and federal agencies. If the plan materializes, water provision and flood damage reduction should also improve across much of South Florida. About $8 billion would be spent on reengineering the water delivery infrastructure of South Florida under direction of the Army Corps of Engineers.

For estuaries bordering the Everglades, Chesapeake Bay, the Gulf Coast, and San Francisco Bay, freshwater stream flows are essential to sustain biotic diversity. When unobstructed, rivers running from land into bays and estuaries carry nutrients critical to all forms of life. In addition, undisturbed estuaries often represent fertile interfaces between saline ocean waters and stream-carried freshwater. The brackish waters created at the interface often support extensive and diverse fisheries. Along

FIGURE 7.25   A fish ladder built to take salmon over the Rock Island Dam on the Columbia. This approach has been only partially successful in allowing salmon to return to spawning areas (courtesy of United States Bureau of Reclamation).

the coasts of Texas and Louisiana, for example, brackish water serves as vital breeding habitat for brown and white shrimp, blue crabs, redfish, and speckled trout.

In the Pacific Northwest, where numerous dams have been built, native trout and salmon populations are in jeopardy—even though great expense and effort have been directed toward eliminating the threats caused by dams (Figure 7.25). In many areas, the obstruction of seasonal floods by dams prevents the renewal of spawning grounds and other fish habitat. Diversions of California's water away from its streams have been largely responsible for the recent decline in trout and salmon populations (Postel 1996). The state has lost 95% of its wetlands. Populations of migratory birds and waterfowl have dropped from 60 million in 1950 to 3 million today (Postel 1996). North America's Great Lakes have fared no better. In 1900, native trout and salmon comprised 82% of the commercial catch. By 1966, less than 1% of the fish caught were native.

In southwestern deserts, rivers through New Mexico, Arizona, and western Texas serve as major corridors for neotropical birds and several endangered birds of prey. The seasonally wet playa lakes and streams can also support diverse life and are especially important to amphibians. The southwest livestock grazing has been a major cause of degradation of vegetation around rivers, streams, and playa lakes as has been previously discussed.

Salinization

A factor further aggravating aquatic and wetland areas is salinization. This process is caused by the return of salt-laden irrigation water to rivers, lakes, and estuaries. This problem is especially acute where soils are saline. Here, irrigation water must exceed the crop water requirement in order for salts to be flushed from the root zone. Soils and water altered by salinization often no longer support the plant communities that native fish and animal species need. Soil salinization reduces the productivity of the nation's farm land, but with proper management it can be greatly reduced (see Brady and Weil 1996).

Who Owns the Water?

The Problem.   In 1889, 3.6 million acres (0.6%) of the nation's 623 million farmland acres were irrigated, principally in California (1 million acres) and Colorado (0.9 million acres) (Guldin 1989). Today, about 45 million acres of farmland are irrigated. The growth of irrigated agriculture was facilitated by publicly supported water development projects that diverted water from major natural water bodies. Support for these projects was maintained by the need to produce ever increasing amounts of food for more people. Thanks to its success, we have been spared hunger and expensive food.

Early on, as settlements and irrigation spread, the need arose for water law to settle disputes. Some states (e.g., New Mexico) were quick to write relatively comprehensive groundwater management statutes. In states such as Texas and California, early water law was revised substantially over the years. However, it still retains significant parts that existed before the rapid development of groundwater resources (Smith 1989).

Historically, only a few primary legal doctrines have governed groundwater use (Smith 1989). The Absolute Ownership Doctrine holds that the water beneath one's land is the property of the landowner. It can be withdrawn with no regard to the effects that may occur on other landowners. This doctrine was developed in England before being transferred to the eastern U.S. It worked reasonably well in areas with abundant water, but was soon found impractical for the more arid West, where the competition for water caused state laws to evolve into the American Rule or the Rule of Prior Appropriation. The American Rule limits a landowner's groundwater right to the amount required for some reasonable and beneficial purpose. The Prior Appropriation Doctrine simply provides that the first landowner to put water to beneficial use, without waste, has a right to continue that use. Most western states have adopted this doctrine. States such as California replaced the Absolute Rule with the Correlative Rights Doctrine. This doctrine recognizes the landowner's right to groundwater but limits that right somewhat by giving other landowners sharing the common source equal, or correlative, rights. The doctrine further states that the amount withdrawn will be put to a beneficial and reasonable use.

The complexities of nature, economics, politics, and sociology continue to frustrate efforts to create the perfect water law. Emphasis on beneficial use often detracts from ecological sustainability because some important ecosystem services cannot at this time be valued monetarily. The emphasis on prior appropriation often fails to coincide with changes in society's values. It thwarts efforts to use water more efficiently. More specifically, by keeping water from the open market economy, prior appropriation laws keep water from flowing to the highest bidder. A big part of the prior appropriation controversy centers around resentment over government subsidies to landowners through dam and water diversion projects that have caused the services of many natural ecosystems to decline. Further resentment arises over restriction on homeowner use of water such as alternate-day lawn watering.

Is There An Equitable Solution?   Most states have passed legislation in recent years to restore stream flows to better meet environmental needs. For example, Florida has mandated substantial stream-flow increases to Everglades National Park. The state of Nevada has mandated increased stream flow to wetlands. California has set in motion numerous projects to increase stream flows for fish and wildlife. In an unprecedented act, the Bureau of Reclamation recently rejected the opportunity to sell electricity valued at $3 million in order to release enough Colorado River water to improve stream bottoms for aquatic life. Many other examples could be cited, but these point out the public's interest in more balanced allocation of water.

The main reason most of these changes came about is the willingness of people to pay either for the benefits they derive directly from environmental improvement or the perceived tourism and related commercial values. Management designed exclusively for biodiversity protection has been much more modest in scope. However, the recent declaration of threatened and endangered status for nine salmonid fish in the

Northwest could become very costly to the public and a powerful test of the willingness of people to pay for biodiversity protection.

Part of the solution to sustainable use of water for all of its diverse applications involves making more efficient use of available water. Farmers have responded by adopting more efficient irrigation techniques. Greater efficiency is being obtained through more careful irrigation scheduling and more sophisticated techniques for tracking soil moisture levels, as well as by modeling crop water use, developing drought resistant crop varieties, and using advanced low-volume irrigation equipment. In the High Plains area, for example, water was frequently sprayed in the open air where it easily escaped to the atmosphere. Now water is applied by systems which roll over the land and place water close to the ground with tubes extending downward from a horizontal pipe above.

Industry is making progress by using water several times before it leaves a plant. Factories continue to find more effective ways to make low-quality water useful through pretreatment technology. As the Environmental Protection Agency (EPA) raised wastewater discharge standards, industry was faced with higher costs to use public wastewater facilities that received their discharge. Many factories found it more economical to treat their own wastewater and in the process learned much about water reuse techniques. More industries are learning how to use brackish water. From 1985 to 1990 the use of reclaimed water grew 30% (USGS 1993). It seems likely that water use per unit of production will continue to decline for most industries.

The public also holds part of the answer as is demonstrated by water use statistics. From 1985 to 1990, people were asked to conserve water in many states. Domestic use increased only 2% while the population grew 4%. In contrast, from 1970 to 1975—when there was less focus on conservation—increase in domestic use was more than double the rate of population growth. Conservation in the 1985 to 1990 period was achieved by numerous approaches and efforts that continue to become more efficient and cost effective. For example, there has been an increase in use of low-volume toilets as well as faucet and shower fixtures that reduce flow rate by decreasing water pressure. Progress toward reducing water applied to home landscapes has been achieved through xeriscaping, which uses plants demanding less water.

Experts agree that the West is in transition from an era of water development to an era of water management and conservation. Said another way, the region is moving away from supply management to demand-side management. Historically, water needs for urban growth and farming in the West have been met by finding more ways to develop potential water resources. Colorado River projects completed decades ago provide an excellent example of how this was done. By constructing dams and diversion structures, the Bureau of Reclamation gathered water, starting with the Wind River Mountains of Wyoming, and delivered it through elaborate systems to southern California. Along the way, dams provide energy and water for crop land, desert cities, and golf courses.

Dam construction now is in decline, however, because of increasing cost, few remaining good sites, and the growing realization that more water must be allocated for other ecosystem services. In its place are increased conservation by different users, more acceptance of water marketing, and the use of water trading strategies (see Chapter 4). For example, some farmers have donated or sold water rights to wildlife advocate groups. Cities have paid for water diversion improvements to avoid diverting more water from natural areas. By providing more efficient conveyance structures that lose less water in transport, more water is conserved for people and nature alike. Above all, such cooperative ventures put financial resources where they are needed as opposed to protracted litigation cases. Solutions are found through cooperation, mutual respect, and the increasing perception of the finite nature of water resources. Such efforts ensure that maximum benefits are obtained from use of the nation's water resources.

MEETING WATER DEMAND IN THE TWENTY-FIRST CENTURY

Water use in the United States will increase substantially over the next 20 years. This is because it is estimated there will be 50 to 70 million more people in the USA by 2020. In order to meet future demands, Owen and Chiras (1995) described seven strategies that include (1) water conservation, (2) reclamation of sewage water, (3) development of groundwater resources, (4) desalinization of seawater, (5) rainmaking, (6) harvesting of icebergs, and (7) diversion of surface water to water-short regions which we will summarize.

Water Conservation

A variety of water conservation practices are available to homeowners, farmers, and industry. Some of the practices available to homeowners include taking shorter showers, minimizing use of the toilet, repairing water leaks, watering lawns during cool hours of the day, and using xeriscaping. Farmers can line their ditches and canals to reduce seepage, use drip and other efficient irrigation systems, and use new computer sensor technology that determines moisture contents of various parts of their fields and subsequent needs for irrigation. Recycling of wastewater is an important way many industries can conserve water.

Reclamation of Sewage Water

Sewage effluent has a 99% water content and can be cleaned through treatment (Perry and Vanderklein 1996). Processed sewage water is used for many purposes, including farmland irrigation, cooling in industrial processes such as steelmaking, and golf course watering.

Development of Groundwater

We have already discussed depletion of groundwater in some detail. New technologies are allowing detection of undiscovered aquifers and easier extraction from those deep in the earth's surface.

Desalinization

Several desalinization plants now exist in the U.S. Although it is an expensive process, it has potential to meet water needs in many coastal areas if future technological improvements occur.

Developing Salt-Resistant Crops

Salinization has made millions of acres of farmland near oceans useless. However, salt-tolerant crops, such as a newly developed strain of barley, show potential to grow on saline soils.

Developing Drought-Resistant Crops

Considerable genetic engineering progress has been made in developing crops that use less water and are more resistant to drought. Sorghum is widely grown in arid portions, and shows potential for increased yields and survival under drought conditions.

Rainmaking

Rainmaking involves seeding clouds with many small crystals of silver iodide that serve as condensation nuclei for raindrops (Owen and Chiras 1995). This technique can potentially increase water supplies in particular areas but at high cost. Other drawbacks are reduction in precipitation in surrounding nontarget areas, lack of control over amount and exact location of precipitation, and possible damage from excess precipitation (flooding, erosion, property damage, and so on). The many legal, political, and environmental problems associated with rainmaking have so far caused a reluctance to apply this strategy.

Harvesting Icebergs

It may be feasible to use water from icebergs in the Antarctic to irrigate crops in desert areas. Under their plan, icebergs would be linked together and pushed to areas where water is needed (i.e., Los Angeles). Warm water from power plants would be used to melt the icebergs. Then the water would be transported inland to meet industrial, domestic, and agricultural demands. The problems are high cost, including adverse environmental impacts (Owen and Chiras 1995). Changes in ocean and coastal water temperatures could greatly affect recreation, climate, and fish populations. Therefore, much study will be needed before there is serious consideration of iceberg harvesting.

Long Distance Water Transport

Long distance water transport involves moving water from where supplies are abundant to where they are scarce using canals and aqueducts. This approach has been used to supply water to desert areas in southern Arizona and California. The California Water Project, the world's most complex and expensive water diversion project, is shown in Figure 7.26. It has been effective in providing water to California's southern desert areas. However, it has also been severely criticized for excessive costs of development, energy, destruction of scenic beauty, and destruction of fish and wildlife habitat. Because of these drawbacks, Owen and Chiras (1995) speculated it may be the last of its kind. In their opinion, more cost-effective and sustainable

FIGURE 7.26   California's State Water Project with major lakes and reservoirs indicated by numbers (from Owen and Chiras 1995).

means of meeting water demands must be used, such as population stabilization; greater water efficiency measures in homes, business, and agriculture; more recycling of water; and recharging of groundwater.

Improved Integration of Water Use

An effective way to make more efficient use of water is to integrate water use. Numerous examples of integrated water use are available. For multipurpose reservoir management, the most beneficial combinations of water level and water transfer have been well researched. After treatment, city wastewater has been used in aquaculture, to grow crops, to water lawns, to provide water for wildlife and fish, and to support water-based recreation. Artificial wetlands are increasingly being used to treat wastewater while they also provide habitat for wildlife and other ecological functions. Thermal effluents from power plants have been used to augment aquaculture and enhance fisheries. Models of water resource systems have been particularly useful in the development of more integrated management approaches.

Water in the Nation's Future

Overall, it seems likely that water withdrawals for agriculture, industrial, and domestic uses will continue to increase as population increases. However, water prices will probably increase and the public will likely respond by employing even more active conservation programs to reduce their per capita use rates. With increased competition for water, new balances will have to be struck in water use between rural and urban areas, especially in the West. Irrigators will likely find it more difficult to compete for water as cities encroach into prime farmlands. Municipal and industrial users can afford to pay much more for water than agricultural users. Ultimately, higher irrigation-water costs will likely be passed on to the consumer of food and fiber goods.

LITERATURE CITED

Aguilar, R. D. and S. R. Loftin. 1991. *Sewage sludge application in semiarid grasslands: Effects on runoff and surface water quality.* In Proc. 36th Annual Water Conf., N. Mex. Water Resour. Res. Inst. Rep. 265, pp. 101–111.

Aguilar, R., S. R. Loftin, T. J. Ward, K. A. Stevens, and J. R. Gosz. 1994. *Sewage sludge application. In semiarid grasslands: Effects on vegetation and water quality.* N. Mex. Water Resources Research Inst. Rep. 285, pp. 1–75.

Allen, S. W. and G. W. Sharpe. 1960. *An Introduction to American Forestry.* 3rd edition. New York: McGraw-Hill Book Company.

Bajwa, R. S., W. M., Crosswhite, and J. E. Hostetler. 1987. *Agricultural irrigation and water supply.* Agricultural Information Bulletin No. 532. Washington, DC: U.S. Department of Agriculture, Economic Research Service. 109 pp.

Binkley, D. and T. C. Brown. 1993. *Management impacts on water quality of forests and rangelands.* U.S. Dep. Agric. Rocky Mtn. For. & Range Exp. Sta. Gen. Tech. Rep. RM–239.

Blackburn, W. H., T. L. Thurow, and C. A. Taylor Jr. 1986. Soil erosion on rangeland. In *Use of cover, soil, and weather data in rangeland monitoring*, pp. 31–39. Symp. Proc., Soc. Range Manage., Denver, CO.

Bouwer, H., D. F. Heermann and B. A. Stewart. 1983. Our underground water supplies: The sometimes dry facts (pp. 448–457). In *Using Our Natural Resources, 1983 Yearbook of Agriculture.* Washington, DC: U.S. Department of Agriculture.

Bowers, W. B., A. Hosford, A. Oakley, and C. Bond. 1979. *Wildlife habitats in managed rangelands in the Great Basin of northeastern Oregon.* U.S. Dep. Agric. For. Serv. Gen. Tech. Rep. PNW–84.

Brady, N. C. and R. R. Weil. 1996. *The nature and properties of soils.* 11th edition. Upper Saddle River, NJ: Prentice-Hall.

Branson, F. A. and J. R. Owen. 1970. Plant cover, runoff, and sediment yield relationships on Mancos shale in western Colorado. *Water Resource Research* 6:783–790.

Branson, F. A., G. F. Gifford, K. G. Renard, and R. F. Hadley. 1981. *Rangeland hydrology.* Range Science Services No. 1. Society for Range Management, Denver, CO.

Clary, W. P., C. I. Thornton, and S. R. Abt. 1996. Riparian stubble height and recovery of degraded streambanks. *Rangelands* 18: 137–140.

Darling, L. A. and G. B. Coltharp. 1973. *Effects of livestock grazing on the water quality of mountain streams.* Proc. Symposium on Water-Animal Relations, Southern Idaho State College, Twin Falls, ID.

DeBano, L. F. 1969. *Observations on water-repellent soils in the western United States.* Symposium on Water-Repellent Soils, University of California, Riverside. Proc. May 6-10, 1969, pp. 17–29.

DeBano, L. F. 1989. *Effects of fire on chaparral soils in Arizona and California and postfire management implications* (pp. 55–62). In N. H. Berg (Coord.), Proc., Symp., Fire and Watershed Management. U.S. Dep. Agric. Pacific Southw. For. & Range Exp. Sta. Gen. Tech. Rep. PSW–109.

DeBano, L. F. and R. M. Rice. 1973. Water-repellent soils: Their implications in forestry. *Journal of Forestry* 71: 220–223.

DeBano, L. F. and L. J. Schmidt. 1989. *Improving southwestern riparian areas through watershed management.* U.S. Dep. Agric., Rocky Mtn. For. & Range Exp. Sta. Gen. Tech. Rep. RM-182.

Dennis, N. 1989. *The effects of fire on watersheds: A summary* (pp. 92–94). In N. H. Berg (Coord.), Proc., Symp., Fire and Watershed management. U.S. Dep. Agric., Pacific Southw. For. & Range Exp. Sta. Gen. Tech. Rep. PSW–109.

Elmore, W. and B. Kauffann. 1994. Riparian and watershed systems: Degradation and restoration. In *Ecological implications of livestock herbivory in the West.* Society for Range Management, Denver, CO.

Gleick, P. H., ed. 1993. *Water in crisis. A guide to the world's freshwater resources.* New York: Oxford University Press.

Guldin, R. W. 1989. *An analysis of the water situation in the United States: 1889–2040.* U.S.D.A. Forest Service Gen. Tech. Report RM–177.

Heady, H. F. and R. D. Child. 1994. *Rangeland ecology and management.* San Francisco, CA: Westview Press.

Heathcoate, I. W. 1988. *Integrated watershed management: Principles and practices.* New York: John Wiley & Sons.

Herdendorf, C. E. 1990. Distribution of the world's large lakes (pp. 3–38). In *Large lakes: Ecological structure and function,* edited by M. M. Tilzer and C. Serruya. New York: Springer-Verlag.

Higgins, D. A., S. B. Maloney, A. R. Tiedmann, and T. M. Quigley. 1989. *Storm runoff characteristics of grazed watersheds in eastern Oregon.* Water Resour. Bull. 25: 87–100.

Holechek, J. L., R. D. Pieper, and C. H. Herbel. 1998. *Range management principles and practices.* 3rd edition. Upper Saddle River, NJ: Prentice-Hall.

Hussey, M. R., Q. D. Skinner, and J. C. Adams. 1986. Changes in bacterial populations in Wyoming mountain streams after ten years. *Journal of Range Management* 39: 369–370.

Jensen, M. E. and J. D. Bredehoeft. 1983. New efficiencies in water use vital for nation (pp. 18–27). In *Using our natural resources, 1983 yearbook of agriculture.* Washington, DC: U.S. Department of Agriculture.

Knight, R. W. 1993. Managing stocking rates to prevent adverse environmental impacts. In *Managing livestock stocking rates on rangeland,* pp. 97–107. TX. Agric. Ext. Serv., College Station, TX.

MacKichan, K. A. 1951. *Estimated use of water in the United States, 1950.* Geological Survey Circular 115. Washington, DC: U.S. Department of the Interior, Geological Survey. 13 pp.

MacKichan, K. A. 1957. *Estimated use of water in the United States, 1955.* Geological Survey Circular 398. Washington, DC: U.S. Department of the Interior, Geological Survey. 18 pp.

Marston, R. B. 1958. Parrish Canyon, Utah: A lesson in flood sources. *Journal of Soil and Water Conservation* 13:165–167.

Miller, G. T. 1990. *Resource conservation and management.* Belmont, CA: Wadsworth Publishing Co.

Nash, L. 1993. Water quality and health (pp. 25–39). In *Water in crisis: A guide to the world's fresh water resources,* edited by P. H. Gleick. New York: Oxford University Press.

Nebel, B. I. and R. Wright. 1998. *Environmental science: The way the world works.* 5th edition. Upper Saddle River, NJ: Prentice-Hall.

Osborn, B. 1956. Cover requirements for the protection of range site and biota. *Journal of Range Management* 9:75–80.

Owen, O. S. and D. D. Chiras. 1995. *Natural resource conservation.* 6th edition. Upper Saddle River, NJ: Prentice-Hall.

Owen, O. S., D. D. Chiras, and J. P. Reganold. 1998. *Natural resource conservation.* 7th edition. Upper Saddle River, NJ: Prentice-Hall.

Perry, J. and E. Vanderklein. 1996. *Water quality: Management of a natural resource.* Cambridge, MA: Blackwell Science.

Platts, W. S. 1981. *Effects of sheep grazing on a riparian-stream environment.* U.S. Dep. Agric. For. Serv. Res. Note INT-307.

Platts, W. S. 1982. *Livestock and riparian-fishery interactions: What are the facts?* In Trans. 47th North American Wildlife and Natural Resources Conference, pp. 507–515. Washington, DC.: Wildlife Management Institute.

Postel, S. 1996. *Dividing the waters: Food security, ecosystem health, and the new politics of scarcity.* Washington, DC: Worldwatch Institute.

Rauzi, F. C. and C. L. Hanson. 1966. Water intake and runoff as affected by intensity of grazing. *Journal of Range Management* 19: 351–356.

Smil, V. 1997. *Cycles of life: Civilization and the biosphere.* New York: Scientific American Library.

Smith, Z. A. 1989. *Groundwater in the West.* New York: Academic Press.

Solley, W. B., C. F. Merk, and R. P. Pierce. 1988. *Estimated use of water in the United States in 1985.* Geological Survey Circular 1004. Reston, VA: U.S. Department of the Interior. 82 pp.

Solley, W. B., R. R. Pierce, and H. A. Perlman. 1993. *Estimated use of water in the United States in 1990.* Geological Survey, circular 1081. Reston, VA: U.S. Department of the Interior.

Speidel, D. H. and A. F. Agnew. 1988. The world water budget. In *Perspectives on water uses and abuses,* edited by D. H. Speidel, L. Ruedisilli, and A. F. Agnew. New York: Oxford University Press.

Supalla, R. J., R. R. Lansford, and N. R. Gollehon. 1982. Is the Ogallala going dry? *Journal of Soil and Water Conservation* 37(6):310–314.

Thomas, A. E. 1986. Riparian protection/enhancement in Idaho. *Rangelands* 8: 224–227.

Thurow, T. L., W. H. Blackburn, and C. A. Taylor, Jr. 1986. Hydrologic characteristics of vegetation types as affected by livestock grazing systems, Edwards Plateau, Texas. *Journal of Range Management* 39: 505–508.

Tiedemann, A. R., D. A. Higgins, T. M. Quigley, H. R. Sanderson, and D. B. Marx. 1987. Responses of fecal coliform in streamwater to four grazing strategies. *Journal of Range Management* 40: 322–329.

U.S. Environmental Protection Agency, Office of Water. 1997. *Water on tap: A consumer's guide to the nation's drinking water.* EPA 815-K-97-002.

U.S. Environmental Protection Agency. 1996. *Environmental indicators of water quality in the United States.* EPA 841-R-96-002.

U.S. Environmental Protection Agency, Office of Water. 1995. *National water quality inventory.* Report to Congress, 1994.

U.S. Environmental Protection Agency. 1984a. *The cost of clean air and water.* Report to Congress, 1984. Report No. 230/05-84-008. Washington, DC: U.S. Environmental Protection Agency.

U.S. Environmental Protection Agency. 1984b. *Nonpoint-source pollution in the U.S.* Report to Congress. Washington, DC: U.S. Environmental Protection Agency.

U.S. Geological Survey. 1993. *National water summary 1990–1991, Hydrologic events and stream water quality.* U.S. Geological Survey Water Supply Paper 2400. Data on trends of selected pollutants found in surface water.

Vallentine, J. F. 1989. *Range development and improvements.* 3rd edition. Provo, UT: Brigham Young University Press.

Wadleigh, C. H. 1968. *Wastes in relation to agriculture and forestry.* U.S. Dep. Agric. Misc. publ. 1065.

Walling, D. E. and B. W. Webb. 1996. Erosion and sediment yield: A global overview. *International Association Hydrological Science* 236:3–19.

World Resources Institute. 1994. *World Resources 1994–95.* New York: Oxford University Press.

CHAPTER 8

Soil: The Basic Land Resource

IMPORTANCE OF SOIL

The most basic of resources is the land and its primary component, the soil. Knowledge of soil characteristics, classification, and management is essential to the natural resource manager. This is because soil is the primary factor determining the potential for crop, timber, wildlife, and forage production of an area within a particular climate. Through history, productive soils have consistently been the cornerstone of the flourishing civilizations. On the other hand, soil destruction or mismanagement was often the root cause of those that declined. Excessive cutting of timber in watersheds, overgrazing by livestock, unsound cultivation practices, and farming of erodible lands are some of the costly mistakes societies have made in managing their soils. In this chapter we provide basic understanding of soil and soil problems. Excellent sources for the topic are provided by Miller (1990), Brady and Weil (1996), and Owen et al. (1998).

Soil is defined as the dynamic, natural body of the surface of the earth in which plants grow (Brady and Weil 1996). It is comprised of minerals, organic materials, and living forms. Soil is often distinguished from a developed *regolith* (unconsolidated rock and materials) by a higher organic matter content, more intense weathering, the presence of horizontal layers, and the presence of living organisms (Figure 8.1). Regolith could be considered potential soil. It performs some of the same resource functions as soils, such as foundations for road and building construction, but is typically less suitable for life-support functions.

Soil characteristics of importance to the natural resource manager include color, texture, structure, depth, pH, organic matter content, and mineral status. The interaction between these seven factors with climate and topography determines the type and quantity of vegetation that an area is capable of supporting. They also determine the nature of support and stability for road, housing, and other construction.

The Soil Profile

Most soils have distinct layers known as horizons. Variation in characteristics and appearance of these layers—referred to collectively as the soil profile—distinguishes the different soil types. In mature or developed soils, the different horizons are quite distinct. These horizons are caused by the interaction of precipitation, temperature,

FIGURE 8.1   Soil profile showing the different layers (courtesy of McKnight 1990, p. 290, p. 348, R12, 36 USDA-NRCS).

and biotic materials through time. Soils in wet, humid areas generally have much more horizon development than those in dry, cold areas.

The first horizon near the soil surface is often an accumulation of fresh or partly decomposed organic matter (Figure 8.1). This thin layer of plant litter is referred to as the O horizon. Next is the A horizon, which is a layer of humus (decomposed organic matter) and mineral matter. The more soluble minerals have been washed from this zone by rainwater. Soil organisms and biotic activity are at a maximum in this zone. Some soils have an E (also called A2) horizon which is a zone of maximum leaching of soluble minerals. The top soil ends and the subsoil begins where B horizon adjoins the A horizon. The B horizon is a zone of accumulation of minerals leached out of the topsoil. Here the original parent materials are greatly altered and clays have tended to accumulate. The true soil ends at the C horizon which is a layer of unconsolidated mineral materials and rock fragments not greatly modified by living organisms. The final zone is an impenetrable layer of bedrock referred to as regolith.

Soil classification centers around the nature of these various horizons. The suitability of land for production of crops, timber, and other human needs primarily involves management of the O and A horizons where the nutrients essential to life are present and readily available.

Soil Formation

Soil formation is initiated by the process of weathering. Two types of weathering occur, which include mechanical and chemical weathering. Mechanical weathering is

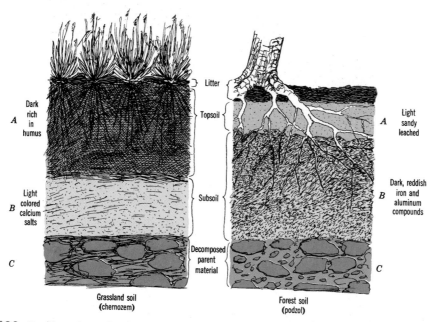

FIGURE 8.2 Profiles of grassland and forest soils. (From R. F. Dasmann, 1976, *Environmental Conservation,* 4th Edition, copyright © 1976, John Wiley & Sons, Inc., New York. Reprinted by permission of John Wiley & Sons, Inc.)

the breaking down of hard rocks by alternate wetting and drying, and by frost action. The process of chemical weathering causes the rocks to be dissolved and structurally weakened primarily by carbonic acid. This acid is formed from the carbon dioxide in the atmosphere and water as follows: $CO_2 + H_2O = H_2CO_3$.

Although this acid is chemically weak, it can dissolve minerals from rocks and bring them into solution as carbonates. Many of these carbonates are associated with important inorganic nutrients required by plants. As carbonates are dissolved the nutrients also are dissolved and become available for plant uptake. Much of the earth's surface is composed of carbonate rock such as limestone and dolomite.

Presently it is well established that climatic conditions and parent material physical and chemical characteristics have considerable influence on weathering. Climate has more effect than any other factor on the rate and kind of weathering that will take place. Weathering rates are generally most rapid in warm, wet, tropical areas. Dry areas are usually characterized by mechanical weathering, which causes only a reduction in particle size. Chemical weathering is the main type occurring in moist areas. This type of weathering may also bring a change in chemical composition.

Soil formation is a function of six factors, which include climate, topography, parent material, biological materials, and time. Climate is probably the most influential of these factors because temperature and precipitation determine the rate of physical and chemical reactions.

Arid and desert soils have little profile development compared to soils of the Tropics. Climate also has a major influence on the living organisms that are associated with a particular area. Living organisms contribute organic matter to the soil. Plant roots strongly influence soil structure. Soil animals such as earthworms play an important role in mixing soil materials. The chemical composition of plant parts affects soil chemical properties. Soils occupied by conifers are usually acid in pH while grassland soils are slightly basic (Figure 8.2). Parent material can control the rate of weathering and the natural vegetation that occupies the area. Limestone, for example, is generally high in bases and can delay acidification in high rainfall areas. Topography influences soil development by modifying climate. This is mainly through its effect on natural erosion. Rolling or steep topography is much more subject to erosion than smooth, flat country. Basins where water collects often have rapid soil accumulation and formation. It may be better to consider time as a dimension rather than a soil-forming factor. This is because time is the period in which the other four factors have to interact. Volcanic soils of the Pacific Northwest show lack of development compared to Southeastern soils with the same precipitation because they are much younger.

Soil Characteristics

Texture.   *Soil texture* refers to the size of the mineral particles comprising the soil. Soil particle sizes, from the smallest to largest, are clay (less than 0.002 mm), silt (0.002 to 0.02 mm), fine sand (0.02 to 0.2 mm), coarse sand (0.2 to 2.0 mm), fine gravel (2 to 5 mm), and coarse gravel (more than 5 mm) (Figure 8.3). Based on the proportions of these particle sizes, a soil can be classed as sand, loamy sand, sandy loam, silt loam, clay loam, or clay.

To a considerable degree, soil texture determines the fertility of the soil. Soils with a high clay content retain nutrients such as nitrogen, phosphorus, and potassium compounds much better than do sands. This is because the small particles have a much greater surface area for attracting and binding nutrients per unit volume than do large particles.

Soil texture also plays an important role in determining the soil moisture content. Water enters coarse, sandy soils much more rapidly than fine clay because there is more space between particles. On the other hand, clay soils retain water much better than do sandy soils. The best balance between moisture infiltration and retention is obtained with loamy soils, which have a mixture of sand, silt, and clay.

In arid regions such as the western United States, sandy soils provide a more favorable habitat for plants, particularly grasses, than do clay soils. This is because the limited precipitation that occurs rapidly infiltrates the soil rather than moving overland. In the higher-rainfall areas (over 900 mm of annual average precipitation) east of the Mississippi River, moisture and nutrient retention by soils has more influence

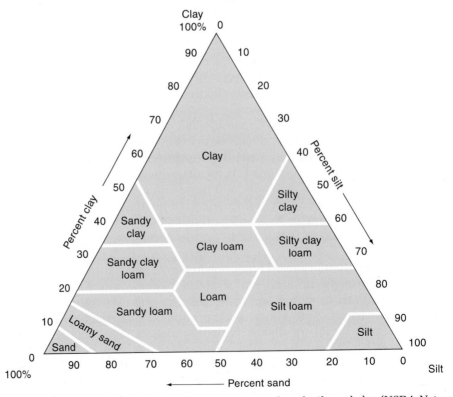

FIGURE 8.3   Soil texture designation based on percentages of sand, silt, and clay (USDA-Natural Resources Conservation Service).

on vegetation productivity than rate of water infiltration. In these areas, soils high in clay are more productive than those high in sand.

Structure. Soil *structure* refers to how soil particles are arranged. Soil structure is quite important because it determines the rate at which water can enter the soil (Figure 8.4). Soil is classified in six basic types of soil structure: platelike, prismlike, blocklike, spheroidal, single grained, and massive. A massive soil structure is tightly compacted with little or no space between soil particles for infiltration and percolation. This structure type often occurs on overgrazed areas or areas compacted by excessive vehicle traffic. Single-grained structure refers to a loose arrangement of individual soil particles that do not stick together, such as sands. Platelike structure refers to soil particles aggregated together into leafy plates. This type of structure can result naturally from parent material or soil-forming forces. Water infiltration is minimized with this type of structure. Prismlike structure refers to soil particles aggregated in vertical columns. This type of structure is most common in arid areas. When the tops are round, this type of structure is referred to as *columnar.* Columnar soils in the West are often associated with a high sodium content and unfavorable soil-water relations. Soils with blocklike structure have a cubical aggregation. The blocks have sharp edges and are found in the subsoil. The spheroidal soils have the best characteristics for water infiltration and storage. The aggregates are found and easily shaken apart. Soils high in organic matter usually have a spheroidal structure. This type of structure is associated with grassland areas.

Depth. Soil depth influences rooted plant productivity because it determines how much moisture the soil can hold. A deep soil in an area with moderate precipitation will often produce more than a thin soil receiving heavy precipitation. Conversely, in some arid areas, grasses on thin, sandy soils underlain with a hard impermeable layer of calcium carbonate (caliche) are more productive than grasses on deeper sandy soils. The impermeable layer restricts moisture to the portion of the soil profile near the surface where it can readily be used by the fibrous roots of the grasses. Depth usually means distance from the soil surface to bedrock or the unconsolidated

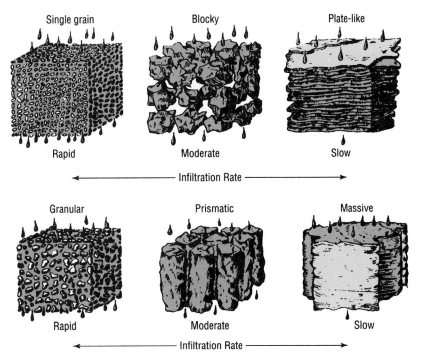

FIGURE 8.4 Note the variety of shapes occurring in soil aggregates and the relative rates at which water can move through them (from Owen and Chiras 1995).

material. Soil depth, in conjunction with texture and structure, largely determines the potential of a site for cultivation.

pH. Soil pH indicates the status of the soil in regard to exchangeable mineral ions. The H in pH is for hydrogen and the pH value expresses whether a soil is basic or acidic. The proportional weight of the free hydrogen ions in the soil solution is given by the actual numerical value. The pH value is a logarithmic expression and is demonstrated as follows: pH 1 = one hydrogen ion per 10,000 ions, pH 2 = one hydrogen ion per 100,000 ions, pH 3 = one hydrogen ion per 1 million ions, and so on. The scale ranges from 1 to 14. A highly acid soil would have a pH of 4, while a highly basic soil would have a pH of 10. Minerals are most available to plants in soils with pH values between 6 and 8. A pH value near 7, which is the pH of pure water, is considered ideal for the growth of most plants. However, many species of plants grow best where soils are either highly acidic or basic.

Most soils in the western United States are slightly basic because they are derived from calcareous parent material and receive low precipitation. In contrast, most eastern soils are slightly acidic. The low rainfall in the West results in little leaching of soil minerals and the associated base losses. *Leaching* refers to the downward transport of dissolved soil minerals as water percolates through the soil. Soils of the Southeast are the most heavily leached in the United States. South America has large areas of soils called Oxisols, which are the most heavily leached soils in the world.

Organic Matter. Soil organic matter represents an accumulation of partially decayed and partially synthesized plant and animal residues. It usually represents a relatively small part of the soil (< 1 to 6%) and is found primarily in the upper 30 cm of soil. Partially decomposed organic matter that has been incorporated into the soil is called humus. Humus provides a constant, although small, supply of nutrients to plants. More important, it serves as a binding site for cations and prevents them from leaching out of the soil profile. Three other important functions of humus include binding of soil particles together, increasing soil moisture-holding capacity, and providing food for microorganisms. For these reasons, it is important to maintain as much humus in the soil as possible. Grassland soils are generally much higher in organic matter than are forest soils. Wetland soils are exceptionally high in organic matter because decay is slowed by anaerobic and toxic conditions.

Fertility. Next to water, fertility is the factor that most limits crop production. On high-rainfall ecosystems (over 1,000 mm of annual precipitation), soil fertility can be a more limiting factor to crop production than precipitation. From the standpoint of plant growth, nitrogen is the most deficient element in most soils. Farmland soils are generally deficient in phosphorus and some in potassium and sulfur. Soils in high-rainfall areas are heavily leached and have low levels of elements required by plant crops, such as copper, cobalt, magnesium, sodium, and zinc. Plants growing on these soils often have inadequate levels of these elements to meet crop requirements. Selenium and molybdenum are two elements that cause both toxicity and deficiency problems for livestock depending on their levels in the soil. Low soil fertility is a major problem limiting forage and livestock production in tropical areas of Africa, Asia, Central America, and South America. Forest soils are generally much more leached and less fertile than grassland soils.

Soil Classification

The soil classification system presently in use was developed by the Soil Survey Staff of the U.S. Department of Agriculture. It is called the "Comprehensive Soil Survey System." The major features of this system are based on the characteristics of soils as they are found in the field. It classifies soils rather than soil-forming processes. Soil names give the major physical characteristics. The levels of classification in this system are as follows: (a) Order, (b) Suborder, (c) Great Group, (d) Subgroup, (e) Family, and (f) Series. This system uses diagnostic surface and subsurface horizons as the basis of soil classification.

The eleven orders of the present soil classification system are discussed below.

Entisols—These are soils found in the Rocky Mountain area of the United States. They are the youngest of soil orders and are lacking in horizon development.

Inceptisols—These soils are slightly more advanced in profile development than Entisols. They are formed from volcanic ash and are found primarily in the Pacific Northwest.

Andisols—These soils are formed on volcanic ash and cinders near volcanic sources (Brady and Weil 1996). They are not highly weathered. They are young soils with unique properties due to their mineral composition and high percentage of organic matter (see Brady and Weil 1996).

Aridisols—These are Southwestern desert soils with little profile development. They are dry more than six months of the year.

Vertisols—These soils have a high content of swelling clay, which shrinks and cracks when wet. They are found mostly in the southcentral United States.

Mollisols—These are the natural grassland or prairie soils. They typically are deep, have a high percentage of organic matter, a high base supply, and moderate profile development.

Alfisols—Alfisols are similar to mollisols but have a higher percentage of organic matter. These soils are associated with the Eastern deciduous forest.

Spodosols—These are leached mineral soils with a distinct organic horizon and a developed B horizon, but with less accumulation of clays than Ultisols. These soils are found primarily in the northeastern and northcentral United States, and generally support coniferous forest.

Ultisols—Ultisols are the most highly leached soils in the United States. They are found primarily in the southeast. These soils are moist much of the time and have a B horizon of clay accumulation, a leached A horizon, and a low base supply. They usually support forest or savannah woodland.

Oxisols—These are the most developed of all soils. They occur in tropical areas and are almost absent from the contiguous United States. Oxisols have the lowest fertility of all orders. They are the least suited for farming.

Histosols—These hydric soils have a very high percentage of organic matter and are often referred to as peat. They are found in diverse locations where wetlands occur.

Different types of vegetation are associated with individual soil orders (Figure 8.5). A map of the major soil orders in North America is shown in Figure 8.6.

We will review the six major soil orders in greater detail. Our discussion of the soil orders follows Brady (1974) and Brady and Weil (1996).

Mollisols.   These are the world's most important agricultural soils. They have moderate profile development and the surface horizons have a granular or crumb structure that does not harden when the soils are dry. This explains their name, which comes from the Latin word *mollis* meaning soft.

Generally mollisols develop under grassland or prairie climatic conditions. They dominate the central Great Plains of the central United States (North Dakota, South Dakota, Nebraska, Kansas, Oklahoma, Texas). However, sizable portions occur further west in the palouse region of Washington, Oregon, and Idaho. Much of the heartland of the former Soviet Union is characterized by mollisols, which also dominate parts of South America (northern Argentina, Paraguay, Uruguay) and Asia (northern China, Mongolia).

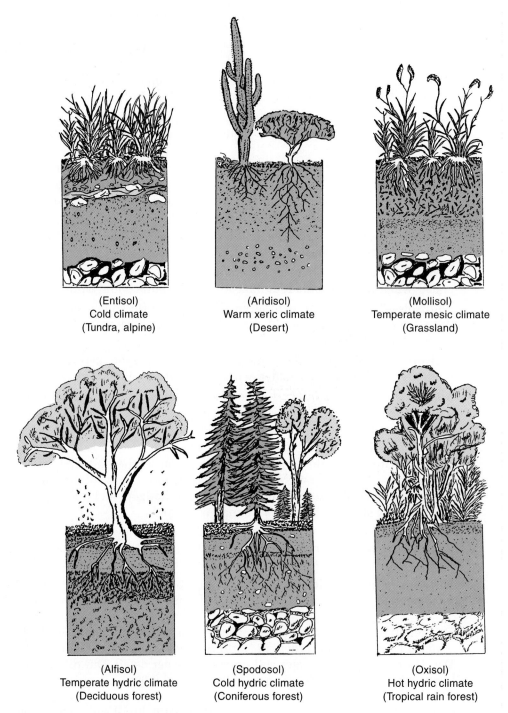

(Entisol)
Cold climate
(Tundra, alpine)

(Aridisol)
Warm xeric climate
(Desert)

(Mollisol)
Temperate mesic climate
(Grassland)

(Alfisol)
Temperate hydric climate
(Deciduous forest)

(Spodosol)
Cold hydric climate
(Coniferous forest)

(Oxisol)
Hot hydric climate
(Tropical rain forest)

FIGURE 8.5 Soil profiles of the major soil orders typically found in six different biomes (adapted from Miller 1990).

Mollisols are considered the very best soils in the world for grains (corn, wheat, sorghum, barley, oats) and many other crops because of their high natural fertility. These soils are characterized by generally moderate rainfall (15–30 inches) that arrives mostly in the spring and summer. Periodic drought (about 1 year in 10) can cause crop failure, but when properly managed these soils can produce bumper crops.

Alfisols.   Alfisols rank second only to mollisols in terms of suitability for crop production. In the United States, these soils occur east of the Great Plains in the Mid-

FIGURE 8.6 Major soil orders of North America (from Owen and Chiras 1995).

western states (Iowa, Illinois, Indiana, Ohio, portions of Minnesota, Michigan, Pennsylvania, New York). In other parts of the world, Alfisols are prominent in central China, much of Europe, southern Africa, eastern Brazil, and southern Australia. Generally Alfisols are characterized by a temperate climate with 30–50 inches annual precipitation fairly evenly distributed throughout the year. Deciduous forests (such as beech, maple, oak, and hickory trees) are the natural vegetation associated with Alfisols.

Alfisols differ from Mollisols in being a bit more leached but less droughty. Early pioneer farming depended heavily on these soils. Corn, potatoes, tobacco, various fruits, various vegetables, and a wide range of forages grow well where Alfisols are properly managed.

Spodosols.   Spodosols are mineral soils differentiated from Alfisols by being colder and more leached. They naturally support coniferous forests (pines, firs). They have distinct A horizons depleted of soluble minerals, but with an accumulation of acidic organic matter at the soil surface. The subsurface B horizon has an accumulation of iron compounds and organic materials. Generally the leached A horizon is light gray or tan while the lower B horizon is a distinct brown. Spodosols characterize the northeastern United States and much of central and eastern Canada. Other large areas include northern Europe, Siberia, and the most southern parts of South America.

Spodosols are naturally infertile. In the early 1900s, many small farms throughout New England and northern Michigan, Wisconsin, and Minnesota were abandoned because of low fertility and a short growing season. Most Spodosols are best suited for timber production. However, when properly fertilized, some Spodosols can be quite productive. Examples are the potato growing region of northern Maine and vegetable producing area in Michigan and Wisconsin.

Ultisols.   Ultisols are highly weathered, acidic soils that developed under warm, consistently moist climatic conditions where annual precipitation is 40 to 70 inches per year. Ultisols are less acidic than Spodosols and less leached than Oxisols. The natural vegetation occurring on Ultisols is forest or savannah woodland. Ultisols dominate the southeastern United States, but are also found in Hawaii and western portions of California, Oregon, and Washington. They are prominent in eastern Australia, southern China, southeast Asia, Paraguay, and southern Brazil.

Although low in fertility, Ultisols can be productive for certain crops when properly managed. Some of these crops include cotton, corn, tobacco, peanuts, and various cultivated forages for livestock (Owen and Chiras 1995). Because of abundant rainfall, a long growing season, and good tillage characteristics, these soils rank third after Mollisols and Alfisols in terms of suitability for agriculture. In the southeastern United States, many areas of Ultisols are hilly and highly erodible. These areas are most suited for wood, forage, and wildlife production.

Oxisols.   Oxisols are the most weathered and wettest of the 10 soil orders. Annual precipitation is typically in excess of 1,200 mm per year. These are the soils of the tropical rain forest, and occur only in Florida in the contiguous United States. The largest areas of Oxisols occur in South America (northern Brazil and neighboring countries) and Africa.

One of the biggest controversies in natural resource management is the conversion of tropical rain forest into agricultural land. Although it would seem these areas could be productive farmland, most agriculture has been unsustainable because of erosion, infertility, and poor soil-tillage properties. Under improved management it may be possible to overcome these problems in certain areas. An alternative approach is to better use natural products from the rain forest such as rubber, cocoa, various tropical fruits, and tree extracts effective in treating human diseases.

The high iron content of Oxisols gives them unique properties. Initially after the trees are removed by burning, Oxisols are moderately productive because of the release of nutrients locked up in the vegetation. However, after a few years the organic matter and nutrients are depleted leaving a high proportion of iron oxides and aluminum. In the hot equatorial sun these materials quickly bake into a hard bricklike substance referred to as laterite. These oxisol "bricks" make great construction materials for buildings because of their hardness and durability (Owen and Chiras 1995). On the other hand, once Oxisols turn into laterite they become almost useless for plant production. Management of the rain forest is one of the great dilemmas that confronts humanity as we enter the twenty-first century.

Aridisols.   Aridisols, as their name implies, are associated with dry, desert-like climates. In the United States they dominate the Southwest (Mojave, Sonoran, and Chihuahuan deserts). In other parts of the world they occur over large regions of China

(Gobi and Taklamakan deserts), the former Soviet Union (Turkestan desert), Africa (Sahara desert), the Middle East, and Australia. In their natural state these soils are dry for over six months of the year and are high in nutrients because they lack leaching. Generally plants are widely spaced in desert areas where wind and erosion are often severe. This results in a layer of stones called "desert pavement" on the surface of many Aridisols (Owen and Chiras 1995).

Aridisols are unsuited for cultivated crops without irrigation. Historically these lands in Africa and the Middle East have been grazed by livestock, particularly sheep and goats, under nomadic herding. Generally productivity is low and rapid human population increase has lead to severe degradation or "desertification" of many desert lands.

When irrigation water becomes available, Aridisols can be productive for high value crops, such as citrus, dates, celery, and nuts, because of their high nutrient content (Owen and Chiras 1995). Over time soluble salts (particularly sodium) may accumulate on the surface of these soils. This process, known as salinization, can be mitigated through management practices involving leaching or flushing, improved drainage, conversion of sodium salts to other compounds, reducing evaporation through improved irrigation practices, and the use of salt-resistant crops.

Soil Erosion

Soil erosion is the detachment and movement of soil particles from one place to another (Miller 1990). Wind and water are the primary forces causing soil erosion. Some geologic or natural erosion always occurs because of these forces. With the exception of desert areas, natural plant vegetative cover and roots keep the loss of soil in balance with topsoil replacement (Figure 8.7).

Soil erosion is categorized into four types: splash, sheet, rill, and gully (see Brady 1974 and Brady and Weil 1996). Splash erosion is the spattering of small soil particles caused by the impact of raindrops on very wet soils. The loosened particles may or may not be removed by surface runoff. Sheet erosion involves wide flow of water moving across a sloping field. This kind of erosion typically occurs when a recently tilled and leveled piece of farmland with no protective plant cover is subjected to heavy rainfall. Rill erosion results from water moving through small channels in the

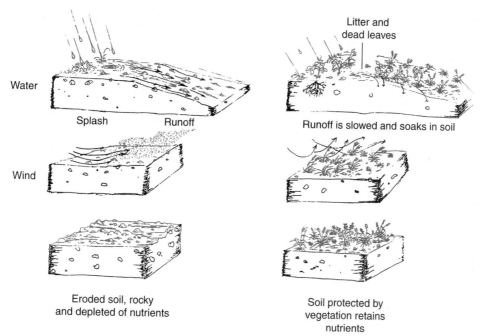

FIGURE 8.7 Vegetation and organic litter protect soil from all forms of erosion (drawing by John N. Smith, based on Nebel 1981).

FIGURE 8.8   Sheet and rill erosion on an unprotected slope. Serious gullying is imminent if protective measures are not instituted soon. Strip cropping or even terracing should be employed if cultivated crops are to be grown. Perhaps better, the field could be seeded and put into the USDA conservation reserve program (courtesy U.S. Soil Conservation Service).

FIGURE 8.9   Severe gully erosion in southern New Mexico as a result of overgrazing by livestock.

soil surface (Figure 8.8). This kind of erosion occurs mostly on farmland that is partially tilled with some vegetative residue or on moderately overgrazed rangeland where plants of low palatability remain to hold the soil in place. Gully erosion occurs where water is concentrated into a high-velocity flow through a large channel that lacks vegetation. This type of accelerated erosion commonly occurs in hilly areas that are heavily grazed or farmed. When human activities such as farming, mining, wood cutting, grazing, and highway construction and other disturbances remove natural plant cover, soil erosion can accelerate to rates that threaten human health and destroy land productivity (Figure 8.9).

Accelerated erosion can severely reduce the productivity of croplands, woodland, and grazing land because the topsoil contains most of the plant nutrients and

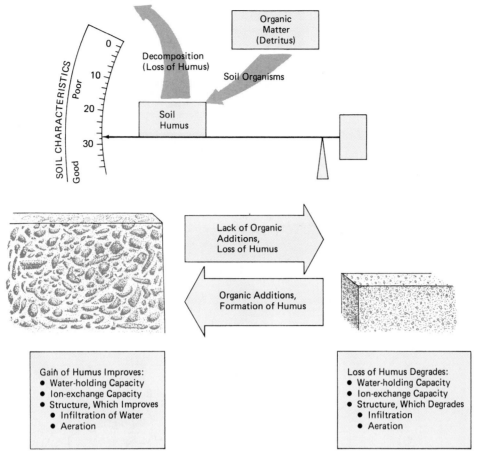

FIGURE 8.10 The amount of humus present in the soil is the result of a balance between addition of organic matter and oxidation of humus. Physical and chemical characteristics of soil improve or degrade according to amount of humus present (from Nebel 1981).

has the greatest water-holding capacity (Figure 8.10). Suspended sediment, which collectively refers to the dislodged soil particles, is the largest source of water pollution. It also causes considerable economic damage by clogging rivers, irrigation ditches, and reservoirs with sediment deposits.

Topsoil is the layer of soil typically moved in cultivation (Brady and Weil 1996). It is a renewable resource because it is constantly being formed by natural processes. It takes somewhere between 220 and 1,000 years to form an inch of topsoil, depending on precipitation and temperature (Miller 1990). Warm, wet, and flat areas have more rapid rates of soil formation than areas where the opposite conditions occur.

Soil erosion problems are greatest on farmlands (Figure 8.11). However, about half of the erosion in the United States is caused by other activities such as grazing, mining, logging, and urban development. Construction sites have greatly increased throughout the United States during the 1990s. The first thing developers and builders usually do on construction sites is to remove all vegetation. Frequently little is done to prevent erosion until the project is complete, which may take several years. Erosion rates on shopping center and highway construction sites can be 100 to 200 times the natural rate.

Soil Erosion in the United States

The United States has been through several alternating periods of both high and low rates of soil erosion on its croplands. These periods have been linked to socioeconomic conditions in the country. During periods of war, demand for agricultural products

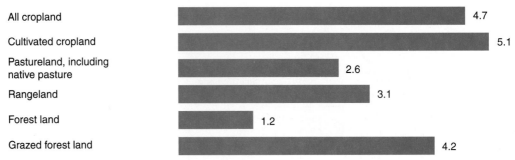

FIGURE 8.11   Average annual erosion rates on various types of land in the United States (courtesy of Natural Resources Conservation Service).

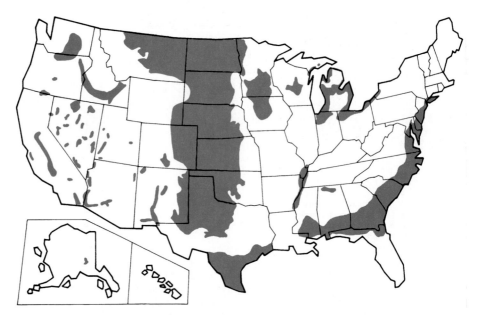

FIGURE 8.12   Cropland areas of the United States with the highest potential for wind erosion (courtesy of USDA-Natural Resources Conservation Service).

accelerates and farmers have strong incentives to maximize production. During these periods, soil conservation practices have been discarded and erosion rates have been excessive. World War I largely created the conditions that led to the infamous "Dust Bowl" era of the 1930s. Because agricultural production was greatly reduced in Europe during the World War I era (1914–1918), prices for most crops skyrocketed. Because the United States was not directly affected by the war, it became the major supplier. Much of the western Great Plains was plowed in areas unsuited for sustained cultivation. Although crop prices decreased somewhat during the 1920s, farmers continued to intensively crop their lands. Generally the land was plowed after each harvest and left bare without strip cropping, crop rotation, or the use of wind breaks. At the same time, heavy livestock grazing destroyed the grass on large areas of rangeland that could not be farmed because of rocky soils on sloping terrain.

Disaster finally struck in the form of concurrent drought and depression in the early 1930s. In 1934, crop prices had dropped over 30% from 1929 levels. It also had the distinction of being the driest year of the century. In the worst portions of the Dust Bowl (Oklahoma panhandle, southwestern Colorado, and southeastern Kansas) (Figure 8.12) hot, dry windstorms caused dust clouds so thick that darkness occurred in midday. Several human fatalities resulted from breathing the dust-filled air. Dead birds and rabbits commonly were found in the wake of these storms. In the spring of 1934, topsoil from the Great Plains was carried across eastern United States to far out in the Atlantic Ocean (Figures 8.13 and 8.14). New York City was literally covered

FIGURE 8.13 A dust storm during the early 1930s in southeastern Colorado. Such storms were the combined result of a natural drought cycle and inappropriate plowing of the western Great Plains. (courtesy of U.S. Government).

FIGURE 8.14 Land in Oklahoma severely damaged by wind erosion during the Dust Bowl era of early 1930s (courtesy of U.S. Government).

with dust. In May 1934, congressional action was taken after dust from the Great Plains leaked into a Washington conference room where legislators were considering programs to deal with the problem (Miller 1990). The USDA Soil Conservation Service (now the Natural Resources Conservation Service) was established in 1935 to promote conservation practices on private lands in the Great Plains. This started an era of improved farmland and rangeland management that lasted until the late 1960s.

The era of improved soil conservation was followed by a period of severe soil erosion that began in the early 1970s. This resulted primarily from mismanagement of the U.S. economy and adverse climatic conditions in several parts of the world. In order to pay for the Vietnam War and finance its social programs, the U.S. government under Presidents Johnson and Nixon decided to print more money rather than borrow it or raise taxes. This devalued the dollar against most foreign currencies, making U.S. farm products relatively inexpensive. In 1972, the Soviet Union started importing food from the U.S. rather than rationing food to make up for production shortfalls. These factors in combination greatly increased crop prices and encouraged plowing of erodible farmlands during the 1970s. However, the 1980s ushered in a new set of economic policies after Ronald Reagan was elected president. The Federal Reserve Board switched from a loose to a restrictive monetary policy, which brought inflation under control and raised the value of the dollar. During most of the 1980s, world climate conditions were favorable for crop production. At the same time, yields of many crops were greatly increased through new technology. By 1984 serious problems were associated with both huge agricultural surpluses and severe soil erosion in many parts of the Great Plains.

In 1985, the U.S. Congress passed a new farm bill (Food Security Act) that had several provisions to retire erodible lands and encourage soil conservation. Over the past 10 years soil erosion has again been contained in the United States. Another potential dust bowl was narrowly avoided. During the early 1980s there were increasing reports of severe dust storms on the western edge of the Great Plains. A severe drought occurred over most of the Great Plains in 1988, but by then the 1985 Food Security Act had returned most of the erodible lands to perennial grasses.

Although estimates vary, USDA-NCRS survey data indicate U.S. croplands have lost an average of about 30% of their original topsoil. This loss may cost the country as much as $30 billion per year. Until the 1985 Food Security Act, about one-fourth of the nation's cropland accounted for about two-thirds of the annual soil loss (Figure 8.15). However, some 30 million acres of these erodible lands have been returned to permanent grasses and soil erosion losses have been cut by over half. Under present policies (1985, 1990, and 1996 Farm Acts), we do not consider soil erosion of cropland to be a serious threat to food security or human health in the United States, although local problems still exist. New farming technologies show potential for overcoming some of the topsoil loss that occurred in the past. We consider the conversion of prime farmland into urban uses to be the most serious threat to sustaining high-quality food production at low prices.

CONTROLLING SOIL EROSION

Generally the key to controlling soil erosion is to maintain a good vegetative cover. Farming practices used to contain erosion include conservation tillage, strip cropping, contour farming, terracing, gully reclamation, windbreak planting, and the retirement of highly erodible lands from cultivation. We will summarize each of these practices, but refer the reader to Brady and Weil (1996) for in-depth coverage of soil conservation techniques.

Conservation Tillage

In order to develop this subject, we will describe conventional tillage and alternative procedures. Our discussion follows Miller (1990) and Brady and Weil (1996). Generally one pass is made over the field with a moldboard plow that inverts crop residues in the upper 15 inches of soil. After a disk is then used to break up the large clods, the field is harrowed to break up the small clods and smooth the surface. After planting, a cultivator is sometimes used to remove weeds that compete for nutrients and moisture and can be a source of contamination when harvest occurs.

An alternative to this system is conservation tillage farming. It is also known as minimum tillage or no-till farming depending on the degree of soil disturbance (Miller 1990). Under this approach the new crop is planted into the residue of the previous crop with only partial disruption of the soil surface (Figure 8.16). With min-

Soil Erosion as a Multiple of the Tolerable Rate (T), 1982

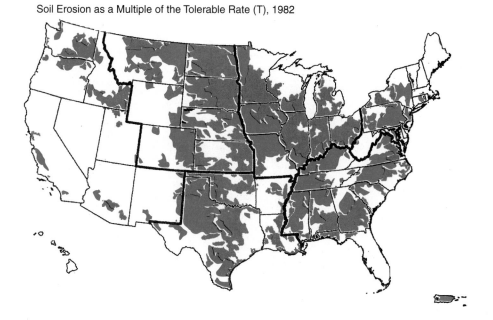

Soil Erosion as a Multiple of the Tolerable Rate (T), 1992

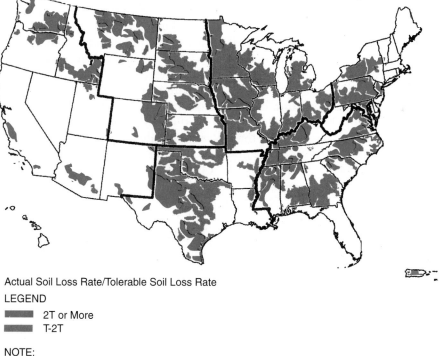

Actual Soil Loss Rate/Tolerable Soil Loss Rate

LEGEND

■■■■■■ 2T or More
■■■■■■ T-2T

NOTE:
Average annual soil erosion by wind and water where cultivated
cropland is greater than 5% of land area.

FIGURE 8.15   Soil erosion as a multiple of the tolerable rate in 1982 and 1992. The area designated in blue is eroding at more than two times the tolerable rate. The area in gray is eroding from one to two times the tolerable rate (courtesy of USDA-Natural Resources Conservation Service).

imum tillage, a special machine is used to loosen but not turn over the soil surface and previous crop residues. With no-till farming, seeds, fertilizers, and herbicides are injected into slits in the unplowed soil.

Conservation tillage greatly reduces erosion, but it also has other benefits. These include decreased soil compaction, soil water loss, and fuel costs associated with

FIGURE 8.16   An example of no-till farming where a specially designed machine applies seeds, fertilizer, and weed killers all at once (courtesy of USDA-Natural Resources Conservation Service).

tillage. Maintaining crop residues also improves habitat conditions for many types of desirable wildlife such as bobwhite quail, pheasants, various species of songbirds, water fowl, and cottontail rabbits. Yields have generally been similar to those under conventional tillage. Although it varies with soil type and crop, more extensive cultivation is needed about every 3–7 years. The main disadvantage is the increased need for herbicide to control weeds. However, this problem can be minimized with improved selection, application, and timing techniques.

In 1992, conservation tillage was practiced on about 35% of U.S. croplands and will probably be used on over half by the year 2000 (Figure 8.17). The USDA estimates that this has reduced the soil erosion rate by about 20–25%. With no-till farming, around 50–100% of the previous crop residue remains on the soil surface compared to 30–45% with conservation tillage and 0–5% with conventional tillage. The primary factor limiting use of conservation tillage is the greater management skill required of farmers.

Strip-Cropping.   Strip-cropping involves the planting of crops in alternating strips to reduce erosion and pesticide and fertilizer demands (Figure 8.18). It is often combined with crop rotation so that a soil-enriching legume crop (alfalfa or clover) is planted after a soil-depleting grain crop (corn, wheat, barley). The annual crop is often referred to as a row crop and the legume crop is referred to as a cover crop. The alternating pattern of row crops and cover crops reduces water erosion and creates barriers to the spread of pests and diseases. In the arid portions of the western Great Plains (Montana, Wyoming, Colorado), a modification is used where strips of wheat or other small grains are alternated with fallow. Every other year the strip is tilled and planted. Most of this land is marginal for crop production, but with alternate-year planting, enough moisture can be stored to ensure a crop. Maintaining the residue during the fallow period greatly reduces erosion and water evaporation losses from the soil. Herbicides are used to retard weeds during the fallow period. On sloping lands, contour strip-cropping in combination with terraces can reduce soil erosion by 50% or more.

Contour Farming.   In contour farming, crops are planted across rather than up and down sloping land (Figure 8.19). This forms a barrier that helps to hold soil and slow water runoff. Contour farming and strip-cropping are often used in combination.

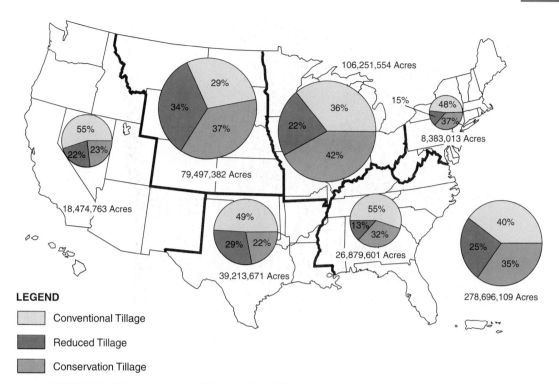

LEGEND

▢ Conventional Tillage

▢ Reduced Tillage

▢ Conservation Tillage

FIGURE 8.17 The percentage of farmland in different regions of the United States under conservation tillage in 1992 (courtesy of USDA-Natural Resources Service).

FIGURE 8.18 A strip-cropping system designed to reduce wind erosion of soil in Montana (courtesy of USDA-Natural Resources Conservation Service).

Terracing. This practice has been used for centuries by societies in Europe, China, India, and South America (Owen and Chiras 1995). In these areas, high human populations relative to the amount of arable land often necessitated farming mountain sides in order to prevent starvation. When terraced, the slope is converted into an alternating series of broad, level benches and sharp or gradual slopes. On gently sloping land (under 10% slope), broad-base terraces are developed to allow farming of the slope (Figure 8.20). On steeper lands (10 to 20% slope), backslope terraces are planted with

FIGURE 8.19   Contour farming of sloping land to protect the soil and grass filter strips to protect water quality in the adjacent stream (courtesy of USDA-Natural Resources Conservation Service).

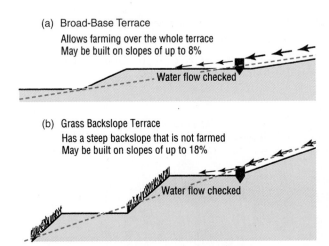

(a) Broad-Base Terrace

    Allows farming over the whole terrace
    May be built on slopes of up to 8%

    Water flow checked

(b) Grass Backslope Terrace

    Has a steep backslope that is not farmed
    May be built on slopes of up to 18%

    Water flow checked

FIGURE 8.20   Ridge terraces. A comparison of the broad-base and grass backslope terraces (from Owen and Chiras 1995).

permanent grass on the slopes. Channel terraces differ from ridge terraces in that a diversion ditch (rather than a flat bench) is dug across the slope to halt and redirect water flow. Although terraces are effective in decreasing soil erosion by reducing the amount and speed of water runoff, they often are incompatible with modern farming methods involving large equipment. In the United States, the approach has been to encourage permanent retirement of most land where terraces would be required. In mountainous developing countries, such as Nepal and Peru, failure to apply terraces on newly cultivated lands is leading to permanent damage from soil erosion.

Gully Reclamation.   Overland flow of water, if not retarded by vegetation, can quickly form gullies (large water-erosion channels) on sloping land (Miller 1990). Recently formed gullies are a major warning signal of a severe soil erosion problem. Generally, retarding erosion on surrounding uplands is more effective than restoring the gully itself. Once this is done, gullies can be reclaimed by seeding quick-growing annual grains such as wheat, oats, rye, and barley. Small temporary dams can be built with hay bales or old tires to check water flow and collect silt in the deeper, more quickly-forming gullies. Once a soil cover is established, perennial grasses and rapidly growing shrubs and trees should be planted to permanently stabilize the soil. Smaller diversion channels can be used to disperse water away from the gully.

Shelterbelt
30-40 Ft. High

FIGURE 8.21 The effect of a windbreak on wind velocity (drawing by John N. Smith).

Windbreaks. Windbreaks or shelterbelts are rows of trees planted perpendicular to the prevailing wind direction to protect soils, crops, homesteads, and more from wind and snow (Brady and Weil 1996). They disrupt the movement of wind and reduce the exposure of soil to its shocking force (Figure 8.21). The shading effect of windbreaks reduces the rate of soil moisture evaporation. Windbreaks are especially effective in reducing erosion in areas with sandy soils such as the Southern Great Plains of Kansas, Colorado, Oklahoma, and Texas. Farm residences in this area are often ringed with windbreaks. They protect the home from wind damage, and reduce the cost of heating in winter and cooling in summer. They provide high-quality habitat for various birds, pest-eating and pollinating insects, and other types of wildlife. Windbreaks add esthetic diversity to plains landscapes that are otherwise uniform and monotonous (Figure 8.22).

After the 1930s Dust Bowl, the USDA Soil Conservation Service implemented an aggressive windbreak planting program on farmlands of the Great Plains. Unfortunately, many of these windbreaks were removed during the 1970s when high crop prices encouraged farming all tillable land. Large farm equipment and irrigation systems became vogue. Following enactment of the 1985 Food Security Act, windbreak plantings have increased in a few areas.

Retirement of Erodible Lands. Land retirement from farm use is an approach to reducing soil erosion that goes back to the drought years of the middle 1930s when the federal government bought out farmers on erodible land in the Great Plains. The National Grasslands administered by the USDA-Forest Service were acquired from these buyouts. In the 1950s, a program called the "Soil Bank" was implemented by

FIGURE 8.22 Windbreaks, or shelterbelts, reduce erosion on this farm in Trail County, North Dakota. They also reduce wind damage, help hold soil moisture in place, supply some wood for fuel, and provide habitat for wildlife (courtesy of USDA-Natural Resources Conservation Service).

the USDA to control soil erosion and reduce food surpluses. It retired about 20 million acres of land from farm use, most of which went back into production in the 1970s when crop prices skyrocketed. The most recent land retirement program was initiated in 1985 when congress passed the Food Security Act. It called for removal of 45 million acres of marginal cropland into a conservation reserve for a 10-year period. This is about 10% of the nation's cropland base. The primary intent was to reduce food surpluses, but soil erosion reduction and wildlife habitat provision were also objectives. Under this program, the farmer contracts with the USDA to withdraw erodible cropland for 10 years and establish a permanent vegetation cover (perennial grasses, shrubs, trees, legumes) that is not to be hayed or grazed unless a drought or some other unusual circumstance occurs (Figure 8.23). The average farmer has received about $50 per acre per year (Knutsen et al. 1998). This program has cost tax payers about 1.8 billion dollars per year. Generally it is considered a success in terms of controlling erosion and improving habitat for wildlife. However, critics consider it too costly and contend it could have been implemented more efficiently. Under the 1996 Food Security Act the Conservation Reserve Program was retained with provisions to encourage further retirement of highly erodible lands and better reward farmers who seeded their lands with plants favoring wildlife.

Maintaining Soil Fertility

In the United States and other developed countries, commercial inorganic fertilizers are primary means for maintaining soil fertility. In developing countries, animal manures and human waste (sometimes referred to as night soil) are still commonly used to fertilize crops. The use of inorganic commercial fertilizers throughout the world has increased by 10- to 15-fold since 1950 and about 20% over the past 10 years (World Resources Institute 1994). They have had the advantages of being easy to transport and apply. It is estimated they have boosted annual food production by 35–40% over the last 50 years. Farmers commonly have their soil tested to determine the proper mix of nutrients. Nitrogen, phosphorus, potassium, and sulfur products are the common forms of inorganic fertilizers. However, other plant nutrients are often added based on soil tests.

FIGURE 8.23 Erodible cropland retired from farming in northeastern Oregon under the Food Security Act of 1985. It has been reseeded with perennial grasses.

The biggest drawback to inorganic fertilizers is that they do not add organic matter (humus) to the soil (Miller 1990). Without periodic additions of organic matter, the soil's ability to hold water and nutrients will decrease. This results from the lower cation exchange properties of soil minerals compared to organic matter, and decreased soil porosity when organic matter is depleted. Inorganic fertilizers also cause water pollution, particularly in areas with sloped land or a water table near the soil surface.

Periodic application of animal manures, green manures, and compost can reduce the problems associated with inorganic fertilizers. Animal manures increase organic nitrogen content, improve soil structure, and stimulate soil microbial growth.

Green manure is green vegetation that is grown to be plowed into the soil to increase organic matter (Miller 1990). Although it often involves legumes such as alfalfa and clover, it can be weeds and grasses in an uncultivated field.

In many areas of the United States, the application of animal manures is impractical today. This is because mixed crop-animal operations of the past have been replaced by more specialized enterprises. Often, manure transportation costs between feedlots for livestock and crop culture are excessive. Therefore some farmers have turned to compost (plant wastes such as cuttings and leaves) as an alternative to manure.

Crop rotation is another means of preventing soil nutrient depletion (Miller 1990). Crops such as cotton, wheat, corn, and tobacco can mine the soil of nutrients if grown year after year in the same field. However, if these crops are rotated with legumes and forage grasses, soil nutrients can be restored and erosion minimized.

THE USDA NATURAL RESOURCES CONSERVATION SERVICE

A primary accomplishment of the NRCS (National Resources Conservation Service) has been the development of a land capability classification system and detailed mapping of soils throughout the United States (Owen and Chiras 1995). Under the NRCS land capability classification system there are eight categories into which land can be placed (Figure 8.24). Class I land is flat, fertile, and well suited for crop

FIGURE 8.24 Classification of land according to capability using the USDA-Natural Resources Conservation Service approach (courtesy of USDA-Natural Resources Conservation Service).

production. Classes II and III are suitable for growing crops but require soil erosion control measures. Classes IV through VII are rangelands and woodlands suitable for grazing and tree production. Class VIII land is suitable only for wildlife habitat, wilderness, and/or recreation due to excessive slope, rocky soils, and/or low fertility. The intent of this system is to help farmers ranchers, foresters, urban planners, and so on best use different types of lands. In the past, however, it helped relegate wildlands to the least productive category. Now significant examples of ecosystems that once occupied intensively farmed areas have become scarce.

The NRCS also helps farmers and ranchers develop conservation plans for their lands. Four steps are involved in development of this plan (Owen and Chiras 1995). They include first conducting an intensive survey of the farm or ranch in which each unit (acre) is mapped on the basis of its capability category. Second, a management plan is developed by the farmer/rancher and a NRCS technician. Under this plan, decisions are made for each land unit regarding use and soil protection measures. Alternative uses might include farming, livestock grazing, tree production, wildlife habitat, home sites, or some combination of these uses. Third, the planned uses and treatments are implemented. The NRCS technician may provide technical and/or monetary assistance (cost-sharing) to the farmer/rancher for certain conservation practices such shelter-belt establishment, terracing, conservation reserve plantings, brush control, and water development. Finally, the farm/ranch plan is monitored for effectiveness from year to year, and updated to incorporate new technologies and government conservation initiatives.

SOIL POLICY

In 1985, the United States Congress passed the Food Security Act which has several provisions that reversed the trend of accelerating erosion during the 1970s. As previously discussed, this act authorized the government to create a conservation reserve by paying farmers to retire their most erodible lands to perennial grasses, legumes, and trees. During the 10-year contract period, the land cannot be farmed, grazed, or hayed. Lands must have an erosion rate three times that of natural soil formation to be eligible. Farmers who violate those contracts are forced to return all government payments with interest.

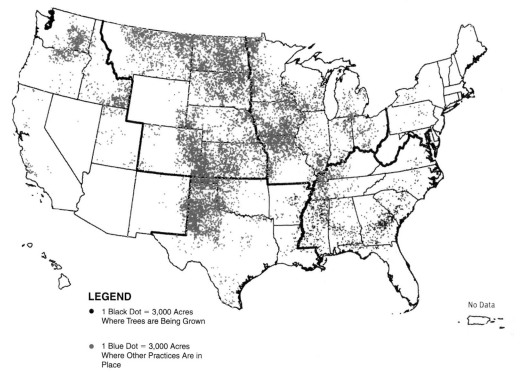

LEGEND

● 1 Black Dot = 3,000 Acres
Where Trees are Being Grown

● 1 Blue Dot = 3,000 Acres
Where Other Practices Are in
Place

No Data

FIGURE 8.25 Conservation Reserve Program (courtesy of USDA-Natural Resources Conservation Service).

By 1995, about 38 million acres or 9% of the nation's farmland had been retired under the conservation reserve program (Figure 8.25). Under the 1996 Farm Act about 20 million acres will be retained in the conservation reserve for the next 10 years. It is generally agreed the conservation reserve has been effective in meeting its goals of reducing soil erosion, improving water quality, and enhancing wildlife habitat. Although estimates vary, the program probably has reduced soil losses by 50 to 60%.

Another part of the farm bill links conservation practices to eligibility for farm subsidies. All farmers must now have a soil conservation plan for their entire farm to receive government benefits. Farmers who agree not to farm highly erodible cropland or wetlands for 50 years can qualify for partial or complete reduction of debts owed to the Farmer's Home Administration. These lands must be converted back to grass and trees or wetland. Farmers who drain or destroy or destroy wetlands on their property are denied federal farm benefits.

In total, soil protection under the 1990 and 1996 Farm Acts has cost the federal government about 2 billion dollars per year. Although there have been some complaints by farmers over excessive regulation and by conservationists over laxity in enforcement of certain provisions (particularly wetland protection provisions), overall it appears these acts have been effective. Even though concern remains over cost, it must be kept in mind that 2 billion dollars represents less than two-tenths percent of the government's 1.5 trillion dollar 1997 federal budget.

Overall we consider the 1985 Food Security Act to be one of the better pieces of legislation that has been developed to protect farmlands from erosion. Modifications in the 1995 Food Security Act better separate highly erodible lands from those with only minor erosion. Only potential lands with definite erosion problems can qualify for renewal of land retirement contracts.

We generally believe that market forces can better solve conservation problems than government interventions. However, soil erosion on farmlands and urban land are among the exceptions to the common rule that market forces are better than government regulations in assuring resource conservation. Here, short-term profit opportunities can

The 1996 Farm Bill's Commitment to Conservation

The 1996 Farm Bill, passed by Congress and signed into law by the President on April 4, 1996, has been heralded as the most progressive environmental farm bill to date. Conservation provisions in the legislation will affect farmers well into the next century. The new provisions build on the conservation gains made by landowners over the past decade. They simplify existing programs and create new programs to address high-priority environmental protection goals. The key provisions:

■ Environmental Quality Incentives program consolidates four existing conservation programs (Great Plains Conservation Program, Agricultural Conservation Program, Water Quality Incentives Program, and Colorado River Basin Salinity Control Program) and directs cost-sharing and technical assistance to locally identified conservation priority areas. Half of EQIP funds are dedicated to livestock-related conservation problems.

■ Wetlands Reserve Program and Conservation Reserve Program are extended through 2002.

■ Farmland Protection Program provides assistance to states that have farmland protection programs to purchase conservation easements.

■ Swampbuster and wetlands provisions from the 1985 and 1990 Farm Bills are modified to provide farmers with more flexibility to meet wetland conservation requirements.

■ Wildlife Habitat Incentives Program helps landowners improve wildlife habitat on private land.

■ Flood Risk Reduction Program provides incentives to move farming operations off frequently flooded land.

■ Emergency Watershed Protection Program allows purchase of floodplain easements.

■ Conservation of Private Grazing Land Initiative offers landowners technical and educational assistance on private grazing land.

■ National Natural Resources Conservation Foundation is created as a nonprofit corporation to foster conservation research, education, and demonstration projects.

■ Conservation Farm Option allows farmers with market transition contracts to consolidate CRP, WRP, and EQIP payments annually, under a 10-year contract, in return for adoption of a conservation farm plan.

■ State Technical Committee membership is broadened to include agricultural producers and others with conservation expertise.

From: A Geography of Hope. 1997. USDA-Natural Resources Conservation Service.

---

often outweigh long-term individual or social welfare from the standpoint of individual farmers or developers. A farmer whose soil gradually erodes away may not be directly impacted by these losses for several years or until after he or she retires or dies. The problems of silt and dust that result from poor management practices are born out by society as a whole rather than strictly by the farmer or developer. For these reasons, coupling conservation compliance to qualification for tax benefits and farm subsidies seems logical. At the individual state level, stiff fines and other penalties are increasingly being used to discourage practices such as leaving soil bare for extended periods by developers and farmers.

Although the United States has made definite progress in controlling soil erosion over the past 10 years, the outlook is somewhat pessimistic in many developing countries (World Resources Institute 1994). In these situations, developed countries could link foreign aid and trade opportunities to application of sound conservation practices. At the same time, slowing population growth and increasing the rate of economic development appear crucial to halting or slowing soil erosion problems in these countries. We refer the reader to Knutson et al. (1998) for in-depth consideration of how government policy influences soil conditions in the United States and other parts of the world.

LITERATURE CITED

Brady, N. C. 1974. *The nature and property of soils.* 8th edition. New York: Macmillan Publishing Co.

Brady, N. C. and R. R. Weil. 1996. *The nature and property of soils.* 11th edition. Upper Saddle River, NJ: Prentice-Hall.

Dasmann, R. F. 1976. *Environmental conservation.* New York: John Wiley & Sons.

Knutson, R. D., J. B. Penn, and W. T. Boehm. 1995. *Agricultural and food policy.* 3rd edition. Upper Saddle River, NJ: Prentice-Hall.

Knutson, R. D., J. B. Penn, and B. L. Flinchbaugh. 1998. *Agricultural and food policy.* 4th edition. Upper Saddle River, NJ: Prentice-Hall.

Miller, G. T. 1990. *Resource conservation and management.* Belmont, CA: Wadsworth Publishing Co.

Nebel, B. J. 1981. *Environmental science: The way the world works.* 2nd edition. Upper Saddle River, NJ: Prentice-Hall.

Owen, O. S. and D. D. Chiras. 1995. *Natural resource conservation.* 6th edition. Upper Saddle River, NJ: Prentice-Hall.

Owen, O. S., D. D. Chiras, and J. P. Reganold. 1998. *Natural resource conservation.* 7th edition. Upper Saddle River, NJ: Prentice-Hall.

United States Department of Agriculture. 1989. *The second RCA appraisal in soil, water, and related resources on federal land in the United States.* Washington, DC: U.S. Government Printing Office.

United States Department of Agriculture. 1994. *Soil erosion by wind.* Natural Resources Conservation Service Agr. Inf. Bull. No. 555.

United States Department of Agriculture. 1997. *A geography of hope.* Natural Resources Conservation Service. Washington, DC: U.S. Government Printing Office.

World Resources Institute. 1994. *World resources 1994–1995.* New York: Oxford University Press.

CHAPTER 9

Ecosystems of the United States

INTRODUCTION

Anyone with even a casual interest in their environment who has done much traveling in the U.S. has observed both subtle and striking variation in forms of land, waters, and vegetation. Combinations of topography, geology, and climate determine the distribution of waters, soils, and vegetation and establish many of the characteristics of ecosystem form and function. Effective management of natural resources depends on understanding the factors that define the different ecosystems. Structural characteristics differentiate ecosystems into three major groups: terrestrial, aquatic, and wetland ecosystems. We will provide a description of the terrestrial, aquatic, and wetland ecosystems of the United States. The reader is referred to Chapter 10 for detailed coverage of forest ecosystems. Our coverage in this chapter is based predominantly on Mitsch and Gosselink (1986), Dasmann (1976), Lerman (1986), Barbour et al. (1980), Miller (1990), Cole (1994), and Holechek et al. (1998).

TERRESTRIAL ECOSYSTEMS

Overview

Grasslands, desert shrublands, savanna woodlands, forests, and tundra are the basic natural land cover types of the world. Precipitation, temperature, and topography interact to determine the type of vegetation that occurs in the area (Figure 9.1). Each of these types is comprised of several plant associations that support a slightly different biota. Activities such as logging, cultivation, grazing, and industrialization have substantially altered the natural biota in all the land cover types. The major land cover types of the world are shown in Figures 9.2 through 9.4.

The term *biome* refers to a set of natural plant and animal communities that exist together and have distinguishing structural characteristics. Broad vegetation categories such as prairie, desert, woodland, grassland, savanna, shrubland, and forest are usually used in naming biomes. All biomes are ecosystems, but not all ecosystems are biomes. While single communities and their habitats are ecosystems, bio-

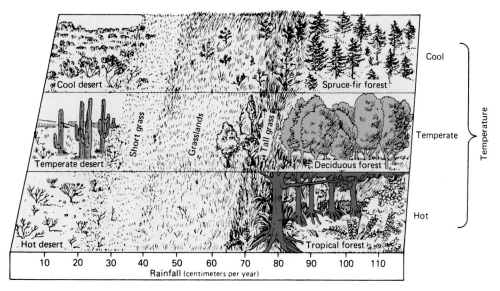

FIGURE 9.1 Average precipitation and average temperature interact to determine the natural vegetation found in a particular area (from Nebel 1981).

mes often include numerous communities and habitats. Thus biomes are larger landscape organizations of basic ecosystem units. The concept of biome is rarely applied to aquatic ecosystems. In this chapter, we will discuss the physical factors that interact to cause the major land biomes of the world, and then provide a more detailed description of the biomes in the U.S.

Grasslands. Grasslands are the most productive lands in the world for farming and grazing. When properly managed they also are an important source of clean water. Grasslands are typically free of woody plants (shrubs and trees) and are dominated by plants in the family Gramineae (grasses). Grasslands occur from sea level to 5000 m but are most common on relatively flat, inland areas at elevations from 1000 to 2000 m.

Grasses, forbs, and shrubs defined: Grasses are distinguished by having hollow, jointed stems; fine, narrow leaves with large parallel veins; and fibrous root systems (Figure 9.5). Many grasslands have a high fraction of sedge (*Carex* sp.) species and rushes (*Juncus* sp.). These grasslike plants have leaves and fibrous roots like true grasses but differ in having nonjointed, solid stems. Forbs are also an important component on many grasslands. Forbs are nongrasslike plants with tap roots, generally broad leaves with netlike veins, and solid nonjointed stems. Shrubs, which are a minor component of most grasslands, have woody stems that branch near the base and have long, coarse roots.

In contrast with shrubs, trees have a definite trunk that branches well above ground. Shrubs and trees have stems that remain alive during dormant periods, but the aboveground parts of most grasses and forbs die back to the crown during winter in temperate areas or during the dry season in the Tropics.

Grassland climate and soils: Grasslands generally occur in areas receiving between 250 and 900 mm of annual precipitation. This precipitation generally occurs as frequent light rains over an extended period (90 days or more). In temperate areas, extended light rains during the summer favor grassland over shrubland because the shallow, fibrous roots of grasses use moisture near the soil surface more efficiently than do the long, coarse roots of trees and shrubs. Winter rainfall with dry summers generally favors shrubland or woodland over grassland in temperate areas because moisture levels near the soil surface are low during the summer period when temperatures are high enough (at least 10°C) for plant growth. During the summer dry period, the

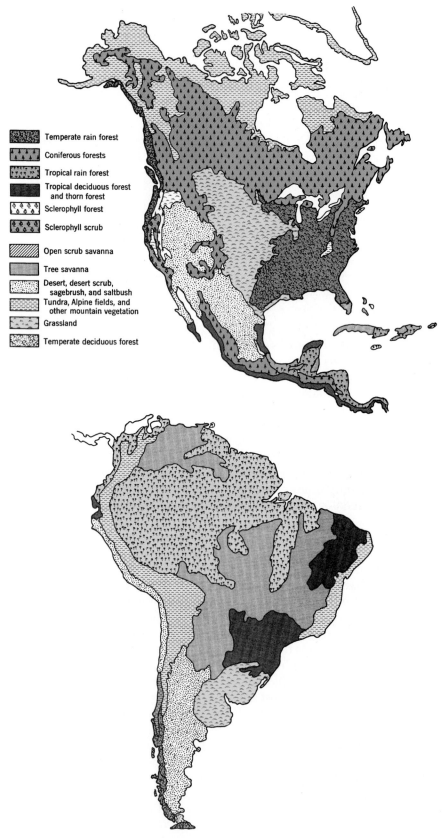

FIGURE 9.2 Major vegetation types of North and South America. "Sclerophyll" refers to woody plants with broad, hard leaves (from R. F. Dasmann, 1976, *Environmental Conservation,* 4th Edition, copyright © 1976, John Wiley & Sons, Inc., New York. Reprinted by permission of John Wiley & Sons, Inc.).

The following are the legend entries for the figure:

- Temperate rain forest
- Coniferous forests
- Tropical rain forest
- Tropical deciduous forest and thorn forest
- Sclerophyll forest
- Sclerophyll scrub
- Open scrub savanna
- Tree savanna
- Desert, desert scrub, sagebrush, and saltbush
- Tundra, Alpine fields, and other mountain vegetation
- Grassland
- Temperate deciduous forest

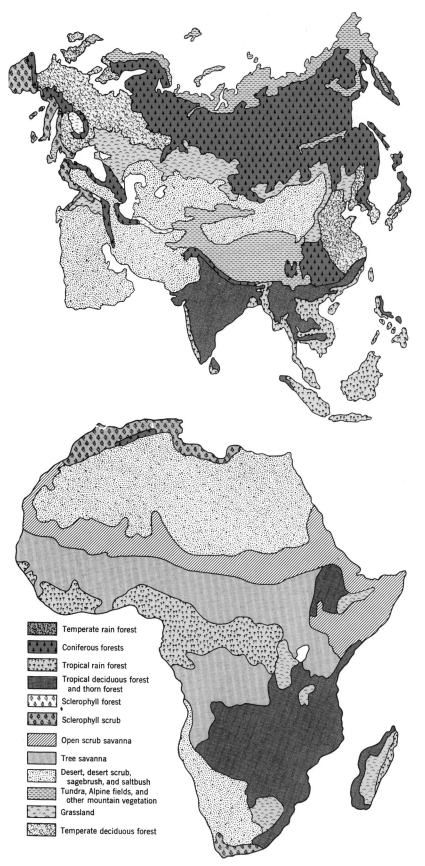

	Temperate rain forest
	Coniferous forests
	Tropical rain forest
	Tropical deciduous forest and thorn forest
	Sclerophyll forest
	Sclerophyll scrub
	Open scrub savanna
	Tree savanna
	Desert, desert scrub, sagebrush, and saltbush
	Tundra, Alpine fields, and other mountain vegetation
	Grassland
	Temperate deciduous forest

FIGURE 9.3 Major vegetation types of Europe, Asia, and Africa (from R. F. Dasmann, 1976, *Environmental Conservation,* 4th Edition, copyright © 1976, John Wiley & Sons, Inc., New York. Reprinted by permission of John Wiley & Sons, Inc.).

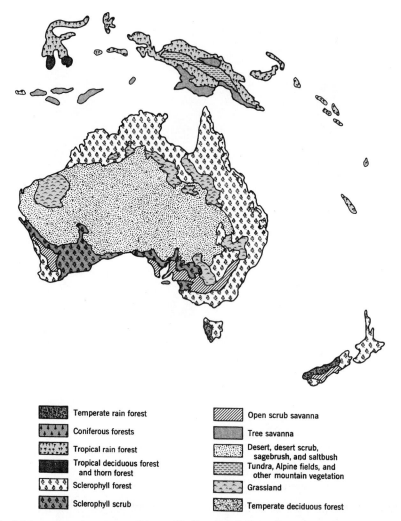

Temperate rain forest

Coniferous forests

Tropical rain forest

Tropical deciduous forest
and thorn forest

Sclerophyll forest

Sclerophyll scrub

Open scrub savanna

Tree savanna

Desert, desert scrub,
sagebrush, and saltbush

Tundra, Alpine fields, and
other mountain vegetation

Grassland

Temperate deciduous forest

FIGURE 9.4   Major vegetation types of Australia (from R. F. Dasmann, 1976, *Environmental Conservation,* 4th Edition, copyright © 1976, John Wiley & Sons, Inc., New York. Reprinted by permission of John Wiley & Sons, Inc.).

long, coarse roots of shrubs and trees can use moisture stored deep (over 1 m) in the soil profile more efficiently than can the shorter, fibrous grass roots. In the Central Valley of California and in the Mediterranean region of Europe and northern Africa, grasslands occur in a wet winter, dry summer type of climate. However, winter and spring day temperatures in these areas are above freezing during most days and are warm enough to permit growth by cool-season grasses (grasses that grow best between 20 and 25°C). Warm-season grasses (optional growth between 30 and 35°C) dominate tropical areas and temperate areas with dry winters and wet summers.

Soils associated with grasslands are usually deep (over 2 m), loamy, textured, high in organic matter, and very fertile (Mollisols). These characteristics make them highly suitable for cultivation. In sandy arid areas (less than 300 mm of annual precipitation), soils (less than 600 mm deep) often support grassland, while deeper soils on the surrounding area support shrubland. The thin soils retain the limited moisture near the soil surface that can readily be used by the fibrous grass roots. In contrast, moisture quickly percolates through the deeper sandy soils and is more readily available to the longer taproots of the shrubs.

In areas with over 250 mm of annual precipitation, heavy clay soils often support grassland while surrounding loamy or sandy soils support forest. The fine clay soils permit less moisture infiltration and retain more moisture near the soil surface than do loams or sands. This results in more favorable conditions for fibrous-rooted than for tap-rooted species.

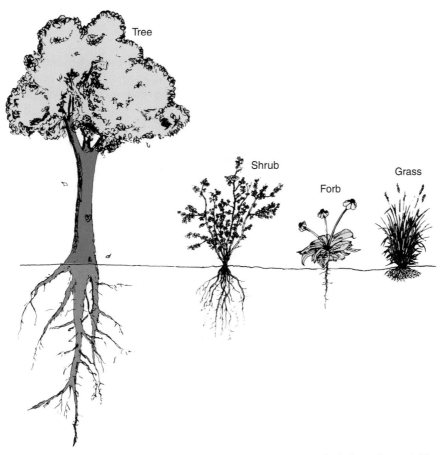

FIGURE 9.5 Rooting depths of grasses, forbs, shrubs, and trees (from Holechek et al. 1998) (drawing by John N. Smith).

Desert Shrublands. Desert shrublands are the driest of the world's rangelands and cover the largest area. Woody plants less than 3 m in height with a sparse herbaceous understory characterize vegetation of this type of rangeland biome. Desert shrublands have been the most degraded by heavy grazing and are the slowest to recover. In some cases, desert shrublands have been created by degradation of arid grasslands by heavy livestock grazing. Desert shrublands provide habitat for numerous unique and rare species, especially where water and stream-side riparian vegetation occurs.

Desert shrublands generally receive less than 250 mm of annual precipitation. The amount of precipitation varies much more from year to year than in the other biomes. In hot desert shrubland areas, precipitation occurs as infrequent, high-intensity rains during a short period (less than 90 days) of the year. This results in long periods where the water content of the soil surface is below the permanent wilting point. This provides highly unfavorable conditions for short, fibrous-rooted plants (grasses). Coarse-rooted plants (shrubs) can collect moisture from a much greater portion of the soil profile than can those with short, fibrous roots near the soil surface. Desert shrub roots extend considerable distances laterally as well as downward. The sparse spacing of desert shrubs permits individual plants to collect moisture over a large area. This explains why they can survive long, dry periods much better than can grasses. In temperate areas with high winter snowfall and dry summers, considerable moisture is available deep in the soil profile during the summer growing season. Shrubs can use this moisture much better than can grasses because of their longer roots.

Sandy to loamy textured soils of variable depth are typical of desert shrublands. Coarse-textured soils permit deep water infiltration and retain little moisture near the soil surface unless there is a restrictive layer. Heavy clay soils and sandy soils with a shallow restrictive layer in desert areas typically show a much higher grass component than do surrounding areas. Desert shrubland soils are mainly Aridisols, although some are Entisols.

Savanna Woodlands.   Savanna woodlands are dominated by scattered, low-growing tress (less than 12 m tall). They have a productive herbaceous understory if not excessively grazed. Heavy grazing usually reduces the understory grasses and increases the density of trees and shrubs. Fire suppression also increases shrub and tree densities on many savanna woodlands. Savanna woodlands often occur as a transition zone between grassland and forest. Shifts toward grassland or forest take place continually in this biome, depending on grazing intensity, fire control, logging, drought, and climate change. Considerable potential exists for conversion of savanna woodlands to grassland when they occur on flat, nonrocky soils over 1 m in depth. Rocky, thin soils favor woodlands in grassland climatic zones because the long, coarse roots of woody plants can grow down into cracks in the rocky layer where moisture is collected. Furthermore, many woody species have long lateral roots that can absorb moisture over a large area of very thin, rocky soil. Without periodic fire most of the wetter portion of the tallgrass type with loamy to sandy soils is quickly invaded by trees and shrubs because considerable moisture reaches that portion of the soil profile below 2 m.

Forests.   Forests are distinguished from savanna woodlands by having closely spaced (less than 10 m apart) trees over 12 m in height. In many areas, forests are managed primarily for timber production and are too dense to have any grazing value. However, they can produce considerable forage for both livestock and wildlife when thinned by logging or fire or when in open stands. Forests also are valued for clean water production, forest wildlife, and aesthetic values. As concern grows about global warming, forests increasingly are valued as storage sinks for carbon.

Forests generally occur in high-rainfall areas (over 500 mm). Under high rainfall, that portion of the soil profile below 3 m has a high water content during most of the year. Much larger quantities of moisture are needed to support the higher biomass of trees compared to grasses and shrubs. Forests occur under as little as 450 mm of annual precipitation in temperate areas of the western United States that have high snowfall in the winter and dry summers. In temperate areas with a fairly even distribution of precipitation during the year, such as the eastern United States, at least 800 mm of annual precipitation is required for forest vegetation. Forest can occur in tropical areas with as little as 500 mm of annual precipitation when soils are coarse textured and the wet season is short (less than 150 days). Coarse-textured and/or thin, rocky soils often favor forest over grassland because they retain low amounts of moisture near the soil surface but store considerable moisture deep in the soil profile and/or rocky crevices. Forest soils typically are less fertile than grassland soils because of greater leaching. Temperate, deciduous forests in most parts of the world are characterized by soils in the order Alfisol. Coniferous forests usually occur on either Spodosols or Ultisols. Oxisols, the most highly leached soils, support tropical forests.

Tundra.   Tundra refers to a level, treeless plain in arctic or high-elevation (cold) regions. Arctic tundra covers about 5% of the world's surface. Large areas of arctic tundra occur in North America, Greenland, northern Europe (the Soviet Union), and northern Asia. Low-growing, tufted perennial plants and lichens dominate this type. Shrubs in the genus *Salix* are the main type of woody plants. The tundra surface is frozen for over seven months of the year. A permanent layer of ice, the permafrost, exists below the surface and restricts trees growing on tundra. Precipitation is low (250 to 500 mm) over much of the area. Strong wind further increases the severity of the area for plant life. Most tundra remains wilderness with limited use by herding and hunting cultures.

Alpine tundra occurs at high elevations (above 2800 m) above timberline in mountainous portions of the western United States and Canada, Europe, South America, and New Zealand. Because of its limited area, rough terrain, high aesthetic value, and short grazing season (less than 90 days), Alpine tundra is of little impor-

TABLE 9.1   Present Extent of the 15 Major Biome Types in the United States

Item	Area (million ha)	Percent in Federal Ownership
Grasslands		
Tallgrass prairie	15[a]	1
Southern mixed prairie	20	5
Northern mixed prairie	30	25
Shortgrass prairie	20	5
California annual grassland	3	6
Palouse prairie	3	15
Desert shrublands		
Hot desert	26	55
Cold desert	73	77
Woodlands		
Piñon-juniper woodland	17	65
Mountain shrubland	13	70
Western coniferous forest	59	62
Southern pine forest	81	6
Eastern deciduous forest	100	7
Oak woodland	16	4
Tundra		
Alpine	4[b]	99

Source: USDA 1972, 1977.
[a]Include sandhills and coastal prairie.
[b]Does not include Alaskan tundra.

tance as a livestock grazing type. Exceptions are Peru and Tibet (China), where alpacas, llamas, and yaks make heavy use of this type.

Biomes of the United States

There are 15 basic land biome types in the United States that are economically important when natural resources and/or total area are considered (Table 9.1). The arrangement of these biomes with movement across the United States is shown in Figure 9.6. Important physical and biological aspects of each type will be discussed using Branson (1985) and Holechek et al. (1998) as primary references.

Grasslands.

Tallgrass prairie: The tallgrass prairie is located in the central United States east of the mixed and shortgrass prairies and west of the deciduous forest (Figure 9.7). Because of its favorable climate and soils, most of the tall grass prairie was placed under cultivation in the 1830s through the 1910s. It is considered to be the best area in the world for growing corn. The major remaining range areas of tallgrass prairie are the Flint Hills of eastern Kansas and the Osage Hills of Oklahoma. These two areas are contiguous and remain because of thin, rocky soils. The Nebraska Sandhills and the Texas Coastal Prairie are considered subunits of the tallgrass prairie, although they are unique in certain vegetational respects. In the northern United States, very little tallgrass prairie remains. The principal areas remaining include the Waubin prairie in Minnesota and Fort Pierre National Grasslands in central South Dakota. Recently, conservation groups have shown interest in buying relict areas of tallgrass prairie, particularly in the more eastern states of Iowa, Indiana, Minnesota, Wisconsin, and Illinois.

The climate of the tallgrass prairie is subhumid, mesic, and temperate. Precipitation ranges from 500 to 1,000 mm. More precipitation is required for tallgrass prairie in the south than in the north because evaporation is higher. Throughout the tallgrass prairie most of the precipitation (75%) occurs during the summer growing season. This is one reason why a grassland is favored. Another reason is that summer drought is periodic and causes considerable mortality of young tree seedlings. Under pristine conditions, most of the tallgrass prairie burned about every three to four years, which further favored the grasses.

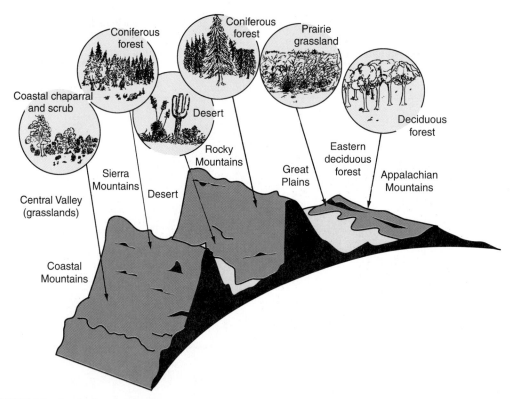

FIGURE 9.6 Gradual transition from one major biome to another between the 34th and 38th parallel crossing the United States. The major factors causing these transitions are changes in average temperature and precipitation (adapted from Miller, 1990 by John N. Smith).

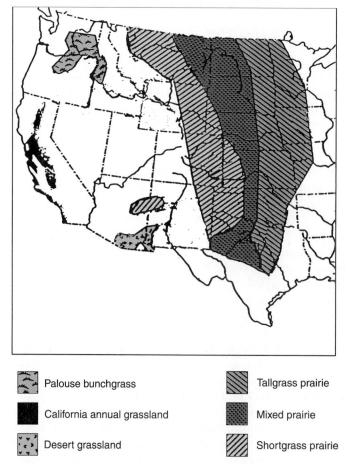

	Palouse bunchgrass		Tallgrass prairie
	California annual grassland		Mixed prairie
	Desert grassland		Shortgrass prairie

FIGURE 9.7 The approximate major locations of the six basic grassland types in the United States (adapted from Branson 1985 and Barbour et al. 1980).

FIGURE 9.8 Tallgrass prairie rangeland in eastern Kansas dominated by big bluestem.

The soils of the tallgrass prairie are primarily Mollisols. These soils are deep, very fertile, and largely free of rocks. They support cultivated grasses such as corn (*Zea mays*) and wheat (*Triticum aestivum*) very well. Generally, the soil profile is not leached enough to impair fertility. The topography of the tallgrass prairie in the north tends to be rolling hills, while it is very flat in the south.

Four grass species characterize the tallgrass prairie (Figure 9.8). These include little bluestem (*Schizachyrium scoparium*), which dominates the uplands, and big bluestem (*Andropogon gerardii*), which dominates the lowlands (Weaver 1954). Yellow indiangrass (*Sorghastrum nutans*) and switchgrass (*Panicum virgatum*) are the other two major grasses. Severely overgrazed areas are dominated by annual sunflower (*Helianthus annuus*), Kentucky bluegrass (*Poa pratensis*), and blue grama (*Bouteloua gracilis*). Ecotones are gradual in the true prairie, although communities can be separated by dominants. Ecotones will shift in wet and dry years. In dry years the upland species will move downslope, while the reverse is true in years of above-average precipitation (Weaver 1954). Several genera of forbs are found in the tallgrass prairie. Leadplant (*Amorpha canescens*) and scurfpea (*Psoralea* sp.) are important legumes with nitrogen-fixing ability. Buckbrush (*Symphoricarpos orbiculatus*) is the primary shrub on the tallgrass prairie. On the eastern edge and along the waterways, several other woody plants are important. These include oaks (*Quercus*), cottonwoods (*Populus*), elms (*Ulmus*), and roses (*Rosa*).

Southern mixed prairie: The southern mixed prairie is the most important of the western range types for livestock production. It extends from eastern New Mexico to eastern Texas and from southern Oklahoma to northern Mexico. The precipitation varies from 300 mm in eastern New Mexico to 700 mm in central Texas. Over most of the area, the frost-free period is at least 180 days. Soils are primarily Mollisols, Entisols, and Aridisols. Because of the wide range of soil and climatic conditions, both productivity and vegetation are variable. Four basic subtypes do occur in the southern mixed prairie (Holechek et al. 1998). These include true mixed prairie, desert prairie, high plains bluestem, and oak savannah. Much of the true mixed prairie and high plains bluestem communities are now under cultivation, although sizable portions still remain as native range in eastern New Mexico and the Texas Panhandle. The desert prairie and oak savannah subtypes exist primarily as native range.

Most of the grasses associated with this type evolved with heavy bison grazing and are relatively grazing resistant. Important grasses occurring over the entire type include blue grama, buffalograss (*Buchloe dactyloides*), little bluestem, various three-awn species (*Aristida* sp.), silver bluestem (*Bothriochloa saccharoides*), vine mesquite (*Pannicum obtusum*), and sideoats grama (*Bouteloua curtipendula*). In the desert prairie, ecotone tobosa (*Hilaria mutica*) and blue grama are the primary grass species. However, as one moves into Texas and climatic conditions become more mesic, silver bluestem and Texas wintergrass (*Stipa leucotricha*) become common.

Areas with sandy, deep soils occur throughout the southern mixed prairie. These areas are characterized by tall grasses, primarily big bluestem, silver bluestem, and little bluestem. In southeastern New Mexico and adjacent Texas these plants grow in association with sand sagebrush (*Artemisia filifolia*) and shinnery oak (*Quercus harvardii*). Collectively, the area supporting these tall grasses is referred to as the high plains bluestem subtype.

In central Texas, the oak savannah occurs on what is known as the Edwards Plateau (Figure 9.9). Originally, this land type supported mostly midgrasses with little bluestem, Texas wintergrass, vine mesquite, silver bluestem, and sideoats grama dominating the composition. However, most of this type has been heavily grazed with three-awn and curly mesquite (*Hilaria belangeri*) replacing the midgrasses.

The southern true mixed prairie occurs primarily in far eastern New Mexico, western Oklahoma, and northwestern Texas. Under climax conditions, this subtype is dominated by little bluestem. Texas wintergrass is a very important species in this type since it is the only cool-season grass of significance to grow in the southern mixed prairie. It provides valuable winter and spring forage when the warm-season grasses are dormant.

Northern mixed prairie: The northern mixed prairie is that portion of the Great Plains extending northward from the Nebraska-South Dakota state line. This type includes the western half of North and South Dakota, the eastern two-thirds of Montana, the northeastern one-fourth of Wyoming, the southeastern part of Alberta, and southern Saskatchewan.

The climate of the northern mixed prairie is characterized by long, severe winters with warm summers. Precipitation ranges from 300 to 650 mm, with two-thirds of it coming during the summer. Most of the other third comes in the spring. Over most of this land type, the peak period of precipitation is June. The average frost-free period ranges from about 140 days in the south to less than 100 days in Canada. The first killing frost usually occurs between September 1 and 10 in the fall and the last freeze generally comes from May 10 to June 10 in the spring.

FIGURE 9.9   Oak savannah rangeland in central Texas. Note browse line on oaks from overuse by whitetail and goats.

The major soils associated with this type are Mollisols. Large areas of Entisols also occur throughout the region. Although much of the soil is suitable for cultivation, severe winters, a short growing season, periodic drought, and low precipitation produce conditions that are not favorable for crop production. Therefore, most of the northern mixed prairie is still rangeland. However, some large areas of productive wheatland occur in central Montana. Much of the northern mixed prairie was farmed in the period between 1900 and 1933, but drought resulted in most of this area being returned to rangeland.

The northern mixed prairie supports the highest diversity of grasses of all the western land types. It has short-, mid-, and tallgrasses as well as both cool- and warm-season grasses. Under climax conditions, the cool-season midgrasses dominate the composition. This type also supports a very diverse composition of shrubs and forbs. Because of the great diversity in vegetation, the northern mixed prairie is one of the best of all types for both wildlife and livestock from a nutritional standpoint. The shrubs and blue grama provide excellent winter feed. However, snow periodically covers the grass in winter, necessitating winter feeding of livestock. Cool-season grasses—such as rough fescue (*Festuca seabrella*), bluebunch wheatgrass (*Agropyron spicatum*) and various bluegrasses (*Poa* sp.)—provide good early spring feed. Green needlegrass (*Stipa viridula*), needle and thread (*Stipa comata*), western wheatgrass, and various forbs provide excellent late spring feed, and little bluestem, blue grama, and sideoats grama provide high-quality summer and fall forage.

Because the northern mixed prairie has a wide variety of shrubs, it supports an abundant and diverse wildlife population. Important shrubs include silver sagebrush (*Artemisia cana*), found on heavy soils in the lowlands; big sagebrush (*Artemisia tridentata*), found on well-drained soils in the more xeric portions; skunkbrush sumac (*Rhus trilobata*) on the hillsides; western snowberry (*Symphoricarpos occidentalis*) and various species of rose (*Rosa* sp.) in the creek bottoms; and fourwing saltbush (*Atriplex canescens*) and black greasewood (*Sarcobattus vermiculatus*) in the more saline areas.

Shortgrass prairie: The shortgrass prairie extends from northern New Mexico into northern Wyoming. Patches of this type are scattered through central Wyoming, western South Dakota, and southern Montana (Figure 9.7). Because of low precipitation, much of this type remains as rangeland.

The climate of the area is characterized by cool winters and warm summers. Annual precipitation ranges between 300 and 500 mm, with 60 to 75% of the precipitation occurs as light rains about evenly distributed over the summer. This type of climate is very favorable to warm-season grasses such as blue grama, which have shallow but extensive root systems.

Soils of this type are largely Mollisols. However, sandy soils in the order Entisol and clay soils in the order Vertisol are scattered through the area. Mid grasses such as little bluestem occupy the sandy soils, while the heavy clay soils are dominated by western wheatgrass. Medium-textured soils support primarily blue grama and buffalograss.

The vegetation of the shortgrass prairie is relatively simple since blue grama and buffalograss comprise 70 to 90% of the composition by weight. The third most important grass species associated with this type is western wheatgrass. Because blue grama and buffalograss evolved with grazing pressure by the American bison, they have morphological and physiological characteristics that make them quite resistant to grazing. An important shrub associated with this type is winterfat (*Ceratoides lanata*). This shrub is palatable to both domestic and wild ungulates. It is an important food of the pronghorn antelope, which reaches its highest numbers in this vegetation type. Scarlet globemallow (*Sphaeralcea coccinea*) is an abundant forb relished by cattle, sheep, and pronghorn. Heavy grazing causes the grasses to be replaced by cactus (*Opuntia* sp.), snakeweed (*Gutierrezia* sp.), Russian thistle, and fringed sagewort (*Artemisia frigida*). During periods of drought, buffalograss tends to replace blue grama.

Past experience has shown that the shortgrass prairie will not sustain cultivation without irrigation. During the drought of the 1930s, severe wind erosion on cultivated land created conditions known as the dust bowl. Vast tracts of cultivated land were abandoned. These tracts were slowly revegetated back to rangeland during the 1940s and 1950s. However, these tracts were again cultivated in the 1970s. In the early 1980s, there was much concern that the dust bowl might be repeated. The 1985 Farm Bill has encouraged the return of these highly erodible lands back to grassland.

Because of flat terrain, good distribution of water, long growing season, and the high nutritional quality of the shortgrasses, this range type is well suited for grazing by both cattle and sheep (Figure 9.10).

California annual grassland: The California annual grassland is found primarily west of the Sierra Nevada Mountains (Figure 9.11). A subtype occurs in Oregon west of the Cascade Mountains as a savannah with an overstory of Oregon white oak (*Quercus garryana*), but with understory species the same as the California annual type.

The climate of the area is Mediterranean, characterized by mild, wet winters and long, hot, dry summers. Rainfall varies from about 200 mm in the southern foothills

FIGURE 9.10   The shortgrass prairie is well suited for grazing by cattle.

FIGURE 9.11   California annual grassland (from Holechek et al. 1998).

to almost 1,000 mm in some areas near the coast. Most of the precipitation comes between October and May, with a peak occurring in January. Little precipitation occurs during the summer and the high evaporation rate quickly vaporizes that which does fall. Summer days are sunny and clear, with maximum temperatures frequently above 40°C. The frost-free period ranges from 200 to 260 days.

Soils of the California annual type are quite variable. Many have prairielike profiles (Mollisols), while others display desert characteristics (Aridisols). In the western Oregon subtype, the soils are almost entirely from volcanic ash (Enceptisols). The more prairielike soils occur near the coast, while the drier types with low organic matter are found in the foothills and uncultivated portions of the interior valley.

The California annual grassland has one of the longest livestock grazing histories of the western range types. Spanish settlements were made in California during the seventeenth century and, by the eighteenth century, much of the valley and coastal areas were being grazed. The original vegetation was composed mostly of cool-season bunchgrasses in the genus *Stipa,* which did not evolve with grazing by large herbivores. Under pristine conditions, the California grasslands were quite beautiful, but they have suffered severe degradation since the arrival of European colonists. Practically all the native perennial grasses are gone. They have been replaced by introduced, cool-season, annual bromes and forbs. These cool-season annuals have nearly ideal environmental conditions in California. They have adequate moisture and temperature for growth and reproduction in winter while their seeds remain dormant during the dry summer period. Today almost 400 species of introduced plants from many lands are found in California and less than 5% of the herbaceous cover is composed of native perennials (Sampson et al. 1951).

The original vegetation of the California annual grasslands was composed of cool-season bunchgrasses dominated by midgrasses such as purple needlegrass (*Stipa pulchra*), nodding needlegrass (*Stipa cernua*), prairie junegrass (*Koeleria pyramidata*), pine bluegrass (*Pos scabrella*), California melicgrass (*Melica imperfecta*), creeping wild rye (*Elymus triticoides*), and California oatgrass (*Danthonia californica*). It appears that introduced annuals may have dominated the area for over 100 years. Numerous fires and overgrazing at an early date probably account for the change from perennial to annual grasses. Today the vegetation is dominated by slender oat (*Avena barbata*), wild oat (*Avena fatua*), soft brome (*Bromus mollis*), ripgut brome (*Bromus rigidus*), foxtail brome (*Bromus rubens*), and little barley (*Hordeum pusillum*) (Figure 9.11).

As one moves up from the valley floor into the Coast Range on the west or Sierra Nevada foothills on the east, conditions become more and more favorable for trees and shrubs. Because of climatic conditions, several evergreen shrubs with small thick leaves known as chaparral occupy the foothill ranges. Important species include pointleaf manzanita (*Arctostaphylos pungens*), wedgeleaf ceanothus (*Ceanothus cuneatus*), hollyleaf buckthorn (*Rhamnus crocea*), poison oak (*Rhus toxicodendron*), chamise (*Adenostoma fasciculatum*), California scrub oak (*Quercus dumosa*), blue oak (*Quercus douglasii*), and interior live oak (*Quercus wislizenii*). Chamise and scrub oak are the dominant chaparral species in California. Other important genera are *Prunus* and *Holodiscus.*

Palouse prairie: The palouse prairie, also referred to as the northwest bunchgrass prairie, has had the highest percentage conversion into farmland of all the western range types. Today it is used primarily for wheat production. The palouse prairie occurs primarily in eastern Washington, northcentral and northeastern Oregon, and western Idaho. Certain areas in northern Utah and western Montana are very similar to the palouse prairie. Although most of the contiguous palouse prairie in Oregon, Washington, and Idaho has been plowed, large tracts remain in these states which support the grasses of this type. The Blue Mountain region of northeastern Oregon and southeastern Washington is an elevated plateau containing a considerable area of open palouse grassland that remains unfarmed because of thin soils and a short growing season. Northcentral Oregon has the largest area of remaining true palouse prairie.

FIGURE 9.12   This palouse bunchgrass prairie is dominated by bluebunch wheatgrass and Idaho fescue in northeastern Oregon. Most of the palouse is now farmed because of excellent soils and climate for wheat production (courtesy of USDA-Forest Service).

Rainfall in the area ranges from 30 to 64 cm annually, with about 65 to 70% falling during the winter months. The months of July and August have the lowest precipitation, although these months are wetter than in the California annual type. The growing season lasts from 140 to 160 days, extending from May 10 to October 10. Winters in the palouse country are relatively mild, and summer temperatures are seldom over 35°C.

The soils of the palouse country are primarily ancient wind-blown dunes in the order Mollisol. They have excellent textural, structural, and chemical properties for agriculture. In many areas the hills or dunes are 100 m deep. Soils on the western edge are heavily mixed with volcanic ash and fall into the order of Inceptisol. The topography of the type is rolling with biscuits and swales. Deep canyons bisect the prairie but are generally unnoticeable as one looks across it.

The major characteristics distinguishing the palouse prairie from other North American grasslands is that the climax vegetation is dominated by bluebunch wheatgrass alone or codominant with either Idaho fescue or Sandberg bluegrass (*Pos sandbergii*) (Figure 9.12). Unlike the central prairie of North America, where the grasses form a sod, the palouse prairie grasses grow in bunches with open interspaces between plants. Like the California grasslands, the grasses of the palouse prairie are almost entirely cool-season bunchgrasses because of the dry summers. With overgrazing, an invader species, downy brome (cheatgrass) (*Bromus tectorum*), replaces the perennial grasses and Sandberg bluegrass increases. In some areas, the unpalatable medusahead rye (*Taeniatherum asperum*) has replaced downy brome after long-term heavy grazing. The palouse prairie supports a wide variety of forbs. Several deciduous shrubs occur in the type. Big sagebrush occupies the more xeric sites and increases with heavy grazing.

DESERTS

Hot Desert

The hot desert is one of the largest western range types. However, because of low precipitation and high temperatures, it rates relatively low from the standpoint of livestock production when compared to other types. This type is found in southern California, Arizona, New Mexico, southwestern Texas, Nevada, and northern Mexico. It contacts piñon-juniper range to the north, chaparral on the west, southern mixed prairie on the east, and the southern Mexico mountains to the south. Elevations for the type range from 925 to 1,400 m.

A desert climate characterizes the type. Precipitation varies from 130 to 500 mm and increases with elevation above sea level. Precipitation occurs primarily during winter and summer with the wettest months in July, August, and September. A much smaller precipitation peak occurs in January. May and June are extremely dry. Summer rains occur as convectional storms caused by solar heating. The winter rains are a result of general frontal movement. Because the storms during the summer are convectional with short duration and high intensity, they are often not very effective. Evaporation generally greatly exceeds precipitation. Temperatures are quite high, with long periods of greater than 38°C expected during the summer months. Bright sunshine is characteristic of the semiarid regions, with some stations in Arizona reporting 90% cloudless days. Relative humidity is generally low. The frost-free period is over 200 days during the year. Many areas may go two to three years without a killing frost. The average last killing frost occurs between February 28 and March 30. The first frost occurs between October 30 and November 30.

Soils of the area are mixed and difficult to categorize, but most are Aridisols. Some areas that are well drained have a calcium carbonate (caliche) layer at the surface or under a shallow profile.

Three general desert areas occur within this type: the Mojave, Sonoran, and Chihuahuan Deserts (Figure 9.13). The Mojave Desert, found in southeastern California, southern Nevada, and northwestern Arizona, is the driest of these three types (Table 9.2). It has the least diversity of vegetation and is primarily a shrubland. The

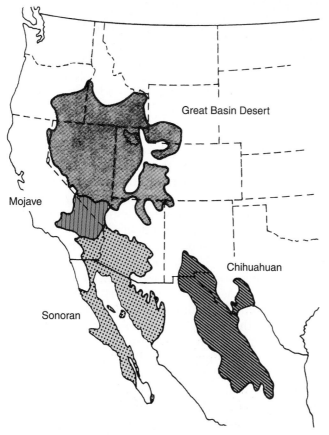

FIGURE 9.13 Deserts of North America. The Great Basin desert is the cold desert (after Shreve 1942 by Blaisdell and Holmgren 1984).

TABLE 9.2   Some Climatic Features of the Four Desert Regions of North America

	Great Basin (Cold Desert)	Mojave	Sonoran	Chihuahuan
Area (km^2)	409,000	140,000	275,000	453,000
Annual precipitation (mm)	100–300	100–200	50–300	150–300
Precipitation in summer (% of total)	30	35	45	65
Snowfall (cm; 10 cm snow = 1 cm rain)	150–300	25–75	Trace	Trace
Winter mean max/min temperatures (°C)	+8/−8	+15/0	+18/+4	+16/0
Hours of frost (% of total)	5–20	2–5	0–1	2–5
Summer mean max/min temperatures (°C)	34/10	39/20	40/26	34/19
Elevation (m)	>1,000	Variable	<600	600–1,400

Source: Barbour et al. 1987.

FIGURE 9.14   Sonoran Desert in southcentral Arizona supports a wide variety of cactus and shrubs because of warm winter temperatures and a biomodal pattern of precipitation (winter and summer peaks).

Chihuahuan Desert, which is found in southwestern Texas and southcentral New Mexico, has less species diversity than the Sonoran Desert in southern Arizona because of colder weather in the winter and less winter precipitation. The Sonoran Desert is quite rich in species because it is virtually frost free and most areas have nearly 300 mm of precipitation. Warmer winter temperatures explain the wider diversity of cactus found in Arizona compared to New Mexico (Figure 9.14).

During early European settlement, the Sonoran and Chihuahuan Deserts were an open grassland or grassland scattered with shrubs (Buffington and Herbel 1965). Much of this area now supports a mixture of shrubs and grasses, with the shrubs more or less dominant. Mesquite (*Prosopis* sp.), catclaw (*Acacia* sp.), and creosote bush (*Larrea tridentata*) are common throughout the area. Several theories have been advanced to explain the increase in shrubs during the past 100 years. Overgrazing, cessation of fire, climatic change, and seed dissemination by domestic animals have all been suggested as possible causes. Probably a combination of these factors explains the increase. Regardless of the cause, brush covers a large area that was originally grassland.

Almost all plants found in the hot desert type are warm season plants. The true climax dominants are almost completely of the genera *Bouteloua, Aristida,* and *Hilaria.* Black grama (*Bouteloua eriopoda*)—which apparently dominated much of the type, particularly southcentral New Mexico—is not highly resistant to grazing under pristine conditions. Intensive grazing in the desert plains results in a disclimax of low-producing perennials with annuals in the genera of *Bouteloua* and *Aristida*. In many areas the grasslands are reduced to almost pure stands of annuals. Broom snakeweed (*Gutierrezia sarothrae*) is an unpalatable half shrub that has severely infested many heavily grazed desert grassland ranges.

Because of low forage production and grazing resistance, livestock grazing is not practical in much of the Mojave and Sonoran Deserts. Because of the dry, sunny climate, large numbers of people have moved into these deserts during the past 20 years. Tourism, wildlife, water, and recreation have become far more important products than forage for livestock. In the Chihuahuan Desert, livestock grazing is and will probably continue to be an important land use. The preservation of open space will probably be the biggest challenge confronting managers of this area in the twenty-first century. The huge influx of people moving into the hot desert over the past 20 years is expected to continue.

Cold Desert

The cold desert, sometimes called the Great Basin, is comprised of the sagebrush shrub steppe and the salt desert (Figure 9.13). These two rangeland types intermingle with each other over vast portions of the intermountain United States. The sagebrush shrub steppe generally occurs at higher elevations than does the salt desert. It is characterized by higher precipitation and less saline soils. Although the frost-free period for both types is typically less than 180 days, the salt desert has warmer temperatures, because of its occurrence at low elevations. Differences in vegetation and managerial components between the two cold deserts warrant a separate discussion of each.

Sagebrush Shrub Steppe. The sagebrush shrub steppe is one of the most extensive of the western range types. There are approximately 39 million hectares of this range type, of which about 65% is in federal control and 35% is in private ownership. Most of the federal land is controlled by the Bureau of Land Management. This land type occurs primarily in Oregon, Idaho, Nevada, Utah, Montana, and Wyoming.

The climate of the sagebrush shrub steppe is semiarid. Precipitation ranges between 200 and 500 mm, with an average of 250 mm. In the northern half about 50 to 60% of the precipitation comes in the late fall, winter, and early spring as snow. Summers are quite dry. In the southern part most of the precipitation also comes in the winter, but slightly more precipitation comes in the summer. The sagebrush grassland occurs between the Cascade-Sierra Nevada Mountains and the Rocky Mountains. Many smaller mountain ranges occur through the area. These mountains result in considerable rain-shadow effect and account for the dryness of this type. Temperatures in the area are extreme, dropping as low as −37°C in winter and climbing up to 38°C in the summer. Most of the area is characterized by about a 100-day growing season. This type occurs at elevations over 1,235 m.

Soils of the area are primarily volcanic materials with much sand and little clay. They are mainly Aridisols. Topography is highly variable. There are many level plains dominated by sagebrush extending to rough foothills. Central Oregon, Idaho, and Nevada have considerable lava.

The sagebrush shrub steppe is characterized by big sagebrush (Figure 9.15). Important grasses of the sagebrush grassland are bluebunch wheatgrass, bottlebrush squirreltail (*Sitanion hystrix*), Idaho fescue, western wheatgrass, Indian ricegrass (*Oryzopsis hymenoides*), needle and thread, and great basin wild rye (*Elymus cinereus*). Originally, the sagebrush grassland had an open stand of big sagebrush, with native wheatgrasses comprising most of the total aboveground vegetation. However, with overgrazing, the big sagebrush becomes more prominent and the invader, downy brome replaces the wheatgrass in the understory.

Associated with this type are several other shrubs which are also important. Black sagebrush (*Artemisia nova*) occurs on rocky and shallow soils in the northern and

FIGURE 9.15   The sagebrush shrub steppe is one of the largest types in the western United States. The high levels of volatile oils associated with the various sagebrushes make them unpalatable to livestock. Sagebrush does provide native big-game animals such as muledeer and pronghorn with important winter feed.

central areas; low sagebrush (*Artemisia arbuscula*) is found on soils with a poor water supply at higher elevations in the northern half; rabbitbrush (*Chrysothamnus* sp.) occupies sandy soils with a low salt content in the central and northern part; antelope bitterbrush (*Purshia tridentata*) occurs on sandy, rocky soils at the higher elevations; and Mormon tea (*Ephedra* sp.) occurs on coarse soils in the central and southern areas.

Salt Desert Shrubland.   The salt desert occurs primarily in the states of Utah and Nevada, although smaller pieces of this type occur in Wyoming, Montana, Idaho, Oregon, Colorado, and New Mexico (Figure 9.13). There are about 35 million hectares of the salt desert shrub, of which about 85% is in public ownership and 15% is privately owned. Most of the salt desert shrub type is controlled by the Bureau of Land Management.

The salt desert occurs in many areas as a mosaic with the sagebrush grassland. The salt desert occurs on lowland depressions of the Great Basin, where drainage is often restricted and the water table is near the soil surface. Evaporation causes salts to accumulate at the soil surface. Sagebrush grassland occupies upland areas free of salts and receiving more precipitation.

The climate of the salt desert is very xeric (it has the lowest precipitation of all types, with the exception of the Mojave Desert) and this condition is made more severe by the high salt content of the soil. Precipitation for this type ranges from 80 to 250 mm, with an average of 120 mm. About one-half the precipitation is snow and one-half is spring or fall rain. Summers are quite dry in much of this type.

Soils of the area are primarily Aridisols. They have varying degrees of alkalinity and salinity. Soils where water collects have the highest salt concentrations. Although both clays and sand occur, sands are more common.

The vegetation of the salt desert is typically characterized by a few species of low, spiny, grayish, and widely spaced microphyllous (small-leaved) shrubs in the Chenopodiaceae and Asteraceae families (Figure 9.16). Shadscale saltbush (*Artiplex confertifolia*) is the shrub species dominating the largest area, with winterfat being the second most common. Important understory species include inland saltgrass, Indian ricegrass, bottlebrush squirreltail, and galleta grass (*Hilaria jamesii*). Much of the salt desert has been severely overgrazed. Vegetation recovery is slow but Yorks et

FIGURE 9.16 The salt desert is dominated by sparse, low-growing, evergreen shrubs. Shadscale and winterfat dominate this salt desert rangeland in central Nevada.

al. (1992) reported meaningful improvement over a 56-year period under moderate livestock grazing pressure.

WOODLANDS

Piñon-Juniper Woodland

The piñon-juniper type is one of the most widely distributed land types in the western United States (Figure 9.17). This type occurs from the state of Washington to 220 km north of Mexico City. Although it occurs on the eastern side of the Cascades and Sierras, most of it is found in the states of Utah, Colorado, New Mexico, and Arizona. North of Utah this type consists entirely of juniper (*Juniperus* sp.) because juniper is more cold-resistant and can tolerate lower precipitation than can piñon pine. This type is found at elevations from 100 to 2300 m. However, in New Mexico it is found as high as 2,400 m. Its upper limits contact Gambel oak (*Quercus gambelii*) and/or ponderosa pine, while the lower limits may contact hot desert, shortgrass prairie, chaparral, or sagebrush grassland. The amount of this type has increased over the past 100 years because of juniper invasion into surrounding grassland.

Piñon-juniper range is commonly referred to as woodland rather than forest since the trees are small and below sawtimber size. The climate of this type is relatively harsh for tree growth. It is characterized by low precipitation, hot summers, high wind, low relative humidity, very high evaporation rates, much clearer weather, and intense sunlight. The annual precipitation varies from 300 to 450 mm, with local areas receiving up to 500 mm. For most of the type, precipitation averages 380 to 420 mm. The frost-free period is variable and ranges from 91 to 205 days.

Soils of this type are poorly developed and are primarily of the orders Entisol and Aridisol. Prior to widespread settlement of the West, piñon-juniper stands were confined largely to the rocky ridges or more level sites with shallow soils. As in many other parts of the West, heavy grazing in the late nineteenth century depleted much of this type. Because of a combination of overgrazing, absence of fires, dissemination of seeds by mammals and birds, and possibly climatic change, trees have encroached on the grasslands and original stands have become more dense (Johnsen 1962). Overgrazing of understory species has reduced protective soil cover and resulted in severe soil erosion in much of this type (Figure 9.18).

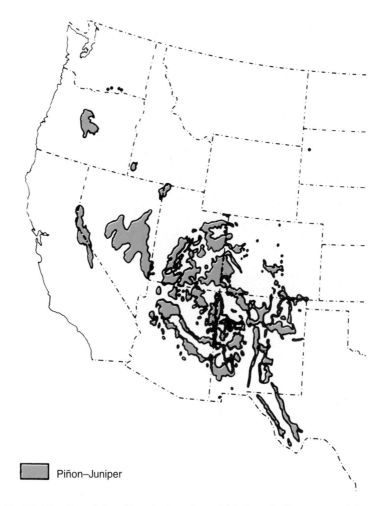

Piñon–Juniper

FIGURE 9.17   Distribution of the piñon-juniper type. This type in Oregon consists mainly of a single tree species, western juniper (*Juniperus occidentalis*) (from Franklin and Dyrness 1969; modified from Clary et al. 1974 and USDA 1936 by Branson 1985).

The understory herbaceous vegetation varies with grazing history, physical features of the site, and density of the trees. Juniper ranges north of Utah have entirely cool-season grasses because of lack of summer rainfall. In Utah, Arizona, Colorado, and New Mexico, this type supports a mixture of warm- and cool-season grasses (Figure 9.18). As the tree overstory increases, perennial grasses and forbs decrease because of shading and increased competition from the trees. Germination of some grasses may be decreased by foliage extract of the juniper.

Land management issues have emerged in the piñon-juniper and the mountain shrub biomes from the subdivision of private tracts into ranchettes, summer homes, and condominiums. This has complicated big game management and reduced incidence of natural wildfires, which have historically occurred at intervals of 10–20 years in these biomes. The proliferation of homes has created strong pressures for fire suppression. This has resulted in a buildup of fuel loads that makes wildfires, when they do occur, far more devastating than those under natural conditions.

Mountain Browse

The mountain browse range type occurs primarily in the Rocky Mountains and Sierra-Cascade Mountains of the western United States. It is a narrow intermittent strip between the uppermost reaches of the grasslands (northern mixed prairie, palouse prairie, sagebrush grassland, California annual grassland) and the coniferous

FIGURE 9.18 An open stand of piñon-juniper in New Mexico with a good grass understory.

forest types. It occurs above the piñon-juniper type in the intermountain area. This type is most prevalent in Colorado, Utah, Oregon, and Idaho.

Climate is intermediate between that favoring grassland and forest. There is not quite enough precipitation for forest. Annual precipitation averages around 460 to 500 cm. Temperatures range from 35°C in the summer to −34°C in the winter. The growing season lasts 100 to 120 days.

Soils are mostly Entisols and Inceptisols. Topography is variable and includes mostly ridges and dry, rocky slopes. Elevations range from 1200 to 2800 m.

The vegetation is dominated by shrubs 1 to 10m tall (most 2 to 4m tall). Important species occurring throughout the type are chokecherry (*Prunus virginiana*) and several species of buckbrush (*Ceanothus*). However, about any western shrub requiring mesic conditions may occur in this type, depending on location. Gambel oak and true mountain mahogany (*Cercocarpus montanus*) are two of the most important shrubs associated with this type in the Southwest. In northern areas, antelope bitterbrush is an important forage species.

Oak Woodland

In the western U.S. oak woodland involves several different species of oak (*Quercus* sp.). There are three basic subdivisions: the shinnery oak type found in southeastern New Mexico, adjoining Texas, and Chihuahua, Mexico; Gambel oak in the central and southern Rockies; and open savannah dominated by tree oaks in California, Oregon, southern Arizona, and central Texas.

The climate of the oak brush ranges is variable depending on location. Oaks are sensitive to winter cold and do not occur much farther north than northcentral Oregon. One species, Oregon white oak (*Quercus garryana*), does occur as far north as southwestern British Columbia. The maximum temperature of the oak brush ranges is about 36°C and the minimum is about −34°C, although oaks may endure even greater extremes during short periods. Soils of the oak brush type are varied but usually well drained. Topography is characteristically rolling uplands and foothills, although the shinnery oak type in west Texas and southeastern New Mexico is quite flat.

As one moves from east to west in the southern Great Plains, the oak species become smaller and smaller as the result of reduced precipitation. In eastern Texas the rather tall-growing post oak (*Quercus stellata*) is dominant; in central Texas and Oklahoma the small tree forms of common live oak (*Quercus virginiana*) are most

FIGURE 9.19 Foothill ranges of the Rocky Mountains are dominated by stands of Gambel oak and other shrubs. Gambel oak is toxic to livestock in the spring, but provides important food and cover for deer.

common; and in western Oklahoma, Texas, and southeastern New Mexico the shinnery oak is the primary species. Oak brush range in Utah and Colorado is composed of Gambel oak (Figure 9.19). A variety of oaks occur in Arizona and New Mexico.

Oak woodland is important to wildlife. The acorns of oaks provide valuable food for many game species, such as the blacktailed deer, whitetailed deer, mule deer, Rocky Mountain elk, collared peccary, wild turkey, bandtailed pigeon, lesser prairie chicken, and bobwhite quail.

Western Coniferous Forest

The western coniferous forest range type is composed primarily of areas dominated by ponderosa pine (16 million hectares) and Douglas-fir (*Pseudotsuga menziesii*) (17 million hectares). This type is found in all of the interior western United States (Figure 9.20). Much of this type exists in several stages.

Ponderosa pine ranges are the largest and most xeric of the true forest types in the western United States. Ponderosa pine is found between the piñon-juniper of various brush ranges and the Douglas fir zone. These ranges occur from 242 to 322 km inland from the Pacific Ocean to Nebraska and the Dakotas and from southern Alberta and British Columbia south into northern Mexico. About 55% of the ponderosa pine subtype is under the jurisdiction of the forest service. Precipitation ranges from 450 to 650 mm, with most as snow in the north and as rain in the south. The growing season ranges from 105 to 140 days, and frost can occur in any month. Soils are primarily Entisols, with Enceptisols on the benchlands and ridges and Mollisols dominating areas with gentle topography.

In the Southwest, ponderosa pine is found at from 2,000 to 2,500 m, while it may occur as low as 1,100 m in northern areas (Figure 9.21). The Douglas fir-aspen zone occurs directly above the ponderosa pine zone. Douglas fir-aspen ranges occur mostly in Colorado, Idaho, Wyoming, Montana, Oregon, and Washington. They occur in isolated areas in Arizona and New Mexico. The bulk of this type is in seral stages, with about 8 million hectares in lodgepole pine (*Pinus contorta*), 5 million hectares in quaking aspen (*Populus tremuloides*), 2 million hectares in western larch (*Larix occidentalis*), and 1.2 million hectares in brush. About 62% of this type is owned by the forest service. This zone receives 640 to 900 mm of precipitation per year, mostly as snow. The growing season lasts from 100 to 125 days. The soils are

Major Forest Types of the United States

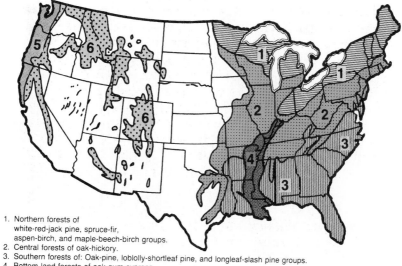

1. Northern forests of
   white-red-jack pine, spruce-fir,
   aspen-birch, and maple-beech-birch groups.
2. Central forests of oak-hickory.
3. Southern forests of: Oak-pine, loblolly-shortleaf pine, and longleaf-slash pine groups.
4. Bottom land forests of oak-gum-cypress.
5. West coast forests of: Douglas-fir, hemlock-sitka spruce, redwood, and some western hardwood groups.
6. Western interior forests of: Ponderosa pine, lodgepole pine, Douglas-fir, white pine, western larch, fir-spruce, and some western hardwood groups.

FIGURE 9.20   Natural forest regions of the United States (from USDA-Forest Service, from Shirley 1973).

FIGURE 9.21   Ponderosa pine is the primary timber tree found on coniferous land in the western United States.

primarily Alfisols, Entisols, and Inceptisols. The topography is rarely level. In the Southwest it is found between 2,500 and 3,100 m in elevation. However, it occurs much lower in the north. Quaking aspen is favored by fire in this type, since it can survive through fire but is intolerant of shade. Douglas-fir, in contrast, tolerates shade but cannot survive fire.

Southern Pine Forest

The southern pine type is one of the largest and most important land types in the United States (Figure 9.22). Primary products from this land type are wood, red meat (cattle), and wildlife. Precipitation for this type averages 1,250 mm or more per year.

FIGURE 9.22   Southern pine ranges are highly productive for wood, wildlife, and livestock. Prescribed burning is used to eliminate excessive accumulation of vegetation and control hardwoods.

The fall period is dry. Precipitation distribution is fairly even during the rest of the year. The frost-free period ranges from 200 to 365 days. Temperatures are high, with a yearly average of 21°C. Soils of this area are highly leached and primarily in the order of Ultisol. Acid soils are a primary limitation to vegetation production. The productivity of this type is the highest of all types, because of warm temperatures and the large amount of precipitation.

The climax vegetation for most of this area is oak-hickory (*Carya* sp.) hardwoods. Presently most of it is in a pine [longleaf pine (*Pinus palustris*), shortleaf pine (*Pinus echinata*), loblolly pine (*Pinus taeda*)] seral stage (Figure 9.22). The pines are grown for lumber and other uses. Open and cutover forests provide abundant forage, particularly in the coastal plain areas (Grelen 1978). Important forage species of the type include grasses in the genera *Andropogon, Panicum, Aristida, Paspalum, Sporobolus,* and *Cynodon.*

Eastern Deciduous Forest

About one-third of the land areas in the conterminous United States is covered by deciduous forest (Barbour et al. 1987) (Figure 9.23). Fingers of deciduous forest extend into the eastern portions of the prairie states (North Dakota through Texas) along the river courses. However, the bulk of it occurs in the central part of the eastern U.S. (Missouri, Illinois, Indiana, Ohio, Kentucky, Tennessee, West Virginia, Virginia). Generally the soils are in the order Alfisol and well suited for cultivation. Where terrain is fairly flat large areas of deciduous forest have been cleared (early through middle 1800s) and are some of the nation's best farmland. This is particularly true in Iowa, Illinois, and Indiana.

Annual precipitation in the eastern deciduous forest varies from 25 inches in the northwest to as much as 75 inches in the southeastern corner. Temperate climate conditions characterize the area, with most of it receiving considerable snow in winter. The precipitation is fairly evenly distributed throughout the year with a tendency toward a summer peak on the western edge. Across most of the deciduous forest, the growing season is five to seven months. Humidity is quite high during the summer months.

The appearance of the deciduous forest changes considerably among seasons (Barbour et al. 1987). In northeastern states, such as Pennsylvania or Ohio, a release of flowering herbs occurs after the snow melts in late March. As the spring pro-

FIGURE 9.23 Eastern deciduous forest in Ohio.

gresses, the trees leaf out—closing the overhead canopy by early June. In July, the early flowering herbs die back and another set of herbs requiring less light begins flowering in August and September. Finally, in early October, the leaves from the different deciduous trees turn into a variety of brilliant colors and begin to drop from the trees. The cycle from leaf formation to shedding in most areas lasts about 180 days extending from early May to early November. Important genera of the eastern deciduous forest trees include the maples (*Acer* spp.), birches (*Betula* spp.), oaks (*Quercus* spp.), hickories (*Carya* spp.), beech (*Fagus* sp.), and basswood (*Tilia* sp.).

Most of the eastern deciduous forest has been heavily altered by farming, logging, and industrialization. These activities have resulted in decreased abundance of some wildlife species while others increased. Whitetailed deer, black bear, ruffled grouse, wild turkeys, woodcock, cottontail rabbits, fox squirrels, and gray squirrels are important game species in the eastern deciduous forest.

TUNDRA

Alpine Tundra

The alpine tundra is the highest land type in altitude. It occurs above the spruce-fir type (see Chapter 10). Alpine ecosystems occupy those mountain areas above timberline that are characterized by short, cool growing seasons and long, cold winters. The vegetation is characteristically dominated by low-growing (20 cm or less in height), perennial, herbaceous, shrubby vascular plants, extensive mats of crytograms [e.g., mosses (*Selaginella* sp.) and lichens (*Cladonia* sp.)], and the complete absence of trees because of permafrost (Figure 9.24). Alpine ecosystems are found primarily in Alaska, Colorado, Washington, Montana, and California, but Oregon, Idaho, Utah, New Mexico, Arizona, and Wyoming have small amounts of this type. Alaska and Colorado are the states with the largest amounts.

The alpine tundra receives 1,000 to 1,500 mm of precipitation, most of which occurs as snow. The alpine tundra is seldom calm, with wind blowing most of the time.

FIGURE 9.24   The alpine tundra type in central Colorado.

The overriding environmental attribute of the alpine tundra is cold temperature. The mean growing season air temperature is often at or near the freezing point. The high winds and low temperatures result in temperature that is very stressful to plant growth. Plants must be adapted to a short growing season and high ultraviolet radiation of high altitudes. Alpine tundra soils range from shallow, rocky, and gravelly Entisols to boggy Histosols.

Compared to floras of other range types, the alpine tundra flora is species poor. Usually, there are no more than 200 to 300 species present, and most of these are common in all alpine areas. Members of the bluegrass (Poaceae) and sedge (Cyperaceae) families occur throughout alpine areas. Additional families with wide alpine distribution are the saxifrage (Saxifragaceae), rose (Rosaceae), mustard (Brassicaceae), buckwheat (Polygonaceae), and pink (Caryophyllaceae). Many of the shrub species are members of the willow (Salicaceae) and heath (Ericaceae) families.

The alpine tundra is a very important source of water in the western United States. Because of esthetics and its fragility, large tracts of alpine tundra have been turned into wilderness areas.

AQUATIC ECOSYSTEMS

Wetland Ecosystems

Overview.   Wetlands form unique ecosystems along hydric gradients between terrestrial and aquatic ecosystems. They are among the most diverse of ecosystem categories (Mitsch and Gosselink 1986). Wetlands range in size from small isolated "pocket wetlands" at groundwater seeps and springs, to huge expanses over thousands of square kilometers, such as the original Everglades of south Florida. Wetlands often are associated with shore zones of lakes, rivers, estuaries, and oceans where they form ecotones that bridge terrestrial and aquatic ecosystems. As a consequence, many wetlands are not easy to identify as discrete ecosystems. All wetlands, however, are associated with water at some point in the annual cycle, often have soils that differ from nearby upland areas, and exclude the many plant species that cannot tolerate flooded soils.

FIGURE 9.25   Bogs are unproductive wetlands with acidic, peaty soils formed from the slow decay of dominant moss and woody shrubs. In the photograph above, the bog is a floating mat extending into and over the pond. Northern spruce forest is in the background (courtesy of USDI-Fish and Wildlife Service).

Wetlands may be permanently covered by water, have permanently saturated soils with no surface water, or be intermittently flooded for only part of the year. Wetlands may be barren of plants (e.g., mudflats), only seasonally inhabited by wetland plants, or permanently and densely inhabited. Wetlands thrive in a wide range of salinity from high marine and arid inland salinity to little more than rainwater concentrations of salts. Wetland plants may be dominated by mosses (*Bryophyta*), sedges (*Cyperaceae*), and woody shrubs in bogs (including muskegs and fens); herbaceous grasses (*Graminae*), sedges, rushes (*Juncaceae*), cattails (*Typhaceae*), and other non-woody plants in marshes; or by larger trees and shrubs in swamps. A great diversity of wetlands occurs in different classes determined mostly by their hydrology, setting in the landscape, water and soil quality, and climate. We refer readers to Cowardin et al. (1979) for detailed coverage of one popular wetland classification scheme. A common way to grossly categorize wetlands is by whether or not they are vegetated. Vegetated wetlands often are sorted into bog, marsh, and swamp classifications.

Bogs.   Also called muskegs and fens, bogs occur in areas of poor drainage and low concentrations of carbonates, calcium, and other nutrients in the surrounding watershed (Figure 9.25). Because of the absence of alkaline carbonates, the water supply to bogs is acidic. Acidity, anoxic soils, and calcium deficiency limit the activity of decomposer bacteria. This typically causes an accumulation of peat, woody material, dissolved organic acids, and other compounds. Sphagnum moss concentrates nutrients and releases hydrogen sulfate (strong acid) through an ion exchange process that contributes to even lower calcium and greater acidity. Because of slow decomposition and low nutrient concentrations, bogs are low productivity ecosystems with a high accumulation of biomass and detrital peat and woody matter. Whereas plant diversity can be quite high in bogs, animal diversity is relatively low compared to marshes and swamps. Low productivity, acidity, dominance of peaty and woody plants that are hard to digest, and deficiency of calcium and other nutrients probably contribute to low consumer diversity. The scarcity of bacteria and fungi in such environments limits detritivore production.

Marshes.   Bogs often grade into marsh communities dominated by herbaceous plants standing in flooded soils or shallow water. Marshes are most characterized by the dominance of emergent and floating herbaceous vegetation rooted in soils that are flooded at least part of the year (Figures 9.26 and 9.27). Although marshes may take root in acidic, peaty sediments, they are commonly encountered in less organically enriched soils and moderately alkaline to alkaline environments. Marshes generally form where ionic concentrations are higher than those that favor bog formation. The diversity of marsh plants tolerating high salinity is rather low. However, they con-

FIGURE 9.26   Salt marshes typically are dominated by cord grasses (*Spartina* sp.) and occur in protected embayments along the coasts, in estuaries, and in arid inland areas of high salinity (courtesy of Virginia Carter and USGS).

FIGURE 9.27   Freshwater marshes often are dominated by diverse herbaceous plant species in the grass, sedge, cattail, and rush families, and many other plant species are incidentally represented (courtesy of USDI-Fish and Wildlife Service).

tribute to extensive salt marshes along the ocean coasts and at inland sites associated with salt formations. Coastal salt marshes often extend well up into estuaries where salinities vary with tidal flux.

Some warm-water marshes are among the most productive ecosystems. This is because of the ideal supply of water, aeration, and sedimentary cycle nutrients in rich soils and plentiful light for photosynthesis above water. Decomposition and mineral cycling is rapid. The vertical stem and leaf structure characteristic of marshes is highly efficient in using light at high stem densities. A diverse array of invertebrate and vertebrate species contributes to and depends on the consumer production. Many primary consumers in marshes are detritivores, which feed on marsh plants once they have died. In marine salt marshes, diverse worms, snails, clams, scallops, oysters, crabs, shrimps, amphipods, and isopods are among the more obvious species encountered. These support numerous marine bird and fish species. Mammalian and reptile species adapted to feeding on marine invertebrates are more restricted.

Swamps. Swamps are characterized by emergent woody vegetation rooted in soils that are flooded at least part of the time (Figure 9.28). Swamps often grade into marshes and bogs. Few trees and shrubs have adapted to permanently flooded conditions because their roots require some aeration. A few species have developed areal extension of roots that function as "breathing tubes," such as the "knees" of bald cypress, which form extensive "blackwater" swamps in the southeastern U.S. Other than the mangroves, most swamp plants have adapted to neutral or acidic soils. Fallen leaves typically contribute to maintenance of acidic sediments and the darkly stained water.

Coniferous swamp species often occur in association with bogs in the Great Lakes region and along the Atlantic coast. Bottomland hardwood swamps often occur adjacent to marshes in the seasonally flooded areas of riverine floodplains. The trees and shrubs in bottomland hardwood swamps survive flooding at certain times of the year, but typically do not tolerate persistent flooding. Although swamp plants are not particularly diverse, swamps often support abundant and diverse consumer populations. This is because of moderate productivity and the complex structural mix of water and islands that typify swamp "landscapes."

FIGURE 9.28 Swamps are dominated by woody plants, such as the bald cypress (*Taxodium distichum*), which occurs throughout the southeastern U.S. (courtesy of USDI-Fish and Wildlife Service).

FIGURE 9.29   Mangrove swamps occur in semitropical and tropical estuaries with waters of variable salinity. They are common in south Florida (courtesy of Richard Frear and the National Park Service).

In semitropical to tropical areas, mangrove swamps (tidal woodlands) frequently dominate the shores (Figure 9.29). Among woody species, only the mangroves have successfully adapted to saline waters. Because they cannot tolerate freezing, yet can survive for long periods at sea, mangroves are circumtropical in distribution along marine coasts and estuaries. Mangroves occur along protected coasts with low tidal flux; often behind barrier islands and in seawater somewhat diluted by freshwater inflows. Mangroves are composed of a number of diverse species from different plant families that have developed parallel adaptations to estuarine conditions. Some of these include waxy leaves to cut freshwater loss, salt pores for secreting salts through the leaves, and specialized roots that extend areal taps to the atmosphere above the anoxic muds. Mangrove species typically attain heights of no more than 10 m and support a few estuarine species in great abundance. Leaf fall from the mangroves is the basis for a detrivore community of oysters, crabs, shrimp, and a variety of fish species, some of which, such as snook and tarpon, are very popular sport species. A variety of large birds (e.g., storks, egrets, herons, spoonbills, ibises) rely on mangrove communities as sources of food and for roosting. Mangroves are the primary refuge for the American crocodile, an endangered species.

Nonvegetated Wetlands. Some important wetlands are not occupied by plants. Mudflats are extensive and important benthic habitats. Many species of worms and clams burrow in the mud of coastal mudflats and provide food resources for shore birds and other predators. Barren zones of high salinity above the tide mark where salts accumulate and create toxic conditions for plants. Numerous temporary shallow ponds that form after intense rains in arid regions support no plant growth. However, many hold the desicated eggs of adapted crustaceans, such as the fairy shrimps (*Anostraca*). They hatch upon inundation and complete their life cycle within a few weeks before the ponds dry up. Although temporary, these shallow ponds can be quite productive for the time they are inundated.

Wetland Functions. Wetlands form zones of habitat convergence for species with both terrestrial and aquatic attributes, such as semiaquatic birds, mammals, and reptiles; some molluscs and crustaceans; and most amphibians. They also support, either directly or indirectly, a diverse array of fully terrestrial and aquatic species. Because of their position between land and sea, wetlands act as areas of exceptional material and energy flux. While they act to trap limiting nutrients, such as phosphorus, they also export carbon and other nutrients in the form of detrital organic matter. They are important in material cycling, especially sulfur. Wetlands also dampen hydraulic energies associated with waves and currents. Therefore, they decrease flood damage and erosion, trap suspended particles in sediment, and contribute to water clarity and offshore productivity. By acting like sponges for retaining water, some wetlands not only dampen floods, but also can contribute to groundwater recharge.

Lake Ecosystems. Lakes seem easy enough to classify and place boundaries on, yet they often grade into wetland, riverine, and oceanic environments in ways that make their definition more difficult than at first thought. Many people think of lakes as distinct from oceans because their water is fresh, but numerous lakes are saline, such as the Great Salt Lake in Utah. Waters like the Caspian Sea in Asia confound those who do not know whether to class it as the largest lake or the smallest ocean. If we define lakes as bodies of inland water disconnected from the oceans and with different salt compositions, then the Caspian Sea must be considered the largest lake. However, freshwater Lake Baikal, also in Asia, is considered the largest lake by volume.

Lakes are "flat water," formed behind a natural or artificial barrier to water flow with surfaces at the same elevation throughout. In contrast, river surfaces follow a slope in elevation and flow in a consistent direction because of the force of gravity. However, on closer inspection, lake surfaces vary in elevation as wind, rain, tributaries, and air pressure changes create currents and a somewhat uneven surface. These variations may be as great as several meters from one end to another in large lakes exposed to strong and consistent winds. Lake Superior often is considered the largest lake based on surface area forming behind a barrier. At the other extreme of size, lakes include very small waters, often referred to as ponds (Figure 9.30). Wetzel (1983), Cole (1994), and Horne and Goldman (1994) provide excellent texts on limnology, the science of lakes and other inland waters.

Lakes may be artificially built impoundments—reservoirs behind dams or excavation lakes dug by mechanical means. Natural lakes were formed in many ways, but most originated from the action of glacial ice. Glacial lakes dot the Northeastern U.S., high mountains of the West and much of the landscape of Canada and Alaska. Floodplain lakes also are quite common as are lakes backed up behind beach dunes along large lake or ocean shores. Other lakes are caused by the gradual splitting of the earths crust associated with plate tectonics. These are among the deepest lakes, like Lake Baikal and Lake Tahoe in the U.S. Many temporary lakes form in wind-eroded depressions, or playas (pans in Africa) in arid and semiarid plains. They typically are seasonal and dry up before the rains or snowmelt returns. Volcanic craters sometimes form lakes, such as Crater Lake in Oregon. A number of other less common origins have been documented and some origins remain a mystery.

FIGURE 9.30   Lakes include a diverse array of aquatic environments typified by impoundment behind a natural or artificial barrier to water flow (courtesy of President's Council on Recreation and National Beauty).

Except at the shallow edges, lake communities are dominated by phytoplankton and zooplankton, many of which are microscopic. Although most plankters have evolved ways to adjust position in the water column, they are, for the most part, at the mercy of wave turbulence. Phytoplankters vary in size; the larger ones being more common in nutrient-rich waters and when it is warm. A predictable succession of smaller to larger species occurs as the seasons warm and cool. A similar succession of zooplankters changes occur, with smaller species dominating in winter and larger ones in summer. However, plankton-feeding fish and invertebrates can markedly influence the sizes of the plankton present because they tend to select the larger plankters for food.

One group of phytoplanktonic bacteria, the Cyanobacteria (blue-green algae), include large species that fix atmospheric nitrogen and produce toxins. Where these nuisance algae dominate in warm nutrient-enriched water, the efficiency of herbivores in consuming phytoplankton is reduced. They add to the accumulation of organic matter that settles to bottom and imparts a bad taste to drinking water. In part because phytoplankton production often is most controlled by phosphorus, water treatment focuses on phosphorus removal to reduce cultural eutrophication.

Lakes exchange water at vastly different rates. Large lakes, like Lake Superior in northeast North America, would take over 140 years to drain. Many small reservoirs would drain in a few days. Lakes with an exceptionally slow exchange rate typically have low nutrient concentrations and low productivity. Deep lakes with high exchange rate, however, may have their plankton production and dependant fish production limited by high flushing rate. Therefore, lakes of intermediate exchange rate appear most productive for most species of sport and commercial fishes. All but a few natural lakes drain from the surface. Many reservoirs drain from depths below the surface. Reservoirs built for irrigation supply typically can be drained from the bottom.

Lakes with a greater amount of settled organic matter than oxygen to decompose it will become anoxic if stratification lasts long enough. This causes numerous environmental changes that affect the aquatic community, often in undesirable ways from the standpoint of resource use. Oligotrophic lakes, with low nutrients and low production, sustain oxygen concentrations at saturation levels determined mostly by water temperature. If the lake is a reservoir with bottom release, hypolimnetic oxygen depletion can extend far downstream in the river below the dam, interfering with productive use.

Although lakes can be thought of as ecosystems separate from their tributaries and watersheds, aquatic ecologists have recognized for many years that the nutrients and

FIGURE 9.31 Streams interact intimately with the landscape they flow through. In forested areas, woody debris is as important as rock and fire sediment in defining the habitat (courtesy of USDA-Forest Service).

sediments entering lakes come from the watershed. When watersheds are low in nutrients and fine-particle soils, as are some sandy regions, lakes remain oligotrophic. Other watersheds are rich in nutrients and are natural nutrient sources for eutrophic lakes. Human activity in watersheds has accelerated nutrient and sediment input through many different routes, but especially through crop culture, fertilization of yards, and domestic waste elimination. This cultural eutrophication often degrades lake resources by causing greater water treatment costs, less desirable fish species, and nuisance algae and plant growths. Toxic contaminants pollute lakes and other waters from agricultural pesticides and industrial and mining sources of various kinds.

Stream and River Ecosystems

Stream Attributes. The key attribute of streams and rivers is consistent downslope flow caused by gravity. Streams vary in size from step-across origins to the largest rivers of the world. Streams flow in channels of their own making during times of high flow and at low or base-flow streams often occupy a small fraction of the channel or dry up entirely. Perennial streams occupy at least some portion of their channel all of the time (Figure 9.31). Intermittent streams leave a channel dry part of the time. As aridity increases, the extent of dry channels contributing to drainage systems increases, forming extensive networks of arroyos in the deserts, which flash flood temporarily during storms. Driven by gravity and the water mass, streams act as an erosive force on the channel. As the channel gradient (slope) becomes steeper and the depth increases, streams become more turbulent and more capable of digging down into the bottom substrate and moving the loosened material downstream.

Flooding is caused by extreme rain or snowmelt events, which increase the depth, turbulence, and erosion in the channel. As the channel is topped and flood water moves over the floodplain, it loses some of its turbulence and capacity for carrying fine sediment, which settles on the floodplain. Floods also erode new channels and leave disconnected channel depressions, which form lakes or wetlands. These often are reconnected during seasonal flooding, sustaining reproductive sites and nurseries for many species of fish and other organisms.

The stream community under dense riparian vegetation depends on the influx of organic matter and its colonization by bacterial and fungal decomposers. This organic detritus is consumed, decomposers included, by relatively large invertebrates such as stoneflies (Plecoptera), caddisflies (Trichoptera), and crane flies (Tipulidae; Diptera). They break the large leaf material down to smaller particles, mostly through generating undigested fecal pellets. These are carried downstream to be consumed by other invertebrates adapted to filtering or otherwise gathering up the fine organic matter. Much of the nutrition provided by detritus is in the form of the colonizing bacteria and fungi. Without the decomposers, the organic matter would be virtually useless calories without sufficient nitrogen and other nutrients to support the needs of the consumers.

Streams and rivers often are classified by their water temperature into cold-water and warm-water streams. In areas of high relief, cold-water streams at high elevations contribute to warm-water rivers at lower elevation. Of course temperature also changes along a continuum and there are intermediate "cool" streams as well, causing variation in the details of this simple temperature-based classification scheme. The variation in temperature also is important. Streams that gain much of their water from groundwater sustain more nearly constant temperatures because of the insulating properties of the earth. A few streams originate in hot springs, which derive their heat from volcanic sources. These typically support few species, but may have luxuriant growths of certain algae and bacteria that tolerate the heat.

Streams also can be classified by their chemistry. Saline streams occur in regions of extreme aridity and geologies of soluble salt deposits. There are limestone streams high in bicarbonate alkalinity and carbonate formations known as travertine. These streams often are rich in inorganic nutrients because the soluble carbonate rocks often are associated with other nutrients. Streams flowing from regions with rock of low solubility, such as granites and schists, usually have low inorganic nutrient fertility. These streams may become acidic blackwater streams stained dark by dissolved organic matter much like bog ponds.

Streams also may be classified by physical shape and structure, including channel gradient, width, and meandering. Related to gradient, bottom types include bedrock, boulder, cobble, gravel, sand, silt, and clay. The shape of the valley is important, varying from steep-sided deep valleys to broad floodplains in shallow valleys. Streams also may be categorized by the degree of artificial channelization that has been imposed on them (e.g., navigation channels, irrigation ditches).

Riparian Communities.   Where floodplains occur and groundwater is close to the surface, a riparian plant community develops different from that on adjacent higher lands (Figure 9.32). Although the soils of floodplains tend to be complex mixtures of cobble, sands, clays, and organic matter because of flooding, nutrient-rich sediments typically are left behind following floods. The combination of permanent water below the surface and nutrient-rich soil makes them especially productive. Riparian communities are most obvious where forest communities extend into areas dominated by grasslands or desert shrub in the upland areas adjacent to the floodplain. These riparian forest communities, however, are not typical of the upland communities in mesic forests. Only certain species become particularly abundant in the floodplain. These species typically are not drought tolerant and tolerate some exposure to flooding, but are not quite as tolerant of wet conditions as the few swamp species described under wetlands. Typical members of riparian communities in the eastern U.S. are box elder (*Acer negundo*), eastern sycamore (*Platinus occidentalis*), eastern cottonwood (*Populus deltoides*), black willow (*Salix nigra*), green ash (*Fraxinus pennsylvanica*), and red maple (*Acer rubrum*). Eastern riparian communities are typically quite dense with limited shade tolerant growth of herbaceous flowering plants. Most grasses and other shade intolerant species are common only in openings where flood events have torn out trees, and along the river margins. In some locations dense impenetrable growth of shrubs become established.

Farther west, box elder is joined by a variety of cottonwood (*Populus* spp.) and willow (*Salix* spp.) species, western sycamores (*Platinus* spp.), and several species of alder (*Alnus* spp.). In the dry Southwest, certain mesquites (*Prosopis* spp.) are common. The

FIGURE 9.32 Riparian vegetation grows in the floodplain adjacent to streams and rivers. It usually depends on groundwater and often on periodic flooding (courtesy of Bureau of Land Management).

introduced salt cedar (*Tamarix chinesis* spp.) has become dominant in many southwestern riparian communities following the failure of native species to reproduce after cutting and grazing. Salt cedar develops exceptionally thick stands. They have a high salt tolerance and concentrate salts in the surface soil as their leaves drop and decay. The salty soils inhibit growth of native plants. Native western riparian communities frequently had substantial herbaceous development of grasses and forbs. These areas can have high understory productivity before the invasion of salt cedar and, in association with shade and water, are often exceptionally attractive to range livestock. Many riparian areas have been greatly altered by grazing and a number of riparian restoration efforts, done mostly by fencing out cattle, has demonstrated that impact.

Many species of birds, amphibians, and mammals are most associated with riparian communities with water in close proximity. Because they form edge communities with the adjacent uplands, the diversity of species typically is highest in riparian areas. Riparian communities in the semiarid and arid parts of the country contribute less than 5% to the total of all vegetation types and support several times that fraction of the diversity. Because riparian communities are more productive, wildlife are usually most abundant there. In many locations riparian areas formed corridors that numerous species travel through under protection of more dense growth than occurs in upland areas. With widespread grazing, crop culture, road development, building construction, dams, levees, and other development, much of the original riparian communities have been altered significantly and greatly fragmented into disconnected patches. The impacts are not entirely well understood, but there is increasing interest in restoring corridors, especially where endangered species might recover as a result.

The River Continuum. Many species of stream organisms have come to depend on the flow of materials carried down river along the river continuum. Other species migrate among the different river and floodplain habitats, and some, like the West Coast salmon species (*Onchorynchus* sp.), move between headwaters and oceans to complete their life cycles. The river continuum concept of Vannote et al. (1980) integrates understanding of geophysical and biological process into a theory of stream and river changes as small tributaries converge into progressively larger rivers. Many stream species have adapted to the continuum of river flow. The interruption of material transport and migratory movement by dams, levees, pollutants, and other barriers has greatly altered stream ecosystems over little more than a century of water resource development. The extent to which this has contributed to the degradation of ecosystem services to humans and especially the endangerment of species, is an active area of scientific exploration. Good introductions to stream ecology include Hynes (1970) and Allan (1995).

Estuarine Ecosystems

Estuaries form in the transition zone (ecotone) where rivers enter the ocean. They are characterized by salinity gradients, predictable tidal flux, and aquatic species that can tolerate sometimes dramatic changes in water elevation and salinity. Because saline ocean water is more dense than fresh water, rivers flow over the seawater forming a saltwater wedge with a relatively sharp boundary. In addition to salinity change, tidal flux often results in large expanses of estuary bottom being exposed to variation in temperature and nutrient as well as salinity (Figure 9.33).

The shapes and dynamics of estuaries are quite variable depending on the coastal land form. In many locations, adjacent estuaries merge contributing to larger estuaries, such as Chesapeake Bay. A rise in sea level following the melting of the last continental glaciers over 10,000 years ago, inundated terrestrial landscapes and extended farther up river valleys. Where river channels were steep and enter directly into the ocean, mixing is rapid and estuary development is minimal. However, where rising sea water inundated flat coastal plains with broad river valleys, mixing became more constrained and formed more extensive estuarine transition zones. Many estuaries have become more shallow from the accumulated deposits of river transported sediments, which settle out on flat gradients. Where the shore steeply rises, as it does along much of the west coast of North America, estuaries tend to be more confined. Glacial scouring of river valleys adjacent to the sea also created deep fjord estuaries.

Formation of coastal barrier islands in shallow coastal environments contributes to constrained mixing and extension of estuarine characteristics. Coastal barrier islands are a prominent land form along the flat mid-Atlantic coastal plain and the Gulf of Mexico, where they baffle the mixing of river and sea and extend estuarine environments along the coastline in lagoons behind the barrier islands. Where coasts are flat and rivers and barrier islands are common, the mix of ocean and freshwater can become quite com-

FIGURE 9.33   Tidal flux in estuaries often exposes large expanses of estuary bottom as shown here in this East Coast estuary where the water is fresh (courtesy of U.S. Fish & Wildlife Service).

plex. Because barrier islands are dynamic and frequently breached and moved by storms, the circulation of coastal plain estuaries is quite variable. Extensive dredging to build and maintain an extensive system of canals and harbors has added to the dynamics. The barrier islands are quite unstable structures, which are continuously moving inland as wind and sea mobilize sand. Large storms often open channels through the islands and alter mixing. Behind the barrier islands, depending on tributary inflow and evaporation, salinities can vary from higher than ocean concentrations to much lower.

Sea grasses of several genera (e.g., *Zostera, Thalassia, Phyllospadix*) have adapted to the protected lagoons behind the barrier islands and to less erosive environments along exposed coasts. The open water lagoons usually link closely with estuaries and to extensive salt marshes. The lagoons often are rich sources of molluscan (oysters, clams, scallops) and crustacean (shrimp and crabs) fisheries as well as productive recreational fisheries for a variety of fin fishes in the sea bass, flounder, and temperate bass families.

Shallow estuaries typically are very productive aquatic environments of limited diversity for a number of reasons. Diversity is limited to species that have been able to adapt to the widely varying and physiological stressful environmental flux associated with the tides. Although many different classes of organisms have adapted, only a few of each class has succeeded. However, the productivity of the successful species makes them particularly abundant and attractive as human resources, including various species of clams, oysters, shrimp, crabs, and fish. Shallow estuaries owe their productivity to plentiful light for photosynthesis, efficient trapping of fine organic sediments washed out of their source rivers, and a fresh supply of nutrients carried in on daily tides. They also are reliably flushed by waste-removing tides, which also sustain high oxygen concentrations.

Land form is critical to determining the productivity of estuaries. Deep fjord estuaries are not nearly as productive because they trap nutrient-rich sediments below the most illuminated waters where photosynthesis is least limited. Shallow estuaries trap sediments at depths where they can be successfully colonized by tolerant algae and vascular plants, such as submerged sea grasses (e.g., *Zostera marina*) and emergent salt marsh grasses (*Spartina* spp.), both of which are consumed directly by few organisms (Figure 9.34). More important, these become substrate for attached algae and both dead

FIGURE 9.34 Salt grass communities, shown here, and submerged sea grass communities are important components of shallow estuarine ecosystems, providing physical habitat and food for many estuarine species (courtesy of USDI-Fish and Wildlife Service).

grass and algae contribute to huge exports of detrital organic matter, which supports many of the bottom invertebrates in estuary and coastal waters and indirectly supports many fish species and birds dependent on estuarine food webs. Crabs, isopods, amphipods, and snails feed directly on the decaying plants. Worms, clams, scallops, oysters, and shrimp filter fine organic particles suspended in the water.

Marine Ecosystems

The Oceans. Although oceans cover nearly three-fourths of the earth's surface, the oceanic depths are among the least well defined of all ecosystems. The different communities and ecosystems are defined as zones, which exhibit gradual transition along gradients determined primarily by depth and proximity to bottom. The *benthic,* or bottom zone, is differentiated from the *pelagic,* or open water zone. The pelagic waters are further divided into the fertile, nearshore waters, or *neritic* zone, and the typically infertile, illuminated waters offshore, the *oceanic* zone. The neritic zone and oceanic zone together compose the photic zone. Below the oceanic zone is the aphotic zone, which includes all water deeper than about 200 m. The benthic communities change along a depth gradient from the intertidal zone nearest shore, through the continental shelf, and along the continental slope into the deep abyssal zone (Figure 9.35).

The *benthos* compose all organisms that live in, on, or just above the bottom. The *plankton* and *nekton* inhabit the water column above the bottom. The benthos frequently use life strategies that sink or anchor them to bottom. In contrast, many planktonic and nektonic species have adapted to open water for completion of their entire life cycles based on the capacity of water density to suspend biomass. Whereas the plankton depend greatly on currents for movement, the nekton are strong swimmers, including most of the fin fishes and all other marine vertebrates. Oceanic depth is an important indicator of physical variables determining the character of oceanic systems. Most importantly, the depth of water determines the amount of light available for pho-

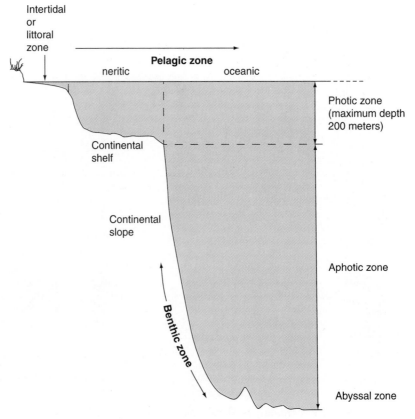

FIGURE 9.35   Major life zones in an ocean ecosystem (from Lerman 1986).

tosynthesis and sight-dependent ecological processes, the amount of heat concentrated (temperature) from absorbed light, and the depth to which heat is mixed by the wind.

Marine organisms are mostly water (over 90% water) and their density usually is quite similar to water. Small variations in body density make the difference between floating upward in the water column or sinking. Many species can control density enough to control their position in the water column with little expenditure of energy. Many of the plankton are especially watery creatures, which can adjust body density through control over salt content, gas vacuoles, fat content, body shape, subtle motion, or other means. Many fish species have gas bladders that allow them to adjust their density. Most bottom fish, on the other hand, have no effective floatation. Many of the benthos take up dense minerals in carapaces and shells that weight them to the bottom. Others adapt behaviorally by wedging themselves into recesses and burrows much of the time. A third group secretes glue-like holdfasts to rock and other solid substrates.

Species are most productive and diverse in the illuminated neritic zone near shore where nutrient concentrations from continental runoff are higher than in offshore waters (Figure 9.36). A diverse array of phytoplanktonic species form the photosynthetic base for the food web. Copepods are common microcrustacean zooplankton that feed on the phytoplankton, but many other species are zooplanktonic, especially in early life stages. Krill is an important larger zooplankton crustacean, which sustains many fish, birds, and whales. Offshore, the oceanic zone is divided into several zones based on depth and light penetration. The epipelagic zone offshore has similar planktonic diversity but the plankton typically are less dense than near shore because of lower offshore productivity. Many of the largest vertebrates (whales, whale sharks) live here; however, where they feed on the plankton by filtration. Just below is a darker environment where many nektonic fish, shrimp, squid, and other species retreat during the day to return to the epipelagic zone to feed under cover of darkness. In the perpetually dark, deep oceans, life exists in continuous cold at very high pressure. Adaptation to the pressure restricts upward movement of the diverse deep sea fishes and invertebrates, which must maintain a pressure equilibrium.

The benthos of the deep oceans live below the disturbance of tides and winds in zones of consistently cool and dark water. The particle size of the bottom sediments tends to become progressively smaller with distance from shore, but is patchy. These are stable environments. The main limitation seems to be food, which comes from detrital fallout and decreases in amount farther offshore as the oceanic zone becomes less fertile. Although the density decreases with depth, the diversity of benthic species increases with depth, probably as a consequence of a long enough evolutionary history to become highly specialized in the stable environment. In specializing, they have evolved several feeding strategies: filtering suspensions from the water, feeding off deposits of particles, absorbing dissolved organic matter, scavenging, and preying on other species. A specialized benthic community inhabits areas around hot volcanic vents where water is warmed locally and bacteria production is high based on chemosynthesis using volcanic elements. A very productive community of tube worms, crustaceans, molluscs, and other species depends on the bacterial production.

Intertidal Ecosystems. The character of the intertidal community (Figure 9.37) in seawater depends firstly on whether the bottom is sandy beach, muds, or consolidated rock and other substrate. On rocky shores, the diversity of marine species increases with consistency of water coverage. In the upper tidal zone the common species are periwinkle snails (*Littorina*) and limpits which scrape algae from the rocks. In the middle intertidal zone diversity increases, but barnacles and mussels often dominate the space by attaching in fixed positions on the rock surface. They are able to filter the water of detrital and planktonic food particles while submersed. Macroalgae, or rockweed (*Fucus* spp.), also shares this space with a variety of crab, snail, starfish, and other predaceous species that feed on the mussels, barnacles, and other crustacean, molluscan, and echinoderm species. The lower intertidal zone is most diverse including a

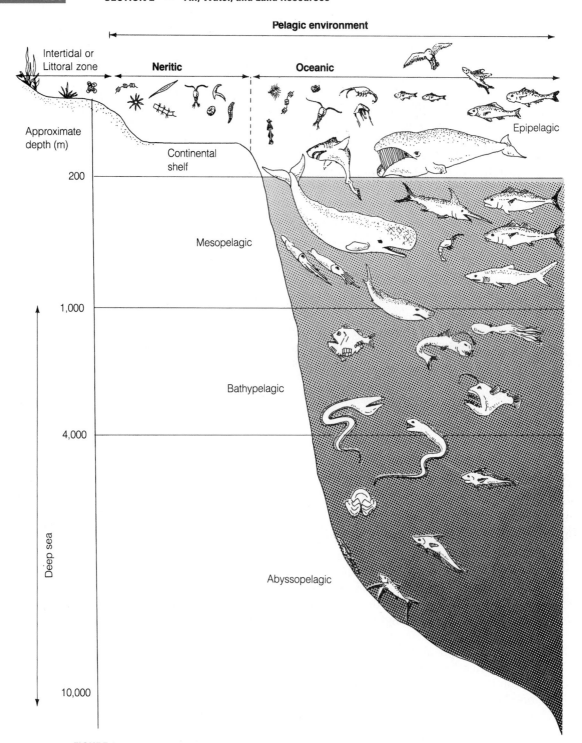

FIGURE 9.36   Stratification of the pelagic environment (from Lerman 1986).

number of kelp and other algae, sea slugs, bryozoans, worms, tunicates, anemones, urchins, crabs, and other crustaceans.

Kelp "forests" inhabit the lower intertidal and subtidal zones of cool western shores of continents. In the U.S., kelp occurs mostly along the California coast. Kelp are dominant macroalgae that can stretch 25 m or more from bottom to surface. Kelp communities also form very productive marine ecosystems, in part because they tend to be most developed in regions of frequent nutrient upwelling from deep, nutrient-rich sources. The upwellings are caused by offshore winds that drive deeper water to

FIGURE 9.37 In the intertidal zone the diversity of living organisms increases with distance off-shore as the amount of time water covers the bottom increases. Here, oysters (*Crassostiea* sp.) are the dominant surface organisms. Many other species are buried in sediment (courtesy of U.S. Fish and Wildlife Service).

the surface. These also are areas of exceptionally high phytoplankton productivity and concentrations of planktivorous fishes, many of which form important fisheries. Kelp communities are sustained by interaction between kelp feeders, such as sea urchins, and predators on sea urchins, such as sea otters. Where this interaction is disrupted, such as it was when sea otters were once overharvested by furs, the herbivores overwhelm the kelp production.

The beach community is less diverse than rocky shore ecosystems because sand is much less stable and there is less primary production to support the community. On the upper beach, a few crustacean species are adapted to feeding on the organic flotsam washed ashore. These include burrowing amphipods and crabs. The intertidal community is more diverse including small burrowing clams, annelid worms, and mole crabs, and very small invertebrates that live in the spaces among the sand grains. These interstitial organisms feed on the fine detritus that is caught in the beach sands from the wave wash. A number of shorebirds are adapted to feeding on beach invertebrates. In the subtidal zone a variety of crustaceans, echinoderms (sand dollars), molluscs, and fish species live on sandy bottoms.

Coral. Coral reefs are tropical marine ecosystems of warm and well illuminated seawater that exist in association with the limestone formations of living organisms, the coral polyps of the phylum (*Coelenterata*) (Figure 9.38). They extract calcium carbonate from the seawater and secrete it layer by thin layer in a slow process that takes thousands of years to replace once the coral dies. The largest coral reef stretches over 1,000 miles along the East Coast of Australia. Not quite so extensive are the coral reefs fringing the tropical Caribbean and Hawaii. Many tropical islands are ringed by reefs. Coral reefs require temperatures above 18° C (64° C) and water over 100 m deep typically is too cold even where light can penetrate to that depth. The light is required by algal symbionts of the coral, which produce food through photosynthesis for themselves and the coral animals. Corals typically are not found near river mouths because the suspended matter absorbs light and the salinity falls below tolerance levels. Because of the

FIGURE 9.38   Coral reefs are formed from biologically secreted calcium carbonate and support some of the most productive and diverse communities on earth (courtesy of the USDI-Fish and Wildlife Service).

coriolis currents, the colder water of continental west coasts is less likely to support corals than the warmer water of east coasts.

Coral reefs are among the most productive and diverse of marine ecosystems. The primary productivity derives mostly from algae living in the coral animals and with encrustation of algae on the dead coral. Little of the primary production comes from phytoplankton. A complex community of herbivores, including the coral polyps, sea urchins, clams, various fish species, and other invertebrates, is fed on by diverse carnivores including sea anemones, starfishes, crabs, carnivorous snails, and diverse fin fishes. At the top are large fish. However, as productive as coral reefs are, they are not important sources of human food resources compared to other marine environments. The diversity results in few species predominating and the coral reef structure resists use of mass harvest techniques. Coral reefs are at risk, however, in many locations because of increased turbidity and toxic contaminants from coastal pollutants.

LITERATURE CITED

Allan, J. D. 1995. *Stream ecology.* New York: Chapman & Hall.

Alonzi, D. M. 1998. *Coastal ecosystem processes.* New York: CRC Press.

Barbour, M. G., J. H. Burk, and W. D. Pitts. 1980. *Terrestrial plant ecology.* Menlo Park, CA: Benjamin Cummings.

Barbour, M. G., J. H. Burk, and W. D. Pitts. 1987. *Terrestrial plant ecology.* 2nd edition. Menlo Park, CA: Benjamin Cummings.

Blaisdell, J. P. and R. C. Holmgren. 1984. *Managing intermountain rangelands—salt desert shrub ranges.* U.S. Dep. Agric. For. Serv. Gen. Tech. Rep. INT-163.

Branson, F. A. 1985. *Vegetation changes in western ranges.* Range Monograph 2. Society for Range Management, Denver, CO.

Buffington, L. C. and C. H. Herbel. 1965. Vegetational changes on semidesert grassland range from 1858 to 1963. *Ecological Monographs* 35:139–164.

Byrd, N. A. and C. E. Lewis. 1976. *Managing southern pine forests to produce forage for beef cattle.* U.S. Dep. Agric. For. Serv. Priv. For. Manage. Bul., Atlanta, GA.

Clary, W. P., M. B. Baker Jr., P. F. O. Connell, T. N. Johnsen Jr., and R. F. Campbell. 1974. *Effects of pinyon juniper removal on natural resource products and uses in Arizona.* U.S. Dep. Agric. For. Serv. Res. Pap. RM-128.

Cole, G. A. 1994. *Textbook of limnology.* 4th edition. Prospect Heights, IL: Waveland Press.

Colinvaux, P. A. 1993. *Ecology 2.* New York: John Wiley & Sons.

Cowardin, L. W., V. Carter, F. C. Golet, and E. T. LaRoe. 1979. *Classification of wetlands and deep water habitats of the United States.* U.S. Fish and Wildlife Service, Washington, DC.

Dasmann, R. F. 1976. *Environmental conservation.* New York: John Wiley & Sons.

Franklin, J. F. and C. T. Dyrness. 1969. *Vegetation of Oregon and Washington.* U.S. Dep. Agric. For. Serv. Res. Pap. PNW-80.

Gay, C. 1965. *Range management. How and why.* New Mexico State Univ. Coop. Ext. Circ. 376, Las Cruces, NM.

Grelen, H. E. 1978. Forest grazing in the south. *Journal of Range Management* 31:244–250.

Holechek, J. L., R. D. Pieper, and C. H. Herbel. 1998. *Range management: Principles and practices.* 2nd edition. Englewood Cliffs, NJ: Prentice-Hall.

Horne, A. J. and C. R. Goldman. 1994. *Limnology.* 2nd edition. New York: McGraw-Hill.

Hynes, H. B. N. 1970. *The ecology of running waters.* Toronto, Ontario, Canada: University of Toronto Press.

Johnsen, T. N. 1962. One-seed juniper invasion of northern Arizona grasslands. *Ecological Monographs* 32:187–207.

Jordan, William R. III. 1997. Ecological restoration and the conservation of biodiversity (pp. 371–398). In *Biodiversity II: Understanding and protecting our biological resources,* edited by Marjorie L. Reacka-Kudla, Don E. Wilson, and E. O. Wilson. Washington, DC: Joseph Henry Press.

Kuchler, A. W. 1964. *Potential natural vegetation of the conterminous United States.* Am. Geogr. Soc. Pub. 36.

Lerman, M. 1986. *Marine biology: Environment, diversity, and ecology.* Menlo Park, CA: Benjamin/Cummings Publishing Company.

Miller, G. T. Jr. 1990. *Resource conservation and management.* Belmont, CA: Wadsworth Publishing Co.

Mitsch, W. J. and J. G. Gosselink. 1986. *Wetlands.* New York: Van Nostrand Reinhold.

Ricklefs, R. E. 1996. *The economy of nature: A textbook in basic ecology.* 4th edition. New York: W. H.Freeman & Co.

Sampson, A., W. A. Chase, and D. W. Hedrick. 1951. *California grasslands and range forage grasses.* Calif. Agric. Exp. Stn. Bull. 724.

Shirley, H. L. 1973. *Forestry and its career opportunities.* 3rd edition. New York: McGraw-Hill Book Company.

Shreve, F. 1942. The desert vegetation of North America. *Botany Review* 8:195–246.

U.S. Department of Agriculture (USDA). 1936. *The western range.* 74th Congress, 2nd Session. Senate Document 199.

U.S. Department of Agriculture (USDA). 1972. *The nation's range resources: A forest-range environmental study.* U.S. For. Serv. Resour. Rep. 19.

U.S. Department of Agriculture (USDA). 1977. *Vegetation and environmental features of forest and range ecosystems.* Forest Serv. Agric. Handbook 475.

Vannote, R. L., G. W. Minshall, K. W. Cummins, J. R. Sedell, and C. E. Cushing. 1980. The river continuum concept. *Canadian Journal of Fisheries and Aquatic Sciences* 37:130–137.

Wetzel, R. G. 1983. *Limnology.* 2nd edition. New York: Saunders College Publishing.

Yorks, T. P., N. E. West, and K. M. Capels. 1992. Vegetation differences in desert shrublands of western Utah's pine valley between 1933 and 1989. *Journal of Range Management* 45:569–578.

The Land-Based Renewable Resources

Vegetation and soil, in their various forms, determine the land-based renewable resources that support forest (Chapter 10), range (Chapter 11), farm (Chapter 12), recreation (Chapter 13), and urban uses (Chapter 14). The land-based renewable resources associated with ground and vegetation may be privately owned or managed under a public authority, which contrasts markedly with the exclusively public ownership of wildlife, fisheries, and endangered species (treated separately in the next section). Very little land is set aside for exclusive non-use (preservation) of soils or vegetation independent of endangered species protection. Even wilderness lands are used for recreation, scientific study, generation of clean water, and a variety of other environmental services. Few land areas are dedicated entirely to one use; instead, multiple use is much more common. In addition to timber production, for example, forested land also supports range, recreation, farming (e.g., Christmas trees), low-density urban development (e.g., ground support for roads, dams, utility lines, homes, saw mills), and a variety of environmental services (e.g., clean water, climate regulation). It is similarly true that land bounded within city limits sometimes supports forest, range, farm, recreational, and even non-use preserves. Also, pristine land converted to a new land use, such as forest to road surface, can be renewed as forest or range when demand is great enough. Several specialty areas of natural resource management are associated with these land-based resource uses including forest, range, farm, outdoor recreation, and urban land-use management. These will be discussed in this section.

The Land-Based Renewable Resources

CHAPTER 10

Forests and Forestry

INTRODUCTION

Over the past decade, forests increasingly have gained public attention. Rain forest destruction and national forest logging controversies are frequent topics during the evening news. Collectively, reports indicate that a growing number of forest-dwelling species are declining from the cumulative effects of deforestation, environmental pollution, and urban encroachment. The public now has much more involvement in deciding how forests will be managed in the United States and abroad. Today's forest managers are better equipped to meet the growing demand for forest products while addressing public concerns over environmental quality. Over the past decade, forestry has been transformed as forest scientists have developed a better understanding of how ecological systems function. The central aim of forestry today is to provide a wide range of ecological goods and services and to manage across broad landscapes as opposed to forest stands. It is no longer acceptable nor scientifically rational to manage forests primarily for wood products and neglect other ecosystem resources.

The key challenge is how to manage natural forests, plantations, and agroforestry systems sustainably while meeting society's growing demands for material goods and social benefits. The U.S. forest sector generates an enormous volume and range of wood products, including lumber, paper, and engineered wood products. They also provide energy, chemical derivatives, forage for livestock, wildlife habitat, various recreational opportunities, and a means of viewing nature apart from humanity's artificial landscapes and stressful lifestyles. In this chapter we will survey the forest resources, examine the ability of forests to meet our material needs, and discuss how forests are managed. In addition, we will review criticisms of forest management and discuss how broader goals can be served by adjusting forest treatments. Numerous concepts will be developed that should enable the reader to more fully understand the technical aspects of forestry issues. From the onset, the reader should recognize that both forests and the science of managing them will continue to change. Forests will continue to adjust to changing environments and management will continue to adjust to society's needs and desires. Important references for this chapter include Sharpe et al. (1976), Spurr (1979), Smith (1986), Perry (1994), and Kohm and Franklin (1997).

TREE STRUCTURE AND FUNCTION

Forest resource management requires understanding some basics of tree structure and function. The differences which set trees apart from other plant life are fairly obvious. Trees are woody perennials, which continue to grow upward and outward over the course of decades, if not centuries. The major parts of a tree are shown in Figure 10.1. Lignin refers to woody material. It is the indigestible component of plant cell walls. Plants lacking lignin in their cell walls are weaker, softer, and unable to support the massive weights attained by trees as they grow upward. Trees elongate as cells divide in the growing tissues of shoot and root tips, known as meristem. This type of growth essentially produces what can be called the primary plant body. The cambial zone (Figure 10.1), found between the inner bark and outer wood, is what enables stems to continue to grow in girth over time. It is called a secondary or lateral meristem. The

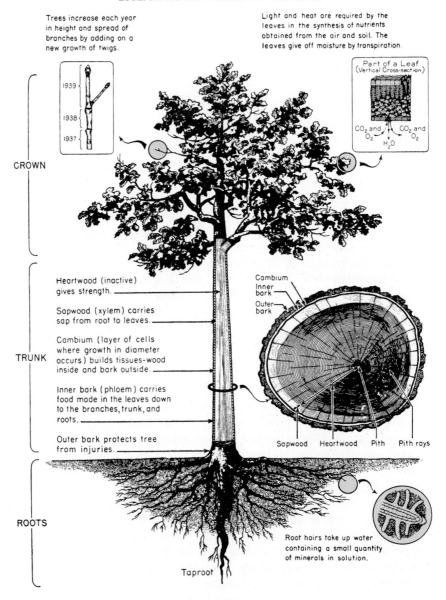

ESSENTIALS OF FORESTRY PRACTICE

Trees increase each year in height and spread of branches by adding on a new growth of twigs.

Light and heat are required by the leaves in the synthesis of nutrients obtained from the air and soil. The leaves give off moisture by transpiration.

Part of a Leaf (Vertical Cross-section)

CO_2 and O_2　　CO_2 and O_2　　H_2O

CROWN

1939　1938　1937

TRUNK

Heartwood (inactive) gives strength.

Sapwood (xylem) carries sap from root to leaves.

Cambium (layer of cells where growth in diameter occurs) builds tissues-wood inside and bark outside.

Inner bark (phloem) carries food made in the leaves down to the branches, trunk, and roots.

Outer bark protects tree from injuries.

Cambium
Inner bark
Outer bark

Sapwood　Heartwood　Pith　Pith rays

ROOTS

Root hairs take up water containing a small quantity of minerals in solution.

Taproot

The buds, root tips, and cambium layer are the growing parts.
The tree takes in oxygen over its entire surface through breathing pores on leaves, twigs, branches, trunk, and roots.

FIGURE 10.1　How a tree functions (with minor modifications, courtesy of USDA-Forest Service).

cambium is a very thin cell layer which produces new xylem cells toward the stem's center and new phloem cells outwardly. Xylem cells comprise tracheids in gymnosperms and tracheids and vessels in angiosperms. In simpler terms, new wood is produced inwardly and new bark outwardly. Each new layer of phloem in the inner bark crushes the previous year's phloem and makes it nonfunctional. In conifers, tracheids mature rapidly, lose their cytoplasm, and become functional in water conduction within a few days to a few weeks. In hardwoods, vessel members (individual vessel cells) connect end to end to form vertically aligned vessels, which provide a pipelike water transporting system throughout the tree.

In temperate zones (i.e., areas where frost occurs), tracheids formed in the spring (called "earlywood") have large diameters and relatively thin walls and are well suited to maximize water supply. Part way through the growing season there is a transition from the production of earlywood to "latewood" tracheids which are typically more lignified, thicker-walled, and have smaller lumens. Latewood is formed as the supply of plant hormones and carbohydrates declines. The structural differences between spring and summer wood make annual growth rings visible in cross sectioned stems. The proportion of earlywood and latewood production affects the density of harvested wood and therefore has practical importance to wood use.

The stems, roots, and leaves of trees serve key functions (Figure 10.1). Roots anchor the tree while absorbing water and dissolved nutrients. The main stem and branches support the weight of leaves and flowers. Through the process of photosynthesis, leaves capture the sun's radiant energy and atmospheric carbon dioxide and convert them into carbon compounds that meet the energy demands of plant growth and tissue maintenance (Figure 10.1). Roots supply the water used in the process.

For the energy demands of nonphotosynthetic tissues to be met, plant sugar must be translocated from leaves to metabolic sinks such as the cambium. The phloem cells located in the inner bark provide the pathway for sugar transport. Water and dissolved mineral transport occur in the outer sapwood. In the heartwood, such cells have become filled with dark chemical substances and are no longer active in water transport. The chemical changes in heartwood makes it more resistant to decay. The cambium also produces parenchyma cells, which line up with xylem cells to form rays. Because ray systems run perpendicular to the longitudinal axis of the stem and bridge phloem cells and xylem cells, they form a critical circulatory system for energy storage. In a sense, the phloem, xylem, and ray systems connect to form a hub and spoke system. This network permits the upward (xylem), downward (phloem), and lateral (ray system) movement of the substances critical for tree survival and growth.

The diverse bark patterns observed in trees are created in part by the presence of a second lateral meristem, the cork cambium. Cork cambia are formed successively in the older portion of the inner bark. They produce outer layers of suberized cork cells which eventually die, are sloughed off, and give trees their characteristic bark patterns.

The wood of different tree species differs based in part on their evolutionary relationships. Two major groups are identified in tree taxonomy. Gymnosperms are plants with naked seeds, not enclosed within a fruit. Angiosperms include broad-leaved flowering trees, such as maples and elms, which are collectively called "hardwoods." Gymnosperms are a rather primitive group of trees whose distant fernlike relatives were present long before angiosperms appeared. Conifers such as pines, cedars, and junipers are cone-producing gymnosperms. Because of their primitive structure, the wood of gymnosperms is much simpler and often easier to saw than the wood of angiosperms. Such trees are also called "softwoods" by foresters. The wood of conifers is largely comprised of tracheids whereas the wood of angiosperms has many more cell types including vessel elements. These fundamental differences have much to do with the way the wood of these groups is used. Paper, for example, requires fibers of uniform type. Pines with their massive store of uniformly shaped tracheids are well suited to paper manufacture.

For trees to survive in cold temperate climates, growth must occur when tissue freeze damage can be avoided. Temperate trees avoid the cold by shutting down growth when daylight periods begin to shorten as winter approaches. An extremely well refined "molecular clock" enables trees to sense the changing seasons. As trees resume growth in the early spring, not all of their main parts begin growth at the same time, nor do they grow at the same rate over the frost-free season. Root growth can precede shoot growth depending on soil temperature and can continue for some time after shoot extension is complete. Growth activity among dormant aboveground parts lag behind root growth. Cambial growth and stem elongation follow. The cambium can remain active for some time after stem elongation is complete. In some species, shoot extension can last for a month or so, but leaves or needles will continue elongation. In fast-growing tree species, shoot extension goes uninterrupted over much of the frost-free season as buds form and elongate in waves of growth activity. By understanding tree dormancy and sequences of tree growth, the forester is able to manage forest regeneration and growth. Some examples are the timing of nursery practices, outplanting, herbicide application, and treatments to enhance seed production.

FOREST PRODUCTS

Forests supply us with a variety of products besides lumber for housing (Figures 10.2 and 10.3). They are a valuable source of food (nuts and fruits), medicines, resins, fuel wood, pulpwood (paper), and decorations (Christmas trees). Most forest lands provide valuable habitats for fish and wildlife and many serve as critical watersheds for large cities. Recreation and livestock grazing are other important forest uses.

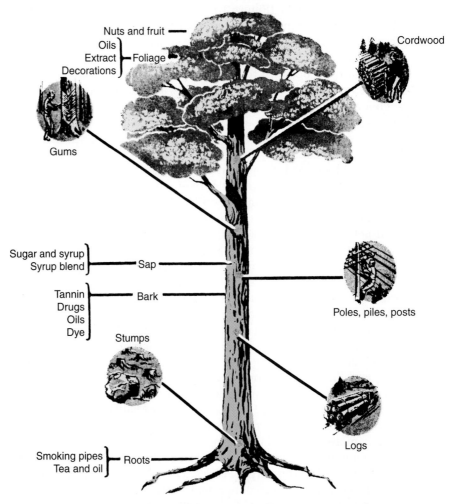

FIGURE 10.2   Some of the useful products obtained from trees (with minor modifications, courtesy of USDA-Forest Service).

FOREST DISTRIBUTION

Forests are areas where trees are the dominant vegetation. In a general sense, they include closed canopy forests, where leaves and twigs of adjacent trees touch, as well as woodlands where trees are more separated and only occasionally overlap. Forest distribution will be discussed first on a global scale (Figure 10.4). At the global scale, the distribution of forest biomes provides a means for presenting a brief overview of the world's diverse forests. For clarity, a forest biome is a major regional ecosystem with its own type of climate, vegetation and animal life. In reality, forest biomes are not sharply separated from other biomes, but merge gradually into one another across

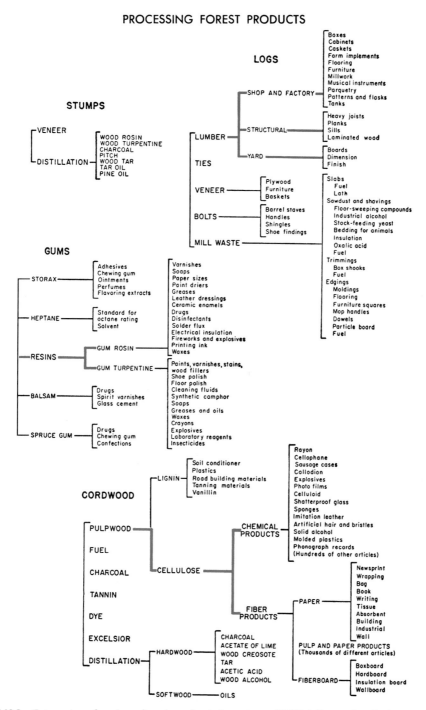

FIGURE 10.3 Categories of various forest products (courtesy of USDA-Forest Service).

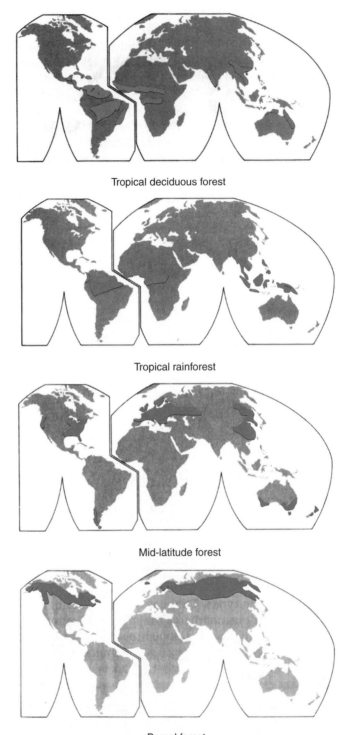

FIGURE 10.4   The general distribution of the major forest types in the world (from McKnight 1990).

ecotones. We refer the reader to Chapter 9 for additional discussion of the forest and woodland biomes.

Boreal Forest (TAIGA)

The boreal forest biome occurs across the high latitudes of North America and Eurasia where it extends from Scandinavia to eastern Siberia (Figure 10.4). These regions are characterized by extreme cold, soils that remain frozen beneath the surface, and a brief (50 to 100 day) frost-free season. Despite the harshness of the climate, the boreal forest is second only to the tropical moist forest in its extent (Pastor et al. 1996). Overall, it accounts for 20% of the world's forest area. Boreal forests are low in tree species diversity because of the harsh climate and their recent formation (glaciers covered the area within the past 10,000 years). Boreal forests are dominated by conifers, especially spruces, firs, and larches. Aspens and birches are important flowering trees. *Picea* (spruces) and *Larix* (larches) commonly occur on poorly drained lowland soils. *Pinus* (pines) are most common on well-drained upland soils, whereas *Populus* (cottonwoods), *Abies* (firs), *Salix* (willows), and certain species of *Picea* occur on the finer-textured upland soils. Boreal forests typically comprise dense stands of relatively small trees, usually less than 30 m in height. Because trees are densely spaced, the forest floor is densely shaded. Fire frequently occurs in boreal forests and strongly influences species composition, nutrient availability, and forest productivity. Forests are slow to recover from disturbance owing to the short growing season and relatively poor soils.

Temperate Deciduous Forests

At their southern boundaries, boreal forests meet temperate forests which have greater tree species diversity and taller stature (Schulze et al. 1996). Temperate broad-leaved deciduous forests occur mostly between 30° to 50° N latitude. Major areas of these forests occur in the eastern U.S., Europe, western Turkey, eastern Iran, western China, and Japan. Temperate forests are much less common in the Southern Hemisphere, but smaller areas can be found in South America, southern Africa, Australia, and New Zealand. Temperate forests are subjected to warm summers and cold winters where temperatures commonly dip to −30 C. The frost-free season is 4 to 8 months. Broad-leaved deciduous forests are concentrated in the eastern portion of North America (see Chapter 9, Figure 9.17). Maples, beeches, hickories, and oaks are among the most representative genera (Figure 10.5). Fire is a major factor in maintaining the diversity of herbs and shrubs on the forest floor. Because they occur in areas long populated by humans, temperate forests have been heavily impacted by unnatural disturbance. Many tree species in temperate forests produce wood of unusually high quality.

FIGURE 10.5
A deciduous forest in Indiana dominated by oaks (courtesy of USDA-Forest Service).

Temperate Coniferous Forests

Temperate evergreen coniferous forests occur in a wide range of climates, and include the montane forests of North America, Europe, and China (see Chapter 9). They also occur on smaller areas of the montane regions of Korea, Japan, Mexico, Nicaragua, and Guatemala. Temperate conifers tend to occur on xeric or infertile soils lacking the water and nutrients needed by deciduous species. Common genera include *Abies, Pinus, Pseudotsuga,* and *Thuja. Pinus* species occur in a wide variety of environments ranging from the hot, arid southwestern United States to cold regions of Scandinavia and Eurasia. Growth rates vary greatly depending on precipitation and soils. Forest management is intensified in regions (e.g., southeastern U.S. and Pacific Northwest) of high production. Even on less productive sites, biomass accumulation (wood) can be substantial over several centuries.

Temperate Mixed Forests

Mixed (deciduous plus evergreen) forests occur throughout the temperate regions where the climates are similar to those described for temperate deciduous and coniferous forests. They are most abundant in the southeastern United States; Europe; and portions of Iraq, Iran, and China. Their occurrence reflects past land use change and local variations in soil conditions. In the southeastern United States, recently disturbed forests, are dominated by conifers. Over time these are replaced by mixed forests, which eventually give way to broad-leaved hardwoods. In the lake states, conifers dominate the drier, less fertile soils. Broad-leaved hardwood forests occur on the more moist soils of finer texture. Mixed forests occur where soil conditions are intermediate.

Temperate Broad-Leaved Evergreen Forests

This group includes broad-leaved sclerophyll and broad-leaved rain forests. The broad-leaved sclerophyll forests occur in areas with a Mediterranean climate, typified by winter rain and summer drought. Broad-leaved rain forests occur in areas with humid, frost-free climates. The sclerophyll forests occur in scattered areas of the United States, around the Mediterranean, and over large areas of Asia. The greatest continuous areas still existing are the Eucalyptus forests of Australia. Temperate broad-leaved evergreen rain forests occur in Japan, Chile, New Zealand, Australia (Tasmania), and in scattered, remnant patches in Asia.

Nothofagus (beech) rain forests occur in stands in New Zealand and southern Chile. The sclerophyll (hard-leaved) forests in the Mediterranean area and the U.S. are dominated by oak (*Quercus*) species. Oak woodlands can be impressive where disturbance has been minimal, but this is rarely the case. In Australia, Eucalyptus can form closed canopy forests up to 60 m in height. Eucalyptus forests are adapted to fire. Large areas of native forests have been cleared since European settlement. Exotic softwood (*Pinus radiata*) plantations have replaced large areas of the native forests in Australia.

Tropical Evergreen Forests

Tropical evergreen forests (rain forests) comprise the largest single forest biome in the world (Figure 10.6). They are found between the Tropics of Cancer and Capricorn in areas where air temperatures are high and change little on a daily or seasonal basis. Rainfall is abundant (more than 1,500 mm per year) and relative humidity is uniformly high. The Amazon Basin supports the greatest area of tropical rain forest, but rain forests also cover the Congo Basin. Tropical forests, highly fragmented by agriculture, are found in Central America. In Asia, they occur along the southeast coast of India, in Sri Lanka, the Malaysian Peninsula, the Indonesian archipelago, Borneo, Sarawak, and Papua New Guinea.

The stature and species composition of tropical rain forests differ according to soil type and elevation, thus making lowland and montane subtypes distinguishable.

FIGURE 10.6 Tropical rain forest in Puerto Rico (courtesy of USDA-Forest Service).

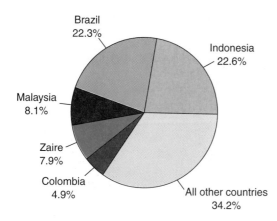

FIGURE 10.7 Percent of tropical rain forest lost (1981–1990) from the top five countries with greatest losses and all other countries combined (Food and Agriculture organization; figure from World Resources Institute 1994).

Rain forest canopies often have three layers. These include a layer of taller trees, which rise above the main canopy layer, and a subcanopy of smaller trees and shrubs.

Rain forest soils are typically infertile, but nutrients are cycled efficiently. Thus much of the nutrient is incorporated in biomass and is removed by intensive logging. Tropical evergreen forests are the most diverse terrestrial ecosystems on earth, with the greatest number of species per unit area. The Amazonian forests contain, for example, more than 2,500 different tree species. Rain forests are the most productive forests in the world in terms of biomass produced. Much of the production is allocated to leaf matter, which forms an energy source for highly diverse insect species.

Tropical rain forests are under constant assault by human activities (Figure 10.7). The most threatening pressures on rain forests differ across the regions (World Resources Institute 1994). In Africa, forests are cleared for agriculture to support a rapidly expanding population. In South America, shifting cultivation is a major factor, but large areas have been cleared for resettlement of immigrants, ranching, modern agriculture, industrial fuel production, logging, and hydroelectric development. Asian forests are being rapidly destroyed by population growth and abusive commercial logging operations.

FIGURE 10.8   Tropical deciduous savannah in Sudan.

Tropical Deciduous Forests

Tropical evergreen forests give way to deciduous forests where rainfall is more intermittent on a seasonal basis. Because deciduous species occur throughout the evergreen forests at least to some extent, the categorical separation has limited practical value. The factor responsible for leaf shedding is moisture availability. The largest areas of tropical deciduous forests are the monsoonal forests of southern and southeastern Asia. They also occupy areas bordering evergreen forests in South America and Africa (Figure 10.8).

The canopies of deciduous tropical forests are typically shorter and more open than those of tropical evergreen forests. In addition, they usually have two rather than three layers. The bottom layer is often occupied by densely growing shrubs that take advantage of the available sunlight and reduced evaporative demand. Tropical deciduous forests are being cleared to make room for grazing and agriculture and are becoming increasingly fragmented.

FORESTS OF THE UNITED STATES

Major Forest Types

The forest vegetation of the U.S. has been classified in numerous ways. One approach has been to map potential vegetation on the basis of climate (e.g., Kuchler 1964). In contrast to Kuchler's method, the Society of American Foresters (SAR) approach identifies forest types that currently exist as opposed to those potentially existing under climax (pristine) conditions (Eyre 1980). For more intensive management purposes, forest vegetation in the western states has been classified according to habitat types as found in Daubenmire and Daubenmire (1968), Franklin and Dyrness (1973), and Pfister et al. (1977).

Hall (1983) combined the numerous forest cover types found in the U.S. to six main groups (Figure 10.9). In effect, forest trees tend to be aggregated into certain groupings or occur in mostly pure stands. SAF calls these identifiable aggregations and pure

Major Forest Types of the United States

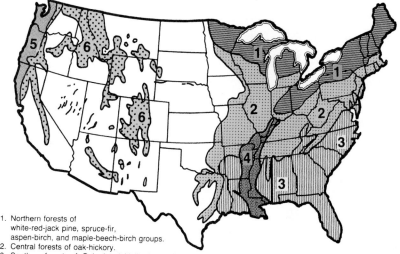

1. Northern forests of
   white-red-jack pine, spruce-fir,
   aspen-birch, and maple-beech-birch groups.
2. Central forests of oak-hickory.
3. Southern forests of: Oak-pine, loblolly-shortleaf pine, and longleaf-slash pine groups.
4. Bottom land forests of oak-gum-cypress.
5. West coast forests of: Douglas-fir, hemlock-sitka spruce, redwood, and some western hardwood groups.
6. Western interior forests of: Ponderosa pine, lodgepole pine, Douglas-fir, white pine, western larch, fir-spruce, and some western hardwood groups.

FIGURE 10.9   The six major forest cover types in the U.S. (courtesy of U.S. Dept. of Agriculture).

stands "forest cover types." In given areas, forest cover types can be transitory or occupy the same area for long periods. Nevertheless, several criteria must be met for a
type to be recognized. Tree crowns should cover at least 25% of the area and the type
must occupy a fairly large area in the aggregate, but not necessarily in a continuous
manner. Many types occur infrequently and merge into others over short distances.

Eastern Forests.   Eastern forest types in the U.S. only occupy about 80% of New England, 50% of the Atlantic Coast States, and only 15% of the Central States where forest land soils are more amenable to farming. The northern forests are dominated by
four cover types: white-red-jack pine, spruce-fir, aspen-birch, and maple-beech-birch.
Figure 10.10 gives the total area of each of these and other forest cover types that are
available for timber harvest (i.e., unrestricted forest land). The white-red-jack pine
type is an important source of pulpwood for the Northeast. The aspen-birch type is
common in the lake states. It is composed of relatively short-lived species that invade
abandoned farms or areas disturbed by fire or logging. The maple-beech-birch type occurs mainly on upland sites in the New England, Middle Atlantic, and lake states regions. This type has expanded in recent years and contains valuable hardwood
species for wood products. Spruce-fir forests grow from New England and the lake
states north into Canada. They occur after long periods without fire. Associates of
spruce and true fir (*Abies*) include white-cedar, tamarack, maple, birch, and hemlock.

The oak-hickory type, the largest of the eastern forest groups, has 127 million
acres available to timber management (Figure 10.10). Much of it occurs on less productive and abandoned farmlands or in mountainous areas.

The southern forest consists of the oak-pine, longleaf-slash pine, and loblolly-shortleaf pine cover types (Figure 10.11). Fire is an important factor determining the extent
of oak relative to pine. Pine forests are fire disclimax communities. Much of the oak-
pine type resulted from selective harvesting of native pine forests. However, before the
late 1980s, many timber industry lands consisting of oak-pine were converted to nearly
pure stands of pine. The longleaf-slash pine type occurs in the southern states and
along the Atlantic Coast with high concentration in Florida and Georgia. The loblolly-
shortleaf pine type tends to grow at higher elevations and farther north. This type accounts for more than one-half of the 96 million acres of cone-bearing forests of the East.
Each of these types have been maintained in the past by planned burning under the

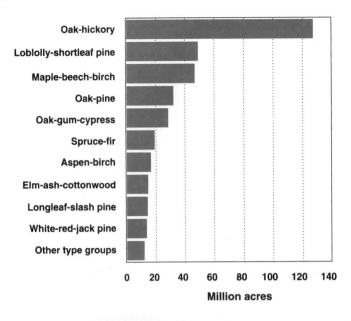

FIGURE 10.10   Forest cover type groups on unreserved forest land in the East (from Powell et al. 1993) Cover types are determined by the dominant species in the overhead canopy (cover).

FIGURE 10.11   Longleaf pine (*Pinus palustris*) savannah in Georgia maintained by periodic burning (courtesy of USDA-Forest Service).

pine canopy. These predominantly pine forests change to hardwood forests if steps are not taken to regenerate pine, for example, through prescribed burning.

Southern bottomland forests of oak-gum-cypress are found primarily along the Mississippi River drainage. They include such valuable species as sweetgum, blackgum, cherrybark oak, tupelo and bald cypress. Common associates include cottonwood, willow, ash, elm, hackberry, and maple. This type is important to the southern hardwood industry, but much of it has been converted to agriculture.

Western Forests.   The western forests can be separated into the highly productive West Coast forest and the less productive western interior forests (Figure 10.9). The West Coast forests consist of the Douglas-fir, hemlock-Sitka spruce, and redwood forest cover types, which tend to be highly productive (Figure 10.12). The Douglas-fir forests found on the Pacific slope are perhaps the most productive softwood forests in the U.S. Common associates of the Douglas-fir type include western hemlock, western redcedar, the true firs, redwood, ponderosa pine, and larch. Western hemlock-Sitka spruce forests are found mainly on the Pacific slope in Oregon and Washington and in coastal Alaska. This type is an important source for lumber, pulpwood, and log exports. Redwood forests are rather localized in occurrence and are found mixed with Douglas-fir, grand fir, and tanoak (Figure 10.12).

(a)

(b)

FIGURE 10.12 (a) Pacific coastal forest dominated by Douglas-fir (*Pseudotsuga menziesii*). (b) Giant Sequoia trees (redwoods) in the Calaveras grove in California (courtesy of USDA-Forest Service).

FIGURE 10.13 Climax ponderosa pine (*Pinus ponderosa*) forest in Arizona (courtesy of USDA-Forest Service).

The tree cover types found in western interior forests include ponderosa pine, lodgepole pine, interior Douglas-fir, white pine, western larch, white fir, Engelmann spruce, and some western hardwood groups (Figure 10.13). Species commonly associated with the ponderosa pine type throughout its range include Jeffrey pine, sugar pine, limber pine, Arizona pine, Apache pine, Chihuahua pine, Douglas-fir, incense-cedar, and white fir. Lodgepole pine, white pine, and western larch types occur where crown fire has destroyed the previous stand and permitted the growth of shade-intolerant species.

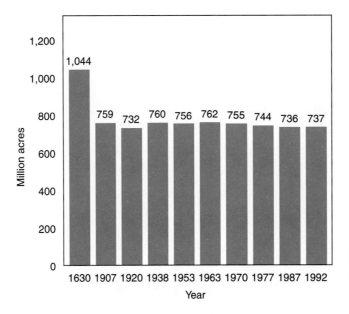

FIGURE 10.14   Trends in U.S. forest area from 1630 to 1992 (from Powell et al. 1993).

The piñon-juniper type consists of ecologically important woodlands scattered throughout the Southwest (Chapter 9, Figure 9.17). The trees lack impressive size, but these woodlands offer vast areas used for wildlife habitat, hunting, grazing, watershed protection, and commercially important products, such as firewood and pine nuts (see Chapter 9, Figure 9.18).

FOREST LAND AREA IN THE U.S.

Today, forests occupy 33% (737 million acres) of the total land area of the U.S. (Powell et al. 1993). The U.S. Forest Service defines forest land as land which is at least 10% covered by forest trees of any size. North American forests have undergone considerable changes from the time early colonists began settling the land. Around 1600, about 49% of the U.S. was forested.

New England, which was 90 to 95% forested when the first colonists arrived, has experienced large declines in forest area (Sedjo 1995). About 307 million acres of U.S. forest land have been converted to other uses since 1630 (Figure 10.14). Land clearing for agriculture has been the primary cause. Between 1850 and 1910, farmers cleared more forest than the total amount cleared in the previous 250 years. Industrialization and railroad construction also consumed wood. For example, between 1870 and 1910, the huge forests of the lake states were cut to supply the material requirements of rapid industrialization. Railroads consumed nearly 25% of the wood used in the 1800s (Sedjo 1995). After 1920, as agricultural production increased per unit area, cropland area stabilized at lower total acreage, and forest production recovered slightly (Figure 10.15). Overall, the area of U.S. forests has been maintained the past seventy-five years (Figure 10.14). However, forest age structure has changed because secondary forests stand today where primary forest stood before. Although old stands were common in colonial America, they were limited by natural disturbances such as insect infestation, disease, and fire, as well as the actions of the native peoples (Sedjo 1995).

THE STATUS OF THE U.S. TIMBER RESOURCE

Our forests range from the highly productive forests of the Pacific Coast to sparse woodland forests of the arid interior West. They include pure hardwood and conifer forests as well as diverse multispecies mixtures. Forests provide diverse goods

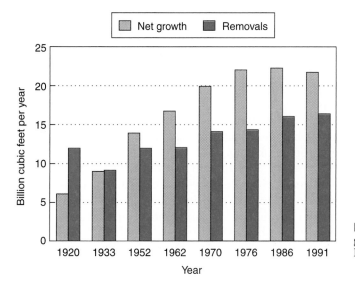

FIGURE 10.15 U.S. timber growth and removals (from Powell et al. 1993).

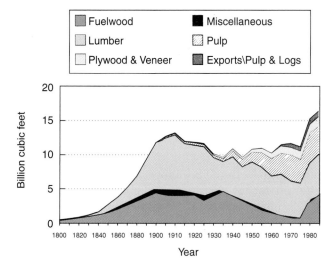

FIGURE 10.16 Domestic production of forest products, 1800–1990 (from Powell et al. 1993).

(Figure 10.16) and services but the importance of timber will increase as other resources and services become more valued.

Commercial Forest Land

By definition, timberland is capable of producing at least 20 ft^3/acre (1.4 m^3/ hectare) of industrial wood per year and is not reserved from timber harvest (Powell et al. 1993). Two-thirds of the nation's forested ecosystems (198 million ha or 490 million acres) are classed as timberland. About 6% (47 million acres) of all U.S. forest land are reserved from commercial timber harvest in wilderness, parks, and other land classifications (Darr 1995). Since 1952, the area of timberland has decreased by 4%, or about 19.3 million acres. This decline has resulted primarily from conversion of public timberland into wilderness or other land uses that do not permit timber harvest.

In the 1920s, timber net growth (accumulated growth after loss to mortality) was about one-half the rate of harvest (Darr 1995) (Figure 10.15). By the 1940s, improved forest net growth rates (partly because of forest protection from fire), as well as declines in harvest rates, resulted in timber net growth and harvest coming into balance (Powell et al. 1993).

Timber Productivity across the Regions

U.S. timberlands cover a wide range of latitudes, elevations, precipitation, and soils, which greatly influence their productivity. Potential productivity measures the ability of a forest to produce wood volume. The Pacific Northwest region has some of the most productive forest land (i.e., land capable of producing 120 ft. 3/acre/year) (Powell et al. 1993). However, large areas in the South covered by loblolly-shortleaf pine and oak-hickory forests are also highly productive. Most of the less productive land is in the West, where forests occur at higher elevations and latitudes, or in piñyon-juniper woodlands, where precipitation is limiting.

Based on the amount of growing stock existing now, timber supplies should meet needs over the next 15 to 20 years. However, beyond 2020 there is considerable concern over wood shortages. By definition, growing stock includes live, sound trees suited for roundwood (timber) products. U.S. timberland contains an estimated 24.3 billion cubic meters (858 billion cubic feet) of timber, in which 92% is in growing stock. From 1952 to 1992, net volume per acre has increased 33% (Powell et al. 1993). Volume growth occurred despite harvests that make the U.S. the world's number one timber producer since World War II (Sedjo 1995). The greatest gain in volume occurred in the South (104%) followed by the North (95%), and the Rocky Mountain region (27%). Recent studies indicate that the forest biomass in the northern Rocky Mountains is 30% greater now than in 1800 (Sedjo 1995). However, the Pacific Coast region has lost about 4% of its timber volume over the last 50 years.

Hardwoods and softwoods supply different types of forest products. Hardwood growing stock is concentrated in the northcentral and southeastern states whereas softwood volume is highest in the Pacific Coast states. Some 57% of the growing volume is in softwoods and 43% hardwoods. About 66% of softwood timber is in the West, and 23% in the South. At least 90% of hardwood volume is in the East.

Timber Removals

Almost all of the timber harvested in the U.S. comes from second-growth forests (those that have regrown after an earlier clearing) or plantation forests. The undisturbed native forests of the U.S. have been almost completely withdrawn from timber harvest. This is because they are now part of park preservation, the wilderness system, or in other set-aside public management (Sedjo 1995). Average timber removals have increased each decade since the 1950s but volume growth has outpaced harvest from the 1950s forward (Figure 10.15). In 1991, almost 55% of all timber removals came from the forests of the South (up from 45% in 1970). Pacific Coast forests provided 23%, the North 17%, and the Rocky Mountains 5% (Powell et al. 1993). Historical timber harvests in the U. S., along with primary uses, are shown in Figure 10.16.

Who Owns the Nation's Forests?

Some 34% of all forest land is federally owned. Most of this land is held by the U.S. Forest Service with lesser amounts under Bureau of Land Management control (Darr 1995). The Forest Service protects and manages 155 national forests which contain 191 million acres (approximate size of Texas). About 18% of this land is designated as wilderness. The area under Forest Service management has remained somewhat constant for many decades (Figure 10.17).

The majority (58%) of U.S. timberland is privately owned. Of the 42% publicly owned, only 18% of commercial timberland is located in the National Forest System. By a wide margin, national forest land is found mainly in Pacific Coast and Rocky Mountain regions (Figure 10.18). Timber corporations own 14% of the nation's timberland area. Much of this corporation land is found in the South. Individual private ownership of forest lands is highest in the eastern U.S.

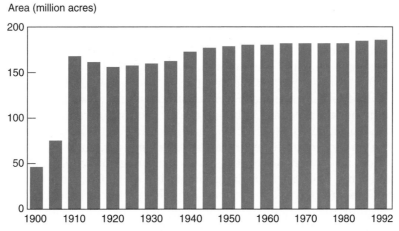

FIGURE 10.17 The total area of forest land under supervision of the USDA-Forest Service from 1900 to 1992 (courtesy of USDA-Forest Service).

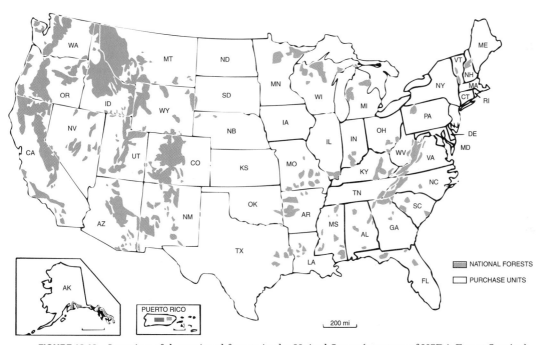

FIGURE 10.18 Location of the national forests in the United States (courtesy of USDA-Forest Service).

FOREST MANAGEMENT

Forest Management Defined

The Society of American Foresters (SAF) defines forestry as "the science, the art, and practice of managing and using for human benefit the natural resources that occur on and in association with forest lands." Forestry, therefore, includes much more than just timber management. It involves watershed hydrologists, wildlife and fisheries biologists, range scientists, and a host of other specialists, in addition to those working most closely with timber resources (silviculturalists). Recreation is a major activity on forest lands. This form of use continues to require ever increasing management on both public and private forests. In 1996, more than 800 million people visited the National Forest System to participate in recreational activities such as: camping, hiking, fishing, hunting, and wildlife viewing.

"Forest management" draws on a family of subjects and is variously interpreted to include mensuration, regulation, valuation, and preparation of working plans. It can also cover financial aspects of forestry (Sharpe et al. 1976). It is the business arm of forest science. Mensuration addresses the measurement of timbered areas, whole forests, single trees, logs, and other pieces and units of forest products. Regulation involves forest inventory and planning activities needed to determine harvest schedules. Valuation puts a monetary value on forest properties and is critical in making sound financial decisions. Although timber production remains a major objective in forest management, other objectives are becoming increasingly important to foresters and the public. In addition to traditional objectives, such as production of game, recreation, and water, other products such as nongame wildlife, biodiversity, esthetics, and forest health are receiving increasing emphasis. As a consequence, assessment of these resources has become as much a part of forest management and public lands as forest mensuration. Forest management, therefore, is an interdisciplinary activity requiring the services of diverse specialists in addition to the traditional forestry focus on tree production.

Silviculture deals directly with managing forest growth and composition. Silviculture draws heavily on ecology, soils, dendrology, and silvics. Dendrology provides a means for classifying and recognizing trees. Silvics addresses such environmental factors as climate, slope, soil, and fire as they relate to tree growth and health. Spurr (1979) proposed three definitions of silviculture, which collectively identify the aims of silviculture: (a) the art and science of producing and tending a forest; (b) the application of silvics in treatment of a forest; and (c) the theory and practice of controlling forest establishment, composition, and growth. Overall, silviculture is at the heart of forestry. It is about decision making, and it enables people to get what they desire from forests whether it be timber or a mix of timber and other goods and services. Done correctly, silviculture avoids forest abuse and has as its principal aim ensuring that forests remain productive for future generations. Overall, the chief purpose of silviculture is to create and maintain the kind of forest that will best fulfill the objectives of the owner.

Criteria Used to Classify Stands

Traditionally, silviculture has been based on the management of forest units called stands. A forest stand is defined as an area of forest whose site conditions, history, species composition, and structure are sufficiently uniform to distinguish it from adjacent areas (Figure 10.19). A stand can contain a few to several hundred acres. Stands can be grouped into forest compartments.

FIGURE 10.19 A stand of jack pine in northern Michigan (courtesy of USDA-Forest Service).

Stands can differ in age, species composition, density and in growth potential as determined by site quality. All of these factors determine how much time is needed for a stand to produce merchantable wood products. The segment of time between stand establishment and harvest is referred to as rotation length.

Age and Size Distribution

Stand structure refers to the distribution of age and/or diameter classes and of crown classes. The distribution of age classes across the entire population of trees is important because trees of near equal age will arrive at harvest size at the same time. Age class distribution can be used to predict whether the flow of timber products from a stand will be continuous or interrupted for unacceptably long periods (Figure 10.20). The latter case might indicate some form of stand treatment to improve the harvest schedule. In even-aged stands, all live trees were established within a narrow period of time so that the difference between youngest and oldest does not exceed a specified period of time. For example, if the stand will require 100 years to reach harvest size from time of establishment, a stand can be considered even-age if all trees are within 20% (or 20 years) of one another in age. In uneven-age stands, there are, by definition, at least three distinct age classes (Figure 10.20). Balanced uneven-age stands contain three or more different age classes with each class occupying about the same area. Age classes are, therefore, spaced at uniform intervals all the way from newly established reproduction to trees near harvest age. Irregular uneven-age stands do not contain all the age classes necessary to ensure that trees will arrive at harvest age at regular short intervals. Uneven-age virgin stands and culled over stands usually are irregular in age distribution.

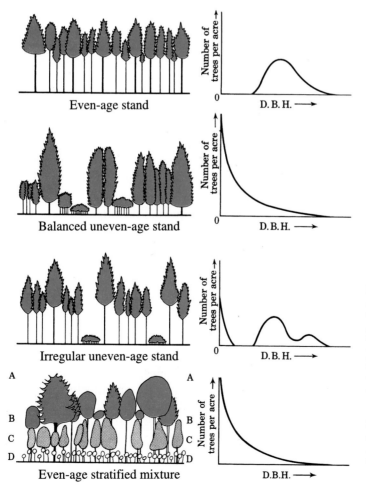

FIGURE 10.20 Typical examples of even-age and uneven-age stands. Trees are of the same species. The diameter at breast-height (D.B.H.) indicates the relative size distribution of each stand condition (Smith, David M. 1986. *The Practice of Silviculture.* New York: John Wiley and Sons. Used with permission of John Wiley and Sons).

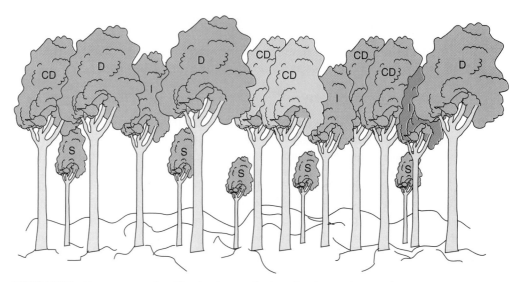

FIGURE 10.21   Tree canopy classification in stands. D = dominant; CD = codominant; I = intermediate; S = suppressed (adapted from Allen and Sharpe 1960).

Simple descriptions of tree sizes in stands can quickly convey the overall structure of a stand. Seedlings or seedling stands, for example, contain young trees only a few feet tall and having stem diameters less than one inch. A sapling stand contains trees at least 4.5 feet tall and up to 4 inches in diameter. Pole timber stands have trees 4 to 10 inches D.B.H. [diameter at breast height (4.5 feet above ground)]. Sawlog stands comprise trees suitable for conversion to lumber and are usually at least 11 inches D.B.H. Finally, an over-mature stand contains trees that have declined in growth rate because of old age and loss of vigor.

Terms are also used to describe the position of individual tree canopies in the stand (Figure 10.21). This indicates which trees have the most and least resources needed for future growth:

    (i) Dominant—Refers to trees that extend above surrounding individuals and capture sunlight from above and around the crown.

  (ii) Codominant—Refers to a tree that extends its crown into the canopy and receives direct sunlight from above but limited sunlight from the sides. One or more sides of a codominant tree are crowded by the crowns of dominant trees.

 (iii) Intermediate—A tree with a crown that extends into the canopy with dominant and codominant trees. These trees receive little direct sunlight from above and none from the sides. Crowns generally are small and crowded on all sides.

 (iv) Overtopped (Suppressed)—A tree that cannot sufficiently extend its crown into the overstory and receives no direct sunlight either from above or the sides. Overtopped trees that lack shade tolerance lose vigor and die (overtopped is synonymous with suppressed).

Species Composition

Stands can be described as mixed or pure depending on species composition. In general, if 80% of a stand is composed of a single species, it can be considered a pure stand. A mixed stand would be one in which two or more species occur and one species accounts for less than 80% of the total stems in the stand.

Classification Based on Density

Density refers to the physical space that the trees take up in the stand. It is a basic measure of the growing space allowed for each tree. One method of describing density is to

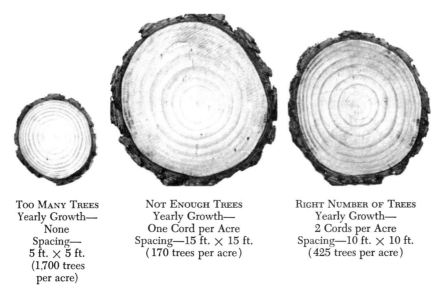

TOO MANY TREES
Yearly Growth—
None
Spacing—
5 ft. × 5 ft.
(1,700 trees
per acre)

NOT ENOUGH TREES
Yearly Growth—
One Cord per Acre
Spacing—15 ft. × 15 ft.
(170 trees per acre)

RIGHT NUMBER OF TREES
Yearly Growth—
2 Cords per Acre
Spacing—10 ft. × 10 ft.
(425 trees per acre)

FIGURE 10.22   Effect of tree spacing and competition for nutrients on annual tree growth as shown by growth rings. These three trees of the same species grew in the same general type of soil and climate and were ten years old when cut (courtesy of USDA Natural Resources Conservation Service).

count the number of trees on a unit area, usually an acre. Density can then be expressed, for example, as the number of stems per acre. This expression provides a straightforward means of describing the density of seedling and sapling stands because most of the trees will be of nearly equal size. As trees get larger, or where stands are uneven-aged, their diameters are less uniform. Therefore, some means must be used to account for this. Such an expression is basal area, which refers to the sum of the square feet of bole area of the individual trees on an acre (e.g., 150 ft²/acre).

Stocking refers to the density of a stand with respect to a standard. A number of approaches have been developed for estimating stocking. Reineke (1933), for example, developed a measure of stocking that equates average tree diameter and number of trees per acre to density. His approach recognizes a concept that is fundamental to the practice of silviculture. This concept is that the total biomass of trees on a site has an upper limit and, as that limit is reached, the competition for light and nutrients among trees will increase. Consequently, some trees will suffer mortality as this natural thinning process occurs.

Stands providing insufficient space for trees to reach merchantable size are "overstocked" (Figure 10.22). There are plenty of individual trees in even-age pine stands, for example, but they are too small for the sawmill to process. In "understocked" stands, trees are so widely spaced that, even with full growth potential realized, crown closure will not occur. With proper density management, growth will be distributed across fewer stems and mortality will not be a significant factor.

Classification Based on Site Quality

Often there is a need to express the productive capacity of a forest stand as it is determined by the ability of a given area to grow trees. The site index is the most commonly-used measure of site quality, or site productivity. It is based on the height attained by the dominant tree in a stand at some base age, usually 50 or 100 years. For trees grown over shorter rotations, the base age can be less. For example, in managing fast-growing southern pine plantations, a base age of 25 years is used.

To determine the site index of a stand, the total height and age of 5 to 8 dominant and codominant trees are measured and compared to the average total stand height and age using base curves (see Figure 10.23). Such a comparison will indicate whether a stand is high, low, or average in productive capacity since other measurements of

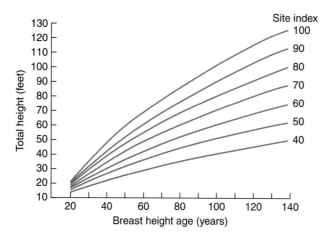

FIGURE 10.23   Site index curves for young-growth ponderosa pine in northern Arizona (from Minor 1964).

yield follow height growth. The total height of dominant and codominant trees in a stand is the best measure of site potential because trees grow taller on good sites than on poor sites. In discussing site index, one might say that it is 100 if trees on a specific site reach a height of 100 feet in 100 years, if that is the base age being used. A site index of 80 would indicate that trees are only 80 feet tall at 100 years. Compared to the first site, this site would be less productive for growing timber. The example of a ponderosa pine site index illustrated in Figure 10.27 assumes that tree heights at different stand ages will differ in a proportional manner from one base site index to another.

We have cited several terms and criteria that can be used to describe stands. When these terms are combined, a fuller description can be obtained for communicating stand information. For example, one could say the area of interest is a fully stocked and even-age, pure stand, or is an uneven-age, mixed, understocked stand on a poor site.

The Concept of Shade Tolerance

The concept of shade tolerance is critical to understanding silviculture and must be addressed before proceeding with stand treatments. "Shade tolerant" species can grow in the shade of others while shade intolerant species grow poorly, if at all, under the same conditions. This fact is critically important for understanding forest succession and for successfully regenerating managed forests (Figure 10.24).

To illustrate the importance of shade tolerance, we will use the example of loblolly pine, a species of major economic importance to southern forests. Loblolly pine cannot grow in stands of larger and older trees that have a high degree of canopy closure. To establish loblolly pine, it is usually necessary to cut all the overstory trees in a designated area to remove the influence of shade.

Additional examples of species needing large amounts of sunlight include ponderosa pine, yellow poplar, and Virginia pine. All of these species are categorized as shade intolerant. In contrast, northern hardwoods such as sugar maple and yellow birch grow well under a relatively closed canopy and can be regenerated without concern for the shade by trees occupying the overstory (Perry 1994). Examples of western species that are shade tolerant include Engelmann spruce, redwood, and western hemlock. Between the extremes of shade tolerant and intolerant species are trees such as Douglas-fir, white oak, and sycamore.

When subjected to increasing levels of light, shade tolerant species will reach their compensation point at lower light intensities than will intolerant species. The compensation point is met when light is sufficient to allow photosynthesis to occur at a rate equal to respiration. Above the compensation point, photosynthesis makes a net contribution to tree growth. Below this point, more energy is being consumed than is produced. Overall, this means that shade tolerant trees can grow under light levels that would result in a net loss of energy in more light demanding species.

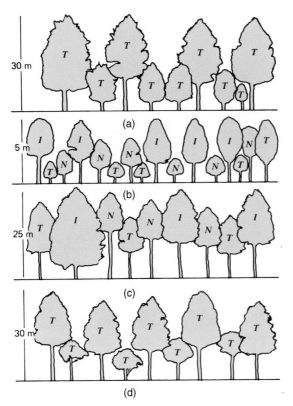

FIGURE 10.24 Schematic view of the distribution of tree crowns by tolerance class and height for four stages of succession initiated by severe disturbance: (a) Before disturbance, (b) Immediately after disturbance (5 years), (c) Midpoint of succession (50–75 years), (d) Full recovery and return to predisturbance conditions (T = tree of tolerant species, N = tree of intermediate species, I = tree of intolerant species) (from Young 1982).

STAND MANAGEMENT

Silvicultural Systems

Stand management is an intensively planned activity involving a silvicultural system. A silvicultural system is a planned program of treatments throughout the life of the stand to achieve stand structural objectives based on integrated resource management goals (see Chapter 5 for planning concepts). A silvicultural system includes integrated harvesting, regeneration, and stand-tending methods or phases. It covers all activities for the entire length of a rotation or cutting cycle. Harvesting techniques (also called regeneration methods because they are the first step in establishing the next stand) are coupled with subsequent intermediate treatments, such as thinning, weed control, to form silvicultural systems. Any reproduction, thinning, or other intermediate method applied without consideration for the other steps needed to manage a stand from establishment to harvest fails to meet the integrative requirements of silviculture systems (Daniel et al. 1979). These are simply treatments imposed independently sometime between the establishment and the harvest phases.

Cutting and Reproduction Methods

The reproduction methods can be categorized into those that produce even- and uneven-age stands from seed and those produced by coppicing. Clearcutting, seed-tree, and shelterwood methods are used to produce even-age stands. The selection method is used to produce uneven-age stands.

Clear-cutting. This method removes the entire stand with one cutting (Figure 10.25). Reproduction is obtained artificially, by natural seeding from adjacent stands, or from the cut trees themselves. In the strict sense, virtually all trees are cut, but at times all trees that cannot be used profitably are left and cutting may resemble a selection thinning. Species that become established naturally with full exposure are assumed to benefit from clear-cutting. Heavy partial cutting can provide the same benefit. This

FIGURE 10.25   A clear-cutting of Douglas-fir where the stand of timber in the background has seeded the logged area (courtesy of American Forest Products Industries).

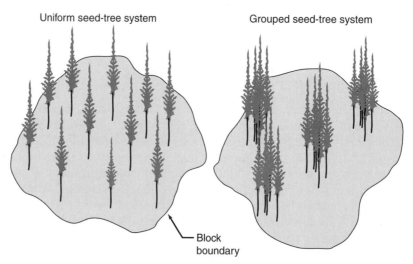

FIGURE 10.26   Seed-tree method (from Ministry of Forestry, British Columbia Forest Service).

method has been applied extensively in North America to Pacific Coast Douglas-fir forests. Douglas-fir is replaced by more tolerant species in the absence of wildfire. Western hemlock, western redcedar, and true firs (*Abies* sp.) are tolerant species.

Seed-Tree Method.   This method calls for removing all trees except a small number of trees left to provide seed for natural restocking. Trees can be spaced to assure uniform seed distribution or grouped to provide wind protection (Figures 10.26 and 10.27).

Shelterwood Method.   This method removes part of the stand to allow room for reproduction. Mature timber is removed in a series of cuttings that extend over a relatively short time (Figures 10.28 and 10.29). The remaining trees shelter the new trees until they are established, then they are removed. This approach prevents the site from being dominated by species that require open exposure to sunlight and encourages even-age reproduction.

FIGURE 10.27
Application of the
seed-tree method on
loblolly pine forest in
South Carolina (cour-
tesy of USDA-Forest
Service, Dan Todd
photographer).

FIGURE 10.28
A shelterwood cut-
ting in eastern
Oregon (courtesy of
USDA-Forest Service,
Roy Filloon photog-
rapher).

Selection Method. This method involves harvesting the oldest or largest trees in a
stand at repeated intervals. The interval between cuts in the same stand is called the
cutting cycle. The entire stand is never completely cut (Figures 10.30 and 10.31).
There is no rotation age at which the mature crop is harvested, as is done under even-
age management. Trees are cut individually or in small groups. Reproduction is an
on-going process because openings are created indefinitely. The age of maturity and
the cutting cycle determine the number of age classes present in the stand. Remaining trees protect seedlings and shade tolerant species are encouraged.

Coppice Forest Methods. This category is addressed separately because regeneration
is derived primarily from sprouts and suckers as opposed to seed (Figure 10.32).
Trees may grow from dormant buds in stumps. The coppice method can be used to
provide browse for herbivores, such as deer.

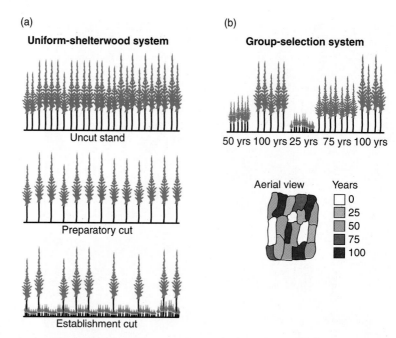

FIGURE 10.29 (a) Uniform-shelterwood and (b) group-shelterwood methods (courtesy of Ministry of Forestry, British Columbia Forest Service).

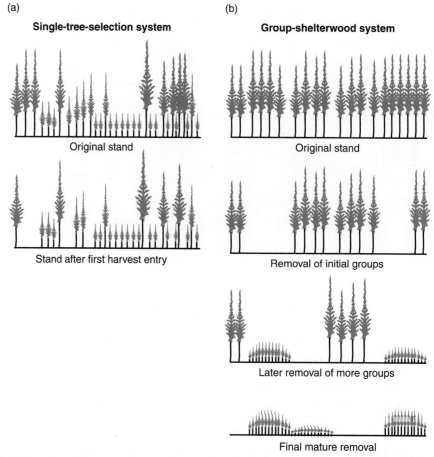

FIGURE 10.30 Single-tree-selection method (a) and group-selection method (b) (courtesy of Ministry of Forestry, British Columbia Forest Service).

FIGURE 10.31 The selection method of cutting applied to a mixed hardwood stand in Michigan (courtesy of USDA-Forest Service, Leland Prater photographer).

Coppice system

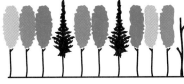

Uncut stand

10 years after harvest

FIGURE 10.32 The coppice method (from Ministry of Forestry, British Columbia Forest Service).

INTERMEDIATE TREATMENTS

An intermediate treatment is any manipulation in a stand applied between regeneration periods. Such treatments are used to ensure the desired composition, stem quality, spacing, and growth performance in a developing stand. Release cuttings cover numerous types of thinning operations done under different stand conditions. Thinning is aimed primarily at reducing stand density to increase growth of residual trees. Commercial thinning involves use of trees that would ultimately die of suppression. Thinnings are called precommercial if the trees cut are too small to be merchantable. In intensively managed stands, trees can be pruned to improve main stem form and reduce knotted wood. Fertilization can accelerate growth but is commonly used only in intensively managed stands. Prescribed burning reduces fire hazard and controls understory competition.

EMERGENCY CUTTINGS OR THINNINGS

Salvage and sanitation cutting represent a separate category of treatments applied on an emergency basis. Salvage cutting removes trees that are dead, dying, or badly damaged, where factors other than competition are involved. Usually, the primary goal is to salvage their timber value before they decay. Sanitation thinning (or cutting) removes trees that have been attacked or are vulnerable to attack by insects or pathogens.

FIGURE 10.33   Harvesting timber on the Lassen National Forest in California (courtesy of U.S. Dept. of Agriculture).

This is done to reduce inoculum or insect populations so that the pest is less likely to spread to other trees.

TIMBER HARVESTING

Timber harvesting refers to cutting, felling, and removal of wood for off-site use (Figure 10.33). Lopping is done after felling. It involves the cutting of branches and tops of large trees so that the resultant slash will lie close to the ground and decay more rapidly. A landing is any place on or adjacent to a logging site where logs are assembled for further transport. Yarding refers to moving logs from the point of felling to a landing. Cable yarding involves the use of a steel cable attached to a powered winch. Variations of this system include high-lead yarding and skyline yarding. A skidder is a rubber-tired machine with a cable winch or grapple used to drag logs out of the forest. Skidding can also involve the use of tractors, horses, or specialized logging equipment to collect felled logs at a landing. Skidding methods vary in their impact on soils and the remaining stands.

TREE PLANTING

Forests can regenerate by natural means from natural seeding or coppicing as discussed. In some cases, such as heavily burned areas, seed trees may not be available. Those areas can be artificially seeded or planted to speed the recovery process. On productive forest land, planting may accelerate early growth and adequate stocking. For whatever the purpose intended, millions of acres have been either seeded or planted over recent decades (Figure 10.34). Planting is usually preceded by some type of site preparation. This may involve mechanical or herbicide treatments to reduce weed competition. On poorly drained soils, the land surface can be altered by forming beds that permit water to drain more freely from the site.

Seedlings are produced under nursery or greenhouse conditions. After a few years of growth, nursery seedlings are lifted from the soil while in the dormant condition and stored under refrigeration until planting. Their roots are bare until they are replanted. Greenhouse grown seedlings are produced in various types of containers.

FIGURE 10.34 Pine trees planted on a hillside in Mississippi (courtesy of U.S. Department of Agriculture).

They require less time (months instead of years) to develop because of the optimal growing conditions provided. Their root systems are removed from the containers before planting. In contrast to bare rooted seedlings, the roots of containerized seedlings remain protected by the growing media. Because they do not lose contact with the soil media, they experience less transplant shock under certain conditions. Over the years, considerable knowledge has been gained concerning which types of seedling planting stock (i.e., "stock types") will perform best under certain conditions.

Overall, successful artificial regeneration requires careful attention to the details of seed source, nursery practices, and planting practices, as well as a thorough evaluation of environmental conditions (Tappeiner et al. 1997). Seed selection zones are used to ensure that future trees are adapted to the planting site. The forest industry has made large investments in genetic improvement of planting stock. Superior tree selections made over many decades have enabled industry to plant seedlings that grow more rapidly and produce better forest products (Figure 10.35).

FOREST PROTECTION

Forest Fire

Fire, insects, and disease are natural components of forests, but cause considerable mortality and growth loss. After the great fires of the early 1900s burned massive areas and took many lives, the U.S. Forest Service and other government agencies joined forces to develop highly effective fire suppression programs (Figure 10.36). In the decades that followed, fire loss steadily declined (Figure 10.37). Fire science was also born as a forestry discipline. Better knowledge of fire behavior and fire danger ratings have greatly reduced loss of life and property from wildfire.

Insects

The threats insect pose to forests are less obvious than fire, but mortality and growth losses can be substantial. The types of insects that threaten forests are highly diverse,

FIGURE 10.35   A mechanical tree shaker comes in handy when it is time to harvest ripe cones from a superior pine tree in a seed orchard (courtesy of U. S. Dept. of Agriculture).

FIGURE 10.36   Fighting forest fires involves a variety of equipment and skills (courtesy of USDA-Forest Service).

but can be grouped by the type of damage they cause. Broad groups include defoliators; sapsuckers; bud, twig, seedling, and root feeders; and cone and seed insects. Bark beetles have been especially destructive over recent decades (Figure 10.38). The southern pine beetle (*Dendroctonus frontalis*) is the most damaging insect in the southeastern United States. It is responsible for killing large numbers of southern pines over wide areas. Other species of the same genus have caused extensive damage to western pines and to spruce found in Alaska.

The larvae of several moths and butterflies damage both conifers and deciduous species, but their greatest damage is to conifers. Spruce budworms infest large areas in the northeastern U.S. The Douglas-fir tussock moth has caused extensive damage

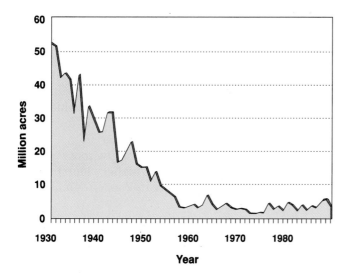

FIGURE 10.37 Trends in U.S. forest land burned by wildfire, 1930–1989 (from Powell et al. 1993).

FIGURE 10.38 The fir engraver beetle attacked this tree in California. The photo shows the egg galleries and larval mines 35 days after attack (courtesy of USDA-Forest Service).

in the northern Rocky Mountains. The Gypsy moth, prevalent in the East, primarily defoliates oaks. However, the larvae also feed on aspen, basswood, paper birch, and willow. Tent caterpillars feed on deciduous species. They live in tentlike silken webs that damage, but generally do not kill infested trees. Insects that feed on cones, seed, and acorns can greatly reduce natural reproduction in some years. These are just a few examples of some better known insect pests.

Insect control is extremely difficult because of the large areas sometimes infested, financial cost of control and potentially adverse effects to the environment. Because insecticides are increasingly viewed as a threat to nontarget beneficial insects, other control measures must be found. The value of natural pest enemies and the promotion of better forest health are receiving much greater emphasis in modern forestry.

Disease

Numerous groups of fungi, bacteria, and viruses damage trees. Fungi are especially damaging, particularly the rusts and those causing cankers, heart rot, wilts, and damping-off. Cankers mark locations where fungi have killed stem tissue. Heart rot, which devalues the wood, is especially damaging to mature stands. Wilts are caused by fungi that plug the water transport system of trees. The most notable example is Dutch elm disease. Damping-off occurs when newly emerged seedlings become infected and die. This problem is most significant in forest nurseries.

Trees can also be damaged by nematodes, which feed on tree roots or invade the vascular system causing wilt. Dwarf mistletoes (parasitic higher plants) cause extensive growth and mortality losses in conifers, especially in Western pines.

Exotic Pests

Pests introduced from abroad have increasingly become a problem. They are especially damaging because native tree species have not developed natural defenses against them as they have against native pests.

Abiotic Factors

Air pollution (see Chapter 6) reduces forest health over vast areas. Particularly troublesome are the acid precipitation, heavy metals, and ozone created by automobiles and industry. Acid precipitation forms high in the clouds where emissions of sulfur dioxide and nitrogen oxides react with water, oxygen, and oxidants. Acid precipitation and other pollutants can travel far from their places of origin, causing stress to forest species with varying abilities to respond. Pollution effects on forest health are complex and difficult to separate from stress-causing factors related to drought, overstocking, and planting species in plantations outside zones where they naturally grow. Acidic deposition can reduce cold hardiness. Ozone absorbed through leaf stomata impairs cells, reducing photosynthesis.

FOREST ECOSYSTEM MANAGEMENT _____

Ecosystem Management Defined

Over the past 25 years, foresters have been increasingly criticized for their forest management practices. Loud protest began in the 1970s when the public became concerned over the large clear-cuts on public lands in the western states. In the public's opinion, clear-cuts threatened the land's ability to recover, displaced wildlife, damaged fish habitat, threatened water quality, and degraded the esthetic quality of forest landscapes (Figure 10.39). Most clear-cuts were done responsibly, but some involved little regard for slope and the fragile nature of forest soils. Across the nation concern grew over the increasing fragmentation of forests by timber cutting and urban sprawl. In the Pacific Northwest, the amount of old growth forest has declined over 50% in the last 60 years (Bolsinger and Wadell 1993). Such destruction is perceived to threaten and endanger native wildlife and fish populations (Marcot 1997). Cutting of old growth forests of the Pacific Northwest has caused one of the most fierce environmental controversies to date. These forests have become symbolic of the conflict between immediate material need and the need to conserve our natural resources for future generations. Forest stewardship concerns go far beyond the spotted owl controversy (see Chapter 4). They can be heard throughout the world, from tropical rain forests to the boreal forests of Siberia. In the U.S., public concerns include the loss of nongame forest wildlife (Bolen and Robinson 1995), the conversion of biologically rich, mixed hardwood forests to pine plantations, the drainage of forest wetlands, and forest health deterioration resulting from fire, exotic pests, pollution, and global warming.

FIGURE 10.39   Land impaired by destructive logging practices (courtesy of USDA-Forest Service).

With the hope of regaining public trust, the U.S. Forest Service and other federal agencies have redefined the practice of forestry by putting greater emphasis on biodiversity, esthetics, and forest health. The name given to this new approach is ecosystem management. Silviculture is being revised in step with policy changes and new laws passed to regulate forest management.

A major change has occurred in how forests are perceived for management purposes. Whereas silviculture previously was practiced at the stand level, the emphasis is now given to forest ecosystems and forest landscapes. A change in management scale was considered necessary to provide greater protection for biodiversity, watersheds, and nature in general. It has provisions for monitoring levels of forest fragmentation. The adoption of geographic information systems (GIS) and computerized decision support systems (DIS) provide the tools for gathering and ciphering the vast amounts of information needed to manage at the landscape level.

The Pacific Northwest will serve as a prime example for discussing how silviculture is being revised in accord with ecosystem management principles. One aim of silviculture has been to develop managed forests that more closely resemble native forests in structural diversity. Forest structure can be viewed on both horizontal and vertical planes. Horizontal structure refers to differences in plant stature that might be encountered over a given area. Where patches of trees meet shrub or grass communities across a landscape, edges (ecotones) are created. Here the horizontal structure is said to be diverse. A landscape having this structure is believed to support more groups of wildlife (Hunter 1990). Vertical structure is well developed where trees and shrubs of diverse heights occur together in multiple canopies. The presence of multiple canopies supports a richer assortment of forest wildlife from small mammals to predatory birds.

Several approaches have been discussed to improve the structure of managed forests. One practice being recommended is to combine thinning with extended rotations. Such proposals usually include a series of silvicultural treatments during development of the stand to ensure creation of specific structural elements.

By extending rotations 50 to 300%, trees are allowed to grow to greater maturity and to develop more structural diversity than they would under conventional (shorter) harvest schedules (Franklin et al. 1997). After some time, trees larger than those harvested today will provide an upper canopy while openings created by natural disturbance will become occupied by smaller trees and shrubs. Longer retention provides numerous additional benefits. Managed forests occupying watersheds will be disturbed less frequently, visual impairment will be reduced, and the quality of the wood harvested will be superior. In older forests, carbon is held in vegetation and soil longer from the atmosphere, thus reducing atmospheric CO_2.

Thinning had not been widely applied in the Pacific Northwest until recently because markets for small wood materials were poor and large volumes of old timber were available. Because thinning usually accelerates understory development and succession, it can improve structural diversity. It can provide even more structural improvements if shade-tolerant conifers are planted in the understory to develop multilayered stands (Tappeiner et al. 1997). Thinnings that promote the development of understory vegetation and small openings can support increased wildlife populations (Curtis 1997). At the same time, thinning can provide income during the intermediate stages of stand development.

Pruning has been advocated and used primarily to increase the amount of clear wood produced in young stands. If pruning is initiated prior to stand closure, removal of lower branches can increase the amount of light reaching the forest floor and favor development of the woody and herbaceous understory. At present, pruning is applied mostly after stand closure has occurred.

Another recommended approach is to retain aggregates of living trees that will "life boat" diverse groups of plants and animals when trees are harvested (Franklin et al. 1997). Aggregates of trees can protect forest microclimate and the forest floor, which should be left undisturbed. In this manner, beneficial fungi, mosses, lichens, and forest animals, including insects and other invertebrates, will survive and spread over time into the surrounding cut over forest area. This approach resembles the shelterwood method mentioned previously. However, the trees left behind must be sound and windfirm so they persist in the forest landscape. Dominant and codominant trees should be used. They will provide critical protection and an energy source for forest fungi and other heterotrophs.

To overcome the adverse effects of forest fragmentation on wildlife, conservation biologists have generally recommended connecting "islands" of intact forests with narrow corridors of intact forest. The area surrounding the islands and the corridors may be sacrificed because of conversion to other uses. However, the retention of forest aggregates over managed landscapes, in conjunction with surrounding unharvested areas (removed from harvest), is a superior approach to maintaining biodiversity (Franklin et al. 1997). The movements of plants and animals will be less impaired and conservation will be better served. Conditions in the landscape matrix—that is, the "sea" surrounding undisturbed forest islands—are the most important factors controlling connectivity in landscapes.

The Clear-cutting Controversy

Clear-cutting typically involves the removal of all trees from a given area in single cutting followed by establishment of a new, even-aged stand. A clear-cut may consist of the whole stand or patches or strips of the stand (Figure 10.40). Nearly 65% of the U.S. is harvested by clear-cutting, which generally requires less cost and planning than other timber harvesting methods. Clear-cutting has other advantages that include less road building, greater timber harvest per acre, more efficient reseeding with improved stock, shorter time requirement for stand regeneration, and efficiency in removal of fire, insect, and storm damaged stands (Miller 1990). If small, irregular areas are cut, clear-cutting can improve habitat for many game species such as deer, elk, turkey, and grouse. Clear-cutting approaches with and without consideration of wildlife are shown in Figure 10.41.

Big block clear-cutting has become an important controversy in the United States, Canada, and countries with tropical forest. It often replaces diverse, uneven-aged stands of old growth with single-species monocultures that lack resistance to disease, insects, and fire (Miller 1990). These monocultures offer poor habitat for most wildlife species. Landscape esthetic quality is severely impaired by large block clearcuts. Severe soil erosion, water pollution, flooding, and landslides have often resulted from large block clearcuts on mountainous (steeply sloping) terrain (see Figure 10.39). Repeated clear-cutting can be quite damaging to the site because of soil

FIGURE 10.40 Patch clear-cuts on the Kootenai National Forest in Western Montana (courtesy of USDA-Forest Service).

erosion and compaction from logging equipment. Public opposition to clear-cutting has led to more emphasis on alternative cutting methods previously discussed (seed-tree, shelterwood selection, coppice methods).

Ecosystem management emphasizes the retention of large trees, snags, and downed logs that are found naturally where disturbance, such as fire or wind, have occurred (Franklin et al. 1977, Tappeiner et al. 1997). These were once considered a liability because they can be hazardous to forest workers and represented a loss from forest production. They are now recognized to provide critical habitat for diverse animals, including cavity nesting birds. Consequently, federal agencies promote their continued existence in forest management plans (Thomas 1979).

Ecosystem Management and Fire

Over past decades, fire danger ratings were standardized and predictions of fire spread and intensity have become increasingly reliable. However, the natural consequences of suppressing wildfire in areas where it has played a major role in shaping forest composition and structure have recently become better understood. In some forests, particularly those in the West, fire suppression has allowed tremendous fuel loads to accumulate, thus raising the probability of catastrophic fire (Figure 10.42). Where fire once occurred at frequent intervals, stand-replacing fires became unusual.

Mature trees survived as fire eliminated the understory and smaller diameter trees. In many areas the stage is presently set for all trees to be killed and for fires to be so hot that soils are damaged. Forests are also threatened by insect infestations, which continue to build in thick, overstocked stands.

In response to the adverse effects of suppression, forest and park managers increasingly view fire as nature's ally and not a force to be eliminated. An increasingly promoted strategy lets naturally ignited fires burn until they extinguish themselves. They are monitored to avoid human risks.

Prescribed fires as well as natural fires have become part of the working tools of today's fire managers, who have an ever-expanding set of land management goals (Figure 10.43). These goals include protection of commercial timber, managing fuels around rural developments (also called the urban interface), and reintroducing or maintaining the natural role of fire in park and wilderness ecosystems (Agee 1997).

(a)

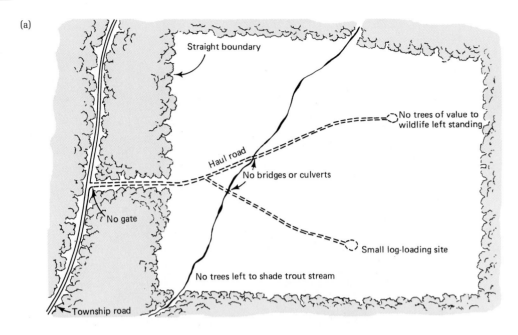

(b)

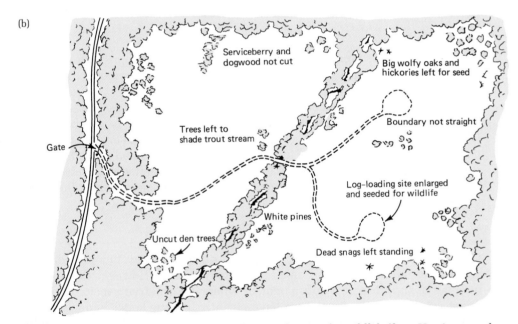

FIGURE 10.41   Clearcut without (a) and with (b) consideration for wildlife (from Hassinger et al. 1981).

Prescribed natural fire use was severely criticized after several allowed fires in Yellowstone National Park required millions of dollars to control. However, on balance, the benefits outweigh the risks in many areas. An approach being used over much of the West is to apply thin-logging and salvage logging to avoid the site deterioration that can result from a wildfire.

Gap Dynamics

Our understanding of succession has grown as ecologists have studied the role of forest gaps. Gaps in forest canopies can differ greatly in size, ranging from a single tree fall to the clearing of large tracts of forests by hurricanes, fire, or humans. These studies increasingly point to the importance of gaps in supporting biodiversity. According to Bazzaz (1996), forests are in a state of continuous change resulting from

FIGURE 10.42 Catastrophic wildfires due to fuel buildup and home construction are becoming an increasing problem on forest lands in the West (courtesy of USDA-Forest Service).

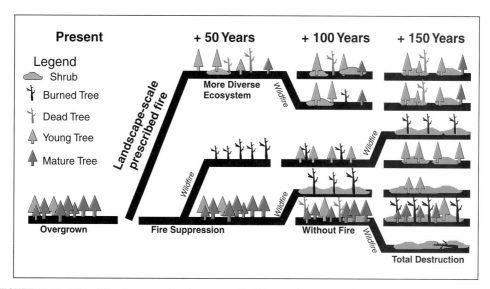

FIGURE 10.43 The risk of not conducting prescribed burns (courtesy of USDA-Forest Service).

the different patches of forest growth that develop in canopy gaps created by disturbance. Depending on gap size, patches can be trees of ages and species different from those dominating the main canopy. Branch gaps, those created when branches of existing trees are lost, might be filled simply by branches of existing trees extending into the open space. Gaps created by the loss of single trees can be filled by neighboring trees. On being released from competition, the small trees in gaps undergo accelerated growth. Large gaps and forest clearings in which much of the resident vegetation is destroyed by fire can be revegetated through seed dispersal from adjacent communities. Overall, the rate of gap creation and filling is determined by several intrinsic and extrinsic factors, and differs greatly among regions.

Forest Succession and Ecosystem Management

We will briefly discuss the relevance of forest succession to forest ecosystem management, referring the reader to Chapter 3 for a more complete discussion of succession. From the work conducted at the Hubbard Brook watershed over several decades, it appears that biomass accumulation following clear-cutting occurs in four stages: (i) reorganization, (ii) aggradation, (iii) transition, and (iv) steady state.

The first stage, "reorganization," marks the beginning of a new stand and lasts for one to two decades. During this time, the ecosystem actually loses total biomass, despite the accumulation of new living biomass in the plants. This stage is characterized by high availability of sunlight, nutrients (decaying roots and above ground litter), and ample water (few plants are competing for it). The time necessary for completion of the reorganization stage varies greatly from one environment to another yet remains critical to ecosystem recovery (Perry 1994). When forest disturbance is severe enough to cause soil loss, the site may revert back (i.e., retrogress) to a former state requiring colonization by more primitive pioneering plants.

During the "aggradation" stage, biomass accumulates to a peak value. The aggradation stage is further characterized by increased tree mortality caused by inter- and intraspecific competition for resources (Landsberg and Glover 1997). This stage marks the point where foresters intervene to reduce competition by thinning the stand. The growth of the remaining trees is accelerated because thinning "releases" them from competition for light and nutrients. The aggrading phase can vary considerably across forest species. The long-lived conifer forests of the Pacific Northwest, for example, accumulate biomass (both living and dead) for hundreds of years, whereas this period is much briefer in forests dominated by shorter-lived trees (Perry 1994).

During the "transition" stage, biomass declines. In this stage, the importance of understory plants is renewed (Landsberg and Glover 1997) because spaces occupied by tree canopies become vacant as some trees die, forming gaps.

The "steady state" arrives when the total amount of biomass fluctuates about the mean, meaning there are periods when more and less than average amounts of biomass have been accumulated. This stage corresponds to the climax stage of forest succession and is characterized by large accumulations of living biomass and coarse woody debris. This stage can have considerable structural diversity (Landsberg and Glover 1997), meaning there are trees of different species and heights (some dead, some alive), and in different stages of decomposition. Consequently, it can support diverse forms of wildlife, including all sorts of fungi, cavity nesting birds, small mammals, and northern spotted owls. However, according to Botkin (1990, 1993) forests will not persist in the climax stage indefinitely as previously reported. With increasing age, the forest becomes deficient in nutrients as it continues to "leak" them. As living biomass is continually lost, the forest can revert to a shrub or grass stage.

The upshot of Botkin's description is that no forest lasts forever. Nature is not static and changeless and there is no such thing as a balance in nature resembling a pendulum that finds its point of rest and remains there.

GLOBAL FOREST PROBLEMS

Two issues concerning forests are of global concern—deforestation and global warming. Deforestation is primarily a problem in countries outside the temperate zone. The conversion of forests in temperate regions to agriculture land has been reversed and forest area is now actually increasing (Figure 10.44). By contrast, the developing countries in the tropics are still in a forest conversion mode (Sedjo 1995). Deforestation remains intense in areas such as Latin America and Africa. Major movements are underway among nongovernmental agencies to reverse forest loss. Deforestation has been the focus of numerous world conferences of one type or another over the past decade.

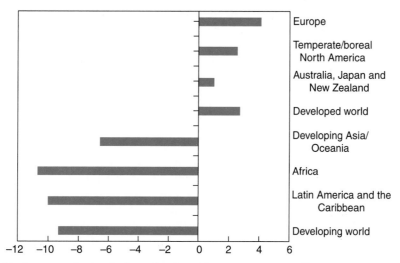

FIGURE 10.44 Losses and gains in forest land area from 1980 to 1990 (from FAO 1997 statistics).

Environmental decline associated with intensively managed forest plantations has received considerable comment, both positive and negative. In response to criticisms, numerous strategies have been developed to make plantations more habitable to diverse wildlife groups. On balance, plantations can be intensively managed to spare natural forests from human disturbance. Forests cover from about 3.4 billion hectares of the earth. To meet world wood demands, 20 to 40% of the growth from these forests must be harvested. High-yield forests, which are now 4 to 8% of Earth's forest area, might meet future world wood demands, depending on technological breakthroughs (Sedjo and Botkin 1997).

Forests could prove to be a close ally as global warming progresses (see Chapter 6). Because forests play a key role in sequestering atmospheric carbon, world forest restoration is considered a high priority by governments and environmentalists alike. The world's forests have been estimated to contain up to 80% of all above-ground terrestrial carbon and about 40% of all below-ground terrestrial carbon (soil, litter, and roots) (Ciesla 1995). About 37% of this carbon is stored in low-latitude (tropical) forests, 14% in midlatitude (temperate) forests and 49% in high-latitude forests (Dixon et al. 1994).

A leading group of atmospheric scientists have predicted an increase in mean global temperature of about 1.5–4.5°C (2.7–8.1°F) in the next century in response to the growing accumulation of greenhouse gases. Temperature changes on this scale would provide natural systems with a warmer climate than any experienced during the last 100,000 years. Some scientists say this change in the global climate could occur many times faster than it did during any previous episode. The rate of change poses a more serious threat than the magnitude of change. If global warming occurs as projected, indirect forest damage could occur over vast regions. This would involve major tree die-offs in many areas and shifts in the potential areas of growth for each tree species. Insect infestations and fire frequency will likely become more frequent. Forests already threatened by pollution, fragmentation, and poor stand condition will face yet another major threat. Sea-level changes alone could pose a serious hazard to low-lying coastal forests. The mangrove forests of the tropical regions are especially vulnerable.

RECYCLING WASTEPAPER

One way to conserve forests is through greater recycling of wastepaper. Up to 50% of the world's wastepaper could be recycled, but only about 30% actually is. The United States, the world's highest consumer of paper, recycles about 30%. This compares unfavorably with most European countries, which recycle about 40%. The key to increasing recycling is requiring people to separate paper from other waste materials (Miller 1990). Otherwise, wastepaper dealers will not purchase it. Another approach is for governments to provide businesses with incentives for recycling. Currently the U.S. federal government is mandated to use recycled paper as much as possible.

REDUCING PAPER USE

Several approaches are available to reduce paper use. These include reduction of overpackaging, more complete use of paper by students (use of both sides), increased use of the Internet for advertising, greater purchasing of food and other goods in bulk quantities, and more use of electronic media in information transfer and storage. Product overpackaging is a major contributor to paper use and waste (Miller 1990). Oversized packages are used to make consumers think they are getting more for their money. According to Miller (1990), product packaging uses about 65% of the paper and 15% of the wood in the U.S.

CONCLUDING REMARKS

Forests play a key role in meeting our needs for material goods, but require careful management to avoid damage to all the other services they provide, including maintaining the earth's biodiversity. The forests of the U.S. and those of other developed countries have actually grown in area as deforestation continues at an alarming rate in countries experiencing a period of rapid economic transition. Key concerns in the U.S. are the condition of secondary forests and conservation of old growth forests. Forest management is in a period of rapid change because of the diverse demands of a changing society. However, its tools and concepts are becoming increasingly refined to meet a broad set of goals. Over the next decade, forests will be increasingly valued and protected in their natural state. At the same time, plantation forestry will provide an increasing share of the world's material timber needs. Global warming poses a serious threat to natural forests because some species will have too little time to adjust. Forests can, however, be used to reduce the threat.

LITERATURE CITED

Agee, J. K. 1997. Fire management for the 21st century (pp. 191–202). In K. A. Kohm and J. F. Franklin, eds., *Creating a forestry for the 21st century: The science of ecosystem management.* Washington, DC: Island Press.

Allen, S. W. and G. W. Sharpe. 1960. *An introduction to American forestry.* 3rd edition. New York, NY: McGraw-Hill.

Bazzaz, F. A. 1996. *Plants in changing environments.* Cambridge University Press.

Bolen, E. G. and W. L. Robinson. 1995. *Wildlife ecology and management.* 3rd edition. Upper Saddle River, NJ: Prentice-Hall.

Bolsinger, C. L. and K. L. Waddell. 1993. *Area of old-growth forests in California, Oregon, Washington.* USDA Forest Service. PNW Station Resource Bulletin PNW-RB-197.

Botkin, D. B. 1990. *Discordant harmonies: A new ecology for the twenty-first century.* New York: Oxford University Press.

Botkin, D. B. 1993. *Forest dynamics: An ecological model.* New York: Oxford University Press.

Botkin, D. B. and E. A. Keller. 1995. *Environmental science: Earth as a living planet.* New York: John Wiley and Sons.

Ciesla, W. M. 1995. *Climate change, forests and forest management.* FAO Forestry Paper 126. Rome.

Connell, J. H. and R. O. Slatyer. 1977. Mechanisms of succession in natural communities and their role in community stability and organization. *American Naturalist* 111:1119–1144.

Curtis, R. O. 1997. The role of extended rotations (pp. 165–170). In K. A. Kohm and J. F. Franklin, eds., *Creating a forestry for the 21st century: The science of ecosystem management.* Washington, DC: Island Press.

Daniel, T. W., J. A. Helms, and F. S. Baker. 1979. *Principles of silviculture.* 2nd edition. New York: McGraw-Hill.

Darr, D. R. 1995. U.S. forest resources (pp. 214–215). In E. T. LaRoe, G. S. Farris, C. E. Puckett, P. D. Doran, and M. J. Mac, eds., *Our living resources: A report to the nation on the distribution, abundance, and health of U.S. plants, animals, and ecosystems.* Washington, DC: U.S. Department of the Interior, National Biological Service.

Daubenmire, R. F. and J. B. Daubenmire. 1968. *Forest vegetation of eastern Washington and northern Idaho.* Washington Agriculture Experiment Station. Tech. Bull. 60. 104 pp.

Dixon, R. K., S. Brown, R. A. Houghton, A. M. Solomon, M. C. Trexlar, and J. Wisniewski. 1994. Carbon pools and flux of global forest ecosystems. *Science* 263:185–190.

Eyre, F. H., ed. 1980. *Forest cover types of the United States and Canada.* Washington, DC: Society of American Foresters. 148 pp.

FAO. 1997. *State of the world's forests.* Rome: Food and Agriculture Organization of the United Nations. 199 pp.

Franklin, J. F., D. R. Berg, D. A. Thornburgh, and J. C. Tappeiner. 1997. Alternative silvicultural approaches to timber harvesting: Variable retention harvest systems (pp. 111–139). In K. A. Kohm and J. F. Franklin, eds., *Creating a forestry for the 21st century: The science of ecosystem management.* Washington, DC: Island Press.

Franklin, J. F and C. T. Dyrness. 1973. *Natural vegetation of Oregon and Washington.* USDA Forest Service General Technical Report PNW-8. 417 pp.

Hall, F. C. 1983. Forest lands of the continental United States (pp. 130–139). In USDA, *Using our natural resources, 1983 Yearbook of Agriculture.* Washington, DC: U.S. Government Printing Office. 671 pp.

Hassinger, J. C., E. Schwarz, and R. G. Wingard. 1981. *Timber sales and wildlife.* Harrisburg, PA: Pennsylvania Game Commission.

Hunter, M. L. Jr. 1990. *Wildlife, forests, and forestry.* Englewood Cliffs, NJ: Prentice-Hall.

Kohm, K. A. and J. F. Franklin, eds. 1997. *Creating a forestry for the 21st century: The science of ecosystem management.* Washington, DC: Island Press.

Landsberg, J. J. and S. T. Glover. 1997. *Applications of physiological ecology to forest management.* New York: Academic Press.

Luken, J. O. 1990. *Directing ecological succession.* London: Chapman and Hall.

Marcot, B. G. 1997. Biodiversity of old forests of the west: A lesson from our elders (pp. 87–106). In K. A. Kohm and J. F. Franklin, eds., *Creating a forestry for the 21st century: The science of ecosystem management.* Washington, DC: Island Press.

McKnight, T. L. 1990. *Physical geography: A landscape appreciation.* 3rd edition. Upper Saddle River, NJ: Prentice-Hall.

Miller, G. T. 1990. *Resource conservation and management.* Belmont, CA: Wadsworth Publishing Company.

Ministry of Forests (B.C. Forest Service), The Province of British Columbia. 1995. *Silviculture systems guidebook.*

Minor, J. M. 1964. *Site-index curves for young-growth ponderosa pine in northern Arizona.* USDA, Rocky Mountain Forest and Range Experiment Station Res. Note 69. 6 p.

Pastor, J., D. J. Mladenoff, Y. Haila, J. Bryant, and S. Payette 1996. Biodiversity and ecosystem processes in boreal regions. In H. A. Mooney, J. H. Cushman, and

E. Medina, eds., *Functional roles of biodiversity: A global perspective.* New York: John Wiley & Sons.

Perry, D. A. 1994. *Forest ecosystems.* Baltimore: Johns Hopkins University Press.

Pfister, R. D., B. L. Kovalchik, S. F. Arno, and R. C. Presby. 1977. *Forest habitat types of Montana.* USDA For. Serv. Gen. Tech. Rep. INT-34. 174 pp.

Powell, D. S., J. L. Faulkner, D. R. Darr, Z. Zhu, and D. W. MacCleery. 1993. *Forest resources of the United States, 1992.* Gen. Tech. Rep. RM-234. Fort Collins, CO: U.S. Forest Service, Rocky Mountain Forest and Range Experiment Station. 132 pp.

Reineke, L. H. 1933. Predicting a stand-density index for even-aged forests. *Journal of Agricultural Research.* 46:627–638.

Schubert, G. H. and J. A. Pitcher. 1973. *A provisional seed-zone and cone-crop rating system for Arizona and New Mexico.* USDA For. Serv. Res. Pap. RM-1045. Fort Collins, CO: Rocky Mountain Forest and Range Experiment Station. 8 p.

Schulze, E. D., F. A. Bazzaz, K. J. Nadelhoffer, T. Koike, and S. Takatsukl. 1996. Biodiversity and ecosystem function of temperate deciduous broad-leaved forests. In H. A. Mooney, J. H. Cushman, and E. Medina, eds., *Functional roles of biodiversity: A global perspective.* New York: John Wiley & Sons.

Sedjo, R. A. 1995. Forests: Conflicting signals (pp. 177–209). In R. Bailey, ed., *The true state of the planet.* New York: Free Press. 470 pp.

Sedjo, R. A. and D. B. Botkin. 1997. Using forest plantations to spare natural forests. *Environment* 39(10):14–20, 30.

Sharpe, G. W., C. W. Hendee, and S. W. Allen. 1976. *Introduction to forestry.* New York: McGraw-Hill.

Smith, D. M. 1986. *The practice of silviculture.* New York: John Wiley and Sons. 527 pp.

Spurr, S. H. 1979. Silviculture. *Scientific American* 240: 76–82, 87.

Tappeiner, J. C., D. Lavender, J. Walstad, R. O. Curtis, and D. S. DeBell. 1997. Silvicultural systems and regeneration methods: Current practices and new alternatives (pp. 151–164). In K. A. Kohm and J. F. Franklin, eds., *Creating a forestry for the 21st Century: The science of ecosystem management.* Washington, DC: Island Press.

Thomas, J. W. (Tech. Ed.). 1979. *Wildlife habitats in managed forests: These Blue Mountains of Oregon and Washington.* Agriculture handbook 553. Washington, DC: USDA Forest Service.

World Resources Institute. 1994. *World resources: 1994–95.* New York: Oxford University Press.

Young, R. A. 1982. *Introduction to forest science.* New York, NY: John Wiley & Sons.

CHAPTER 11

Rangeland and Range Management

RANGELANDS DEFINED

The term rangelands refers to uncultivated land that provides forage and browse for large animals (Figure 11.1). Most rangelands in the world are characterized by low precipitation, thin soils, rugged topography, and/or cold temperatures. Generally, rangelands are managed using extensive (ecological) practices that center around controlling the kinds and numbers of grazing animals and their timing, frequency, and distribution of use. Unlike pasturelands, rangelands receive little in the way of routine agronomic inputs such as cultivation, irrigation, and fertilization. Most rangelands support native vegetation, but areas do exist in the western United States where nonnative plants such as crested wheatgrass have been established and maintained without periodic cultiva-

FIGURE 11.1   Rangeland in New Mexico grazed by cattle.

tion. Forage refers to standing plant material (grasses, forbs, shrubs) that is edible by livestock and wildlife. Grazing is the consumption of grasses and forbs, whereas browsing is the consumption of edible leaves and twigs from woody plants. Herbivory by grazing and browsing animals is the primary focus of rangeland management.

Historically the primary contribution of rangelands to humans has been forage for livestock. However, in countries such as the United States, where food supplies are abundant, water, wildlife, and recreation are overtaking forage as important rangeland products. Most rangelands, when properly managed, can provide all these products without being degraded. The concept of managing rangelands for a variety of uses and products is referred to as multiple use. Our discussion of range management follows Holechek (1994) and Holechek et al. (1998). Other textbooks that provide valuable information on range management include Vallentine (1990) and Heady and Child (1994).

RANGELAND MANAGEMENT DEFINED

Range management involves the manipulation of grazing by large herbivores so both plant and animal production will be maintained or increased (Holechek 1994a). The art and science of range management developed during the early twentieth century in the western United States when it became evident that declining forage production, severe soil erosion, and heavy livestock mortality were caused by poorly controlled livestock grazing. Manipulation of the intensity, timing, and frequency of livestock grazing were the primary cornerstones of early range management. However, since the 1960s the scope has broadened to also include manipulation of fire, wildlife, human activities, and various other factors. Still, the distinguishing feature of range management is the control of livestock grazing. Presently range management is defined as the manipulation of rangeland components to obtain the optimum combination of goods and services for society on a sustained basis (Holechek et al. 1998).

Basic Concepts

Range management is based on five key concepts (Holechek et al. 1998):

1. Rangeland is a renewable source.
2. Energy from the sun can be captured by green plants, which can only be harvested by the grazing animal.
3. Rangelands supply humans with food and fiber at very low energy costs compared to those associated with cultivated lands. Ruminant animals are best adapted to use range plants. Unlike human beings, ruminants have microbes in their digestive systems that efficiently break down fiber, which is quite high in most range plants.
4. Rangeland productivity is determined by soil, topographic, and climatic characteristics.
5. A variety of "products" are harvested from rangelands, including food, fiber, water, recreation, wildlife, minerals, and timber.

Generally forage production increases and more of the forage can be consumed by grazing animals without damage to the range as precipitation, temperature, and soil depth increase. The flat, humid ranges in the Great Plains and Southeast are much more productive and resistant to grazing than the arid rangelands in the intermountain West. The failure to recognize that different management practices are required for humid compared to arid areas has been a major contribution to rangeland degradation (Figure 11.2).

FIGURE 11.2 The humid, tallgrass prairie range in central Oklahoma shown at the top can withstand a grazing intensity of around 50% use of forage. In contrast, grazing intensity on the arid Chihuahuan desert New Mexico range in the bottom picture must be kept around 35% or less to avoid degradation.

TYPES OF RANGELAND

Climate, soils, topography, and human activities interact to cause differences in the types of rangeland that occur throughout the world. Basically rangelands can be categorized as grassland, desert shrubland, savannah woodland, forest, and tundra biomes (see Chapter 9). These biomes will be briefly discussed following Holechek et al. (1998).

Grasslands. Grasslands are the most productive in terms of forage for livestock and wildlife. They occur on about 24% of the world's land surface, and are dominated by plants in the family *Gramineae* (grasses). Grasses are characterized by a fibrous root system; hollow jointed stems; and fine, narrow leaves. Forbs, also important on grasslands, differ from grasses by having broader leaves, taproots, and solid, nonjointed stems (Figure 11.3). Shrubs and trees, a minor component of grasslands, have thick,

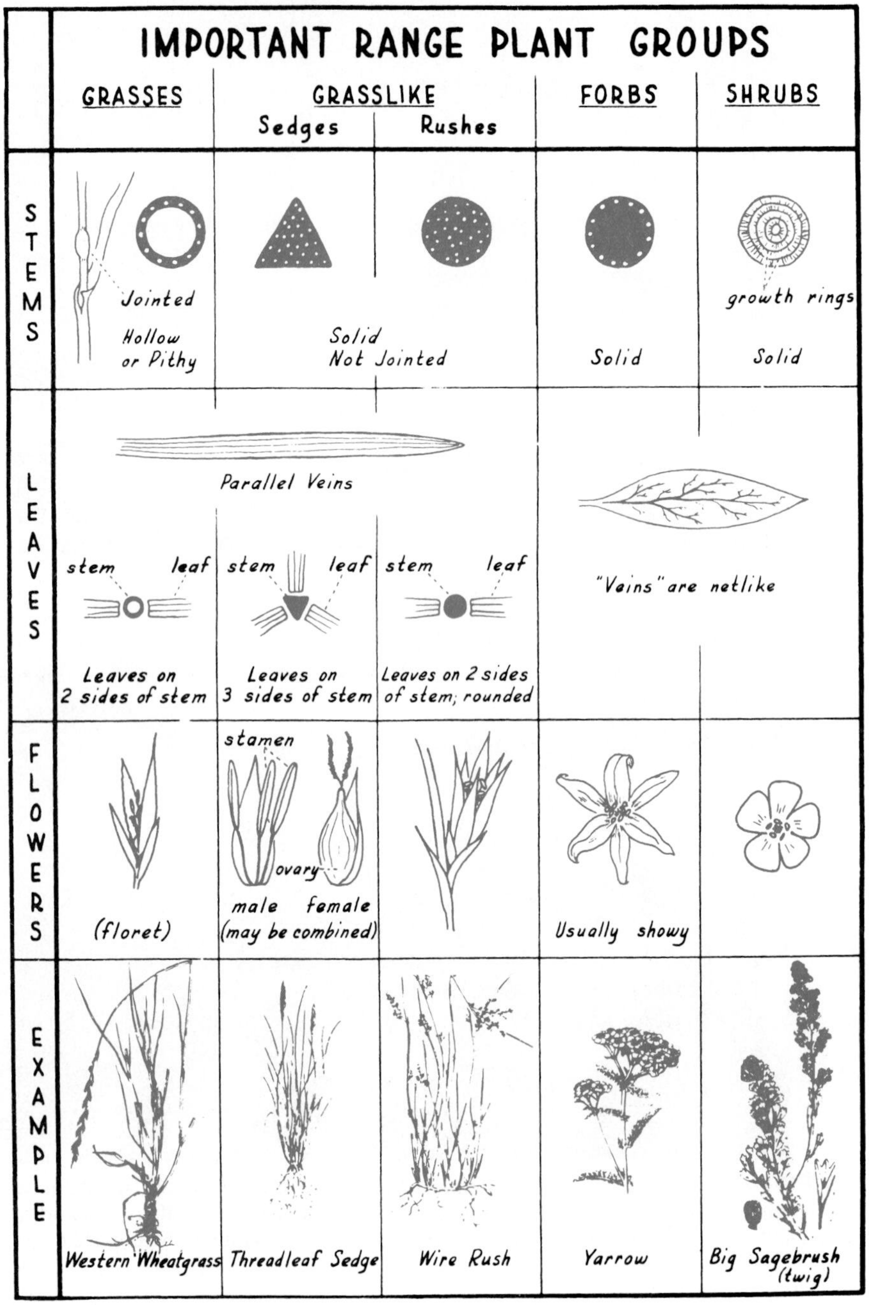

IMPORTANT RANGE PLANT GROUPS

	GRASSES	GRASSLIKE		FORBS	SHRUBS
		Sedges	Rushes		
S T E M S	Jointed / Hollow or Pithy	Solid / Not Jointed		Solid	growth rings / Solid
L E A V E S	Parallel Veins / stem leaf / Leaves on 2 sides of stem	stem leaf / Leaves on 3 sides of stem	stem leaf / Leaves on 2 sides of stem; rounded	"Veins" are netlike	
F L O W E R S	(floret)	stamen / ovary / male female (may be combined)		Usually showy	
E X A M P L E	Western Wheatgrass	Threadleaf Sedge	Wire Rush	Yarrow	Big Sagebrush (twig)

FIGURE 11.3   Characteristics of the major range plant groups (from Gay 1965).

woody above- and belowground parts. Shrubs differ from trees by branching at ground level rather than from a trunk above ground.

Climatically grasslands are favored by light, frequent rains during the summer, and dry winters. This maintains moist conditions during the growing period (summer) in the soil surface where most grass roots occur. Loamy and clay-loam soils favor grasslands because they slow moisture infiltration deep into the soil profile and keep it near the surface. Both the climate and soils of the Central Great Plains of the U.S. are highly favorable for grasslands (see Chapter 9).

Desert Shrublands.   Desert shrublands occur under the driest climatic conditions, and comprise about 30% of the world's rangeland. Generally precipitation is erratic with infrequent storms of high intensity. Soils are often sands or sandy loams. This climate-soils combination causes the soil surface to be dry for extended periods creating unfavorable conditions for grasses. However the coarse, deep-rooted shrubs that have the capability to collect moisture from large portions of the soil profile, are well adapted to desert conditions.

Forests. Forests account for about 30% of the world's rangelands. Most forests are characterized by high annual precipitation because large amounts of precipitation are needed to support the greater biomass of trees compared to grasses and shrubs. These wetter conditions cause greater leaching and lower fertility of forest compared to grassland soils. The clearing and burning of tropical forests to convert them into grazing land has been a major environmental concern in South American countries (see Chapter 9). There is doubt about the capability of soils in the tropical rain forests to sustain production of forage and crop plants.

Savannah Woodlands. Savannah woodlands are transition areas between grasslands and forests. They occupy about 8% of the world's rangelands and are typically characterized by scattered, low-growing trees with an understory of grasses. Many of the more humid grasslands will revert into savannah woodland without periodic fire.

Tundra. Tundra involves cold, treeless areas with low-growing grasses and shrubs in Arctic or high elevation regions. Because the deeper portions of the soil remain frozen throughout the year (permafrost), trees do not grow well on the tundra. About 5% of the earth's surface is occupied by tundra.

HISTORICAL PERSPECTIVE

Range management as a profession had its beginning in the United States in the 1890s. In this period, biologists such as Jared Smith first defined the problems of uncontrolled livestock grazing on western rangelands. Some of the problems Smith identified included replacement of desirable forages with unpalatable plants, compaction of soil by livestock, high loss of soil during periods of torrential rain, increases in prairie dogs and jackrabbits, and reduction in grazing capacity. Smith (1899) was the first to recommend basic practices that included control of livestock numbers, range rest periods, water development, brush control, and range seeding. These have become cornerstones of modern range management.

Generally, published accounts of livestock grazing problems and scientific studies on rangelands do not exist from other parts of the world prior to 1900. Therefore many consider the science of range management to be established in the United States. However, pastoral tribes in Asia and Africa have grazed livestock on rangelands for several thousand years. These herders used grazing rotation systems similar to some of the more sophisticated systems used in the United States today. Therefore, the concept of controlled livestock grazing on rangelands did not entirely originate in the United States. We refer the reader to Chapter 2 for a discussion of the history of rangeland policy in the United States.

As we move more into the twenty-first century, western rangelands will undoubtedly become increasingly important in providing open space, wildlife, recreation, and water for the American public. If present trends continue, livestock grazing will be deemphasized on the public rangelands in the West but will receive greater emphasis on Great Plains and eastern rangelands. During the past 15 years, many ranches in the West have subdivided into ranchettes of 40 acres or less and livestock are no longer produced. This trend will undoubtedly continue, and many more ranches will be subdivided. Public pressure has increasingly forced government agencies to emphasize recreation and wildlife on federal lands. As urbanization in the western states continues, this pressure will probably increase. The efficiency of livestock production in the West compared to the Great Plains and the East is much lower (Holechek and Hawkes 1993). The future use of rangelands in the United States will depend on what happens in other countries as well. As long as the world population grows at a rapid rate, pressure to increase livestock in developing countries also will grow. However, great opportunities exist for many of these countries to derive income from other uses of rangelands as well as livestock if appropriate government policies are implemented. This will be discussed later.

RANGELAND ECOLOGY

Grazing Effects on Range Plants

The most fundamental concept in range management centers around how excessive grazing influences the long-term status of palatable forage plants. Scientific study has established that excessive grazing of these plants causes their death and replacement by a succession of plants with lower productivity and often lower palatability; some are even poisonous (Figure 11.4).

Under moderate or light grazing levels the poisonous, unpalatable plants are at a competitive disadvantage because they invest part of their products from photosynthesis in poisonous compounds (e.g., alkaloids, oxalates, and glycosides) and appendages (e.g., spines, thorns, and stickers) that discourage defoliation rather than contribute to growth. In contrast, the palatable plants use their photosynthetic products mainly for growth in the form of roots, leaves, stems, rhizomes, stolons, and seeds. Under excessive defoliation levels the photosynthetic capacity of the palatable plants is reduced to the point that they are unable to produce enough carbon compounds for maintaining roots systems, regeneration of leaves, respiration, and reproduction. Over time, they are replaced with unpalatable plants that have had lower rates of defoliation.

As excessive grazing continues, the palatable plants tend to be replaced by a succession of plants that increasingly are lower in palatability, lower in productivity, and often more poisonous. This process is referred to as retrogression. When defoliation is reduced to a correct rate, the palatable plants again have the competitive advantage and succession occurs back to the original or climax vegetation.

FIGURE 11.4   This fence line in New Mexico shows the effects of long-term proper grazing (left) and excessive grazing (right) on range vegetation. Productive, palatable perennial grasses dominate the range on the left, while the excessively grazed range is dominated by poisonous, unpalatable forbs and shrubs.

The driving environmental forces in succession are moisture (rainfall) and temperature. In wet, humid range types such as the southern pine forest in the southeastern United States or the tallgrass prairie in the eastern Great Plains, recovery after retrogression is both rapid and predictable. The climax plants will usually again dominate the site within 5 years if severe soil erosion has not occurred. In the drier range types such as the Chihuahuan desert, the palatable plants tend to be less resistant to grazing. Here retrogression can occur within just a few years under excessive grazing, but recovery, if it occurs, is a slow process often requiring 20 or more years. On some sites with severe soil erosion, only minor improvements have been observed after 20 or more years even under complete elimination of livestock grazing. In many areas, secondary succession to the original climax does not occur; rather a new and stable climax occupies the site.

RANGELAND CONDITION AND TREND

Range vegetation naturally shows considerable fluctuation through time because of climatic variability. However, several long term range research studies reviewed by Holechek et al. (1999) show that sound grazing management practices maintain the desirable plants at much higher levels than poorly controlled grazing. Those plants that naturally occupy the site in the absence of severe disturbances—such as excessive grazing, fire, and cultivation—are collectively referred to as the climax. On grassland ranges, these plants are generally the most palatable and productive. However, in natural forests, desirable forage grasses and forbs are most plentiful when the trees are completely or partially removed by fire or logging.

Rangeland condition refers to the state of health of the range and is usually judged by the amount of climax vegetation that remains. J. E. Dyksterhuis in 1949 developed a system known as "the quantitative climax approach," which has been widely used to evaluate range condition. With this system, rangeland condition scores depend on the amount of remaining climax vegetation. Ranges with 76% or more of the climax are classified in excellent condition, those with 51 to 75% are in good condition, 26 to 50% are in fair condition, and 0 to 25% are in poor condition. This system has been widely applied by government agencies, such as the USDA Natural Resources Service, and works well on native grasslands.

Under the quantitative climax approach, plants are placed into categories of decreasers, increasers, or invaders depending on their response to grazing (Figure 11.5). Decreasers are the highly palatable climax species that steadily decline under excessive grazing. Increasers are those plants that are part of the climax and initially increase as the decreasers are eliminated. Increaser I plants are distinguished from increaser II plants based on their moderate palatability and eventual decline under excessive grazing. Increaser II plants are unpalatable and in many cases poisonous; therefore, they steadily increase under excessive grazing. Invaders are those species that were unimportant members of or not part of the original climax vegetation. Invader I plants are distinguished from invader II plants by having some grazing value. Many invader plants are annuals or biannuals while decreasers and increasers are long lived perennials.

In recent years, soil stability and suitability of the existing vegetation for the intended use have become criteria for assessing range condition, as well as the amount of climax vegetation that remains. On most ranges, limited amounts of increaser and invader plants are advantageous because they provide food and cover for certain wildlife species and often grow when decreaser plants are dormant and low in nutritional value. Although if heavily consumed, increaser and invader plants are often toxic, moderate or small amounts in livestock diets can be nutritionally advantageous. Ranges with a mixture of different plant types generally meet the needs of livestock and wildlife better than pure grasslands or shrublands. On most grassland and desert ranges, about 60% remaining climax (good condition) gives an ideal mix of

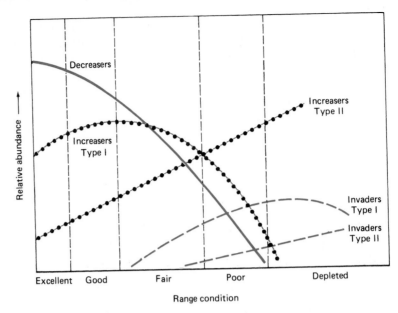

FIGURE 11.5 Changes in abundance of species groups for different range condition classes of the USDA-Natural Resources Conservation Service.

plants for livestock and wildlife and will ensure soil stability. This type of range is sustainable under sound grazing management.

Range trend refers to the rate and direction of change in range condition. Many range managers consider knowledge of trend to be the most important aspect in evaluating the effectiveness of different range management practices. Generally an upward trend is considered to be succession toward the climax or the desired plant community while a downward trend would be retrogression away from the climax or desired plant community.

In evaluating trend, climatic influences must be separated from those associated with grazing or other management practices. Exclosures (ungrazed areas) are an important tool for making these separations. Generally, grazing management is considered to be sound if vegetation improvements on the grazed area equal or exceed those in the exclosure. However, improper management would be signaled where a grazed area reveals less recovery during wet periods and more rapid retrogression in drought periods than range in an exclosure.

RANGE ANIMAL ECOLOGY

Early range research and management was concerned mostly with maintaining important range forage plants. Starting in the 1950s, increasing emphasis was placed on the nutritional needs of range livestock, their foraging behavior, and interaction with other range animals. Most recently, attention has focused on desirable rangeland wildlife species (e.g., pronghorn, mule deer, elk, white-tailed deer, and various upland gamebird species) as a result of increasing public willingness to pay for hunting and viewing these animals. Animal ecology is now as much a part of range management as plant ecology.

Comparative Digestive Systems

Knowledge of the role of different forages in meeting the nutritive needs of different range ungulates is critical for good range management. This knowledge is particularly useful for decisions regarding brush control, range seeding, grazing management, and forage allocation to different ungulates. We will discuss the role of grasses, forbs, and shrubs in meeting the nutritional requirements of range ungulates following Holechek (1984) and Holechek et al. (1998).

Forage Selection by Different Ungulates

Range ungulates can be divided into three groups based on their foraging habits. These groups include the grazers which consume grass-dominated diets; the browsers, which consume primarily forbs and shrubs; and the intermediate feeders, which use equal amounts of grasses, forbs, and shrubs.

The Grazers. Cattle, elk, bighorn sheep, mountain goats, musk oxen, and bison are North American ungulates classified as grazers. However, on some ranges these ungulates, with the exception of bison and musk oxen, do consume large amounts of forbs and shrubs. This occurs primarily when green grass is unavailable. These ungulates avoid shrubs high in volatile oils (e.g., junipers, rabbitbrush, and various sagebrushes) because they lack mechanisms to reduce the toxic effects of these substances.

The Browsers. Moose, pronghorn, mule deer, domestic goats, and white-tailed deer feed primarily on forbs and shrubs throughout the year regardless of location. With the exception of domestic goats, these ungulates experience digestive upsets if forced to consume diets dominated by mature grass. This group of ungulates consumes a limited amount of grass in the spring when it is green and forbs and shrubs are unavailable. However, dry mature grass is almost completely avoided. The smaller ruminants in this group can consume large amounts of forages high in volatile oils because their small, pointed mouth parts enable them to select the portions of these plants with the lowest levels of volatile oils. In addition, the small ruminants chew their food to a much greater extent than large ruminants or monogastric animals. Apparently, fine chewing of plants high in volatile oils results in release of these substances as gasses and greatly reduces their assimilation by the animal's digestive system. If assimilated at high levels, the volatile oils found in many sagebrushes, rabbitbrushes, and junipers can be toxic to the animal.

The Intermediate Feeders. Domestic sheep, burros, and caribou are classified as intermediate feeders. These animals have the greatest capability to adjust their feeding habits to whatever forage is available. Domestic sheep are probably better adapted to the forage resource in the intermountain West than any other ungulate because they will readily use grasses, forbs, or shrubs depending on availability. The primary problem with domestic sheep is that their short legs and relatively large body make them very susceptible to predation.

Comparative Nutritive Value of Grasses, Forbs, and Shrubs

On ranges of the western United States, the primary nutritional constraints on ungulate productivity are inadequate concentrations of energy, protein, phosphorus, and vitamin A in the diet. With a few localized exceptions, mineral deficiencies other than phosphorus are not a problem.

Various studies on forage nutritive quality show different forages provide different levels of critical nutrients at different times of the year. Therefore, ranges with the widest diversity of plant species provide the best nutritional conditions for domestic or wild ungulates when yearlong grazing is practiced.

The Grasses. Grasses typically have lower crude protein, phosphorus, and lignin concentrations and higher total fiber and cellulose concentrations than do forbs and shrubs. Digestibility of grasses is generally less than forbs and shrubs. At comparable growth stages, cool-season grasses are higher in crude protein, phosphorus, and digestibility and lower in fiber than warm-season grasses. Plant fiber is digested more slowly than the cell contents. The high cellulose (digestible portion of fiber) concentration and high ratio of cellulose to lignin (indigestible portion of fiber) makes grasses best suited to large ruminants such as cattle or cecum digestors (horse) that have low nutrient requirements per unit body weight. Leaves of grasses are nutritionally superior to stems. For this reason, short grasses are nutritionally superior to mid and tall grasses particularly during dormancy. Grasses are usually the component of the forage resource available in the greatest quantity unless overgrazing has been severe. Their high availability makes them important to large ruminants that have a high total forage requirement.

The Forbs.   Compared to grasses or shrubs, actively growing forbs have higher levels of crude protein, phosphorus, and digestibility, and lower levels of fiber. Leaves from deciduous shrubs are similar to forbs in nutrient content. When dormant, forbs, and deciduous shrub leaves rank intermediate between grasses and evergreen shrubs in nutritive quality. Because of their low fiber levels, forbs and deciduous shrub leaves break down quickly in the rumen and permit higher intakes than grasses or evergreen shrub leaves during active growth. Forbs and deciduous shrub leaves are critical dietary components for small ruminants that require low fiber diets, such as white-tailed deer and pronghorn.

Because many forbs are poisonous, range managers have often collectively dismissed them as undesirable. Studies with cattle, sheep, and goats show that poisonous forb problems are minimal if the range is in good condition (a high diversity of palatable plants) and grazing intensity is moderate.

The Shrubs.   Evergreen shrub leaves and buds from deciduous shrubs have higher crude protein, phosphorus, carotene (Vitamin A), and digestibility levels and lower fiber levels than grasses and forbs when forage is dormant. Woody material from shrubs is highly lignified and very low in nutritive value. Therefore, grazing animals select for leaves, buds, fruits, and young twigs with low lignification. Ruminants with small mouth parts such as goats or pronghorn can select against woody material much better than cattle or elk. However, evergreen shrub leaves do provide an important crude protein, phosphorus, and carotene supplement to cattle and elk on many ranges when grasses are dormant. Deciduous shrubs with broad leaves, such as snowberry and ninebark, are heavily used by cattle during periods of drought. Many evergreen shrubs (e.g., oaks, sagebrushes, and junipers) have volatile oils or tannins that bind up proteins, reducing the nutritive value of the forage. Goats, deer, pronghorn, and other animals with small mouthparts use these plants most efficiently.

Digestive Systems.   Ungulates have two basic types of digestive systems that include the rumen and cecum systems (Figure 11.6). Both systems evolved to enable ungulates to digest plant fiber (plant cell walls) by microbial (bacteria and protozoa) fermentation.

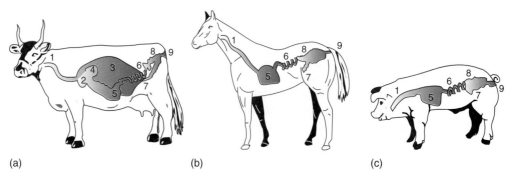

(a)   (b)   (c)

FIGURE 11.6   Stylized representation of the digestive anatomy and arrangement of (a) ruminant, (b) postgastric fermenter, and (c) non-ruminant herbivores. 1, Esophagus; 2, Reticulum; 3, Rumen fermentation compartment; 4, Omasum; 5, Stomach (abomasum); 6, Small intestine; 7, Cecum fermentation compartment; 8, Large intestine; 9, Anus (adapted from Huston and Pinchak 1991).

The fermentation processes are quite similar in both the rumen and cecum. The systems differ in that the rumen is an enlarged portion of the digestive tract that food must pass through before entering the true stomach. The cecum occurs as an enlarged portion of the large intestine that food enters after passing through the true stomach.

From the standpoint of digestive effectiveness, the rumen system has two advantages over the cecum system. First, the process of rumination (regurgitation and rechewing of forage) results in substantial reduction of particle size that provides more surface area for microbial digestion. Because food must be broken down to fine particle size to leave the rumen, retention of fiber is longer than in the cecum. This results in more complete digestion of fiber in the rumen than in the cecum because fiber digestion is a time dependent process. Second, microbes are passed from the rumen into the abomasum where they are digested and then assimilated, providing the animal with an important source of protein. Little microbial protein is assimilated by cecum digestors because microbial fermentation occurs after the food has passed through the stomach. However, research shows horses will ingest their feces when their diet is low in protein, which partially compensates for the inefficient use of microbial protein associated with cecum digestion.

The primary advantage of cecal digestion of fiber is that forage material can pass easily out of the cecum without any great reduction in particle size. Although fiber digestion in cecum digestors is less complete than in ruminants, cecum digestors compensate by consuming a much greater amount of forage. They do not have to break fiber down to a small particle size to pass it out of their system.

As might be deduced from the previous discussion, cecum digestors can subsist on lower quality diets than ruminants. However, they must have a greater forage supply because they use the forage less efficiently. This explains why horses can survive on coarse, mature grasses better than cattle.

Large ruminants can subsist on higher fiber diets than small ruminants because they have lower nutrient requirements per unit body weight. Therefore, a large portion of the diet is typically composed of highly available forage, such as grasses for bison and cattle, or woody browse for moose. The small ruminants, such as white-tailed deer and pronghorn, must consume diets dominated by leafy material and fruits from forbs and shrubs that have high levels of crude protein, phosphorus, and digestibility, and low levels of fiber. These animals can afford to be selective for these materials because they have a low total forage demand.

Domestic sheep subsist better on grass-dominated diets than other small ruminants, such as white-tailed deer or pronghorn. This is facilitated in sheep by a relatively large rumen size in relation to their body weight.

Use of Nutritional Knowledge in Management

Until the 1980s, range management practices were often directed toward replacing forbs and shrubs with pure stands of grasses. The vast acreages of crested wheatgrass in the Great Basin and lovegrasses in the Southwest stand in support of the above statement. During active growth, pure stands of grass provide good forage for cattle or, in some cases, elk and mule deer, but poorly meet the nutritional requirements of all ungulates during most of the year. Research in the Great Basin and the Southwest show that inclusion of palatable forbs and shrubs in seeding mixtures with grasses can greatly improve livestock performance during forage dormancy and provide better habitat for small wild ungulates and other wildlife species than pure stands of grasses.

Range condition is usually based on the density and production of native, palatable, perennial grass. A better criterion may be the diversity of palatable forage species. Using this criterion it may be desirable to manage for up to 20% of the yearly forage production in the form of palatable annuals. Because many annual grasses and forbs grow only in periods when perennials are dormant, a mix of perennials and annuals tends to sustain the longest period of nutritious forage.

A large number of studies involving both domestic and wild ungulates in North America are consistent in showing that forage selection changes tremendously within and between years. The nutritive quality of various forage species also fluctuates greatly within and between years. The greater the degree of forage selection a range provides domestic or wild ungulates, the more likely they will meet their nutrient needs.

Animal Suitability for Different Rangelands

Different animal species are adapted to different types of rangelands. These differences are reflected in the influence of rangeland on the animal and the effects of the animal on rangeland.

Water requirements, topography, predators, climate, parasites and disease; poisonous plants; type of forage; and economic costs and returns are the important considerations in selection of type or types of livestock for a particular range.

Cattle and horses, because of their large size, are best suited for flat, well-watered areas that are dominated by grasses. European cattle (*Bos taurus*) such as the Hereford and Angus breeds, are well adapted to cooler climates while the African cattle (*Bos indicus*), such as the Brahman breed, do well in hot, humid areas.

Sheep and goats are better adapted to hot, humid climates and rugged terrain than cattle or horses. Unlike cattle and horses, they can be watered every other day, rather than daily, and still perform well. Because of their small size and surefootedness, sheep and goats make better use of steep terrain than large animals.

In some areas, pests and diseases make it difficult to raise domestic livestock. Large areas in Africa are unsuitable for cattle because of the tsetse fly. Game ranching (raising of native ungulates for meat) shows potential as a sound alternative to domestic livestock based on evaluations by Dr. David Hopcraft in Kenya.

In some parts of the western United States, sheep ranching has been more profitable than cattle ranching. The price of lambs has usually been greater than that for calves, and wool further adds to returns from sheep. However, cattle do have the advantages of reduced susceptibility to predation and they require less labor than sheep. The cowboy persona associated with cattle has given them a special appeal.

Common-use grazing is the intentional use of rangelands by more than one type of animal. This involves some combination of animals such as cattle and sheep in the mountains of Utah or cattle and white-tailed deer in central Texas. Common-use grazing often has a number of advantages, such as better distribution of animals, harvesting more of the available plant species, and reduction in risk from diversification. The disadvantages include greater labor requirements, more handling facilities, and conflicts with wildlife when certain combinations of domestic animals are grazed at excessive stocking rates.

RANGELAND MANAGEMENT

Range management centers around control of livestock grazing. The primary principles in controlling grazing are proper livestock numbers, proper livestock distribution, proper kinds of animals, and a proper system of grazing. A secondary objective in range management involves range vegetation manipulation with tools other than large animals. Practices used to achieve this objective include seeding, fertilization, and control of undesirable plants with fire, herbicides, mechanical means, or insects. Rapid increases (2- to 10-fold) in forage production can occur within short time periods (1 to 5 years) using these intensive practices. Forage increases from vegetation manipulation through grazing management generally occur more gradually, but risks and costs are much lower.

Importance of Correct Stocking Rate

Selection of the correct stocking rate is the most important of all grazing management decisions from the standpoint of vegetation, livestock, wildlife, and economic return.

Although this has been the most basic problem confronting ranchers and range managers since the initiation of scientific range management early in the twentieth century, specific approaches to this problem were generally unavailable until the late 1980s (see Holechek 1988). It is generally agreed that there is no substitute for experience in stocking rate decision on specific ranges. However, procedures now available will give reasonable estimates for most ranges in the United States and should work in other parts of the world.

Carrying or *grazing capacity* is a term commonly used when discussing stocking rate. It refers to the maximum stocking rate possible year after year without causing damage to vegetation or related resources. Although actual stocking rates may vary considerably between years because of fluctuating forage conditions, grazing capacity is generally considered to be the average number of animals that a particular range will sustain over time. In most cases, ranches are bought and sold on the basis of their grazing capacity.

Stocking rate has more influence on vegetation and livestock productivity than does any other grazing factor (Table 11.1). When all North American studies are averaged, annual herbage production has increased by 35% when livestock use was reduced from heavy to moderate. An average increase in forage production of 28% resulted from switching from moderate to light use. On some ranges, forge production was actually less under light grazing than under moderate grazing. Herbage production on most ranges can be substantially increased by switching from heavy to moderate or light grazing intensities. This is particularly true for grassland ranges.

In the more arid shrubland ranges of the Southwest and intermountain regions, light grazing can be a useful means of improving forage production during the early stages of range deterioration when the desirable forages are still present but in low vigor.

Commonly, grazing capacity is expressed as the stocking rate, which is the number of animal units placed on a given area for a particular time period. An animal unit is a measure of how much forage a mature nonlactating bovine weighing about 1,000 lbs will remove over a 1-year period. Various studies show ruminant animals eat about 2% of their body weight per day on a dry matter basis. Therefore, a 1,000-lb cow requires

TABLE 11.1 Influence of Grazing Intensity on Winter Sheep Production at the Desert Experimental Range in Utah and Cattle Production at the Southern Great Plains Experimental Range in Oklahoma

Sheep—Desert Experimental Range, Utah[a]		
	Excessive Grazing	Moderate Grazing
Utilization of forage (%)	68	35
Ewe weight change (fall to spring) (lb)	+1.1	+9.3
Average fleece weight (lb)	9.68	10.63
Lamb crop (%)	79	88
Death loss (%)	8.1	3.1
Lambs weaned per ewe (lb)	67.0	77.0
Net income (3,000 head flock) ($)	5,072	10,390
Net income per ewe ($)	1.69	3.45
Net income per acre ($)	0.14	0.39

Cow/Calf—Southern Great Plains Experimental Range, Oklahoma[b]		
	Excessive	Moderate
Acres/cow	12	17
Estimated forage use (%)	62	44
Calf crop weaned (%)	81	92
Calf weaning weight per cow (lb)	314	424
Calf weaning weight	388	461
Net returns per cow ($)	9.00	29.44
Net returns per acre ($)	0.70	1.88

[a]*Source:* Hutchings, S. S. and Stewart, G. (1953). *Increasing forage yields and sheep production on intermountain ranges.* U.S. Department of Agriculture Circ. 925.
[b]*Source:* Shoop, M. C. and McIlvain, E. H. (1971). Why some cattlemen overgraze and some don't. *Journal of Range Management* 24:252–257.

about 7,300 lb of forage per year, 610 lb per month, and 20 lb per day. Horses and donkeys eat about 50% more than ruminants and have dry-matter intakes near 3% body weight per day. The following represent some accepted animal unit equivalents.

Animal	Weight (lb)	Animal unit equivalent
Cow	1,000	1.00
Steer	750	0.75
Bull	1,200	1.20
Sheep	150	0.15
Goat	100	0.10
Deer	150	0.10
Elk	700	0.70
Bison	1,800	1.80
Donkey	700	1.05
Pronghorn	120	0.12
Horse	1,200	1.80

Grazing intensities that can be applied to different ranges depend on season, range condition, climate, soil, and topographic features. On flat, wet, humid ranges, such as the tallgrass prairie, where forage production (dry matter basis) exceeds 2,000 lb per acre, 50% of the forage can be removed without damage to the plants. On semiarid ranges, such as the shortgrass prairie, 40 to 45% can be removed. On desert ranges, no more than 30 to 35% removal is recommended. Exceeding these levels not only damages the range under most conditions, but usually reduces livestock productivity and economic returns.

During a drought, livestock operators need to be prepared to reduce livestock numbers or provide feed supplements to prevent overgrazing. Failure to destock during and following drought periods has been one of the most serious causes of rangeland degradation in the United States and other parts of the world. For more detailed discussions of how to set stocking rates, the reader is referred to Holechek (1988) and Holechek and Pieper (1992).

Improving Livestock Distribution

Uneven use of rangeland by livestock has been and continues to be a major problem confronting range managers. On many ranges, improvement will occur without reduction in livestock numbers if practices to secure more uniform use are implemented. Distribution problems are most severe in arid or desert areas and in mountainous terrain. These conditions characterize most of the area west of the Rocky Mountains in the United States. Factors causing uneven use of rangelands include distance from water, rugged topography, diverse vegetation, wrong type of livestock, pests, and weather. Practices commonly used to improve livestock distribution include water development, fencing, trail building, salting, controlled burning, fertilization, and changing types or kinds of animals grazed.

Distance from Water. Poor water distribution is the chief cause of poor livestock distribution on most ranges. In arid regions of the world, water is in short supply and poorly distributed. Where watering points are infrequent, large sacrifice areas around watering points have often occurred. The heavy use of vegetation around watering points is well documented (Figure 11.7).

Although livestock will travel great distances to water, this is not in the best interest of the animal or the range resource. Travel increases animal energy expenditure that otherwise would go into production (weight gain, milk production) and takes away from grazing and resting time.

If animals must travel large distances between water and available forage, a series of trails will be created that gradually become larger and more numerous. These trails become water channels that cause severe erosion. The proportion of rangeland in livestock trails is a good indicator of grazing severity.

FIGURE 11.7 The zone of degradation around a livestock watering point in south-central New Mexico. Rotation of access to watering points has been effective in improving these degraded areas.

In arid areas, water point spacing of about 3 to 4 miles apart has proven most practical when development cost and efficient use of the range are both considered. In the more humid ranges, 2- to 3-mile spacings are practical because more grazing capacity is gained relative to watering-point cost.

Failure to adjust stocking rates for distance from water has been an important cause of rangeland degradation and poor livestock performance. Various range studies show that with cattle, areas over 2 miles from water should be deleted in grazing capacity estimates. A 50% reduction in grazing capacity is required for the zone 1 to 2 miles from water, but no adjustment is necessary for the zone within a mile of water.

Salt blocks and supplemental feed in strategic places can also be useful for improving livestock distribution.

Topography. Rugged topography is the second most important cause of poor livestock distribution on rangelands. The reluctance of livestock to use steep slopes is not entirely undesirable because these areas are often fragile and valley bottoms can better withstand grazing. However, in many cases slopes serve as barriers to the use of benches and ridgetops above valley bottoms.

Livestock vary considerably in their willingness to use steep terrain. Heavy animals, such as mature cattle or horses, have difficulty in traversing steep, rocky slopes. Cattle make little use of slopes over 20% (Figure 11.8). Because of their smaller size, greater agility, and surefootedness, sheep and goats use these areas more readily. Research shows slopes up to 45% are well used by sheep. Beyond this point, sheep use drops off rapidly. This applies to deer and goats as well.

Grazing Systems

Specialized grazing systems have been a major focus of range researchers and managers since the 1950s. During the 1950s and 1960s, deferred-rotation systems received considerable attention. Rest-rotation grazing was heavily applied on public lands in the intermountain West during the 1970s. Both deferred and rest-rotation systems are still being used, particularly in mountainous areas. In the 1980s, short-duration (cell) grazing became the newest fad in grazing systems. Through the years, impressive claims have been made for each new system regarding increased stocking rates and livestock production. However, actual research has shown that

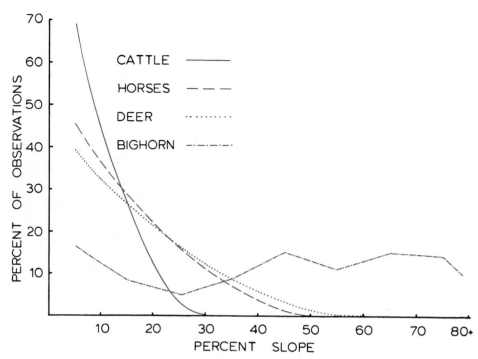

FIGURE 11.8   Relationship of slope gradient to the percent of observations of cattle, feral horses, deer, and bighorn sheep (from Ganskopp and Vavra 1987).

specialized grazing systems generally give modest (10 to 30%) to no increase in grazing capacity over season-long or continuous systems (Holechek et al. 1999).

Deferment, rest, and *rotation* are terms that receive constant use when grazing systems are discussed. *Deferment* involves the delay of grazing in a pasture until after the growth period at the time of seed maturity in the key forage species. This permits the better forage plants to gain vigor and reproduce. In *rest* systems, the range receives no use for a full year, rather than just during the growth period. This gives plants a longer period to recover from past grazing influences and provides wildlife with a pasture free from livestock use during the critical dormant period. A disadvantage of both deferment and rest is the need for increased grazing load on other pastures during the critical growth period. It is questionable whether periodic nonuse during critical periods compensates for periodic heavy use. However, deferment and rest do provide plants on sacrifice areas with some opportunity for recovery. On mountainous areas, season-long or continuous grazing generally results in degradation of areas convenient to livestock. Rotation grazing involves the movement of livestock from one pasture to another on a scheduled basis. It is the critical feature of all specialized grazing systems. The main advantage of rotation is the provision of primary grazing areas and key forage plants with periodic nonuse during the critical growing season. Systems with deferment and rest typically involve livestock rotations. Short-duration (rapid-rotation, cell, time-control) grazing is distinguished from other specialized systems by the greater frequency of nonuse and grazing rotations during the growing season.

Considerations in Grazing System Selection.   Grazing systems commonly used in the United States and other parts of the world include continuous, deferred-rotation, rest-rotation, short-duration, Merrill three-herd/four-pasture, high intensity-low frequency, best-pasture, and seasonal-suitability. Climate, topography, vegetation, kind or kinds of livestock to be grazed, wildlife needs, watershed protection, labor requirements, and developments (fence and water) are important considerations involved in grazing system selection. Specialized systems have been most useful where:

1. Terrain is rugged
2. Wildlife is an important consideration

3. Water distribution is poor

4. Poor distribution of precipitation over the range occurs within years (the south-western United States)

5. Carefully timed grazing is necessary to prevent tree damage

6. The vegetation has low grazing resistance

We will discuss the conditions where the various grazing systems give best re-sults, updating the discussions of Holechek (1983) and Holechek et al. (1998).

Continuous Grazing. Continuous grazing involves grazing a particular pasture throughout the grazing season year after year. Grazing systems other than continuous are commonly called specialized systems because scheduled moves of livestock from one pasture to another are involved.

The primary problem associated with continuous grazing is that livestock have preferred plants and areas for grazing. These plants and areas receive excessive use even under light stocking rates. The preferred areas generally occur where water, for-age, and cover are in close proximity. These are often the most productive parts of the pasture.

Continuous grazing at a stocking rate that gives moderate use (40% removal of current year's growth) has given good results in the flat shortgrass prairie country of the Great Plains where watering points are usually not farther than 2 miles apart and differences in palatability between forage species is minimal. The shortgrasses, blue grama, and buffalo grass are the primary forage species. They evolved with heavy grazing by bison and are quite grazing resistant. Precipitation in the shortgrass coun-try occurs as several light rains throughout the summer months. Therefore, consid-erable opportunity exists for regrowth after defoliation. The flat nature of the terrain and the close proximity of watering areas minimizes the tendency of livestock to con-gregate and linger in the most convenient areas. Sacrifice areas can be reclaimed by temporary fencing to allow vegetation recovery. Since most of the sensitive areas oc-cur around water, control of access to watering points can be used to provide sensi-tive areas with periodic nonuse.

Continuous grazing has given superior results to specialized systems in the Cali-fornia annual grassland type when use was moderate and practices such as salting, fencing, and water development were used to obtain proper distribution. Annual grasses need only to set seed year after year to maintain themselves, unlike perennial grasses that must store carbohydrates. Differences in palatability between most of the annual grasses are small.

In both the shortgrass and California annual grassland types, livestock perform-ance has been better under continuous grazing than under specialized systems. This is explained by the fact that continuous grazing allows livestock to exhibit maximum forage selectivity and minimizes livestock disturbance due to gathering, trailing, and quick change in forage quality.

Deferred-Rotation Grazing. Deferment involves delay of grazing until seed maturity of the important forage species is completed (Figure 11.9). Rotation is the movement of livestock from one pasture to another on a scheduled basis. Initial research on this system was conducted by Arthur Sampson in the Blue Mountains of northeastern Oregon in the early 1990s. The system he studied involved dividing the range into two pastures. Each pasture received deferred grazing every other year. Vegetation re-sponse under this system has been superior to continuous grazing on bunchgrass and mountain ranges in the northwestern United States and on tallgrass ranges in the eastern Great Plains. Deferred-rotation grazing provides a better opportunity for pre-ferred plants and areas to maintain and gain vigor than continuous grazing. It works best where considerable difference exists between palatability of plants and conven-ience of areas for grazing. On mountain ranges, stringer meadows and riparian zones will often received excessive use by cattle even under extra light grazing intensities while surrounding uplands will receive light or no use. The deferred rotation system

Year 1	A	B
June 15–August 15	Grazed	Ungrazed
August 16–October 15	Ungrazed	Grazed

Year 2	A	B
June 15–August 15	Ungrazed	Grazed
August 16–October 15	Grazed	Ungrazed

FIGURE 11.9   A one-herd, two-pasture deferred-rotation grazing plan used on the Starkey Range in northeastern Oregon.

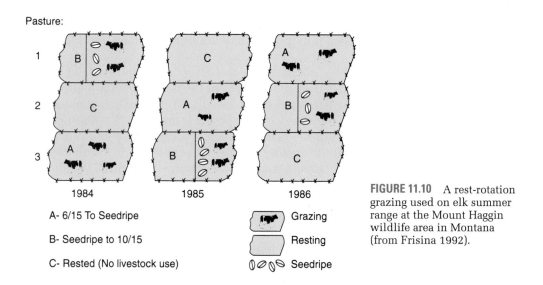

Pasture:

1984   1985   1986

A- 6/15 To Seedripe

B- Seedripe to 10/15

C- Rested (No livestock use)

Grazing

Resting

Seedripe

FIGURE 11.10   A rest-rotation grazing used on elk summer range at the Mount Haggin wildlife area in Montana (from Frisina 1992).

provides forage species on the lowland sacrifice areas with the opportunity to store carbohydrates and set seed every other year.

Rest-Rotation Grazing.   Rest-rotation grazing was developed by Gus Hormay of the forest service in the 1950s and 1960s. This system is unique in that one pasture receives 12 months of nonuse while the other pastures absorb the grazing load. Presently most rest-rotation schemes involve three or four pastures. Various sorts of rotation schemes are used on the grazed pastures (Figure 11.10). The problem with rest-rotation grazing is that the benefits from rest may be nullified by the extra use that occurs on the grazed pastures.

From a multiple-use standpoint, rest-rotation grazing with moderate stocking of grazed pastures has a number of advantages. Small mammals and upland game birds are provided with a good vegetative cover throughout the year on at least part of the area. In eastern Montana, waterfowl production was increased threefold when rest-rotation was compared to continuous grazing (Mundinger 1976). The rest pasture provided waterfowl with the heavy cover they need for good nesting success. Big game animals, such as deer and elk, that generally prefer to avoid livestock are provided with an area free from disturbance and have maximum forage selectivity in the ungrazed pasture. From an esthetic standpoint, the public prefers to see a certain amount of the range ungrazed. The major drawback with rest-rotation grazing is that livestock numbers may have to be reduced to prevent excessive use on the grazed pastures. In areas with large big-game populations, the rest pasture may receive heavy use by big game.

The Merrill Three-Herd/Four-Pasture System. In the early 1950s, Leo B. Merrill in south-central Texas developed a grazing system involving three herds and four pastures. With this system, each pasture is grazed continuously for a year, and then given a four-month period of nonuse. The period of nonuse in each pasture has occurred during all times of the year at the end of a four-year cycle. This system has given good results where effective precipitation and plant growth can occur at any time during the year. It also works well where common use of the range by more than one grazing animal is practiced. In Texas some combination of cattle, sheep, goats, and whitetailed deer graze most ranches. Each grazed pasture is assigned a different type of livestock. Every four months the types of livestock are interchanged between pastures. White-tailed deer prefer the pasture receiving nonuse by livestock. The Merrill system has been the best studied of all the specialized grazing systems. In Texas, it is superior to continuous grazing from the standpoint of livestock, forage, and wildlife production. This system should be effective in areas where forage plants have the potential for growth throughout the year and common-use grazing is practiced. The Merrill system can be modified for areas with seasonal precipitation by dividing the ranch into four pastures and providing each pasture with growing season nonuse, once every four years.

High Intensity-Low Frequency Grazing. High intensity-low frequency (HILF) grazing differs from rapid-rotation grazing in that periods of grazing are typically longer than 2 weeks and nonuse periods are over 60 days. This system requires at least three pastures per herd of livestock. The stocking rates must be light to moderate under this system or severe declines in livestock production and excessive defoliation will occur. This system, like short-duration grazing, works best in flat terrain with an extended growing season and forage species of similar palatability. An important objective of the HILF grazing system is to force use of old, coarse, unpalatable but grazeable forage. Because this system has negative impacts on animal nutrition and forces heavy plant defoliation, it has been largely abandoned in favor of short-duration grazing, which allows greater animal selectivity, permits lighter levels of defoliation, and prevents more of the forage from maturing prior to grazing.

Seasonal-Suitability Grazing. Seasonal-suitability grazing involves partitioning a range into pastures based on vegetation types or conditions classes. The best pasture from a nutritional standpoint is used for each season of the year. When seeded pastures are used in conjunction with native range, this system is often called complementary grazing. In areas such as the Southwest, where local rains can cause considerable difference in forage availability between pastures, livestock are moved to pastures where green forage is available. Under these conditions the grazing scheme is called the "Best Pasture System."

Short-Duration Grazing. Short-duration grazing was developed in Zimbabwe by Allan Savory. This system typically involves a wagon wheel arrangement of fences with water and livestock handing facilities located in the center of the grazing area (Figure 11.11). It is recommended that no fewer than eight pastures (paddocks) of equal grazing be built that radiate as spokes from the central area where the water is located. Each paddock is given a short, intensive period of grazing followed by a long period of nonuse. Ideally, the grazing period of each paddock would be 5 days or less followed by 7 weeks of nonuse. The high stock density is supposed to improve water infiltration into the soil as the result of hoof action, reduce selectivity so that more plants are grazed, improve the leaf area index, and give more even use of the range. It is claimed that these benefits permit stocking rates to be substantially increased compared to continuous grazing.

From a theoretical standpoint, rapid-rotation grazing should work best in flat grassland areas that have an extended period of plant growth (at least 3 months), small differences in plant species palatability, and at least 12 inches of average annual precipitation. These conditions apply to most of the prairie region east of the

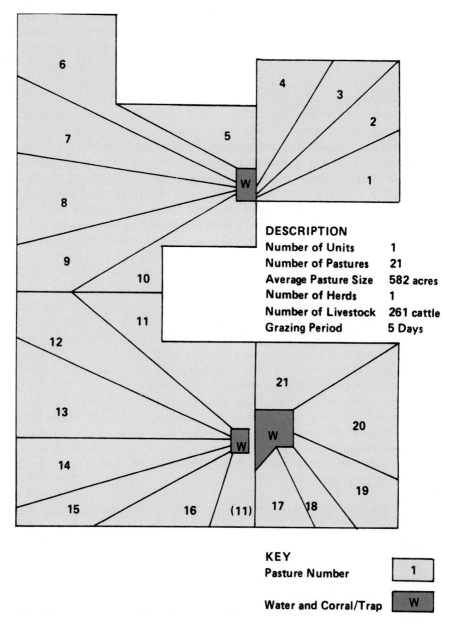

FIGURE 11.11 Short-duration grazing system on a small cow-calf ranch in southeastern New Mexico (from Fowler and Gray 1986).

Rocky Mountains. In the drier, more rugged parts of the western United States, fencing and water distribution problems make rapid-rotation grazing less feasible. Concentrating livestock in the early part of the grazing season has the potential for severe trampling of plants and soil compaction on mountain and sagebrush ranges where much of the effective precipitation comes from snow melt and heavy, early spring rains. These areas typically have growing seasons that are under three months. If plants are heavily defoliated, there is little opportunity for regrowth. Differences in plant palatability are great and livestock will starve before they will accept many of the shrubs such as big sagebrush or rabbitbrush.

There are many potential problems with short-duration grazing on desert ranges in the Southwest. The biggest of these problems is that a tremendous amount of fence must be built per paddock to accommodate a reasonable number of livestock (100 head of cows) because of the sparse nature of the vegetation (good condition desert grassland ranges typically produce 250–350 pounds of forage per acre compared to about 800 pounds per acre for shortgrass prairie). In most years, growth of the perennial grasses occurs in less than a 60-day period, which minimizes the positive aspects of repeated light defoliation. Precipitation that does occur is often from one or two intense thunderstorms. Concentration of animals, therefore, has the potential for severe soil compaction. Another problem is that most of the desert grasses are very low in grazing resistance unlike the prairie grasses that evolved with bison grazing. One failure to move cattle at the correct time under the rapid-rotation system could severely damage grasses, such as black grama or Arizona cottontop. Lowlands dominated by tobosa grass and big sacaton are productive because they collect water runoff from uplands and have deeper soils. On these ranges short-duration grazing may be practical.

Although much controversy surrounds short-duration grazing, many ranchers have found the holistic management principles developed and taught by Allan Savory appealing and useful. Rancher ingenuity in applying these principles varies tremendously, which explains in part why some claim success and others failure. The reader is referred to Savory (1999) for a complete discussion of holistic management.

Grazing Systems for Riparian Zones. Continuous grazing is most damaging to streamside areas (riparian zones) and wetlands because livestock congregate and linger there to use the convenient forage, water, and cover. Riparian zones are the most important part of the range from the standpoint of wildlife, water quality, esthetics, and forage productivity. Many managers and researchers have concluded that the only means of restoring and maintaining these valuable areas is complete livestock exclusion. However, this alternative is unacceptable to ranchers. Recent studies have shown improvement of riparian zones may be possible without complete livestock exclusion (Clarey 1999) (Figure 11.12).

Replacing cattle with herded sheep is a workable solution in some areas where livestock operators graze both species or can switch from one species to another without economic hardship. Herding of sheep permits careful control of grazing timing, frequency, and intensity on riparian zones.

Researchers in Oregon have found that fencing and delayed grazing of riparian zones on mountain rangelands can be beneficial to wildlife, livestock, and vegetation. Their scheme involves restriction of cattle to upland areas until late summer when the gates are opened to the riparian zones and meadows. By this time nesting birds and small mammals have completed critical activities associated with reproduction. The

FIGURE 11.12 This riparian area in southcentral Arizona has responded well to rest-rotation grazing and conservative stocking.

growing season is over so impacts on vegetation are minimal. The intensity of grazing can be controlled by the time at which cattle are permitted access to the riparian zone and are removed from the zone. Livestock performance under this strategy has been found to equal or exceed season long use of the riparian zone. Problems associated with gathering cattle for removal in the fall are greatly minimized because the cattle are concentrated on a small area of flat terrain with good visibility. The only drawback to this scheme is the cost of fencing. The reader is referred to Elmore and Kauffman (1994) for a detailed review of livestock management strategies for riparian zones.

Specialized Grazing Systems and Stocking Rate.   Unfortunately, managers on both public and private rangelands have held the belief that the stocking rate could be largely disregarded if some miracle specialized grazing system was applied. However, actual research has shown overwhelmingly that no specialized grazing system alone will counteract the long-term effects of overstocking. A careful analysis of grazing management research shows the stocking rate has had far more impact on range condition, forage production, livestock production, and economic returns than any specialized grazing system (Table 11.2) (Holechek et al. 1999).

Across all studies financial returns have averaged about 40% more under moderate stocking rates than under heavy stocking rates (Holechek et al. 1998). The monetary benefits of moderate grazing have increased as the length of study time has increased. This is because degradation of soil and water resources under excessive grazing occurs gradually.

On the arid western rangelands net profits have consistently been higher under moderate, continuous grazing compared to rotational grazing systems. However, on the more humid ranges of Texas, the Great Plains and the southeastern United States, specialized grazing systems have often been advantageous from both vegetation and monetary standpoints.

RANGELAND LIVESTOCK PRODUCTION

In the past, "range" has been strongly associated with livestock grazing in the western United States. The traditional use of North American rangelands has been to produce meat for human consumption. However, this use—particularly in the 11 western states—is now being challenged not only by the nonranching public, but also by ranchers as they find it difficult to make a profit. Historically, annual returns on capital investment for range livestock enterprises have been low (2 to 6%) compared to

TABLE 11.2   Rangeland and Cattle Production Characteristics for Different Grazing Management Strategies on the Fort Stanton Range in New Mexico

	Fort Stanton Experiment Range—Cow/Calf Operation		
	Moderate Continuous	Heavy Continuous	Best Pasture Rotation
Duration of study (years)	10	10	10
Average annual precipitation (inches)	15	114.8	15.1
Average forage production, lb/acre	740	607	819
Total palatable forage production in 1974 drought, lb/acre	235	103	379
Range condition	Good	Fair	Good
Acres/animal unit	67	54	54
Forage use (%)	40–45	50–55	50–55
Calf crop (%)	93	91	85
Average calf wean wt./acre (lb)	6.2	7.0	6.4
Death losses (%)	<2%	<2%	<2%
Total 10-year accumulated value ($)	388,575	358,744	382,380

Source: Adapted from Pieper et al. 1991 and Holechek 1998.

returns from investments in industry (10 to 30%). In New Mexico, annual net returns on capital investment in the best of years has averaged less than 4% (Holechek et al. 1998). In most years the overall return is a little over 1%. Compared to the peak in the late 1970s, land prices in the late 1990s for ranches in the western United States have dropped 10%, operating costs have increased about 60%, and prices paid for livestock have increased about 11% (Workman and Evans 1993, Holechek et al. 1998). Based on this scenario, the economic situation for range livestock production in the western United States has been highly unfavorable. Presently, many ranchers face an economic climate that severely threatens their survival. However, there is some other optimism due to recent opportunities for income from enterprises such as recreation, raising non-traditional animals (such as ostriches, red deer, bison, and llamas), producing plants for landscaping, and fee hunting and fishing operations.

Importance of the West

The 11 western states support about 20% of the nation's cattle and 50% of the nation's sheep. Cattle numbers are highest in the Great Plains and Southeast, with the 11 western states having the highest sheep numbers. It is often unrecognized that costs associated with cattle ranching are lower in the Great Plains than in the West (Holechek and Hawkes 1993). This is because the much lower amounts of land required per animal unit reduce property taxes, fuel costs, and labor costs and permit better herd management in the Great Plains than in the West. Supplemental feed costs are less in the Great Plains because droughts are less frequent and severe. Also, active forage growth occurs for a longer period in the Great Plains. Although forage is generally low in quality in the Southeast and the natural vegetation is woodland, the high quantity of forage produced, flat terrain, long growing season, and ready availability of water make livestock production fairly efficient there.

In the most arid portions of the West where forage production is below 100 lb/acre, livestock grazing probably will be unprofitable in most years (Holechek 1992). This is because the fixed costs per animal unit (fences, roads, watering points, property taxes, and insurance) become excessive. Livestock management becomes increasingly difficult when animals are scattered over large areas. In sparsely vegetated areas the energy the animal spends seeking forage can become greater than the energy value of the forage the animal is able to consume. Stoddart and Smith (1943) stated that when more than 180 acres are required to support an animal unit for a year, grazing becomes uneconomical.

Livestock Management During Drought

Drought is a fact of life for stockmen using most rangelands. As aridity increases, severity and frequency of drought tend to increase (see Chapter 6). Climatic records show that ranchers can expect drought conditions to prevail about 3 years out of every 10 years on most western rangelands. Learning to live with drought is the major challenge confronting ranchers in most of the western United States and many parts of Africa. Advance planning is the key to drought survival (Thurow and Taylor 1999).

Inexperienced ranchers often underestimate the degree to which drought can reduce forage production. Studies from a variety of range types show drought can reduce forage production by more than 50% of the annual average (Table 11.3). The best way to deal with drought is to apply conservative stocking rates (Holechek 1999, Thurow and Taylor 1999). Under drought conditions, heavy grazing reduces forage production much more than moderate grazing. Partial confinement and dry-lot feeding of livestock can be advantageous under drought conditions. This allows ranchers to avoid liquidation of livestock when prices are low followed by repurchase after the drought when prices are high because of restocking by other ranchers. Another problem with liquidation is that disease-free, high-quality herds are usually difficult to obtain. Animals that have experience on a particular range perform better than those brought in from another area.

TABLE 11.3 **Herbage Production (Pounds per Acre) on Heavily and Moderately Grazed Shortgrass Prairie in Colorado During a Drought Year Compared to the 5-Year Average**

Grazing Intensity	Drought Year	5-year Average	Drought Year as Percentage of Average
Excessive: 54% use of forage	312	595	52
Moderate: 37% use of forage	577	766	75
Light: 21% use of forage	609	817	75

Source: Klipple, G. E. and Costello, D. 1960. *Vegetation and cattle responses to different intensities of grazing on shortgrass ranges of the central Great Plains.* U.S. Department of Agriculture Technical Bulletin 1216.

Poisonous Plant Problems

Annual livestock losses to poisonous plants average between 2 and 5% on western ranges. Heavy livestock losses to poisonous plants are generally linked with excessive grazing that reduces availability on palatable, nonpoisonous species and increases those that are poisonous. Studies from experimental ranges at a variety of locations show annual poisonous plant losses of livestock to be under 2% when stocking rates were moderate.

A few poisonous plants, such as larkspur, may be consumed by livestock when they are stocked moderately on good range. Although larkspur is quite toxic to cattle, it has little effect on sheep. Delay of grazing until larkspur dries is an effective management practice. Grazing sheep instead of cattle on range with larkspur is another approach.

Most poisonous plants are only toxic for a short period of time, usually during active growth. Many poisonous plants initiate growth earlier than nontoxic grasses, and in this period they may be preferred by livestock because they are green and succulent. Delay of livestock grazing until green growth of perennial grasses is readily available is one of the most effective ways to minimize poisonous plant losses on mountain rangelands. Livestock that are born and reared in a particular environment are less prone to ingest poisonous plants than those brought in from another area.

CONTROLLING RANGELAND VEGETATION

The manipulation of range vegetation by practices other than control of grazing, such as with herbicides, mechanical control, and fertilization, received great emphasis in the United States in the period from the mid-1950s to the early 1970s. However, in recent years, grazing management has received increased attention, while management practices geared toward causing major changes in range vegetation have been deemphasized. This has partially resulted from the high supply of meat and low prices for livestock in the 1980s and 1990s that has made cost reduction management more attractive to ranchers than practices that may substantially increase rangeland forage productivity, but at great cost and risk. Reduced land prices in the 1990s have made the purchase of more land an attractive alternative to vegetation manipulation when increased grazing capacity is desired. Pressure from environmental groups coupled with reduced government spending on conservation has greatly curbed brush control projects. Based on present trends, we do not foresee a shift back toward heavy emphasis on manipulation of range vegetation in the western United States until well into the next century, if then. However, vegetation manipulation is the only practical way to increase forage for livestock and to improve wildlife habitat on some ranges. Furthermore, for some ranchers this may still be the best way to increase economic returns.

The risk associated with revegetation is high because the practice may not give the desired effects even when properly done. Control of unwanted plants, revegetation, and/or fertilization can cause production to increase two- to tenfold within 1 to 3 years. High management inputs are required once these risky, costly practices are used if the land manager wishes to realize a reasonable return on his or her investment.

Few, if any, land managers use intensive practices exclusively on a unit of rangeland. Rather, some combination of beneficial practices is applied whereby both intensive and extensive principles are used. The use of various practices changes with time as dictated by economic, political, and social conditions, or as improved technology becomes available. Land managers must be flexible and innovative in planning operations on a range unit. Practices that work on one range unit may be entirely unsuited for the range unit next to it.

Methods used by range managers to control unwanted plants have included spraying with herbicides, controlled burning, mechanical control, and biological control by releasing animals or insects that will consume the plant.

During the 1950s and 1960s, herbicides such as 2,4,5-T were widely used to kill invading woody plants. However, they are now seldom used because of their high cost and possible threats to human health.

Controlled burning has become a more popular method of controlling undesired plants. It is considered to be ecologically natural and involves much lower cost than herbicides. The disadvantages of burning are air pollution, soil erosion, and the possibility of fire getting out of control. Generally burning causes shifts from shrubs and trees to grasses and forbs. When well planned and executed, burns can have minimal negative impacts on the environment, and can be quite beneficial to livestock and wildlife.

Some introduced predatory insects have given impressive biological control of unwanted plants. Goats can be used in suppressing certain shrubs. Biological control is less expensive than other methods, but finding and establishing predatory species on the unwanted plant has been difficult.

Mechanical control of unwanted plants by root plowing, shredding, bulldozing, or chaining has been little used in recent years because of high cost and severe soil disturbance. This approach is most practical in the wetter rangeland types where taller brush species occur and severalfold increases in forage are possible after their removal.

After unwanted plant control, proper management of livestock is essential. Generally 1 to 2 years of nonuse after control is recommended to give desirable plants opportunity to become established.

GOVERNMENT POLICY

To the early settlers in the United States, land resources seemed unlimited. Until the 1930s, government policies were concerned with disposing of the vast land resource. These policies were oriented toward converting undisposed land into farmland. It was generally unrecognized that large areas of the West were unsuited for farming and suited only for grazing or other use of lower intensity. The vastness of the land resource coupled with the failure of government to recognize that most of the West could not sustain farming resulted in tremendous land damage and economic loss.

Prior to the Forest Reserves Act (1891) and the Taylor Grazing Act (1934) (see Chapter 2), there was no control over how federal lands were used. If one person did not exploit the forage in an area, his or her neighbor probably would. This is commonly referred to as the "tragedy of the commons" (see Chapter 2). This resulted in a constant struggle in the intermountain area to get on spring ranges as soon as the snow melted and to stay on them as long as possible. This competition between stockmen to gain and maintain use of the best forage resulted in total disregard of prudent grazing practices.

The Taylor Grazing Act of 1934 triggered the era of land retention by the federal government. It placed under management the vast, arid areas of the West that were suffering from severe overgrazing. With the passage of this act, the government formally recognized that unsold arid western lands were unsuited for cultivation and that grazing on these lands must be controlled. Prior to the Taylor Grazing Act, little attention was given to the custodianship of unsold western lands.

Importance of Federal Lands

The issue of livestock grazing on federal lands has been quite controversial during the 1990s (Wuerthner 1990). Therefore we have included a brief discussion of its importance (Holechek and Hess 1994). At present 30,600 permittees graze cattle on federal lands. This is about 2% of the nation's ranchers or about 7% of the ranchers in the 11 western states. Bureau of Land Management and Forest Service rangelands provide forage for about 10.8 and 7.7 million animal unit months (AUMs), respectively, for a total of 18.5 million AUMs. This represents 1.54 million animal units (AUs) or 3.76% of the nation's beef cattle herd (41 million AUs). At an average fair market value of $80 per AUM, the total value of federal land grazing permits is roughly 1.48 billion dollars.

Although federal rangelands provide only a small part of total livestock forage requirements, they are seasonally important in the production process. Around 22% of the yearling cattle in the United States spend a portion of their lives on federal rangelands, which play an even bigger role in sheep production. They support about 20% of the nation's stock sheep from which about 21% of the nation's wool is shorn.

It is doubtful that discontinuation of federal land grazing would have much impact on the price of meat to the consumer. Increases of beef production on private lands in the Great Plains and Southeast would likely compensate for any reduction on federal lands in the West.

Discontinuation of federal land grazing would severely harm some local economies. Negative impacts on wildlife populations would be likely if private land holdings associated with federal land grazing are subsequently subdivided into ranchettes. Further, many water points on federal lands would no longer be maintained. Livestock grazing is an important use on federal lands. However, situations do exist where livestock grazing on public lands should be discontinued because of conflicts with other uses or because low forage productivity makes grazing financially unsound.

Over the last 50 years considerable progress has been made in managing and improving the nation's rangelands (Table 11.4). Despite this progress, range improvement has not been rapid enough to satisfy many ranchers, conservationists, and environmentalists.

Presently, as in the past, the biggest problem on public rangeland continues to be the overobligation of grazing privileges. This problem is exacerbated by government programs, mainly cost subsidies, that encourage ranchers to maximize livestock numbers on private and public rangelands, often at the expense of land conservation.

The federal government has had various cost subsidies for range improvement and private lands since the 1950s. These programs have shared from 50 to 100% of the cost of water developments, brush control, grazing systems, and range seeding with the operator. Many conservationists have objected to these programs, claiming

TABLE 11.4   Comparative Percentages of Bureau of Land Management Rangelands in Excellent, Good, Fair, and Poor Condition Between 1936 and 1992

Year	Excellent	Good	Fair	Poor
1936	1.5	14.3	47.9	36.3
1966	2.2	16.7	51.6	29.5
1975	2.0	15.0	50.0	33.0
1984[a]	5.0	31.0	42.0	18.0
1992[a]	5.0	34.0	38.0	13.0

Source: USDI 1984, 1993.
[a]Less than 100% total because some lands have not been rated as to range condition.

that they have rewarded ranchers for bad management, they were often harmful to wildlife, their costs were higher than the benefits, and they contribute to an over-supply of livestock and depressed prices. The largest of these projects, the Vale Rangeland Rehabilitation Program, sheds light on these arguments. Our source of information for the following discussion is Holechek and Hess (1994).

The Vale Program

Before 1963, the Vale District was characterized by severe degradation from over-obligation of grazing privileges. Livestock numbers were estimated to exceed grazing capacity by 40%. The majority of the allotments were both small and communal (shared by two or more ranchers). Low levels of private investment because of small ranch size, unwillingness of permittees to invest in communal allotments, and un-compromising protection of permitted livestock numbers explain why stocking remained too high and why range improvement was neglected.

When finally faced with the prospect of livestock reductions, ranchers, local politicians, and the Bureau of Land Management (BLM) joined forces to pressure Congress into funding massive range reclamation (1963 Vale Rangeland Rehabilitation Program). This federal program provided for widespread spraying, plowing, and seeding of overgrazed rangelands along with fencing and water development—all at taxpayer expense. Between 1963 and 1985, a total of about 18 million dollars was spent on the Vale project. At the beginning of the program (1963) there were about 332 permittees, but by 1985 the number had dwindled to 184. Roughly $97,000 had been spent for every permittee remaining on the Vale district in 1985. In 1992 dollars (adjusted for inflation), this amounts to $56 million in total, or $304,348 per remaining permittee. About 750,000 acres were involved in the Vale project (119,000 seeded, 170,000 sprayed, 461,000 native range). Looking at it another way, the project incurred $24/acre actual cost and $75/acre inflation-adjusted (1992) cost.

In terms of grazing capacity, the Vale project was considered to be capable of handling 285,000 AUMs in 1963, though the actual number on the area was 400,000 AUMs. In 1986, the estimated grazing capacity was near 485,000 AUMs, but the actual number on the area remained at 400,000 AUMs. Seedings best sustained grazing-capacity increases while the sprayed areas had declined to about the same productivity as untreated range at the beginning of the program. It is of particular interest that average grazing capacity on untreated native rangelands increased about 40% between 1963 and 1986. This was attributed to reduced stocking, water development, fencing to facilitate grazing systems, season of use changes, and more favorable precipitation patterns.

There was no definite trend in numbers of most wildlife species over the course of the Vale project. It does appear that pronghorn benefitted from water developments and crested wheatgrass seedings that included alfalfa. Sage grouse were declining in the area in 1986, but this could have been partially due to wildfires in the early 1980s. Riparian areas in 1986 generally were in fair or poor condition for fish and wildlife, with the exception of those fenced off from livestock.

The bottom line on the Vale project is that the BLM created about 200,000 AUMs of forage at a cost of $90/AUM in absolute dollars of $280/AUM in 1992 inflation-adjusted dollars. This represents 3.5 times the present fair market value of BLM grazing permits in most areas ($80/AUM). There is now uncertainty regarding whether the 200,000 additional AUMs created between 1963 and 1986 can be sustained over the next 30 years. After detailed economic analysis using several scenarios, it was concluded that livestock benefits alone were not sufficient to justify the Vale project.

The option of destocking Vale program rangelands from 400,000 to 250,000 AUMs through government purchase of overobligated grazing privileges from permittees apparently was never considered. Our estimates indicate that this could have been done at about 16% of the final cost of the program. Although long-term benefits of conservative stocking have never been evaluated on Oregon sagebrush ranges, research from other range types indicates they could be substantial from vegetation, livestock, financial, and wildlife standpoints. The impressive improvement in grazing capacity (35–40%) on the untreated native range in the 1963–1986 period of the Vale project supports the destocking approach in conjunction with low-cost management practices (e.g., fencing for grazing systems, water development, season of use changes).

One aspect of the Vale program completely overlooked is opportunity cost. We hold the view that monetary resources are scarce. Public benefits realized from the Vale program (most of which accrued to 184 permittees) entailed public benefits lost to the rest of society—benefits that would have occurred had Vale project resources been allocated to ranches in other areas or, for that matter, to the retirement of the national debt.

Policy Changes

A variety of rangeland policy changes have been advocated by reformists that favor greater reliance on market forces (Hess 1992, Holechek 1993, Holechek and Hess 1996). Discontinuation of the cost subsidies previously discussed is an important part of most rangeland policy reforms. This would allow the marketplace to punish ranchers who use unsound grazing management practices.

The other aspect of rangeland policy reform centers around permitting ranchers and environmentalists on public lands to resolve their differences through the marketplace rather than by increased regulation. Holechek (1993) recommended some changes in federal rangeland policy that might benefit ranchers, environmentalists, range conservationists, and the taxpayers. They include the following:

1. Option of grazing privilege purchase from permittees in heavily urbanized areas
2. Promotion of conservative stocking rates with conservative grazing fees as an incentive
3. Active management of custodial allotments by BLM
4. Allow permittees to exercise allotment vacancy if grazing fees are paid
5. Integration of aesthetic values and wildlife needs into brush control projects
6. Fees for recreational use of all federal rangelands

Purchase of Grazing Privileges.   Some tracts of federal rangelands near large urban centers such as Los Angeles, Tucson, Phoenix, Denver, Reno, Las Vegas, and Salt Lake City would better serve society if livestock grazing was discontinued. Not only are financial returns to the operator low or negative from grazing these areas, but the costs of assessing and mitigating the externalities are high. Similar problems may exist on 8 to 10% of federal rangelands.

Presently the effect of livestock grazing on riparian habitats is a major concern. Although new grazing management technology can ameliorate some of the adverse effects, there probably will be tremendous pressure to eliminate livestock grazing along streams in areas heavily developed for recreation, such as Jackson Hole, Wyoming; Bend, Oregon; and Bozeman, Montana. Even carefully controlled grazing has some negative impacts on trout, and recreationists using the mountain streams do not like the fences and dung associated with livestock.

Heavy recreational use adversely affects the efficiency of livestock production. Just a few of the negative influences include increased vandalism, increased livestock death loss from road hits, increased livestock handling problems from failure to close gates, and reduced weight gains from livestock avoidance of watering points and preferred feeding areas when occupied by recreationists.

The most practical and equitable solution when major conflicts develop between recreation and livestock grazing may be for the government to purchase the grazing permit from the rancher at fair market value on a willing seller/willing buyer basis.

This would involve a cost of up to $80 million based on a value of $70 or $90 per AUM. Many ranchers on these lands would welcome the option of selling their grazing privileges. They recognize that grazing is not financially effective but lack opportunities to sell their permits as a reasonable price. This type of program may seem costly, but it must be viewed in the context of established public policy on equity rights and the future costs of administration, monitoring, and litigation as these conflicts increase. Perspective may be gained by remembering that the federal government is paying farmers about $2 billion a year not to farm CRP lands.

Application of Conservative Stocking Rate. Most Bureau of Land Management lands are still stocked on the heavy side of moderate with the goal of around 50% use. A 50% use level may work in the flat, humid regions of the Great Plains and Southeast because of their high productivity and high adaptability of the plants to grazing. However, in most cases it causes range destruction in the rugged, arid ranges of the West. Research shows stocking rates that involve a 30 to 35% forage use level will enhance range recovery, maintain adequate food and cover for wildlife, protect soil resources, and give the highest long-term economic returns with the least risk on nearly all the western range types. Procedures are available that make conservative grazing strategies easy to apply.

Present federal rangeland policies actually create a disincentive for ranchers to apply conservative stocking rates on their allotments because they are billed for the number specified on the permit, even if grazing a lower number of livestock. Some ranchers on federal lands by choice do apply conservative stocking rates and pay for the permitted AUMs of forage left unused. Ranchers who apply a conservative stocking strategy to their allotments could be given reduced grazing fees ($1.80 per AUM) without having their permit reduced; while those stocking at a heavier rate could be assessed higher fees ($3.00 to $4.00 per AUM). Ranchers using the conservative grazing strategy might also be given preference for water developments, brush control, fences, and other range improvements. This would create some capitalistic incentives for ranchers to better manage federal lands by providing a tangible reward to those ranchers who leave an optimum amount of residue for soil protection and wildlife.

Allow Allotment Vacancy. Under present federal policy, if a grazing permit is not exercised within two years, the right will be transferred to another qualified applicant who will graze the allotment. This reduces the opportunity for conservative grazing management or rest by the permit holder. In addition, this blocks certain organizations, such as the Nature Conservancy, from buying grazing privileges from ranchers on federal lands and then removing livestock. It would be progressive for the BLM to change this policy so both permit holders and private organizations could temporarily or permanently reduce or retire livestock grazing on areas with extraordinary wildlife or recreational values. Purchase of the grazing privileges should be negotiated with the permit holder.

Active Management of Custodial Allotments. In the early 1980s the BLM developed a policy that resulted in placement of their rangelands into three categories based on condition. The maintenance (M) category was given to allotments judged to be in acceptable condition; the basic goal was to sustain present management. Allotments in a deteriorating condition judged to have potential for recovery were given the improvement (I) category, and management was geared toward improvement through better grazing management, water development, brush control, and other management. Allotments with low forage production potential were given the custodial (C) category, and management on these allotments is based on maintenance of the present condition. Much of the destructive grazing that now is occurring on BLM lands in the Southwest is on custodial allotments. Consistently these allotments produce under 100 pounds of forage per acre, which makes monetary losses to the operator highly probable. Both high fixed costs and low livestock productivity account for these losses.

Part of the rationale behind the custodial category was to minimize the difficulties of managing the smaller, more scattered parcels of BLM land and those parcels that are interposed with state and private lands. However, in actuality a number of allotments with three or more sections (in some cases over 50 sections) have been given the custodial category. The low condition or potential of these areas is not a justification for practices that are ecologically and economically unsound.

Better Designed Brush Control.   Brush control on federal lands is widely regarded second to destructive grazing as a major concern of environmentalists. Failure to consider wildlife and esthetics often remain as problems with these projects. Under present economic conditions, it is difficult to rationalize the need for brush control projects on federal lands if the only goal is more forage for livestock. However, these projects become more justifiable if wildlife habitat improvement, soil stability, and esthetics are other benefits. We believe brush control on federal rangelands generally should involve small tracts and more often employ the practices of strip spraying, variable rate spraying, skipping riparian corridors, and use of irregular treatments.

It would seem prudent to suspend major brush control initiatives where threatened or endangered species are involved until the impacts are fully understood. Ameliorating the impacts of poorly planned brush control projects on threatened wildlife could prove to be expensive to taxpayers in the future.

Fees for Recreational Use.   Recreational use of rangelands in the U.S. has risen sharply in recent years, a result of the rapid human population increase in the 11 contiguous western states combined with more affluence and leisure time. In some areas, such as southern California, central Colorado, southern Arizona, central Idaho, and central Oregon, recreation has become a more important use of federal rangeland than grazing.

Recreational use of rangelands includes a variety of activities and the economic values of many, such as camping, hiking, and water skiing, are difficult to quantify. Some forms of recreation, such as off-road vehicle travel and trail biking, can be as destructive as overgrazing if uncontrolled. Litter is also another problem associated with recreational use.

To prevent range destruction in the future, much greater regulation of recreation on federal rangelands will be necessary. It seems only reasonable that recreationists using federal land, as well as ranchers, be charged a fee for their activities.

RANGE MANAGEMENT AND THE FUTURE

The approach of the twenty-first century brings both challenges and opportunities to range managers in the United States, and other parts of the world. New technologies will probably cause further increases in livestock productivity and reduce worldwide prices. Those producers who survive will be most effective at reducing production risks and costs, and most successful applying the technologies that are most efficient in expanding supply.

Primary range management challenges confronting range managers in the twenty-first century were discussed by Holechek et al. (1998). They include:

1. Sustaining ranching as an occupation and way of life.
2. Preservation of open space (Figure 11.13).
3. Prevention and resolution of social conflicts over usage and management of rangeland resources.
4. Preservation of threatened and endangered species.

FIGURE 11.13   Subdivision of privately owned rangeland into small tracts for housing and other uses caused considerable reduction in open space in the western United States in the 1990s.

5. Expansion of supply of rangeland products as follows:

   ■ livestock products
   ■ recreation
   ■ wildlife
   ■ water
   ■ esthetics

These issues are discussed throughout the book so we will not elaborate further here. Rather, we will focus on range management opportunities.

The biggest past and present obstacles to successful range management have been the difficulty of integrating climatic, biological, financial, and political information into a management framework. New low-cost/user-friendly computers are now coming on the market that give ranchers the capability to solve this problem. Already, grazing management software programs (APSAT) have been developed for Texas rangelands that allow ranchers to make seasonal stocking rate adjustments (Kothmann and Hinnant 1992). Within just a few years, software programs will be available that help with decision making on brush control, forage allocation to livestock and wildlife, watering point spacing, and supplemental feed allocation.

The Internet (computer information highway) now provides ranchers and range managers instantaneous databases on different aspects of range management that would take several days to collect from libraries and other sources. This kind of information should make arid-land ranching a much more efficient, less risky enterprise than it has been in the past. It should change it from a natural resource to a technological-based operation.

Other uses of computers and sophisticated electronic equipment will include corral systems that monitor animal condition and sort animals for supplemental feeding, livestock ear tags that facilitate record keeping, and automated feeding systems that provide livestock with the correct amount of supplement (Holechek et al. 1998).

There are also some promising developments involving bioengineering that should positively affect rangelands in the next 20 years (Holechek et al. 1998). Instead of chemicals for noxious plant control, new herbicides will be developed from plant allelopathic toxins that do not contaminate the soil or water and do not release residues harmful to humans. Genetically engineered rumen microflora will provide range livestock with increased potential for converting range forage into useful nutrients. Genetic engineering will also improve range plants and animals. Energy used on the ranch will be largely generated by solar and wind power with the sale of the surplus to urban areas. Robotics will be used to reduce ranch manual labor requirements. Enterprise diversification into stock and bond markets, recreation, wildlife, and nontraditional animals are ways that ranchers can reduce risk and improve financial returns. Better information is needed on the long-term outcome of grazing management and range improvement practices. We believe the key to ranching survival in the twenty-first century will be the capability to effectively use biological and financial knowledge. The world trend toward free markets, less government intervention, and rapid technological advance in agriculture will make the acquisition and application of knowledge more important than ever before.

Increasing demand for outdoor recreation, water, wildlife, and open space on a shrinking land base provides a formidable challenge for future range managers. However, it provides opportunities as well. Historically the problem of increasing demands on limited resources has created incentives for technology that greatly improved both human and natural resource conditions.

LITERATURE CITED

Clary, W. P. 1999. Stream channel and vegetation responses to late spring cattle grazing. *Journal of Range Management* 52:218–227.

Dyksterhuis, J. E. 1949. Condition and management of rangeland based on quantitative ecology. *Journal of Range Management* 2:104–115.

Elmore, W. and Kauffman. 1994. Riparian and watershed systems, degradation and restoration. In *Ecological implications of livestock herbivory in the West*. Society for Range Management, Denver, CO.

Fowler, J. M. and J. R. Gray. 1986. *Economic impacts of grazing systems during drought and nondrought years on cattle and sheep ranches*. New Mexico Agric. Exp. Stn. Bull. 725.

Frisina, M. R. 1992. Elk habitat use within a rest-rotation grazing system. *Rangelands* 14:93–96.

Ganskopp, D. and M. Vavra. 1987. Slope use by cattle, feral horses, deer, and bighorn sheep. *Northwest Science* 61:74–81.

Gay, C. 1965. *Range management: How and why*. Las Cruces, NM: New Mexico State Univ. Coop. Ext. Circ. 376.

Hanselka, C. W. and L. White. 1986. Rangeland in dry years: Drought effects on range, cattle, and management. In R. D. Brown, ed., *Livestock and wildlife management during drought*. Cesar Kleberg Wildlife Research Institute, Texas A&M University, Kingsville.

Heady, H. F. and R. D. Child. 1994. *Rangeland ecology and management*. San Francisco, CA: Westview Press.

Hess, K. 1992. *Visions upon the land*. Washington, DC: Island Press.

Hibbard, B. H. 1924. *A history of the public land policies*. New York: Macmillan.

Holechek, J. L. 1983. Considerations concerning grazing systems. *Rangelands* 5:308–311.

Holechek, J. L. 1984. Comparative contribution of grasses, forbs, and shrubs to the nutrition of range ungulates. *Rangelands* 6:245–248.

Holechek, J. L. 1988. An approach for setting the stocking rate. *Rangelands* 10:10–14.

Holechek, J. L. 1992. Financial benefits of range management practices in the Chihuahuan desert. *Rangelands* 14:229–284.

Holechek, J. L. 1993. Policy changes on federal rangelands: A perspective. *Journal of Soil and Water Conservation* 48:166–174.

Holechek, J. L. 1994a. Rangeland grazing. *Encyclopedia of Agricultural Science* 3:535–547. New York: Academic Press.

Holechek, J. L. 1994b. Financial returns from different grazing management systems in New Mexico. *Rangelands* 16:237–284.

Holechek, J. L., H. Gomez, F. Molinar, and D. Galt. 1999. Grazing studies: What we've learned. *Rangelands* 21(2):12–16.

Holechek, J. L. and J. Hawkes. 1993. Desert and prairie ranching profitability. *Rangelands* 15:104–109.

Holechek, J. L. and K. Hess. 1994. Free market policy for public land grazing. *Rangelands* 16:63–67.

Holechek, J. L. and K. Hess. 1995. The emergency feed program. *Rangelands* 17:133–136.

Holechek, J. L. and K. Hess. 1996. *Market forces would benefit rangelands.* Forum for Appl. Res. and Public Policy 2:5–15.

Holechek, J. L. and R. D. Pieper. 1992. Estimation of stocking rate on New Mexico rangeland. *Journal of Soil and Water Conservation* 47:116–119.

Holechek, J. L., J. Hawkes, and T. Darden. 1994. Macroeconomics and cattle ranching. *Rangelands* 16:118–123.

Holechek, J. L., R. D. Pieper, and C. H. Herbel. 1998. *Range management principles and practices.* 3rd edition. Englewood Cliffs, NJ: Prentice-Hall.

Huston, J. E. and W. E. Pinchak. 1991. Range animal nutrition. In R. K. Heitschmidt and J. W. Stuth, eds., *Grazing management.* Portland, OR: Timber Press.

Hutchings, S. S. and G. Stewart. 1953. *Increasing forage yields and sheep production on intermountain ranges.* U.S. Dep. Agr. Circ. 925.

Klipple, G. E. and D. F. Costello. 1960. *Vegetation and cattle responses to different intensities of grazing on shortgrass ranges of the central Great Plains.* U.S. Dept. Agric. Tech. Bull. 1216.

Knutson, R. D., J. B. Penn, and W. T. Boehm. 1990. *Agricultural and food policy.* 2nd edition. Englewood Cliffs, NJ: Prentice-Hall.

Kothmann, M. M. and R. T. Hinnant. 1992. APSAT: An operational level grazing management model. *Proc. West. Sec. Amer. Soc. Anim. Sci.* 43:354–357.

Mundinger, J. G. 1976. Waterfowl response to rest-rotation grazing. *Journal of Wildlife Management* 40:60–68.

Pieper, R. D., E. E. Parker, G. B. Donart, and J. D. Wright. 1991. *Cattle and vegetational response to four-pasture and continuous grazing systems.* New Mexico Agr. Exp. Sta. Bull. 576.

Savory, A. 1999. *Holistic management: A decision making framework.* Washington, DC: Island Press.

Shoop, M. C. and E. H. McIlvain. 1971. Why some cattlemen overgraze and some don't. *Journal of Range Management* 24:252–257.

Smith, J. G. 1899. *Grazing problems in the southwest and how to meet them.* U.S. Dep. Agric. Div. Agros. Bull 16:1–47.

Stoddart, L. A. and A. D. Smith. 1943. *Range Management.* New York, NY: McGraw-Hill.

Thurow, T. L. and C. A. Taylor, Jr. 1999. Viewpoint: The role of drought in range management. *Journal of Range Management* 52:413–419.

United States Department of Agriculture (USDA). 1993. *Agricultural statistics.* Washington, DC: United States Government Printing Office.

United States Department of Interior (USDI). 1984. *Fifty years of public land management.* U.S. Department of Interior, Bureau of Land Management.

United States Department of Interior (USDI). 1993. *Public land statistics.* U.S. Department of Interior, Bureau of Land Management.

Vallentine, J. R. 1990. *Grazing management.* San Diego, CA: Academic Press.

Workman, J. P. and S. G. Evans. 1993. Utah ranches—An economic snapshot. *Rangelands* 15:253–255.

Workman, J. P., S. L. King, and J. F. Hopper. 1972. Price elasticity of demand for beef and range improvement decisions. *Journal of Range Management* 25:337–340.

Wuerthner, G. 1990. The price of wrong. *Sierra* 25:38–48.

CHAPTER 12

Farmland and Food Production

INTRODUCTION

The capability of modern agriculture to meet the food needs of a human population that now numbers 5.6 billion and has increased by fivefold over the last 100 years is one of the greatest technological accomplishments of mankind. Since Englishman Thomas Malthus published his 1798 essay on the human population limits, a succession of like-minded observers have warned of limits to human population growth. Many have predicted that unchecked human population growth will ultimately outgrow the capacity of humans to feed themselves. So far, however, they have consistently underestimated the capability of technology to solve problems of food scarcity and natural resource depletion. To achieve that feat, however, agriculture now must use more of the earth's land, water, soil, plant, animal, and energy resources than any other human activity. It has also been a major cause of environmental degradation, including pollution and loss of biodiversity.

In developed countries technological breakthroughs involving improved crop varieties, irrigation techniques, pesticides, better tillage methods, and fertilizer usage have led to huge food surpluses, which are permitting large areas of marginal lands to be retired from farming. Other technologies are reducing environmental degradation problems previously associated with agriculture. Serious problems of food shortages and environmental degradation still plague many developing countries, and non-point sources of U.S. agricultural pollution remain problematic (see Chapter 7). However, we are optimistic that these challenges can be met if socioeconomic changes are made that encourage the technological developments and applications that have contributed to the food surpluses in the more developed countries. We see the present global trend toward a slowdown in the human population growth rate, rapid technology increase, and more democratic, free-market sociopolitical systems as encouraging. In developing this chapter, we have relied heavily on Miller (1990), Schiller (1994, 2000), Owen and Chiras (1995, 1998), and Knutson et al. (1995, 1998).

A BRIEF HISTORY

Over the course of world history agriculture has undergone four revolutions, each of which closely correspond to improvements in the human condition (Laetsch 1979). Our discussion of these revolutions closely follows Bolen and Robinson (1995). A more detailed discussion is provided by Chrispeels and Sadava (1977). About 10,000 years ago the domestication of wild plants initiated the first period. This happened in separate parts of the world such as China, South America, and Europe. Some of the first plants domesticated were wheat, rice, corn, oats, and potatoes, but gradually many others were added. The domestication of plants permitted the development of organized civilizations. Because people were freed from daily procurement of food, plant culture permitted specialization of labor. It changed societies from nomadic to sedentary. With permanent settlement and a reduction in the amount of time needed for acquisition of food, other important aspects of human life could develop, including art, crafts, education, religion, government, science and advanced technologies.

Exploration and discovery were the basis for the second agricultural revolution. This era dates back to the thirteenth century when Marco Polo journeyed from Europe to China on a trading mission in quest of spices. A highlight of the exploration era was Columbus' landing in America in 1492. Throughout the 1500s, a series of European explorers including Balboa, Magellan, Cabot, and Coronado traversed both the oceans and major land areas of the world accumulating previously unknown knowledge of the world's geography, wealth, and people. Land and sea trade routes were established for the exchange of agricultural products and other commodities. New World (North and South America) plants such as corn and potatoes were introduced to Europe. Sugar, coffee, tobacco, and a variety of spices became key items of trade among widely scattered countries.

The third major agricultural revolution resulted from industrialization that started in the eighteenth century with the inventions of the steam engine and the cotton gin. These inventions led to both improvements in mass transit and the substitution of machines for strenuous human and animal labor. Steamboats in the early 1800s and railroads in the mid-1800s greatly reduced the time and cost of human travel. By the 1890s, agricultural goods could be readily transported by railroad across nearly all portions of the United States. A variety of machines were widely used in farming such as reapers, combines, balers, and drills. Farm mechanization continued quite rapidly after World War I with the advent of the tractor. The replacement of labor with machines greatly boosted the supply of agricultural goods, which lowered food costs. This caused a large scale reduction in the number of farms in the United States as average farm size increased and marginal lands were abandoned.

We are now going through a fourth revolution in agriculture based on biotechnology breakthroughs. Advances in chemistry and genetics are translating into improved crop varieties, fertilizers, herbicides, and insecticides that give productivity increases far beyond those of the Industrial Revolution (Figure 12.1). Basically plant geneticists are improving on nature. Highly adapted hybrid crops and livestock are being developed for each of the specific climatic regimes, soil types, and pests that occur in different parts of the world. Agricultural technology appears to be winning the race with the growing human population by boosting food and fiber yields to previously unimagined highs.

MAJOR TYPES OF AGRICULTURE

About 30 crops provide over 80% of the world's food supply although some 80,000 plants are edible (Miller 1990). Wheat, rice, corn, and potatoes account for over half of the world's annual production of food. Livestock products such as meat, milk, eggs, and cheese are much more expensive than plant products. This is because about 90% of the energy in plants is lost when plants are consumed by livestock rather than

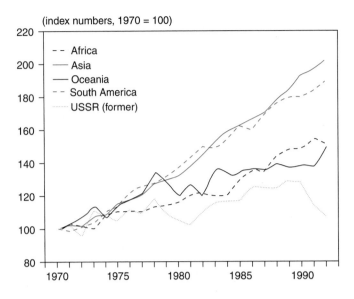

(index numbers, 1970 = 100)

FIGURE 12.1   Index of food production by region, 1970–1992. Food and Agriculture Organization of the United Nations (FAO), Agrostat PC, on diskette (FAO, Rome, 1993) (from World Resources Institute 1994).

consumed directly by people. Therefore, people in developed countries, where incomes are high, typically can afford to consume much more meat and other animal products than those in developing countries. Exceptions to this generalization occur in parts of Africa, the Middle East, and South America where herders depend primarily on livestock for their sustenance.

Basically two types of agricultural systems are used to produce crops and livestock—industrialized and subsistence agriculture (Figure 12.2). Industrialized agriculture is widely practiced in the developed countries, where specialized operators grow a single crop or one type of livestock on large areas using high inputs (fertilizer, pesticides, mechanized cultivation) (Miller 1990). In developing countries, the common form of food production is still subsistence agriculture involving small plots of land tended by individual farmers with primitive equipment pulled by draft animals. Various modifications of these two basic forms of agriculture are applied in both developing and developed countries. Using the intensive approach, large plantation farms grow coffee, sugar, bananas, and cacao in many tropical or semitropical developing countries such as Kenya, Colombia, and Brazil. On the other hand, organic farms using minimal inputs of inorganic fertilizer and pesticides are becoming common in some developed countries including the United States.

THE GREEN REVOLUTION

The basic options for increasing food production are (1) increasing yields from existing cropland and (2) bringing more land under cultivation (Miller 1990). Most of the increase in world food production since 1960 has been brought about by huge crop yield increases made possible by the "green revolution."

The green revolution centers around genetic selection of highly adapted plant varieties that are responsive to large inputs of inorganic fertilizer, pesticides, and irrigation water (Chrispeels and Sadava 1977). The first green revolution plants were short-stemmed varieties of wheat resistant to lodging (bending) developed in the United States in the 1950s. Later in the 1970s, the same approach was used to create fast-growing dwarf varieties of rice for tropical climates in Asia. The short and stiffer stalks of the new varieties of wheat and rice enabled them to produce larger seed heads without toppling over. With appropriate water, fertilizer, and pesticide inputs, three- to fivefold increases in yields over traditional varieties were possible. Another advantage was the faster growth of green-revolution plants. Multiple cropping (two or three consecutive crops within a year) became possible on the same unit of land.

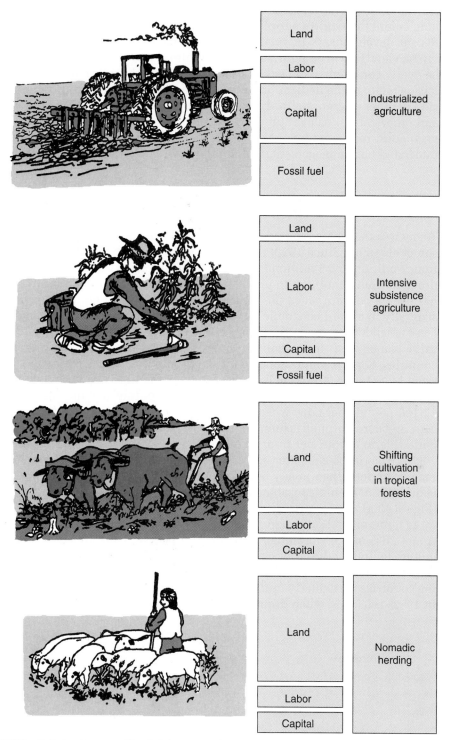

FIGURE 12.2 Relative inputs of land, labor, capital, and fossil fuel energy into major types of agricultural systems (drawing by John N. Smith, adapted from Miller 1990).

Since 1960, most of the increase in world grain output has resulted from green revolution techniques (Miller 1990). Although the green revolution is considered a great success story, it does have its critics. The primary argument against the green revolution is that it requires large fossil fuel input for cultivation, irrigation, and manufacturing of pesticides and fertilizers. Agricultural use of fossil fuels has increased about six- to eightfold since 1950 (Miller 1990; World Resources Institute 1994). It now takes twice as much oil to produce a ton of grain as it did 40 years ago.

On the other hand, it seems probable that new energy sources will be developed long before the world's supply of fossil fuels is depleted (see Chapters 19 and 20). Recent developments in solar and nuclear energy lend support to the more optimistic view. However, policy related to future research investments cannot afford to be overly confident about easily mastering the challenges that lie ahead. Other criticisms relate to the dedication of very large areas to single crops without interspersion of other crops or natural areas. This increases the probability of pest outbreaks and other natural catastrophes. In addition to negative environmental effects from pesticides, recent evidence that pests adapt rapidly to pesticides has given rise to integrated pest management, which relies more on natural ecological processes.

AGRICULTURE PROBLEMS IN THE UNITED STATES

Today less than 1% of the workforce in the United States engages in farming (Knutson et al. 1998). Since 1920, the number of farms in the United States has dropped from 7 million to 2.5 million (Figure 12.3). At the same time the farm population has dropped from 32 million to under 10 million. While the reduction in farm numbers has been a concern since the nineteenth century, it has no doubt contributed substantially to overall economic development and the rise in living standards. The release of labor from food production is one of the primary contributors to technological advance and industrialization in the United States and other developed countries. Some examples of agricultural productivity since the early 1950s summarized by Schiller (1991) include the following: annual egg production has jumped from 183 to 243 eggs per laying chicken, milk output has increased from 5,400 to 12,100 pounds per cow, wheat output has increased from 17.3 to 35.3 bushels per acre, and corn output has jumped from 39.4 to 102 bushels per acre. During the same period, farm output per hour of labor, has increased even more phenomenally by 700%. Americans eat better and at the same time spend a lower percentage of their income on food than in any other country. People in most parts of the world spend about 40% of their disposable income for food compared to 13% in the United States (Figure 12.4).

In terms of total sales, agriculture is now the biggest industry in the United States. About 20% of the workforce, or 25 million people, work in some part of U.S. agriculture including the growing, processing, or marketing of food. Agriculture exports from the United States play a critical role in meeting food needs of several developing countries particularly during periods of sociopolitical unrest such as in Somalia, in 1992–1993, and when drought conditions occur in India or China.

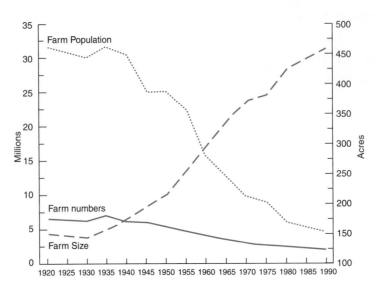

FIGURE 12.3 Farm population, farm numbers, and farm size for selected years 1920–1990 (from Knutson et al. 1995) (Agricultural Statistics, Washington, D.C.: USDA, various issues).

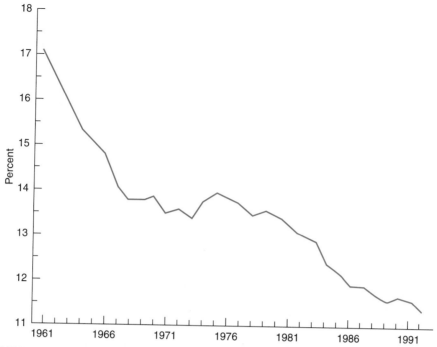

FIGURE 12.4 Share of consumer income spent on food (from Knutson et al. 1995) (Denis Dunham, Food Cost Review 1992, AER-672 Washington, D.C.: ERS/USDA, September 1993, p. 46).

FIGURE 12.5 This picture of a family farm in West Virginia conveys an image of prosperity. However, a variety of factors including urban sprawl, rising production costs, declining exports, and regulation now seriously threaten this type of operation (courtesy of U.S. Department of Agriculture).

Although its successes are many, American agriculture confronts a number of problems that are discussed in detail by Schiller (1991), Knutson et al. (1995), and Owen et al. (1998) (Figure 12.5). These include soil erosion (see Chapter 8), increasing regulation, declining exports, urban sprawl, rising production costs, restricted water supplies, and atmospheric pollution. We will briefly discuss these issues following Owen et al. (1998).

Land Uses Occupying Over 50% of Area

LEGEND
- ▨ Developed Land
- ▨ Cropland
- ▨ Rangeland/Pasture
- ▨ Forest Land
- ▨ Mix of Two or More Land Uses
- ☐ Federal Land/Water/Non-Sampled Area

FIGURE 12.6   Dominant uses of private land in the United States in 1992 (courtesy of USDA-Natural Resources Conservation Service).

Increasing Regulation

As discussed in Chapter 8, farmers must increasingly comply with federal, state, and local regulations regarding use of pesticides, use of herbicides, tillage of soil, management of wetlands, and plowing of erodible lands. Although many of these regulations are needed to protect environmental quality and human welfare, they greatly increase farming costs and can make U.S. farm products less competitive with those of other countries. Past misuse of fertilizers and pesticides has polluted groundwater supplies in some heavily farmed areas. Certain insecticides used by farmers have adversely impacted wildlife populations. Some pesticides have contaminated soils and adversely affected desirable soil bacteria that take nitrogen from the air and fix it into a form (nitrate) needed by crops.

Urban Sprawl

In the period from 1982 to 1992 the United States lost about 3 million acres (there are 382 million acres total) of its farmland to urban sprawl based on USDA Natural Resources Inventory. This rate of conversion was lower than in previous decades (1950s to 1980s) (Figure 12.6). What makes this problem critical is the best watered, most level, and most productive soils generally occur near cities. A tremendous increase in shopping malls, airports, factories, and highways has occurred around almost every major city during the past 40 years (see Chapter 14). This problem is made more severe by the unpredictable way in which much urban sprawl has occurred (Figure 12.7) (see Chapter 14).

Rather than developing lands around cities in tightly organized fashion, as in much of Europe, the United States has permitted random development patterns in

FIGURE 12.7 Urban sprawl into farmland near Phoenix, Arizona.

Letting Farmland Stay Farmland

Conversion of farmland to nonfarm uses may change land use irreversibly and alter the character of an area. It also may weaken the local agricultural economy. In heavily developed areas, loss of even a few acres of remaining farmland suggests for many the end of a way of life and separation of people from their roots.

Fifteen states, mostly in the Northeast, have enacted laws and appropriated funds to pay farmers willing to keep their land in an agricultural use. Easements stay with the land even after its sale, guaranteeing that farmland stays farmland.

Since the mid-1970s, farmland preservation laws have protected nearly 420,000 acres of farmland at a cost of almost $730 million—about $1,750 an acre. Funding for the programs has come mostly from sale of bonds and levy of sales, property, and other taxes. An additional $195 million was available early in 1996 for further purchases—$107 million in New Jersey alone.

Among the leaders in farmland protection are Maryland, which has spent about $125 million to purchase easements on 1,117,000 acres of farmland, and Pennsylvania, which has spent more than $150 million to protect almost 75,000 acres. Massachusetts and New Jersey have each spent more than $80 million to protect 35,907 acres and 27,924 acres, respectively.

The Federal Agriculture Improvement and Reform Act of 1996 established a Farmland Protection Program with a funding level of $35 million. The program will help states with farmland protection programs purchase conservation easements. Prior to the end of the 1996 Federal fiscal year, $15 million were made available under the new program in 17 states through 37 individual programs. An estimated 150 to 200 farms will be signed up under these various programs.

From: A Geography of Hope. 1997. USDA-National Resources Conservation Service.

---

which adjoining farmlands are broken up and cut off by an interspersion of homes, business, and public works (also see Chapter 14, Figures 14.2 and 14.15). This greatly compromises the efficiency of farms that remain because of restrictions on movement of equipment, pesticide use, fertilizer use, and raising of livestock such as hogs, chickens, and dairy cattle. This pattern is particularly severe in New England, the mid-Atlantic states, Florida, and parts of the far West.

Rising Production Costs

Since the early 1970s, the costs of fuel, fertilizer, and pesticides have been rising faster than prices for most farm crops (Figure 12.8) (Knutson et al. 1998). This is more a problem of overproduction of farm products rather than an excessive increase in

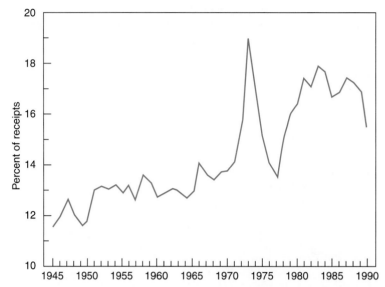

FIGURE 12.8   Farm cash expenses account for an ever-increasing share of farm cash receipts including government payments (from Knutson et al. 1995) (Economic Research Service, Economic Indicators in the Farm Sector: National Financial Summary 1990 and 1991, Washington, D.C.: USDA, November 1991 and January 1993.).

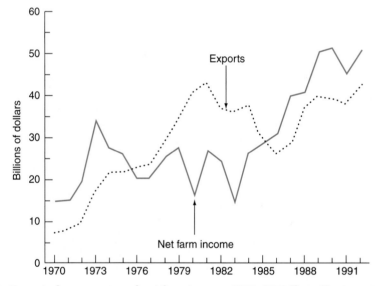

FIGURE 12.9   Boom in farm exports and net farm income, 1970–1979 (from Knutson et al. 1995) (Council of Economic Advisers, Economic Report of the President, Washington, D.C.: Executive Office of the President, February 1992, p. 409).

the cost of farming inputs. Generally farm product prices have lagged behind the consumer price index since 1980. This is also true of energy, fertilizer, and pesticide prices, but to a lesser extent. The consumer price index measures the average annual changes of a standard group of needs by a typical household.

Declining Exports

Although farm exports from the United States grew rapidly in the 1970s, this trend reversed in the 1980s (Figure 12.9). Several factors including a strengthening U.S. dollar, higher interest rates, increased regulation, increased labor costs, changes in international trade laws, and higher farm input costs have tended to reduce the competitiveness of U.S. farm products with those of several other countries. Argentina,

Australia, Brazil, Canada and other countries have greatly expanded their output of several agricultural products. Their production costs are generally lower than costs in the United States. China and India, once major food importers in the 1970s, have become nearly self-sufficient because of the green revolution. For small grains such as wheat, American farmers do have a cost advantage over European countries. However, the Europeans since 1974 have generally chosen to protect their farmers with high subsidies and excessive tariffs on agricultural imports.

Atmospheric Pollution

Crop yields near major cities and highways are adversely impacted by carbon dioxide and other pollutants from vehicles using fossil fuels. The exact costs of this pollution are difficult to estimate and depend on the area and type of crop. However, losses could be as high as $5–10 billion in years when there is little or no surplus production.

Restricted Water Supplies

In the Southwest and central Great Plains, the amount of water available for irrigation is rapidly dwindling. The greatest concern is the depletion of the Ogalalla aquifer in the plain states of Texas, Oklahoma, New Mexico, Kansas, and Nebraska (see Chapter 7). In California and Arizona, rapid urbanization and industrialization are making increased demand on limited water supplies previously used by agriculture.

Excess Capacity

In developed countries (the United States and western Europe), the biggest problem confronting farmers has been excess capacity (Knutson et al. 1995). In free-market economies, such as the United States, farm prices tend to trend downward over time because of rising productivity and low elasticity of demand. Price elasticity of demand refers to the relationship between quantity of a product demanded and price. Products with high elasticity of demand, such as automobiles and computers, show large increases in demand as prices fall. However, many agricultural products generally show inelasticity of demand in which demand is fairly constant whether prices rise or fall. This means that small increases in supply relative to demand can cause sharp drops in prices.

During the depression of the 1930s, the United States government began using price supports and cost subsidies to raise farm prices. These policies led to increased farm output requiring higher exports, government food purchases, and farmland set-aside programs to avoid a buildup of food surpluses. With the exception of the 1970s, market surpluses have characterized American agriculture since World War II. This problem has been made more severe by heavy subsidy of agriculture in western European countries (France, Spain, Germany, England, and so on) since the early 1970s. In response to the oversupply of farm products, developed countries have dumped these products on world markets at below true production costs. This has hindered agricultural progress in those developing countries that often have a competitive advantage in agricultural production of certain products such as sugar, beef, wheat, rice, and peanuts because of less regulation and lower labor costs. During the mid-1990s, the developed countries began shifting their agricultural policies towards lower price supports and the retirement of marginal farmlands through cash payments. Direct income payments to farmers decoupled from agricultural production are being considered among the ways to preserve smaller family farms that cannot compete with large corporations who practice industrialized farming.

AGRICULTURE IN DEVELOPING COUNTRIES

Subsistence agriculture is still the primary livelihood for much of the world's population. In its various forms this type of agriculture occurs on about 55% of the world's cropland. It is the primary type of agriculture in Africa, China, India, and South

America. Generally subsistence agriculture involves small plots of family-owned land (mostly under 10 acres) on which a variety of crops are grown using little resource supplement.

Some of the farming techniques used in subsistence farming include intercropping, polyculture, and agroforestry. These will be briefly discussed following Miller (1990).

Intercropping.   This involves growing two or more different crops in alternating rows on the same plot. The strategy is generally to combine high-protein nitrogen-fixing crops, such as soybeans, with energy-rich but nitrogen-depleting grains, such as millet or sorghum.

Agroforestry.   This is a variation of intercropping in which crops and trees are planted together. With this approach, trees producing fruits or nuts might be planted around grains, legumes, or a fiber-producing plant such as cotton. In some cases, acacias or other nitrogen-fixing trees are planted for fertilization and are fuel wood.

Polyculture.   This highly complex form of intercropping involves interspersion of a large number of different plants. In Thailand and other South Asian countries, these plants might involve fast growing grains and vegetables, moderate growing bananas, and slow growing tubers, such as cassava. The soil is protected year-round and the diverse root systems at varied depths efficiently capture nutrients and moisture. By creating a diversity of habitats for natural predators and fully using the soil profile, the need for pesticides and herbicides is nearly or completely eliminated. Polyculture systems also provide subsistence families with food throughout the year and crop diversity reduces the risk of crop failure caused by unfavorable weather conditions. In addition, labor requirements are more evenly distributed during the year.

The trend in the more progressive developing countries is to integrate subsistence agriculture with more modern farming techniques. No country has done this better than China. Although it has less farmland than the United States, it must feed over four times as many people (1.2 billion). Over the past 20 years, China has largely eliminated malnutrition by doubling its grain production and increasing per capita food consumption by 50%. This has come about from a combination of political reforms and technological applications (Miller 1990). The political reform has involved shifting from a centrally planned, state-controlled agricultural system to a market-oriented system where individual families make the ground decisions. Also important have been a more equitable distribution of farmland and a comprehensive program designed to reduce the growth rate in the human population. Some of the technological applications have involved use of scientifically developed hybrid crops, increased use of commercial inorganic fertilizer, fish farming using intensive pond management, biological control of insects using birds (ducks and geese), frogs and insects, increasing use of organic wastes (human and animal wastes, garbage, crop residues, and more), and increased regulation of urban sprawl. India, the Philippines, Chile, and Korea are other developing countries that have substantially increased their agricultural productivity.

Although there are reasons for optimism about food production in developing countries, it is important to recognize that per capita food production in most of Africa has actually declined over the past 20 years (Figure 12.10). Here the future is uncertain because of rapidly rising human populations, political instability, disease outbreaks, and failure to apply technology. Miller (1990) provides an overview of the worsening food situation in Africa. He summarizes the problems as rapid population growth (most African countries have human population doubling of 20–30 years), extended drought conditions lasting nearly 20 years, poor food distribution systems, frequent wars within countries (Rwanda and Sudan provide recent examples), severe underinvestment in agriculture, severe loss of topsoil from overgrazing and deforestation, and price fixing that discourages farmers from expanding output (governments keep food prices low to prevent urban unrest).

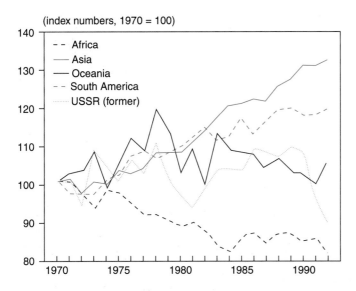

(index numbers, 1970 = 100)

FIGURE 12.10 Index of per capita food production by region, 1970–1992 [Food and Agriculture Organization of the United States (FAO), Agrostat PC, on diskette (FAO, Rome, 1993)] (from World Resources Institute 1994).

Famine problems have been greatest in the central portion of Africa known as the Sahel, which is on the edge of the Saharan Desert. The basic problems are poor preparation for droughts, which characterize the area, and unstable socioeconomic systems, which discourage investment, savings, and infrastructure development. Because of high currency inflation rates and unsound banking systems, farmers and herders have little incentive to convert food surpluses in good years into currency that can be used to buy supplies needed in hard times. Instead they retain their crops and animals and practice a bartering system that involves exchanging a portion of their crop or a few animals for other goods when needed. During droughts, herd levels are too high and over half the animals often perish from starvation. A large amount of stored crop production is lost to weather, insects, and fungus. Unless this system is improved, the prospects for much of Africa would appear bleak at best.

Developing countries with the brightest agricultural prospects are in Latin America and Southeast Asia (Figures 12.10 and 12.11). Chile and Argentina have been particularly successful in increasing agricultural output and economic growth. Success in both countries has involved a series of free market reforms that privatized state run companies, encouraged foreign investment, stabilized the currency, removed trade barriers, reduced unnecessary regulations, lowered service on borrowed foreign capital, and permitted movement towards a more democratic government.

Latin American countries have some important advantages over African countries that are worth mentioning. These include higher rainfall, less cultural division, and a common language (Spanish). Perhaps the biggest problem in Africa is tribalism. The native peoples of Africa are fragmented into a series of tribes with different languages and customs. This has been a major barrier to economic development because of the constant quarreling among various groups. National unity has been a prerequisite for any country to make rapid progress in improving the human condition.

PESTICIDE CONTROVERSIES

One of the biggest controversies in modern agriculture centers around the use of pesticides (weed, rodent, and insect killers). Without question pesticides have played a crucial role in keeping world food supplies ahead of population growth. Since World War II, pesticides have also prevented millions of deaths and untold human suffering from insect-transmitted diseases such as malaria (from mosquitoes), bubonic plague (rat fleas), typhus (fleas and lice), and sleeping sickness (tsetse fly) (Miller 1990).

It is estimated that 40 to 50% of the world's potential food supply is lost to pests (Miller 1990, Pedigo 1999). Without pesticides, these losses could be 30 to 50% higher.

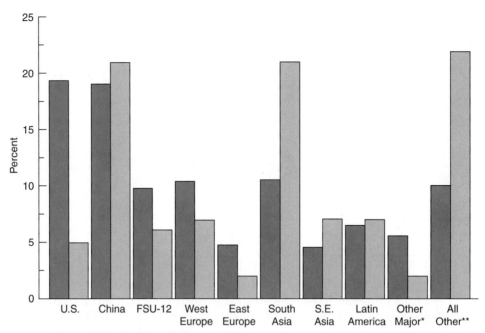

FIGURE 12.11 Share of world crop production and population by region, 1990–1992 average (from Knutson et al. 1995) [World Agricultural Production (Washington, D.C.: USDA/ERS, March 1993) and The World Factbook: 1990 (Washington, D.C.: Central Intelligence Agency, 1993)].

Present pesticides are an efficient, cost-effective, and seasonally safe means of controlling weeds, insects, and rodents when properly used. Other advantages include ease of shipping and handling. Pesticide companies are now developing pesticides that are safer to use and less ecologically harmful than those used over 30 years ago.

We will explore the argument for and against pesticides following Miller (1990) and Pedigo (1999). The main argument against pesticides is that eventually insects and weeds develop genetic resistance to them. This is particularly true with insects that have quick generation turnovers and produce large numbers of offspring. In tropical areas, insects can become resistant to a specific pesticide in less than 3 years while in temperate areas it takes about 5 years. Another part of the problem is that pesticides usually kill off most of the pest's natural predators. Genetic resistance can lead to a cycle of ever-increasing costs for pesticides that are less and less effective.

Pesticides also can kill desirable wildlife species and contaminate air and water resources. Several major fish kills in the United States have been caused by misuse of pesticides. DDT, a fat-soluble, slowly degradable pesticide, interferes with reproduction in birds such as the bald eagle. This pesticide, now banned in the United States, can be increasingly concentrated in fatty tissues with upward movement through the food chain. DDT in fish is concentrated at about 10 times the level of the plankton they consume (Miller 1990). Eagles that eat these fish further concentrate the DDT. Because of this process, referred to as biological amplification, dilution is not always a solution to reducing the pesticide hazard to human health. The recent improvement of southern bald eagle status, now listed as threatened instead of endangered, is associated in part with the reduction of DDT and similar compounds in eagle habitat.

There is considerable concern that human health could be threatened by long-term, low-level exposure to pesticides. However, the Federal Food and Drug Administration requires that pesticides be intensively tested for harmful effects on humans before approved for use on farmlands. Many of the pesticides now in use can cause cancer in humans if used at extremely high levels over an extended period of time. However, a 1987 National Academy of Sciences report estimates that, at worst, about

20,000 cases of cancer a year are caused by the pesticides in food. Some of the foods most likely to contain pesticide residues include potatoes, oranges, lettuce, carrots, celery, apples, grapes, and tomatoes.

Critiques of intensive pesticide use point out there are farming alternatives that are practical and cheaper. These can be categorized into cultivation methods, plant breeding, biological pest control, allelopathic plants, insect sterilization, and insect hormones. These will be briefly discussed following Miller (1990) and Pedigo (1999).

Cultivation Methods.   This approach basically involves using diversified farming instead of monotypic farming. Rather than planting one crop year after year over large areas, a variety of crops are grown in rotation systems. Hedges can be planted within and around fields to create a habitat for pest predators. Planting times can be manipulated so that major crop pests are deprived of their food source for long periods and starve to death. In addition to crop rotation, intercropping and agroforestry can be used to further increase diversity.

Biological Pest Control.   With this approach, diseases, parasites, and natural predators of a problem pest are identified and released to obtain population reduction. Worldwide there have been about 300 successful biological pest control projects. In the northwestern United States, chrysolina beetles have been effective in controlling an aggressive weed called St. John's wort. In Australia, the prickly pear cactus has been successfully controlled with the Argentine moth bopper. A viral disease, myxomatosis, has been effective in controlling wild rabbits in Australia. In Africa, mealybugs, which cause considerable loss of cassava crops, have been controlled by wasps. Ducks, geese, and chickens can be useful in controlling crop weeds and insects. However, they also consume certain vegetable crops. Goats have been effective in controlling some unwanted shrubs on rangelands.

Allelopathic Plants.   Many plants produce substances toxic to (allelopathic) other plants that may be competitors. Through selective breeding, crop varieties of various grains can be developed that suppress weeds. Some combinations of crops planted together can be resistant to insects.

Insect Sterilization.   This approach involves the use of radiation to sterilize large numbers of male insect species. They are then released and allowed to mate with wild fertile females. Screwworms, a costly livestock and wildlife parasite in the southwestern United States causing several millions of dollars in damage, were largely eliminated by use of sterilized males. Problems with male sterilization are high cost, prevention of reinfestation, and the difficulty of producing enough nonsterile males to ensure low reproductive success.

Insect Hormones.   Chemicals extracted from the insect or made in the laboratory are strategically applied during the insect's life cycle to cause death or reproductive failure. A variation of this approach is to use insect extracts to attract the pests into traps with poison or to use attractants to help predators find pest insects. There have been some successes with hormonal growth disruptors and attractants. However, they are costly, require careful timing, and are not immediately effective.

Integrated Pest Management.   Increasingly, chemical cultivation and biological methods of pest control are being combined to optimize efficiency, cost, and environmental sustainability. This approach is known as integrated pest management (IPM). IPM programs can reduce pest control costs by 50% or more, reduce pesticide use by over 50%, and sustain or increase crop yields. Among the disadvantages with IPM are that it requires much greater knowledge than alternative methods and often is more labor-intensive. In the United States there has been a gradual shift toward IPM in the 1990s. It is used to some degree on about 35–40% of the nation's farmland.

Other countries applying IPM include China, Brazil, and Indonesia. We refer to Pedigo (1990) for a more complete discussion of insect pest management.

SUSTAINABLE AGRICULTURE

Sustainable agriculture is popular terminology for a set of agricultural practices that are an alternative to modern monotypic, industrialized farming. Many farmers and ranchers have been part of the sustainable agriculture movement. This approach substitutes polycultures for monocultures and minimizes use of fossil fuels, inorganic fertilizer, pesticide use, and irrigation water (Miller 1990). Its philosophical roots come from a group of environmentalists and conservationists concerned that topsoil depletion and pesticide contamination under industrial farming will lead to degradation of ecosystems functions and services.

Although most features of the sustainable agriculture movement are sound under specific conditions or when used in moderation, an immediate complete switch to this type of farming would sharply lower food production and raise good prices. Breakthroughs involving genetic engineering, tillage, integrated pest management, and improved irrigation systems have greatly reduced the environmental problems associated with conventional agriculture. In the United States the need to crop fragile, erodible soils has been largely eliminated by the increases in yields associated with the high input approach. In our view, the best industrial and subsistence practices can be integrated into sustainable agricultural practices that will best meet the food needs of society and maintain environmental integrity as we move into the twenty-first century.

FARMLAND POLICY

In most years, the United States government spends 15 to 25 billion dollars on farm subsidies (Figure 12.12). Since 1980, these subsidies have cost taxpayers about 350 billion dollars (Schiller 1994). With this much money, the government could have bought out every farmer in 40 states. The high cost of farm programs is of considerable concern to tax payers. Evidence indicates many farm programs are

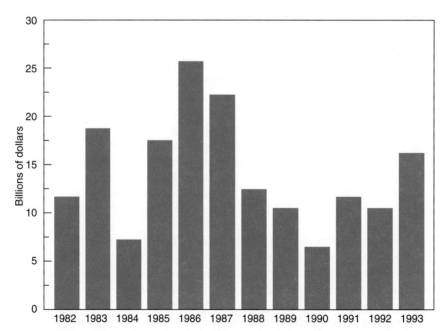

FIGURE 12.12 Cost of government programs for agriculture, 1982–93 (from Knutson et al. 1995) (Economic Research Service, Agricultural Outlook, Washington D.C.: USDA, various issues).

ineffective in halting the decline in farm numbers or reducing food surpluses. A view growing among conservatives is that less government intervention in farming would be better for both farmers and food consumers.

Agriculture is one of the most competitive of all industries (Schiller 2000). This is because there are few barriers to entry. It is easier to get into farming than most other businesses although, in order to have a profitable operation, farmers need increasingly larger acreages of land, more expensive equipment, and more credit. Because of low barriers to entry, farmers can quickly expand their output during periods of high crop prices usually caused by adverse climatic conditions. For this reason profits are usually short-lived and farmers must continually invest wisely and cut costs in order to stay in business.

The huge farm production increases made possible by technology have greatly benefited consumers but created a problem for American farmers (Schiller 2000, Knutsen et al. 1998). This problem exists because the demand for food is relatively inelastic. In other words, the amount of food that people can eat is limited, and well-fed consumers do not increase food purchase by very much when harvests are abundant. Therefore, a small increase in food supplies often leads to large drops in food prices. For example, in the periods from 1910–1914 and 1993–1995, the ratio of farm to nonfarm prices fell 50%. This ratio would have fallen even farther without foreign demand for U.S. farm products and government farm subsidies. Over the long run the ability of U.S. agriculture to produce has greatly outpaced the growth in the human population and the quantity of food demanded.

Another problem is that farmers also must confront abrupt price swings caused by changes in annual climatic conditions and world political upheaval (Schiller 1994, Figure 12.13). Favorable weather results in abundant harvests but low prices. On the other hand, droughts, freezes, insect infestations, and outbreaks of crop diseases can drastically reduce harvests but sharply push up prices. Unlike most other industries, farmers cannot quickly increase supplies when climatic adversity occurs. A year or more is required for farmers to respond by planting more land, or increasing their livestock herds when food prices go up. This lag intensifies price swings and makes it difficult for farmers to take advantage of higher prices. When corn prices are high because of unfavorable weather, farmers will plant more corn the following year. After a year of high corn prices it is probable that better weather and increased

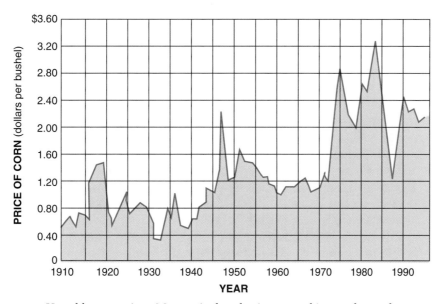

FIGURE 12.13 Unstable corn prices. Most agricultural prices are subject to abrupt short-term changes. Notice how corn prices rose dramatically during World Wars I and II, then fell sharply. Poor harvests in the rest of the world increased demand for U.S. food in 1973–74 (from Schiller 1994) (U.S. Department of Agriculture).

crop acreage will cause corn prices to drop. It is very difficult for individual farmers find it very difficult to avoid these booms and busts in prices.

Prior to the 1930s' depression, the U.S. government had little involvement in agriculture. Generally the farm sector was prosperous because of expanding foreign demand, a rapidly human population, and a slow pace of technological advance. The 1910–1919 period was particularly favorable for American farmers because of World War I which depressed farm output in Europe (Figure 12.14). After 1920, international trade encountered increased restriction, and European demand for American farm products declined. At the same time there was considerable improvement in agriculture efficiency mainly from mechanization. In the depths of the depression in 1932, farm prices had fallen 75% below 1919 levels (Schiller 1994). Average farm income fell from about $2,600 to $850. Farmers pressured the U.S. Congress to take action on their problems. The main objective of the resulting law (Agricultural Adjustment Act of 1933) was to increase farm product price levels to those of the 1909–1914 period, which was considered fair or parity. The basic approach was to establish a price floor in which the government compensates farmers for the difference between the market and the parity price. The consequences of this policy have been surpluses that must be eliminated by government food purchase and giveaway, export sales, or restrictions on supply (Figure 12.15). Since the 1930s, the government has also subsidized farmers through acreage set-asides, marketing orders, import quotas, crop insurance, and cost subsidies. These will be briefly discussed.

Acreage Set-Asides.   The surest way to increase farm prices without causing a surplus is to pay farmers to take land out of production (Schiller 1994, Knutson et al. 1995). Legislation passed in 1936 and 1938 was the basis for allotments and quotas designed to limit the supply of specific crops. Farmers have always responded to these restrictions by more intensively farming allotment acres. Quotas are often applied with allotments. They restrict the quantity of a commodity a farmer is allowed to sell. Discontent with the regimentation associated with allotments and quotas led to voluntary land retirement programs that paid producers to take land out of production. Land productivity has been used as the basis for payments. There have been three

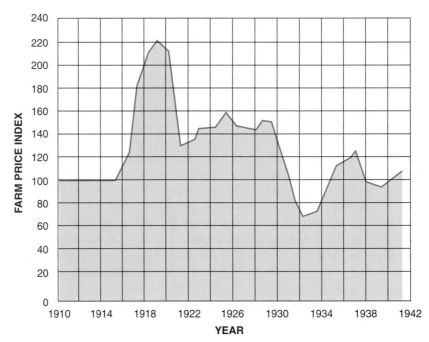

FIGURE 12.14   Farm prices, 1910–1940 (1910–1914 = 100). Farm prices are less stable than non-farm prices. During the 1930s, relative farm prices fell 50%. This experience was the catalyst for government price supports and other agricultural assistance programs (from Schiller 1994).

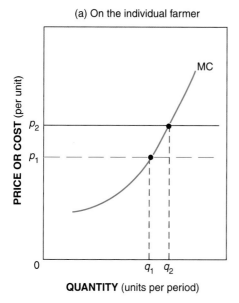

(a) On the individual farmer

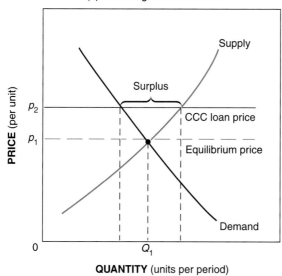

(b) On the agricultural market

FIGURE 12.15 The impact of price supports. In the absence of price supports, the price of farm products would be determined by the intersection of market supply and demand (see Chapter 4). In this case, the equilibrium price would be p_1, as shown in part b. All individual farmers would confront this price and produce up to the point where MC = p_1, as in part a. Government price supports raise the price to p_2. By offering to buy (or "loan") unlimited quantities at this price, the government shifts the demand curve facing each farmer upward. Individual farmers respond by increasing their output, resulting in a market surplus (part b) (from Schiller 1994).

major land retirement programs since the 1930s. These include the soil bank, the payment-in-kind, and the conservation reserve programs. Our discussion of these programs follows Knutson et al. (1995, 1998).

The soil bank, first implemented in 1956, paid farmers to retire land from production for a 10-year period. By 1960, nearly 30 million acres were in the soil bank. Under this program whole farms could be retired from production. This created economic hardship for some rural communities because of a drop in commerce. The large increases in crop prices during the 1970s caused most of the soil bank lands to go back into production.

Drops in farm prices because of inflation control in the U.S., a rising dollar, and increased foreign agricultural output resulted in huge grain surpluses by 1983. To deal with this problem, USDA announced its PIK (payment-in-kind) program which paid farmers with grains for 80 to 95% of their farm yields in return for a 1-year retirement of their land from production. Slightly over a third (82 million acres) of the eligible cropland was removed from production in 1983 (Figure 12.16). This program was effective in temporarily lowering grain surpluses and raising farm prices, although it was expensive.

FIGURE 12.16   Quantities of farmland removed from production, 1983–1993 (from Knutson et al. 1995) [Economic Research Service, Agricultural Resources: Cropland, Water, and Conservation, AR-30 (Washington D.C.: ERS/USDA, May 1993), p. 4].

The Conservation Reserve Program, or CRP, was part of a 1985 farm bill that had dual objectives of reducing farm commodity oversupply and controlling soil erosion. It differed from the soil bank program in that only land classified as highly erodible could be retired for a 10-year period. In order to minimize adverse economic consequences to rural communities, only 25% of the land could be retired in any county. By 1995, about 35 million acres had been retired at an annual cost of 1.8 billion dollars per year. Like the soil bank program, payments to farmers are based on land productivity. Although this program is considered costly, it has been praised for its benefits associated with soil and wildlife conservation. Its effectiveness in reducing oversupply of certain crops has been more controversial.

Marketing Orders.   Although the production of individual farmers is too small to raise market prices by withdrawing output, they can increase market prices through collective action (Schiller 1994). Marketing orders set minimum prices that processors must pay for agricultural commodities. Since the depression, the federal government has allowed farmers to limit the quantity of produce they market. The surpluses that result from market orders are wasted (consumed by no one) by the individual farmer. Huge amounts of oranges, lemons, nuts, raisins, and various fruits have been unnecessarily grown since the 1980s. In addition to waste, this practice adds unnecessarily to environmental degradation.

Import Quotas.   Restrictions are placed on the amount of specific farm products that can be imported from other countries (Schiller 1994). Sugar, dairy products, certain grades of beef, peanuts, and cotton can generally be produced more cheaply in other countries than in the U.S. Therefore, import quotas have been imposed that protect U.S. farmers. Import taxes (tariffs or duties) on foreign agricultural products are also used to restrict their importation to the U.S.

Cost Subsidies.   Cost subsidies involve direct or indirect benefits that farmers receive from the federal government, which help to reduce farm production costs (Schiller

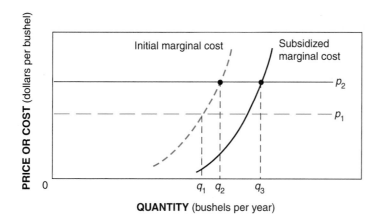

FIGURE 12.17 The impact of cost subsidies. Cost subsidies lower the marginal cost of producing at any given rate of output, thereby shifting the marginal cost curve downward. The lower marginal costs make higher rates of output more profitable and thus increase output. At price p_2, lower marginal costs increase the desired rate of output from q_2 to q_3 (from Schiller 1994).

1994). These benefits include irrigation water from federally funded reclamation projects, cash payments to cattle ranchers for feed in drought years, and partial payments for a variety of practices that include drainage, brush control, land leveling, water development, fencing, and more. These various subsidies lower production costs and encourage more output than would otherwise occur (Figure 12.17).

Crop Insurance. Most private insurance companies generally do not offer adequate crop insurance to farmers because of excessive risk (Knutson et al. 1995). Therefore, federal crop insurance (FCI) is set up to extend insurance to all crops and give producers a choice in the level of coverage. The intent was for the federal government to subsidize 30% of the cost of this program. The problem with FCI is that participation has been highest in high risk, nonirrigated, low rainfall areas, and low where the opposite conditions prevail. Because this program has run at a one-half-billion-dollar annual loss during the early 1990s, there is general agreement that the program needs to be improved.

Farm Programs in Other Countries

The United States does not stand alone in protecting agriculture with federal subsidies. Most other developed countries subsidize agriculture to a greater degree than the U.S. (Figure 12.18). This is particularly true in France, Germany, Switzerland, Italy, Spain, Great Britain, and Japan. The reasons for farm subsidies include a national desire to maintain a healthy farm sector, maintain rural esthetics and open space, provide food security in case of war, and maintain strong political pressure to help certain special interest groups involved in agriculture (Knutson et al. 1995). It is estimated that the various subsidies and agricultural protection program costs the average consumer in Europe 200 dollars and the figure is thought to be even higher in Japan. Some of the more costly subsidies from a consumer standpoint involve sugar in the U.S., rice in Japan, vegetables in the Netherlands, and grain in Germany and France. In all these cases, other countries can produce these products at a much lower cost (have a comparative advantage), but because of government protection, inefficient producers are kept in business.

A major drawback to agricultural subsidies and protectionism in developed countries is the adverse impact they can have on developing countries (Foster 1992). Many developing African and Latin American countries have a comparative advantage in world trade in production of basic goods such as food and textiles, fossil fuels, and minerals. Their low labor costs and favorable climatic/soil conditions make them efficient producers of sugar, coffee, sorghum, peanuts, and cotton as well as various livestock products. In order to generate currency to purchase the complex goods necessary for economic development, they need to be able to sell their basic goods in world markets. However, they typically confront an international system of quotas and tariffs that reduces their opportunities to sell what they can produce at comparative advantage. On the other hand, the developed nations faced with huge surpluses of subsidized farm products habitually dump excess food on world markets at less than production costs.

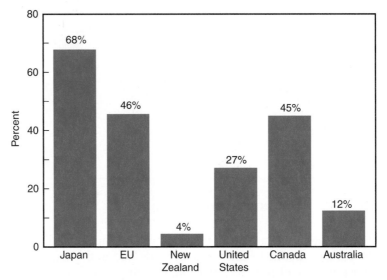

FIGURE 12.18   Levels of government intervention in agriculture, by country, 1989–1992 (from Knutson et al. 1995) [Agricultural Policies, Markets, and Trade: Monitoring and Outlook 1993 (Paris, France: OECD/OCDE), p. 169]

Food relief is another challenging aspect of subsidized oversupply of food in developed countries. Providing food relief to developing countries experiencing famine is commonly considered humanitarian. An alternative view concludes that it does more ultimate harm than good. Massive food aid can depress local food prices and reduce food production incentives. It may also cause migration from farms to cities and create a disincentive for investment in agricultural development. Many experts in foreign development now believe food aid should only be provided on a temporary basis when some natural disaster occurs.

Trends in Current Policy

During the mid 1990s, the United States and other developed countries have been in the process of moving toward free trade and away from agricultural subsidies and protectionism. Another proposed change in agricultural policy in the United States is the decoupling of agricultural payments to farmers from level of farm production. These changes will be briefly discussed following Knutson et al. (1995).

Trade Policy.   GATT (General Agreements on Tariffs and Trade) is the main global forum for various nations to negotiate trade agreements. It involves 92 member governments, and its main goal is to liberalize and expand trade by reducing trade barriers. As a whole progress has been greater in liberalizing trade of industrial products than agricultural goods. This is because of increased agricultural subsidy and protection in the developed countries since World War II. Nevertheless there have been some breakthroughs since Uruguay hosted negotiations in 1987. These have included reductions in import barriers, reductions in direct and indirect subsidies, and minimizing the effects of unnecessary health and sanitary regulations on trade. Some of the key provisions were to reduce the 1986–1990 level of domestic subsidies by about one-third to convert quota barriers on trade to tariffs, then to phase out the tariffs and provide increased access for products barred from specific countries such as rice in Japan.

NAFTA (North American Free Trade Agreement), passed in 1993, is the most important recent trade agreement affecting the United States, Canada, and Mexico. Under NAFTA, most trade barriers among the three countries will be eliminated over a 15-year period. Basically NAFTA phases out tariff and nontariff barriers to trade, establishes rules for investment, strengthens intellectual property rights, and sets up procedures for settlement of trade disputes.

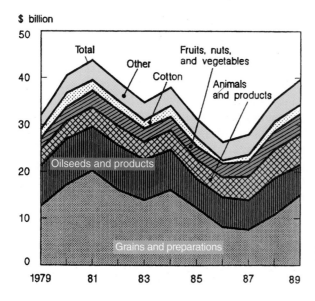

FIGURE 12.19 Value of U.S. agricultural exports by commodity (from Knutson et al. 1995) [Extension Service, Agricultural Chart Book 1990, Agric. Handbook 689 (Washington D.C.: USDA, April 1990), p. 81]

Free trade is in the interest of all countries because different countries have comparative advantages in producing different goods. Columbia is well suited for coffee production, Caribbean countries can best produce sugar, Australia and New Zealand are highly efficient producers of wool and mutton, and the United States is an efficient producer of grains and soybeans (Figure 12.19). Comparative advantages results from the ability of a country to produce and export a specific good more efficiently (lower input costs) than other countries. When free trade is maximized, different countries can specialize in producing goods where they have comparative advantage. Each country benefits because they have access to more goods at lower cost. Although the country as a whole benefits from free trade, some companies and workers in inefficient industries may be harmed in the short run because of increased competition. At the same time, free trade has often forced existing industries to become more efficient or competitive in producing goods for consumers. The U.S. government has had various trade adjustment programs to ease the hardship when economic losses occur because of trade agreements.

Disengaging. There has been increasing pressure to greatly reduce or eliminate government involvement in agriculture (Knutson et al. 1995). However, various farm groups resisted these efforts, claiming that it would reduce their incomes and cause a reduction in farm asset values. One alternative that is receiving consideration is decoupling. It involves the separation of income payments to farmers from market prices and production. Basically under this approach all farm programs that alter a farmer's production decisions are eliminated. Decoupling to some extent transforms farm subsidies into entitlement or welfare payments. However, it largely avoids the free market distortions that result from production subsidies. To some extent, land retirement initiatives such as the Conservation Reserve Program are a form of decoupling.

ELIMINATING GOVERNMENT INVOLVEMENT IN AGRICULTURE: THE NEW ZEALAND CASE

The case for continued government assistance to farmers centers around farming being different from other kinds of enterprises (Schiller 2000, Knutson et al. 1995). The argument is basically that unpredictable weather and other crop growing conditions make farming an unstable enterprise. Therefore, government payments to farmers are vital to protect the agricultural sector from periodic disaster.

However, many economists and congressmen now question whether farm program benefits exceed costs to taxpayers, and some actually think they do more harm than good. Some farms in the U.S. growing unsubsidized crops—such as potatoes, fruits, and vegetables—are doing quite well. New Zealand provides some of the most convincing evidence against farm subsidies.

In 1986, New Zealand ended its farm subsidy programs because of economic stagnation and huge government budget deficits (Merline 1995). Despite dire predictions otherwise, few farmers went bankrupt or committed suicide.

New Zealand's farmers actually prospered after the elimination of subsidies (Merline 1995). Output of some agricultural commodities fell, but production efficiency noticeably improved. There were definite environmental benefits from ending subsidies in New Zealand. Marginal land was taken out of production, and usage of fertilizer and pesticides became more judicious as their true market cost became more apparent. Jobs in the farming sector dropped, but this was merely a continuation of a trend in place long before the elimination of subsidies. Farm organizations are generally pleased with the agricultural revival in New Zealand, and there has been no movement for a return to subsidies. Farmers in New Zealand now widely believe that they are better off by being independent of the government.

According to a recent study by the USDA Economic Research Service, farmers in the U.S. who do not participate in government farm programs appear to be doing better than those who do. Some 63% of nonparticipating farms and 60% of participating farms in government programs had a favorable financial position. About 4% of nonparticipating compared to 7% of participating farms were vulnerable to insolvency (bankruptcy).

One problem with government farm programs not always recognized as such, is that they change constantly. New farm bills are passed every 5 years and often amended. This makes long-term planning difficult and may cause farmers to focus more on qualifying for government programs rather than on improving operation efficiency. Therefore, farm programs may contribute more to instability than stability in the farming industry.

Conclusion

In conclusion we believe that governments in both developed and developing countries will increasingly move away from farm subsidies. Not only are farm subsidies costly, they distort international trade, discourage production efficiency, and may encourage environmental degradation. Countries such as New Zealand and Chile that have recently eliminated farm subsidies are experiencing improved economic growth and greater productivity. Decoupling of farm subsidies from crop prices and land productivity is an alternative to supply-oriented subsidies where the goal is to maintain farm numbers. Land retirement programs, such as the Conservation Reserve in the United States, are forms of decoupling that have benefits in terms of reducing crop oversupply, reducing soil erosion, and increasing wildlife populations. However, this approach is costly and in the long run may be no more effective than if market forces were allowed to operate. We believe minimal government intervention with reliance on market forces will prove to be best from the standpoints of increasing agricultural productivity and maintaining a healthy farmland environment.

LITERATURE CITED

Bolen, E. G. and W. L. Robinson. 1995. *Wildlife ecology and management.* 3rd edition. Upper Saddle River, NJ: Prentice-Hall.

Chrispeels, M. J. and D. Sadava. 1977. *Plants, food, and people.* San Francisco, CA: W.H. Freeman and Company.

Erlich, P. 1968. *The population bomb.* New York: Ballantine Books.

Foster, P. 1992. *The world food problem.* London: Lynne Rienner Publishers.

Knutson, R. D., J. B. Penn, and W. T. Boehm. 1995. *Agricultural and food policy.* 3rd edition. Englewood Cliffs, NJ: Prentice-Hall.

Knutson, R. D., J. B. Penn, and B. L. Flinbaugh. 1998. *Agricultural and food policy.* 4th edition. Englewood Cliffs, NJ: Prentice-Hall.

Laetsch, W. M. 1979. *Plants: Basic concepts in botany.* Boston, MA: Little, Brown & Co.

Merline, J. 1995. Can farming service markets? *Investor's Business Daily,* March 9, 1995, 1–2A.

Miller, G. T. 1990. *Resource conservation and management.* Belmont, CA: Wadsworth Publishing Co.

Owen, O. S. and D. D. Chiras. 1995. *Natural resource conservation.* 6th edition. Englewood Cliffs, NJ: Prentice-Hall.

Owen, O. S., D. D. Chiras, and J. P. Reganold. 1998. *Natural resource conservation.* 7th edition. Englewood Cliffs, NJ: Prentice-Hall.

Pedigo, L. P. 1999. *Entomology and pest management.* 3rd edition. Upper Saddle River, NJ: Prentice-Hall.

Schiller, B. R. 1991. *The economy today.* 5th edition. New York: McGraw-Hill.

Schiller, B. R. 1994. *The economy today.* 6th edition. New York: McGraw-Hill.

Schiller, B. R. 2000. *The economy today.* 8th edition. New York: McGraw-Hill.

World Resources Institute. 1994. *World resources: 1994–95.* New York: Oxford University Press.

CHAPTER 13

Outdoor Recreation

INTRODUCTION

Outdoor recreational opportunities in the United States are the envy of the world. Few, if any, other countries provide such vast, multifaceted landscapes and associated opportunities to enjoy their recreational benefits (Figures 13.1, 13.2, 13.3, and 13.4). Relatively early in the settlement of the United States, a change developed which replaced an attitude of conquest and exploitation of natural areas to one of respect and a desire to maintain such areas in perpetuity. This transformation initially required great effort on the part of a few to change the prevailing national attitudes toward nature and wilderness. Americans in the millions now take for granted the ability to board a plane, car, train, or boat and readily reach a destination where nature can be viewed and enjoyed in silent contemplation or wild abandon. However, controversy has resulted over the appropriate allocation of resources for recreation on public lands.

Outdoor recreation is a component of the world's largest industry, tourism. Recreation—the pleasurable use of leisure time—is highly popular and greatly stimulates the world's economy. In 1994, travel and tourism generated $3.4 trillion of the global gross output in goods and services, it created employment for 204 million people (1 in every 9 workers), and it accounted for 11% of consumer spending in the world. In the United States, tourism generated $820.1 billion in 1994, and it sustained over six million jobs. Only auto and food sales rank above tourism as a retail industry.

In this chapter we will discuss the development of outdoor recreation mostly on public lands in the U.S. Recent trends and challenges regarding public land recreation will be addressed. Recreation based on the management of fisheries and wildlife is addressed in more detail in Chapters 15 and 16. Also see Chapter 14 for outdoor recreation on urban lands. Rangelands will receive special emphasis because recreation there, versus other uses, has been especially controversial in recent years. We refer the reader to Jubenville (1978) and Zinser (1995) for more complete coverage of outdoor recreational management.

FIGURE 13.1 A family camping and picnicking in the Ouachta National Forest in Arkansas (courtesy of U.S. Dept. of Agriculture).

FIGURE 13.2 An angler enjoys fishing in the Pisgah National Forest in North Carolina (courtesy of U.S. Department of Agriculture).

What is Recreation?

Recreation is, after all, an antidote for seriousness. Therefore most people would rather put recreational management concerns out of their minds. However, the quality of contemporary life in the U.S. would not be the same without recreation managers who do take it seriously. Nine out of ten Americans participate in some form of outdoor recreation and many rural economies in the U.S. and elsewhere depend on it. Anticipation of satisfying outdoor recreation has become an integral part of contemporary life.

FIGURE 13.3 Water-related recreation, such as sailboating, is particularly attractive to outdoor enthusiasts (courtesy of Jack McCaw).

FIGURE 13.4 Farmlands, recreation, and beauty go hand in hand. Bird hunters in Virginia enjoy a nice autumn day (courtesy of U.S. Department of Agriculture).

The word recreation derives from the french *recreare,* which means to restore, refresh, or to "create anew." Recreation is a partner of leisure time. In centuries past, only the wealthiest had time to "play" and, for many perhaps, recreation had more to do with escaping boredom than with respite from hard work. One of the major advances made by humankind has been the freeing up of time for a more pleasurable existence—one that balances fulfilling work with rejuvenating recreation. In keeping with the relatively recent escalation of leisure time in human lives, recreational management is one of the more recently evolved specialty areas in natural resources management.

Contemporary recreation usually restores energy, interest, and creativity through some form of diversion from a work routine. Although it is most valued by participants for its immediate rewards, recreation may be more valuable for the increased work productivity that follows. Regardless of whether outdoor recreation is active or passive, physical or intellectual, intense or relaxed, regular or sporadic, organized or happenstance, it usually involves a change in human behavior and environment. For most people it requires translocation from indoors to some distant point outdoors. Travel mode and distances are critical factors in choice of recreational activity and outdoor recreational planning.

No matter the form or place, effective outdoor recreation restores health and motivation—it "creates anew" both physical and mental function. Even if the quality of a recreational site is unexcelled, satisfaction can be devastated by poor or uncomfortable access. Thus comprehensive recreational management is necessarily integrated with transportation and travel management. One of the most prominent changes in the twentieth century is the degree to which outdoor recreation has become a part of most lives in the U.S. and other developed nations. Management of natural resources for outdoor recreation has become one of the most important emphases of public land and water management. It is also of growing interest and importance in the private sector.

The Importance of Outdoor Recreation

We rarely think about how important recreation has become to us during work breaks, after work, and on weekends, holidays, and vacation. Regardless of what we call it, we now take for granted the need for recreation in all forms. Studies in 1960 and 1987 showed a consistent 89 to 90% of the U.S. population participated in outdoor recreation away from the home (TFORRO 1988). Labor agreements often emphasize paid time off for rest and renewal. Newer work environments often are designed to provide short "getaways" from the grind. Many larger employers provide outdoor recreational outlets, such as parklike grounds, picnic tables, exercise areas, and pleasant walkways. Windows are important to many employees in part because they provide brief escape to the "outdoors" and seem to stimulate mental productivity. Professional meetings often are set up in attractive locations and environments designed to provide outdoor recreational opportunities as a work incentive. Cities vie for tourism dollars associated with business and purely for pleasure. Whole industries have grown up in support of recreational opportunities. Recreational tourism, including a mix of indoor and outdoor opportunities, has become one of the world's biggest businesses.

The growth of large corporate employers during the industrial and information age created conditions leading to more leisure time and a greater need for recreation. When most people worked for themselves, primarily on family farms, holidays and vacations were rare because business required continuous attention. Technology and mass production techniques greatly increased per capita production and average real income. Yet the work of family businesses was more diverse and less mind numbing than the specialized routines required of many company jobs. Work concessions were won by organized labor. It soon became apparent that productivity was especially high when workers were well rested and refreshed. This typically required shorter work days and work weeks.

By the 1940s, most employees of large companies expected a 40-hour workweek and employment benefits including a week or more of paid vacation. After World War II, family participation in outdoor recreation boomed with the baby boom. Time-off benefits rose to a higher priority level in employment negotiations. As the information age emerged, many jobs became more sedentary, mentally more demanding, and often harder to leave behind "at the office." Even though the amount of reported leisure time has actually decreased in recent years, for a variety of reasons (e.g., more work commuting time), the demand for outdoor recreation has been sustained (TFORRO 1988). On-the-job boredom and physical inactivity continue to be the most debilitating aspects of routine work in many organizations. Outdoor recreation is one of the more effective prescriptions for many people.

Attributes of Outdoor Recreation

Although, in the strictest sense, outdoor recreation means simply any restorative behavior done outdoors, it usually connotes a different form of activity than those commonly done inside. Benefits from outdoor recreation depend heavily on availability and quality of outdoor environment.

Outdoor recreational management centers around providing ideal outdoor environments for a wide variety of recreational activities. While provision of appropriate landscaping, site access, shelter, vegetation management, interpretive services, and facilities can greatly improve upon outdoor recreational satisfaction, the original choice of site often is keystone to management success. Many of the present locations accessible to either private or public recreational use were chosen decades ago and more often than not for reasons other than recreational use. Therefore, much of what recreational resource management does involves making the best of less than optimum recreational conditions.

Important management variables are natural topography, climate, ecosystem productivity and diversity, other human uses, and land-use authorities. The attributes and diversity of topography are critical for supporting diverse forms of outdoor recreation. The distribution and proportions of land and water determine the diversity of recreational activities possible at a managed site. Rugged topography is critical for certain types of recreation, such as mountain and rock climbing, skiing, back-country hiking, wilderness camping, whitewater navigation, and all-terrain vehicle use. Other recreation requires somewhat less rugged terrain including most field sports, ice skating, jogging, snow sledding, cultivated gardens, bicycling, beach activity, and waterside access to docks, piers, marinas, and boat ramps. Diverse climates facilitate alpine skiing, waterskiing, ice fishing, whitewater rafting, fall hunting, swimming, and sunbathing on sandy beaches. Without a diversity of natural ecosystems, many forms of outdoor recreational opportunity would be diminished or defunct, such as hunting, fishing, nature photography, and bird-watching.

Resource Conflicts and Resolution

The history of recreational use of public resources in the U.S. is in part a study in resource use conflicts and their resolution. The amount of public land and water made accessible and developed for recreational use in the natural state is unexceeded anywhere else. Outdoor recreational activity has increased with vastly improved quality of life and increased diversity of recreational opportunities. However, other uses of public lands and waters have come into increasing conflict with recreation use, including timber, range, water supply, and flood management. Even different recreational groups can find themselves in conflict. At overcrowded sites, for example, those who would enjoy themselves in silent contemplation often find themselves at odds with those who prefer wild abandon in noisy group play. Recreational management has come to include making up rules, enforcing laws, and otherwise managing the conflicts that can arise in multiple-use situations.

The diversity of viewpoints relative to the appropriate limitations on the recreational use of public lands has stirred controversy. Has enough public land been set aside in the right places for present amount and distribution of recreational demand? Should private development be encouraged over public development? Should there be a cap on the total amount of land excluded from private ownership and development? How much should other authorized uses of public lands be allowed to diminish the value of recreational use and vice versa? To what extent should private concessions and other partnerships between public and private development be promoted in public policy? To what degree is recreational development compatible with emerging concerns for biodiversity protection and ecosystem management?

Recreational management often is wedged in with other natural resource management. In a multiple-resource management setting, it frequently is the last management added when budgets are fat and the first to be cut when budgets are lean. Attitudes toward recreation often exhibit less respect for recreational resources than other resources even when economics strongly suggests otherwise. Although the reasons for this are complex, some of the most basic probably extend back to a time in European and American history when having fun was viewed with great suspicion.

HISTORICAL PERSPECTIVES

Federal Management

The federal government through its various land management agencies, principally the Forest Service, Bureau of Land Management, National Park Service, and Fish and Wildlife Service, manages almost one-third of the land area of the United States. In addition, the water resources agencies contribute greatly to water-based recreation on inland reservoirs and waterways; especially the Army Corps of Engineers, Bureau of Reclamation, Natural Resources Conservation Service and the Tennessee Valley Authority. The acquisition of most federal lands resulted from a series of federal laws that transferred large blocks of land from the public domain to federal management. While the largest percentage of federal ownership within a state in the eastern and central states does not exceed 8% (with the exception of South Dakota with 17%), the percentage of federal ownership in the western states ranges from 35% in Washington State to 85% in Nevada. Most of this federal land acquisition occurred in the nineteenth century (see Chapter 2).

Early Attitudes

The initial philosophy of the American public toward natural resources was one which perceived superabundance and fostered conquest over nature. The wilderness was to be subdued and whatever wealth it contained was to be extracted for personal and national gain. For most people, little time was available for recreation and religious leaders often disapproved of recreational pleasure. The Spanish, French, English, and Russian interests in the New World were intent on exploiting living and nonliving resources. In the eastern U.S., the fishery and logging industries were an important economic impetus. For most of the immigrant masses, however, land was cleared of timber for conversion to agriculture. The rich fertile soil, which once supported luxuriant and seemingly endless stands of forest, was the resource most sought. As the U.S. borders moved farther west, the same pattern continued—only it was the extensive grasslands that were plowed and grazed, and the original ungulate fauna replaced by domestic sheep, cattle, and horses from the Old World. The federal government acquired large tracts of public domain land mostly west of the Applachian Mountains. While much of this land was sold at low price or given away (see Chapter 2), a large fraction was retained under federal management. Lands east of the Rocky Mountains quickly became privately owned. By the end of the nineteenth century, most remaining public land was located in and west of the Rocky Mountains region.

The philosophical and intellectual basis in America for the preservation of nature for its enjoyment was introduced by Ralph Waldo Emerson and Henry David Thoreau before the Civil War. Nature study and tourism started to become popular. Thoreau proposed small, unspoiled parks where people could seek nature. However, it is landscape architect Frederick Law Olmsted whom we owe most for the first concepts of national and state park systems, and the basis of modern park planning. He believed humans exposed to undisturbed park settings benefited psychologically and that parks should be established for the enjoyment of all levels of society. Olmsted laid the foundation of modern park planning through his management plans for Central Park in New York City and the Yosemite State Park in California. Yosemite State Park was established by an act of Congress in 1864 on most of what is now Yosemite National Park "upon the expressed conditions that the premises shall be held for public use, resort, and recreation and shall be held inalienable for all times." This state park was followed by the creation of New York State reserves at Niagara Falls and in the Adirondock and Catskill Mountains. In March 1872, President Ulysses S. Grant signed the bill that created what was to become Yellowstone National Park and which initiated the national park system. Several state and county park systems and numerous city park systems were established between 1880 and 1890. The next major growth in outdoor recreation occurred after World War II when leisure time increased and mass-produced automobiles and the expanded road system facilitated travel.

Recent Trends

The most influential modern document on outdoor recreation land-use planning was prepared by the Outdoor Recreational Resources Review Commission (ORRRC), which was established by an act of Congress. Their report, published in 1962, was a comprehensive survey of outdoor recreation in America and made outdoor recreation recommendations relative to "what policies and programs should be recommended to ensure that the needs of the present and future are adequately and efficiently met." The ORRRC noted the lack of outdoor recreation facilities near metropolitan areas and that the present recreation lands were not meeting the needs of Americans. They recommended the creation of a Bureau of Outdoor Recreation in the Department of the Interior. The Bureau was created in 1962 and was the first federal agency specifically assigned to formulate a nationwide outdoor recreation plan. Outdoor recreation greatly increased in the 1960s and 1970s. Federal expenditures for recreation increased from $75 million in 1960 to $1.4 billion in 1980. Outdoor recreation stabilized or moderately increased in the 1990s on federal lands (Table 13.1).

The next outdoor recreation report, titled *American Outdoors: The Legacy, the Challenge,* was prepared by the President's Commission on Americans Outdoors (1987). The major findings reported were as follows:

1. "Americans place a high value on the outdoors; it is central to the quality of our lives and the quality of our communities.

2. "Outdoor recreation provides significant social, economic, and environmental benefits. Because these benefits are difficult to assess in dollars, recreation and resources protection suffer in competition with other programs for public and private dollars.

3. "High quality resources—land, water, and air—are essential to fishing and boating, camping and hiking, skiing, and bicycling, hunting and horseback riding, and every other outdoors activity.

4. "Quality of the outdoor estate remains precarious. People continue to misuse and abuse resources and facilities. We are becoming aware of more pervasive long-term threats such as toxic chemicals, water pollution from nonpoint sources, groundwater contamination, and acid precipitation.

5. "We're losing available open space on the fringe of fast growing urban areas and near water.

6. "Wetlands and wildlife are disappearing.

7. "Wild and free-flowing rivers are being dammed, while residential and commercial development is cutting off public access to rivers in urban areas.

8. "With more people doing many different things outdoors, competition for available lands and waters is increasing; to accommodate these pressures we will have to better manage what we have.

TABLE 13.1   Visitation Time Spent at Federal Recreation Areas (Millions of Hours)

	1980	1985	1990	1991	1992
Forest Service	2,819	2,705	3,157	3,346	3,452
U.S. Army Corps of Engineers	1,926	1,721	2,280	2,306	2,306
National Park Service	1,042	1,298	1,322	1,344	1,390
Bureau of Land Management	68	246	518	540	563
Bureau of Reclamation	407	289	280	280	269
Tennessee Valley Authority	87	79	10	13	14
All Areas	6,367	6,403	7,567	7,829	7,995

9. "The quality of recreation services delivery is inadequate. Though some services are improving, much remains to be done.

10. "Inadequate funding for staff, development of facilities, and maintenance limits recreation use of some public lands.

11. "People in central cities have a harder time experiencing the outdoors.

12. "Barriers to investment prevent the private sector from reaching its potential as a recreation provider.

13. "Resources management and recreation programs offered by public and private providers are not coordinated as well as they should be.

14. "The liability crisis is limiting our opportunities to enjoy the outdoors.

15. "We don't have a good overall picture of what we have; we lack systematic monitoring of resource conditions and public needs."

FEDERAL RECREATIONAL MANAGEMENT

National Forest Service

The national forests are the most important providers of public outdoor recreation in the United States (Table 13.1). Lands within this system encompass 191,453,354 acres and include national forests, national grasslands, and portions of the National Wilderness Preservation System, national recreation areas, wild and scenic rivers, and wildlife protected areas. The evolution of the national forest system has been one of de-emphasizing commodity values (principally timber production and livestock grazing) and emphasizing amenity values (principally recreational opportunities) (Figure 13.5). This reflects the changing attitudes of the American public followed by appropriate legislative mandates.

In 1886, Congress established the Division of Forestry under the Department of Agriculture to regulate timber cutting. However, it was the Forest Reservation Act of 1891 which gave the president of the United States the power to establish public lands as timber reservations. By 1905, over 100 million acres had been placed in reserves. The Forest Management Act of 1897, known as the Organic Act, instituted the basic philosophy and structure of the national forest system. The act emphasized the

FIGURE 13.5   Forest service lands provide vast areas for enjoying outdoor recreation (courtesy of Jack McCaw).

need to furnish a continuous supply of timber and to protect the forests for secure, favorable water flows. This law clearly established that the forest reserves were to be managed for utilitarian purposes in contrast to national parks which were based on preservationist policy. The Weeks Act of 1911 authorized the purchase of deforested private lands in the eastern United States to protect watersheds. This enabled the establishment of national forests throughout the United States. By 1930, 20 million acres had been added to the national forest system in the East.

Although recreation use of national forest continued to increase, Congress refused to appropriate funds for recreational purposes. This was because recreation was viewed as inappropriate on national forest lands. High demand for lumber following World War II and increasing recreation in the national forests ultimately created conflict between the two uses. Some recreationists, including advocates for wilderness areas, viewed logging as a threat to their interests. The Forest Service sought to resolve this conflict through the passage of the Multiple Use, Sustained Yield Act (MUSY) of 1960 which mandated a multiple use of national forest lands. Under this act, the uses designated for national forests broadened to include recreation, timber, livestock grazing, watershed protection, and provision of wildlife and fish habitat.

This legislation was the result of social pressures following World War II, which called for a change in commodity-oriented use of natural resources. National conservation organizations—such as the Audubon Society, Wilderness Society, and Sierra Club—exerted political pressure to ensure that amenity values of natural resources were given equal standing.

The next goal of amenity users was wilderness legislation. Wilderness areas were first established beginning in 1919 at the suggestion of Arthur Carhart, an employee of the Forest Service. In 1924, through the efforts of Aldo Leopold, the Forest Service designated the Gila Primitive Area within the Gila National Forest. By 1937, primitive areas encompassed 14 million acres. However, these areas were not fully protected by law. In 1964, President Lyndon B. Johnson signed the Wilderness Act, which gave national legal status to wilderness protection.

The act established 54 areas totaling 9.1 million acres on Forest Service land. The wilderness system has since expanded to other federal lands and now includes 579 areas totaling 96,056,113 acres. Alaska has the largest percentage of wilderness area within a state totaling 57,408,796 acres or 59.8 percent of the total area.

The National Forest Management Act of 1976 also affected recreational use of federal lands. The act expanded multiple use to include wilderness, instituted a formal public review process which incorporated the public in decisions of national park planning, and required that the national forest prepare fifty-year multiple-use plans for national forests. It required that planning be interdisciplinary involving the biological, physical, and social sciences.

National Park Service

Americans can clearly identify with the National Park System because of the spectacular geologic and scenic features within the system (Figure 13.6). The National Park Service Act of 1916 placed under one agency the previously designated 14 national parks and 21 national monuments. The mission of the Park Service was "to conserve the scenery and the natural and the historic objects therein and to provide for the enjoyment of the same in such a manner and by such means as will leave them unimpaired for future generations."

The National Park Service is charged with managing a variety of natural, historic, and cultural resources under several designations and sometimes in cooperation with other agencies, such as the Bureau of Land Management. These include national parks, national recreation areas, national monuments, national preserves, national seashores, national lakeshores, national memorials, national historic sites, national historic parks, national battlefield parks, national rivers, and national parkways. The primary national parks and national monuments are shown in Figure 13.7. National

(a)

(b)

FIGURE 13.6 The National Park System in the United States includes spectacular geologic and scenic areas. Photo (a) shows the Grand Canyon in Arizona (courtesy of the National Park Service). Photo (b) shows White Sands National Monument in southcentral New Mexico (courtesy of Bill Walker).

parks consist of large areas that protect natural and cultural heritage. There are 47 national parks encompassing 47.4 million acres. National monuments protect a nationally significant resource. There are 81 national monuments encompassing 6.3 million acres. With few exceptions, national monuments are less than 1,000 acres in size. However, the largest monument, Escalante National Monument in Utah, is 1.7 million acres. National preserves are restricted to two units in the lower 49 states and 10 in Alaska. Hunting, trapping, and fishing, and the extraction of minerals and fuels are allowed in preserves under restricted conditions. Visitations were less than

FIGURE 13.7   Primary national parks and national monuments in the United States (data from U.S. Geological Survey).

1 million in 1994. The 20 national recreation areas are usually associated with dam impoundments in the West and encompass 3.4 million acres. Lake Mead National Recreation Area (1.5 million acres) in Nevada and Glenn Canyon National Recreation Area (1.2 million acres) are the largest units, comprising 80% of the total acreage of the national recreation areas. The other units are much smaller.

Other important protection legislation included the Wild and Scenic Rivers Act of 1968. This act gave protection to rivers in their free-flowing state and brought to a halt the emphasis placed on dam construction. It provided that select "wild, scenic, or recreational rivers which, with their immediate environments, possess outstandingly remarkable scenic, recreational, geologic, fish and wildlife, historic, cultural, or other similar values, shall be preserved in free-flowing condition, and that they and their immediate environments shall be protected for the benefit and enjoyment of present and future generations" (Figure 13.8). The National Trails Systems Act of 1968 mandated "trails be established (i) primarily near the urban areas of the Nation and (ii) secondarily within established areas more remotely located." These trails are

FIGURE 13.8   Over 7,300 river miles in over 720 rivers have been designated as wild, scenic, or recreational rivers (courtesy of USDA-Forest Service).

administered by the National Park Service, Forest Service, and the Bureau of Land Management.

Federal Water Resources Agencies

Among federal agencies, the Army Corps of Engineers is second only to the Forest Service in providing recreational opportunities, mostly at the 461 reservoirs it manages for navigation, flood damage reduction, hydropower and other purposes. The Corps provides more boating access and related recreational service than any other agency. Through development of the Intracoastal Canal, riverine navigation projects, and harbor maintenance the Corps also has provided for thousands of miles of recreational boating in addition to the primary commercial navigation purpose.

Among other important federal providers of water-based recreation the Bureau of Reclamation and the Tennessee Valley Authority stand out. Most of the Bureau's 355 storage reservoirs are located where water-based recreation is highly valued, in the arid West. In 1991 it provided over 23 million days of recreation activity (Zinser 1995). The Tennessee Valley Authority has provided recreational management at its managed sites on the Tennessee River since 1969 (Zinser 1995). The last estimate of total recreational use, in 1980, indicated over 70 million recreation visits per year occurred at all sites under its management.

OTHER RECREATIONAL MANAGEMENT

State and Local Government

State parks hosted 666 million visits on over 10,000,000 acres of land by 1986 (TFORRO 1988). Although these state and local opportunities add up to a small fraction of the total public land and water area under federal authority, over 88,000 sites existed in 1986, distributed mostly near cities. This proximity to population concentrations makes them much more densely visited than most federal recreation lands and dwarfs them in total provision of recreational use. However, these recreational areas tend to complement rather than compete with the federal recreational lands. In contrast with longer-term recreation common on federal lands, most use is day use on state and local government lands.

Private Recreation Opportunities and Tourism

Much outdoor recreational management takes place on private lands. However, the opportunities and use provided on a private basis are not as well documented as for public lands, but are very important, especially where there is little public land. Thus private lands have tended to complement public land recreation rather than compete with it. Many concessions on public lands are operated privately under contract with public agencies including campgrounds, ski areas, marinas, hotels, restaurants, stores, and services. A need exists for updated data on recreational use of private lands to help maintain proper public recreational investments. The most complete data are over three decades old. In 1965, over 1 billion visits were estimated for private holdings, either free of charge or seeking profit from recreational use (TFORRO 1988). An additional 800 million visits took place under the auspices of clubs or on private lands not seeking profit for recreational use. As of 1988, about 785 million acres of private land were available for public recreation either free or for a fee (TFORRO 1988).

RECREATIONAL CHALLENGES ON PUBLIC LANDS

In this section we will discuss some of the major recreation challenges and approaches to management on public lands in the United States. We will emphasize public rangelands in the western United States because of their large area and the intense controversies that have recently occurred over how they will be used (Wuerthner 1990, Jacobs 1991). Much of our discussion follows Holechek et al. (1998). We refer the reader to Chapter 10 for additional discussion of recreation conflicts with other uses on national forest lands and Chapters 7 and 16 for discussion of water resources conflicts.

Importance of Public Land Recreation

Recreational use of public lands in the United States has accelerated tremendously during the past 30 years. Much of the increase has resulted from the rapid human population increase in the 11 contiguous western states, combined with rising affluency and leisure time. In some areas, such as southern California, central Colorado, western Arizona, central Texas, and central Oregon, recreation has become a far more important use of rangeland than livestock grazing. If present trends of urbanization continue in the 11 western states, it appears likely the recreational value will exceed livestock grazing value on most rangelands within the next 15 to 20 years.

Recreational use of public lands includes a variety of activities (Table 13.2). The economic values of many of these activities, such as camping, hiking, and waterskiing, are difficult to quantify. Some forms of recreation, such as off-road vehicle travel and trail biking, can be as destructive as overgrazing if uncontrolled.

Subdividing Private Grazing Lands

The sale of recreational homesites has had the most severe impact of all the wildland recreational uses in the western U.S. Literally thousands of recreational homesites are being sold every year in each of the western states (Figure 13.9). Although the total area lost each year is relatively small [about 1 million acres (404,694 ha)] compared to the total land base, this trend is having a major impact on the local range livestock industry.

TABLE 13.2   Types of Rangeland Recreational Uses

Activities with Minor Impacts on Rangeland	Activities with Major Impacts on Rangeland
Hiking	Hunting
Camping	Horseback riding
Fishing	Trail biking
Skiing	Motorcycling
Boating	Jeeping
Rock hounding	Dune-buggy riding
Tubing	Home building
Canoeing	
Mountain climbing	
Relic hunting	
Bird-watching	
Picnicking	

Source: Holechek et al. 1998.

Large-scale conversion of agricultural and rangeland to housing is occurring in the mountain valleys of the intermountain West. These lands are critical in terms of providing forage for livestock during the winter when snow makes grazing infeasible at the higher elevations. Without winter-base property, grazing of surrounding summer range, typically in federal ownership, becomes impractical. The people who build homes in the mountain valleys generally come from the urban areas and use these second homes primarily for vacation purposes. They heavily use the surrounding summer ranges for recreational pursuits. Under these conditions, recreation replaces livestock production as the primary use on the remaining federally owned rangeland.

The subdivision of mountain valleys can severely limit big-game populations, particularly mule deer and elk. These animals are directly affected by the loss of important winter ranges. Indirectly, they are subjected to greater stress on remaining ranges as the result of increased human activity that goes with subdivisions, such as snowmobiling, woodcutting, and hiking.

To maintain esthetic values, clean air, wildlife populations, and other amenities that make mountain valleys so desirable, restrictions on private land use may be necessary. Otherwise, the very qualities that make these areas so attractive will be destroyed.

Practices that provide economic incentives to landowners to retain their land in agriculture show some promise. In Oregon, for example, state legislation has reduced

FIGURE 13.9   Thousands of recreational homesites are sold every year in the western United States, severely impacting esthetic, wildlife, and recreational values in many areas.

property taxes for land kept in agriculture. This coupled with zoning laws that restrict subdivision in farming and ranching areas have kept a good balance between development and agriculture. Similar approaches to those used in Oregon will probably be applied to other western states as they become more urbanized. However, some question whether government should benefit comparatively few land owners and recreationists at substantial cost to the tax-paying public. Such decisions require careful analysis of the benefits and costs (see Chapter 4).

Agriculture on the Urban Interface

Large zones of rapid urbanization have occurred around many western cities in the 1990s (see Chapter 14). The problems of farming and ranching on the urban-rangeland interface are considered by Huntsinger and Hopkinson (1996). Many people moving to rural areas do not understand rural customs and activities just as many rural people do not understand urban sensibilities and behavior. Farmers and ranchers on the urban-rangeland interface often complain about marauding dogs, vandalism, trespass, carelessness with gates and fences, increased liability costs, and introduction of exotic plants (Hart 1991; Forero et al. 1992). In turn, suburban and resort residents often complain about the use of prescribed burning, predator control, and weed control. Stray livestock can cause property damage and vehicle accidents. Commuter traffic may be blocked by slow ranch vehicles. The new neighbors also may complain about livestock odors and the threat of pollution from pesticide or fertilizer applications.

For these and other reasons, better planned development is critical to ranchers as well as society at large. Huntsinger and Hopkinson (1996) believe environmentalists should be more tolerant in their rangeland appearance expectations while ranchers should rethink their concept of property ownership. They mentioned that debates over the ecological impacts of grazing, wildlife management imperatives, and riparian zone restoration become moot when grass is replaced by concrete.

Huntsinger and Hopkinson (1996) point out that sustaining rangeland ecosystems in the future will be as much a social-economic process as an ecological one. They provide examples from Marin County, California, of how successful planning and alliance building between ranchers and environmentalists prevented development of a large farming-ranching area. Near San Francisco a combination of zoning, conservation easements, tax relief for farmers and ranchers, community leadership, and recognition of the heritage value of rural lifeways all played a part in the success. They noted that a similar pattern was emerging in other parts of the West. They believe carefully planned ranching can play a vital role in conserving biodiversity in many threatened range landscapes. This planning process should include ecological, social, and economic factors. The land market is driving the conflict over scarce, desirable land. Where the market does not adequately reflect all costs, the government needs to assure otherwise through benefit-cost analysis (see Chapter 4). Although environmentalists and ranchers differ in many beliefs, Huntsinger and Hopkinson (1996) point out that both groups generally want to preserve open space and share a love of nature. When environmentalists were given more exposure to ranchers and their way of life, many came to see ranching as compatible with their goals.

Scenic Beauty and Range Management

The impact of range management practices on scenic beauty in the United States has been an important concern in recent years because of the public's increased sensitivity to the environment and its greater role in the decision-making processes on public lands. A study of 241 dispersed recreationists on the Malhuer National Forest in eastern Oregon evaluated the public's response to range management activities (Sanderson et al. 1986). Photographs of selected ecosystems, range management practices, and management intensities were used to elicit attitudes of dispersed recreationists concerning management of the range resource. Recreationists were broken into categories of fishermen, hunters, and campers.

Recreationists responded most favorably to photos showing environmental scenes where livestock were grazed but range management practices were least visible. Here, livestock numbers were within grazing capacity of the pasture and a minimum of fences were used. No methodology was used to distribute livestock evenly through the pasture. Extensive management of the environment and livestock was rated second. Here, the goal was full livestock use of forage facilitated by range management methods that promote even distribution of livestock over the range. Intensive management of the environment and livestock was ranked least appealing. Here, the goal was to maximize forage production for livestock consistent with environmental constraints, including multiple use. Cultural practices were used to increase forage production and practices that improve livestock distribution were fully applied.

Fishermen rated aesthetic quality represented in photographs lower than campers. They considered grazing near riverbanks, herbicide spraying, alteration of upstream vegetation, and improved river access to recreationists as unacceptable management practices. They reacted more favorably toward fences than did other groups because they considered fences vital to maintenance of the fishery by excluding cattle.

Hunters gave the photographs higher ratings than did campers or fishermen. This appeared to have been due to the high hunting success in the area. Hunters generally responded adversely to practices restricting access, such as road closure and establishment of more wilderness areas.

Campers, in general, felt that cattle were more appropriate for the mountain meadow and mountain grassland ecosystems than in the forest. They tended to consider open spaces as areas for grazing and forest as areas for camping. Generally, they did not mind seeing cattle from a distance on open areas, but they did not want cattle present near camping sites in the forest.

This study indicated several relationships between range management activities, recreationists, and the scenic qualities of rangeland. These are summarized as follows:

1. Recreationists do perceive differences in the visual quality of rangelands.
2. Different types of recreationists differ in their perception of visual quality.
3. The greater the familiarity with national forests, as measured by the number of prior visits, the greater the willingness to accept intensive management.
4. The public, in general, is not aware of the requirements for efficiently managing forest-range environments for increased forage.

The last relationship merits discussion. Many recreationists asked why cows were grazing in a national forest and why certain portions of a stream were fenced off. There was a failure to understand that the fences were to control livestock. Recreationists were asked if they had heard the term "multiple use." The majority (84%) were unfamiliar with the term. When participants were asked if they knew what the concept "more intensive management of the range resource" meant, 91% of the users had no idea of its meaning. About 1% properly defined it as the application of techniques to improve the quality and quantity of range forage. The remainder believed that it meant the practice of grazing a greater number of cattle per unit of land. Based on this study, it appears that the public generally tolerates range management activities, but generally has a poor understanding of what they involve.

More recently, visitor perceptions about cattle grazing on national forest land were evaluated in Colorado (Mitchell et al. 1996). The number of visitors indicating range livestock added to their stay (34%) was no different than the number stating a negative relationship (33%). Visitors in dispersed campsites where livestock roamed tended to be more critical of grazing than those in developed campgrounds where cattle were excluded. The authors speculated this difference might derive from consequences of camping where cattle had previously been. This study indicated that if livestock are kept out of developed campgrounds and adjacent riparian areas used for fishing, there is little objection to them. It was concluded that if possible, livestock should be managed so they are away from dispersed camping areas during times of high recreational demand.

Public Opinion and Management of Federal Rangelands

Public opinions on grazing issues have become an important concern of federal rangeland managers. A survey conducted by Brunson and Steel (1994) showed U.S. residents place little confidence in livestock, mining, and energy extraction groups. They tend to believe more protection should be given to wildlife and fish resources. In another survey, Brunson and Steel (1996) found only weak opinion differences on range issues between the eastern and western United States. However, there were definite attitude differences between urbanized areas and rural regions, which could eventually result in political power loss for range constituencies in the western United States.

This survey was consistent with Sanderson et al. (1986) and Brunson and Steel (1994), which all indicate the need to better inform the public on range management issues. At the same time many new and divergent interpretations of multiple-use management on public lands are being advanced that must be considered by modern range managers. Many range scientists have come to believe that systems are needed that provide more sustainable, environmentally sensitive range-livestock production as alternatives to traditional systems (Box 1995, Heitschmidt et al. 1996, Vavra 1996). Meeting this need will be an important challenge for range scientists and managers in the twenty-first century. Many ranching operations are marginally successful and any additional management cost may threaten their continuance, unless they take innovative steps.

Recreation and Ranching

Advantages to Ranchers.   The urbanization of the western United States might be turned to advantage by the ranching industry. Returns from livestock on western ranches as a percentage of capital investment have been low (they have averaged about 1% to 3% during the past 25 years). However, the increased human population in the West provides considerable opportunity for ranchers who are willing to diversify their enterprise. The potential of fee hunting has been discussed previously (Chapters 11 and 15). Other forms of recreation, such as packing trips, horseback riding, sight-seeing trips, dude ranching, and fishing, are proving to be highly lucrative in certain areas. Many ranches have built special ponds stocked with fish and charge either daily fishing fees or so much per fish. They often also rent cabins and provide meals to vacationing tourists. In some cases, these enterprises bring in far more net income than the sale of livestock. It appears that recreational enterprises will increasingly displace livestock as the main source of income from ranches in many parts of the western United States.

Disadvantages to Ranchers.   Ranchers using public rangelands generally view recreation negatively, often with good reason. Increased human disturbance can reduce animal performance. Increased incidences of vandalism, fire, and losses of livestock from theft, traffic collision, and shooting usually accompany heavier recreational use. In some cases, camping occurs around livestock watering points and livestock not used to human activity may be reluctant to water when this occurs.

Managing Recreation Costs on Public Lands

Many segments of the public now believe that federal agencies should charge fees for recreational use of public grazing lands. These fees could then be used to mitigate some of the negative effects of recreation on livestock production. In reality, many recreational uses—particularly those forms involving unregulated off-road vehicle travel—can affect the range more severely than livestock grazing or logging (Figure 13.10). To prevent destruction in the future, regulation of recreation on public lands will be increasingly necessary. It seems only reasonable that recreationists using pub-

FIGURE 13.10 Damage to public rangelands in southern California from uncontrolled travel by four-wheel-drive vehicles.

lic land be charged a fee for their activities just as ranchers are charged for grazing. However, many recreationists believe this grazing fee is too low and management is ineffective for protecting recreational interest.

Many of the basic principles used in controlling livestock grazing can also be used in controlling recreational use (Heady and Vaux 1969). Proper stocking rate can be modified to proper numbers of people or vehicles on a particular piece of rangeland for a particular recreational activity. Proper distribution and timing of use is just as important with most recreational activities as it is with grazing. This is particularly true for off-road vehicle travel.

Off-road vehicle travel is a recreational pursuit that can have considerable impact on rangeland soils and vegetation. This was evaluated on a northern Great Plains range in southeastern Montana (Payne et al. 1983). These researchers studied the influence of different numbers of trips at different times of the year on soil and vegetation characteristics using typical off-road vehicle (four-wheel-drive Chevrolet Blazer). When soils were dry, they found that up to eight trips in the same tracks did little long-term damage to living plants or the soil. However, above eight trips there was a strong likelihood of damage carrying over into following years. Damage to both soils and vegetation was much more severe with wet than with dry soils. It was recommended that drivers be encouraged not to follow other tracks when vegetation is actively growing because one to two trips had a negligible effect on soils and vegetation compared to repeated trips (eight to 32). However, the recommendation is reversed for dry, mature vegetation (winter use), because breakage became an important factor. If the range was not being reserved for winter use, they recommended that drivers be encouraged not to follow other tracks. When soils are wet, off-road travel should not occur because of severe damage to both soils and vegetation.

In a Nevada study, the comparative influences of motorcycle, four-wheel-drive truck, and no traffic on infiltration rate and sediment production were studied (Eckert et al. 1979). In this study infiltration rates were decreased and sediment production was increased by traffic. The four-wheel-drive truck had a much greater impact than did the motorcycle.

As a consequence of such studies, off-road vehicle travel on federal lands in the United States has become highly regulated during the past 10 years. It seems likely that this form of range recreational use will be even more restricted in coming years if recent trends in public use of federal land continue.

Conflict Resolution in Multiple-Use Decisions

In recent years, public land management agencies have become heavily burdened with resolving controversies among various interest groups regarding natural resource use and management. Capability in conflict resolution has become of critical importance in managing public lands. Coordinated Resource Management Planning (CRMP) has been widely used in the western United States to deal with multiple-use conflicts. CRMP brings together public and private interests to resolve multiple-use and land management conflicts through reasoned-scientific analysis, bargaining, and compromise (Anderson 1991).

The participants in CRMP have a common goal in correcting past land management mistakes and making informed decisions on management alternatives. It has been effective in resolving many conflicts on public rangelands (Anderson 1991, McClure 1992, Grizzle 1992). However, quantitative decisions based on economic trade-offs are usually complex because many natural resource products on public lands are free and their values vary among interest groups (Anderson 1981). However, other than biodiversity values, most natural resource values can be quantified either through direct market means or indirect, non-market valuation (see Chapter 4). CRMP guidelines that have developed from 40 years of experience are available from Cleary and Phillippi (1993).

Another approach called "dispute resolution" is discussed by Torell (1994). It involves bringing together parties in disagreement to participate in joint decision-making processes that seek win/win solutions and avoid litigation. An objective third party is commonly recruited to assist in resolving conflicts. A study of environmental disputes found that 78% of the cases where alternative dispute resolution techniques were used resulted in settlement (Bingham 1986).

OUTDOOR RECREATIONAL MANAGEMENT

Interfacing People With Resources

The goal of outdoor recreational management is to provide optimum conditions for people to obtain outdoor recreational opportunities. Outdoor recreational managers usually are responsible for the management of a site or a group of sites linked to a specific geographic area, which may be privately or publicly owned. Site managers must understand principles of landscape and habitat management pertinent to recreation and other uses of the area. They must also be informed of user needs with respect to resource qualities, access, facilities, interferences, costs, and availability of compatible and synergistic activities. Before all else, recreational management requires careful planning. The main difference between managers and biologists in natural resource management for recreation is the understanding of recreational management principles. Fish and wildlife "management" is especially likely to underestimate the importance of recreational management principles and overemphasize the importance of biology.

Recreational Planning

Recreational management begins with development of a management plan with goals, inventory information, analysis and forecasts of use, management objectives, a prescription for implementation based on the most cost-effective alternative, and an evaluation of management effectiveness (see Chapter 5 for general planning principles). Most recreational management plans combine geographic and event plans. They usually emphasize identification of resource use areas, access to those areas, placement of supportive facilities, various landscape modifications for improving resource utility, compatibility with other uses, timing of events, and specific locations of organized events. Recreational management planning starts with choosing the location of the site to be developed and managed. If, as is increasingly the case, the plan is for a site that already has been identified and developed for recreational use, the conditions left by previous site managers become important constraints for future planning possibilities.

Management plans are guided by the mission and goals of the management organization or group of organizations involved in a coordinated or integrated planning process. Information for planning usually is gained through one of several approaches. A common approach is to use existing studies of resource condition at the site and recreational activities. This is an inexpensive and efficient approach as long as the conditions at the site are similar and sites have not greatly changed. Reliance on generic studies, or studies of sites elsewhere, often risks being at variance with conditions at the site. Management targets, such as numbers and locations of campsites, docks, trails, roads, boat ramps, toilets, food concessions, and more may be misidentified. Old surveys may be out of date because both the landscape and human preferences changed more rapidly than expected.

Recreational sites rarely exist outside some larger context of recreational opportunities. A site developed 20 years ago in a rural setting may now be surrounded by suburban growth. While previously prized as a primitive wildland recreational resource, the site now may have resource attributes more compatible with homes and traffic in close proximity. The site may have been the only location with water-based recreation and camping and now is surrounded by new reservoirs and campgrounds under the same or different management authorities. It is important to know if the targeted site is a high priority destination of recreational users, only average, or gets used only when preferred sites in the vicinity are filled. For private development, what the competition is going to do will affect the income of all competing parties.

Management objectives are identified based on resource supply and demand relationships and the interactions of recreational use with other uses. First, a general vision of the desired future condition of the site is developed, which often requires integration with other land management goals and objectives. An effective approach is to develop several alternative plans for converting the vision to reality and then selecting the most cost-effective plan. Objectives are specified as clearly as possible. They often include specifications for landscape shaping, planting, flooding, draining, other habitat modifications, and construction in the detail needed for contractors to bid on and carry out the necessary work.

Recreational Economics

A chronic dilemma for natural resource managers is judging the extent to which they should budget and invest in recreational improvements relative to other resource development. For example, if a public reservoir serves multiple uses, including recreational fishing, boating, and swimming, how much should recreation dictate the water level maintained in the reservoir over hydropower generation and water supply? What are the tradeoffs? It is virtually impossible to compare numbers of fish caught to kilowatts of energy produced or gallons of water delivered without transforming those outputs into some comparable measure of value, such as money (see Chapter 4).

While it is relatively easy to estimate the market value of a kilowatt of energy or a gallon of water delivered, recreational value may be more problematic depending on whether the site is private or public. If the site is privately owned, access is controlled, a fee is charged for recreation, and the profit made from recreation at any particular water level can be readily compared to the profits obtained from hydropower and water supply. Then the optimum water level for profit maximization from all three services can be estimated through use of economic planning tools.

If the site is public, the approach taken to estimate benefits from recreation depends on the site attributes and public policy. Where access can be controlled and a user-pay policy exists, management emphases can be decided on the basis of the total revenue derived from water, electricity, and recreational uses. Localized areas around lakes and reservoirs often are amenable and especially attractive as recreation sites. Where access cannot be controlled, there is no practical way to establish a recreation market. The benefit derived from recreation must be estimated indirectly from surveys of user willingness to pay. The approach preferred by many economists is the travel-cost method (Loomis 1993). In this approach, the willingness of users to pay a "price" for recreating at the site is based on the travel costs incurred. The effect water level or other site management has in increasing or decreasing the willingness to use the site for recreation can then be compared to effects on other services provided at the site.

Models have been developed for estimating recreational use based on site qualities under different management strategies (e.g., Ward et al. 1997). An important variable in estimating the willingness to recreate at a site is the availability of substitute sites with the same kind of recreational services. Substitute sites may be visited instead when they have similar quality but cost less to visit or when they offer better recreation at similar travel cost. Such models allow assessment of complex interactions resulting from management in a system of sites spread throughout a region. They can facilitate coordination of different public authorities for the most efficient provision of recreation. Cole and Ward (1996), for example, demonstrate this capacity for public sportfishing.

Integrative Management Planning

The primary elements in the recreational management system are the natural resources, the visitors, and the management plan (Jubenville 1978). High-quality recreational experiences that satisfy visitors require careful integration of these components (see Chapter 5 for more detailed planning concepts).

Integrative management planning for outdoor recreational use takes into consideration the plans of other agencies and private enterprise so as to avoid over- or under-development of opportunities relative to anticipated demand. Integrative management planning typically is regional in perspective. The average distances traveled to participate in outdoor recreation often exceed two hours. With high speed highways in the vicinity of a population center, the region may easily include 30,000 square miles of alternative weekend recreational opportunities. Within that region, a wide variety of federal, state, county, municipal, and private authorities are likely to provide recreational opportunities. Although outdoor recreation is an increasingly important aspect of quality of life for Americans and people in other nations, past planning has not emphasized integrated approaches.

Outdoor recreation is highly interactive with transportation access to the recreational resources and with transportation planning. Transportation choices may include plane, train, and boat. However, once within a region, the automobile is by far the preferred means of travel to outdoor recreational locations. Sight-seeing by car is one of the most popular outdoor recreational activities (TFORRO 1988). Sight-seeing commonly includes specific target locations as travel stops or an array of possible stops any of which might be suitable as whim and physical need dictate. Regional metropolitan or state planning often revolves around transportation needs. Outdoor recreation is an important consideration. In many U.S. locations, the main travel use of rural roads is for recreational purposes and the main source of outside income is based on recreational demand.

Recreational Resource Managers

The recreational resource manager must have broad understanding of the needs of the visitor, the site's natural resources, and the services needed to make the social and resource environments satisfying (Jubenville 1978). Therefore, the primary components of recreational management are visitor, natural resources, and service management. Visitor management involves controlling distribution of use, safety, and access to information and education. It requires understanding of visitor perceptions, needs hierarchy, and style of participation. Natural resource management is the protection and enhancement of the landscape, soils, waters, vegetation, geological formations, and other natural resources for recreational purposes. Service management takes many forms, but emphasizes provision of access to resources and basic facilities needed to meet basic visitor needs.

Visitor Management

People seek different things from recreation, but all need food, shelter, and safety appropriate to the recreational experience. Most desire being part of a recreational partnership or group. Some recreators spend a lot on equipment and supplies usually in response to some perceived recreational status or competency level they seek to attain. Others desire a very personal sense of recreational satisfaction often independent of group dynamics.

Differences in behavioral style are important. For instance, a participant may be motivated to participate in hunting for the meat, a trophy, or mastering the technology. Camping attracts a wide variety of recreational types ranging from wilderness camping to transient vehicular camping. Perceptions of experience can vary. Therefore, feedback between users and managers is important in assessing the perception of the experience. Was it too crowded or the experiences not intense enough or too sporadic? Was the information service sufficient? Were there unexpected, threatening hazards? Were opportunities diverse enough to satisfy the whole group?

A primary strategy for managing visitors is to control their distribution. Controlling access and use of sites is a critical management responsibility because crowding affects user satisfaction and landscape condition. Use may be redirected or dispersed in a number of ways. Regional and area planning is the most effective means for controlling visitor access and distribution. Recreational use zoning is an effective way to sort out compatible and incompatible uses. This can be facilitated through information services and types of access and facilities ranging from highly developed roads, food services, and utilities, often on the edges of public lands, to primitive and wilderness conditions more typically located in the site interior (Figure 13.11).

Improper original location of access and facilities will make any subsequent visitor management more difficult. Intensively used sites may require restriction or even site closure. While limited use requires expensive supervision, site closure is the last resort. Regional and on-site information provision is the most important avenue to visitor management. Where economically justified, visitors should be provided information in advance about any change in services and/or access to recreational locations.

An effective means of rationing use is through access fee charges, a means often advocated by resource economists. Some people object to this, believing in principle that public lands belong equally to all. However, the largest cost for outdoor

FIGURE 13.11 A well-designed and highly accessible state park in Oregon, which accommodates both day-trips and overnight campers (courtesy of U.S. Government).

recreation typically is associated with travel costs and equipment. Fees are a minor consideration for most.

Natural Resource Management

Intensity of Use.   On large federal and state public lands, much recreation management must be integrated with other specialty areas including forestry, range, wildlife, and fisheries management. A number of management strategies are applied depending on purposes and circumstances. Recreational use may be dispersed or concentrated on developed sites to protect the natural resources from degradation and erosion. Intensively used sites may be culturally treated through fertilization and irrigation to increase their natural resistance to human use. Area use may be rationed by establishing capacity limits on continuously used sites and by closing and rotating site use. Once sites are degraded, they may be naturally restored or enhanced through cultivation techniques. Certain intensively used areas may be hard-surfaced in esthetically compatible ways. Unintentional destruction by visitors sometimes is a problem, which often is corrected through information and education, or through simple prevention of access, such as vehicle barriers. Law enforcement has become important at most intensively used outdoor recreational areas.

Vegetation.   Vegetation management of turf, shrubs, and trees is an important aspect of recreational resource management in moderate- to high-use areas. Turf may be improved by thinning shade trees, irrigation, fertilization, and other treatments. For special use areas, such as golf courses and various field sports, more intensive and specialized turf management practices are applied. Trees and shrubs often need to be monitored for disease and trimmed or removed to prevent spread. Vegetation also needs to be managed for fire control through encouraging certain species, thinning, and creation of firebreaks. The natural history of fire in the region needs to be incorporated into the management. Monitoring is done with a variety of scientifically developed techniques using photography, fixed point surveys, plots, line transects, and other approaches. Typical attributes measured include the percent of vegetative cover and a ranking of plant vigor. Recreational managers typically rely on a diverse array of professional support services such as the Cooperative Extension Service associated with Land Grant Universities, state departments of conservation, and state soil testing labs.

Access.   For areas managed to highlight natural beauty, access to visual resources is an important part of recreational service. Placement of road, trail, and waterway access with vista views that do not degrade the value of the resource is an important planning activity at national parks and other public lands. This planning often requires classifying landscapes in terms of dominant features and their arrangements and with respect to the qualities users seek in visual resources. On the ground, vegetation often needs to be managed not only to allow viewing, but to frame it with complementary foreground, much as art is presented in a museum.

Hazards.   Outdoor experiences include hazards, some of which are necessary for a fully satisfying recreational experience. Among natural hazards are cliffs, landslides, avalanches, electrical storms, high winds, desiccation, falling trees and branches, dangerous wildlife, disease, freezing conditions, deep water, currents, tidal changes, flooding, and natural obstructions in water, trails, and roads. The most common anthropogenic hazards include old mine shafts and mine high walls, abandoned buildings, wells, poor road design, and the behavior of other visitors. While many hazards are accepted and anticipated, others are not tolerated. A hazard-free recreational experience is viewed by many participants as sterile. However, hazards caused by human engineering and mismanagement are much less likely to be accepted than hazards associated with nature. Acceptance of hazards in one situation does not necessarily carry over to other situations depending on the age, experience, and attitudes of participants.

Hazard management objectives, which need to be first defined, depend on the recreational services demanded of the site. Hazards next are identified and evaluated for their degree of threat and mitigation strategies are adopted appropriate for the conditions. A basic decision is whether to manipulate the use of the site or the site's resources. In some cases, facilities may require relocation to reduce hazard encounters. Provision of information in appropriate locations and form is a basic management tactic. The recreation manager has to consider liability as well as recreational quality in seeking the optimum level of hazard management, especially in private provision of recreational opportunity.

Information Service Management

Information services inform recreation participants of opportunity availability, special events, use restrictions, other sources of information, and hazards. Regional services are aimed at resource users before they reach the site. The information provided helps them to choose among site alternatives. The most comprehensive services, such as catalogs of recreational campgounds, often include both private and public recreational opportunities with detailed summary descriptions of all facilities and services provided, as well as important restrictions. Other forms of regional information may be provided through mass media (typically expensive), public hearings, and group contacts. The Internet offers opportunities for providing much more detail on public use areas.

Visitor centers are the most common means for providing information at large recreational areas, such as national parks. Decentralized information stations often are used at local campgrounds, trailheads, boat ramps, and other staging points for recreational activity. These may be little more than an information shed or box that protects maps, pamphlets, and notices from the weather. The most explicit information is provided by large and unavoidable signs. In some locations, radios and other gadgetry have been used, but remain rare because of expense and frequent failure. The design of information packages is a critical aspect of communication. Avoidance of information overload is important and loose paper often becomes litter.

Interpretive service is a special category of information service designed to elicit more from the recreational resources. Interpretive services are most commonly employed where natural and cultural history are primary objectives of the recreational use. At many large sites, reference publications often are provided for the more serious student. Interpretive services usually inform visitors about what they may otherwise overlook at the site; often inform about management goals, such as preservation of natural, cultural, and historical features; and sometimes inform about the organization and its mission.

FUTURE DEMANDS FOR OUTDOOR RECREATION

Several factors impact the future demands for outdoor recreation in the United States. Demographic changes in the American population have implications for managing outdoor recreation, in particular the increase in minority ethnic groups and increases in the older age groups. Racial minorities will significantly increase population percentage and total numbers. During the period between 1990 and 2025, minorities (principally those of Spanish and African origin) will increase from 24 to 35% of the population and will continue increasing well into the next century. The greatest growth will occur in the Spanish-descendent population. The Hispanic population increased 44% from 1980 to 1990. About 22 million Hispanics now live in the United States, with 13 million in the southwestern United States. The Hispanic population in the U.S. is projected to increase to 31 million in 2000 and 81 million in 2050, when the non-Hispanic white population will constitute 53% of the U.S. population.

Relative to the age of the U.S. population, the median age increased from 23 in 1900 to 30 in 1980, and is expected to reach 41 by the year 2025. The percentage of individuals 65 and older will increase from 12.6% in 1990 to 21% in 2025, with a

FIGURE 13.12   There is growing demand by joggers, hikers, and joy walkers for safe and esthetically pleasing natural areas within or adjacent to urban areas (courtesy of USDA-Forest Service).

marked increase after 2010. Also, urban residents continue to increase with only moderate rates of growth in rural states such as Montana, Idaho, and Wyoming. The increasing urbanization of the American population will require additional recreation facilities near urban areas, and oriented more toward an aging society.

Cordell et al. (1990) projected that the recreational activities with the greatest increase in near-future demand will be downhill skiing, cross-country skiing, pool swimming, backpacking, visiting prehistoric sites, running and jogging, and day hiking (Figure 13.12). By 2040, the most popular recreational activities will be sight-seeing, walking for pleasure, pleasure driving, pool swimming, picnicking, day hiking, family gatherings, bicycle riding, photography, stream/lake/ocean swimming, wildlife observation, visiting historic sites, and camping.

Relative to social, economic, and environmental implications of demand-supply comparisons, the following are projected by Cordell et al. (1991):

1. "The social characteristics of selected multicounty communities across the United States can be compared with the available recreation opportunities to yield information on social imbalances. In general, Americans who are elderly, less educated, part of a racial minority, economically disadvantaged, disabled, or living in cities have fewer opportunities to participate in resource-based recreation than do others.

2. "The uneven distribution of opportunities can have adverse social effects including reduced family stability, more crime and juvenile delinquency, less opportunity for social bonding, more social conflict, and slower ethnic and cultural assimilation.

3. "Increased economic opportunities for the private sector are projected for several categories of recreation. These include investments in developed recreation areas and the provision of associated goods, services, and information. Increased government revenue generated by user fees is expected to be offset by higher management costs for dispersed recreation.

4. "Impacts on natural systems from most outdoor recreation and wilderness uses are minimal compared to more consumptive uses such as lumbering or mining. Recreational impacts such as soil compaction and erosion are generally local in nature and the greatest damage occurs during the initial use of an area.

5. "Outdoor recreation and wilderness use can benefit natural systems through improved esthetic quality, greater environmental awareness, and preservation of natural systems. For example, demand for water opportunities has generated pressure in governments and industry to improve water quality in rivers, especially near urban areas."

The obstacles hindering attainment of opportunities in outdoor recreation include the following:

1. "A major problem is the imbalance between recreation and wilderness land distribution (mostly in the West) and the population distribution (mostly in the East).

2. "Private landowners are often hesitant to provide access to their land for public use without economic incentives or protection of the uses for which they own the land.

3. "Insufficient funding, information, cooperation, and coordination among agencies contributes to problems in reducing the recreation-wilderness supply-demand gap (Cordell et al. 1990)."

These challenges are some of the same ones confronting recreation development throughout its history. Particularly important in the future will be the inclusion of private landowners in providing additional opportunities for recreation participants. Recreation is a basic component of integrated natural resources management. The inclusion of recreational components in integrated ecosystem planning involving state, federal, and private entities are essential to the sustainable use of natural resources and the provision of recreational facilities for future generations.

LITERATURE CITED

Anderson, E. W. 1991. Innovations in coordinated resource management planning. *Journal of Soil and Water Conservation* 46:411–414.

Anderson, T. L. 1981. *Multiple use management: Pie slicing vs. pie enlarging.* Workshop on political and legal aspects of range management. Jackson Hole, WY: Natural Resources Council.

Bingham, G.. 1986. *Resolving environmental disputes, a decade of experience.* Washington, DC: The Conservation Foundation, p. 43.

Box, T. W. 1995. A viewpoint: Range managers and the tragedy of the commons. *Rangelands* 17:83–85.

Brunson, M. W. and B. S. Steel. 1994. National public attitudes towards federal rangeland managements. *Rangelands* 16:77–81.

Brunson, M. W. and B. S. Steel. 1996. Sources of variation in attitudes and beliefs about federal rangeland management. *Journal of Range Management* 49:69–75.

Chavez, D. J. (Tech. Coord.). 1994. *Proceedings of the second symposium on social aspects and recreation research.* General Technical Report. PSW-GTR-156. USDA Forest Service, Albany, CA: Pacific Southwest Research Station. 186 pp.

Cleary, C. R. and D. Phillippi. 1993. *Coordinated resource management guidelines.* Denver, CO: Society for Range Management.

Cole, R. A. and F. A. Ward. 1996. Sustaining reservoir fisheries and angler benefits through comprehensive management analysis. In L. E. Miranda and D. R. Devries, eds., *Multidimensional approaches to reservoir fisheries management.* Bethesda, MD: American Fisheries Society.

Cordell, H. K., J. C. Bergstrom, L. A. Hartman, and D. B. K. English. 1990. *An analysis of the outdoor recreation and wilderness situation in the United States: 1989–2040.* General Technical Report RM-189. Fort Collins, CO: Rocky Mountain Forest and Range Experiment Station, USDA Forest Service.

Dwyer, J. F. 1994. *Customer diversity and the future demand for outdoor recreation.* General Technical Report RM-252. Fort Collins, CO: Rocky Mountain Forest and Range Experiment Station.

Eckert, R. W. Jr., M. K. Wood, W. H. Blackburn, and F. F. Peterson. 1979. Impact of off-road vehicles on infiltration and sediment production of two desert soils. *Journal of Range Management* 32:394–398.

Forero, L., L. Hutsinger, and W. J. Clawson. 1992. Land use change in three San Francisco Bay Area counties: Implications for ranching at the urban fringe. *Journal of Soil and Water Conservation* 47:475–480.

Grizzle, G. 1992. Coordinated resource management planning in New Mexico. *Rangelands* 14:272–274.

Hart, J. 1991. *Farming on the edge.* Berkeley: University of California Press.

Heady, H. F. and H. J. Vaux. 1969. Must history repeat? *Journal of Range Management* 22:209–210.

Heitschmidt, R. K., R. E. Short, and E. E. Grings. 1996. Ecosystems, sustainability, and animal agriculture. *Journal of Animal Science* 74:1395–1405.

Holechek, J. L., R. D. Pieper, and C. H. Herbel. 1998. *Range management principles and practices.* 3rd edition. Upper Saddle River, NJ: Prentice-Hall.

Huntsinger, L. and R. Hopkinson. 1996. Viewpoint: Sustaining rangeland landscape: A social and ecological process. *Journal of Range Management* 46:167–173.

Jacobs, L. 1991. *Waste of the West: Public lands ranching.* Lynn Jacobs. P.O. Box 5874, Tucson, AZ.

Jubenville, Alan. 1978. *Outdoor recreation management.* Philadelphia, PA: W. B. Saunders.

Loomis, J. 1993. *Integrated public lands management: Principles and applications to national forests, parks, wildlife refuges, and BLM lands.* New York, NY: Columbia University Press.

McClure, N. R. 1992. Ranchers and resources reaping benefits of CRM. *Rangelands* 14:249–250.

Mitchell, J. E., G. N. Wallace, and M. D. Wells. 1996. Visitor perceptions about cattle grazing on national forest land. *Journal of Range Management* 49:81–86.

Outdoor Recreation Resources Review Commission. 1962. *Outdoor recreation for America.* Washington, DC: U.S. Government Printing Office.

Payne, G. F., J.W. Foster, and W. C. Leininger. 1983. Vehicle impacts of northern Great Plains range vegetation. *Journal of Range Management* 36:327–331.

President's Commission on Americans Outdoors. 1986. *Report and recommendations to the President of the United States.* Washington, DC: U.S. Government Printing Office.

President's Commission on Americans Outdoors. *Americans outdoors: The legacy, the challenge.* 1987. Washington, DC: Island Press.

Sanderson, H. R., R. A. Meganck, and K. C. Gibbs. 1986. Effects of overland flow on water relations, erosion, and soil water percolation on a Mojave Desert landscape. *Soil Science Society of America Journal* 53:1567–1572.

Task Force on Outdoor Recreation Resources and Opportunities. 1988. *Outdoor recreation in a nation of communities: Action plan for Americans outdoors.* Superintendent of Documents. Washington, DC: U.S. Government Printing Office.

Torell, D. J. 1994. Viewpoint: Alternative dispute resolution in public land management. *Journal of Range Management* 47:70–74.

United States Department of Agriculture (USDA). 1980. *Report of the Forest Service: Fiscal year 1985.* Washington, DC: United States Department of Agriculture, Forest Service.

United States Department of Agriculture (USDA). 1992. *Report of the Forest Service: Fiscal year 1992.* Washington, DC: United States Department of Agriculture, Forest Service.

United States Water Resources Council (USWRC). 1978. *The nation's water resources, 1975–2000.* Second National Water Assessment. Vol. I. Summary. Superintendent of Documents 052-045-00051-7. Washington, DC: U.S. Government Printing Office.

Vavra, M. 1996. Sustainability of animal production systems: An ecological perspective. *Journal of Animal Science* 74:1418–1423.

Ward, F. A., R. A. Cole, R. A. Deitner, and K. Green-Hammond. 1997. Limiting environmental program contradictions: A demand systems application to fishery management. *American Journal of Agricultural Economics* 79:803–813.

Wuerthner, G. 1990. The price is wrong. *Sierra* 25:38–48.

Zinser, C. I. 1995. *Outdoor recreation.* New York: John Wiley and Sons.

CHAPTER 14

Urban Land-Use Management

THE URBAN LANDSCAPE

Resources and Services Expectations

Chances are good that you are now located in an urban building or vehicle engineered according to human design. A large majority of people live in towns and cities with more than 2,500 people. Virtually all live in an engineered environment of buildings, vehicles, and supporting infrastructure. We inhabitants of the urban landscape take for granted the provision of diverse services provided from natural resources and human ingenuity.

We expect safe water out of our taps, clean air upon opening a window, heat when the thermostat is turned up, light to read by, music or a movie at the flick of a switch, and routine garbage removal. We want ready and affordable access to food, clothing, furnishings, building materials, and other goods provided from natural resources usually produced somewhere well beyond the city limits. Provision of these basic goods and services is normally taken for granted and happens as a consequence of well-designed urban land use and associated natural resource management.

Chances also are good that you very recently participated in an outdoor activity of some kind—perhaps no more than a commute to work, a shopping trip, a basketball game in a backyard, or an outing at a nearby park or nature reserve. Your choice of activity would have been decided by your activity needs and preferences, the time you had available, proximity to suitable sites, available modes of travel, expense, and quality of the environment in terms of health, safety, and esthetics. Most of us take for granted the design of urban land use and how much it affects the quality of our daily lives. The ease of getting to desired destinations outside the home and workplace contributes importantly to our quality of life. Chronic worry and stress on congested and unsafe streets and walkways is a common urban complaint.

Fries (1997) suggests that Americans have a long history of ambivalence or even hostility toward cities, as evidenced by Thomas Jefferson's (an architect among other things) likening of cities to sores on the human body. Despite their doubts, Americans have steadily moved from rural to urban landscapes (Figure 14.1). How well these habitats serve humans function depends on how effectively individuals and private

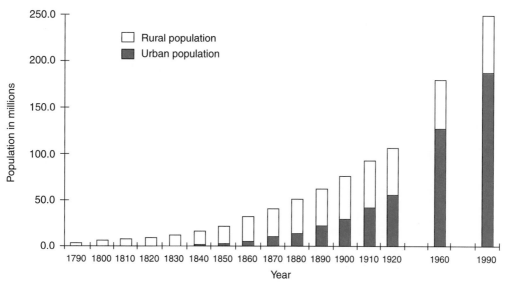

FIGURE 14.1 Urban growth in the U.S. from 1790–1990 (from Judd and Swanstrom 1994 and with permission of Cullingworth 1997).

and public institutions interact in planning land resource use in the urban environment (Fowler 1992, Kaiser et al. 1995). The provision of basic goods and services safely, in a nurturing environment, depends on city and natural resource managers working closely with private enterprise. We refer the reader to Kaiser et al. (1995), Garvin (1996) and Cullingworth (1997) for additional reading about urban planning.

The Urban Ecosystem

Material and Energy Flux. Like other ecosystems, cities have structures and functions that depend on management of energy and material resource inputs and waste outputs (Bridgeman et al. 1995, Hough 1995). Cities are the ultimate destination of most extracted natural resources. The structures of cities are developed from inputs of materials including mineral, wood, oil (e.g., plastics, asphalt), metal, natural life forms, and other physical resources. Cities function from reliable and continuous inputs of food, energy, electricity, fuel and solar energy, water, air, and land resources. Efficient functioning requires dampening the extremes associated with flood, drought, snow, storms, fire, and other disturbances. Resource conversion to useful products also results in waste products, which are concentrated where resource use is most intense—in the cities. Wastes are eliminated in gaseous, liquid, and solid forms into air, water, and earth. Improper elimination threatens physical and mental health and renewal of needed natural resources. Waste elimination solutions are found in more efficient resource use, waste assimilation, and waste dispersion.

The quality of ecosystem services supporting urban life depends on how well ecosystem structure and functions are managed. The urban landscape operates fundamentally like any other ecosystem, but is at the high end of the spectrum of human influence and need. That influence and need extends deeply into rural environments and sources of urban natural resources. The problems of natural resource supply and demand concentrate in urban landscapes more than any other type of ecosystem. Much of what natural resource managers do in urban landscapes is to regulate use, prevent abuse, and contribute to design of urban form and function. Managers of urban natural resources often complement the activities of environmental engineers who develop artificial approaches to augment natural ecological services, such as waste treatment facilities.

FIGURE 14.2   Urban areas increasingly extend into farm, forest, range, and wild recreational lands (courtesy of U.S. government).

Extent of Urban Impacts.   Urban impacts extend influence far beyond city limits. Although the urban land use occupies only about 50% of the total land surface, roads and utilities thread into the more extensive forest, range, farm, and wild recreational lands (Figure 14.2). A high flight over most anywhere in the U.S. outside Alaska reveals an interconnected network of villages, towns, cities, and megalopolises linked by highways, railroads, navigational waterways, and utility right-of-ways. Urban infrastructure extends like roots deep into the land's resources. The miles of roads on U.S. forests alone greatly exceeds that of the entire interstate highway system. These extensions of travel and utility corridors from the urban ecosystem enable the extraction of agricultural, mineral, forest, and other natural resources and outdoor recreational use. The interspersion of urban resource demand and rural resource supplies contributes significantly to transportation needs and associated environmental impacts.

Whereas land classified as urban area is only 5% of the land surface, the extensions of urban development into other urban and rural areas via roads, pipelines, waterways, atmosphere, utility right-of-ways, transmission lines, and railroads also extend urban demands and impacts far beyond the city limits (NRC 1992, Hough 1995, Landsberg 1981). The natural resource effects of urban development are especially intense, not only as sources of resource demand and destinations for extracted resources, but also because of their widespread impacts on air, water, and land resources (Hough 1995) (Figure 14.3). In Chicago, for example, 45% of the land surface is impervious concrete and other hard surfaces (NRC 1992). This impervious surface alone dramatically changes the dynamics of water movement and energy flow through the urban ecosystem, altering the quality of land, water, and air, often far from the urban setting. The urban "metabolism" generates much of the carbon dioxide and other gasses now believed to influence the radiation and thermal balance of the biosphere (Houghton et al. 1996).

FIGURE 14.3 High fractions of impervious material are an aspect of densely populated urban areas and cause increased amount and intensity of contaminated storm runoff (courtesy of U.S. government).

Integrating Ecological Services into Urban Design. Environmentally sensitive urban design is the most fundamental approach to managing natural resources in cities to sustain a high quality of urban life (Fowler 1992, Roseland 1998, Beatley and Manning 1997). To do that effectively, the ecological ser-vices provided by healthy ecosystems need to be identified, protected, and restored wherever it is cost-effective. Much of past urban development proceeded in ignorance of ecosystem services, sometimes requiring expensive engineering alternatives. Where ecological services are insufficient, environmental engineering can be adapted to natural process to increase efficiency of those ecological functions most in demand. Many of the engineering procedures used for environmental improvement attempt to improve upon natural processes. Sewage treatment plants, for example, collapse the natural process of decomposition into a smaller space where it is enhanced through artificial mixing and sedimentation. However, spatial efficiency comes at a high cost in energy requirements, which contribute to other waste problems. Wherever more natural ecosystem services can be used, they generate a cost-savings.

Ecosystem services are most cost-effectively provided when land, water, wetland, and atmospheric systems remain intact. The ecological functions supporting services of urban ecosystems often are overwhelmed by over use and abuse of the landscape, which is common in poorly managed urban landscapes. Careful land use planning is the key to improved provision of needed urban services. Clean water, for example, is provided by natural filtration and sequestering of contaminants and prevention of erosion by vegetated watersheds and wetlands. Those areas that form recharge zones for groundwater aquifers serving cities often are particularly sensitive to disturbance. Coastal wetlands, river floodplains, and barrier beaches can serve to protect urban ports and shores from storm and flood damage (Hudson 1996). Coastal wetlands also generate organic detritus, which supports valuable coastal fisheries. By infiltrating water, uncompacted soils free of impermeable concrete or other hard surface reduce damaging storm run off (Hough 1995, Perry and Vanderklein 1996). Natural vegetation absorbs noise and carbon dioxide (an important greenhouse gas), regenerates oxygen, dampens climate extremes, and forms habitat for diverse wildlife-based recreation and other outdoor recreation. Where wild communities remain intact and are managed carefully, diverse wildlife often provides natural controls of noxious pest species that flourish in habitats with low biological diversity. However, rapid and poorly planned urban sprawl, with insufficient consideration of all costs, continues in most parts of the United States. It is a particular problem in the Southwest and Pacific coastal states.

Urban Form and Function

Urban form largely determines urban function and quality of life for human habitats (Figure 14.4). The best urban form optimally links private residences, workplaces, stores, and other service structures through a public infrastructure of roads, walkways, railways, waterways, sewerage, electric and fuel utilities, recreational areas, and greenways. It functions to sustain safe, healthy, and positively stimulating environments. As urban form materializes, it both provides and constrains choices for the next generation. Much of urban form carries over functions that may have been more appropriate for past culture. The history of urban development is informative because it reveals how cities can become prisoner to their past, unable to adapt, and finally dysfunctional.

The rise of modern urban structure began thousands of years ago in basic urban form that persists to this day (Mumford 1961, Morris 1994). Urbanization accelerated during the industrial revolution as people moved from sparsely populated agricultural land to concentrate at new employment centers. The cities that resulted were compact because most people had to walk to work. Also, land speculation was an early profitable investment in the U.S. More money was made from subdividing into closely spaced small units. Archeology reveals that even the earliest cities tended to rely on geometric design with an emphasis on straight lines and sharp angles that form an open grid for maximum access in and out of neighborhoods (Ward-Perkins 1974, Morris 1994).

The rise of modern urban structure appears to have begun with trade centers at crossroads thousands of years ago establishing a basis for urban form that persists to this day. The most natural extension of streets from crossroads is a grid of streets running parallel to the crossroads. This design left the greatest flexibility for access to the main roads of the city. While perfect grids were rare in older cities, continuously curving streets occurred mostly where topography demanded it. This was typically

FIGURE 14.4 Urban form greatly influences the quality of life of most Americans (courtesy of U.S. government).

along main arteries, and was uncommon in neighborhoods until the automobile made walking nearly obsolete. Fries (1997) traces back colonial town and city design to the classical grid design of Greek and Roman periods. In particular the town square, or *agora,* served as a community common space—a precursor of public space in U.S. cities and towns. Cities changed rapidly in form as people relied more on automobiles and less on walking and mass transportation. One result was much greater urban sprawl and more interface with rural ecosystems.

Human communities occupy the most modified ecosystems, which can grade into adjacent rural and wilderness ecosystems in ways compatible with, or in sharp disruption of, ecosystem structure and function. Well-designed and maintained parks, gardens, and yards soften the harder edges of urban existence (Hough 1995, Nasar 1998). Natural parks can be designed into the urban landscape to thread through and extend to the urban fringes (Figure 14.5). When designed well, these natural spaces can provide recreation as well as contribute to other ecological services such as cleaner air, cleaner water, and less soil erosion. However, excessive urban sprawl with its intense impact on natural process has led to widespread degradation of natural functions. The main causes of dysfunction are environmental contamination, environmental fragmentation, and widespread conversion of watersheds to inert and impermeable structure (asphalt and concrete).

The high contact area with landscapes makes travel and transport corridors (roads, waterways, railroads, airports) a major source of abiotic and biotic contaminants for land, air, and water. Roads are especially problematic because their impact is so extensive. Travel corridors are major routes for contamination by exotic plants, petroleum products, ozone, noxious aerosols, dust, sediment, diverse spilled chemicals, and litter (Hough 1995). In addition, they are one of the main ways that wildlife habitats are split into nearly useless fragments.

Integrating Urban Form with Natural Functions

Much of the history of urban development in the twentieth century has been in search of urban form more compatible with natural function. Cities have adapted mostly by extending outward from the stark artifice of city centers into rural outskirts through suburbanization. A main motivator has been a desire to achieve a more parklike environment where the best of nature is captured and the worst is eliminated or controlled. The results have not always been satisfying because of competing demands for space and costly demands for infrastructure to serve urban sprawl (Figure 14.6).

FIGURE 14.5 A strip of woods along a river, long ago dedicated as a park, provides both open space relief and flood water storage in a densely settled area (courtesy of U.S. government).

FIGURE 14.6 This photo demonstrates both the loss of space and the high amount of infrastructure needed to serve urban sprawl (courtesy of U.S. government).

Parks, parkways, and parklike yards of suburban developments have been a central theme in connecting the city center to a more natural urban world. The best of park planning is multipurpose. Parkways along flood-prone stream drainages, for example, can serve as corridors for highways and for floodways when needed. The tight association of narrow green strips with highways has greatly reduced their value for recreation and wildlife use. However, they can absorb highway sound, reduce side glare and distraction, protect river banks from erosion, and possibly provide some aesthetic comfort for travelers. Where compatible, some parkways also provide paths for bicycles, joggers, and walkers. However, high-speed traffic in narrow parkways often forces all other parkway uses into secondary and less satisfying roles. Numerous parkways, built in a more leisurely time, are no longer well adapted to satisfying multiple use because the demands of the automobile culture now take precedent.

Depending on planning, roads and other infrastructure can be designed to fit more naturally and with less negative impact on ecosystems. The first urban parkways were built during the late nineteenth century to connect larger park openings for the recreational traveler in a horse carriage or on a bicycle. As the U.S. became more wealthy, more attention was placed on the esthetics of urban form. Two architects were particularly influential in urban planning (Fowler 1992). Frank Lloyd Wright emphasized a need for individual space in residences and natural integration with nature. He influenced many suburban developers of suburbia starting in the 1930s. Others emphasized revitalization of the downtown areas with tall high-rise residences surrounded by greenways. U.S. cities initiated a metamorphosis in form during the decades around the turn of the century—a time of growing resentment for wasteful resource exploitation. At the same time, belief that efficient utility of resources was compatible with natural beauty and culture became popular. This concept of efficiency was closely associated with the emerging conservation movement. It assumed an elegance that extended into architectural design of buildings and land-

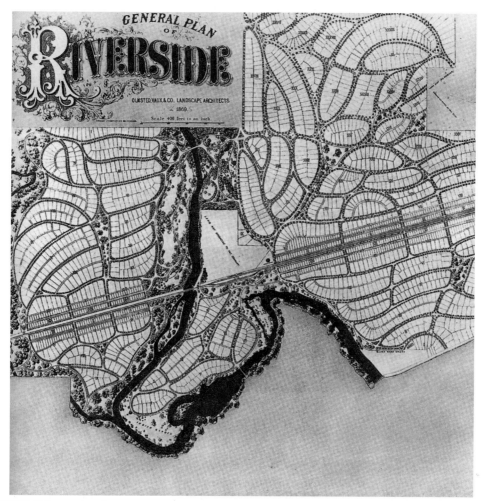

FIGURE 14.7 Frederick Law Olmsted, an early landscape architect, was very influential in establishing contemporary concepts of parks and parkways. Many modern suburban developments have adopted Olmsted's concepts and propagated them across metropolitan landscapes. The plan shown above was proposed by Olmsted and partners in 1869 (courtesy of the National Park Service).

scapes. When transformed into roads, the new concept of efficiency emerged in curving boulevards, which replaced the grid with more aesthetically pleasing links between locations. Often, the parkways connected larger parks and public structures, such as museums, theaters, and colleges. Although they were not the most time-efficient travel corridors, these curving parkways served as models for the endlessly reiterated curved mazes of later suburban developments.

Parks have played an important role in U.S. urban form since early in the last century (Garvin 1996). Among the earlier and most influential park designers was Frederick Law Olmsted, who created Central Park in New York and many other city parks. It was built to be a mix of functional playgrounds and naturalistic environments on land that was almost totally landscaped and artificially planted. Within a decade, the results at Central Park were impressive enough to establish both Olmsted and city parks as fixtures in numerous U.S. urban plans.

Many of the early landscape architects, including Olmsted, became city planners (Garvin 1996). They developed a new urban architecture, later led by Frank Lloyd Wright and Corborsier, which sought to spread out and blend more with the urban landscape (Figure 14.7). The influence of these innovative home and landscape architects extended into the parklike development of suburban communities a century later—a process that has come to haunt the natural ecosystem planners.

Some city parks in the U.S. are modeled after the highly structured and geometric patterns of many European parks. However, the largest U.S. urban parks promote natural landscapes of wild vegetation or naturalistic landscapes of undulating topography, lawns, gardens, and thinned out woods based on nature. Water is typically a central element in most large parks. Like parkways, parks often are managed for many purposes. Water, for example, is an important natural element that often is integrated into parks for flood and pollution control purposes as well as recreation and esthetics. In recent years, cities like Portland, Oregon, Austin, Texas, and the greater New York region have developed plans that protect natural land from suburban development (Beatley and Manning 1997). A mix of strategies are being applied that include purchase and reservation of natural areas for watershed and low density uses, such as lands protected in the New Jersey pine barrens for groundwater recharge and other low intensity uses. New Jersey is considering zoning plans for protecting remaining rural lands for traditional uses.

History shows that without careful land-use planning in urban landscapes, costly adjustments often are needed to sustain necessary functions. The costs have to great extent been borne by the taxpayer in the form of new and rehabilitated public infrastructure. Despite the obvious need, urban planners often are considered as necessary evils in the U.S. and elsewhere (Fowler 1992), where regional planning often is viewed as an infringement on private property rights. However, it is a necessary one given the tendency for the marketplace to overlook long-term community needs. As a consequence, the role of natural resources management in urban planning often is contentious and political.

DEVELOPMENT OF URBAN INFRASTRUCTURE

Public and Private Partnership

Private Independence. The most highly valued land and water resource use is for private development of residences and businesses. In contrast with wild lands and waters, which are predominantly owned by the public, ownership of urban and agricultural landscapes is mostly private. Because of its high social value, demand for urban development usually establishes the prevailing land prices around existing centers of urban development. Much of the nation's richest farmland already has become urbanized, and much more is likely to be if the pattern of recent urban growth continues unabated (see Chapter 12). Similarly, natural ecological services other than food production typically require large land areas, such as for producing clean water, absorbing flood water, and providing wildlife and fish habitats. Starting with the earliest attempts, urban design in the U.S. has been an exercise in providing an appropriate balance between private development and public service needs (Judd 1988).

Most Americans have placed a high value on economic, religious, and political independence. Rights to private property and freedom of movement are important ideals. The promise of plentiful land and the opportunity to live on it and develop it were important reasons for immigrants to come to the U.S. The typical American dream was, and continues to be, independent land ownership centered around home and family, freedom for personal beliefs and expression, opportunity for wealth accumulation, and mobility. By the mid-twentieth century, the American dream for the average citizen was a privately owned, appliance-filled family home with a big yard, a car in the garage, and enough personal security, wealth, and leisure time to enjoy them. The Federal Housing Act of 1949 extended this dream nearly to the status of a right by declaring its intent to provide a decent home and environment for every American (Garvin 1996).

Public Interdependence. American private well-being, however, also depends on publicly owned community infrastructure—most notably parks, roads, and water resource systems. The town square often was, and still is, common-use space in the geographical center of town life. Town centers frequently became the first city parks,

FIGURE 14.8 Public buildings, historic structures, and a landscaped plaza are united to form a community focal point in Alexandria, VA (courtesy of U.S. government).

which were surrounded by buildings dedicated to key businesses, town government, education, and religion (Figure 14.8). As people gained more opportunity for relaxation and recreation, parks became more important public resources.

Early Americans accepted travel and even came to seek it for pleasure once invention produced a comfortable means. The concept of parkways emerged, blending public transportation corridors with public recreation and creating an esthetic sense of connection and community (Garvin 1996). American taste for pleasurable travel extended well beyond the city limits to enjoy some of the most diverse and awe-inspiring terrain in the world. Infatuation with the "great outdoors" grew with the improvement of the automobile. Similarly, Americans supported public development of navigation, water storage, and flood control in its waterways. Later, it would accept with little hesitation public water treatment systems for protection of public health and the environment.

Public and Private Coordination. The value of private real estate usually depends on the extent to which publicly funded and managed infrastructure is provided in coordination with private development. Public lands typically sustain the engineered infrastructure required for contemporary urban functions—public buildings, streets, highways, parking lots, airports, railroads, waterways, dams, walkways, bridges, tunnels, utilities, water supply, sewerage facilities, and public recreational facilities. The value of private real estate has had much to do with how the public infrastructure was developed. However, provision of infrastructure also has had major impact on real estate value and on ultimate public welfare. Too frequently, special interest wealth influenced infrastructure without concern for the public interest.

Poor planning coordination in the past has contributed to many of the urban resource management problems of the present. Many of these problems will require decades to solve even under the best of circumstances. Much past development of private land and public service has progressed incrementally with minimum concern for cumulative regional impact. Past and present urban expansion is the consequence of planning that integrated public subsidy with private investment. Tension between national and local interests in the U.S. continues to make regional planning a strenuous process (Fowler 1992, Kaiser et al. 1995), but it remains the most promising way to sustain and improve upon quality of urban life.

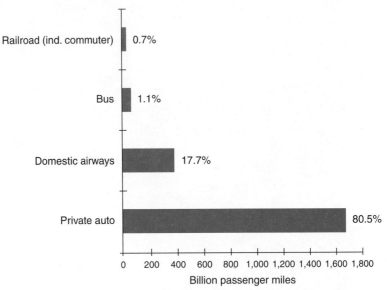

FIGURE 14.9   In 1992, the automobile exceeded all other forms of passenger transport in the U.S. (with permission from Cullingworth 1997).

Transportation

The development of modern urban infrastructure has been dominated by public policy in support of transportation. Many of the most impressive inventions and developments by Americans have responded to American predilection for travel. The means we have chosen for travel have shaped the modern urban landscape at least as much as the buildings in which Americans live, work, learn, shop, govern, pray, and play (Figure 14.9). More than any other form of transportation, the automobile has come to dictate the design of most U.S. cities (Jakle 1990, Langdon 1994, Roseland 1998).

Where urban environments were more consolidated, as in Europe, there was less space available for the demands of the automobile. But for many U.S. cities, congested, noisy, and exhaust-polluted streets became the norm (Figure 14.10). Many European countries were already densely settled before cars became commonplace. Their larger cities are older than in the U.S. Therefore, historical appreciation was great enough to prevent extensive remodeling around the automobile. As a consequence, Europeans have relied more on a network of mass transport within and between cities. Some of the larger cities of the U.S. continued to build on their investment in railroad mass transportation. The first subway systems were constructed late in the nineteenth century in Europe and the U.S. Later, insufficient resource wealth also dictated an emphasis on mass transportation systems in various other countries.

The automobile was invented at a time in U.S. history when a frontier mentality and unfettered entrepreneurial spirit still motivated independent thought and action. Coordinated planning and design processes received little consideration. Americans had long been a mobile society that solved local resource depletion problems simply by moving to unexploited resources—usually westward and off the beaten path. The U.S. populace was still more rural and small village than big city. Many of the first buyers of automobiles were farmers seeking more frequent connection to cities. Farmer's market and fresh "truck crops" quickly became a staple of city life. These were a predecessor of the modern supermarket that now supplies various foods from sources thousands of miles away.

By World War I, a sizable fraction of U.S. society could afford a car, but prices remained out of reach of the majority. That was changed by Henry Ford and other automakers through the development of mass production techniques (Jakle 1990). Cars, trucks, tractors, and many other manufactured items became much more affordable. Private enterprise also developed more questionable strategies for construction of in-

FIGURE 14.10 Congestion in downtown Pittsburgh in 1937. Congestion was a major problem facing cities at that time (courtesy of Carnegie Library in Pittsburgh).

frastructure based around the internal combustion engine. General Motors purchased electric trolley companies and then removed the trolley tracks and replaced them with internal combustion busses (Jakle 1990). Clearing the streets of trolleys made automobile travel much easier and encouraged more auto commuting farther from the city center (Figure 14.11).

As more people bought cars, industry successfully lobbied for public investment in highway transportation designed for internal-combustion traffic. The electric trolleys were gradually displaced. Because of uncoordinated local control, highway development created a patchwork of roads of variable quality. The federal role in influencing highway system development was justified based on the need for improved interstate commerce. The Federal Highway Act of 1916 was the first major involvement of federal government in highway expansion since early nineteenth-century investments in a national highway westward. By offering matching funds for highway construction, this act coerced states to create state highway departments with authority to oversee and replace local road construction. Car ownership boomed during the economically expansive period following World War I and the lure of suburban development increased.

Utilities

The wealth that powered suburban development also influenced local and state governments to invest in the infrastructure along the public highways. Our modern concept of utility networks grew out of that convenience. The government subsidized private enterprises by providing access through the public lands associated with roads. Electric and telephone lines usually were strung up along the highways at much less cost to private utilities than would have occurred without public highways. As the need for traffic lights and suburban electrification increased, electric services increasingly became an indispensable part of the urban infrastructure. Some private water supply utilities also began to provide treated water along highway right-of-ways, followed by natural gas lines mostly after World War II.

FIGURE 14.11   Chicago 1909. Electric street cars, although fuel efficient, impeded automotive traffic. General Motors bought up street car companies and converted them to busses with the intent of favoring the internal combustion engine (courtesy of Chicago Historical Society).

Street flooding was another problem that had to be addressed. The right-of-ways beneath highways also were convenient for draining storm water runoff. At first, water supply and waterborne waste removal were locally provided through wells, latrines, and septic waste fields. Some private water supply services developed where urban development occurred in locally deficient areas. But local worry about dependable water supply and waterborne disease grew into widespread public health concerns. As a consequence, local governments gradually assumed most water supply and waste removal services. Sewerage pipes were added to the infrastructure buried beneath the streets. Garbage removal initially depended on individual disposal responsibility and hired services. Garbage dumps began to degrade the landscape as people dumped virtually anywhere to get rid of solid wastes. As public health and esthetic concerns grew, public management of solid wastes became the rule.

Housing Subsidies

The depression of the 1930s slowed urban expansion less than might have been the case without government intervention and plentiful oil resources. In response to the depression's impact on housing construction, the Federal Housing Administration Act of 1934 increased mortgage insurance and made mortgages much easier to obtain (Cullingworth 1997). Before the act, a mortgage usually covered only one-third to one-half of the house cost and had to be refinanced every five to 10 years. The act enabled house purchases with mortgages up to 30 years with only a 5% down payment by the buyer. No down payment was required from World War II veterans. Housing starts in the suburbs increased rapidly in the midst of the depression years and accelerated after the war. Cheap fuel sustained the use of cars throughout this period. Burgeoning oil, trucking, and intercity bussing industries joined the automotive industry to continue to influence government investment in the highway system. In addition to other public works, such as dams, highway building was also accelerated during the depression years using public funds. It encouraged employment while building an improved infrastructure for future expansion of private enterprise.

FIGURE 14.12 In no other city are superhighways more a part of the landscape than in Los Angeles, where they relieved congestion at the expense of air quality (the haze in the background is smog) and fragmentation of neighborhoods (courtesy of the Environmental Protection Agency).

The Superhighways

During the depression, the government became much more active in building public works and often focused on project objectives with single-minded intensity. Science and technology were widely accepted as the way to a more satisfying future. Much private land was claimed and purchased for eminent domain needs pertaining to public transportation. Robert Moses, in New York City, was one of the more effective public transportation managers of the time (Caro 1974). Like most other regional planners focused on one land use, he advocated development of roads for transportation benefits, without in-depth consideration of their ultimate effect on other aspects of lifestyle (curiously, Moses never drove himself) (Figure 14.12).

The new superhighways were designed and developed through close association of automotive engineers and executives with highly focused planning objectives. For many decades, the typical superhighway systems were designed as if to discourage mass transportation and pedestrian traffic. New highways frequently sliced through neighborhoods creating isolated urban backwaters and worsening decay (Muller 1990) . They concentrated urban noise and air pollution. In addition to providing the means for suburban expansion and more rapid traffic flow, they encouraged urban conditions that stimulated even greater urban flight to the suburbs (Figure 14.12).

During the Cold War following World War II, federal support of highways was justified for the rapid evacuation of nuclear targets and to facilitate movement of troops and retaliatory missiles. This initiative received enthusiastic support of the auto, oil, trucking, and bussing industries. The Interstate Highway Act of 1956 subsidized further urban decentralization along interstate routes. This act was followed by the Instate Highway Act, which funded and coordinated minimum access "beltways" around city centers. The beltways were needed to relieve the interstate highways from the escalation of traffic that was occurring on them. Most metropolitan areas of that time favored growth, and beltways provided additional access to existing suburbs and a stimulus to develop more suburbs.

In a typical expression of public works and private land development, the beltways reshaped land development almost overnight. Business complexes sprang up

FIGURE 14.13 The development of high-occupancy vehicle (HOV) lanes to encourage less traffic congestion has had disappointing results. Urban demands make car pooling a time-consuming and demanding task that is not a very attractive alternative to slower traffic in low-occupancy lanes (courtesy of the Environmental Protection Agency).

around the beltway forming nuclei for new "edge" cities (Garvin 1996). Much of the public supported these massive engineering projects—especially those developed to relieve the traffic-choked highways first built in the 1920s. Many of those earlier highways had become magnets for commercial development and cross traffic leading to subdivision housing. Thus a reinforcing feedback between public highway development and private suburban development began and continues to dominate city planning and function. From 1950 to 1992, the number of cars in the U.S. quadrupled. Now about 80% of all passenger traffic is in cars. Most travel to work is by car, and most commuters travel alone. To reduce commuter traffic, large amounts of money have been invested in special travel lanes for high-occupancy vehicles in the larger metropolitan areas of the U.S. with the intent of reducing traffic (Figure 14.13). The effect overall has been disappointing, however. Commuting contributes only about 20% to total automotive travel, with similar contributions each from pleasure driving, shopping, and personal needs.

URBAN GROWTH AND DECLINE

Urban Growth

The industrial revolution of the eighteenth and nineteenth centuries accelerated the urban growth concentrated around city-center industry and business. This continued until after World War II into the present. As cities continued to grow, they diversified into identifiable neighborhoods and districts, which began to replace the city center as discrete places of residence and sources of specialty services (e.g., financial, theater, warehouse districts). However, most central cities continued to thrive until after World War II despite the nearby neighborhood decay. Starting in the early twentieth century, sky scraper construction and rail transportation rejuvenated city centers by multiplying useful work space in the city center. It facilitated efficient access and networking among growing businesses. The city center continued to be a hub where activities converged in the "controlled chaos" of a "people center" dependant mostly on walking (Goldfield 1977).

U.S. cities attracted people from many other parts of the world as well as rural America. The influx of mostly low-income people overwhelmed housing and urban infrastructure (Cullingworth 1997). The consequences of overcrowding and under-employment caused city residents who could afford alternative housing to move to suburbs, taking needed tax revenue with them. Federal legislation in support of more affordable mortgages encouraged urban flight. Mortgage restrictions discouraged purchase of older houses, especially in urban neighborhoods with risky futures. They diverted prosperous people, who might have otherwise stayed, to new housing in the suburbs. As urban decay advanced, disease, fire, and crime progressively crippled older parts of the city center (Judd 1988).

Urban Decline

Residential neighborhoods varied in health according to income, with the poorest neighborhoods typically remaining on the industrial and warehouse fringes around the city center. While most neighborhoods were well represented, the poorest were least represented and ended up in the least tolerable ghettos. As they became more wealthy, a succession of people moved from poorer neighborhoods to newer neighborhoods nearer the city edge or beyond. The poorest remained trapped in decline and grew in number, especially as the tenant farmer system of the South gradually collapsed during the first half of the twentieth century. During that time, many unskilled people moved off the land to the cities in search of scarce employment (Lemann 1991).

The Role of the Automobile

The automobile was a "catalyst" for decline of many city centers, starting before the depression and rapidly accelerating after World War II (Goldfield 1977). Those cities most able to sustain a healthy city center, such as New York, had developed efficient mass transportation systems early in the century. Cities more dependent on the automobile suffered more severe and persistent declines. The private car required too much space to maintain areas dedicated to walking and mass transport. Car exhaust also overwhelmed assimilation by the urban atmosphere. Tall buildings trapped fuel exhausts, especially where natural atmospheric inversions caused stagnant air (Goldfield 1977). Air quality worsened even in cities with little industry like Washington, D.C., and Los Angeles.

Starting in the 1930s and reaching a peak in the 1960s, neighborhoods and city centers were increasingly split up by limited access highways. Helped by federal funding, parts of some cities became "spaghetti bowls" of multistoried freeways (Muller 1990). It became more expensive for large businesses to attract people to downtown locations. They began to leave for the new satellite cities growing rapidly along major intersections with superhighway beltways. Before the war, suburbia remained dependent on the large city centers for employment, shopping, and many services. With the establishment of satellite cities along the beltways (Figure 14.14), a new form of centralized retailing came into its own—the shopping mall (Cullingworth 1997, Garvin 1996).

The first shopping malls were built during the depression years, but they did not ascend to banality until after the superhighway systems were built in the 1960s. During the Cold War, the Department of Defense invested heavily in technical products including aerospace, computer, communications and other products of technical firms which grew up around interstate highways, especially the beltways around large cities (Cullingworth 1997). Satellite cities attracted local business centers and

FIGURE 14.14   Satellite, or "edge," cities grew up rapidly after the interstate highway system was developed. Tysons Corner, VA, pictured here in 1989 on the beltway around Washington, D.C., was little more than a country store in 1935 (courtesy of Fairfax County Public Library Photographic Archive).

finally light industry, especially in aeronautics, computers, communications, and other "high-tech" employment of the information age. These industries attracted employees from the older industrial centers, which were in decline as a consequence of cheaper labor associated with international competition. Defense spending supported the cost of moving workers from old industrial city centers to new beltway locations. Many heavy industries closed, leaving blighted "brownfields" of rusting abandoned buildings and debris with few prospects for desirable land development.

Urban Sprawl and Downtown Renewal

Suburban Sprawl.   Urban decentralization was especially problematic for cities. It also has created problems in the suburbs, which grew twice as fast as the central cities during the 1980s (Cullingworth 1997). In many localities, suburbia sprawled over most of the land between city centers and the edge and began rapidly extending into rural environments beyond. Many of these metropolitan areas have coalesced into a sprawling "megapolis" where the city landscape sprawls over hundreds of square miles (e.g., Phoenix, Arizona; Los Angeles, California; Atlanta, Georgia; and from New York City to Washington, D.C.). Most "suburbanites" sought a more restful, cleaner, safer, and convenient environment. Their dream has been temporarily realized, but at considerable environmental cost (loss of open space, congestion, pollution, high local taxes, and so on).

Goldfield (1977) called the lure of suburbia the "Golden Fleece of the pastoral ideal." Suburbanites paid in part for their urban independence through utter dependency on cars and associated noisy, congested, esthetically distasteful thoroughfares with numerous costly side effects. Roads and traffic became a major source of air pollution, suburban and urban fragmentation, and competition for needed mass transportation and urban redevelopment resources. Kaiser et al. (1995) summarized this as follows: "Low density sprawl with its leapfrog development pattern is seen to be not only ugly and wasteful of land but also fiscally inefficient, environmentally harmful, socially isolating, and exclusionary." Numerous books have been written on the social and environmental problems associated with urban decentralization (e.g., Langdon 1994, Kunzler 1996). Sprawl is typically regional, extending beyond the limits of urban jurisdiction, and even state jurisdiction (Figure 14.15). This makes city governments unsuited managers of sprawl. Some state governments have tried to address the problem with mixed success (Kaiser et al. 1995).

Among the more pernicious problems in suburban environments is the sense of isolation that often befalls people in the midst of thousands of people. Except for the local side streets and cul de sacs, suburban arteries often are designed for two to three lanes of traffic with auto speeds of up to 50 miles per hour, harkening to an era when mass evacuation was on the minds of Cold War planners (Kunzler 1996). Because crossing lights often are spaced far apart, the roads become threatening barriers to pedestrians, forcing reliance on vehicles. Children need to be chauffeured or bussed

FIGURE 14.15 Urban sprawl displaces irrigated farmlands in Utah (courtesy of U.S. government).

most places. Many suburban areas have limited outlets, forcing people to drive the same routes repetitiously and farther than necessary past monotonously similar buildings (Langdon 1994). Much driving is done alone because people are so spread out. Retail services have become centralized in large shopping malls or multiservice stores, which can only be reached comfortably by vehicle.

The protection of public open space and the provision of improved infrastructure is becoming increasingly problematic as land prices increase and the public resists further conversion of private land to public control. Local governments have won constitutional approval for zoning use of private lands (Cullingworth 1997), which provides a valuable land-use planning tool. Zoning has often promoted urban sprawl and exclusionary policies resulting in large minimum-size lots and wide separation of homes from services (Cullingworth 1997). Public investment in infrastructure is becoming more costly per person as the distance from service centers increases and the density of people served decreases. Much of the existing highway infrastructure is aging to a point requiring major repair and rehabilitation—a costly process in both taxes and driver time. As concern for private property rights has grown, it is becoming increasingly difficult for private lands to be condemned and purchased for rehabilitation of public infrastructure. Older suburban developments and edge cities are beginning to show signs of decay as the process of rehabilitation and renewal congeals. The wealthier people seek solutions by moving farther out or, increasingly, back into renewed city centers.

The public is now less willing to support the private real estate interests of the most wealthy through infrastructure development. Municipalities are increasingly shifting the burden of new suburban infrastructure to private developers as the public balks at more taxes (Cullingworth 1997). There is increasing pressure to convert minimum-access "freeways" to tollways as the technology allows boothless toll collection. Some proponents expect tolls will reduce use and encourage more environmentally compatible forms of transportation or uses of time. It might also increase the competitive advantage of smaller local services from large centers. The rush to suburbia has shown signs of slowing in recent years, but the economic forces behind further development remain intense. The wealthiest, at least, will continue to move to the fringes, repeating a pattern that started well over a century ago.

Urban Renewal. Urban renewal and rehabilitation are measures taken to counteract urban decay. The basic strategy of urban renewal has been to eliminate urban blight through destroying old structure and rebuilding on the cleared land. The strategy of rehabilitation is preserving the best of the old urban structure and renovating it for new use.

Urban renewal started with the Housing Act of 1949, which sought to provide affordable decent housing and environment for every American. It fell short of aspirations in large part because of resistance by private real estate and development interests (Garvin 1996). Because of delays, it took nearly 20 years for the act to accomplish its goals. Subsequent amendments established qualification rules for commercial properties. It eventually subverted the act's primary purpose as property values escalated and low-income inhabitants could not afford the housing. Early urban renewal translated into whole-scale physical destruction rather than restoration of what was left of decaying neighborhoods. In some locations, renewal got no farther than clearing the land for private redevelopment, which was reluctant to invest in many locations. This was because of a perceived lack of interest by middle-class homeowners who, for the most part, wanted a suburban location (Cullingworth 1997). In other locations, government subsidized the construction of large high-rise structures surrounded by open, parklike areas. Without a sense of neighborhood and community function, and with high unemployment rates, some of these buildings became high-rise ghettos permeated and surrounded by crime and devastation. Even though many renewal projects were quite successful, the disappointing results of some highly publicized renewal projects have cooled government support.

Cities have learned that successful redevelopment requires a complete integration of social function with the structural development. It is important to create a mix of residences and community services that provide the physical support for neighborhood rebirth. Private enterprise is favored, with government playing more subtle roles in subsidizing desired rehabilitation and reconstruction. Zoning to recognize historic value of architecture is one approach to rehabilitation that has had success in sustaining or reviving urban neighborhoods. However, before the Tax Reform Act of 1976, federal tax laws discouraged rehabilitation. In some locations, municipal and state incentives have encouraged renovation (Cullingworth 1997). To be thoroughly successful, neighborhood businesses such as local stores, restaurants, entertainment, and personal services need encouragement. Safe parks, playgrounds, and other open spaces are important components for success. Judicious investment in new mass transportation routes has helped gain middle class interest in complete neighborhood renovation. Problems remain, however, in placing people of lower income into the process of private restoration.

Urban rehabilitation of city centers began to take shape in the 1950s and 1960s in cities such as Minneapolis. The focus was on downtown revitalization and mass transportation links from neighborhoods to downtown areas (Goldfield 1977). In many cases, they restricted auto travel or eliminated it entirely from downtown pedestrian malls, while providing necessary parking on the periphery. Other successes include more effective bus or rail mass transportation, such as in Washington, D.C., and attractive downtown centers, such as the river walk at San Antonio, the old-town Bourbon Street area in New Orleans, the wharfs in San Francisco, and the river fronts in New York and Philadelphia. In the last two decades, the construction of public aquariums has been an especially effective way to rehabilitate once decaying urban waterfront areas, such as at Baltimore, Boston, Camden, Chattanooga, Monterey, New Orleans, and Tampa.

URBAN LAND-USE PLANNING

Historic Land-Use Development

Land-use planning in the U.S. has been and continues to be mostly a local process (Cullingworth 1997). If city persistence is a clue, the effective urban planning has always taken best advantage of the natural resources and trade made accessible through transportation. The ghost towns of the U.S. are typically at a dead end or along a little used highway. Most large cities are strategically situated between major resource supply and demand centers. Chicago, for instance, grew rapidly as it received grain and livestock by rail, which were processed and distributed using the Great Lakes and the Mississippi River system. East Coast cities shipped products overseas. Much city planning from the beginning took place to assure someone or some group obtained greater wealth based on trade, land speculation, or land development. For many centuries, London was the world's leading trade center. It has an excellent harbor for trade, in part because it is in close proximity to major European markets.

Trade in Boston, Providence, New York, Philadelphia, Baltimore, Charleston, Savannah, New Orleans, and San Francisco soon grew to rival that of European cities. Other inland cities became established on rivers and lakes large enough to serve navigational commerce, such as Cincinnati, Saint Louis, Memphis, Chicago, Detroit, Buffalo, and Pittsburgh. Strategic naval location also was a factor in cities like Anchorage, Honolulu, Norfolk, San Diego, and Seattle. Some of the oldest cities in the world were established along overland routes in Asia and North Africa, such as the "Silk Road" (really a network of caravan trails) linking the Mediterranean to China. In the U.S., some of the oldest cities also were built on inland trade routes along the Rio Grande at El Paso, Albuquerque, and Santa Fe. But these land-route cities were later eclipsed by cities with railroad linkages, improved water supplies, modern highways, and improved pest and disease control (e.g., mosquito-borne diseases). Air conditioning encouraged city growth where it would not flourish otherwise. Advances in those and other areas have resulted in the greatest recent growth of inland cities in the southern and western U.S., including Atlanta, Dallas, Denver, Houston, Jacksonville, Los Angeles, Miami, and Phoenix.

The extent to which urban planning has become a formal and comprehensive process varies tremendously across the nation, depending much on local history and attitude toward government. Back to the early colonial governments, urban planning has been dominated by a wariness of strong central government (see Chapter 2). Americans were more disdainful of "big government" controls than had been the case in Europe, where urban landscapes are more commonly governed by regional and national regulations (Cullingworth 1997). What appeared to be an inexhaustible land supply promoted a laissez faire approach to land-use regulation and planning driven by private enterprise. That attitude is most persistent in contemporary U.S. politics where population densities are relatively low and per capita land supply is abundant, such as in the Rocky Mountain states.

The major evidence of extensive government influence in urban planning before the twentieth century shows up in government capitals. More commonly, local planning was accomplished more through reaction to undesirable trends than through proactive prevention of such trends. Local urban regulations started immediately following settlement to assure dependable commerce and safe conditions. Regulations were developed, for example, to prevent devastating fire, mired streets, depleted wood, depleted wildlife and fishery resources, and to assure that local food needs would be met. As time passed, new cities learned lessons from older cities regarding fire, flood, mud, disease, and other resource misuse. In addition, local governments came to depend more on state and federal government for disaster relief, especially after the Great Depression of the 1930s. Such aid relieved local governments of more careful land-use planning in the paths of flood, fire, earthquake, hurricane, tornado, and other natural events.

The few attempts at early urban planning were typically associated with a strong civic leader. One of the more successful was James Oglethorpe, who designed Savannah, Georgia, around two dozen neighborhood parks laid out in central squares to maintain a more natural urban environment (Garvin 1996, Fries 1997). William Penn designed colonial Philadelphia to be composed of one-acre lots with single homes, orchards, and vegetable gardens. But the design was viewed impractical by landowners who quickly subdivided the land into close rows of houses on minuscule lots. Reliance on walking and maritime trade were major factors in determining the high density of early U.S. cities along their waterfronts (Muller 1990).

Washington, D.C., as the seat of the federal government, came closest in the U.S. to being completely planned in response to the recommendation of President Washington. The French planner, L'Enfant, envisioned a basic grid superimposed on radiating boulevards and traffic circles linking houses of government and monuments (Fries 1997). The design created many odd-shaped plots of lands and intersections, which may have slowed city development as transportation changed. However, while at first viewed as an embarrassment, the U.S. capital has since become a world attraction because of its monuments, museums, and unique urban design (Zelinsky 1990).

Other early federal influence on urban planning stemmed from the way the public domain lands were organized for transfer to private ownership. The practicality of organizing land ownership and development on a grid was institutionalized by the U.S. federal government soon after the nation formed. The national grid formed square-mile townships composed of 640 acres each. These township boundaries defined the basic urban pattern for many Midwestern cities.

Most existing urban lands in the U.S. developed without a comprehensive plan. They expanded incrementally on private lands at the margin of the existing city. Long before nationhood, the colonies emphasized the importance of private land holdings as the basis for land control. Exceptions were made for necessary infrastructure, such as roads and central fenced areas for grazing dairy animals. The old European concept of farm villages, built around the commons and surrounded by agricultural fields, was faithfully transferred to many colonial towns (Figure 14.16). The concept of the town-square was transferred westward as people migrated from the original colonies. In the Southwest, it is manifested as the central plaza.

In contrast with Europe, where the largest cities became centers of civilization and bases for government and public policy, religion, and culture, most large U.S. cities became seats of private enterprise and entrepreneurial spirit (Mumford 1961). As urban growth began to be influenced more strongly by the automobile in the early 1900s, the first urban regional planning agencies were formed in Los Angeles and Chicago. Urban regional planners grew influential enough to encourage comprehensive planning initiatives at the federal level during the Great Depression. They emphasized developing sustainable and self-contained cities that linked comfortably

FIGURE 14.16 This landscape pattern of villages surrounded by farmland in the northeastern United States was typical of many communities until the 1960s when rapid urban expansion began to occur (courtesy of U.S. government).

with the natural environment. This was, however, eliminated before any substantial progress was made. The first zoning laws also originated during the early 1900s and, although contested, have withstood the test of constitutionality in providing for the general welfare. Although zoning is a means for organizing district conformity, and not planning *per se,* it has been used in numerous ways to meet planning objectives (Cullingworth 1997).

Gaps materialized routinely between the development of private lands and public infrastructure before laws were adopted to facilitate integration. Even so, gaps still remain common as evidenced by incomplete superhighways, typically placed in areas of dense population with diverse stakeholder concerns, such as in Memphis, Boston, Baltimore, Washington, D.C., and central New Jersey. Private developments of wood frame houses continue to be built immediately adjacent to fire disclimax ecosystems on public lands. Although local governments have learned from past mistakes, the learning process has been a slowly evolving one.

Perhaps most has been learned about planning transportation networks and much urban planning now centers on them. When the nation was young, roads often "evolved" more than they were purposefully designed and engineered. The fringe population in early cities often were isolated during bad weather. Bridges, if built at all, were frequently washed away during floods. Topography was a prevalent consideration in early urban development, which had to conform to the shorelines of adjacent waters, steep topography and drainage channels. However, the desire to maintain an urban grid network of streets on a reasonably flat foundation often required substantial restructuring of the landscape, which sometimes took years to complete. Such areas often were badly engineered and created other problems, usually related to drainage and inadequate natural earth foundation. These problems showed up especially at times of "natural catastrophe" during flood, storm, earthquake, and fire. Despite all that has been learned, the mistakes of the past continue to catch up with the future and often are repeated.

CONTEMPORARY URBAN LAND-USE PLANNING

Planning Professionals

Contemporary land-use planning is a highly developed discipline with a large professional membership. Much has been learned and published about urban planning. According to Kaiser et al. (1995), effective land-use planning considers all interests while attempting "to integrate land use, transportation, public service, environmental protection, hazard mitigation, public finance, historic preservation, and other related functions into a comprehensive plan that accounts for the connections among these areas." Much of the experience of the regional planning process has been gained through urban land-use management. Kaiser et al. (1995) liken urban land-use planning to a "big-stakes game of serious multiparty competition over an area's future land use pattern." Planning helps create the rules of the game from existing law and agreement of stakeholders in the planning process. Although regional urban planning has been a highly contentious process with mixed success, comprehensive regional planning is gaining influence. The growth of environmental, transportation, and social restrictions at all levels of government has encouraged comprehensive planning (Cullingworth 1997). The alternatives have proven too costly in the long run.

One of the basic objectives of urban land-use planning is that benefits will accrue to some degree to each of the many stakeholders. In providing for those interests, it is the intent of planners to generate a livable environment. This is not an easy task, given the fragmentation of planning authorities, marked differences in attitudes and goals among different planning authorities, and rapid urban growth in many areas. Contemporary urban-use planning now deals routinely with tensions among environmental, market, and social-use activists. Increasingly, the intent is to "strike a balance" among the interests in managing land-use changes through sustainable development. However, there remains great regional variation in commitment to land-use plans that will sustain benefits far into the future.

Nongovernment Stakeholders

In countries such as the U.S., where much of the land is privately owned, private interest is the primary motivator of changes in land-use patterns (Kaiser et al. 1995). Among those interests are stakeholders in three main categories: market-value interests, neighborhood resident interests, and environmental interests. Market value interests are motivated by profit and are concerned about land-use decisions that will affect their investments. These interests range from the local home owner concerned about resale value to the large land development organizations concerned about profit for themselves and their shareholders.

Many urban problems exist because of the incremental approach to planning that private land control promotes in its interactions with public services. Business planning tends not to be comprehensive; rather it "externalizes" the side effects of its planning process, relegating them to someone else's problem bin. In typical fashion, the real estate entrepreneur speculates on land that has the best potential for future development. A stable regulatory and comprehensive planning environment may be desirable but is difficult to bring about because land development itself typically creates a need for adaptive change. Provision of appropriate transportation, utilities, recreation and other public facilities usually makes the difference between mediocre and exceptional development success. Therefore, the influence of real estate interests on public infrastructure planning remains very important.

The preferences and attitudes of other stakeholders also changes, adding to the dynamics of long-term planning. Rapid changes in the urban environment after World War II caused a relatively rapid change in social and environmental policy in the 1960s and 1970s, which has greatly influenced urban land-use planning. Neighborhood stakeholders are concerned about sustaining or improving the social values of their neighborhood that make it a satisfying place to live. Most environmental stakeholders are concerned about the protection or restoration of those ecological services that support a healthy, interesting, and esthetically pleasing environment. The aim of some environmentalists is to stop and even roll back further urban growth, but most groups target more specific concerns. Whereas few people seem willing to pay much for purely nonuse ecological protection for inherent values—so called existence values—the groups fostering protection of these values have grown rapidly in recent decades.

The challenge of urban planning is to plan land use to benefit all stakeholder interests as much as possible within the conflicting demands placed on land use. Although many interests have been included in most recent urban land-use planning, one of three interest areas usually dominate. Thus the development of certain communities is obviously dominated by real estate markets while other development is dominate by urban environmental concerns and by neighborhood protection, restoration, and rehabilitation.

Government Role in Planning

Local government is the primary arbiter for resolving conflict among competing stakeholder interests. However, local governments have only partial authority over regional planning problems. The effectiveness of local urban planning very much depends on the actions of federal and state governments. The diversity of approaches among local governments often subverts any comprehensive planning process that might leap ahead of private land speculation. In small towns and cities, urban sprawl is of little concern. However, the larger cities often extend over different municipal, county, and state governments requiring cumbersome intergovernmental planning commissions.

Regional planning has been most successful where federal and state agencies provide assistance or regulation, as for river-basin water supply, navigation, federal highway, and environmental needs. Respecting states rights, the federal government in the U.S. has done relatively little to regulate urban sprawl, leaving it to local interests to "muddle through." However, the lack of government coordination has aggravated regional land-use problems. For example, federal housing policy inadvertently promoted urban decentralization and associated problems in part to protect the mortgage lenders' interests and to ensure tax collection. The federal investment in superhighway transportation in support of national commerce and security purposes also facilitated local shortsighted planning that contributed to undesirable attributes of urban sprawl. More recently, the federal government has attempted a more positive effect, but usually indirectly through environmental, social, housing, and transportation regulation. The general principle here is to establish national standards for protecting the public interest and allow local flexibility in meeting those standards.

Federal government effects urban change through regulation, providing grants in aid, withholding grants, and through development of national infrastructure. Federal government has been influential by setting and enforcing national environmental standards, providing transportation funding and regulation, and by supporting community development and public housing. In recent years its influence has moved more toward setting standards that must be met and away from direct funding of corrective actions. This has been transferred to the responsibility of states and local government (Kaiser et al. 1995). The Coastal Zone Management Act of 1972 is one example of how federal legislation influences regional planning when natural resources are being severely threatened by reckless development. The Clean Air Act (1970) and Clean Water Act (1972 and 1977) are outstanding examples of influence through regulation and funding for municipal treatment systems. Federal government continues to use grants in aid to encourage local decisions that are in the national interest. The Intermodal Surface Transportation Efficiency Act of 1991, for example, encourages links between land-use plans and transportation plans, environmental protection, energy conservation, and maintenance of transportation infrastructure.

The states are important players whose roles in urban land-use planning has increased in recent years (Kaiser et al. 1995). The states split out into three categories. The first group of states has enough interest in the problems associated with urban sprawl to enact legislation that controls urban growth. This group includes some of the more crowded and fastest growing states as well as states with strong ecological values such as Hawaii, Florida, Oregon, Washington, New Jersey, Vermont, California, Rhode Island, Maryland, and Georgia (Cullingworth 1997). A second group is debating problems, considering about what to do with them, and developing legislation. They include Colorado, Massachusetts, New York, North Carolina, Virginia, West Virginia, and Pennsylvania (Kaiser et al. 1995). Most of these states have legislated some form of comprehensive planning law with goals that municipalities must recognize in their own planning process. However, the effectiveness of these laws varies greatly among states and many are undergoing amendment as the forces of private and public interests vie for satisfactory balance. Even though comprehensive regional planning has grown in influence, there are many forces operating contrary to successful planning processes. Most large, metropolitan areas are continuing to become more socially fragmented as suburban flight continues. Most states either perceive no problems with urban growth, are dominated by philosophies that discourage interference with market-driven development, or are just beginning to feel the consequences of unplanned growth.

Environmental concerns have encouraged a regional integration approach to protecting natural resources while providing for sustainable development. Early guidelines for environment integration into land use were offered by McHarg (1969), who developed a coherent approach to land suitability analysis in land-use planning. His approach examined basic land and water attributes as foundations for environmental development and protection. The carrying capacity of the urban habitat has been analyzed in an approach to estimating limits to growth (Clark 1981). Planning for environmental concerns has resulted in numerous land preservation or regulated-use decisions to establish and maintain nature preserves, watersheds, groundwater aquifers, flood plains, and buffer areas around pleasing vistas, parks, and other protected lands. Recently the comprehensive concept of sustainable development has emerged (The President's Council on Sustainable Development 1996, Roseland 1998). One goal is to identify the level and intensity of land use that can be sustained without environmental damage while providing for present and future economic and social needs. In this view, the costs of protecting and restoring environmental services associated with ecosystem stability and diversity are most likely to be successfully borne by a healthy economy and high quality of life.

MANAGING THE URBAN ECOSYSTEM

Integrating Urban Ecosystem Services

The most challenging part of natural resource management may exist in the urban landscape. U.S. urban ecosystems, including interlinking travel and utility corridors, are among the primary sources of biosphere pollutants and fragmentation of wildlife habitats. Natural ecosystems provide many ecological services that are influenced by how well the urban landscape is designed. Unimpaired land, water, and air provide services that are indispensable for a high quality of life. They provide the oxygen, carbon dioxide, and water molecules essential to all life and assimilate waterborne, gaseous, and solid wastes (Daily 1997). Well-planned natural environments can help assimilate urban wastes in air, water run off, and terrain, and dampen extremes of wind, temperature, and humidity.

Contemporary natural resource concerns often concentrate on the interfaces between urban and wild landscapes where environmental stress is especially acute. Natural resource professionals are involved in the management of public space which provides the supporting infrastructure for the urban landscape—the highways, pathways, waterways, public railways, greenways (natural vegetation corridors), and recreational areas. The urban ecosystem often dominates natural resource development far beyond city limits. Water resources development, for example, typically is oriented either to agricultural or urban landscape needs. Waterway development and maintenance for navigation is as much a travel corridor activity as highway and railway development and maintenance. Most national expenditure for flood damage reduction is justified to protect urban property. Highways, railways, and utilities reach into and through otherwise wild lands to sustain connections among urban centers and to reach remote natural resources destined for urban use. This is the arena of regional planners, landscape architects, highway and waterways engineers, environmental engineers and regulators, public health professionals, and urban forest, wetland, fisheries, wildlife, parks, and recreation managers. They usually work in interdisciplinary planning teams to coordinate their areas of specialization into effective sustained use of natural resources.

The urban environment frequently dominates the regulatory attention of the Environmental Protection Agency, Army Corps of Engineers, Fish and Wildlife Service, and other government agencies concerned with protecting and restoring the quality of air, water, wetland, wildlife, and other natural resources. These activities are carried out under the auspices of numerous federal and state authorities. Some of the more important ones include the Water Resources Planning Act of 1962, Solid Waste Disposal Act of 1965, Clean Air Act of 1970, Coastal Zone Management Act of 1972, Federal Water Pollution Control Act of 1972, Endangered Species Act of 1973, the Water Resources Development Act of 1986, and the Intermodal Surface Transportation Efficiency Act of 1991. The Federal Water Pollution Control Act of 1972 is especially aggressive legislation, which seeks protection and restoration of the physical, chemical, and biological integrity of all aquatic habitats. These and other laws form the policy basis for managing urban natural resources.

Land

Esthetics. Much of the livability in cities is based on the appeal of urban land development, or, as Nasar (1998) refers to it, the city evaluative image. Although the functional performance of the city as a place to live is greatly influenced by the arrangement of structural landmarks, paths, districts, edges, and nodes, the feelings and meanings linked to those structures also are important. The convenience of linkages of courthouses, libraries, schools, museums, monuments, and religion centers by paths (roads and walkways) to residential and business districts is of basic importance. Also important is the degree of transition from one area to another (edge) and the appropriate provision of activity centers (nodes), such as shopping malls, community centers, and entertainment and recreational outlets.

Nasar (1998) found that the "likability" of urban settings depends on a number of criteria including naturalness, upkeep, openness, historic significance, orderliness, and complexity. Most people respond positively to an urban environment that incorporates open vistas, natural areas, and artificial structures in a way complex enough to develop interest, but not so complex as to confuse and threaten. They also like neat and orderly neighborhoods with a sense of continuity among structures including identifiable belonging to a historic period. Esthetic evaluative techniques need further improvement, but show promise for providing needed input information to planning process. Desirable urban land-use patterns needs to include esthetic appeal. This means integrating natural ecosystems and vegetation management into urban landscapes (Figure 14.17).

Transportation. Some of the most heavily used public resources exist in transportation right-of-ways. Congestion alone is estimated to cost $40 billion annually in lost benefits (Cullingworth 1997). Transport and utility corridors linking urban ecosystems form extensive ecological edges and barriers. They may be developed and managed to promote the protection of resources, such as habitat for endangered species, watershed, and wetlands, or they may be haphazardly developed and mismanaged to social detriment. Transportation right-of-ways typically create sharp changes in vegetation and landform, often fragmenting wildlife habitat. They also form conduits for toxic spills, dissolved contaminants, and suspensions of eroded particulates into streams, lakes, wetlands, and coastal zones.

FIGURE 14.17 A public park that is integrated into a primary business district in Oakland, California (courtesy of U.S. government).

Vehicular collision with wildlife escalates where human travel corridors intersect with wildlife travel corridors, posing threats to both wildlife and people. Design that provides safe alternatives for both drivers and animals is an important aspect of contemporary right-of-way management. Development of natural corridors, or "green belts," in urban areas for recreation and other ecological services intensifies the need for careful management of undesirable encounters between nature and humanity. Storms, floods, and other natural stresses frequently exert the greatest damage along travel corridors. The management of natural resource degradation along transport, utility, and natural corridors starts with comprehensive planning based on understanding of the many interfaces between the urban ecosystem and more natural ecosystems.

Natural resource professionals associated with transportation often are involved in meeting resource protection requirements. Through auspices of the National Environmental Protection Act (1969), any projects funded with federal money need to determine the environmental impact and seek the least environmentally degrading choice within similar cost ranges. The Endangered Species Act of 1973 and similar state laws require planning around listed federal and state endangered species. The Clean Water Act of 1972 and 1977 established provisions for protecting national waters and wetlands. It requires biologists, hydrologists, and soil scientists familiar with ecosystem processes to assure regulations are met on all highway development. All unavoidable wetland damage must be compensated through creation or restoration of similar wetlands. Some states have even more stringent laws.

The Intermodal Surface Transportation Efficiency Act of 1991 requires the states to prepare their own transportation plans and integrate environmental considerations into the plan. The plans must address issues such as congestion management, land-use effects of transportation, consistency among transportation and land-use plans, and environmental, energy, and social effects of transportation development. One important section of the act restricts federal funds for highway development where the Clean Air Act is violated. Methods for reducing automobile traffic have to be proposed. Through that act, highway funds are available for mitigating impacts on wetlands and other wildlife habitat, preservation of historic sites, air quality improvement, highway beautification, construction of bicycle and pedestrian facilities, and alternative forms of ground transportation. To get the money, a 25-year transportation plan must be developed.

Urban Outdoor Recreation and Natural Aesthetics.   Urban outdoor recreation includes many of the natural resources services usually associated with park development and management. A short list of recreational opportunities includes development and management of facilities for public swimming, boating and marinas, bicycling, hiking, dog walking, fishing, camping, golf, tennis courts, baseball, football, basketball, lawn bowling, skating, running, archery, horseback riding, motorized off-road vehicles, sight-seeing, nature study, picnicking, and lounging. Many larger park systems attempt to provide most of these opportunities in addition to maintaining heavily used traffic arteries.

The primary challenge for recreational managers is how to develop limited space for diverse use in a safe environment. Many outdoor activities are inherently dangerous to nonparticipants who find themselves in the path of a vehicle, golf ball, boat, arrow, skater, bicycler, or nervous dog. Recreational planners work closely with landscape architects and construction engineers to place tracks, paths, roads, courts, rinks, links, diamonds, lawns, restrooms, benches, waterways, service buildings, vegetation screens, earth barriers, bridges, various service concessions, and other structures in ways that minimize collision of uses while maximizing user satisfaction as a whole. Facilities need to be developed with enough challenge to satisfy skilled users while remaining safe enough for novices. Regulation and direction are unavoidable necessities. Signs need to be obvious yet unobtrusive, informative but concise, and universally understandable for polyglot users.

A variety of specialists are involved with recreational management including urban foresters, horticulturists, fisheries and wildlife managers, environmental scientists, educators, sports trainers, law enforcement specialists, architects, and engineers. They typically provide services for a network of parks and related recreational facilities. Recreational managers often are involved in urban planning. Because air and water quality are especially critical in park settings, they often are trained in environmental assessment.

Urban Wildlife Management. In barren city centers and industrial areas before the turn of the last century, virtually all vegetation was trampled, cleared away, burned for fuel, or otherwise used. Wildlife were reduced to a few highly tolerant undesirable species. Most of these species evolved with people in Eurasia, where they ate their grain and castoffs and nested in or on their buildings. Prominent among them were rock doves (pigeons), house sparrows, starlings, house mice, and Norway rats. They were often joined by feral domestic cats and dogs, which adapted to the frequently abundant rodent and bird populations.

These few species of urban wildlife flourished especially before cars and trucks replaced horses, whose feces contained digested grain that formed the basis of a food chain. Waste grain and other foods also were common at granaries, wharfs, railroad yards, stables, markets, and city dumps. As medical understanding of disease transmission improved, especially for such worrisome diseases as bubonic plague, urban wildlife became more of a concern. Cities began to spend public money in education and control programs. Even after WWII, urban wildlife management frequently connoted public health management.

As open land and natural vegetation has become more integrated into the urban environment, some native forms of wildlife have adapted to the urban scene. Certain wildlife species have always been well adapted to the farm-village habitats of humanity wherever some semblance of natural habitat persisted. However, before hunting was regulated, many species were less common near human habitation than now. Rapid suburbanization following World War II resulted in increased encounters between people and wildlife, especially where large areas of natural vegetation remained among homes and developments. In addition to the Eurasian imports, suburban backyards frequently are visited by robins, house finches, mockingbirds, chickadees, titmice, cardinals, crows, grackles, mourning doves, and more.

Where wild parklands enter city centers, especially along waterways, many other species may be commonly encountered. These include rabbits, squirrels, skunks, weasels, opossums, raccoons, fox, coyotes, deer, muskrats, various bats, turtles, snakes, lizards, various frogs and salamanders, small rodents, and a great diversity of insects and other invertebrates. Where developments have included boat slips and ponds in otherwise suitable habitat, alligators, water snakes, and aggressive turtles are increasingly encountered. Some of the most desirable homes back up to parks and other wild space. In such areas, even wilderness wildlife such as bears, mountain lions, and moose may be encountered.

Urban wildlife have become very abundant in some locations—a boon to many and a nuisance to others. A number of factors have contributed to their increase. Because "green space" is desirable, especially when it threads through cities and connects to wilder lands at city edges, a greater diversity of urban wildlife has become more common. Large areas of "slum" housing were cleared and never redeveloped in some locations. Abandoned industrial sites have colonized with natural vegetation following the closure of many factories. Lethal applications of pesticides are less commonly used now and hunting is outlawed in most urban and suburban locations. Most cities require tight control of dogs.

In addition to the interest they add, urban wildlife cause accidents, damage yards and homes, transmit disease, inflict bodily harm, and frighten some people. Large animals, such as deer and alligators, cause traffic accidents which are sometimes lethal. Beaver have become common once again after near total extirpation in the eastern

U.S. Skunks, raccoons, and possums are frequent visitors to Fido's food bowl on the back porch, where they dump garbage cans and sometimes harm small pets. Even larger pets are sometimes killed and consumed by coyotes, mountain lions, and alligators on suburban margins. Burrowing muskrats break through pond liners and drain them. Moles, muskrats, and mice burrow into lawns. Rodents, deer, and rabbits eat gardens. Children try to catch animals of all kinds, sometimes getting bit in the process. Rabies, although a relatively rare threat, is a continuing concern. Canada geese defecate all over park and yard lawns. Birds are sucked into jet aircraft and occasionally cause a crash. Poisonous snakes are encountered in new developments with a lot of wildland. Even harmless snakes remain a phobia for many people.

The urban wildlife manager uses many tools to develop compatibility of wildlife and human needs. Among the most effective is appropriate design of urban space, which uses knowledge of wildlife and habitats to confine wildlife to desirable areas. Education and, where necessary, regulations are important means for assuring that people behave responsibly with respect to waste, pet, and yard management affecting wildlife pest problems. As a last resort, wildlife management requires direct control by removal of offending wildlife using safe and ethical methods.

Land Wastes.   One of the more vexing problems of urban areas is solid waste disposal (Hough 1995). Solid wastes have increased at a rate greater than the growth in consumer purchasing power. Much of the solid waste problem derives from a life style that places a premium on time savings. Because most people in households now work outside the home there has been a disproportionate increase in pre-prepared and packaged foods. Elimination of disposable diapers alone is estimated to cost $4 billion/yearly, nearly equal to their entire market value (Cullingworth 1997). In addition to the metal, plastic, paper, cardboard, wood, garden, and food wastes generated by the average household, industrial and infrastructural wastes require environmentally compatible disposal. Debris from demolished buildings, roads, and utilities needs to be recycled or eliminated. Fly ash from coal-fired power plants, slag wastes from various industrial processes, and sewage treatment sludge all need disposal. The primary ways to dispose of materials include land dumping, ocean dumping, incineration, and recycled use.

Environmental concerns have greatly altered the way solid wastes are managed. The burning dumps that commonly blackened urban fringes through the 1960s are no longer tolerated under modern environmental laws protecting land, water, and air. Most dumping in marine environments is prohibited except where materials might contribute to development of desired structure, such as artificial reefs. Some landfills used to bury chemical wastes now require expensive treatment or isolation under superfund legislation. Landfills need to be managed to assure groundwater and surface water quality is not degraded. Restored land use must be compatible with the conditions in the landfill (Freedman 1995, Cullingworth 1997). Space suitable for landfills is rapidly disappearing in many urban areas. The NIMBY (Not In My Back Yard) phenomenon is common. Some eastern municipalities ship garbage to western rural locations where space is sold for landfill use. Even rural disposal has become controversial in many locations.

Incineration is safe for uncontaminated organic matter. One of the major problems with incineration of sewage sludge is contamination by toxic metals such as mercury, which can be dispersed through the atmosphere (Freedman 1995). Use of sewage sludge for agricultural soil enrichment is controversial because microbial copper, zinc, cadmium, and nickel contaminants could exist (Freedman 1995). Yard and park organic matter contributes much to solid waste and often is intensified by overuse of fertilizers (Hough 1995).

Recycling has greatest utility where there are associated savings. Recycling is especially profitable for metals, glass, and to a lesser extent for paper products. It also makes sense to recycle cuttings and leaf litter from yards and parks. The old automobile junkyards that once dotted the landscape have for the most part disappeared as the scrap became valuable for metal recycling. Aluminum cans and whole glass containers are much less commonly encountered in litter along highways because of their recycling value. However, for the most part, the bulkiest nonorganic matter remains problematic. In some urban areas that have instituted recycling programs, material is accumulating in storage areas faster than demand can reuse it. However, major advances have been made in recycling and more can be expected in the future.

Hazardous waste is the most difficult form of land disposal problem to manage. Discovery of contaminants leaking into basements from an old waste pit at Love Canal in Niagara Falls, New York, precipitated wider investigation of such conditions around the country. There was, it turned out, no easy way to determine how many such sites existed. Congress passed what has been popularly known as the Superfund Act in 1980 to deal with this problem. It authorized the EPA to develop a priority list starting with 400 sites. Now there is careful documentation of hazardous waste production, use, and disposal designed to avoid future problems. One of the earliest results of the Superfund Act was closure of many landfills too costly to comply with federal law. Instead wastes were treated with the aid of funding under a 1984 amendment. One of the critical provisions of this law is that those responsible for improper hazardous waste would also pay for its cleanup. Since cleanup costs can reach billions of dollars, this has led to contentious and costly legal battles. One of the loopholes in present law is that household wastes are excluded from the hazardous waste category. This makes hazardous materials typical of most households a problem for waste management.

Water

Water Supply. Water supply is as important to city ecosystem function as it is to natural ecosystems. The influence of urbanization on resources far beyond city limits is no where clearer than in managing water supplies for navigation and consumption. Water is used for many domestic, industrial, and outdoor purposes consuming hundreds of gallons per capita per day. Water is needed for cooking, cleaning, drinking, industrial processes, food preparation, irrigation of lawns and gardens, fire fighting, cooling, waste removal, transportation, recreation, and sustaining natural aquatic ecosystems. Clean local sources of water are valuable because water purification, or importing it from afar, are expensive (Perry and Vanderklein 1996). Thus water management has been a dominant theme of urban resource management nearly as long as people have inhabited cities.

Water management remains inadequate in many locations of the world. Over half of the world's population may be affected by waterborne diseases resulting from poor water supply management (Perry and Vanderklein 1996). Even though the worst problems are in underdeveloped countries, water supply continues to rank high among urban resource management problems in industrial nations. Water uses for transportation, waste dilution, and consumption become particularly intense in city areas. Therefore, the acquisition of clean water usually requires that cities go to remote sources. Most sources of water contamination are local, but in regions of sprawling suburbia and other intense land use, contamination may originate in remote parts of watersheds. The cumulative effects of rural and urban water contamination have increased costs of treatment and acquisition of clean water. Among the more difficult problems caused by navigation is contamination from oil and derivatives, most of which occur from normal traffic. Occasionally disastrous large spills occur, such as the disastrous *Exxon Valdez* spill in Alaska coastal waters, which had widespread economic and ecological impacts. Millions of gallons of fuel are transported annually on railroads and roads to urban destinations and threaten spills into lakes, rivers, and estuaries.

Water dirtied by inorganic and organic sediment needs to be treated by settling and filtration through beds of fine sand or other media (see Chapter 7). Disinfection is done by chemical oxidation, usually with chlorine or ozone. Culturally eutrophied water often has a fishy or musty taste. This is difficult to remove through physical and chemical engineering and prevention in the first place often is more cost-effective. Water that is contaminated with metals, nitrates, arsenic, and synthetic compounds requires additional expensive treatment through chemical precipitation and specific filtration and absorption media. Increasingly, contamination from organic compounds, such as gasoline and oil, is being treated with bacteria that decompose the contaminants to harmless elements.

Urban water management is greatly influenced by federal and state laws. The federal Clean Water Acts of 1972 and 1977 have been especially important for establishing the ultimate objectives of water quality management and regulation. The acts seek "restoration and maintenance of the physical, chemical, and biological integrity of the nation's waters." The 1977 act established protection of wetlands, which now requires management for no net loss of wetland function and value after centuries of draining and filling. Some cities have integrated wetland protection with water treated by managing the natural capacity of wetlands to trap nutrients and other wastes (Figure 14.18). Because many urban areas are situated near water and wetlands, the Clean Water Act will have an important impact on future land use and urban development. These acts have had positive influences, mostly through regulation and financing massive development of waste treatment plants. Still, many plants are inadequately funded for optimum operation (Cullingworth 1997, Adler 1993).

Groundwaters: Because it has been naturally filtered and only rarely contaminated by natural sources of pollutants, unimpaired groundwater is especially valuable wherever it is close to the surface and in large aquifers. However, many groundwater aquifers have been contaminated, especially near the ground surface (Perry and Vanderklein 1996). Drilling and pumping from deeper groundwaters are more expensive. Demands on local groundwater and surface water have exceeded supply in many large metropolitan regions, requiring conservation practices, recycling, and importation of clean water from outside city limits. Overpumping at rates faster than new water can percolate into the aquifer results in lowered water tables, higher pumping costs, and eventual use of the entire available aquifer. Overpumping near ocean coasts results in salt water intrusion, which is costly to desalinate. It can cause land subsidence leading to building and infrastructural damage. Other sources of salinity contamination of groundwater include improper use of deicing salts and

FIGURE 14.18   Created wetlands in urban areas are increasingly being used to provide recreation, wildlife habitat, and water treatment facilities, often with several integrated services (courtesy of U.S. Dept. of Agriculture).

agricultural irrigation in dry environments. Overpumping also can dry up wetlands and springs that serve diverse needs, including support of threatened and endangered species. Unlike federal regulation of surface waters, groundwater regulation is left up to the states.

Chemical contamination of groundwaters in urban areas can come from many sources (see Chapter 7). Common ones are from septic tanks, industrial waste disposal lagoons, urban landfills, mines, buried storage tanks for chemical wastes, fuels from gas stations and home heating, and chemical spills of diverse kinds. Hydrocarbon compounds are the largest type of pollutant because they are so widespread in our internal combustion culture. Nitrate fertilizer seepage from local lawns or nearby agricultural use also is a source of groundwater pollution. Sewage, gas, and other pipes collapse and leak, especially in older urban areas. Bacterial contamination from broken pipes, old septic systems, past landfills, and graveyards is a major problem in some cities. Abandoned and poorly maintained wells can be a route of contaminated flow into groundwaters. Aquifer protection is an important concern for many urban areas. They usually approach these problems through setting standards for quality, regulating land management practices, and forcing polluters to pay cleanup costs. Especially sensitive aquifers include those that lie very close to the surface, are in karst formations of fissured limestone, and areas of volcanic rock.

Surface waters: Another cheap source of clean water is from groundwater that has emerged through springs into streams and lakes. Many urban areas long ago contaminated their local surface waters. Land purchase and water transport to the city was cheaper than intensive clean up. Because much of New York City is on impermeable rock with little groundwater, it has purchased many acres of undisturbed watersheds and water supply reservoirs. Urban water-supply watersheds and reservoirs usually are protected from any degrading use and managed for compatible uses, such as wildlife refuges.

But many urban areas have little recourse other than to use waters that have been affected by local land use. Thus waste treatment required in the Clean Water Act contributes to water treatment savings in downstream communities. Threats to continuous supply of urban water during droughts have increased as regional urban development has intensified. Another reason for investing in groundwater for urban use is its dependability during drought. For those many areas of the country that must augment municipal water supplies, the protection of water yield from watersheds is critical.

In arid lands of the western U.S., cities compete for water with agricultural users. Water law in most western states treats water as a private good associated with irrigatible land (see Chapter 7). Cities have bought control of watersheds or the purchase of water rights mostly from agricultural users. Los Angeles has been particularly dependent on water produced elsewhere. It began buying and diverting water from the Owens River valley some 200 miles away early in the twentieth century. Resulting problems with dust pollution in what used to be Owens Lake may ultimately result in a return of some water to control dust pollution. This follows a California Supreme Court decision with respect to Los Angeles' right to acquire water rights and divert water that would damage the Mono Lake ecosystem, a unique wildlife habitat. In so doing, the court said all water rights remained subject to review and revision to assure the public trust is protected (Getches 1993).

Waste Assimilation, Transport, and Contamination. Untreated wastes used to be dumped directly into streams, lakes, and ocean coastal waters. Systems of drains were developed in many cities simply to collect sewage for dumping directly to natural waters without treatment. With growing medical recognition of waste hazards in the last century, urban sewage systems began to appear in larger cities. In more rural areas, individual septic systems were used to treat wastewater. However, as rural population density intensified, leakage and contamination from individual systems became a widespread problem, requiring extension of centralized treatment systems into the urban sprawl. By the 1920s and 1930s, there was growing scientific awareness of the extent of water contamination from organic wastes. Many miles of urban and industrial waters were devoid of intolerant organisms including virtually all fish. The first treatment plants were designed to remove organic waste and to chlorinate for control of pathogens, but much of the organic matter remained in the wastewater. Because organic waste treatment focused on oxygen restoration, the Biological Oxygen Demand (usually referred to as BOD) was used as the gauge for treatment need and success.

After World War II industrial growth and increasing use of chemicals for various purposes, including fertilizer applications, contributed to rapidly worsening water quality. Conditions in some urban river settings were so bad from oil pollutants that they caught fire—the most infamous example occurring in downtown Cleveland, Ohio. Inorganic nutrients remained a problem in treated sewage and in urban runoff, contributing to cultural eutrophication (nutrient enrichment). This caused excessive water "blooms" of phytoplankton, many of which were foul tasting, somewhat toxic, and further depleted oxygen. Following the Clean Water Act of 1972, the U.S. embarked on a mission of complete cleanup, but was most successful with point sources of such pollutants. The EPA administered a construction program for sewage treatment plants with greater capacity to remove BOD and inorganic nutrients, such as phosphorus and nitrogen. In some locations additional treatment is needed to protect pristine environments such as Lake Tahoe in Nevada. Wetlands have been shown to be effective in waste water "polishing" (Mitsch and Gosselink 1986). Although there has been much improvement in waste removal from natural waters, contamination still remains and many of its sources are urban.

Urban Runoff.   Surface runoff from rain and snow in over half the urban areas is drained away from the sewage treatment systems to natural drainages (Perry and Vanderklein 1996). Runoff from streets, parking lots, roofs, and lawns is contaminated with a variety of nutrient and toxic materials. Lawn pesticides and fertilizers, waste oil and gasoline, road salts, vehicle emission particles, various solvent cleaning agents, inorganic suspended solids, heavy metals, and various pathogens all are flushed into storm drainage systems. Combined storm water and sewage systems exist for about 20% of the U.S. population compared to more than half in Europe. These systems are rarely built with the capacity to contain excessive storm runoff. The treatment plants often spill untreated sewage. Whenever it rains over 0.75 inches in New York City, for example, 500 million gallons of untreated sewage is flushed into the receiving waters (Perry and Vanderklein 1996). The most common approach to managing urban runoff is detainment in settling basins and to encourage infiltration into the soil. These need to be large enough to hold water for a day or more to have significant settling effect (Perry and Vanderklein 1996). Percolation basins also risk groundwater contamination if not carefully placed. Another strategy gaining attention is to reduce the extent of continuous impervious surface by using porous concrete, grass swales in place of concrete gutters, and more natural material spaced intermittently with concrete surfaces.

Waste Heat.   Thermal pollution occurs when heated water is released to the environment in a way that reduces the functioning of aquatic ecosystems. The largest source of waste heat is from cooling systems of steam-generated electric power plants. Cooling water effluent is either mixed rapidly into large water bodies or is enclosed in cooling ponds to avoid exposure of natural ecosystems. Cooling water ponds often require barriers of various kinds to prevent fish impingement and death. Attempts have been made to use heated water for aquaculture and other low-grade uses of heat, but with marginal success. Cooling ponds are more readily included in some forms of recreational use, such as fishing and waterfowl management.

Flood Damage Control and Navigation and Reclamation. The U.S. Army Corps of Engineers (the Corps) is the lead agency authorized to promote river, harbor, and intracoastal navigation for interstate shipping. This authority extends back nearly two centuries when snag removal began mostly in the Mississippi River and tributaries. Water transport remains the most cost-effective means for moving bulk extractions of natural resources, such as coal and grain, to urban destinations. The Corps has developed and operates an extensive system of locks, dams, and levees that promote navigation deep into the heartland along the nation's waterways. In association with that system, it oversees sediment dredging from thousands of miles of river, harbor, and intracoastal canal. It also oversees modification of navigable waters and wetlands to assure no further net loss of wetland functions and values (Clean Water Act of 1977).

The Corps became the lead agency in flood prevention as a natural complement to its navigation responsibilities. A system of over 500 flood-control reservoirs and flood-protection levees was developed over less than a century. In addition, the Corps sustains a coastal-flooding control program including construction and maintenance of both artificial and natural protective structures. Most of the larger urban areas of the U.S. are affected in some way by flood-prevention engineering provided by the Corps. In recent years, the Corps has begun to emphasize more nonstructural solutions to flood damage, such as relocation of people out of particularly troublesome flood areas (Figure 14.19). Beatley and Manning (1997), however, note that trade-offs of moving out of the floodplain need to be carefully evaluated and negative effects mitigated in the settlement area. Related to this is an increasing emphasis on more natural solutions to flood damage reduction, such as wetland and barrier beach restoration. Environmental improvement became equally important in the Corps's mission during the past decade. The Corps increasingly seeks integration of natural process with engineering to provide navigation, flood-damage reduction, and other environmental benefits to both urban and rural environments. For example, the Corps is leading a number of watershed studies designed to improve environmental services and administers funding to restore ecosystems in partnership with local project sponsors.

The term reclamation has come to mean transforming natural lands for a more productive use as defined at the time. Many cities located on rivers, lakes, estuaries, and oceans have expanded over wetlands (Hudson 1996). Although much of that expansion appeared warranted at the time, at least some of it seems regrettable, given concerns about climate change and earthquakes in many areas. Recent earthquakes in Japan (Kobe) and the San Francisco Bay area were exacerbated because of the effect earthquakes have on the liquefaction of filled wetlands and water bodies. If

FIGURE 14.19 These poorly located homes were subjected to periodic flooding. In minimally developed areas especially, relocation away from the floodplain is now being advanced by the U.S. Army Corps of Engineers as an alternative to physical control methods (courtesy of U.S. government).

global warming materializes as many scientists now suspect, rising water levels could be especially troublesome to urban development over low-lying filled areas (Nasar 1998). Whereas the U.S. has attempted to resist further net loss of wetlands, through the Clean Water Act of 1977, further filling for urban expansion is anticipated and will be an increasingly contentious area of natural resource management.

Aquatic Resources. Recreational fisheries is a specialized area of urban natural resource management. Urban fisheries are unique only because they are so intensively used. They typically rely on continuous stocking to sustain the catch. Rainbow trout and channel catfish are the most common species stocked in urban waters. However, many urban anglers fish in adjacent large rivers, estuaries, and coastal waters for wild fish. Where intense enough, fishing may require special restrictions so as not to deplete the fishery. Urban fisheries programs by and large are maintained by state agencies who coordinate with municipal recreational programs.

Environmental problems often are of greater concern than overfishing. Many fish species are contaminated with metals and synthetic organics associated with urban wastes. In coastal waters, the U.S. Environmental Protection Agency and the National Marine Fisheries Service interact with state agency programs to assure minimum contamination and healthy populations. Inland, the U.S. Fish and Wildlife Service complements the National Marine Fishery Service. Certain shellfish, such as oysters, are especially likely to concentrate microbial contaminants from incompletely treated fecal wastes. Contaminants also may limit reproductive success of certain species. For coastal anadromous species, dams remain a barrier to successful reproduction in many locations. Many commercial and recreational species are associated with wetlands at one or more stages of their lives. This is one of the reasons why wetlands now receive added protection among the diverse ecosystems in the U.S. The Clean Water Act requires compensatory mitigation of unavoidable wetland degradation or destruction.

Air

Sunlight, Wind, Heat, and Humidity. Like natural ecosystems, cities absorb sunlight via concrete, roofing material, tar, asphalt, metal, stone, wood, open water, bare soil, and through the photosynthesis of vegetation in lawns, gardens, and natural areas. Cities form geographic heat islands (Bridgeman et al. 1995), which also influence rainfall patterns (Landsberg 1981). Water can store higher quantities of heat without raising temperature as much as most other materials. Therefore, cities with lots of natural vegetation and water remain cooler than cities without. Trees may reduce the need for electricity by 200 billion kilowatt-hours per year in the U.S. by reducing air-conditioning needs (Committee on Science, Engineering, and Public Policy 1992). Depending on form, urban structure can substantially reduce cooling breezes that flush warm air from hot city landscapes or inversions of cold air during winter. Stagnant air accumulates exhausts from city functions including heating and cooling systems, vehicles, and manufacturing processes (Landsberg 1981, Bridgeman et al. 1995). Other city forms can funnel moving air into high winds between tall buildings. Wind movement over water also increases cooling by evaporation as it replaces humid air with drier air. Because natural vegetation is mostly water and air, lawns and gardens maintain cooler temperatures and greater humidity at the earth's surface than mineral surfaces (asphalt and concrete). Stores and restaurants in Phoenix, Arizona, create a curtain of fine mist around outdoor walks and patios to cool and humidify them. Rising hot air from cities with little natural vegetation may add to convection and result in more rain downwind.

Elementary understanding of urban climate suggests it might be managed by improved integration of naturally vegetated and aquatic environments into urban design. An urban ecosystem in a forested biome typically elevates mean temperature several degrees above that of surrounding natural forest. In contrast, a suburban area in a desert biome may be cooler than the natural ecosystem. Nonvegetated urban areas typically have lower humidity than vegetated areas (Landsberg 1991, Bridgeman et al. 1995). Integration of vegetation and water may provide humidity and air movement closer to optimum levels for human activity than surrounding drier or wetter ecosystems. By supplying water to desert urban landscapes, the resulting trees, gardens, lawns, and surface water cool and humidify the urban landscape. Because of the space dedicated to concrete, asphalt, and other building materials in a wetter urban landscape, the city is warmer but possibly less humid than surrounding forest ecosystems, depending on airflow through the city.

Auxiliary Forms of Energy.   Solar energy provides light for an average of half a day—an inconvenience for an animal that requires about eight hours of sleep and cannot see well in the dark. Also, the cavelike structures that humans typically inhabit often are dark even during daylight. Until the invention of electric light by Edison about a century ago, auxiliary light was provided by an open flame. Flames enclosed in furnaces and stoves have been effective means for heating buildings and water, generating steam, and cooking foods. They continue in such service fueled mostly by natural gas, coal, oil, and other fossil fuels, which must be imported from remote sources by the city. Industrial and home use of oil, wood, and coal for heating caused massive emission of incompletely burned particulates until conversion to natural gas mostly during the last few decades.

Open flames are inefficient sources of light and risk fire much more than electric light, which was adapted to city illumination soon after invention. Electricity also was an effective power supply for household appliances. Electric fans were among the more important inventions for encouraging greater evaporative cooling. Two of the most important appliances are electric refrigerators and air conditioning. They did not become widely affordable until after World War II. They accelerated growth of the southern cities in the U.S. as urban growth slowed in the northern cities. Before World War II, the icebox was a primary means for storing food. Electricity is now so widely used to power appliances, it is difficult to imagine household life without it. The urban demand for electricity has increased exponentially since its invention.

Atmospheric Contamination.   Boorstin (1973) recounts an exuberant promotion of the purity of Los Angeles' air in 1874. How things have changed! Before improvements in the 1960s and 1970s, air quality in many urban areas often was unhealthy and sometimes deadly (Figure 14.20). It was not uncommon in some of the industrial cities of the northeastern U.S. for street lights to remain on most of the day because of the darkness caused by aerial particulates (Freedman 1995). The widespread burning of coal especially contributed to particulate pollution as well as to gaseous emissions. Coal heated homes and work places, provided fuel for electric power generation, and was the main energy source for heavy industry and the railroads. Particulate conditions have much improved in the U.S. as employment shifted away from heavy industry to light industry; cleaner atomic energy, fuel oils, and gas replaced coal; and regulation of atmospheric emissions became more technically feasible and effective. The particulate atmospheric problems of a half century ago in the U.S. still persist in developing countries. Heavy industry is now concentrated in these countries and they continue to rely on dirty fuels with little regulation of emission.

The demand for cleaner sources of power for electric generation and heating, such as wind and solar energy, was thought to be incentive for altering the scope of cities, returning them to walking size (Goldfield 1977). However, recent history indicates no such trend. The emphasis instead has been on centralizing electricity production in large power plants and cleaning up their emissions. Other conservation measures include extending natural gas supply plumbing to households and workplace, and emphasizing energy conservation through building codes and economic incentives. Power plants, factories, and automobiles are the major sources of air quality problems.

FIGURE 14.20 Pittsburgh at 11:00 A.M. in 1945 darkened by industrial air pollution. Clean air legislation following World War II eliminated this extreme form of particulate air pollution (courtesy of Allshevy Conference in Community Development).

The major gases produced are carbon monoxide (CO), carbon dioxide (CO_2), sulfur dioxide (SO_2), hydrogen sulfide (H_2S), ammonia (NH_3), nitric oxide (NO), nitrous dioxide (NO_2), nitrous oxide (N_2O), ozone (O_3), and various hydrocarbons. Carbon monoxide ranks highest among U.S. toxic emissions. It is a highly toxic by-product of fossil fuel combustion, which is oxidized rapidly to carbon dioxide and rarely becomes life threatening in the open environment. Carbon dioxide is generated from combustion of organic matter in any form and from respiration of plants. It is also absorbed by plants and fixed in biomass. Thus, widespread removal of plant biomass contributes to the reservoir of carbon dioxide in the atmosphere. Global emissions of carbon dioxide from fuel combustion have increased by a factor of more than 40 since 1860 (Freedman 1995). The reduction in forest biomass and combustion of carbon dioxide have about equally contributed to the observed increases (see Chapter 6).

The accumulation of carbon dioxide is believed to be causing the greenhouse effect (heat increased in the atmosphere), resulting in rising global temperatures (Freedman 1995). Other gasses, mostly from urban sources, may also contribute to the greenhouse effect. There remain many unanswered questions about the link between greenhouse gasses and temperature change on the earth's surface. Depending on the extent to which the earth actually warms from future green house gas emission, sea levels, agricultural production, natural fire occurrence, biodiversity, and many other earth attributes could be effected. Carbon dioxide can in theory improve agricultural production; therefore, all of the impacts are not necessarily negative and the implications are an active area of inquiry.

Sulfurous gasses contribute to acid rain and are toxic enough to slow plant growth. The sulfur gasses are emitted from fossil fuel combustion. Anthracite coal has a particularly high concentration and western softer coals have lower contents. Oil products have somewhat lower concentrations. Power plants and manufacturing are particularly important sources of sulfur gasses although some sulfur is removed from smokestack effluents to prevent formation of acid rain. Concentrations of sulfurous gasses average about 500 times greater in urban areas than in remote locations

FIGURE 14.21 Although improved since the 1960s, when this picture was taken, smog remains an urban problem most associated with automotive emissions. Cities close to higher elevations are most likely to accumulate smog from automobile and other sources because atmospheric circulation is reduced by topography and thermal inversion (courtesy of U.S. government).

(Freedman 1995). Because of high emission stacks and changing winds, the effects of emission gasses are typically dispersed over relatively large areas.

The largest sources of nitrogen gasses and hydrocarbons are associated with natural outgassing processes. However, the internal combustion engine has added to this atmospheric emission and causes especially high concentrations in urban areas (Figure 14.21). Nitrogen gasses can cause some toxic response in plants, but usually at higher concentrations than observed even in dense urban areas. Although various hydrocarbons emitted from both natural and human sources have little direct effect on plants or animals, they react in the presence of sunlight to form ozone, which is much more toxic to plants. Gaseous pollutants are less threatening to animals and humans if the atmosphere circulates well. Irritation and health threats to vulnerable people increase sharply when thermal inversions occur and emissions become trapped. Local smog is especially a problem for cities near mountains from which cool air drains and circulates to form thermal inversions.

When electricity is used, it degrades to heat and contributes to the warming of the urban environment above background levels. Although the technology now exists to convert solar energy to electricity (Chapter 20), alternative sources remain cheaper and more widely used at this time, including internal combustion, steam, wind, or water power. Of those, natural resource availability and economics has favored steam generation. One advantage to steam generation is that it can be located almost anywhere there is a source of cooling water. Many older plants were located close to urban users to reduce loss of electricity through line transmission. Unlike wind and water power, steam generation requires a massive supply of cooling water to condense and regenerate the steam in the cycle that powers the generators. The waste heat is eliminated either directly to aquatic ecosystems, or to a closed system with or without large evaporative coolers. Electric utilities have heavily invested in the reduction of thermal pollution and beneficial use of waste heat. It is sometimes used for winter heating. Some power plant cooling systems have been integrated into other urban services, such as urban sport fisheries and aquaculture. There remain many more potential applications of thermal heat.

Artificial power production requires massive material input in the form of fuels and elimination of material by-products. Numerous metals also contaminate the emissions of fossil fuel power plants, waste incinerators, and internal combustion engines of automobiles and other sources (Freedman 1995). Toxic metals such as cadmium, chromium, copper, nickel, vanadium, and zinc concentrate near heavily used roads and may be dispersed some distance. Before lead was removed from most gasoline in the late 1980s, it was dispersed to remote locations before being washed out of the atmosphere by rain. Mercury is a particularly troublesome toxin because it can be magnified in human food webs to quite high concentrations. It is a natural contaminant of coal, and is distributed through the manufacturing processes. Mercury and other metals accumulate in sewage sludge, some of which is incinerated.

REGIONAL PLANNING CHALLENGES

Urban resource management has historically relied on the assimilative capacity of land, water, and air to absorb the impact of urban waste products in the form of waste gasses, liquids, and solids. In recent decades, it has become increasingly clear that the carrying capacity of the urban environment is insufficient for assimilating wastes. As pollutants degraded the quality of natural resources both locally and in other urban areas, the need for state and federal direction grew and environmental laws rapidly proliferated. The strength of federal environmental law came about in part because the development interests in the states saw pollution control as a local competitive disadvantage unless the laws were universally applied.

The environmental laws have been based mostly on the effectiveness of regulating emissions to sustain resource quality. Although there has been substantial success in the U.S., there remains substantial need for further improvement. Ozone concentrations, for example, are nearly universally in violation of the Clean Air Act (Cullingworth 1997). The high acid rain sources, mostly power plant emissions of the north central U.S., heavily contribute to acidification of waters in the northeastern U.S. and Canada. They have relatively little local effect because of naturally alkaline environments that buffer their effects. A continuing debate is over who should pay for the additional costs of burning low-sulfur coal. In water pollution, about one-third of the nation's waters do not meet standards. Integrating the operations of 59,000 water supply utilities, thousands of local governments, diverse agencies in 50 state governments and at least 27 federal agencies in water policy is a forbidding challenge (Smith 1995).

In closing, we believe improvement in atmospheric conditions will be made through several complementary strategies including regulation, greater conservation of materials use, recycling, and land-use planning. These strategies will encourage less fossil fuel use and provide for greater ecological services in air, water, and land quality. Those strategies need to be approached through regional planning processes that identify who will benefit and who should bare the costs of corrective actions (see Cullingworth 1997).

LITERATURE CITED

Adler, R. 1993. *The Clean Water Act: 20 years later.* Washington, DC: Island Press.

Beatley, T. and K. Manning. 1997. *The ecology of place: Planning for environment, economy and community.* Washington, DC: Island Press.

Boorstin, D. 1973. *The Americans: The democratic experience.* New York: Random House.

Bridgeman, H., A. Warner, and J. Dodson. 1995. *Urban biophysical environments.* New York, NY: Oxford University Press.

Caro, R. A. 1974. *The Power Broker: Robert Moses and the fall of New York.* New York, NY: Vintage.

Clark, J. 1981. The search for natural limits to growth. In Judith de Neufville, ed., *The land use policy debate in the United States.* New York: Plenum Press.

Committee on Science, Engineering, and Public Policy. 1992. *Policy implications of greenhouse warming.* Washington, DC: National Academy Press.

Cullingworth, B. 1997. *Planning in the U.S.A.: Policies, issues and processes.* New York, NY: Routledge.

Daily, G. C., ed. 1997. *Nature's services: Societal dependence on natural ecosystems.* Washington, DC: Island Press.

Dramstad, W. E., J. D. Olson, and R. T. T. Forman. 1996. *Landscape ecology principles in landscape architecture and land-use planning.* Washington, DC: Island Press.

Fowler, E. P. 1992. *Building cities that work.* Montreal, Quebec, Canada: McGill-Queens University Press.

Freedman, B. 1995. *Environmental ecology: The ecological effects of pollution, disturbance, and other stresses.* 2nd edition. New York, NY: Academic Press.

Fries, S. D. 1997. *The urban idea in colonial America.* Philadelphia, PA: Temple University Press.

Garvin, A. 1996. *The American city: What works, what doesn't.* New York: McGraw-Hill.

Getches, D. H. 1993. Water resources: A wider world (pp. 124–161). In Lawrence J. MacDonald and Sarah F. Bates, eds., *Natural resources policy and law: Trends and directions.* Washington, DC: Island Press.

Goldfield, D. R. 1977. Recovering the american city: Minneapolis as a case study (pp. 456–497). In J. Cairnes, Jr., K. L. Dickson, and E. E. Herricks, eds., *Recovery and restoration of damaged ecosystems.* Charlottesville, VA: University Press of Virginia.

Hough, M. 1995. *Cities and natural process.* New York, NY: Routledge.

Houghton, J. T., L. G. Meira Filho, B. A. Callander, N. Harris, A. Kattenberg, and K. Maskell, eds. 1996. *Climate change 1995: The science of climate change.* New York, NY: Cambridge University Press.

Hudson, B. J. 1996. *Cities on the shore: The urban littoral frontier.* New York, NY: Pinter.

Jakle, J. A. 1990. Landscapes redesigned for the automobile (pp. 293–310). In M. P. Conzen, ed., *The making of the American landscape.* New York, NY: Routledge.

Judd, D. R. 1988. *The politics of American cities: Private power and public policy.* 3rd edition. Glenview, IL: Scott Foresman.

Judd, D. R. and T. Swanstrom. 1994. *City politics: Private power and public policy.* New York: HarperCollins.

Kaiser, E. J., D. R. Godshalk, and F. Stuart Chapin Jr. 1995. *Urban land use planning.* 4th edition. Urbana, IL: University of Illinois Press.

Kunzler, J. H. 1996. *Home from nowhere: Remaking our everyday world for the twenty-first century.* New York, NY: Simon and Schuster.

Landsberg, H. 1981. *The urban climate.* Volume 28, International Geophysics Series. New York, NY: Academic Press.

Langdon, P. 1994. *A better place to live: Reshaping the American suburb.* Amherst, MA: University of Massachusetts Press.

Lemann, N. 1991. *The promised land: The great black migration and how it changed America.* New York, NY: Alfred A. Knopf, Inc.

McHarg, I. 1969. *Design with nature.* Garden City, NJ: Doubleday.

Mitsch, W. J. and J. G. Gosselink. 1986. *Wetlands.* New York, NY: John Wiley & Sons.

Morris, A. E. J. 1994. *History of the urban form: Before the industrial revolution.* New York, NY: Longman Scientific and Technical and John Wiley and Sons.

Muller, E. K. 1990. The Americanization of the city (pp. 269–292). In M. P. Conzen, ed., *The making of the American landscape.* New York, NY: Routledge.

Mumford, L. 1961. *The city in history: Its origins, its transformations, and its prospects.* New York: Harcourt Brace (Penguin edition).

National Research Council. 1992. *Restoration of aquatic ecosystems.* Washington, DC: National Academy Press.

Nasar, J. L. 1998. *The evaluative image of the city.* Thousand Oaks, CA: Sage Publications.

Perry, J. and E. Vanderklein. 1996. *Water quality: Management of a natural resource.* Cambridge, MA: Blackwell Science.

Roseland, M. 1998. *Toward sustainable communities: Resources for citizens and their governments.* 2nd edition. Stoney Creek, CT: New Society Publishers.

Smith, Z. A. 1995. *The environmental policy paradox.* 2nd edition. Englewood Cliffs, NJ: Prentice-Hall.

Steiner, F. 1991. *The living landscape: An ecological approach to landscape planning.* New York: McGraw-Hill.

The President's Council On Sustainable Development. 1996. *Sustainable America: A new consensus for prosperity, opportunity, and a healthy environment for the future.* Washington, DC: U.S. Government Printing Office, Superintendent of Documents.

Ward-Perkins, J. B. 1974. *Cities of ancient Greece and Italy: Planning in classical antiquity.* New York, NY: George Braziller.

Zelinsky, W. 1990. The imprint of central authority (pp. 311–334). In M. P. Conzen, ed., *The making of the American landscape.* New York, NY: Routledge.

SECTION 4

The Wild Living Resources

The wild living resources of U.S. landscapes include all of the organisms that contribute to what we know as wildlife, fisheries, and natural biodiversity. They are characterized by life in an undomesticated and mostly uncontrolled wild state. Whether they are highly mobile or not, wild living resources live independently of human husbandry and culture. Unlike the land-based soil and vegetation resources, which can be privately owned, most wildlife and fishery resources of the U.S. are public resources managed for an increasing variety of public interests. Chapter 15 treats wildlife resources and their management and discusses the often confusing and incomplete distinctions made among species' assemblages comprising wildlife, fisheries, and biodiversity. In practice and convention, if not in professional declaration, wildlife are mostly terrestrial organisms and fisheries, treated in Chapter 16, include mostly aquatic organisms. Threatened or endangered life forms contributing to biodiversity become part of the public trust once they are protected under provisions of the Endangered Species Act, which is the most important legislation protecting biodiversity in the U.S. Although wildlife and fisheries include species that contribute to biodiversity, treated in Chapter 17, the concept of biodiversity is an all inclusive one that extends to life processes of all kinds and at all organizational levels. Biodiversity management is now closely linked with ecosystem management, a concept developed in the last section. Wild living resources are intimately connected to the land through its soil, water, plants, and other features. These combine to create essential habitats, without which no wild organisms can live independently. Because habitat often is privately owned and used, the fate of wildlife, fisheries, and biodiversity frequently depends on public regulation of private land and water resource use. This critical link between life and its habitat places the management of habitats supporting wildlife, fisheries, and biodiversity in the center of a growing political storm over the sometimes divergent needs of public and private interests. Very little natural resource management takes place today without in-depth consideration of the wild living resources.

CHAPTER 15

Wildlife Conservation and Management

INTRODUCTION

In recent years, concern for wildlife has grown throughout the world. Sustaining global wildlife resources may be among the biggest challenges confronting humanity in the twenty-first century. In this chapter, we will identify and discuss what we consider to be primary wildlife challenges and conflicts. Past and present approaches to wildlife management will be addressed, along with the historical perspective. In this chapter we emphasize wildlife habitat sustainability and management and refer the reader to Chapters 3, 7, 10, 11, 12, 14, 16, and 17 for other relevant information. We also refer the reader to Scalet et al. (1996), Bolen and Robinson (1999), and Anderson (1991) for complete and detailed coverage of wildlife ecology and management.

WILDLIFE VALUES AND CONFLICTS

The Controversial Resource

Wildlife conservation and management programs have long aroused political and socioeconomic controversy. Differing human attitudes and disparate values attached to wildlife create contentious issues and management challenges. Surveys show that most Americans derive positive benefits from wildlife of some type through observation, photography, hunting, trapping, feeding, and other recreational or commercial use. Some of those same people, however, complain about health threats and property damages from some of the same species in a different setting. Some species are consistently reviled by most of the public. Other species have been elevated to valued status by official listing under the Endangered Species Act. Some Americans question the rationale of stopping construction projects worth millions of dollars to save obscure mice and toads, or banning timber harvest on thousands of acres of western forests to protect an owl. Cattlemen and government agencies have engaged in long and costly court battles over the reestablishment of endangered wolves in Yellowstone National Park (Figure 15.1). Yet, despite continuing attempts to weaken or do away with the Endangered Species Act, the majority of the public resists any modification. The animal

FIGURE 15.1   Restoration of wolves in areas where they were extirpated has been contentious because of the differing values attached to wolves by sectors of the American public (courtesy of Lynn Starnes).

rights movement goes beyond population protection to protection of the individual animal. They oppose the use of animals in research, the fur industry, recreational hunting, or other activities that might result in animal pain, suffering, and death.

Although most Americans no longer depend on wild animals for their livelihood, wildlife continues to be very much a part of their cultural, political, judicial, and economic life (Kellert 1996). The common controversies caused by wildlife derive from an incomplete understanding of wildlife management needs by people who are both positively and negatively affected by wildlife. This incomplete understanding derives from both economic and ecological ignorance and sometimes results in policy inconsistency. Wildlife are controversial in part because human relationships with wildlife have changed rapidly and unevenly across society. Conflicts can even arise between groups holding positive but different wildlife values such as hunters and animal rights groups, as well as between groups in conflict over the benefit and the damage associated with wildlife. Recent suburban sprawl into wildlife habitat has established a new awareness, respect, and fear for the damage wildlife can cause (see Chapter 14). Because wild animals are a publicly owned resource in the United States, the resolution of conflicts over them typically involves government employees charged with their management (Grosse 1997).

Wildlife Authority

Most of the laws now pertaining to hunting and other wildlife-related regulations are set by the respective states, which have jurisdiction over resident wildlife. State wildlife agencies are usually directed by a commission of one to several members, usually appointed by the governor. The commissioners oversee the administrative organization of the state wildlife agencies, which are headed by directors typically hired by the commission. The states set seasons, limits, and license fees for harvesting game birds, mammals (including furbearers), and fish.

Federal agencies have regulatory powers over migratory birds, managing national refuges, coordinating endangered species programs, administering federal aid to states, and negotiating international wildlife agreements. The lead federal wildlife agency is the U.S. Fish and Wildlife Service within the Department of the Interior. One of its important responsibilities is managing the 452 federal refuges located in 49 states with a total area of 92 million acres. Refuges had 30 million visitors in 1996 (Wildlife Management Institute 1996). The National Marine Fisheries Service within the Department of Commerce is the major agency responsible for conducting research on and managing marine vertebrates (fishes, whales, and seals, among others), and marine invertebrates. It is the lead agency in managing offshore wildlife resource development. Most of the wildlife on other public lands are the responsibility of the Bureau of Land Management (Department of the Interior), the U.S. Forest Service (Department of Agriculture), and the Department of Defense.

However, state regulations regarding harvest and other take also apply to federal public lands (Grosse 1997, Bean and Rowland 1993).

WHAT IS WILDLIFE?

The Professional Concept of Wildlife

Wildlife in the broad sense includes all wild animals. This definition excludes domestic animals—although feral animals (domestic animals that survive without the direct assistance of humans), such as horses and burros that roam freely in public lands, are a management concern of wildlife biologists. Some biologists extend the concept of wildlife to plants; however, the usual usage restricts the term wildlife to terrestrial vertebrates (most mammals, birds, and reptiles) and semiaquatic vertebrates (amphibians, waterfowl, walruses, otters, muskrats, beavers, and seals, among others), whereas fisheries refer to aquatic animals harvested for food, sport, and other uses such as fish, crustaceans (shrimp, crayfish, lobsters), mollusks (squid, clams, oysters), marine turtles, and cetaceans (whales, dolphins, and porpoises). In practice, wildlife management has been reserved for terrestrial ecosystems while fisheries management addresses aquatic ecosystems. The two areas of management focus overlap at the water's edge where many wildlife species use aquatic resources. Wildlife typically refers to those species that reproduce on land and always depend on atmospheric sources of oxygen.

Early government concepts of wildlife were closely associated with commercial and sport hunting values. Some state agencies continue to be called "game and fish" departments, reflecting the continuing influence of sport hunting in the wildlife concept. However, the Endangered Species Act, passed in 1973, greatly changed how wildlife is defined in professional practice and in the public vernacular. This law defined the term fish or wildlife as any member of the animal kingdom, with certain exceptions. Insects classed as pests with "overwhelming and overriding" risks to humanity were exempted from the act. States have also had to expand their protection authority to a broad inclusion of many species. Some states have assumed management authority over all wild vertebrates and an increasing number of invertebrates and wild plants. However, domestic animals and privately owned animals usually are exempt, with certain exceptions. Privately owned elk, other deer, and other native animals, normally free-ranging, continue to come under a wildlife authority in many state agencies. Owners must have a permit to keep them captive and sell them for meat and other uses. Exotic animals that have become naturalized in the wild and are considered game species, such as the ring-necked pheasant, are typically placed under a wildlife (or game) authority. The Wild Horses and Burros Act of 1971 extended federal protection to "wild free-roaming horses and burros" because a large portion of the public viewed them as part of our national heritage despite the damage they might cause to native wildlife habitats (Grosse 1997). Thus the legal definition of wildlife has changed dramatically over the past century and is likely to change further in the future.

The Public Concept of Wildlife

The sense that a species must have a net positive value to be "classified" as wildlife remains part of the generally accepted public concept of wildlife. It explains why many people do not consider pests and vermin as wildlife animals. This differentiation, based on the negative and positive values assigned by people, is the source of many wildlife controversies. Coyotes, for example, are viewed by many as a valued part of our natural and cultural heritage while others see them only as livestock killers. While many people exclude insects, most will include butterflies and other esthetically pleasing invertebrates among wildlife. These are among the invertebrates most valued by hobbyist collectors and fit the positive esthetic value typically associated with wildlife (Kellert 1996).

The wildlife concept has evolved from one almost entirely associated with the positive values of wildlife as game to a much more inclusive one. The public concept of wildlife is continuing to evolve and the boundaries determining inclusion and exclusion continue to change. For most people, wildlife conjures positive feelings associated with a "natural" setting, even including naturally planted backyards. Professional attention has strongly leaned toward natural settings as well, reflecting where the public has directed funding for management. As species once thought to be useless or vermin became more widely recognized for their ecological and social values, they gained wildlife status. At some point in the future, all wild organisms may eventually be valued for the roles they play in ecosystem processes. Then they will be managed collectively to sustain their ecological services as well as their commercial and recreational values. It is doubtful though that all wildlife species will be equally valued. Conflicts will continue to arise over their management because of the different values assigned to wildlife by different sectors of the public.

MANAGEMENT PHILOSOPHY: WILDLIFE CONSERVATION

The prevailing management philosophy is wildlife conservation, which advocates the sustainable use of native wild animals and their habitats for present and future generations of human societies. Whereas individual populations under private control may not be managed sustainably, such as some populations managed for meat or hide production on commercial farms (e.g., alligator farms, venison farms), government-managed wildlife populations typically are managed for sustainable use. In public management, even when native species are viewed as pests, total eradication is rarely contemplated. Rather the species is managed to reduce damage while sustaining population viability. On the other hand, exotic species (animals introduced in areas where they do not naturally occur) may be considered for eradication, especially where they have altered the natural ecosystem in undesirable ways.

Wildlife conservation is a social process that focuses on maintaining and improving the productivity of wildlife habitat. Wildlife conservation involves planning, research, population and habitat assessment and manipulation, education, public relations, administration (logistics), establishing and enforcing wildlife law (e.g., wildlife harvest regulations), and other forms of management. Wildlife management involves the application of scientifically based knowledge to sustain production of wild animals for recreational, commercial, and protection purposes (e.g., endangered species). Wildlife management often involves manipulating populations or their habitats to achieve human objectives, such as providing hunting, fishing, viewing, or wildlife-product opportunities, protecting the options for use by future generations, or reducing animal damage. Wildlife management may be private or public, targeted for the select few, or broadly directed toward a large sector of the public. It applies population and habitat research results to the management of wildlife, habitats, and human use.

Wildlife professionals in the earliest years of management were concerned principally with game birds and mammals—those wild vertebrates that provided hunting recreation and required a state or federal license or permit to harvest. Wildlife professionals now manage a much greater variety of wildlife including nongame, endangered, exotic, and pest species (Figure 15.2). The enlargement of concern arose as urban development spread into more natural landscapes, more harmful human actions impacted wildlife and their habitats, and people sought less consumptive use (e.g., hunting and trapping) and more nonconsumptive use (e.g., feeding, photography) of wildlife.

The measurement of management effectiveness in satisfying conflicting demands depends on estimation of costs and benefits associated with wildlife. Effective man-

FIGURE 15.2   Threatened and endangered nongame animals, such as the bald eagle shown here, now receive protection and management consideration in the U.S. (courtesy of Jack McCaw).

agement of wildlife resources improves human health and nutrition, economic prosperity, leisure enjoyment, and future options for wildlife-based goods and services. Management that provides the greatest total benefit for the costs depends on determining who and how much each public segment will benefit from and pay for wildlife protection, enhancement, and damage control. The inability to convert all wildlife values into easily compared monetary terms underlies many of the controversies in wildlife management.

Values of Wildlife

Wildlife have esthetic, social, commercial, educational, ecological, and recreational values. During most of the history of human existence, wildlife and plants provided all of the necessities for human survival. It has only been within the last 10,000 years that humans began to depend on agriculture, industrial development, and fossil fuels as sources of food, fiber, fuel, and medicine. Previously, fishing, hunting, and plant gathering were the only human activities involving natural resource use. Indeed, there are still human populations, even in North America, whose primary source of meat protein is from wildlife populations.

Many of the controversial aspects of wildlife conservation are rooted in a dynamic history of wildlife-societal relationships and the diversity of values that have emerged out of that history. In the United States, wildlife is now less valued for its commercial consumptive uses—that is, for meat, hide, fur, oils, down, and other goods—than for its recreational consumptive and nonconsumptive uses. Trapping has greatly declined since World War II as fewer people think it justified for what they perceive to be luxury goods. Recreational hunting has become less popular as more people question the ethics of killing for pleasure and fewer young people find the opportunity to participate in this American heritage (Dunlap 1988).

While some wildlife benefits are easy to assess, others are more difficult, and some are not yet possible to estimate and to compare in monetary terms. Commercial benefits are easily estimated through the market place. Recreational and esthetic benefits can be estimated through indirect economic methods that estimate public willingness to pay. It is more difficult but possible to assign value to wildlife that perform ecological

FIGURE 15.3   Some species of wildlife, such as this great horned owl, are rarely seen, but they have great nonconsumptive value for many outdoor recreationists and perform important ecosystem services (courtesy of New Mexico Department of Game and Fish).

services, such as natural regulation of agricultural pests (Figure 15.3). In contrast to benefits, wild animals often entail costs in the form of damages they cause to property, human health, and management programs, such as endangered species. Negative impact costs also differ in ease in which they can be estimated. Much more research is needed on wildlife economic benefits and costs (Johnson 1987, Kellert 1996).

Wildlife economic values continue to increase. The National Survey of Fishing, Hunting, and Wildlife-Associated Recreation (U.S.D.I. 1997) reported that over 77 million United States residents 16 years or older participated in some type of wildlife-related recreation in 1996. During that year, 62.9 million residents participated in at least one type of nonconsumptive recreation activity in which wildlife enjoyment was the primary purpose. About 22% (12.8 million males and 1.2 million females) were consumptive users (hunters). Americans spent $101 billion on wildlife-related recreation. Trip costs totaled $30.0 billion, equipment costs summed to $60.4 billion, and other items added another $10.8 billion. Hunters and fishers spent a total of $71.9 billion. They spent $20.5 billion on trips (food, lodging, and transpor-tation), $43.7 billion on equipment, and $7.7 billion on related goods and services (magazines, membership dues, contributions, land leasing, land ownership, licenses, stamps, tags, and permits). The 62.9 million nonconsumptive participants spent $9.4 billion on trip-related expenses, $16.7 billion on equipment, and $3.1 billion on magazines, membership dues, and contributions to conservation or wildlife-related organizations.

Esthetic, ethical, and cultural values can be more important than other values. Individuals who spend time in the outdoors, be it wilderness areas or in their backyards, gain great psychologically rewarding experiences by simply viewing wildlife. Contact or communing with nature is mentally refreshing and contributes to our mental well being. Wildlife may even have an existence value in that some people seem to gain satisfaction from simply knowing it exists, independent of any potential use. Human ties to the natural world are evident in the use of animals as state, national, and societal symbols. The animals used are usually wild forms. Thus animals remind us of our national origins and heritage and establish a common bond among citizens. However, some of the symbols remind us of past profligacy because they are now locally extinct (e.g., grizzly bear in California, jaguar in Arizona). Wild animals also provide great educational benefits by creating awareness and respect for nature and natural processes (Kellert 1996, Dunlap 1988).

HISTORICAL AND LEGISLATIVE PERSPECTIVES OF WILDLIFE CONSERVATION

Foods and Other Goods

Early History. Although wildlife management emerged relatively recently among the scientifically based natural resource management specializations, professional roots extend back to medieval Europe. In this period feudal lords directed game wardens to prevent the taking of game by unauthorized individuals. "Poachers" were typically poor serfs seeking protein nourishment and clothing that competed with the feudal lords' concern with preserving recreational "sport" hunting. The two different views of wildlife as sources of "necessary" goods versus sources of "luxury" services influenced the later development of wildlife-use ethics and wildlife law in the United States.

The relationship of Native Americans and wildlife before European colonization was complex and diverse. However, many of the tribes developed a spiritual tie to the wildlife they hunted and used. Most killed animals were in some way used for food, clothing, housing, utensils, decoration, religion, talismans, and trade. Resource dependency frequently resulted in a respect for the spirit as well as the embodiment of wildlife. Even wildlife species that threatened competition or direct harm, such as wolves, were personified and accorded respect (Hughes 1996).

For prehistoric Native Americans the contemplative experience associated with hunting or trapping wildlife seems to have been more religious than recreational. Recreational sport seems to have evolved in humanity as a primary use of wildlife only after food and other necessities were assured through means other than wildlife. The Cro-Magnon cave art of the last glacial advance indicates a similar spiritual tie to wildlife in the early ancestry of Europeans. But by the time Europeans colonized North America, their relationship to wildlife had changed. For most colonists, wildlife were simply resources in support of livelihood—a gift from God to be subdued and used the same as any inanimate resource might be. Wildlife conservation rarely crossed the minds of early European settlers who believed God would provide for the faithful, especially in the New World of unbelievable wildlife bounty. The possibility of wildlife extirpation was unthinkable. Because there seemed to be plenty for everyone, a personal claim to ownership seemed unnecessary. In 1633, a traveler through southern New England recorded the bountiful wildlife resources in southern New England and its widespread use. The supply of waterfowl, turkeys, deer, and fishery resources seemed endless. However, by the mid-nineteenth century, Henry David Thoreau had recounted the decline of forests and wildlife in his native New England (Bolen and Robinson 1999, Cronon 1983, Matthiessen 1987).

Wildlife Resource Depletion. Overexploitation of species for food and other goods was a major source of wildlife depletion during European colonization. By the nineteenth century, the first of a number of species extinctions took place. The great auk, a large penguin-like bird common on oceanic islands of the northern Atlantic Ocean, was slaughtered for its flesh, oil, and eggs. It was the first species extirpated in North America. The Labrador duck followed in 1875. The passenger pigeon was doomed by unregulated market hunting in conjunction with railroad expansion, which made rapid shipping to distant game markets feasible. The last passenger pigeon died in the Cincinnati Zoo in 1914. Shore birds, such as the Eskimo curlew and golden plover, were heavily harvested by market hunters for their feathers, as were snowy egrets in the south. Fashionable ladies' hats were adorned with egret feathers (Figure 15.4). Many of the waterfowl species became scarce (Trefethen 1975).

Wildlife in the West fared no better. The emigrants who settled in the Midwest and Great Plains overhunted game birds and mammals, such as elk, bison, prairie chickens, and pronghorn. Game meat of all kinds was shipped east on railroads for market and restaurant sale (see Chapter 2). Hostility also played a role in the decimation. A bill presented to President Grant for protection of bison from market

FIGURE 15.4   In the late 1800s, many wildlife species, such as the egret in the photo, now protected by law, were subjected to unregulated market hunting that nearly resulted in their extinction (courtesy of Jack McCaw).

FIGURE 15.5   Large mammals were not the only wildlife extirpated in the Great Plains. Prairie dogs once occurred in the millions in grasslands but were eradicated because they were viewed as incompatible with livestock husbandry (courtesy of New Mexico Department of Game and Fish).

hunters was pocket vetoed with the intent of starving plains Indians into submission. Settlers converted much of the prairie to cropland and livestock range, degrading habitat suitability for many prairie species (Figure 15.5). Only the establishment of Yellowstone National Park in 1872 saved the bison from extinction in the United States (Trefethen 1975, Wagner 1978).

Personal consumption of wildlife contributed significantly to the decline of some species. However, unregulated market hunting was blamed for the most devastating damage in the last half of the 1800s. Early environmentalists such as Henry David Thoreau, and later John Muir, called attention to the accelerating changes in the landscape and wildlife populations. However, wealthy sportsmen were most instrumental in outlawing market hunting and establishing the system of state fish and wildlife agencies that now enforce most wildlife laws throughout the nation (Reiger 1975). The first state agencies for game protection were established before the Civil War. Most states and territories had "game departments" by the early twentieth century. The game laws particularly restricted market hunting, which steadily declined in the United States. The last significant market hunting, for alligator hides and curios, decimated alligator populations. The alligator recovered rapidly once it gained protection in various states and under the Endangered Species Act. This was also true for early game species after the market carnage was stopped at the turn of the century. Until 1900, the federal government played little role in wildlife management, other than to create national parks where wildlife protection was authorized, but not always strongly enforced. The first federal law to significantly strengthen enforcement of state game protection laws was the Lacey Act of 1900. The Lacey Act made it a federal offense to transport illegally killed game across state lines and greatly contributed to the demise of market hunting (Grosse 1997).

During most of the twentieth century the commercial development of wildlife goods has been limited to game farm and ranch production where animal husbandry was overseen by state agencies. The commercial production of wildlife for meat and hides has continued. Important commercial species include elk, other deer, pheasant, quail, mink, fox, nutria, and alligators. Fur farming has declined since fur demand for fashionable clothing diminished. The trapping of wild furbearers also has declined markedly and may be entirely outlawed in individual states as antitrapping groups gain influence from an increasingly sympathetic public. However, trapping remains an important source of income for certain rural communities in Canada and Alaska. Hunting remains an important source of protein in subsistence cultures. The commercial production of wildlife goods from native species has become quite specialized and is a now a small part of the estimated economic value of wildlife. However, the personal use of wildlife for food and fiber remains a significant part of the recreational hunting experience (Bolen and Robinson 1999).

Milestone Federal Wildlife Legislation

1900. Lacey Act. Prohibited the transportation of illegally taken wildlife, fishes, plants, and other organisms across state borders and prohibited the importation of certain exotic species.

1913. Migratory Bird Act. Federal government assumed regulatory powers over migratory birds.

1918. Migratory Bird Treaty Act. Provided for coordination between U.S. and Canada in managing migratory birds and later amended to include other nations.

1931. Predatory Mammal Control Program. Authorized the Department of Agriculture to study and control predatory mammals causing damage to crops and livestock.

1934. Migratory Bird Hunting Stamp. Required that waterfowl hunters purchase a duck stamp and monies generated be spent on wetland conservation programs.

1934. Fish and Wildlife Coordination Act. Authorized the Department of the Interior to assure the welfare of fish and wildlife in water development programs initiated or licensed by federal agencies.

1935. Creation of the Cooperative Wildlife Research Units. The units conducted research and established graduate programs in wildlife science at state universities.

1937. Federal Aid in Wildlife Restoration Act. Imposed a federal excise tax on sporting arms and ammunition with proceeds to be distributed to states for wildlife-related projects.

1969. National Environmental Policy Act. Required federal agencies to submit environmental impact statements describing potential negative effects of any major project on the environment (including wildlife) before the project begins.

1972. The Marine Mammal Protection Act. Established protection of marine mammals under the authority of the Department of Commerce.

1973. Endangered Species Act. Initiated a list of endangered U.S. species. The 1973 version directed federal agencies to protect and restore endangered species and their habitats.

1985. Food Securities Act. The Conservation Reserve Program established a voluntary program for landowners to improve soil and water resources including fish and wildlife habitats.

1986. The North American Waterfowl Management Plan. International agreement between the U.S. and Canada for restoring waterfowl habitat and populations across North America.

Recreation

The Rise of Recreational Use.   Expenditures on wildlife-based recreation now greatly exceed their material value. The amount people pay for equipment, travel, supplies, and other hunting costs far exceeds the average food, hide, and other value of the harvested goods. Theodore Roosevelt was more influential than any other President in initiating federal policies favoring recreational use of wildlife over commercial use. Roosevelt was an ardent sportsman and leading member of the Boone and Crockett Club, a sportsman's group influential in establishing the laws that eliminated market hunting of wildlife. Roosevelt added 148 million acres to the national forest system, which were open to legal sport hunting except for game refuges established after 1916. Roosevelt established the first national wildlife refuge by executive order in 1903. He convened a White House conference in which governors were invited to discuss methods of implementing conservation measures. By the time Roosevelt left office, unregulated market hunting for most species had become a thing of the past. The nation was committed to the recovery of its reduced game populations (Trefethen 1975).

After nearly half a century of shaping sportsmen's ethics through sporting magazines, wildlife conservationists, led by George Bird Grinnell, established greater value in the experience than in the goods that followed from a successful hunt. It had become quite obvious that wildlife populations could not sustain the demands for meat, hides, and other goods exerted in the late nineteenth century (Kimball and Johnson 1978). If their populations were to be sustained, they needed to be reserved for limited sporting recreation and personal use of meat and other goods. "Game hogs" were severely criticized in the sporting magazines and commercial hunting was condemned. However, the widespread decimation of game noticeable to most hunters probably was the main motivator for the change in hunting attitude and behavior.

After the states had passed laws severely limiting market hunting, personal recreation harvest was further restricted for numerous species. Hunting was increasingly limited to a fall season for most species with meager daily bag limits. For deer, buck-only laws were instituted. An important federal law, the Migratory Bird Act of 1918, established federal jurisdiction over migratory birds and more coordinated management and regulation. J. N. "Ding" Darling, a wildlife administrator, cartoonist, and conservationist, led a sportsmen's push for the Migratory Bird Hunting Act of 1934. It required purchase of a stamp by duck hunters and authorized use of the revenues for preservation of waterfowl habitat. More recent federal laws have provided monetary incentives to farmers for managing their lands favorably for waterfowl, upland game birds, and other recreational wildlife (Dunn et al. 1993).

FIGURE 15.6 Elk, once almost extirpated, now are numerous in the West. This large herd of elk is being aerially censused. Aerial censusing is an efficient technique under certain circumstances, such as in winter, when some large mammals congregate in large mixed-sex herds (courtesy of New Mexico Department of Game and Fish).

Early Management Successes. Although once widespread in the eastern U.S., moose, elk, bison, mountain lions, and black bears were greatly reduced or, as in the case of elk and bison, extirpated (Figure 15.6). White-tailed deer were much scarcer than now, especially near population centers. Turkeys, prairie chickens, quail, various shorebirds, and several species of waterfowl were growing scarce. The passenger pigeon and heath hen were about to become extinct. The public right to hunt was in jeopardy (Trefethen 1975).

The recovery of most game species is a management success story owed primarily to influential recreational hunters who also became formidable conservationists and politicians influenced, in particular, by George Bird Grinnell (Reiger 1975). They convinced the voting public and state and federal governments that recreational use had to prevail over wholesale consumption of meat, hide, and other goods if game species were to survive through the twentieth century. As a consequence of their influence, and that of dedicated management professionals in government agencies, there have been dramatic recoveries of most over-hunted game populations. Their recovery was mediated by effective enforcement of game laws, reintroduction programs, refuge establishment, and habitat improvement. A more controversial aspect of recovery was the "control" of large predators, such as wolves and mountain lions, which was strongly supported by livestock owners concerned about stock losses. William T. Hornaday, director of the New York Zoo, in 1913 wrote *Our Vanishing Wildlife, Its Extermination and Preservation,* in which he condemned all sources of excessive wildlife killing by people and by nonhuman predators. His views were influential in early game management practice.

Once scarce, white-tailed deer greatly increased from less than 1 million to nearly 30 million. They are now the most numerous big game species in North America (Figure 15.7). In Texas alone, with a population of over 3 million white-tailed deer, hunters harvested about 452,000 whitetails in 1993–1994. In the western United States, all ungulate populations have stabilized or increased, including mule and white-tailed deer, elk, moose, pronghorn antelope, bighorn sheep, and mountain goats. Elk increased from less than 100,000 in the early 1900s to an estimated 960,000 in 1995. Small mammals and furbearers, including eastern tree squirrels, cottontail rabbits, muskrats, Virginia opossum, raccoons, beavers, mink, red and gray foxes, bobcats, and coyotes also have stable or increasing populations. Coyote populations are particularly adaptable and continue to expand geographically despite widespread

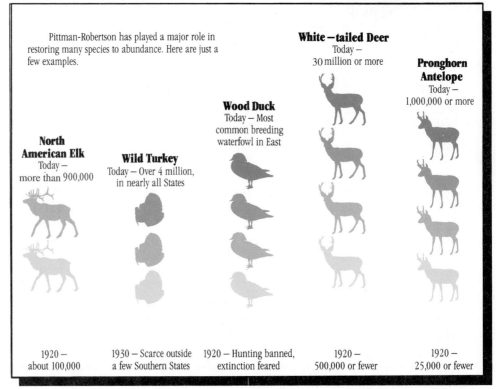

FIGURE 15.7   Examples of wildlife recovery from overharvest in the U.S. over the past 80 years, 1920–2000 (courtesy of U.S. Fish and Wildlife Service).

control programs. Beavers, at one time at the verge of extinction, now are widespread and common (Council for Wildlife Conservation and Education 1992).

Among upland game birds, turkeys have increased significantly since the turn of the century and now number over 4 million because of reestablishment programs and favorable landscape changes. Bobwhite quail numbers have declined in some areas because of intensive agricultural practices that have eliminated habitats. Forest game birds, such as blue grouse and ruffed grouse, have been stable or increased (Flather and Hoekstra 1989).

Animal Damage Control

In colonial North America, the first wildlife targeted for eradication usually were those species that conflicted with human economic interests. As early as 1630, the Massachusetts Bay Colony placed a bounty on wolves, which were extirpated in much of New England by 1800. Before European colonization, large numbers of Carolina parakeets ranged throughout the eastern U.S. deciduous forests from New York to Florida and west to Nebraska and Texas (Matthiessen 1987). They fed on agricultural crops, causing considerable damage. As a consequence they were persecuted by farmers and, by 1914, they were extinct.

FIGURE 15.8 Many wildlife species, such as jackrabbits, have been targeted for control because they are perceived as incompatible with livestock husbandry. However, large increases in jackrabbit populations can be the result of poor grazing management (courtesy of New Mexico Department of Game and Fish).

In the early period of game management, predators were believed to be an impediment to the restoration of game to former abundance. During the 1800s, more than half of all Americans lived off the land. Because most kept small livestock, predators were widely condemned as thieves and killers. The largest predators threatened human life and, therefore, many people thought they should be completely eradicated. Once state game agencies were established, predator control became nearly as important as enforcing game harvest regulations. States hired professional hunters to kill bears, wolves, coyotes, foxes, mountain lions, bobcats, skunks, weasels, eagles, hawks, crows, and other "varmints." Wolves and grizzly bears declined to extremely low numbers south of Canada. Anyone could shoot or trap predators with impunity and often be rewarded for it.

Seeking an alternative to costly paid hunters, many states revived an old practice of paying bounty to anyone who provided physical evidence of a killed predator. After years of costly payments, studies conducted in Michigan on coyotes and in Pennsylvania on weasels, showed no signs that these programs reduced predators or increased the game animals the predators were blamed for decimating. Bounties also were subject to fraud. After thorough discredit bounties were eliminated as a wildlife management tool (Bolen and Robinson 1999, Allen 1984).

Valid government programs remained to control coyotes in the western United States, crop-eating birds on farmlands, and nuisance species in suburban and rural areas. In 1931, the Predatory Mammal Control Program was created in the Department of Agriculture to study and control mammalian damage. It initiated federal animal damage control, which has remained a necessary and sometimes controversial activity ever since (Figure 15.8). Coyotes have remained controversial because they can cause significant livestock losses. The use of lethal control measures, including poisons, were thought most effective. The poisons too often caused unintentional death to other wildlife and had questionable effect on coyote population abundance. The use of a predator control toxin known as 1080 was banned by Executive Order in 1970.

Other management problems originated from the intentional and accidental introduction of numerous species. When successful, exotic species more often than not have competed with native species and complicated wildlife management. The rock dove, English sparrow, starling (intentionally released), house mouse, and Norway rat (accidentally introduced) are now considered urban pests. The total number of exotic species released in the United States is lengthy, but relatively few have survived. Many were introduced to supplement game hunting. Ring-necked pheasants, first successfully introduced in Oregon and later in other states, flourished and are now widely accepted and valued. However, there are indications they compete with certain endangered native species (Bolen and Robinson 1999).

FIGURE 15.9 Exotic species have been introduced accidentally and purposefully. Large mammal species, such as oryx introduced from Africa for sport hunting purposes, have the potential to cause damage to native species and habitats. This oryx has been fitted with a radio collar (note antennae) to monitor its movements.

After World War II, the demand for hunting jumped and with it a demand for exotic species (Figure 15.9). Numerous exotic deer, sheep, and antelope species, successfully introduced on fenced-in private lands in Texas, have escaped. The nutria, a South American wetland rodent, was introduced for its fur in Louisiana, where it escaped from captivity. It now is a widespread wetland nuisance in warmer regions of the Southeastern United States. Other isolated populations of exotic species occur elsewhere in the United States. Government agencies for the most part now disparage exotic introductions, considering them more of a risk than a benefit.

Management Perspectives in the Twentieth Century

Wildlife management in the first half of the twentieth century emphasized the establishment and enforcement of game laws and the destruction of game predators (Bavin 1978). Wildlife legislation was largely a reaction to changing perceptions of the American public toward wild animals. The most influential wildlife law following the Lacey Act was the Migratory Bird Treaty Act of 1918, which initiated protection of migratory birds, outlawed market hunting, and established regulations on season lengths and bag limits. It also facilitated the coordination of hunting seasons and regulations for harvest of migratory birds among the states and among countries, starting with Canada. The treaty was later expanded to include Mexico, Japan, and the Soviet Union. Except for migratory birds, however, wildlife management remained under state control (Grosse 1997).

Another management emphasis was the reintroduction of wildlife to areas where they had been locally extirpated, or into an entirely new range. Elk, turkey, bighorn, pronghorn antelope, bison, whitetail deer, black bears, beaver, and many other native species were transferred from refugial populations. Many introductions were successful, hastening the recovery of game species. After World War II, the restoration of game species had reached a point where habitat was the major limiting factor. In some cases, game animals were causing damage to private property and threatening human health.

The personnel of the early state game departments were typically few and often were expected to cover huge areas to assure that harvest regulations were enforced. They also were charged with determining the population status of diverse species. To contend, they enlisted the aid of volunteer wardens from among sportsmen's groups and other government agencies. Many of the rangers in the United States Forest Service became game wardens on the lands they managed. Among them was a forester named Aldo Leopold, who would lead in the development of the new discipline of wildlife science and write the first wildlife management book, *Game Management* (Leopold 1933).

Using his forestry background, Leopold emphasized the importance of habitat management as the basis of wildlife production. He recognized that once populations returned to the "carrying capacity" of the land, only changes in habitat could alter the number of wildlife produced. Realizing that most wildlife species require a mix of different vegetation and habitat attributes, he was the first to point out the importance of the arrangements and relative amounts of different plant communities in determining wildlife production.

Leopold's (1949) book, *A Sand County Almanac,* published a year after his death, is a classic that established a land ethic based on a respect for natural process. Leopold emphasized that wildlife was a renewable resource that could be replenished through sound knowledge of its relationships to its habitat. He believed humans had an ethical responsibility to respect the integrity, stability, and beauty of natural communities. His holistic approach integrated management of wild animals, agricultural practices, and other natural resources on public and private lands. *A Sand County Almanac* is recommended reading for an understanding of human relationships and ethical responsibilities with respect to the natural world. Leopold established one of the first university programs in wildlife science at the University of Wisconsin.

During the 1920s and 1930s, a small group of wildlife professionals began to coordinate their influence on legislation and eventually formed the Wildlife Society in 1937. Through the Fish and Wildlife Coordination Act of 1934, federal law established that wildlife resources be managed cooperatively by federal agencies. The Federal Aid in Wildlife Restoration Act of 1937 levied an excise tax on sporting arms and ammunition to raise money to support wildlife research, land acquisition, and habitat management (Trefethen 1975).

Ding Darling was politically very influential. In 1934, he became the director of the government bureau that was the predecessor of the Fish and Wildlife Service. He had much to do with originating duck stamps as a source of funds for purchasing wetland habitat for waterfowl. In addition, he was the main political force behind creation of the Cooperative Wildlife Research Units at state universities. These became the main source of professional wildlife researchers and managers during the period following World War II when interest in hunting and other wildlife-based recreation escalated.

During the 1960s, human interest and awareness in the welfare of the environment and wildlife increased. Rachel Carson's book, *Silent Spring,* warned of the serious damage to humans and wildlife caused by the indiscriminate use of persistent pesticides such as DDT. Numerous conservation groups led by the National Audubon Society, Wilderness Society, Sierra Club, Nature Conservancy, and the National Wildlife Federation joined with other sectors of the American public to lobby the federal government for improved environmental legislation. The Wilderness Act of 1964 directed the Forest Service, Fish and Wildlife Service, and the National Park Service to take action to preserve wilderness areas. The Endangered Species Act of 1973 and subsequent amendments directed federal agencies to protect and restore endangered species and provided expanded funding toward this goal. Research on predator, prey, and wildlife habitat relationships greatly expanded the database for managing wildlife populations and habitats (Bolen and Robinson 1999, Bean and Rowland 1993).

After years of fluctuating increases following the drought years in the 1930s, duck populations declined more than 30% during the decade between 1970 and 1980. They dropped from 44 million birds in 1972 to 28 million in 1980. This was due largely to declining waterfowl habitat. In 1986, the United States and Canada initiated a joint waterfowl conservation plan named the North American Waterfowl Management Plan

FIGURE 15.10 Plantings around prairie potholes preserved under "The North American Waterfowl Management Plan" have contributed to recovery of waterfowl and several other bird species (courtesy of the Natural Resources Conservation Service).

FIGURE 15.11 Ducks and geese have increased significantly since 1990. There are 30 species of ducks, geese, and swans in the United States (courtesy of New Mexico Department of Game and Fish).

(Figure 15.10). The goals were to protect and create 6 million acres of wetland habitat and increase waterfowl numbers to 62 million breeding ducks with a fall flight of 100 million birds (Figure 15.11). The fall flight in 1986 was only 73 million birds. By 1997, duck populations had significantly increased and only two species remained below the goals of the plan. The fall flights in 1997 and 1998 numbered over 100 million birds. Canada goose populations have increased to the point that they are considered a nuisance in some areas (Ankney 1996). Those mammal and game bird species that have significantly increased do well in habitats with early to mid-successional stages, usually associated with human-induced changes.

The North American Waterfowl Management Plan: Waterfowl for the Future

Waterfowl are a precious natural resource, important to hunters, bird-watchers, and naturalists. But numbers of some of our most popular duck species are at an all-time low. Millions of acres of wetlands have already been destroyed by agriculture, urban development, and industry, and more are threatened each year. *Waterfowl need help.*

What's being done? The United States and Canada have joined forces to reverse the decline in certain populations of ducks and geese. In 1986, U.S. and Canadian officials signed a far-reaching document: the **North American Waterfowl Management Plan.** Today, that historic agreement has become an innovative international partnership in wildlife conservation. The plan has inspired cooperation between federal, provincial, and state governments as well as private conservation agencies in the two countries.

Work has already begun. The plan has established headquarters in both the U.S. and Canada. At the grassroots level, partners have formed **"joint ventures,"** composed of representatives from public and private organizations. *In the U.S., six priority joint ventures are underway:*

*Atlantic Coast	*Central Valley
*Gulf Coast	*Lower Great Lakes-St. Lawrence Basin
*Lower Mississippi Valley	*Prairie Pothole

In Canada, there are four joint ventures:

*Arctic Geese	*Black Duck
*Prairie Habitat	*Eastern Habitat

Partners in the joint ventures are carrying out specific tasks—**developing economic incentives to change land-use practices, striking agreements with private landowners, improving water management, and sponsoring research studies.** Land acquisition is only one facet of the joint venture initiatives.

The plan, with its 15-year horizon to 2000, **establishes specific objectives to restore duck populations to the levels of the 1970s.** It aims for breeding populations of 62 million that should produce a fall flight of 100 million birds. Attaining these objectives would mean that 2.2 million hunters could harvest about 20 million ducks annually. **The plan also lists population objectives for geese and swans.**

Habitat management objectives are equally ambitious. **The plan targets critical waterfowl breeding, staging, and wintering areas in both countries.** In the **Prairie Pothole Region,** cooperators intend to protect and improve 1.1 million acres of habitat. In the **Lower Mississippi Valley and Gulf Coast,** the objectives total 686,000 acres. **California's Central Valley** has an objective of 80,000 acres. Along the **Atlantic Coast,** 50,000 acres will be protected and improved. In the **Lower Great Lakes-St. Lawrence Basin,** more than 10,000 acres have been identified. In Canada, the objective is to protect and improve 3.6 million acres in the prairie Provinces, 60,000 acres in the Lower Great Lakes-St. Lawrence Basin, and 10,000 acres in the Atlantic Coast.

Although the plan focuses on waterfowl, other species will benefit too. Waterfowl habitat offers shelter for a variety of other waterbirds, shorebirds, songbirds, small mammals, and resident game species. Wetland communities also produce invertebrates and plants that are important foods for fish and wildlife. These ecosystems act as pollution filters, floodwater stores, and erosion controls.

The price tag on this unprecedented habitat protection and enhancement program is an estimated $1.5 billion. Federal and state governments will need help securing this sum and are looking to the private sector for support. The plan calls for $1 billion to be spent in Canada, 75% of which is scheduled to come from U.S. sources.

So far, the North American Waterfowl Management Plan has involved Canada and the U.S., but both countries are pursuing a working agreement with Mexico.

Source: The North American Waterfowl Management Plan, U.S. Fish and Wildlife Service.

An excellent conservation program that benefits wildlife and private landowners is the Conservation Reserve Program administered by the United States Department of Agriculture. This program was first established by the Food Security Act of 1985. The objective of the program is to conserve and improve soil and water resources by taking erodible land out of crop production. Cooperating landowners agree to establish and maintain permanent vegetative cover that cannot be grazed or hayed but on which hunting is permitted. These revegetated lands are especially important for wildlife because they provide habitats, nesting areas, and winter cover in intensively farmed areas (Figure 15.12). Federal employees in the USDA Natural Resources Conservation Service and state biologists assist landowners in selecting appropriate species of plants, their arrangement, and their proportions for maximum wildlife-based benefits (Dunn et al. 1993). Chapters 8 and 12 provide more information on the Food Security Act of 1985 and the Conservation Reserve Program.

Threatened and Endangered Species

The Endangered Species Act of 1973 resulted in many changes in wildlife management. Before the Endangered Species Act, attempts to prevent extinction from any cause other than hunting was sporadic and of questionable effectiveness. Most early wildlife management by the state agencies focused on regulation of harvest and predator control. Loss of wetlands encouraged federal interest in creating and restoring wetlands for waterfowl through legislation passed in the 1930s. Extinction was prevented for a number of these species. By the 1960s, it had become clear that a growing number of species were at risk of extinction for reasons other than harvest. Habitat loss and fragmentation was the most frequent threat to these species. The majority of these species were not game animals.

In cooperating with the Fish and Wildlife Service, most of the state wildlife agencies assumed authority over state threatened and endangered species. For many states this significantly widened a nearly single-minded focus on game management to include nongame species. For the federal land management agencies, which had never had an explicit game management authority, the new Endangered Species Act

FIGURE 15.12   An upland game-bird hunter on conservation reserve land in northcentral Oregon. The Conservation Reserve Program, initiated in 1985, has been effective in reducing soil erosion and restoring farmland wildlife species such as ring-necked pheasants, bobwhite quail, and cottontail rabbits.

quickly redirected attention to low profile wildlife species. Through its emphasis on protecting support ecosystems, the Endangered Species Act began to shift management emphasis from individual species with game value to the entire complex of species inhabiting the ecosystem (see Chapter 17). The concept of wildlife has broadened as the concept of ecosystem management has developed.

RESPONSIBILITIES OF A WILDLIFE MANAGER

Changing Emphasis

Today's wildlife managers must have a broad awareness of natural resources coupled with more specialized knowledge in wildlife management (Figure 15.13). The agencies that are responsible for the sustainability and management of public wildlife provide the data needed to make knowledgeable management decisions. Many professionals continue to concern themselves with the management of a select group of species, typically valued for their game qualities, such as deer, turkey, or waterfowl. However, after the Endangered Species Act was passed, management has made a gradual transition toward a more comprehensive approach to wildlife conservation. This is based on ecosystem management, which incorporates socioeconomic, cultural, political, and biological processes into management decisions. This transition has accelerated in the past decade as leading natural resource scientists have questioned the single-species focus used in meeting objectives of the Endangered Species Act (Marcot et al. 1994).

In the ecosystem approach, individual wildlife species are viewed as biodiversity components in an interactive process that includes all the intended and unintended effects of humans (see Chapter 17). While certain wildlife species continue to be emphasized, they no longer totally dominate public wildlife management. The ecosystem approach is inherently interdisciplinary. Wildlife managers routinely work with range, forest, park, urban, fisheries, and other resource managers and scientists. This transition is incomplete and uncertain, with federal agencies, especially the U.S.

FIGURE 15.13 Wildlife professionals must not only have a broad knowledge of natural resources management, but also a deep appreciation for nature and dedication to wildlife management for the public good (courtesy of New Mexico Department of Game and Fish).

Forest Service, moving into a leadership role (see Chapter 23). Many state agencies still lag behind, some even attempting to retrench in their traditional roles emphasizing game conservation through regulation setting and enforcement. Regardless of this shift toward more systems analysis, the basic elements of wildlife management remain the same: population assessment and management, habitat assessment and management, resource demand assessment and management, resource user satisfaction, management coordination, planning and administration, education, and research. Their underlying responsibility is to approach management through science and objective assessment of resource supply and demand. This requires continuous review of the scientific literature and reexamination of the effectiveness of past approaches. Many aspects of these responsibilities are similar to those summarized in more detail for fisheries management in Chapter 16.

Population Assessment and Management

Three of the highest priorities in wildlife management are determining where species occur (distribution), how many there are (population size), and why they occur where they do (the combination of biotic and abiotic factors that constitute an animal's suitable habitat). In evaluating, monitoring, and managing wildlife, biologists review previous information, conduct new population estimates to obtain data on distribution, movements, growth, and reproduction; estimate population numbers; determine the harvestable quota for hunted species; classify and quantify vegetation cover and its nutritional value; and determine what organisms eat by observation and stomach and fecal analysis (Figure 15.14).

More so than in the past, wildlife professionals attempt to assess the diverse interactions among predators, prey, and competitors. They are now much more likely to consider secondary and tertiary impacts of such interactions. They are less likely to assume that "empty" niches occur, which can be occupied by exotic species without harm to native wildlife (see Figure 15.9). Studies of interactions between native and exotic game species, for example, have revealed previously overlooked compe-

FIGURE 15.14   Wildlife biologists use a variety of capture techniques. These sandhill cranes have been netted. They can then be weighed, measured, and banded to study migration routes. Many other species are similarly captured and used to reestablish populations in areas where they have been extirpated (courtesy of New Mexico Department of Game and Fish).

tition (Mungall and Sheffield 1994). Even though population biologists responsible for endangered species and game management often work independently, their cooperative interaction is essential.

Habitat Assessment and Management

Habitat disturbance, reduction, and destruction are the most serious problems facing wildlife populations. Therefore, wildlife professionals must constantly evaluate habitat quantity and quality. Management may involve the reduction of domestic and wild animals grazing an area, such as along streams to restore riparian vegetation, burning of living or dead vegetation, or altering forest harvest practices in ways that provide for wildlife needs (Morrison et al. 1998, Bookhout 1994, Payne and Bryant 1994).

Whereas wildlife managers are allowed almost total control over wildlife habitat in wildlife refuges, most have to influence the actions of other public or private land managers. Wildlife managers work closely with private landowners to maintain and improve habitats, such as shelter belts, buffer strips, wetlands, streams, and uncultivated plots that provide necessities for wild populations (Burger 1978). Urban wildlife management is becoming an increasingly important aspect of wildlife management through park management, nature centers, backyard management, and animal damage control.

Usually, habitat management is designed to mitigate the impacts of human-induced changes in ecosystems, such as from agriculture, animal husbandry, forestry, mineral extraction, energy development, urban development, water usage, and waste disposal. Many wildlife professionals are hired by the public land and water management agencies authorized to manage all public resources for sustainable public benefit and multiple use.

Resource Demand Assessment and Management

The "human dimension" also requires assessment and management. Wildlife managers often establish check stations where hunters submit animals for weighing, measuring, and examining for parasites and general health. Professionals can also obtain information from the participants themselves such as the age, sex, background of hunters, distances traveled, and other aspects. This information aids recreation specialists in planning future opportunities. The most common interface between representatives of the wildlife profession and the public is through law enforcement personnel. Their social skills are crucial in establishing a positive relationship between the public and the agency (Fazio and Gilbert 1981).

Development of regulations and law enforcement often are the primary means by which wildlife agencies manage wildlife use. However, evidence suggests that education can shape public opinions and greatly improve compliance with wildlife regulations. Law enforcement is most effective when the vast majority of the public accepts the need for regulation and voluntarily complies. Even before the state agencies established a warden-based approach to wildlife management, public education through sporting magazines, clubs, and other groups was quite effective in calling attention to obvious management needs. However, the closer proximity of most Americans to the land and its wildlife made such education easier in the past than today, when much of the public is removed from the land. Education effectiveness today depends greatly on how wildlife professionals reach the mass media and school programs.

A continuing challenge for many state agencies is how to most effectively administer law enforcement as well as other programs with limited management funds. Demands on agencies, especially with respect to nongame management surveys and public education, have multiplied as hunting has declined in recent years. Meanwhile, the hunting public continues to be very influential because they contribute most to management funding (Bavin 1978).

Surprisingly few studies have been conducted to assess the effectiveness of law enforcement (Bolen and Robinson 1999). Much of present law enforcement justification is based on the apparent success of early management practice, which emphasized law enforcement, widespread introductions, and predator control. While the latter two strategies have been well evaluated, greatly reduced in scope, or even reversed, a traditional approach that integrates law enforcement with other agency activities continues in many of the smaller agencies. This seems to be partly a matter of agency size. Many of the larger agencies have treated law enforcement as a specialty area, mostly separate from other management activities. Where wildlife professionals must split their responsibilities, the duties of law enforcement tend to make it the primary activity. This can interfere with other important management operations.

Resource User Satisfaction

Because the public owns wildlife, their involvement in establishing management goals is critical. Public support for goals should be obtained prior to implementing management plans. Public support is necessary to appropriate funds for wildlife and fisheries programs. Therefore, public attitudes toward wildlife management programs have to be measured and proper actions taken to deal with unfavorable public opinion.

One of the more difficult and sometimes controversial areas of wildlife management is the extent to which the public, through commissions of lay representatives or otherwise, should influence agency conduct. The skills of professional wildlife managers, especially among the upper ranks of agency administration, are of paramount importance in maintaining the proper balance of public and professional input in determining what is scientifically most justified. Wildlife professionals often need to become legal scholars as well as tactful "politicians" as they explain the obligations of wildlife conservation policy and make objective arguments for change in existing laws and regulations (Bean and Rowland 1993, Fazio and Gibert 1981).

Wildlife professionals, like other natural resource professionals, are continuously pressured into joining coalitions with influential special interest groups. Because positive policy changes often require involvement of special interest groups, this creates an ethical challenge to the wildlife manager. The relationship of wildlife professionals with special interest groups is one of the more pervasive and chronic problems associated with wildlife management. This tends to be true in all areas of natural resource management.

Education And Research

Most wildlife professionals are to some extent involved in research and education, but only a fraction specialize in formal education and research processes. In universities, research and education usually are joint responsibilities of faculty, especially where graduate programs exist. A critical aspect of most graduate education is the application of the scientific process in original research. This information is the basis for sound wildlife management. Faculty in wildlife programs are almost universally required to have earned a Ph.D. This requires scholarly and specialized research into diverse areas of the wildlife profession. The most effective teachers of wildlife management also have a comprehensive general knowledge of wildlife management history, ecology, economics, social sciences, management planning, administrative processes, and interactions of wildlife management with other natural resource management areas.

Other research specialists work in research divisions within the wildlife management agencies. The wildlife scientists in the Cooperative Wildlife Research Units, now located in most states, hold faculty positions in state universities. Here they work cooperatively with other faculty in research programs. They emphasize graduate student involvement. These units now are administered by the Biological Resources Branch of the U.S. Geological Survey (USGS).

The National Park Service, U.S. Forest Service, and U.S. Fish and Wildlife Service conduct some research internally and fund more research outside, often through the USGS research units. Specialized federal research laboratories address issues in forest management, wildlife population management, contaminants ecology and management, water resources management, and other diverse areas associated with wildlife (Bolen and Robinson 1999, Anderson 1991).

CONTEMPORARY CONCEPTS IN WILDLIFE MANAGEMENT

Managing Wildlife Supply and Demand

Wildlife management is focused on managing supply and demand for wildlife resources. Ecology is the primary science supporting wildlife management. It contributes basic knowledge about the factors controlling the abundance, productivity, and supply of wildlife resources. The social sciences, including economics, are the primary sources of knowledge about the human dimension in wildlife management, especially those factors determining the demand for wildlife-based goods and services. Decisions about which resources to promote through management involves interaction among disciplines. Management is most effective when the disciplines are well integrated (Marcot et al. 1994, Morrison et al. 1998).

To be effective, wildlife managers must assess the demand for the different resources—that is, what people would like, if they could have it—and explain to the public when necessary why they cannot always have what they want. Wildlife managers manage both the supply of and the demand for wildlife. The supply can be managed by changing the habitat in ways that are more or less supportive of targeted wildlife. Habitat can be improved where wildlife is desired and purposefully degraded in locations where wildlife are not desired. The wildlife supply can be depressed or enhanced directly by various removal and stocking techniques, or indirectly by adjustment of regulations when the populations are harvested. Demand for a particular wildlife species can be adjusted through regulation and its enforcement, education, and provision of substitute wildlife goods and services. In the latter strategy, for example, harvest limits on one species may be relaxed in compensation for greater restrictions of another species' harvest.

The Role of Public Input

Historically, wildlife agencies relied greatly on political input for determining the demand for wildlife resources and politics remain a part of the process to this day. Because wildlife are defined by law as public property, government agencies are involved with virtually all aspects of wildlife use including oversight of private commercial operations based on wildlife. As a consequence, most wildlife management is conducted either as a public service or in response to public oversight in private service. State agencies typically are responsible to an appointed or elected commission, which is supposed to represent the public interest in wildlife management. Commissions typically are authorized to regulate the harvest of specified wildlife species and to carry out specific management actions such as habitat development; wildlife introductions, stocking, and removal; animal damage control; provision of access to public lands; refuge management; and public education. Agency personnel present program proposals to the commission, which may or may not be accepted. Often, for example, regulations are set as a consequence of both professional and public inputs to the commission through informal process and through formal public hearings.

The result is not always optimal for the sustainability of wildlife or for the benefit of all of the public. Those commissions that accept public input through scientifically based survey procedures are more likely to accurately assess public demand. Despite a trend toward more use of scientific evidence, a political process that responds to pressure from special interests continues to be a factor in state and federal management. The most effective managers are those who manage based on clearly set principles, are respectful of the rights of others, can clearly and objectively articulate the relevant scientific information, and are tactful and patient with those who are not easily persuaded.

Categorizing Wildlife for Management

Wildlife are categorized by important management attributes. Big game (large mammals such as hoofed mammals and bears) and small game (squirrel, rabbits, red fox, mink, raccoon, and others), upland game birds (turkey, quail, pheasants, and others inhabiting terrestrial areas), waterfowl (ducks, geese, swans), shorebirds (small wading birds such as plovers, snipes, and avocets), raptors (hawks and owls), songbirds, small mammals, predators, and furbearers are important wildlife categories. Furbearers are those wildlife with pelts of commercial value that are not usually hunted for sport purposes. Like hunted species, they are harvested. However, their use is consumptive. Consumptive use and nonconsumptive use differ in approach and intensity of management needed. Nongame species are those noted for purposes other than hunting or commercial uses, such as songbirds, avian predators, most reptiles, most rodents, most amphibians, and certain predatory mammals. Some nongame animals may be killed without limit, such as most rodents and coyotes (in many states). Such nongame species have been viewed in the past mostly as pests with little positive value. As the ecological value of all species becomes more scientifically demonstrable, regulations are likely to tighten even on pest species.

Habitat Management

The wildlife ecologist collects habitat data and predicts what will happen under a given set of circumstances to both the habitat and the dependant populations. Wildlife ecologists can theorize regarding the long-term consequences of management actions or lack of actions based on data gathered in a scientific manner. In managing the habitat of wildlife populations, wildlife managers confront a host of challenges because of specific habitat requirements for each species. For example, species inhabiting a forest may favor certain variables such as a specific tree species composition, age of trees, tree density and diameter, and presence or absence of a vegetation understory. The carrying capacity of habitat for each species differs because of their unique demands. Management of habitats for a number of wildlife species usually means some condition less than optimum for each species needs to be identified (Bookhout 1994, Patton 1997).

Habitat preservation, creation, and restoration are the key elements in sustaining wildlife populations. A key goal is population sustainability at some level determined by human demand for wildlife and by ecological limits. A major objective of habitat management is to identify and mitigate those factors in the habitat that limit wildlife populations. Providing these necessities at the proper time and the proper amounts is the key to successful wildlife management. To survive and reproduce, terrestrial wildlife populations must be provided with food, water, and shelter, such as escape terrain. Habitat management is an indirect approach but the benefits are greater and longer lasting than any other practice.

Habitat Supports Food Sources.   The habitat supports a biotic community including food organisms. Food must be available in sufficient quantity and suitable quality to provide all the needs of each species. Young may require different and more nutritious foods than adults. Young nesting birds require a diet high in insects because they provide the high protein content necessary for rapid growth. In temperate regions with freezing temperatures, animals may require a high caloric diet in winter to offset loss of body heat. During fall most large mammals build up fat reserves that can comprise up to 20% of their total body weight. They rely on fat reserves during days of inclement weather when food is difficult to obtain.

Water.   Most terrestrial species require free water obtainable from standing bodies of water or dew. Some may require water only during certain times of the year, such as during the dry season. Some can obtain their water requirements from vegetation with high water content, such as cacti. Small quantities of widely available water better meet the needs of wildlife than a few areas with large quantities of water. Water becomes more of a management tool where it is scarce in arid and semiarid habitats. Careful provision of water where it is limiting may add substantially to the habitat carrying capacity.

Habitat As Cover.   Wildlife use different types of cover for shelter to escape predation; feed in security; rest; protect themselves from cold, heat, wind, and precipitation; reproduce; and nest and breed. Cover is provided by the physical structure of earth and vegetation (Figure 15.15). Especially important attributes of wildlife cover are its dimensions, opacity, effect on air circulation, and resistence to destruction by natural forces.

Rock, soil, and snow provide important cover for burrowing, dens, and retreats. As the structure of the burrowing medium varies from optimum, the amount of energy and time that must be devoted to burrowing diverts from other life processes. Many species sustain burrows year round for many life functions while others maintain them part of the year, usually for winter protection, hibernation, and breeding. Availability of snow of the right depth and quality may be critical for survival of overwintering birds and mammals because it provides insulation from severe cold. For desert wildlife, burrows provide insulation from extreme heat and cold as well as a more humid retreat for water conservation. Some species rely on other species to provide burrows. Prairie dog burrows, for example, may be occupied by burrowing owls, snakes, other rodents, and black-footed ferrets. The availability of suitable burrow habitat may be the limiting factor for dependent wildlife, but artificial burrows are only rarely provided through wildlife management because it is expensive.

Vegetation provides cover needs for most wildlife. Vegetation structure is the primary objective of habitat cover management. The density and form of vegetation determines the visibility of many species of wildlife. Just as important, it modifies the

FIGURE 15.15   Healthy wildlife habitats provide food, water, space, and cover for a variety of species. Note the diversity of plants and how these desert mule deer blend in with their background (courtesy of New Mexico Department of Game and Fish).

microclimate of animals and especially air circulation. Endothermic wildlife, which generate their own body heat (birds and mammals), require much less energy to sustain body temperatures where retreats from wind and precipitation are readily available. The tangle of stems and branches in grassland and forest habitats foils the larger predators of small prey species. Den trees with rotted cavities provide retreats and nest space for woodland mammals, birds, and reptiles. Downed and rotting logs are important habitat resources for salamanders and snakes.

Habitat Arrangements.   Animals need to move freely among areas with food, water, and cover within their home range. The home range is an area in which an animal obtains its requirement for survival and reproduction. Animals neither occupy nor use habitats randomly. The arrangement of habitats providing food, water, and cover is as important as the total amount of habitat. The juxtaposition (proximity) and interspersion (mix) of needed habitat components, such as feeding areas and escape cover, are important in the animal's daily movements. Figure 15.16 shows high-quality farmland habitat in Wisconsin. Note the irregular fields with odd areas and strips of trees that provide travel corridors.

The size of the area necessary for providing the habitat needs of an organism varies greatly. A cottontail rabbit can live its entire life within 1.25 acres. A wolf can cover an area of 25 sq. miles or more while a male mountain lion may require over 100 sq. miles. Migratory animals may have two home ranges, such as many waterfowl that have a summer and winter home range separated by hundreds of miles.

The Concept of Edge.   The more diversity that exists in a habitat the greater the probability that it will provide the necessary requirements for a variety of wildlife. This is the reason why wildlife tends to be more abundant along the edges of adjoining habitats (Adams 1994). This edge effect can occur in natural situations, as where a forest meets a grassland or it can be human-induced, as where agricultural lands adjoin

FIGURE 15.16   Farmed landscapes can support diverse wildlife populations if interspersed with natural vegetation. This farmland in Wisconsin supports healthy populations of deer and numerous other wildlife populations (courtesy of U.S. Department of Agriculture).

wooded areas or fence row vegetation (Figure 15.17). Creating the edge to suit the needs of desired wildlife is one of the most important concepts in habitat management.

Edges usually support a greater number of species because the greater variety of habitats provides more physical *niches* for occupancy by different species. The variety of plants present, including grasses, forbs, and shrubs, produce a variety of fruits, seeds, and nuts. They support insect populations and other invertebrates essential in meeting the dietary needs of a host of wildlife species. They also create different structural patterns offering a variety of cover types to different species (Figure 15.18).

High edge development is more likely to provide the different food and cover needs of animals in different seasons. Also, it is more likely than monotonous habitat to provide all of the diverse needs of species with complex life cycles. The habitat needs for breeding, feeding, loafing, and rearing young for most wildlife species typically are better met in areas of high edge development than where large continuous stands of vegetation occur (Martin 1992). On the other hand, a few species are adapted to large expanses of the same habitat (i.e., spotted owl) and are at a disadvantage in areas where there is high edge development. Most of these species are associated with climax forest communities (Reese and Ratti 1988). Too much edge can be as undesirable as too little edge if biodiversity is to be sustained (see Chapter 17).

Wildlife Associations With Successional Communities.   Community succession is a basic concept in wildlife habitat management (see Chapter 2). Succession refers to the sequence of communities that replace one another on a given area. If, for example, a forest area is cleared of vegetation by fire or other natural or human-induced change, a series of vegetation types occur. This begins with pioneer stages (low seral) of initial vegetation growth (small forbs and grasses), which are replaced by more mature communities (midseral and high seral stages composed of shrubs and trees) until the original dominant vegetation community or climax vegetation reestablishes itself; in this case, a forest. Wildlife can be classified according to the successional stage they prefer.

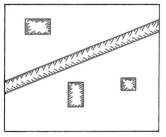

Habitat diversity with little interspersion and poor edge effect. Four plant communities (letters) meet just once in this case.

Straight ditch and regular sided patches of vegetation (shaded areas) add little edge effect.

Large ponds (shaded areas) have relatively little shoreline edge per surface area of water.

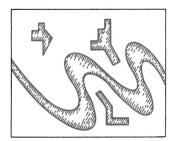

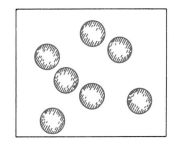

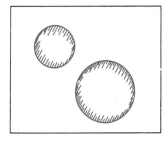

Habitat diversity with good interspersion and edge effect. Four plant communities now have many more contact points, without reducing the total area occupied by each.

Meandering ditch and irregularly shaped patches greatly increase the amount of edge for wildlife use.

Numerous small ponds create a large proportion of edge with no loss of total surface area.

FIGURE 15.17   Schematic diagrams comparing low interspersion and edge effect (top row) with high interspersion and edge effect (bottom row) (reproduced from Bolen and Robinson 1999).

FIGURE 15.18   Wild turkeys in central Texas taking advantage of edge effect. They use the more open areas for feeding and the more heavily wooded areas for escape cover.

Wildlife biologists use this knowledge to modify habitats to accommodate and increase the numbers of specific species being managed.

Wilderness or climax wildlife are those species that prefer undisturbed habitats (Table 15.1). Species in this category occur in wilderness areas, national parks and national forests and include the grizzly bear, polar bear, mountain goat, caribou, musk ox, wolf, mountain lion, lynx, and wolverine. These species are most harmed by human land use. Therefore, suitable habitat usually occurs far removed from human impacts. The grizzly bear, caribou, and mountain sheep are in this category because they fail to thrive in close contact with humans. In the case of grizzly bears, it is because they prey on livestock and can be dangerous to humans. Bighorn sheep are susceptible to diseases transmitted by livestock. Mountain goats and polar bears require such rough or extreme habitats that they are automatically wilderness species. Mountain lions are more likely than other wilderness species to move into areas where suburbs abut wilderness. Here they can occasionally threaten human life.

Forest and range (midstage in vegetation succession) wildlife consist of species inhabiting lands compatible with forestry and livestock interests (Figure 15.19). These respond well to human-induced changes in habitats. Many common game species are in this category, such as white-tailed deer, mule deer, elk, moose, turkey, ruffed grouse, scaled quail, black bear, pronghorn antelope, several species of tree squirrels, and alligators (Figure 15.19). White-tailed deer, turkey, raccoons, and some

TABLE 15.1   **Examples of North American Mammals and Birds Associated with Climax and Seral Communities**

Climax Communities		Seral Communities	
Mammals	Birds	Mammals	Birds
Caribou	Prairie chicken	White-tailed deer	Bobwhite quail
Mountain goat	Spotted owl	Mule deer	Mourning dove
Bighorn sheep	Spruce grouse	Pronghorn antelope	Ruffed grouse
Bison	Ptarmigan	Elk	Scaled quail
Musk ox	Mearns quail	Prairie dog	Red-tailed hawk
Grizzly bear	Meadowlark	Jackrabbit	Lark bunting
Polar bear	Aplomado falcon	Coyote	Loggerhead shrike
Wolf	Cassins sparrow	Fox	Brownheaded cowbird

FIGURE 15.19 White-tailed deer are woodland species that respond well to human-induced habitat changes. In this case logging has increased edge effect and food supplies for both deer and moose (from Hunter 1990).

grouse do well in landscapes interspersed with forest patches and cultivated areas. White-tailed deer, squirrels, and raccoons are among the forest species most likely to invade suburban habitats of cities. Traveling by waterways into backyard boat slips and parks, alligators have become a nuisance and even life threatening in some southern suburban locations. Mule deer, pronghorn, and elk tolerate moderate grazing by cattle. Most of these species do best when the stages of vegetational succession are below climax (Wagner 1978).

Farm wildlife are adapted to the earliest stages of vegetation succession and exhibit a high tolerance for habitat modifications, including urban habitats. They typically have small home ranges. These species present the least habitat problems because they do well in agriculturally developed areas. They do require interspersed areas of natural vegetation free of cultivation for escape cover, loafing, and raising young. Examples of these species are bobwhite quail, mourning doves, ring-necked pheasant, cottontail rabbit, fox squirrel, opossum, and red fox (Burger 1978).

Habitat Fragmentation. Modern forestry, farming, and urban development have led to lower wildlife populations because of large-scale conversion of the pristine habitat to early successional stages or introduced plant communities. Habitat fragmentation is the process by which habitats are reduced to small and disconnected units. The ratio of edge to habitat area increases dramatically and favors those species that flourish where there is high edge development. One good example of an apparent edge effect is the increase of cowbirds, an edge species, at the expense of forest species. Cowbirds are nest parasites that lay their eggs in the nests of smaller bird species. Because of size, the young cowbirds have an advantage over the young of parent birds and often are the only survivors in the nest. Cowbirds have been increasing in abundance in the eastern U.S., especially in areas where deciduous forest is fragmented by agriculture and suburban development. Roads and rights-of-way contribute to fragmentation and may be associated with observed declines of some species (Adams 1994, Robbins 1991).

FIGURE 15.20   Large fields without buffer strips, windbreaks, or other natural vegetation planted to the same crops year after year such as this Washington wheat field provide little habitat for wildlife. The trend in recent years has been away from this type of farming because of problems relating to soil erosion, pesticide requirements, and eradication of wildlife habitat (courtesy of U.S. Department of Agriculture).

Mechanized agriculture has enabled the clearing of vast areas of land eradicating needed cover and food resources for many wild animals. Mechanization, with its use of large farm machinery, greatly increased the size of plowed fields (Figure 15.20). It often eliminated native vegetation that existed between the smaller fields. These uncultivated strips provided food and cover for songbirds, small mammals, and game animals such as cottontails and bobwhite quail. Another detrimental factor for wildlife resulting from mechanized agriculture was the shift toward cultivation of a single crop such as soybeans. The practice of crop monoculture eliminates the diversity of plants needed by most species of wildlife. It also increases the probability that the few species that respond positively to the habitat change will become pests requiring expensive control. Grain-feeding blackbirds, for example, have become extremely abundant in many agricultural areas and present an animal damage control problem of impressive proportion (Bolen and Robinson 1999).

Large tracts of pure pine plantations, large single-species grass pastures, expansive cuts of mature timber, and other monocultural practices typically provide poor wildlife habitat because they decrease the biodiversity of species contributing to food and cover variation. One of the primary strategies of wildlife management is to provide a diversity of vegetation communities in a given area to meet the needs of several species. Considerable research is underway to define the optimum areas of habitat and the amount of habitat interconnection to maximize or optimize species diversity. Habitat fragmentation is also discussed in Chapters 3, 16, and 17.

Habitat Management Strategies

Habitat management typically is expensive. It is most justifiable when provision of a small amount of habitat will have a large impact on wildlife survival or birthrate.

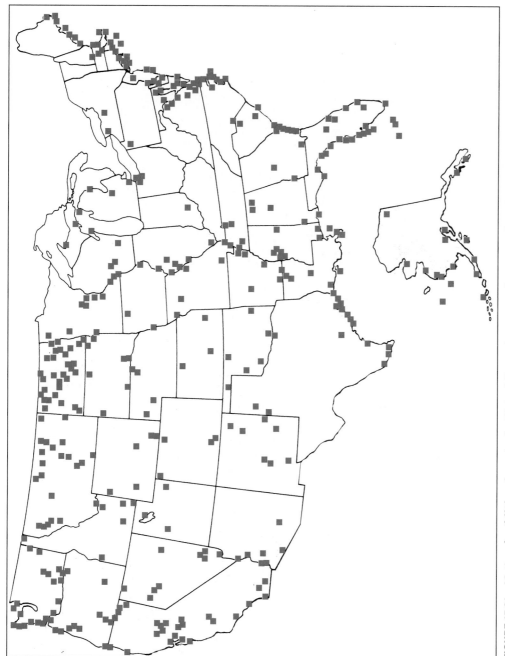

FIGURE 15.21 National wildlife refuges in the United States (courtesy of USDA-Fish and Wildlife Service).

FIGURE 15.22   Federal rangelands under control of the Bureau of Land Management and U.S. Forest Service are now managed to provide habitat for wildlife as well as livestock. The proper juxtaposition and interspersion of habitat components is critical for meeting the needs of diverse wildlife populations, such as these mule deer (courtesy of New Mexico Department of Game and Fish).

Nesting boxes for birds and the creation of ponds and wetlands for waterfowl are examples of artificial habitat manipulation that can be highly effective. Wildlife refuges operated by state agencies and the Fish and Wildlife Service typically emphasize protection or creation of scarce habitat (Figure 15.21). A high proportion of refuges are wetland habitats critical for wildlife reproduction and resting during migrations. Other extensive refuges are in wilderness areas where conflicting use demands are minimal, such as in Alaskan tundra and in the Nevada desert.

Forest and range habitats are too dispersed and too much in demand for other uses to justify extensive conversion into refuges. Early in the century, forest refuges were established on U.S. Forest Service lands, mostly as retreats from overhunting. They were later eliminated as their effectiveness as habitat refuges was shown to be minimal. Instead, the needs for forest wildlife were incorporated into a multiple-use approach to habitat management. In that approach, forest management for timber, watershed, and recreational uses is balanced against critical habitat needs for wildlife. For example, the location, size, shape, and intensity of timber harvests are considered for their impact on wildlife habitat as well as timber revenues. Similarly, public rangelands are now managed to provide forage for wildlife as well as livestock (Figure 15.22). Attention paid by the Bureau of Land Management and Forest Service to water and riparian resources for wildlife has been most critical in recent years. The cover, water, and food values associated with riparian areas make them both critical and scarce habitat embedded in larger range landscapes (see Chapters 9 and 11). Because multiple use often provides greater total human benefit than single use, its application often justifies higher management costs on public lands.

Managing Wildlife Populations

Because the states have complete jurisdiction over the take of nonendangered resident wildlife, state wildlife agencies have the primary responsibility for wildlife population management. Compared to the federal government, state land control is relatively minor. For wildlife, it is generally limited to scarce habitat acquisition that can be justified for special wildlife needs—most usually for wetland species management. However, in densely populated eastern states, some public lands have been purchased specifically to allow hunting access. This is because private owners generally prohibit hunting on their lands. For many states where public land is scarce, management of wildlife is primarily through laws regulating wildlife use, especially wildlife harvest.

Harvest Management. A basic premise of wildlife management is that populations can be managed for sustained resource use while individual animals cannot. Death comes to all animals in one form or another, including the death incurred by human harvest. Harvest refers to the take of wildlife for sport, subsistence, or commercial purposes. Harvest involves killing or otherwise removing wildlife from populations by shooting, trapping, snaring, or other means. Whether or not an animal is killed, the population effect is the same when it is removed from the population—it contributes to the population loss rate (see Chapter 3 for details on population dynamics). Wildlife is a renewable natural resource, meaning that mortality can be replenished by natality (birthrate) if the population is not overexploited and if required habitats are provided. Sustaining a wildlife population at a specified number requires balancing the birthrate (natality) and the death rate (mortality). Such management can result in increased densities.

Wildlife managers must objectively determine whether harvest through hunting or other means decreases population abundance below a targeted sustainable number. Based on many studies, wildlife agencies have determined that removing "a surplus" of animals from some populations is not detrimental to their long-term sustainability. Studies of quail have shown that population numbers fluctuate regardless of population size. A certain portion of the population will die each year from disease, active predation, and other causes and hence hunting a population in the fall removes those animals that would normally die of other causes during the winter. Of course, regardless of the cause of death in a natural setting, the dead animal usually is consumed by other animals.

When humans harvest wildlife, they necessarily alter the flow of energy through the ecosystem and decrease the productivity of other species. Responsible wildlife management assures that this effect does not cause decline in diversity and extinctions. To avoid this, a larger ecosystem view of effected populations is required rather than the traditional focus on the viability of the harvested species.

Present agricultural practices and urbanization have greatly modified the natural functioning of wild populations (Flather et al. 1994). The elimination of key predators, such as wolves, has been particularly important. Under these conditions, some species of animals such as whitetailed deer overpopulate areas, leading to depletion of their food resources and eventual decrease in the quality and abundance of the population. Deer in a borderline nutritional state associated with over-population have enough food for subsistence only and are highly vulnerable to die-offs caused by disease or other stress. Disturbance of ecosystem controls, such as predation, often result in more erratic population fluctuations with longer periods of high and low population abundance than what occurred before the disturbance. It also often results in habitat degradation by overly abundant prey species.

In urban areas, overpopulation of wildlife leads to high incidence of animal-caused damage (see also Chapter 14). The increase in elk in some urban areas of Alberta, Canada, and moose in New England has led to human injuries. These problem elk and moose have had to be removed. In the eastern United States, large populations of black bears and white-tailed deer in some areas have caused concern because of damage to property and human well-being. A large source of human injury is wildlife collision with automobiles. The increased incidence of Lyme disease, a virus spread to humans by tick bites, has been aided by the high populations of deer and other wildlife which harbor the ticks. The threat of rabies has increased with urban sprawl. This is because the abundance of garbage increases the incidence with which pets and humans encounter potentially rabid animals, such as raccoons and skunks.

Under such circumstances, managing wild populations without hunting or some other form of lethal removal is difficult and often prohibitively expensive. However, alternatives such as animal trapping and transfer frequently are used where hunting is not acceptable. Animal transfers are expensive and often only shift problems because resident wildlife already occupy the available habitat. So far methods used to induce infertility so as to avoid outright killing have proven impractical (too costly). While an increasingly vocal minority are repelled by any human killing of wildlife, the ecological and economic realities lead the majority to accept the most cost-effective safe and humane control available—and that is usually some form of killing.

The elk herd in the Grand Teton region of Wyoming provides a rural example of human-induced changes that have resulted in overpopulation of ungulates (hoofed mammals such as deer). The original wintering range of elk is now occupied by cattle ranches and agricultural development. The elk spend summers in national parks, but in winter they seek food in private lands, which causes conflicts with the landowners. A hunting program has been established to remove excess animals. Bison often wander outside the boundaries of Yellowstone National Park. They also come into conflict with private landowners. Because this population of bison has been shown to be possible transmitters of brucellosis to domestic livestock, they have been shot when they leave the park (Bolen and Robinson 1999). Elephant populations in Africa have caused similar conflicts because they destroy agricultural crops.

There is increasing evidence that remote sources of habitat change can have major effects on wildlife abundance. For example, the increasing probability of continued climatic warming has many implications for wildlife management. A good example of how remote effects can combine to cause population changes is provided by a recent study of snow geese (Figure 15.23). Because of changes in habitat and harvest management, snow goose populations have reached densities that are degrading their northern tundra nesting and rearing habitat in Canada. Most of the population increase has come as a consequence of much improved winter habitat in the south-central U.S. where there is a history of increasingly mild winters and conversion of habitat area to agricultural production. Grains, which are consumed by the geese, have improved their reproductive condition and increased their egg clutch size once they reach the nesting grounds. The changes in winter temperature may be a consequence of global warming caused by atmospheric management policies. A combination of decreased interest in hunting and conservative harvest limits set by the states has prevented increased mortality from balancing the increased natality. Researchers

FIGURE 15.23 Snow goose populations in North America have increased to levels that are now endangering nesting habitat. Global warming may be a factor in their increase (courtesy of Jack McCaw).

predict that the population will eventually crash when the food resources for goslings are totally overwhelmed. The dimension of management problems encountered with snow geese shows how important a total ecosystem perspective is in wildlife management.

Harvest strategies have shifted from maximum sustained to optimum sustained yield. Initially, the objective in harvest management was to maximize sustained yield over a long period of time without decreasing future harvest. This concept was based primarily on the production of goods, such as from market hunting and commercial fisheries (see Chapter 16 for further development). Maximum sustained yield was a difficult management objective to attain because potential yields often vary annually, depending on ecological conditions. Only when managers have complete data on population structure and size, which is often not the case, can a reasonably precise estimate of maximum yield be made. Misjudgements frequently result in overharvest and prolonged reduction of allowable population yield. A philosophy of optimum sustained yield replaces maximum sustained production of specific goods, such as meat. This allows balancing all goods and services of benefit to society over the long run and allows for error in estimating sustainable yield.

Stocking and Reintroduction.   The reintroduction of game species had been for the most part a highly successful aspect of management in the early years of state agency programs. After World War II, reintroduction had become mostly a thing of the past as most of the extirpated populations were restored wherever suitable habitat remained. During those early years, another common activity was the continuous stocking of game birds in areas that underwent high natural or hunting mortality (Allen 1984). However, scientific study soon revealed that stocking usually was too costly for the human benefit produced. Except for some intensively hunted private lands, routine stocking was discontinued.

Following passage of the Endangered Species Act, reintroduction into suitable habitat became one of the approaches for recovery of endangered species populations. However, recent attempts to reintroduce wolves in Yellowstone National Park, eastern Arizona, and in isolated locations in the southeastern U.S. have been quite controversial. Discussion of grizzly bear reintroduction also generates controversy. Although habitat protection and restoration are emphasized in endangered species management, reintroduction will remain an important part of recovery programs where suitable habitat exists.

COMMERCIALIZATION AND WILDLIFE MANAGEMENT

Wildlife commercialization (private sector economic gain from exploitation of wildlife resources) is a controversial issue. Some contend that commercialization, in the form of paid hunting and the sale of meat and wildlife by-products, threatens the public ownership of wildlife and our whole system of wildlife management. Others contend that only the wealthy nations, such as the U.S., can afford the luxury of limited wildlife use mostly for recreation. The commercial value of wildlife is a reality in many parts of the world, including the U.S. In some cases commercial development is the only impetus to conserve wildlife resources.

Although some countries raise significant income from ecotourism and other nonconsumptive use of wildlife, such as nature viewing in national parks, these activities alone do not guarantee the sustainability of wildlife. In many parts of the world, wildlife must be able to generate economic gains for the general populace. Welfare conservation, the dependence on foreign aid or domestic government funding to support conservation projects, is only a short-term solution. In Africa, large areas are being used to produce wildlife for harvest as a means for economically justifying land use where wildlife is the principal product.

FIGURE 15.24 Hunting leases on private land are becoming more important as a source of income to landowners. Modern agriculture and hunting are generally compatible. In the photo grain stubble has been left to provide cover for pheasants and good shooting for hunters in a private club (courtesy of U.S. Dept. of Agriculture).

Where wildlife remains nonmarketable in economically underdeveloped regions of the world, rural populations and landowners will not participate in conservation programs. Wildlife habitats will continue to be degraded in favor of agricultural development and wildlife will continue to be viewed as an impediment to economic gain. Wildlife use can be particularly appropriate where other forms of land use can be detrimental, such as in marginal lands easily eroded and not favorable to agricultural development or livestock grazing. Wildlife commercialization can enhance biodiversity by maintaining natural habitats outside protected areas while enhancing rural economic development.

Commercialization of wildlife in the United States is legally practiced in a variety of enterprises. In leasing programs, hunters purchase exclusive rights to hunt on a private property for an agreed upon time. Leases are available for big game, waterfowl, game birds, such as pheasants and quail, and sportfishing (Figure 15.24). In some cases, income from hunting enterprises exceed that of livestock production. Property owners can set quotas for the number of animals collected, but must abide by state and federal laws that restrict the methods of hunting, bag limits, and seasons, with some exceptions.

In Texas, where many exotic species are available for hunting on private land, the state has no laws governing exotic harvest and the landowner is given full discretion. In California, a special program known as Ranch for Wildlife permits landowners to develop a wildlife management plan detailing the wildlife and fish enhancement programs. If approved by the California Fish and Game Department, the landowner is licensed to operate a private wildlife program, which allows the landowner to charge for hunting and fishing under special game seasons and limits approved by

FIGURE 15.25 Wildlife can be a lucrative source of income to private landowners. Bighorn sheep hunts have been auctioned for several hundred thousand dollars. Some landowners in arid areas of Sonora, Mexico have abandoned livestock production and devoted their efforts into more profitable wildlife enterprises (courtesy of the Bighorn Institute).

the department. In some cases, depending on game population numbers, seasons and game bags may be more liberal than statewide regulations. This and other programs stimulate private participation in wildlife conservation and permits the landowner the flexibility needed to manage wildlife and fishery resources for increased income. It also increases hunter participation of hunting license sales and sustainable management of wildlife and fishery resources on private lands. In Texas ranches with quality trophy buck populations, seasonal leases per acre range from $5 to $7. A 4- to 5-day trophy deer hunt can be priced from $2,000 to $5,000. Landowners in Sonora, Mexico charge $50,000 for a desert wild sheep hunting permit (Figure 15.25). In desert areas such as Sonora, where cattle production is highly risky because of the uncertainty of forage production, big game hunting enterprises are an important source of supplemental income. In some cases, landowners have abandoned cattle production and are concentrating on wildlife ranching enterprises.

Game ranching (production of wildlife on open grazing lands) and game farming (intensive husbandry of wildlife for commercial purposes), often using exotic species, are commercially viable enterprises in several countries. In the U.S., fallow deer have been farmed in at least two states. Game farms can produce large numbers of deer in a small area. One Texas farm has 3,500 fallow deer on about 400 acres. A pivot irrigation system provides for the growth of fresh forage, and diets are supplemented with high protein and energy foods. In New Zealand, European red deer on game farms exceed 1 million animals. The sales of red deer antlers and meat now exceed $300 million.

Texas has imported more species of exotic ungulates (hoofed mammals) than any other state or country. Some of these exotics and their hybrids, such as ibex (Asian wild goat) and domestic goat crosses, are used in game ranching operations. In 1963, a state survey revealed 13 exotic species totaling 13,000 head. By 1988, there were 67 species totaling 164,257 animals on 486 ranches. Six species had established wild free-roaming populations: nilgai antelope, blackbuck antelope, and axis deer from India, sika deer from Japan, fallow deer from Europe, and aoudad or Barbary sheep from northern Africa. In some ranches, exotics became more numerous and were considered to have greater priority than native species because of the high prices some hunters are willing to pay for hunting trophy animals. Game ranchers have come to realize that native species such as trophy white-tailed and mule deer command higher prices than common exotics. Nonetheless, exotics will remain a viable income option for Texas ranchers (Mungall and Sheffield 1994).

Marketing can be a year-round option because exotics in most counties are exempt from state regulations. They can be incorporated in multispecies grazing systems with livestock enterprises. Negative aspects of exotic game ranching include transfer of diseases to native wildlife and livestock and competition for forage resources with native species. Some states, viewing the potential dangers of exotics, have prohibited game farming and the use of exotics in game ranching.

CHALLENGES AND TRENDS IN WILDLIFE MANAGEMENT

Wildlife managers in the U.S. have been especially successful in establishing a system of laws and enforcement that have sustained game populations and prevented extinction of several species. Public land management has greatly improved over the past century, integrating wildlife with other forms of land use. Wildlife management has been less successful in protecting and restoring habitat for nongame wildlife, especially on private lands, and protecting wildlife from habitat contaminants of various kinds (Flather and Hoekstra 1989). As a consequence, numerous wildlife species are threatened or endangered.

Many wildlife species require large tracts of land to maintain viable populations. Mountain lions, for example, require hundreds of square miles to acquire the necessary resources for survival and reproduction. Where land is private, public agencies have not been as effective as certain nongovernment organizations, such as the Nature Conservancy, in maintaining habitat. Migratory birds can travel thousands of miles during migration to and from their nesting grounds. This requires the coordinated management of federal and state wildlife agencies and those of other countries, and private landowners (Agee and Johnson 1988).

However, even coordination among federal agencies is difficult and unqualified success stories appear rare. One of the better examples of apparent success is the North American Waterfowl Management Plan, which has significantly advanced coordination of wetland habitat development across nations along migratory pathways. All too often, federal agencies do not coordinate their management plans, resulting in different goals, which sometimes are counterconstructive to wildlife management programs and wasteful economically (Wright 1992). A recent movement toward what is called ecosystem management is an attempt to better integrate agency activities (see Chapter 23).

The ever-increasing competition for natural resources between the needs of wildlife-based benefits and other benefits is a constant concern. Water diversion and groundwater pumping for irrigated agriculture and human and industrial use will mean less water for wildlife. Wildlife managers will also have to deal with the toxic trace elements and pesticides from agricultural and industrial contamination (Langner and Flather 1994). They continue to pollute ecosystems even after the Federal Water Pollution Control Act of 1972. But the water pollution threats are worse in other parts of the world. Pollution of inland seas and coastal estuaries, combined with overfishing, is diminishing the world's fish catch. The reduction in marine fish is causing increased concern for marine mammals and birds also dependent on fish.

Millions of acres of undeveloped lands are annually converted to urban, transportation, and specialized commercial uses (see Chapter 14). Industrial expansion, urban sprawl, road construction, water resource development, and other factors have often eroded the quality and quantity of habitat to the detriment of wildlife species (Flather et al. 1994). Only the collective cooperation of resource managers, landowners, concerned conservation organizations, and general public can reverse the continued decline of wildlife habitats. Progress in recent decades has shifted from an emphasis on protection of environmental resources, including wildlife, to restoring degraded ecosystems of significant environmental value.

FIGURE 15.26 The availability of accessible wild landscapes with abundant and diverse wildlife populations is a national heritage that will require the cooperation of the public and private sector to maintain its esthetic and economic values for future generations.

Most progress has been made on public lands. However, of the 2.26 billion acres of land in the U.S., 60% is privately owned. The percentage is much higher in the eastern U.S. than in the West. Private land ownership and control still remains a paramount pursuit of most Americans. Obviously, private lands must be included if programs to benefit wildlife and other renewable natural resources are to succeed. Proper policy incentives for wise land use and wildlife conservation must be provided to landowners or they will be unwilling to make necessary monetary investments. It is difficult to convince farmers that they should not drain a wetland, clear a shrubland, or plow a grassland if it does not translate into more cash returns and less financial disparity in their budgets. Similarly, it is difficult to convince real estate developers that provisions for wildlife should be included in regional plans if they do not translate into profit. Changes in knowledge about the costs of uninformed and unregulated private land development start with education of each citizen.

Natural resources managers, farmers, ranchers, agriculturists, and land developers must understand the needs and concerns of each other, integrate planning for more effective management—and compromise when necessary for the larger good (Marcot et al. 1994). This requires an investment in objective research and diversification of wildlife conservation and management concepts. Diversified wildlife management can be accomplished through flexibility and cooperation.

Two major types of management are practiced in North America: protective management and multiple-use management. Protective management refers to land that is locked into parks and preserves and is dedicated to a single use or highly compatible multiple uses (Figure 15.26). Multiple-use management is commonly practiced by most agencies involved with wildlife resource management. In this system, the benefits and values of various resources are balanced against one another, including timber, livestock grazing, mining, energy extraction, and wildlife resources. In the past, multiple-use management often resulted in independent resource management proposals and resolution of conflicts that became evident as the implications of different proposals become clear. In such management, most of the planning process is conducted independently and then aligned at relatively late planning stages. As Chapter 23 discusses, recent analyses of the management process reveal a need for more integration of wildlife programs with those for timber, rangeland, water, fishery, recreation, wilderness, and mineral development through an integrated planning process. Educational and technical assistance programs that aid private landowners to improve habitats should be encouraged in order to promote the enhancement and sustainability of wildlife resources.

Although the majority of the public is "pro" wildlife, they are woefully uneducated about the values of wildlife resources, the dynamics of natural populations and ecosystems, and the costs of management that sustain wildlife. They typically relate more to the welfare of individual animals than to the sustainability of populations in ecosystems. There is growing sentiment against any killing of wildlife and a general shift toward emphasis on nonconsumptive wildlife recreation. This has contributed to a gradual decline in funds derived from hunting licenses over the past two decades. Because that was a primary source of funds for management and research, including numerous nongame issues, supportive research and management funding is not keeping pace with need. Ecological and social education needs to be made more effective in primary and secondary schools if the public is to become more knowledgeable about funding needs for research, management, and land acquisition. Even though the state and federal agencies responsible for wildlife and fisheries management are staffed by knowledgeable, dedicated personnel, it will require the cooperative efforts of all citizens to ensure that the animals and their habitats are sustained. The professional resource managers of the future also will need to be better communicators and educators of the public as well as technically competent in managing habitat and populations.

LITERATURE CITED

Adams, L. W. 1994. *Urban wildlife habitats: A landscape perspective.* Minneapolis, MN: University of Minnesota Press.

Agee, J. K. and D. R. Johnson, eds. 1988. *Ecosystem management for parks and wilderness.* Seattle, WA: University of Washington Press.

Allen, D. 1984. *Our wildlife legacy.* Rev. ed. New York, NY: Funk and Wagnalls.

Anderson, S. H. 1991. *Managing our wildlife resources.* 3rd edition. Upper Saddle River, NJ: Prentice-Hall.

Ankney, C. D. 1996. An embarrassment of riches: Too many geese. *Journal of Wildlife Management* 60:217–223.

Bavin, C. R. 1978. Wildlife law enforcement (pp. 350–364). In H. P. Brokaw, ed.,*Wildlife and America.* Washington, DC: U.S. Government Printing Office.

Bean, M. J. and M. J. Rowland. 1993. *The evolution of wildlife law.* 3rd edition. Westport, CT: Praeger.

Bolen, E. G. and W. L. Robinson. 1999. *Wildlife ecology and management.* 4th edition. Upper Saddle River, NJ: Prentice-Hall.

Bookhout, A., ed. 1994. *Research and management techniques for wildlife and habitats.* Bethesda, MD: The Wildlife Society.

Burger, G. V. 1978. Agriculture and wildlife (pp. 89–107). In H. P. Brokaw, ed., *Wildlife and America.* Washington, DC: U.S. Government Printing Office.

Council for Wildlife Conservation and Education. 1992. *The hunter in conservation.* Washington, DC: Wildlife Management Institute.

Cronon, W. 1983. *Changes in the land: Indians, colonists and the ecology of New England.* New York, NY: Hill and Wang.

Dunlap, T. R. 1988. *Saving America's wildlife.* Princeton, NJ: Princeton University Press. 222 pp.

Dunn, C. P., F. Stearns, G. R. Guntenspergen, and D. M. Sharpe. 1993. Ecological benefits of the Conservation Reserve Program. *Conservation Biology* 7:132–139.

Fazio, J. R. and D. L. Gilbert. 1981. *Public relations and communications for natural resource managers.* Dubuque, IA: Kendall/Hunt Publishing Co.

Flather, C. H. and T. W. Hoekstra. 1989. *An analysis of the wildlife and fish situation in the United States: 1989–2040.* Ft. Collins, CO: USDA Forest Service General Technical Report RM-239.

Flather, C. H., L. A. Joyce, and C. A. Bloomgarden. 1994. *Species endangerment patterns in the United States.* Ft. Collins, CO: USDA Forest Service General Technical Report RM-241.

Grosse, N. J. 1997. *The protection and management of our natural resources wildlife and habitat.* 2nd edition. Dobbs Ferry, NY: Oceana Publications.

Hornaday, W. T. 1913. *Our vanishing wildlife, its extermination and preservation.* New York, NY: New York Zoological Society.

Hughes, J. D. 1996. *North American Indian ecology.* El Paso, TX: Texas Western Press.

Johnson, F. R. 1987. Wildlife benefits and economic values (pp. 219–228). In H. Kallman, chief ed. *Restoring America's wildlife, 1937–1987.* Washington, DC: USDA, Fish and Wildlife Service.

Kellert, S. R. 1996. *The value of life.* Washington, DC: Shearwater Books and Covelo, CA: Island Press.

Kimball, T. L. and R. E. Johnson. 1978. The richness of American wildlife (pp. 3–17). In H. P. Brokaw, ed. *Wildlife and America.* Washington, DC: U.S. Government Printing Office.

Langner, L. L. and C. H. Flather. 1994. *Biological diversity: Status and trends in the United States.* Fort Collins, CO: USDA Rocky Mountain Forest and Range Experimental Station, General Technical Report RM-244.

Leopold, A. 1933. *Game management.* New York, NY: Charles Scribner's Sons.

Leopold, A. 1949. *A sand county almanac.* New York, NY: Oxford University Press.

Marcot, B. G. M., M. J. Wisdom, H. W. Li, and G. C. Castillo. 1994. *Managing for featured, threatened, endangered, and sensitive species and unique habitats for ecosystem sustainability.* Portland, OR: USDA, Forest Service Pacific Northwest Research Station General Technical Report PNW-GTR-329.

Martin, T. E. 1992. Landscape considerations for viable populations and biological diversity. *Transactions of the North American Wildlife and Natural Resources Conference* 57:283–291.

Matthiessen, P. 1987. *Wildlife in America.* Rev. ed. New York, NY: Viking-Penguin.

Morrison, M. L., B. C. Marcot, and R. William Mannan. 1998. *Wildlife-habitat relationships.* 2nd edition. Madison, WI: University of Wisconsin Press.

Mungall, E. C. and W. J. Sheffield. 1994. *Exotics on the range, the Texas example.* College Station, TX: Texas A&M University Press.

Patton, D. R. 1997. *Wildlife habitat relationships in forested ecosystems.* Rev. ed. Portland, OR: Timber Press.

Payne, N. F. and F. C. Bryant. 1994. *Techniques for wildlife habitat management of uplands.* New York, NY: McGraw-Hill.

Reese, K. P. and J. T. Ratti. 1988. Edge effect: A concept under scrutiny. *Transactions of the North American Wildlife and Natural Resources Conference* 53:127–136.

Reiger, J. F. 1975. *American sportsmen and the origins of conservation.* New York, NY: Winchester Press.

Robbins, C. S. 1991. Managing suburban forest fragments for birds (pp. 253–264). In D. J. Decker et al., eds., *Challenges in the conservation of biological resources: A practitioner's guide.* Boulder, CO: Westview Press.

Scalet, C. G., L. D. Flake, and D. W. Willis. 1996. *Introduction to wildlife and fisheries.* New York, NY: W. H. Freeman.

Trefethen, J. B. 1975. *An American crusade for wildlife.* New York, NY: Winchester Press.

U.S.D.I., Fish and Wildlife Service and U.S. Department of Commerce, Bureau of Census. 1997. *1996 national survey of fishing, hunting, and wildlife-associated recreation.* Washington, DC: U.S. Government Printing Office.

Wagner, F. H. 1978. Livestock grazing and the livestock industry (pp. 121–145). In H. P. Brokaw, ed. *Wildlife and America.* Washington, DC: U.S. Government Printing Office.

Wildlife Management Institute. 1996. *Outdoor News Bulletin* 50:3.

Wright, R. G. 1992. *Wildlife research and management in the national parks.* Urbana, IL: University of Illinois Press.

CHAPTER 16

Fishery Conservation and Management

THE FISHERY RESOURCE

Fisheries exist wherever aquatic animals are actually or potentially harvested or captured by humans. Most harvest of finfishes (true fishes) is for food or other goods. However, some sport fisheries are nonconsumptive and many fisheries are for animals other than finfish. A catch-and-return sportfishery is one example of a nonconsumptive fishery. The least consumptive fisheries "capture" fish on camera or in viewing. Fishery capture more usually relies on various nets, traps, hooks, lines, spears, harpoons, arrows, snares, poisons, anesthetics, clubs, and even bare hands. Some types of fishing have much in common with hunting except they occur in aquatic habitats. Fisheries based on use of harpoons and arrows are examples.

Modern fishery science and management in the U.S. originated from fish culture over a century ago. Fish culture, now more commonly known as aquaculture, involves intensive management of waters to produce fish for commercial markets and for public management of sport fisheries. From aquacultural roots, the American Fisheries Society has grown to be the principal professional organization in the U.S., now representing diverse aquatic interests and concerns. It is involved with all areas of consumptive and nonconsumptive use, including preservation of fishery stocks for future resource renewal (Figure 16.1).

Since antiquity, aquatic resources have held great meaning for diverse cultures. Fishing implements are common among the artifacts of prehistoric cultures. Early Egyptian, Greek, Assyrian, Chinese, and Roman art and writing depict fishing by net, spear, rod, and line (Radcliffe 1921, Waterman 1975, Figure 16.2). The world's aquatic resources are now valued for food and other goods, for recreational experience, and for their wondrous biodiversity. The roots of the word "fishery" reach back to when most aquatic animals were casually classified as "fish."

Excepting birds, fisheries include resource populations from all vertebrate classes and many phyla of invertebrates. Vertebrate fisheries include the true fishes, certain turtles and frogs, porpoises, and whales. Among invertebrate animals, the crustaceans (crabs, lobsters, shrimp, crayfish) and molluscs (clams, oysters, scallops, mussels, squid, octopus, snails) are among the most important fisheries in the U.S.

FIGURE 16.1   The Coho salmon is an important recreational and commercial fish species that needs to immigrate upstream as adults to spawn. Although some natural barriers can be by-passed, high dams have contributed to declines by impeding passage (courtesy of USDI-Fish and Wildlife Service).

FIGURE 16.2   Depiction of Egyptians fishing about 4,000 years ago by various means, including rod and line (courtesy of USDI-Fish and Wildlife Service).

and elsewhere. These two phyla contain most of the shellfish species. Sponges, jellyfishes, corals, segmented worms, and sea urchins make up the remainder of marine invertebrate fisheries. Freshwater invertebrate fisheries include crayfish, freshwater shrimp, and freshwater mussels.

While the finfishes dominate aquatic diversity (finfish species outnumber the total of all other vertebrate species), weight harvested, and total economic yield, the most commercially valued category is invertebrate—the shrimps. Marine turtles, certain marine mammals, and certain salmoniform fishes are high-profile threatened and endangered species among the many aquatic species at risk. Important types of freshwater fish, marine fish, and shellfish are shown in Figures 16.3 and 16.4.

The first human interest in fisheries probably was for subsistence food and clothing. Subsistence fishing evolved into commercial food and recreational trade, now supporting employment income estimated at nearly $30 billion annually in the U.S. alone. Recently, with increased recognition of widespread degradation of aquatic resources, the concept of protecting resource options for future potential use has become a predominant concern in fishery management. This chapter follows this lineage of management concern starting with fishing for food and other goods, then fishing for recreation and, finally, biodiversity concerns. The chapter ends with a summary of different occupational activities in fishery resource management.

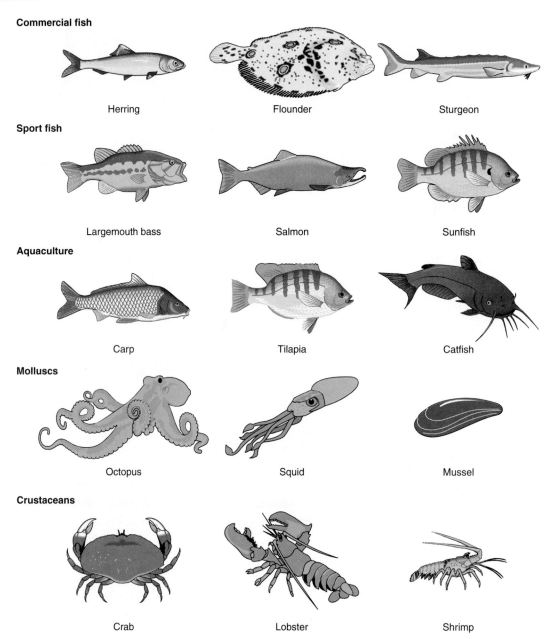

Commercial fish

Herring   Flounder   Sturgeon

Sport fish

Largemouth bass   Salmon   Sunfish

Aquaculture

Carp   Tilapia   Catfish

Molluscs

Octopus   Squid   Mussel

Crustaceans

Crab   Lobster   Shrimp

FIGURE 16.3   Some important examples of vertebrates and invertebrates that contribute to fisheries, including species emphasized in aquaculture.

FISHING FOR FOOD AND OTHER GOODS

Cultural Importance

Aquatic resources have provided important sources of protein, clothing, medicines, fuels, jewelry, animal feeds, and other goods for many cultures of the world (Miller and Johnson 1989). Subsistence fishing remains an important theme in sustaining the cultural integrity of remaining nonindustrial societies (Miller and Johnson 1989). For many other societies, fishing has become an important basis for commerce and monetary income. Commercial fishing was a well-established enterprise by the time of the Roman Empire (Radcliffe 1921).

Fisheries provide more protein for people than terrestrial animals (Miller and Johnson 1989). Until recently, use of unexploited fisheries kept pace with growth of the human population. However, the limits of the world's aquatic food resources are

Marine reptiles

Sea snake   Marine iguana   Leather back turtle

Marine mammals

Sperm whale   Manatee   Walrus

FIGURE 16.4   Some important marine mammals and reptiles, a few species of which have composed fisheries (e.g., sperm whales and sea turtles). Note: manatees are marginally marine species inhabiting both estuarial and totally fresh waters.

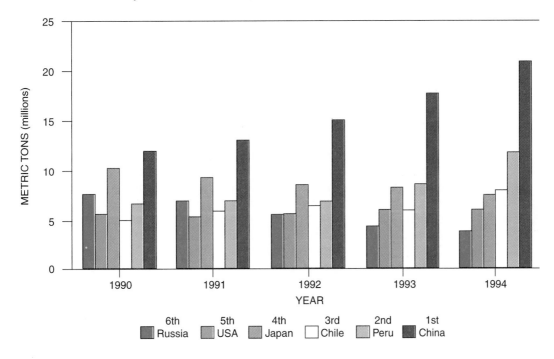

FIGURE 16.5   The world's commercial catch for each of the leading fishery nations during the early 1990s (from National Marine Fishery Service 1995).

now being reached, even in the vastness of the open sea (McGinn 1997, Johnson 1989). The world's commercial catch by the leading fishery nations is shown in Figure 16.5.

Early Fishery Allocation

The contemporary means by which food fisheries are allocated is based on a tradition of resource use that started over thousands of years ago in Europe and Asia. Monarchs held claim to territory for themselves and their subjects, reserving the most desirable resources for royalty. In medieval Europe all land initially belonged to "the crown" for use at its discretion. Private lands came from grants of crown land to independent ownership. Most of these lands were granted to aristocracy who then controlled access and fishery use of inclusive waters. Larger, navigable waters were more freely used for transportation and commerce including fishing. Commerce corridors became common-use areas for all citizens early in the evolution of nations from kingdoms and estates.

Early food fisheries remained close to land but eventual fishery depletion encouraged expansion into deeper waters where gear was limited mostly to long handlines holding hundreds of baited hooks. Crews from western Europe fished off the North American shore soon after Columbus landed (Waterman 1975, Kurlansky 1997). Conflicts over fishing rights grew as oceanic fishing expanded, contributing to the growth of national navies, larger ships, more far-reaching exploration, and new territorial claims. International tension has continued to accompany oceanic fishing throughout its long history.

North American Fisheries

Native American fisheries were well established when Europeans first colonized the Americas (Figure 16.6). Much of the early exploration financed by England was motivated by natural resources and the fishery resource of the northeastern Atlantic coast was found to be exceptionally bountiful. The Pilgrims came to New England in part to fish for food and commerce (Kurlansky 1997). By the eighteenth century, commercial fisheries supported numerous coastal economies in colonial North America. Most North American commercial fishing remained close to the East Coast until the nineteenth century, when it expanded rapidly along all coasts, inland, and over the open seas. Whale, finfish (mostly cod and haddock) became mainstays for employment in New England and Maritime coastal communities. Lobster fishing later became established there in the nineteenth century. The steam engine gradually replaced sails during the nineteenth and early twentieth centuries, extending daily reach from home ports. Fisheries thrived along the lower Atlantic and Gulf Coasts based on shrimp, crab, clam, oyster, shad, sea basses, and other finfishes.

After early desimation of fur seals and marine otters on the West Coast by Russian crews, salmon, halibut, herring, and oyster fisheries contributed significantly to local economies of newly established Pacific Coast communities. Fishing industries became established on the Sacramento and Columbia rivers and other West Coast locations soon after canning technology was developed in the middle 1800s. Later in the nineteenth century, salmon were efficiently caught in wheel nets, which cycled and dumped migrating salmon into attached barges. Halibut and salmon depletion resulted in formation of the International Halibut Commission in 1923 (Bell 1970) and the Pacific Salmon Commission in 1937 (Larkin 1970).

The Great Lakes and larger interior rivers also supported sizable settlements based on freshwater fisheries started in the early nineteenth century. Lake sturgeon populations plummeted rapidly followed by whitefish, chubs, and lake trout late in the nineteenth century. Although inland fisheries were mostly based on finfishes, significant shellfisheries developed for crayfish, freshwater mussel shell, and pearls in midwestern and southern rivers (Pennak 1978).

Technological Revolution

The late nineteenth and early twentieth century was a period of exceptional technical innovation that transformed marine fisheries. Internal combustion engines, hydraulic systems, electric generators, refrigeration systems, telephones, radios, and plastics were invented. These advances in technology initiated profound change in twentieth-century lifestyle, condition of the aquatic environment, and fishing efficiency. Steam power and electrical refrigeration were especially important because they allowed marine fisheries to extend far from shore. Harvest grew many fold, greatly benefiting numerous cultures. However, some fish populations began to decline from the combined effects of habitat deterioration and overfishing.

Fishery habitat problems increased rapidly as the U.S. industrialized in the late 1800s. Problems were caused by water project engineering, sediment from watershed abuse, industrial contaminants, and, after World War II, applications of artificially developed fertilizers and pesticides. Although environmental conditions have improved because of environmental laws, major habitat problems still remain. In addition to impeding fish migration, dams, canals, dikes, and other water control struc-

The manner of their fishing.

FIGURE 16.6 A sixteenth-century artist's depiction of native North Americans fishing with spears and a fish trap (upper left) (courtesy of USDI-Fish and Wildlife Service).

tures changed water quality, altered nutrient movements, and modified river flow characteristics. Environmental regulation of the newly emerging oil, plastics, and other chemical industries was weak at best. They released toxic materials directly into public waters. Soil erosion accelerated as agricultural technology expanded.

The once impressive inland fisheries of the Great Lakes suffered from damaging invasions of lampreys and alewives caused by placement of locks and canals in the Great Lakes (Smith 1970). Salmon abundances continued to decrease as dams accumulated on West Coast spawning rivers. Numerous spring and riverine habitats were threatened or eliminated by groundwater pumping and irrigation diversion. Many

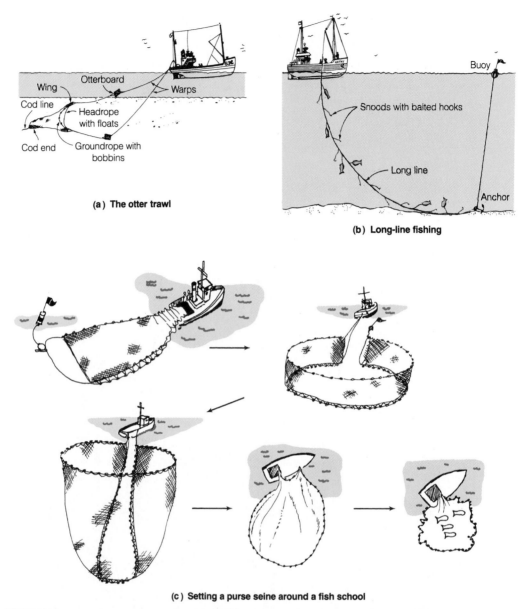

FIGURE 16.7   Common methods of commercial fishing (from Lerman 1986). The otter trawl (a) is towed just above bottom to capture shrimp and bottom finfish. Long-line fishing (b) requires the baiting of many hooks tied along lines extending thousands of feet. Purse seining (c) involves setting a long, deep net around and under a school of fish, then drawing the catch in.

lake, riverine, and estuarine fisheries declined from sedimentation, pollution, and eutrophication (see Chapter 3).

After World War II, marine fisheries rapidly expanded because of technical advances including sonar, improved marine communications equipment, and synthetic materials for larger and more efficient gear (Figure 16.7). By the 1970s, commercial fishers from diverse nations were using huge factory boats and fishing gear. Fish were harvested, processed, frozen, and stored for return to port and sale. The fraction of fish traded in the international market increased from under 10% before World War II to 20% in 1950 and over 40% at present (Miller and Johnson 1989).

Technological advances affected size of catch much more than selectivity of catch. Large masses of nonmarketable species, called by-catch, were killed and dumped overboard (Wise 1984). As the most desired commercial species were depleted, new species were selectively harvested using trawls, seines, gill nets, pound nets, and various other means. An effect of efficient harvest of marketable fish and mortality of by-catch was a reduction in food resources for fish-feeding fishes, mammals, and birds (Northridge 1984).

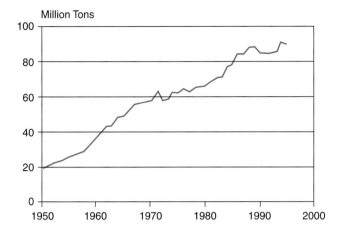

FIGURE 16.8 World fish catch, 1950–1995 (from McGinn 1997).

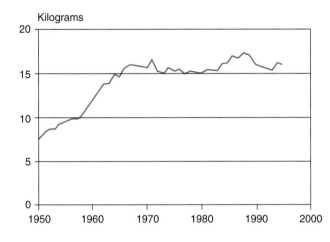

FIGURE 16.9 World fish catch per person, 1950–1995 (from McGinn 1997).

Reaching Food Fishery Limits

Many advances were made in understanding ecological process during the 1950s and 1960s. One of the most basic discoveries was that aquatic communities retained only 5 to 15% of biomass energy originally fixed in photosynthesis with each transfer of energy from one feeding level to the next (see Chapter 3). Based on estimates of plankton production, ecological effeciency, and the species judged suitable for harvest, Rhyther (1959) estimated that sustainable yield of food fisheries worldwide was about 100 million tons. Present trends (Figure 16.8) suggest that sustained yield is approaching that limit. The per capita harvest has already leveled off and may decline in the future (McGinn 1997) (Figure 16.9).

Soon after World War II (WWII), fishery professionals realized that many specific marine finfisheries had been or were becoming overexploited even as the total fishery yield continued to increase (Miller and Johnson 1989). Sardine populations "crashed" and the viability of several whale species was questioned (Ahlstrom and Radovich 1970). This exploitive trend was an outcome of what Hardin (1968) called the "tragedy of the commons" (see Chapter 2). This occurs when renewable resources are used in an area held in common by many people without clear responsibility for the future of the resources. If one person does not exploit the resource, another probably will. There is an inclination to exploit rather than to regulate use for resource sustainability. The need for more harvest control was clear. Starting in the 1970s, fishing nations declared sovereignty over larger fishing territories, usually extending them 200 miles offshore. Each nation established and enforced its own regulations for fish harvest method and amount. The 200-mile limit also forced more competition for limited resources within the U.S. territory, increasing tensions among commercial, recreational, and biodiversity interests.

Following WWII, the yield from wild inland freshwater fisheries of the U.S. and elsewhere decreased and aquacultural production increased. The Great Lakes fisheries reached an economic low in the 1950s (Smith 1970). Introductions of West Coast salmon into the Great Lakes established an attractive recreational fishery. However, commercial food fisheries remain a tenuous enterprise to this day. Technological improvements favoring oceanic exploitation and alternatives to fishery products also contributed to inland commercial fisheries' decline. Development of large electric freezers permitted open ocean fisheries to better compete with small inland fisheries. Freshwater mussel fisheries declined rapidly once superior synthetic materials were invented to make buttons and related items (Pennak 1978). The world's commercial fisheries are now about 10% freshwater and 90% marine.

By the early 1990s the commercial catch in the Great Lakes had become a very small part of the total. The commercial harvest in Alaska exceeded all other U.S. harvests in weight (Figure 16.10). The high harvest in the Gulf includes diverse species dominated by shrimp and menhaden, a herring used for animal feed. The Chesapeake Bay is especially recognized for its shellfish and New England for cod and lobster.

World harvest continued to climb through most of the 1970s and 1980s, reaching a value of $26 billion ($6 billion in the U.S.) by 1988 (Miller and Johnson 1989). However, the rate of increase in worldwide harvest began to decrease after 1970 and now appears to be leveling off (McGinn 1997), as predicted over three decades ago by Rhyther (1969) based on ecological understanding. Even though total harvest in 1995 reached an un-

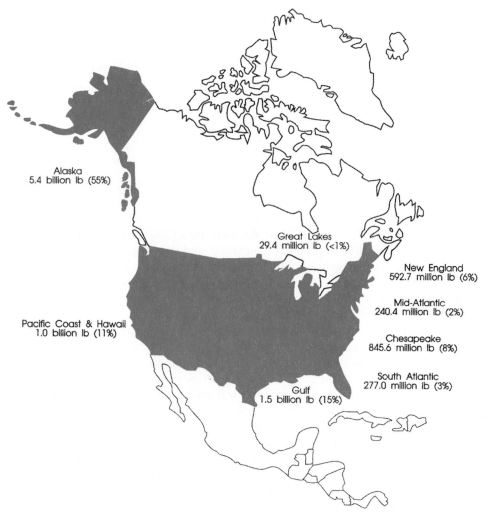

FIGURE 16.10   U.S. commercial fishery landings by region in 1994 (from National Marine Fishery Service 1996a).

precedented peak of 112 million tons (McGinn 1997), signs of decline now appear in many stocks of the world's marine fisheries. The National Marine Fishery Service (1996) recently declared over half of the stocks in U.S. coastal waters to be overexploited.

Despite the 200-mile limit, the U.S. and Canada have been forced to restrict or eliminate harvest for many of their most profitable wild commercial fisheries until they recover. Harvest restriction has resulted in local economic hardship and higher prices for prime seafood. The means of controlling commercial operator entry into the fishery is being reevaluated as are the government subsidies provided to commercial fishers by both governments. The impacts of repeated trawling over huge areas and fishing miles-long gill-nets is now appreciated after years of denial. Appropriate management should allow greater total sustained yield than was possible under previous fishery conditions. By-catch remains an inefficient waste of important resources, which may be reduced by improved gear. Unconventional species, such as krill, may eventually contribute more to future fisheries, if economical harvest techniques and demand develop (Wise 1984). Because krill and other abundant unconventional species also provide food for many other species, the ecological implications of such harvest are being considered.

The results of a long-term study along the Alaskan coast recently added evidence to the growing probability that commercial fishing is contributing to complex changes in ocean systems (Estes et al. 1998). Sea otter populations recently declined sharply over much of the Alaskan coast after a century of recovery from near extinction. The decline is likely to be due to predation by orcas (killer whales) following a population collapse of the orca's preferred foods—seals and sea lions. The seal and sea lion collapse is linked to the loss of their fish prey, which has been associated with commercial fishing and climate change. Because sea otters consume sea urchins, which eat kelp, the kelp-based communities are in decline.

The Aquacultural Potential

There has been much discussion about the potential of aquaculture to replace wild fishery resources. Fish farming has been an important practice in Asian cultures for thousands of years. Aquaculture was described in Chinese writing over 2,000 years ago (Miller and Johnson 1989, Bowen 1970). It is most important in Asia, where in 1994, 87% of the world's aquacultural yield occurred. Aquaculture outside Asia was initiated in the early nineteenth century. Food markets for catfish and trout developed rapidly in the U.S. following WWII (Bowen 1970).

Aquaculture is now the fastest growing sector of commercial yield, providing for nearly 19% of the total world fish consumption (McGinn 1997, Figure 16.11). Aquaculture of food fishes has increased in importance as wild fisheries, harvest has leveled. Outside Asia, aquaculture concentrates on product quality more than quantity. In the U.S., trout culture began for the market soon after the Civil War (Scott 1875). Warm-water culture was less successful except for catfish culture, which thrived in the southern U.S. after WWII when the interstate transportation system was completed and truck refrigeration was perfected. More recently, West Coast salmon have

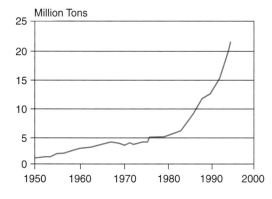

FIGURE 16.11   Aquacultural production for the world (from McGinn 1997).

contributed to U.S. aquacultural production. Crayfish, tilapia, freshwater shrimp, and oysters are cultured at lower levels of production.

Elsewhere in the world, carp and other large minnow species are widely cultured in Asia and Eastern Europe. Some tropical species show promise for growing product acceptance in Africa and Asia. Mariculture (marine aquaculture) has grown rapidly in recent decades, especially for salmon, oysters, shrimp, and other high-quality products. Atlantic salmon are cultured in Scandinavia. Asian mariculture has been an important source of regional protein for centuries. Shrimp are now the major species exported from Asia. A special area of aquaculture is the tropical fish pet trade. In 1986 world imports were $68 million (Miller and Johnson 1989).

As technology improves, aquaculture should contribute substantially more to human protein consumption, especially where labor and environmental costs are low. However, aquaculture is not a panacea for overexploitation of wild fishery resources of high quality (Wise 1984). The products must be of high enough quality to warrant prices required for profit. Aquaculture in the U.S. requires high facilities investment, high operations costs, and high cleanup costs to protect the environment. Only a small fraction of aquatic species are likely to prove suitable for aquaculture in the foreseeable future.

RECREATIONAL FISHING

Cultural Significance

Most of the value of recreational fishing is in the experience, rather than in food value. Fishing appears to have been a common form of fun and relaxation for many cultures. Rod, reel, and line were first depicted in a thirteenth-century Chinese painting (McClane 1974). The first extensive account of recreational fishing has been attributed to Dame Juliana Berners in an essay written in 1496 and entitled "A Treatyse Of Fysshinge Wyth An Angle." Recreational fishing was common among the European affluent well before Isaac Walton wrote the first edition of his famous treatise on sportfishing in 1653 (Walton 1676). His book was to become the philosophical basis for sportfishing among the educated, as signaled by its title: *The Complete Angler or the Contemplative Man's Recreation.*

Although food fishing was common in colonial North America, sportfishing was neither common nor frequently advocated. An exception was Captain John Smith, who grew rich from commercial fishing for cod (Kurlansky 1997) and admitted to fishing for fun (Waterman 1975). Leisure time was a rare luxury for the working-class society. The Puritans disparaged pleasurable use of time. Presaging certain twentieth-century concerns, some religious leaders claimed that pain caused as a consequence of sport was sacrilege.

Attitudes toward relaxation and pleasure gradually changed as human welfare improved. By the 1840s, many Americans admitted they enjoyed the sport as well as the harvest for food. Unlike Europe, even the "working class" owned land adjacent to fishable water, or knew someone who did. In keeping with European tradition, large waters were open to common use including fishing. Many people fished both for food and for a break in their daily routines (Figure 16.12).

Growth of Sportfishing in the U.S.

After the U.S. Civil War, tourism based on sportfishing grew at the remaining wilderness areas. The rapidly growing railroad network facilitated sportfishing. Tourist "destinations" grew up around wilderness resorts in the "north woods" of Maine, New York, Michigan, eastern Canada, the Rocky Mountains, and in Florida. The first National Parks were set aside from remaining wilderness lands. Unspoiled stream, pond, estuary, and coastal fishing remained close to large cities like New York even though pollution was becoming more common (Scott 1875). By the early 1900s,

FIGURE 16.12   Eel spearing at Setauket, Long Island, is shown in this painting by William S. Mount, 1845 (courtesy of U.S. Department of Interior).

much of the western U.S. had been declared public land with open access to anglers and other recreationists.

Technology had much to do with the expansion of sportfishing in the U.S. It freed up leisure time for the working classes. Following the invention of the internal combustion engine, all forms of sportfishing continued to grow in popularity as automobiles, roads, and boat motors improved and became more affordable. The first molded plastic fishing lures and rod parts were developed. The costs of fishing equipment dropped as manufacturing methods improved.

After WWII, families looked forward to a promising and enjoyable future. They demonstrated their optimism by expanding their family sizes and by recreating more outdoors. Sportfishing was the ideal outdoor family pastime for the babyboom generation, especially when combined with boating and other outdoor recreation. Sportfishing gear further improved and anglers became better educated by the new invention, television. Development of the interstate highway system provided easy access to many more fishing sites. Fishing intensity increased rapidly at many locations (Figure 16.13).

The combination of growing participation, more sophisticated participants, and more effective gear increased angler efficiency and the need for improved sportfishery management. Starting in the 19th century, anglers formed clubs for comradery and policy influence. Effective coalitions were formed in the Izaak Walton League, Trout Unlimited, the Sportfishing Institute, and other organizations, often with support of the sportfishing industry. In early response to growing demand, Congress passed the Dingell-Johnson Act in 1950, which provided federal aid for sportfishery management. It was administered by the U.S. Fish and Wildlife Service and authorized sales taxes on fishing equipment and supplies. Revenues were distributed to the states according to a formula based on licenses sold (60%) and geographic area (40%).

Government also invested heavily in water-based public works programs, which generated a 40% increase in the surface area of fishable water excluding Alaska and the Great Lakes (Jenkins 1970). The public water projects were constructed for numerous primary uses, such as for flood control, irrigation, navigation, and electricity generation. Policy required management for multiple use including recreation. The large reservoirs and canals were designed and operated by the Bureau of Reclamation, Army Corps of Engineers, and various public utilities. The Soil Conservation Service (now the Natural Resources Conservation Service) constructed nearly one million small reservoirs,

FIGURE 16.13   Anglers fishing at an especially good location in the Northwest illustrate the increasing intensity of fishing that occurred after World War II at many locations (courtesy of U.S. Fish and Wildlife Service).

mostly on private lands. Through these diverse water management agencies and the U.S. Fish and Wildlife Service, the federal government formed intricate management relationships with state management agencies and private groups and individuals.

The post war expansion of reservoir habitat, highway access, and management revenues helped fishery managers keep up with burgeoning sportfishing interest and increasing fishing efficiency. After reaching a low in the 1960s, water quality began to improve following comprehensive legislation in the early 1970s under the lead of the U.S. Environmental Protection Agency. Previously degraded habitat once again became suitable for sportfishing. Provision of amenities like campgrounds, boat ramps, and sanitary facilities encouraged even more sportfishing interest. At the same time fish habitat expansion peaked, leveled, and then began its present slow decline. To keep up with recreational demand, the Dingell-Johnson Act was amended in 1984 (Wallop-Breaux Amendment) to authorize expanded tax revenue collection with special provisions for improved boating access, services, and aquatic education.

The Limits to Sportfishing

With important help from massive reservoir construction, recreational fisheries management in the U.S. has sustained generally good sportfishing. The number of anglers in the U.S. has increased over the past 40 years (Figure 16.14). In 1991, anglers in the U.S. alone spent $25 billion on sportfishing, rivaling the world trade in commercial fishing. However, there are indications that limits to quality sportfishing are being reached even as demand for more sportfishing grows.

Most suitable reservoir sites have been developed. Numerous existing water projects are viewed as more of a liability than an asset because of undesirable environmental effects on biodiversity. Some of the most used sportfishery management strategies are now questioned because of undesirable effects on biodiversity (Williams et al. 1989). Many reservoirs are filling with sediment and more water is being diverted to uses incompatible with sportfishing. In coastal waters, competition with food fisheries could become a limiting factor. Deteriorated water quality remains a limiting factor despite water quality improvements in the past several decades. The net result is the beginning of a decline in total habitat surface area and increased competition with other resource uses. Although remaining habitat can be upgraded to higher quality with improved environmental regulation, it will need more intense and smarter management to keep pace with demands from competing interests. One of the most important competing interests is protection of biodiversity.

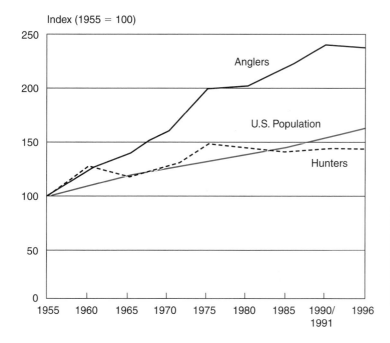

Index (1955 = 100)

FIGURE 16.14 Relative increase in sportfishing participation compared to hunting and the total U.S. population from 1955 to 1996 (courtesy of United States Dept. of Interior).

FISHERIES BIODIVERSITY ISSUES

A Growing Concern

Long before the emergence of widespread environmental awareness in the 1960s, individuals and agencies became concerned about loss of biodiversity associated with aquatic resources. Fur seals and sea otters were decimated during the eighteenth and nineteenth centuries. Precipitous declines and extinction of unique finfish species were especially evident in the upper Great Lakes. Here a diverse assemblage of salmonid fishes known as chubs were negatively impacted by overfishing and invasion by the marine lamprey through locks and canals (Smith 1970). Desert fishes were growing scarce because of habitat alteration and water diversion (Minckley and Douglas 1991).

Although relatively few marine fish and invertebrates are at risk, marine mammals and turtles have been exceptionally vulnerable to overharvest. Several species of whales have been declared endangered. Until devices were required to allow sea turtle escapement, many turtles were trapped and drowned in fishing trawls (Figure 16.15). Accumulated evidence indicates that huge drift nets set in the oceans are a hazard to many large species, including endangered whales (Federal Register 1997), and drift nets have been prohibited in many locations.

A high fraction of native freshwater fish species became endangered or extinct over the past century. Williams et al. (1989) listed 254 living fish species and subspecies of inland waters within the U.S. as threatened or endangered. Nehlsen et al. (1991) identified risk to 214 salmonid stocks on the Pacific Coast. In Canada and the U.S., 48% of all crayfish (Taylor et al. 1996) and 72% of mussel species (Williams et al. 1993) are at risk of extinction or already extinct. Causes include introductions of exotic species, habitat changes, pollution, and river system modification for flood control, irrigation, and navigation purposes (Figure 16.16).

Growing concern about genetic loss through extinction resulted in passage of the Endangered Species Act of 1973. The Act initiated revolutionary change in the practice of aquatic resource management. The primary intent of the Endangered Species Act is the conservation of species for their diverse human-based values (see Chapter 17). This anthropocentric basis for the law is consistent with the U.S. Constitution and a century-long tradition of utilitarian conservation based on future

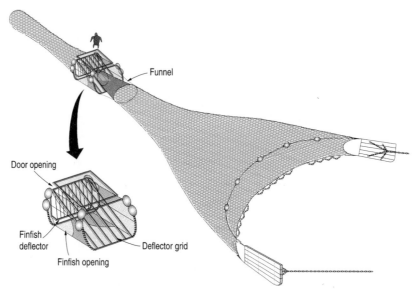

FIGURE 16.15   The turtle excluder device (TED) has been demonstrated to substantially reduce accidental catch and killing of endangered marine turtles (courtesy of USDI-Fish and Wildlife Service).

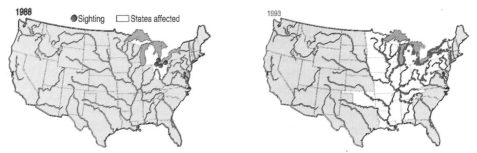

FIGURE 16.16   The zebra mussel is an exotic species introduced accidentally from Eurasia. Once introduced, it spread rapidly and adds to the risk of endangerment of natural species (courtesy of USDI-Fish and Wildlife Service).

human needs. In practice, the protection of "species" has included unique populations within species, called stocks by fisheries professionals. Especially important at this time are threats to the diverse salmonid stocks of the U.S. West Coast. The U.S. Fish and Wildlife Service and the National Marine Fishery Service are lead organizations in protection of aquatic species.

Biodiversity Values

The construction and popularity of numerous public aquariums during the past decade reflects great public interest in the biodiversity of the world's waters. However, aesthetic, scientific, and educational value make up only a part of the value held in the genetic diversity of aquatic life. Diversity also has ecological value to the extent that it contributes to maintaining important ecosystem functions needed for human welfare. There are increasing concerns that various fish-eating mammals and birds are threatened by the commercial harvest of their foods or are contaminated with materials in their food that may reduce their reproduction rates.

Perhaps most importantly, protection of global biodiversity protects the options of future resource users (Bishop 1987, Loomis and White 1996). Once a species is extinct, there is no use option left. Unlike the self-centered focus of conservation in the early twentieth century, the Endangered Species Act enlarged the conservation perspective to *potential* resource use. The broad definition of use here includes all consumptive and nonconsumptive use, including the costs to preserve species simply to know they continue to exist.

Society has shown increasing inclination to pay for protection of what has been called "intrinsic worth" held in living organisms. The "existence value" (Bishop 1987) of this concept of intrinsic worth, independent of human value, is an important motivator for species protection (Deacon and Deacon 1991, Holmes 1991). Some argue that intrinsic value is priceless, and therefore should not be considered among competing interests in determining human benefits. However, the biocentric ethic behind this approach to valuation has yet to become well established in human societies. Economic methods to estimate values associated with endangered species and biodiversity have been difficult to develop, but are becoming more available and reliable (Loomis and White 1996).

FISHERY SCIENCE AND MANAGEMENT

Early Management Emphases

Knowledge of fishery management had begun to accrue by the time medieval European landowners closed their waters and lands to public hunting and fishing. Like hunting, the early sportfishing in inland waters was reserved for "the crown" and anyone receiving land grants including water. Fishing prohibition enforced by wardens was designed to discourage subsistence fishing by hungry peasants. The perceived need for regulating access to small waters indicates the ease with which inland streams and ponds could be depleted of their aquatic resources.

Diverse Native American fisheries existed before Europeans colonized the New World. Not all Native Americans fished, but the traditions of certain cultures were intricately linked with the animals they caught. They promoted resource sustainability through personal self-regulating belief systems, usually based on spiritual respect for the harvested animals. Europeans brought new fishing methods and an exploitive culture somewhat less respectful of the animals as "brother beings." They relied on a social system of written laws, which were established in response to problems, usually as a last resort.

Local food fishery depletions caused protective regulation among East Coast settlements as early as the seventeenth century, soon after European colonization (Stickney and Johnson 1989). It was based in a legal tradition established in Europe centuries before, but was decided more democratically through town government instead of by the lord of the manor. Only local and spotty management remained in effect, however, until after the U.S. Civil War.

Linnaeus's development of an organized classification of life forms spurred a growing scientific interest in natural history and species inventory. A prolonged period of natural resource exploration and inventory followed and grew as travel improved with better boats, roads, and the railroad. Much of the earliest investigation was financed privately. Government financing grew slowly. A number of initial surveys were conducted during exploration of new transportation routes, starting with the Lewis and Clark Expedition in 1804–06, and continuing throughout the century. Other studies were conducted through colleges and museums. The Philadelphia Academy of Science was an early leader (McHugh 1970) followed by Harvard College, which initiated marine surveys sponsored by the U.S. Coastal Survey (McClane 1974). Many of the scientists associated with museums and educational institutions contributed to the development of modern fishery science and management late in the nineteenth century.

Modern Fishery Management

Concerns.   By the time of the U.S. Civil War, fish stocks were locally depleted near many settled areas and popular tourist resorts. The Commission of Fish and Fisheries, later to

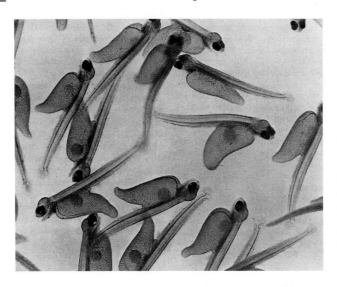

FIGURE 16.17   Fish stocking of fry (shown here), small "fingerlings," or larger sizes has been an important management strategy for over a century (courtesy of USDI-Fish and Wildlife Service).

become the U.S. Bureau of Fisheries, was created in 1871 by Congress to address the issue of declining fisheries (Thompson 1970). The Commission initiated inventories of fishery resources, concentrating on marine resources, which continued over the next several decades. Except for the Great Lakes, inland inventories were left more to the discretion of individual states, educational institutions, and private interests.

Much of the loss of sportfisheries was blamed on indiscriminate harvest methods, such as poisons and gillnets, which were used for market fishing. Between 1857 and 1871, ten states formed fish commissions to regulate fishing and to reduce impacts of dams (Thompson 1970). By 1875, New York had outlawed use of poisons and required fish sluices to be built in dams where migratory fishes occurred (Scott 1875). For certain species, seasons were closed to sportfishing and gear was limited to hook and line angling. Other states and Canadian provinces enacted similar laws. By the turn of the century, most states and provinces had authorized commissions and supporting agencies to establish and enforce game and fish regulations.

Stocking and Introduction.   Following the lead of European advances, stocking of fish fry and small "fingerlings" was advocated as an effective way to recover and sustain fisheries (Scott 1875), and is retained today for certain fisheries (Figure 16.17). Hatcheries initiated large scale production in the U.S. just after the Civil War. Fish were cultured to stock depleted habitats and to serve as a base for new species introductions. By 1870, about 200 individuals were operating private hatcheries in the U.S. They met to form the American Fish Culturist's Association, the precursor of what was to become the American Fisheries Society in 1884 (Thompson 1970).

The fish culturists were effective activists. Scott (1875) suggested: "Every farmer should have a trout preserve" for the hatching of eggs, stocking, and sale to others. In addition to trout, some 28 other species were successfully spawned, hatched, and stocked with fish fry into various waters where stocks of Atlantic salmon, American shad, and striped bass had declined (Bowen 1970). These anadromous species, which migrated from the sea to spawn, were severely depleted in northern coastal streams where industrial mills built numerous dams. The fish culturists spurred federal legislation to assess and improve fishery conditions.

In 1873, the Commission on Fish and Fisheries was authorized by the U.S. Congress to support aquaculture, fry stocking, and introduction of desirable species. Over 90% of the Commission's activity was devoted to aquaculture, an emphasis that continued for the next six decades at a federal facility near Woods Hole, Massachusetts.

Local anglers, fish culturists, and entrepreneurs sought to "improve" native fisheries by widely introducing European species to U.S. waters. The Federal Fish Commissioner promoted and facilitated widespread introduction of carp and other exotic species, especially along the railroads, which burgeoned after the Civil War. In addition to the Eu-

FIGURE 16.18 Many fish species were introduced to waters they did not previously inhabit. After 1973 the Endangered Species Act slowed the process down because of negative effects on rare native species. However, accidental introduction remains a big problem (courtesy of USDI-Fish and Wildlife Service).

FIGURE 16.19 Sleuthing the source of fish kills like the one shown here is an important aspect of fishery management associated with pollution prevention (courtesy of USDI-Fish and Wildlife Service).

ropean brown trout, many native sport species were widely introduced to new locations. They included brook trout, rainbow trout, largemouth bass, northern pike, muskellunge, Atlantic salmon, and channel catfish. The eggs of striped bass and American shad, both Atlantic Coast natives, were successfully introduced into California coastal waters soon after the transcontinental railroad was completed in 1869 (Figure 16.18).

Habitat Quality Issues. By the end of the nineteenth century, water pollution was intense in settled watersheds. Causes included poor livestock management, poor logging practices, untreated sewage, slaughter houses, and flour, saw, pulp, textile, metal, and industrial mills.

The American Fish Culturist's Association lobbied for legislation to curb pollution and provide fish ladders around dams on coastal rivers, but emphasized stocking and regulation of indiscriminate harvest as the primary needs (Scott 1875). The first comprehensive statewide and federal pollution laws were enacted about the turn of the century, outlawing certain types of waste dumping into public waters. Investigation of fish kills from pollutants is essential to preventing further pollution and kills (Figure 16.19).

Public Land Management.   Expansion of public sportfisheries became a popular management goal for newly created land management agencies. The first national parks and state reserves were set aside before 1900, and the national forests followed soon after. Stocking to sustain recreational fishing was widely promoted even in national parks. The negative effects of stocking on native aquatic communities were unknown at the time. The conservation of genetic information never received consideration because the genetic basis of evolution was yet to be elaborated. Wide availability of managed public lands and waters, particularly in the West, helped encourage continued growth of sportfishing as human population, means of travel, and leisure time increased.

Scientific Management.   Most of the main research themes of fishery biology were initiated before World War I (McHugh 1970). The earliest research concentrated on classifying the different species of fish which started in the U.S. soon after Linneas developed a satisfactory classification system. Research also grew out of aquacultural needs for understanding the early life stages of fish, starting early in the nineteenth century in Europe. Nutritional, health, and predation studies were conducted in association with fish culture studies. The hypothesis that most fish mortality occurs during the first few weeks of life was first proposed in 1914. It focused attention on the suitability of habitat for the earliest life stages.

In addition to exploratory studies on distribution and life history, the first successful studies of fish movement by marking fish and recapturing them was conducted by the Commission of Fish and Fisheries in 1873. Age and growth studies, the foundation of modern fish population dynamics, were initiated about the turn of the century. Studies of fish population dynamics started with Petersen's (1903) concept of a sustained population at equilibrium between fish death rate (including fishing mortality) and fish birthrate. The earliest physiology and behavior studies were reported in 1913 because of concern for water pollution effects. The first research into differences among fish stocks within a species was based on fish morphology.

In 1918, the American Fisheries Society began a campaign to develop fisheries science curricula in universities, much like those previously developed for agricultural colleges (Carlander 1970). By then, a school of fisheries had been established at the University of Washington and courses had been started at Cornell University. Before then, most training had been done by apprenticeship under early fish culturists with little formal education. Fishery science grew slowly at first and was almost entirely biological in orientation. Based on membership in the American Fisheries Society, the number of fishery professionals remained steady from 1920 to the end of World War II (Benson 1970).

After WWII, the number of professionals increased rapidly through the 1960s (Benson 1970) and has slowed during recent years. Numerous college programs were developed (Carlander 1970). The growth of fishery science was closely associated with development of federal aid for sportfisheries management, growth of food-fish aquaculture, Sea Grant, and environmental and endangered species legislation. These new sources of funding became available mostly between 1950 and 1984, the period of most rapid professional growth.

Fishery education became highly focused on biological processes to the extent that the title "fishery biologist" was nearly synonymous with fishery scientist (Figure 16.20). The social side of fisheries received little educational attention and remains underdeveloped to this day. This educational orientation had much to do with the development of principles guiding fishery management.

Management for Biodiversity.   Until the Endangered Species Act, the attention of fishery management agencies was directed mostly at consumptive use, in keeping with a long tradition established in the last century. The federal act was quickly followed by companion state laws that authorized federal fund expenditures, state enforcement, and listing of additional state-protected species. Numerous fish species were included on the original lists of threatened and endangered species. In this way, the authorites

FIGURE 16.20 Fishery management education was at first highly focused on fishery biology but increasingly brings to it a human dimension. Many managers now must educate the public about their professional duties (courtesy of the Fish and Wildlife Service).

FIGURE 16.21 Crayfish are important invertebrate members of freshwater ecosystems. Management of crayfish is needed to assure rare species are sustained and common ones are not allowed to invade areas where they can do damage to other crayfish species or to desirable natural aquatic vegetation (courtesy of USDI-Fish and Wildlife Service).

of state and federal fishery management agencies were substantially expanded and diversified, resulting in a greater diversity of human interests served by each of them.

Because consumptive resource management activities sometimes conflicted with the intents of the Endangered Species Act, the new authority caused significant changes in agency goals and operations. Introductions of exotic organisms, continuous stocking of nonnative stocks, increased reservoir surface area, and various pollutants contributed to native stock declines. Some exotic species are superior competitors and predators primarily responsible for reduced populations of certain native species (Williams et al. 1989). In addition to introduced sportfish, bait-bucket transfer introduced highly competitive minnow species. Certain crayfish species were introduced to control aquatic plants, which interfered with sportfish harvest (Figure 16.21). They also competed with native crayfish (Williams et al. 1993). Through inbreeding, stocking programs have homogenized the native stock diversity of a number of species. The great increase in sportfishery habitat, associated with reservoirs and water diversion, also reduced and fragmented pristine habitat for many native species (Minckley and Douglas 1991).

New Management Principles

Harvest. The underlying rationale for fishery science and management is the assurance of fishery resource sustainability for human benefit. This concept was distilled with greater understanding of the relationship between fishery harvest and fishery renewal

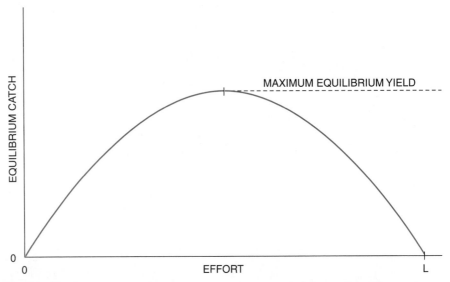

FIGURE 16.22   The theoretical average catch at equilibrium with fishing effort. The maximum equilibrium yield is indicated halfway between no fishing and fishing to catch extinction. The maximum equilibrium yield is the maximum sustained yield (from Roedel 1975).

rates. The management principle of fishing for maximum sustained yield evolved out of quantitative understanding of fishery population dynamics in the 1930s (McHugh 1970). It postulated that the maximum yield of a stock could be sustained indefinitely at some stock number less than the unfished stock number. For a number of stocks, the maximum sustained yield appeared to exist when fishing reduced the original stock number by half (Figure 12.5). Assuming no interaction among stocks, the maximum yield sustained by all fisheries would result from the sum of the individual stocks. The concept of maximum sustained yield proved to be an inadequate guiding principle.

Over time it became clear that fishing changed stock quality by reducing average age and size of fish in the remaining population. Because management based on maximum sustained yield did not consider changes in yield quality, it was inconsistent with the economic value of the fisheries. In the early 1940s, a discourse began between biologists and economists in a debate about the main objective of fishery management (McHugh 1970). Was it to be a biological target of maximum sustained yield of fish biomass or was it to be an economic target of maximum sustained human welfare based on optimum sustained yield?

Management based on the principle of optimum sustained yield considers social factors in addition to total yield, such as the quality of catch and the fishing experience. It holds that the maximum value sustained from a fishery occurs at some optimum yield, which typically is less than the maximum sustained yield (Figures 16.22 and 16.23). An intellectual shift toward the concept of optimum sustained yield occurred following a 1928 symposium (Roedel 1975). However, management of both food and recreational fisheries has had difficulty moving away from the biological concept of maximum sustained yield. Partly responsible is the past biological training in the fishery profession, which has emphasized resource quantities over economic qualities. Another reason is the greater complexity involved in determining optimum sustained yield.

Maximum sustained yield as expressed in Roedel (1975) emphasizes food and recreational fishery values. However, the concerns legislated in the Endangered Species Act moved biological forces in the fishery profession toward the practice of optimum sustained yield. This shift is based less on quantified economic theory and more on the professional imperative to insure species viability is sustained through time.

Ecosystem Management.   The great complexity of managing for diverse aquatic resource needs, including compliance with the Endangered Species Act, has led the responsible federal agencies and many state agencies toward a more holistic approach to management. This new approach, often referred to as ecosystem management, emphasizes the

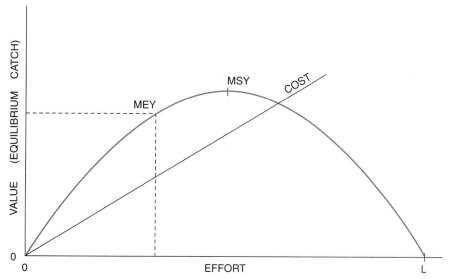

FIGURE 16.23 The theoretical maximum economic yield, which occurs at an optimum sustained yield less than the maximum sustained yield. The difference is caused by a number of factors, such as the average size of fish caught (from Roedel 1975).

close association of ecological and social processes (Adhoc Committee on Ecosystem Management, Ecological Society of America 1995). It seeks to resolve conflicts among various social sectors with stakes in aquatic management decisions. Most agencies concerned with aquatic resource management have used a watershed approach. This is because watersheds are the common means by which aquatic ecosystems are bounded.

Watershed abuse has lead to increased nutrient, sediment, and pesticide concentrations. The need for more comprehensive understanding is widely recognized as social demands on resources trend more toward collision with resource sustainability goals. Industrial development, including atmospheric contamination of rain and snow, has altered the basic chemistry of watersheds and contributed to toxicity and food web contamination. Because they are near the end of the chain of these effects, estuaries often are drastically affected by contaminants. Many ecosystems also suffer from widespread purposeful and accidental introduction of highly competitive exotic species. These exotics have contributed to the decline of numerous native species, which evolved separately from the exotics and were not adapted to live with them. Human-caused environmental impacts tend to favor fewer species with wide environmental tolerances.

The stakeholders affected by the management of watershed ecosystems include many diverse groups whose actions might be regulated because they affect water quantity and quality. These stakeholders include people associated with timber harvest, crop culture, urban development and maintenance, range grazing, mining, transportation networks, utilities operations, and virtually all other human activities, both private and public. Seeking maximum sustained benefit out of optimum provision for each of these diverse and often conflicting interests is the central challenge of fisheries management. In this context, the following section describes some of the common activities of future fishery professionals.

THE FISHERY PROFESSIONAL

Changing Expectations

Aquatic resource professionals consistently have focused their efforts on sustaining aquatic resources. The range of their focus has greatly widened as the complexity of resource management has grown. If formally trained at all, most of the first fishery professionals were trained as fish culturists and ichthyologists. Only a handful of

U.S. universities offered fisheries programs before WWII (Carlander 1970). Now the number offering several specialized courses in fisheries exceeds 100.

The concerns of aquatic resource professionals has diversified and expanded beyond the narrow focus on harvestable stocks. However, "fishery" remains firmly entrenched in professional titles and optimum sustainable yield (or potential for yield) remains a central management concept. As the complexity of management for diverse interests has grown, fishery professionals increasingly find themselves in the company of professionals from other disciplines, such as oceanography, limnology, hydrology, forestry, range science, soil science, wildlife science, economics, and political science. The profession now relies on interdisciplinary team approaches to solve many pervasive, critical, and complex problems. In the U.S., the lead fisheries professional organization remains the American Fisheries Society. It functions to enhance communications among fishery professionals and policy makers through technical publication, conferences, lobbies and other services. It supports numerous sections representing geographical regions and subdisciplines. The society is in the forefront of changing expectations for fishery resource management.

The activities of many contemporary fishery scientists are only remotely connected to fish catch and harvest. Many are concerned about sustaining aquatic populations and ecosystems for other reasons. Scuba diving, beach combing, and aquarium keeping, for example, are pastimes that need not be consumptive for enjoyment of aquatic resources. As more is learned about the role of different species in sustaining aquatic functions, species are increasingly managed for their ecological value. A prevalent contemporary employment focus is on protection of threatened and endangered species. One of the more rapidly growing aspects of the fishery profession in recent years has been the "human dimension," including sociology, economics, ethics, law, and conflict resolution.

Fishery professionals typically are engaged in one or more of several core activities. These include: (1) bioassessment and management; (2) habitat assessment and management; (3) resource demand assessment and management; (4) resource user satisfaction assessment; (5) management effectiveness assessment; (6) management planning, coordination, and administration; (7) original research; and (8) education. Some aquatic resource professionals are specialists; particularly those engaged in original research. Others are generalists, participating in numerous activities. Fishery professionals also may become involved in various support activities such as development of new habitat and law enforcement. However, these support services often do not require a fishery education as long as fishery professionals guide operations.

Bioassessment and Management

Bioassessment. The status of all kinds of aquatic species are assessed through surveys of distribution, abundance, age and size structure, birthrate, growth rate, death rate, nutritional status, and health (Nielsen and Johnson 1983, Schreck and Moyle 1990, Kohler and Hubert 1993). Analytical skills are prerequisite, including at least an elementary understanding of mathematics and statistics. Some fishery professionals become biometricians, but most work with statisticians to accomplish work objectives. Most professionals in this area have earned an M.S. or Ph.D. degree.

The main objective of most bioassessments is to identify changes in resource status as a consequence of management, habitat modification, change in harvest, or suspicion of species decline and endangerment. Scientific surveys have adapted methods used in commercial harvest as well as new methods, such as electrofishing. Formerly commercial fish and sportfishes have been the main objects of such surveys. Now diverse vertebrate, invertebrate, plant, and microbial populations may be surveyed because they are ecological indicators of ecosystem productivity, sustainability, and condition.

Management. Past management of aquatic populations has relied on introductions of new species, reduction of less desirable species, and routine stocking of desirable species. Stocking of popular sportfish species remains the backbone of biomanagement, although its effect on genetic diversity of native stocks is an increasing concern

(Winter and Hughes 1997). Aquatic species are now rarely introduced to new habitats because of their potential negative impacts on native fish. Reduction of abundance of certain competitor and predator species to favor desirable species may have merit in some circumstances when the cost-benefit ratio is favorable. Such management may be pursued to protect endangered species as well as to increase recreational or commercial fish species.

Habitat Assessment and Management

Assessment. Habitat assessment involves measurement of habitat extent and suitability for various species. It relies on the sciences of limnology and oceanography (Wetzel 1983, Cole 1994, Horne and Goldman 1994). Different aquatic communities are broadly associated with habitat features based on temperature, salinity, pH, flow velocity, plant nutrients, bottom condition, water clarity, the dependability of oxygen concentration, and the concentrations of toxic materials. Although environmental quality has generally improved since the low point in the 1960s, it remains problematic in numerous waters.

Habitat assessment also examines the amounts and arrangements of habitats needed for the main life stages from eggs to reproducing adults. This is especially critical for management decisions. Habitat assessment often requires interactions with hydrologists, chemists, engineers, and others familiar with the relationships among climate, watersheds, and aquatic habitats. The dimensions of habitat assessment have increased as understanding of ecosystem process has improved.

Management. Management of aquatic habitat may be as simple as adding bottom structure, such as artificial reefs or spawning gravel for protective cover or reproduction (Kohler and Hubert 1993) (Figure 16.24). It also may involve engineering more complex structures including dams and fish ladders over dams, or structures that stop the passage of unwanted species. On the other hand, more old engineered structures are being modified or removed to restore natural connections among habitats. Many

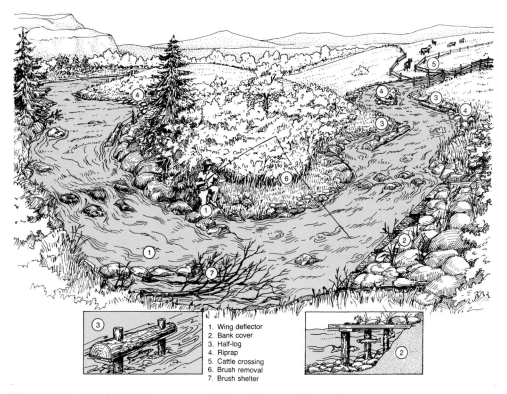

1. Wing deflector
2. Bank cover
3. Half-log
4. Riprap
5. Cattle crossing
6. Brush removal
7. Brush shelter

FIGURE 16.24 Fish habitat improvements (from Owen et al. 1998).

degraded habitats can be cost-effectively aerated to increase oxygen concentration, fertilized to increase productivity, or dredged to remove sediment or aquatic plants.

In its most complex form, habitat management integrates the management of whole watersheds. Sustaining an optimum balance of surface water and groundwater contributions to aquatic habitats, controlling erosion of sediment and nutrients, sustaining optimum habitat shapes and sizes, and controlling pollution are all parts of the management process (see Chapter 7). Habitat management requires close coordination and cooperation among people responsible for management of various parts of watershed ecosystems. Because habitat degradation is the major cause of aquatic resource decline, habitat management often is the key to recovery.

Resource Demand Assessment and Management

Assessment.   Aquatic resource demand typically is evaluated by monitoring resource use. It includes assessing the amount of fishing (the fishing effort) and fish death caused by harvest and by-catch. Hooking mortality of returned sportfish is an important consideration. In on-site surveys, the condition of both harvested and returned fish is assessed, including sizes, ages, and general health (Sigler and Sigler 1990, Schreck and Moyle 1990, Kohler and Hubert 1993). Differences in the characteristics of caught fish from those left in the habitat are examined for characteristics and trends that could indicate reduced sustainability or other important change.

In sportfisheries, the resource demand sometimes is estimated indirectly through angler reports of what they harvested. This method provides an inexpensive and comprehensive means for assessing fishing activity. But because of less control over accuracy, it usually is used in combination with on-site assessments by professionals (Pollock et al. 1994).

Management.   Resource demand is managed mostly through setting regulations and through public education. Regulation setting requires knowledge of fish population behavior, population dynamics, and renewal rates as well as knowledge about the acceptability of the regulation to the user public and the difficulty entailed in enforcement (Kohler and Hubert 1993). Regulations may close habitats for specified times, restrict the numbers and sizes harvested, or restrict the fishing methods used.

Resource users often are educated about the rationale behind regulations, which sometimes encourages self-monitoring in place of regulations. Education may by used to improve fishing efficiency while reducing unnecessary mortality. Many management programs report where the sportfishing is currently most rewarding and provide helpful hints on how anglers can be more successful at underused sites. Education also may be used to develop demand for underused species and reduce demand for overused species.

Resource User Satisfaction

Knowledge of resource user satisfaction helps to assess management effectiveness. Satisfaction is an expression of the value obtained from resource use. A high satisfaction index implies highly rewarding use. Various levels of sophistication are used to assess resource user satisfaction, but most depend on well-designed mail, telephone, or on-site surveys.

Satisfaction surveys become especially useful when satisfaction estimates can be compared to the management costs. That kind of information allows managers to redistribute their management revenues toward the most satisfying use. In that way they maximize benefits relative to management costs.

One of the recurrent questions associated with benefit-cost analyses using satisfaction indices is the comparability of satisfaction indices among various activities. Is the basic value of a day of trout fishing on a remote wilderness lake the same as a day of angling for sunfish at the local pond? Answering this or similar questions requires an economic estimate of angler willingness to pay for their fishing. Therefore, fishery science increasingly has an economic orientation.

Management Administration

Much of what modern aquatic resource professionals do involves coordination with other professionals and with the public (Kohler and Hubert 1993). This aspect of the profession demands good communication and interpersonal skills as well as a thorough knowledge of aquatic resource ecology and sociology. As more species become scarce and their demand increases, conflicts over resource use and allocation will intensify.

Conflict resolution is becoming a critical part of public service because of growing and diverse demands for limited resource supplies. Often at the heart of conflict is inadequate knowledge regarding how various stakeholders will be affected by implementation of natural resource decisions. Economic analysis, although imperfect, can help quantify the various degrees of benefit gain or loss by various stakeholder groups.

Much of what a fishery professional does involves teamwork and coordination with people of diverse educational backgrounds. Communication skills are critical. In addition to clear, complete, and concise technical writing skills, fishery professionals need strong and public speaking and social skills. They need to be able to work effectively with people having diverse values. They also need to be careful planners, considering alternative pathways to desired ends. A critical aspect of future aquatic resource management will involve the development of attainable goals for future ecosystem condition and the planning of strategies to reach these goals.

Original Research

Original research is conducted to develop new understanding of ecological, social, and managerial processes. It is the means by which assessment techniques are improved. Original research in fisheries is conducted mostly in research universities and in government laboratories. Most original research is led by a "principal investigator" who has earned a Ph.D. in a highly specialized area. The principal investigator often supervises others who actually carry out the specialized activities involved in research. These research technicians frequently have earned an M.S. degree, but some find positions with a B.S. degree. They often have relatively little to do with research design or communication of results, but are experts on specific methods. Research in universities usually is conducted with graduate assistants because education of future researchers is a primary mission.

Many of the easiest research problems have been well addressed, if not entirely solved. Therefore, research is becoming more interdisciplinary and team-oriented, as the dimensions of primary problems become better known. To be successful, team researchers need to develop communication and interpersonal skills in addition to the specialized skills required in their specialties. This usually means learning more about the other disciplines.

Education

Most fishery professionals have some opportunity to educate other professionals and the lay public. However, relatively few teach in universities or other formal educational environments. The most effective educators are themselves broadly enough educated to understand the diverse needs and interests of students.

In the fisheries professions, university education is closely tied to original research. With careful planning, research and teaching can merge to produce greater strengths in both areas. The best educators serve as models for their graduates. They are enthusiastic, optimistic, and adaptive while remaining realistic about professional challenges. One of the greatest challenges for fishery educators is to develop and use their specialized skills in a more integrated and interpersonal learning environment. They play a key role in teaching students, by example, how to work effectively in groups of diverse people.

FUTURE FISHERY ISSUES

A quick review of the predominant issues in fishery resource management shows recurrent themes in a variety of aquatic ecosystems settings. They include the following issues and concerns.

- The sustainability of aquatic resources is threatened by growing human demands on aquatic ecosystems and by inadequate past management. Fishery professionals need to address diverse challenges to maximize human benefit while sustaining resource renewal.

- The negative effects of environmental change in fishery ecosystems have been cumulative and interactive. Predictive understanding and more effective management requires a more holistic ecosystems approach.

- Understanding of the ecosystem process, including cultural influences, remains too qualitative to accurately predict the beneficial consequences of costly "corrective" management proposals. Adaptive management, which incorporates research into the management process, is a suitable approach for analyzing management effectiveness and improving benefits estimation.

- The negative effects of overfishing have continued to worsen in many locations despite warning and regulation. This has occurred mostly because (1) resource demand exceeds optimum sustained yield, (2) short-term emphasis on employment outweighed long-term sustainability, and (3) by-catch mortality was poorly controlled.

- There is a need for accelerated assessment of biodiversity, especially in regions of rapid environmental change, and improved protection of resource options before they become threatened with extinction.

- All relevant private and public benefits and costs associated with fishery management have not been estimated objectively and thoroughly enough for sound decision making. Greater investment is needed in benefits estimation and cost accounting in all areas of fishery management, including biodiversity.

- A professional belief that recovery of ecosystem "integrity" is the most appropriate means for obtaining optimum sustained benefit has gained considerable "scientific" credence. However, little empirical information is available to determine appropriate recovery targets. This requires a greater investigation of the relationship of biodiversity to ecosystem structure and functions.

- There needs to be improved education of fishery professionals in the U.S. Agencies need to be encouraged to work in an integrated way toward the ideal of improved general public welfare.

These fishery issues need to be addressed by a combination of research, technical, policy, education, and public relations strategies. Students who elect to enter the fisheries profession can look forward to personal involvement in at least some of these issues over the next several decades.

LITERATURE CITED

Ahlstrom, E. H. and J. Radovich 1970. Management of the Pacific sardine (pp. 183–193). In N. G. Benson, ed., *A century of fisheries in North America.* Washington, DC: American Fisheries Society.

Bell, F. H. 1970. Management of Pacific Halibut (pp. 209–221). In N. G. Benson, ed., *A century of fisheries in North America.* Washington, DC: American Fisheries Society.

Benson, N. G. 1970. The American Fisheries Society, 1920–1970 (pp. 13–24). In N. G. Benson, ed., *A century of fisheries in North America.* Washington, DC: American Fisheries Society.

Bishop, R. C. 1987. Economic values defined (pp. 24–33). In Daniel J. Decker and Gary R. Goff, eds., *Valuing wildlife: Economic and social perspectives.* Boulder, CO: Westview Press.

Bowen, J. T. 1970. A history of fish culture as related to the development of fishery programs (pp. 71–93). In N. G. Benson, ed., *A century of fisheries in North America.* Washington, DC: American Fisheries Society.

Cairns, M. S. and R. T. Lackey. 1992. Biodiversity and management of natural resources: The issues. *Fisheries* 17(3):6–10.

Carlander, K. D. 1970. Fishery education and training (pp. 57–70). In N. G. Benson, ed., *A century of fisheries in North America.* Washington, DC: American Fisheries Society.

Cole, G. 1994. *Textbook of limnology.* 4th edition. Prospect Heights, IL: Waveland Press.

Deacon, C. and J. E. Deacon. 1991. Ethics, federal legislation, and litigation in the battle against extinction (pp. 109–121). In W. L. Minckley and J. E. Deacon, eds., *Battle against extinction: Native fish management in the American West.* Tucson, AZ: The University of Arizona Press.

Estes, J. A., M. T. Tinker, T. M. Williams, and D. F. Doak. 1998. Killer Whale predation on sea otters linking oceanic and nearshore ecosystems. *Science* 282:473–475.

Federal Register 1997. *Endangered fish or wildlife: Special prohibitions.* North Atlantic Right Whale Protection. Proposed rule; temporary closure of fishery, National Marine Fisheries Service. Vol. 62, No. 212, 59335.

Hardin, G. 1968. Tragedy of the commons. *Science* 162:1243–1248.

Holmes, R. III. 1991. Fishes in the desert: Paradox and responsibility (pp. 93–108). In W. L. Minckley and J. E. Deacon, eds., *Battle against extinction: Native fish management in the American West.* Tucson, AZ: The University of Arizona Press.

Horne, A. J. and C. R. Goldman. 1994. *Limnology.* 2nd edition. New York, NY: McGraw-Hill.

Jenkins, R. 1970. Reservoir management (pp. 173–182). In N. G. Benson, ed., *A century of fisheries in North America.* Washington, DC: American Fisheries Society.

Johnson, F. G. 1989. Fisheries: Harvesting life from water (pp. 2–9). In F. G. Johnson and R. G. Stickney, eds., *Fisheries: Harvesting life from water.* Dubuque, IA: Kendall/Hunt Publishing Company.

Kohler, C. C. and W. A. Hubert, eds. 1993. *Inland fisheries management in North America.* Bethesda, MD: American Fisheries Society.

Kurlansky, M. 1997. *Cod: A biography of the fish that changed the world.* New York, NY: Penguin Putnam.

Larkin, P. A. 1970. Management of Pacific salmon of North America. (pp. 223–236). In N. G. Benson, ed., *A century of fisheries in North America.* Washington, DC: American Fisheries Society.

Lerman, M. 1986. *Marine biology.* Menlo Park, CA: Benjamin/Cummings Publishing Co.

Loomis, J. B. and D. S. White. 1996. Economic values of increasingly rare and endangered fish. *Fisheries* 21(11):6–11.

McClane, A. J., ed. 1974. *McClane's new standard fishing encyclopedia.* New York: Holt, Rinehart and Winston.

McGinn, A. 1997. Global fish catch remains steady (pp. 32–33). In Linda Stark, ed., *Vital signs: The environmental trends that are shaping our future.* New York: W.W. Norton & Company.

McHugh, J. L. 1970. Trends in fishery research. In N. G. Benson, ed., *A century of fisheries in North America.* Washington, DC: American Fisheries Society.

Miller, G. T. 1990. *Resource conservation and management.* Belmont, CA: Wadsworth Publishing Co.

Miller, M. L. and F. G. Johnson. 1989. Fish and people (pp. 10–23). In F. G. Johnson and R. R. Stickney, eds., *Fisheries: Harvesting life from water.* Dubuque, IA: Kendall/Hunt Publishing Company.

Minckley, W. L. and M. E. Douglas. 1991. Discovery and extinction of western fishes: A blink of the eye in geologic time (pp. 7–18). In W. L. Minckley and J. E. Deacon, eds., *Battle against extinction: Native fish management in the American West.* Tucson, AZ: The University of Arizona Press.

National Marine Fishery Service, National Oceanic and Atmospheric Administration. U.S. Department of Commerce. 1996b. *Fisheries of the United States, 1995.* Silver Spring, Maryland.

National Marine Fisheries Service, National Oceanic and Atmospheric Administration. U.S. Department of Commerce. 1996b. *Our living oceans: Report on the status of U.S. living marine resources, 1995.* Silver Spring, Maryland.

Nehlsen, W., J. E. Williams, and J. E. Lichatowich. 1991. Pacific salmon at the crossroads: Stocks at risk from California, Oregon, Idaho, and Washington. *Fisheries* 16:4–21.

Nielsen, L. A. and D. L. Johnson, eds. 1983. *Fisheries techniques.* Bethesda, MD: American Fisheries Society.

Northridge, S. P. 1984. *World review of interactions between marine mammals and fisheries.* FAO Technical Paper 251. Rome, Italy: Food and Agriculture Organization of the United Nations.

Owen, O. S., D. D. Chiras, and J. P. Reganold. 1998. *Natural resource conservation.* 7th edition. Upper Saddle River, NJ: Prentice-Hall.

Pennak, R. W. 1978. *Freshwater invertebrates of the United States.* 2nd edition. New York, NY: John Wiley & Sons.

Petersen, C. G. J. 1903. What is overfishing? *Journal of Marine Biology Association.* (U.K.) 6:587–594.

Pollock, K. H., C. M. Jones, and T. L. Brown. 1994. *Angler survey methods and their applications in fisheries management.* Bethesda, MD: American Fisheries Society Special Publication 25.

Radcliffe, W. 1921. *Fishing from the earliest times.* Unchanged reprint of the edition. Chicago, IL: Ares Publishers.

Rhyther, J. H. 1959. Potential productivity of the sea. *Science* 130:602–608.

Roedel, P. M., ed. 1975. *Optimum sustainable yield as a concept in fisheries management.* Proceedings of a Symposium. Special Publication No. 9. Washington, DC: American Fisheries Society.

Schreck, C. B. and P. B. Moyle, eds. 1990. *Methods for fish biology.* Bethesda, MD: American Fisheries Society.

Scott, G. C. 1875. *Fishing in American waters.* New York: The American News Company.

Sigler, W. F. and J. W. Sigler 1990. *Recreational fisheries management, theory, and application.* Reno, NV: University of Nevada Press.

Smith, S. H. 1970. Trends in fishery management of the Great Lakes. In N. G. Benson, ed., *A century of fisheries in North America.* Bethesda, MD: American Fisheries Society.

Taylor, C. A., M. L. Warren Jr., J. F. Fitzpatrick Jr., H. H. Hobbs III, R. F. Jezerinac, W. L. Pflieger, and H. W. Robison. 1996. Conservation status of crayfishes of the United States and Canada. *Fisheries* 21:25–38.

Thompson, P. E. 1970. The first fifty years—the exciting ones (pp. 1–11). In N. G. Benson, ed., *A century of fisheries in North America.* Bethesda, MD: American Fisheries Society.

U.S. Fish and Wildlife Service and U.S. Bureau of the Census. 1993. *National survey of fishing, hunting, and wildlife-associated recreation.* Washington, DC: U.S. Government Printing Office.

Walton, I. 1676. *The complete angler or the contemplative man's recreation.* Reprinted in 1985 with modernized spelling. London: Harrap Limited.

Waterman, C. F. 1975. *Fishing in America.* New York: Holt, Rinehart and Winston.

Wetzel, R. G. 1983. *Limnology.* 2nd edition. New York: Saunders College Publishing.

Williams, J. 1962. *Oceanography.* Boston, MA: Little, Brown and Company.

Williams, J. E., J. E. Johnson, D. A. Hendrickson, S. Contreras Balderas, J. D. Williams, M. Navarro-Mendoza, D. E. McAllister, and J. E. Deacon. 1989. Fishes of North America, endangered, threatened and of special concern. *Fisheries* 14:2–20.

Williams, J. D., M. L. Warren Jr., K. S. Cummings, J. L. Harris, and R. J. Neves. 1993. Conservation status of freshwater mussels of the United States and Canada. *Fisheries* 18:6–22.

Winter, B. D. and R. M. Hughes. 1997. Biodiversity position statement—corrected version. *Fisheries* 22(3):16–23.

Wise, J. P. 1984. The future of food from the sea (pp. 113–127). In Julian L. Simon and Herman Kahn, eds., *The resourceful earth.* New York: Basil Blackwell.

CHAPTER 17

Biodiversity and Endangered Species Management

THE DILEMMA

The myriad of life forms inhabiting the earth have been a constant source of wonder and wealth to human societies. The diversity of life is the result of the evolutionary process that began some 2.5 billion years ago when the first forms of life appeared on the earth. *Homo sapiens* is a late arrival; our direct human ancestors appeared only a few million years ago in Africa and from there spread throughout the world. Recent evidence indicates that humans reached North America over 30,000 years ago from Asia via the Bering Land Bridge, which connected Siberia and Alaska (Hopkins et al. 1982). Humans evolved the skills and technology necessary to fully exploit the earth's natural resources, often with catastrophic consequences for life on earth. By the end of the twentieth century, the earth's natural bounty had been degraded to an alarming degree by overexploitation, uncontrolled technology, and continued human population growth and resource demand. This degradation has forced human societies to evaluate their impacts and to develop corrective resource management policies.

Following years of neglect, biodiversity protection and restoration is now recognized in law and convention as necessary for the long-term survival and economic viability of human societies. Humanity is developing an ethic for more responsible interaction with all living organisms, which places a higher priority on sustainability of renewable resources than on immediate economic gain (Kellert 1996). Maintaining biodiversity depends heavily on the few who disproportionately control most of the earth's resources. However, economic wealth in and of itself should not be blamed for poor conservation ethics. Wealth enables education, in-depth reflection, and influential action. In addition to self-interest, the drive for wealth is the engine of investment in scientific understanding, universal education, and the democratic process. Without some degree of wealth, ignorance, despair, and self-serving urgency tend to dominate human action. However, the unconstrained rush for wealth is the main enticement away from the practice of more prudent conservation ethics. The poor man's understandable pursuit of economic gain is collectively as damaging to resource sustainability as the avarice of an unprincipled corporation executive. Solutions to problems resulting from deficient environmental ethics worldwide are most promising if they include an improved distribution of the wealth derived from the world's natural resources—including its biodiversity.

BIODIVERSITY

What is Biodiversity?

Biodiversity refers to the variety within and among living organisms. Biodiversity is quantified in terms of genetic variation within species, the number of species or kinds of living organisms inhabiting an area, the number of communities and other biotic (living) assemblages, and the variety of ecosystem functions and structures (DeLong 1996, see relevant parts of Chapter 3). Many of the world's living organisms have yet to be identified. Our knowledge of the world's biodiversity is based mostly on larger species, such as vertebrates, and economically important forms. However, most of the invertebrates remain undescribed. A conservative estimate of the number of species in the world is 5 to 10 million, but less than 2 million have been discovered and named (Wilson 1992). The richest areas of biodiversity occur in the forested Tropics, although coral reefs and oceans also support a rich but still poorly documented diversity. The Tropics support as much as 50% of the species in the world although it comprises less than 10% of the global land base (Reid and Miller 1989). Tropical biodiversity is highly vulnerable because of human population pressure, resulting in rapid deforestation (Figure 17.1). About 80% of the human population is concentrated in South America, Africa, and southern Asia.

Conservation biology, founded in the 1970s, united several disciplines including genetics, physiology, biogeography, population biology, and wildlife conservation with the common objective of halting the decline of biodiversity. The ultimate goal of conservation biology is to ensure the perpetuity of natural process, including the diversity of habitat dynamics and evolutionary adaptations. Important strategies include restoring degraded ecosystems and sustaining the integrity of ecological and evolutionary processes. Ecological processes are the basis for sustaining life and evolution, which affords organisms the continuing ability to adapt to short-term and long-term changes in

FIGURE 17.1 Protected areas, such as this wildlife refuge in India, are important for maintaining biodiversity. Many of the plants and animals found in such refuges are rare in surrounding areas.

the environment. Comprehensive overviews of conservation biology include Cox (1997), Meffe and Carroll (1997), Primack (1998), Wilson (1992, 1988), and Reaka-Kudla et al. (1997). *Conservation Biology,* the bimonthly journal of the Society for Conservation Biology, is the major source of technical articles on the subject.

Environmental degradation and the loss and fragmentation of habitats in conjunction with overexploitation, human population growth, and exotic introductions are major factors contributing to the extinction of species and loss of biodiversity. Healthy terrestrial and aquatic ecosystems and their component species are among the ultimate sources of the wealth of nations. It has become evident that the decline in biodiversity threatens the well-being of humanity on a global scale (Morrell 1999). Ecosystem services are the processes through which ecosystems produce the exhaustible natural resources necessary for sustaining human societies and perform life-support functions such as nutrient cycling. However, they are seriously threatened in many areas by the poorly planned, uncontrolled, and ignorant actions of humans.

The Need for International Cooperation

The decline of many species can only be halted by cooperative international management programs. For example, migratory songbirds, which nest in the U.S. and Canada and winter in the New World Tropics, are declining because of habitat loss, habitat fragmentation, and other impacts in both the U.S. and tropical countries. Only a strategy that strives toward global sustainable development (development that meets the needs of present generations without compromising the ability of future generations to meet their own needs) can resolve this threat. The United States is a signatory of the Convention on Biological Diversity. Recognizing the values of and threats to biodiversity, many nations have agreed to initiate a concentrated international effort to formulate strategies to curb the loss of biodiversity and to institute programs that sustainably and equitably use living resources (Grosse 1997).

The Escalating Loss of Biodiversity

Concern for declining biodiversity developed because of the high rate of extinction of animal and plant species due to the actions of humans. In the past century, almost five species per year are known to have gone extinct. However, this is probably a conservative estimate because many species that may have become extinct have been unrecorded. It is estimated that as many as 25% of the world's organisms could become extinct by the year 2015 if current trends continue. Humans have caused the extinction of entire species, greatly reduced numbers of organisms, and greatly modified and decimated habitats through mining, unrestricted logging, urban expansion, dam construction, overgrazing of rangelands, recreational activities, unregulated hunting and fishing, introduction of exotic animals, pollution, and agricultural development. Extinctions have been a natural phenomenon, such as during the ice ages when there were major widespread climatic and consequent vegetation changes. However, extinction rates have accelerated since human populations developed advanced technology (Reid and Miller 1989, McNeely et al. 1990).

An alarming example of habitat degradation resulting in the decline of a native fauna is the case of native fish in the United States (Flather and Hoekstra 1989, Flather et al. 1994, Langner and Flather 1994). The estimated 500,000 adult Atlantic salmon that migrated up 34 New England river systems in precolonial times has dwindled to about 7,000 adult salmon in 16 river systems. Most of these are from hatchery stock. Salmon in the western U.S. have also drastically declined because of hydroelectric development, commercial harvest, habitat degradation, and hybridization with exotic species. The stocks of commercially important fishes in the Great Lakes were depleted by 1900. Four of these species are now extinct, two are endangered, one is threatened, and two are considered rare. Large percentages of fishes from many of the major river systems are extinct, endangered, threatened, or declining. In the Colorado River system alone, 34% of native fishes are extinct, endangered, or threatened.

FIGURE 17.2 Unknown to most people, many animal species are important in maintaining natural ecosystems. Bats, such as this species from the American Tropics, are the second largest order of mammals. Some species are important pollinators and disseminators of seeds.

Biodiversity Services and Value

Genes, species, populations, communities, and ecosystems form the bases of a network of interactive, interdependent, and easily disrupted associations. When maintained in a healthy condition, they can provide sustainable natural resources and economic viability for human populations. Wild organisms interacting in ecosystems provide many services vital to the well-being of humans. The breakdown of toxic materials is one essential service. About 90% of the 12 billion tons of organic wastes, of which 130 million tons are human excreta and 2 billion tons are manure produced by livestock, are decomposed by the interactive complex of soil and water invertebrates. These include earthworms, protozoans, nematodes, fungi, and bacteria. Microorganisms aid in disintegration and neutralization of more than half a million tons of pesticides, 4 million tons of soaps and detergents, 18 million tons of ammonia, 12 million tons of alkalies, and 4 million tons of acids produced annually by United States agriculture, industry, and homes. Of the more than one-half million tons of pesticides applied to crops in the U.S., less than 1% reach the target pests. Their disintegration by soil organisms is vital to reducing pollution. New varieties of microorganisms are being developed to aid in facilitating breakdown of the 7 million barrels of oil spilled into the environment annually (Pimental 1981, Daily 1997b).

Wild organisms perform many other valued services. An estimated 15 billion tons of nitrogen, vital for plant growth and production of high protein foods worth an estimated $3.5 billion, is biologically fixed by soil microbes. Cross-pollination by insects, for which no substitutes are available, is essential in the production of 90 United States crops worth nearly $4 billion. The value of 9 additional crops, valued at $4.5 billion, is enhanced by insect pollination. Thousands of wild plants also depend on insects and other animals for pollination (Figure 17.2). Naturally occurring and introduced predators, such as insects and parasites, are essential in controlling pest insect populations of forestry and agriculture crops worth billions of dollars (Pimental 1981, Daily 1997b).

The integrity of natural ecosystems and their associated biodiversity are critical in the storage and cycling of global material resources. Intact communities sustain

FIGURE 17.3   Eland, shown here, have been recently domesticated from wild stock. Many wild animal species have the potential for becoming sources of new domesticated animals.

healthy root systems that resist soil erosion and nutrient loss. The biomass of intact forests stores a large fraction of the biosphere's carbon, which is now accumulating instead in the atmosphere, where it appears to be causing climatic warming.

Intact watershed ecosystems store water and release it at moderate rates, thereby reducing flood impacts on ecosystems and human communities downstream. Upland watersheds combine with the unimpeded functioning of floodplain ecosystems to absorb and transform flood impacts into useful forest and wildlife productivity. Further, they protect human property and health. In arid parts of the world, riparian ecosystems provide a disproportional high fraction of the natural production and biodiversity.

Resource Option Value.   Genetic diversity is a natural library of information waiting to be "read" for the scientific insights that are held within (Wilson 1988). The genetic diversity contained in wild organisms is the basis for future food, medicine, and other resource development with an inestimable contribution to the future world economy. Sustaining genetic diversity preserves options that would otherwise be denied.

The genes of many wild forms related to domestic crop species possess resistance to parasites and pests. Their genes can be used to impart resistance to cultivated forms. New crops with greater tolerance to heat, drought, and salinity can also be developed from wild forms and aid in greater crop production and greater human welfare. The benefits of genetic diversity were understood by pre-Columbian southwestern cultures. They planted several varieties of maize to assure successful crops under a variety of weather conditions.

Wild animals provide a wealth of genetic diversity for creating new species of domestic animals. They can be crossed with domestic animals to produce new, more adaptable and productive animals from wild forms (Figure 17.3). All of the major domestic terrestrial animals produced in North America originated in the Old World, except for the turkey. However, other native species could be developed to advantage for use in North America, such as the bison. Because bison coevolved with North American native plants, they are better adapted than domestic sheep and cattle for making efficient use of native rangeland plants. Bison also are better adapted to the climatic regimes where they evolved.

The world depends on less than ten animal species (sheep, cattle, chickens, and pigs, among others) for most of the world's terrestrial animal food production. Less than twenty crops account for 90% of the world's plant food production. None of these species are native to North America north of Mexico. Fish protein accounts for as high as 60% of the animal protein consumed by people on a regional basis in parts of the world other than the U.S. Thousands of wild plants and animals could potentially be used to produce new domesticated forms better adapted to local conditions. They are more productive, and require less fertilization, pesticide application, and tillage. Because of the high reliance on a few domesticated species for food, human populations are highly vulnerable to crop devastations. Preventing the further loss of species is a global priority because wild or semiwild forms of potential economic importance could become extinct, resulting in great future economic losses (McNeely et al. 1990).

Medicines are another category of beneficial goods derived from wild organisms. Poisonous marine vertebrates and invertebrates provide antimicrobial, antiviral, cardioactive, and neurophysiological substances. Many plants, fungi, and other life forms also contribute massively to our store of medications. These substances are essential in providing better health services and they generate billions of dollars for the pharmaceutical industry. Primates are used for medical research, including the search for a cure for AIDS. Many animals and plants have medicinal properties that have yet to be developed or are unknown. Wild animals also provide oils, ornamental by-products, fuel, feathers, and pets worth billions of dollars in international markets (Daily 1997a, 1997b).

ENDANGERED SPECIES

What Are Endangered Species?

Endangered species are at risk of extinction and require protective management to avoid extinction. A species is considered endangered if its death rate consistently exceeds its birthrate; if it is incapable of adapting to natural or human-induced environmental changes; if its habitat is threatened by destruction or serious disturbance; or if its survival is threatened by the unwanted introduction of other species through predation, competition, disease, or environmental pollution. A threatened species is scarce throughout its range and will probably not become extinct as long as conditions are managed to remain stable and favorable.

There were 902 (349 animals and 553 plants) species listed as endangered and 233 (117 animals and 116 plants) listed as threatened by the United States Fish and Wildlife Service (USFWS) as of April 1998. Only eight have been removed from the list since 1973 and 18 species have been upgraded to threatened status.

The main source of information on endangered species is the International Union for Conservation of Nature and Natural Resources (IUCN), a nongovernmental conservation organization headquartered in Gland, Switzerland. The *IUCN Red List of Threatened Animals,* a publication that is periodically updated, presents information on the current status of globally threatened vertebrates and invertebrates, exclusive of microorganisms. The 1996 edition lists 5,205 species that are threatened with extinction. The listed species are those experiencing significant population declines, those restricted in distribution, those with very low population numbers, or those exhibiting any combination of these characteristics. A large percentage of the vertebrates are threatened with extinction, including 25% of mammals, 11% of birds, 20% of reptiles, 25% of amphibians, and 34% of fish. Countries with the largest number of endangered mammals are Indonesia (128), China (75), and India (75), all of which also support extremely high human populations. Countries with the largest number of known endangered birds are Indonesia (104), Brazil (103), and China (90). Countries with the largest number of endangered fishes are the United States (123), Mexico (86), and Indonesia (60).

In the United States, the federal agency charged with maintaining a list of endangered species is the Fish and Wildlife Service. Of the species listed as threatened or endangered, 10% were considered to have increasing populations, 38% were considered declining, 31% were estimated to have stable populations, 19% had unknown status, and 2% were believed to be extinct. The greatest percentage of species with declining populations or unknown status were invertebrates (81%), fish (51%), and amphibians and reptiles (79%). This indicates that smaller species and aquatic species may be at a greater risk of extinction. Forest ecosystems were associated with the greatest number of listed species (312), followed by rangeland (271), water (244 dominated by fish and mollusks), and barren land (beaches, dry salt flats, exposed rock, areas disturbed by mining, and other habitats) (176 species). Wetlands supported fewer (155) threatened and endangered species than other habitats. However, wetlands—which comprise only 5% of the contiguous U.S.—support a disproportionally high number of listed species (30% of listed animal species and about 15% of listed plants) (Flather et al. 1994). Of all the habitats, riverine habitats hold the highest concentrations of endangered species in the U.S.

Of the factors contributing to species endangerment, habitat losses and fragmentation associated with intensification of land and water use were the most important. Interspecific (between species) interactions, especially those associated with introduced species, and human overuse associated with the harvest, collection, or commercial trade of species are second and third in importance. Agricultural development was a prominent factor for endangerment among most species, particularly mammals and birds. The primary factor impacting fishes and birds was human-introduced exotic species. Commercial exploitation and collecting were particularly detrimental to snails, reptiles, crustaceans, amphibians, and plants (Wilcove and Bean 1994).

Causes of Extinction

The five main factors contributing to species extinction are habitat modification, natural causes, unregulated hunting, introduced predators, and nonpredatory exotics. Natural causes are part of the evolutionary process. Animals become overspecialized and are unable to adapt to environmental changes or cannot compete with other evolving organisms. Natural catastrophic events—such as sudden climatic changes, earthquakes, or volcanic eruptions—can also be involved in extinctions. From a management standpoint, human-caused extinction is the greater concern and these causes are discussed in more detail below.

Habitat Modification. Habitat modification by humans is the primary threat to biodiversity (Figure 17.5). Such modification occurs through logging, draining of wetlands, agricultural and industrial development, reservoir construction, urban sprawl, pollution, and other changes (Flather et al. 1994). Of the recent extinctions involving habitat loss, humans are probably responsible for over 90%.

The most significant factors noted in the *Red List* (IUCN 1997) are habitat reduction, fragmentation, and degradation. Natural systems are being replaced by human-dominated landscapes resulting from unrestrained economic development in conjunction with human population increases. Once widespread wildlife populations are being fragmented into small, isolated subpopulations (Figures 17.4 and 17.5). For those animals listed as threatened or endangered, habitat loss or degradation have been identified as one of the main reasons for listing of 75% of mammals, 44% of birds, 68% of reptiles, 58% of amphibians, 55% of fishes, 52% of crustaceans, 47% of insects, 95% of clams and oysters, and 50% of snails. Unplanned and unrestricted exploitation of many of the larger terrestrial vertebrates and marine fishes in the *Red List* has resulted in severe population declines.

Increasing human development impacts around rivers, lakes, and oceans are continuing to change the nature of aquatic areas and their aquatic forms. Extensive stream channelization, construction of dams and reservoirs, and conversion of wetlands have been principal components in altering aquatic habitats (see also Chapter 7). Forty fish

(a)

(b)

FIGURE 17.4 Suburban sprawl (a) and dams (b) have been two primary factors associated with habitat fragmentation and species endangerment in the U.S. and other parts of the world (courtesy of U.S. government).

FIGURE 17.5 Asiatic lions are now restricted to a lone population fragment in the Gir Forest of western India. Lions originally occurred from southern Africa to India. Populations have been extirpated from the region between India and central Africa.

FIGURE 17.6 In many countries, species are exploited without regard for their continued existence. This snow leopard skin was being offered for sale in a Chinese market even though protected within the country.

forms, including 27 species and 3 genera, became extinct in North America, and of these, 9 became extinct after 1964 (Williams and Miller 1990). The American Fisheries Society listed 364 North American fish species and subspecies as either threatened (114 species), endangered (103 species), or of special concern (147 species) (Cairns and Lackey 1992).

Unregulated Hunting and Fishing.   Excessive and unregulated hunting includes hunting for commercial purposes, such as in the cases of the great auk and passenger pigeon, and those targeted for control because they transmit diseases or are pests, such as the Carolina parakeet. Demand for specific animal parts for folk medicines and ornamentation, mostly in Asia, has caused the excessive killing of rhinoceros, tigers, and other animals (Figure 17.6). Harvest exploitation has been specifically identified as one of the main reasons for the threatened status of listed species among 13% of mammals, 7% of birds, 31% of reptiles, and 68% of marine fishes.

Introduced Predators.   Introduced predators can be highly destructive because native faunas have not evolved defenses to cope with the new predator, especially in small, isolated areas such as islands. Some introduced predators, such as the Indian mongoose in the Caribbean Islands and Hawaii, were released to control pests, such as rats and snakes, but instead concentrated on the more vulnerable native prey species. Introduction of Nile perch into African rift valley lakes has led to the disappearance of large fractions of the native fish.

Even the seemingly harmless domestic cat is a wildlife conservation dilemma. There were an estimated 60 million pet cats in urban and rural regions of the United States alone in 1990. This estimate does not include cats that are semiwild or free ranging. The combined total was estimated at 100 million cats. Over a billion small mammals and hundreds of million of birds are killed by rural cats alone each year (Bolen and Robinson 1999). This estimate does not include those wild animals, such as songbirds, killed each year by urban and suburban cats.

FIGURE 17.7   Exotic species can pose significant problems. The rock dove or feral pigeon shown in this photo has become abundant throughout the U.S. It carries diseases that can be harmful to humans and native birds.

Nonpredatory Exotics.   Exotic organisms can be detrimental to native species by displacing native animals, altering ecosystems, introducing foreign diseases and parasites, hybridizing, and diverting funds from native wildlife programs for control and eradication. Nonpredatory exotics include introduced carriers of diseases and parasites that can infect native species, which have no or low resistance (Figure 17.7). Even the Native American human population was vulnerable to diseases introduced by Europeans, such as small pox, resulting in the deaths of many thousands of people.

The introduction of exotics also has been a serious threat in aquatic systems. Hybridization of exotic forms, such as cutthroat and rainbow trout, with native forms, such as Gila trout, have genetically altered native species. Competition and diseases associated with exotics, plus the alteration of ecological relationships between species, threaten entire aquatic ecosystems.

Exceptional Vulnerability of Island Species.   Island forms are particularly vulnerable to exotic predators because they evolved in isolation free from predators and human interference. Feral cats have been particularly destructive when introduced on islands previously devoid of land predators. One of the best examples of dramatic island change following human colonization is Hawaii, where unique plant and animal life has been greatly diminished.

In the Hawaiian Islands, as many as 98 bird species became extinct between A.D. 400 when Polynesians first colonized the islands and the first European contacts in 1778. Another 17 species became extinct in Hawaii after European settlement. These extinctions were due to the clearing of forests for agricultural development, alteration of native vegetation, exotic plant introduction, hunting, predation by introduced species (rodents, exotic birds, domestic cats, and the Indian mongoose), and introduced avian diseases. Half of Hawaii's endemic (distribution restricted to the island) birds have become extinct or been greatly reduced since 1880 (Royte and Jones 1995). Similar extinctions have occurred on many other land masses including New Zealand, Australia, and Madagascar where, after colonization by European cultures, domestic species such as pigs, sheep, goats, and rabbits degraded the native vegetation (Reid and Miller 1989).

ENDANGERED SPECIES
POLICY AND MANAGEMENT

Legislation and Treaties

Endangered Species Act. The strongest and most direct legislation pertaining to endangered species in the United States is the Endangered Species Act of 1973. It gave broad powers to the U.S. federal government in regulating actions affecting endangered animals and plants. The Endangered Species Act made it unlawful to be in possession of endangered species anywhere in the United States, provided a formal structure for listing and management of endangered species, required all federal agencies to use their existing authorities to conserve listed species, and provided a means for citizens to bring suit against any federal agency for failure to meet its obligations as specified in the legislation (Grosse 1997).

The Endangered Species Protection Act of 1966 authorized the Secretary of the Interior to develop a list of species facing extinction. It also sponsored research and authorized acquisition of terrestrial habitats to protect endangered species. The Endangered Species Conservation Act of 1969 added protection for federal lands and activities. The Endangered Species Act of 1973 superceded the two previous laws and, with subsequent amendments, has become the federal law directing the identification of threatened and endangered species. It regulates the taking and capture of all listed animals and plants, including foreign species, as well as their interstate and foreign commerce. It provides for designation of critical habitats, land acquisition, and financial assistance to states and foreign governments. In addition to public lands, it extended federal authority over private land use practices that could potentially affect endangered species. Projects that are a threat to an endangered species or its critical habitat must be altered to remove the threat or canceled. The act directs the government to become involved in international efforts to protect species by conducting law enforcement investigations and prohibiting the importation of endangered and threatened species. The U.S. Fish and Wildlife Service is responsible for land and freshwater species and the National Marine Fisheries Service in the Department of Commerce is responsible for marine species (Grosse 1997).

The USFWS may nominate a species for listing, but an individual or a private organization may also petition to initiate the listing process. The act provides protection for species threatened or endangered by one or more of the following factors: (1) present or threatened destruction, modification, or curtailment of the species' habitat or range; (2) overuse for commercial, sporting, scientific, or educational purposes; (3) disease or predation; (4) inadequacy of existing regulatory mechanisms for preventing decline or degradation of habitat; and (5) other natural or man-made factors affecting its continued existence. Threatened species are those likely to become endangered within the foreseeable future.

The term endangered species as defined in the law includes any species, subspecies, or population that is in danger of extinction throughout all or a significant portion of its range. This allows for listing of unique forms in portions of a species' distribution. For example, it was possible to specifically designate the subspecies of mountain sheep in southern California as endangered. This subspecies is not endangered in other parts of its distribution and this mountain sheep species occurs over a wide area of the United States. Individual states may also designate endangered and threatened populations under authority of state laws (George et al. 1998).

The Secretary of the Interior and Secretary of Commerce have the power to initiate actions that promote recovery of species. In order to restore a species to a nonendangered status, a recovery plan must be initiated for each species listed. Recovery plans are usually developed by wildlife biologists and resource specialists with expertise on the species being considered, or with other relevant expertise, and who form recovery teams. Recovery teams make recommendations on the steps necessary to remove a species from the endangered list. Removal from the list means the species is no longer in need of protection and therefore considered recovered. Conclusive evidence of extinction also leads to removal from the list.

Recovery plans have been developed for many species. They may include captive breeding programs, such as for the California condor, peregrine falcon, whooping crane, Apache trout, masked bobwhite, black-footed ferret, and Mexican wolf. Once propagated, species are released in the wild, although not always successfully. Harvest prohibition has been most successful for species that had been previously overharvested, such as sea otters, fur seals, beavers, snowy egrets, and American alligators. Translocation (moving animals from one area to another) programs were important in reestablishing populations once locally exterminated by hunting, including white-tailed deer, turkeys, elk, pronghorn antelope, bison, and bighorn sheep. However, most of those transplants occurred before genetic variation was understood. Unique populations were in some cases lost, such as Merriam's elk in the southwestern U.S. Maintenance or development of proper habitat and acquisition of essential habitats are other management strategies for rehabilitating endangered species. The creation of the Arkansas National Wildlife Refuge in 1937 on the Texas coast was of vital importance in saving the whooping crane from extinction. Among the animals the Fish and Wildlife Service has recently downlisted or plans to delist in the near future because they have recovered to stable, viable population sizes are the bald eagle, American peregrine falcon, brown pelican, island night lizard, and Ash Meadows amargosa pupfish.

The Endangered Species Act of 1973 precluded consideration of economic impacts. However, in 1978 it was amended to include the possibility of appeal to an Endangered Species Committee of high-level federal administrators, who might decide that the costs of protection were too high for the perceived benefits. Appeal to this committee, sometimes referred to irreverently as the "God Squad," has often been debated but rarely made.

Habitat Conservation Plans

An amendment to the Endangered Species Act in 1982 allows a Habitat Conservation Plan to be developed by a landowner whose land use has been identified as critical habitat of a threatened or endangered species (Beatley 1994). In exchange, the U.S. Fish and Wildlife Service issues an incidental take permit. The plan must be approved by the Fish and Wildlife Service and is developed under the oversight of a committee of informed stakeholders. This amendment was passed to contend with the increasing conflicts that developed among environmental groups, developers, and land owners over endangered species issues. The intent of Habitat Conservation Plans is to allow land use while assuring protection of the species despite some incidental loss to the population. The plan must meet certain standards. It must spell out the steps for minimizing incidental take, the impact that would result from incidental take, alternative plans and why they were not adopted, and the source of funding for carrying out the plan. For a plan to be approved, the applicant must assure minimum effect of incidental kill, funding to carry out the plan, and convincing evidence that the incidental take will not have a negative impact on species viability.

Beatley (1994) reviewed the progress of habitat conservation planning with respect to urban development in three of the fastest growing states: California, Texas, and Florida. He provided a sample of species that have been addressed through Habitat Conservation Plans, including fringe-toed lizard (California), American crocodile (Florida), desert tortoise (California), several kangaroo rats (California), Key Largo woodrat (Florida), Buena Vista lake shrew (California), San Joaquin kit fox (California), several butterflies (California and Florida), several songbirds (California and Texas), and several plants (California and Texas). Strategies typically include reservation of land for habitat protection, habitat restoration, and management providing special needs (e.g., removal of exotic species and competitor control).

CITES. The most influential international approach to endangered species conservation is the Convention on International Trade in Endangered Species (CITES), which was signed in the U.S. in 1973. International trade can be an important factor in the decline and extinction of some species if trade promotes overhunting for furs or hides, for food products, for exhibition, for sport, for scientific experimentation, or for other purposes. The convention addressed the protection of species from the standpoint of how trade impacted the status of a particular species in the wild in its native country.

Overexploited, imported, or exported species are eligible for protection under the convention (Grosse 1997). The participating parties provided for appendices listing endangered and potentially endangered species. Stringent controls were directed at regulating trade activities of endangered species. All trade of such species, their parts and derivatives, including manufactured products, require permits from both the importing and exporting countries. As of 1997, 134 countries had signed the treaty. CITES aided in curtailing the rampant trade in a host of endangered species, including cheetahs, other spotted cats, tigers, elephants, rhinos, reptiles, parrots, and many other birds. Nonetheless, trade in endangered species continues, at an estimated $5 billion to $10 billion per year, and requires the persistent concerted efforts of the international community to limit its destructiveness.

Management Needed to Maintain Biodiversity

Halting further degradation and fragmentation of existing natural habitats and ecosystems is essential for the conservation of biotic resources. Protection of intact natural areas and restoration of degraded habitat are key components of protecting biological diversity from further erosion. A fundamental cause of habitat degradation is human population growth beyond the capacity of landscapes to support biodiversity. Limiting human population growth is essential for sustaining biodiversity. Development of more efficient uses of natural resources without degrading biodiversity values is also a fundamental strategy. In the United States, wetlands have declined 53% (221 million acres in the 1780s to 104 million acres today). The surface area of wetlands in the 48 contiguous states has been reduced from 11% to 5%. The conversion of forests and grasslands to agricultural uses has also greatly diminished the native fauna and flora. Only 1% of the original 10 million km^2 of tallgrass prairie in the eastern Great Plains remains in natural vegetation. Certain forest types have substantially decreased, particularly bottomland hardwoods of the lower Mississippi Delta and riparian forests in the semiarid to arid West. These forest types supported communities rich in biodiversity and produced timber products in high demand (Flather and Hoekstra 1989).

Because federal lands provide habitats for a large proportion of native wildlife, federal agencies must be major participants in conserving biodiversity. Of the 660 listed species for which data are available, 24% occurred on U.S. Forest Service lands, 17% occurred on lands managed by the Bureau of Land Management, and 26% occurred on Department of Defense Lands. Noss and Cooperrider (1994) provide guidelines for managing forests, rangelands, and aquatic ecosystems. Reserve networks and monitoring programs are important features of a national biodiversity conservation strategy.

Most federally owned lands occur in the West. Lack of federally owned land limits the government's role in conserving diversity in the eastern U.S. Even in the West, many elements of diversity occur outside of federally protected lands. Species that require very large tracts of undeveloped land occur principally on federal lands. However, these protected areas—primarily national parks and national forests—may not be sufficient in size and interconnection to ensure the long-term viability of large mammals, such as grizzly bears, gray wolves, Florida cougars, and wolverines. In areas where little public land exists, habitat fragmentation is a major problem even for small species.

Except for habitat conservation planning, the Endangered Species Act discourages private landowners from creating, restoring, or enhancing habitats for endangered species. Private landowners fear greater regulatory actions if endangered species are found on their properties. Incentives for landowner participation in managing endangered species could include the option of partially deferring estate taxes and allowing participants to claim tax deductions for costs associated with habitat management practices. A conservation strategy that entices more participation of private landowners is essential and vital in curtailing the loss of biodiversity (Wilcove et al. 1994). We refer the reader to Anderson and Leal (1991) for consideration of alternatives that would reward landowners for protection of threatened and endangered species habitat.

Although the federal government administers endangered species law, nongovernment organizations have played key roles in the management of endangered species. Outstanding examples of such organizations include the Nature Conservancy, Audubon Society, Sierra Club, Defenders of Wildlife, Environmental Defense Fund, World Wildlife Fund, and the National Wildlife Federation. The Nature Conservancy has been especially active in maintaining biodiversity through private land acquisition. It has taken the lead in developing an ecosystem approach in purchase and management of land. Several nongovernment organizations also have played important roles in habitat conservation planning. Also important is the public outreach and educational aspect of these organizations through magazines and other media.

ENDANGERED ECOSYSTEMS

One of the declared purposes of the Endangered Species Act is to "provide a means whereby the ecosystems upon which endangered species and threatened species depend may be conserved." The primary strategy used to attain this end has been identification of critical habitat and its protection through regulation and habitat conservation planning. The focus is on critical habitat determination based predominantly on the patterns of past and existing distribution of the threatened species and its associated habitat. One recent effective approach is the use of Geographical Information Systems (GIS) to map land uses, individual population distributions, and habitat as indicated by remote sensory signatures of vegetation and geological features. Using an overlay approach, those areas of greatest unique biodiversity and threatening existing or potential land uses can be identified for priority attention. However, overattention to a pattern approach has been criticized as insufficient without careful attention to the preservation and restoration of processes that sustain diversity (Smith et al. 1993, Risser 1995).

The prevalent forces determining the interactive dynamics among species within ecosystems often originate from outside a community including storms, fire, atmospheric changes, flooding, drought, off-site changes in habitat of migratory species, and colonization (invasion) by new or already established species. The evolutionary role of these forcing factors can be clearly seen in the invasion of island ecosystems by new species. The slow rate of natural invasion has resulted in the unique biodiversity of many island ecosystems. However, the accelerated rate of invasion caused by humans has greatly reduced that unique biodiversity. Invasion in itself is not the problem; it is the extent that the rate and intensity varies from historic conditions. The periodicity and intensity of forcing factors from outside ecosystems have been profoundly important in determining the unique character of ecosystems. Ecosystems are most endangered where the rates and intensities of those forces have been greatly modified either by accident or by human design.

Ecosystem processes are adapted to these influential forces. The unique character of ecosystems, and of many of the unique species within them, often depends on the extent to which the forces continue to operate as they have during past evolutionary history. Noss and Cooperrider (1994) have pointed out the importance of identifying endangered ecosystems as a precursor to the prevention of individual species endangerment. Certain wetland and prairie ecosystems are among the most deserving candidates of endangered ecosystem status. In prairie ecosystems the deprivation of fire as a natural forcing factor has been critical in determining biodiversity and the ability of an ecosystem to recover from drought. In wetlands and rivers, the periodicity and intensity of flooding is often a critical forcing factor for sustaining ecosystem processes. Of course, both wildfire and unrestrained flood are threatening to human property and health. They require control where those benefits have higher priority than protection or restoration of natural ecosystems.

Future recovery of existing endangered species is likely to emphasize the importance of protecting and restoring fundamental ecosystem processes, including restoring those forces that were most influential in sustaining the unique attributes of ecosystems. This will be challenging for natural resource managers because past political decisions about land ownership and control have placed little emphasis on maintaining natural ecosystems and their processes. The floodplain communities of large rivers, for example, are most often privately owned and influenced by artificial river control and hydropower structures. Similarly, only small remnants of tall grass prairie remain in isolated islands of protected land in otherwise developed or developing landscapes.

Past natural resource management tactics, such as flood and fire prevention, usually will have a higher priority to society than sustaining pristine ecosystem processes. However, there often are alternatives that can reduce threats to biodiversity while sustaining the most crucial artificial forcing factors. Research on biodiversity maintenance is increasingly turning its attention to these large landscape processes. Conflicts between the public interest and individual interest in property protections and development will continue to be a fundamentally divisive issue.

LITERATURE CITED

Anderson, T. L. and D. L. Leal. 1991. *Free market environmentalism.* Boulder, CO: Westview Press.

Beatley, T. 1994. *Habitat conservation planning: Endangered species and urban growth.* Austin, TX: University of Texas Press.

Cairns, M. A. and R. T. Lackey. 1992. Biodiversity and management of natural resources: The issues. *Fisheries* (Bethesda) 17:6–10.

Cox, G. W. 1997. *Conservation biology.* 2nd edition. Dubuque, IA: Wm. C. Brown. 362 pp.

Daily, G. C., ed. 1997a. *Nature's services: Societal dependence on natural ecosystems.* Covelo, CA: Island Press. 392 pp.

Daily, G. C. 1997b. Introduction: What are ecosystem services? (pp. 1–10). In G. C. Daily, ed., *Nature's services.* Covelo, CA: Island Press.

DeLong, D. C. Jr. 1996. Defining biodiversity. *Wildlife Society Bulletin.* 24:738–749.

Flather, C. H. and T. W. Hoekstra. 1989. *An analysis of the wildlife and fish situation in the United States: 1989–2040.* USDA Forest Service General Technical Report RM-178. Fort Collins, CO: Rocky Mountain Forest and Range Experiment Station. 146 pp.

Flather, C. H., L. A. Joyce, and C. A. Bloomgarden. 1994. *Species endangerment patterns in the United States.* USDA Forest Service General Technical Report RM-178. Fort Collins, CO: Rocky Mountain Forest and Range Experimental Station. 24 pp.

George, S., W. J. Snape, and M. Senatore. 1998. *State endangered species acts: Past, present, and future.* Washington, DC: Defenders of Wildlife and Albuquerque, NM: Center for Wildlife Law. 133 pp.

Grosse, W. J. 1997. *The protection and management of our natural resources, wildlife, and habitat.* Dobbs Ferry, NY: Oceana Publications.

Hopkins et al., eds. 1982. *Paleoecology of Beringia.* New York, NY: Academic Press.

IUCN. 1997. 1996 IUCN red list of threatened animals. Gland, Switzerland: IUCN. 368 pp.

Kellert, S. R. 1996. *The value of life: Biological diversity and human society.* Washington, DC: Island Press. 263 pp.

Langner, L. L. and C. H. Flather. 1994. *Biological diversity in the United States.* USDA Forest Service General Technical Report RM-244. Fort Collins, CO: Rocky Mountain Forest and Range Experimental Station. 24 pp.

Marcot, B. G. M., M. J. Wisdom, H. W. Li, and G. C. Castillo. 1994. *Managing for featured, threatened, endangered, and sensitive species and unique habitats for ecosystem sustainability.* USDA Forest Service General Technical Report PNW-GTR-329. Portland, OR: Pacific Northwest Research Station.

McNeely, J. A., K. R. Miller, W. V. Reed, R. A. Mittermeier, and T. B. Wemer. 1990. *Conserving the world's biological diversity.* Gland, Switzerland: IUCN. 193 pp.

Meffe, G. K. and C. R. Carroll. 1997. *Principles of conservation biology.* 2nd edition. Sunderland, MA: Sinauer Associates.

Morrell, V. 1999. The variety of life. *National Geographic* 195:6–31.

Noss, R. F. and A. Y. Cooperrider. 1994. *Saving nature's legacy: Protecting and restoring biodiversity.* Washington, DC: Island Press.

Pimental, D. 1981. *Biological diversity and environmental quality.* (pp. 44–47). In Proceedings of the U.S. Strategy Conference on Biological Diversity. Department of State Publication 9262, Washington, DC.

Primack, R. B. 1998. *Conservation biology.* 2nd edition. Sunderland, MA: Sinauer Associates.

Reaka-Kudla, M. L., D. E. Wilson, and E. O. Wilson, eds. 1997. *Biodiversity II.* Washington, DC: Joseph Henry Press.

Reid, W. V. and K. R. Miller. 1989. *Keeping options alive, the scientific basis for conserving biodiversity.* New York, NY: World Resources Institute. 128 pp.

Risser, P. G. 1995. Biodiversity and ecosystem function. *Conservation Biology* 9:742–746.

Royte, E. and C. Jones. 1995. Hawaii's vanishing species. *National Geographic* 188:3–37.

Smith, T. B., M. W. Bruford, and R. K. Wayne. 1993. The preservation of process: The missing element of conservation programs. *Biodiversity Letters* 1:164–167.

West, N. E., ed. 1995. Biodiversity on rangelands. Logan, UT: College of Natural Resources, Utah State University. 114 pp.

Wilcove, D. S. and M. J. Bean, eds. 1994. *The big kill: Declining biodiversity in America's lakes and rivers.* Washington, DC: Environmental Defense Fund. 275 pp.

Wilcove, D. S., M. J. Bean, R. Bonnie, and M. McMillan. 1994. *Toward a more effective Endangered Species Act for private land.* Washington, DC: Environmental Defense Fund. 20 pp.

Williams, J. E. and R. R. Miller. 1990. Conservation status of the North American fish fauna in fresh water. *Journal of Fish Biology* 37:79–85.

Wilson, E. O., ed. 1988. *Biodiversity.* Washington, DC: National Academy Press. 521 pp.

Wilson, E. O. 1992. *The diversity of life.* Cambridge, MA: Belknap Press of Harvard University Press. 424 pp.

The Mineral and Energy Resources

Some of the same basic mineral fertilizer and solar energy resources required for ecosystem functions are managed directly to meet human needs. Chapter 18, on minerals, reveals how, through technology and exploration, people have extracted and refined minerals for many applications. Minerals are the basis for developing much of the infrastructure that forms the artificial habitat of humanity, including rock, sand, gravel, lime, clays, and other basic earth resources used in concrete, tile, building stone, brick, mortar, and glass. The metals, many of which are nutrients required for life or toxic in excessive concentrations, are indispensable parts of infrastructure, machinery, household appliances, and other material goods. Chapter 19 describes development and management of nonrenewable energy resources of atomic and fossil fuels. Fossil fuels include oil, coal, and natural gas. Chapter 20 addresses the renewable energy resources, including solar, wind, water, and biofuels. The development and management of minerals and energy have many implications for sustainable economic development. Their impact on ecosystem functions and quality of air, water, and land resources may be more important than their eventual depletion. For that reason, policies governing development of minerals, energy, and biotic resources are closely linked.

CHAPTER 18

Mineral Resources

INTRODUCTION

The purpose of this chapter is to provide an overview of the earth crust resources that supply our material needs. To comprehend the nature of our dependence on these resources, we will discuss their origin, classification, properties, abundance, and use. Our focus will be nonfuel minerals (including metal minerals and nonmetallic resources mined from the earth). Coal, oil, oil shale, sand tar, and uranium will be discussed in Chapter 19 on nonrenewable energy resources. The availability of these resources is often determined by cyclic processes working over time scales beyond our immediate comprehension. Our use of these resources is of critical importance in determining the sustainability of global and local economies. In this chapter, we provide an overview of mineral availability, mining, and environmental problems and refer the reader to Kesler (1994) for more comprehensive coverage of these subjects.

GEOLOGICAL FOUNDATIONS

The Realm of Minerals and Rocks

Minerals and rocks provide us with a host of material goods, many of which escape our attention. To understand their origin, we must provide a framework of definitions and classification suited to our needs. By definition, a rock is a naturally occurring solid containing one or more minerals. A mineral is a naturally occurring inorganic solid with specific physical properties and a consistent internal structure. In composition, minerals can range from those containing a single element, such as native copper, to those comprising several elements. A mineral can be nonmetallic such as salt, gypsum, or sand, or metallic such as aluminum or copper. Strangely, water in its frozen form satisfies this definition owing to the crystalline structure of ice, whereas liquid water does not. Also, coal is routinely listed in tables of economic minerals, but in fact is not a mineral. It more correctly is an organic sedimentary rock.

For identification purposes, geologists rely on several mineral properties. These include color, crystal form, specific hardness and density, luster, and cleavage. Crystal form reflects atomic structure of the mineral and refers to the assemblage of flat faces that make up the surface. Hardness indicates resistance to scratching or abrasion, and determines which materials can be used to efficiently cut hard metals or drill into rock. Cleavage refers to the manner in which minerals split along planes determined by their crystal structure. For example, cubic minerals such as halite often display three mutually orthogonal cleavage directions. Mica has a perfect basal cleavage in one direction and splits into thin sheets. Feldspars commonly show two strong cleavages.

To understand the chemical and physical properties of minerals, one must examine the ways the atoms they contain bond to one another. Bonding refers to the way electrons are shared or released by atoms. It can occur in three basic ways, which include ionic, covalent, and Van der Waal's bonding. Ionic bonding is a linkage formed by transferring or shifting electrons from one atom to another. Ionically bonded minerals are moderately hard, have high melting points, and dissolve easily in water (for example, halite or NaCl). Metal atoms are joined by shells of electrons overlapping them. This results in a solid swarm of electrons that are free to move about and which enable metals to conduct electricity and transmit heat efficiently. Because of the dense arrangement of their electrons, metals can be twisted or beaten into thin sheets without breaking, and can be drawn into thin wire. Covalent bonds are formed by sharing electrons and create the strongest types of chemical bonds. Diamonds are extremely hard owing to the covalent bonds in their structure. Covalently bonded minerals do not readily dissolve in water, have high boiling points, and do not conduct electricity. The weakest of the chemical bonds is Van der Waal's bonding maintained by the weak residual charges between sheets of atoms. Minerals with this type of bonding, such as the micas, can be split into sheets using only a fingernail. The individual sheets are held by covalent bonds and are stronger. Like mica, most minerals exhibit more than one kind of bonding.

Concerning classification, all minerals can be grouped into either silicates or nonsilicates. Because the two most abundant elements in the earth's crust are oxygen and silicon, silicates (silicon-oxygen compounds) form the most abundant group of minerals. Most minerals in Earth's crust are silicates. Silicate minerals are further classified according to the structural arrangement of the silicate anions. The feldspar silicate group is most abundant and represents a family of closely related minerals having sodium, potassium, or calcium in addition to silicon and oxygen. Ferromagnesian silicates are rich in magnesium or iron. Soil scientists are especially concerned with the mineralogical properties of clay minerals found in the mica group. In soils, clay minerals can greatly affect moisture and nutrient retention. Talc, found in any drug store, is a member of the same group.

Nonsilicates include minerals that can be grouped by common chemical characteristics. These include oxides, sulfides, sulfates, carbonates, halides, and native elements. Native copper, graphite, and diamond are examples of native elements. Gypsum is calcium sulfate, hematite is iron oxide, and dolomite is one of the carbonate forms. Lead is obtained from galena, a sulfide mineral. Other examples of nonsilicates and silicates, and their sometimes complex mineralogical structures and elemental compositions, are discussed more fully in geology texts (see Hoeffer 1995).

The formation of economic minerals can be understood by examining the fundamental structure of Earth, including how crust is created and destroyed by plate tectonic activities. This information is necessary to understand how rocks and minerals are created and recycled over eons of time. Our understanding of Earth's dynamic geological processes has grown considerably in recent decades. Earth usually is depicted as consisting of four concentric layers differing in composition and density (Figure 18.1). These layers include Earth's inner core, outer core, mantle, and crust. Within Earth's inner core, the pressures are so great that the iron and nickel contents are solid despite high temperature. In the outer core, temperature and pressure are so balanced that matter is molten and exists as a liquid. Because their contents are believed to be similar, it is added pressure that separates the inner and outer core layers.

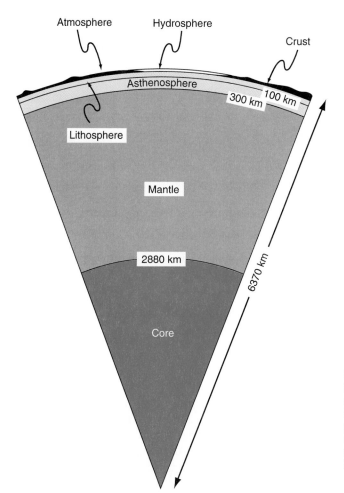

FIGURE 18.1 Distribution of the lithosphere, hydrosphere, and atmosphere (except for that part in the hydrosphere, the biosphere is too small to be shown at this scale) and major divisions of the solid earth (from Kesler 1994).

Geologists divide the mantle and crust into three distinct regions. The mesosphere ("intermediate or middle sphere") exists from the core-mantle boundary to a depth 350 km beneath Earth's surface. In this region, rock is so highly compressed it exists as a solid, even though the temperature is high. Within the upper mantle, from a depth 350 km to 100 km, is a region where the balance between temperature and pressure is such that rocks have relatively little strength. In this region, the asthenosphere ("weak sphere"), igneous rock is easily deformed. Above the asthenosphere is the region composed of upper mantle and crust called the lithosphere ("rock sphere"). Here rocks are stronger and more rigid. It is rock strength, not rock composition, that results in regional distinctions. The crust, or rigid outermost layer of Earth, consists of continental crust and oceanic crust. The crust varies in thickness, from about 56 km (35 miles) under the continents to 5 km (3 miles) under the ocean basins. Earth's crust consists mainly of alumino-silicates. The granitic materials of the continental crust are less dense than the basaltic materials of oceanic crust.

Plate Tectonics

The theory of plate tectonics has become widely accepted and is said to have revolutionized our thinking about some of the most fundamental issues in geology. According to this theory, the lithosphere is broken into plates which can slide over the asthenosphere. This movement is possible because the asthenosphere is soft and flows easily. The mechanism by which Earth's plates move involves the formation of convection cells, which develop as hot mantle material. This rises from the earth's interior, moves laterally, then cools, and begins to sink (Figure 18.2). The energy driving this movement comes from hot convection currents in the mantle. Sources of the

Plate Tectonics and Mineral Deposit Environments

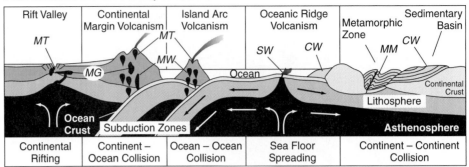

FIGURE 18.2 Plate tectonic environments showing their relation to ore-forming magmatic and hydrothermal processes (from Kesler 1994).

earth's interior heat are residual heat resulting from the earth's cooling from a molten state, heat produced by the decay of radioactive isotopes, and heat produced by crystallization of the core's liquid portion.

Plate Boundaries

Actions occurring at the margins of plates explain how new lithosphere is produced and how rocks are recycled. Along divergent plate boundaries, or spreading centers, plates are moving apart. Spreading centers can involve oceanic to oceanic plate margins and continental to continental plate margins. Along these margins, basaltic lava eruptions and earthquake activity are common and new lithosphere forms. The mid-Atlantic ridge marks the location of ocean-to-ocean plate margins where the lithosphere is being pushed upward by rising hot mantle. Along such ridges, convection currents bring up hot rock from deep in the mantle and the asthenosphere becomes hot enough to begin melting. The magma that forms beneath the ocean ridge rises upward to the top of the lithosphere where it cools and hardens to form ocean crust. This new oceanic crust moves away at right angles from both sides of the ridge (horizontally), carrying the continents with it. The rift valleys of Africa mark the locations of divergent boundaries where fragments of continental crust are moving apart.

Convergent plate boundaries are areas where two plates move toward each other, forcing either development of a subduction zone or a collision zone. In a subduction zone, one plate capped with less dense continental crust will override the denser oceanic plate, which sinks along the contact margin. The entire oceanic plate becomes increasingly hotter as it descends deeper into the earth. As the plates grind against each other, temperatures increase even more. Where the subducting plate comes into contact with the asthenosphere, melting occurs and magma is created. As the less dense magma makes its way upward, it either erupts on the earth's surface and solidifies as extrusive igneous rock, or solidifies within the earth to become intrusive igneous rock. New metamorphic rock is created where rock near the subduction zone does not melt because of the pressure present. In collision zones, continental crust is not recycled into the mantle because this crust is lighter and less dense than the contents of the mantle beneath. In short, as continental plates move toward each other, their crust is too buoyant to be dragged downward on top of the sinking lithosphere. Instead, collision zones push up massive mountain ranges, such as the Himalayas. Transform plate boundaries are areas where two plates are sliding past each other, resulting in earthquake activity but no volcanic activity. California's San Andreas fault is an example of this type of boundary.

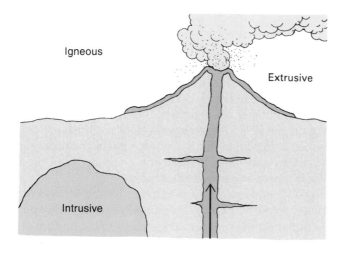

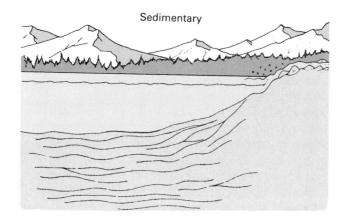

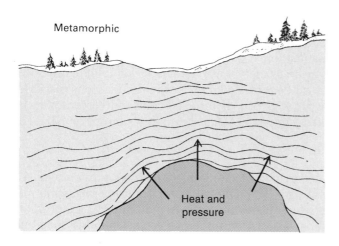

FIGURE 18.3 The three basic types of rocks include igneous rocks formed from cooling magma, sedimentary rocks formed from consolidation of deposited particles, and metamorphic rocks produced when heat and pressure act on preexisting rocks (from McKnight 1990).

The Rock Cycle

Plate tectonics offers an explanation of how rocks are created and destroyed as they are subjected to extreme pressures and heat (Judson et al. 1987, McKnight 1990, Kessler 1994). A more complete explanation of rock genesis is to be found in the rock cycle, which calls upon tectonic activities and above-ground factors including the actions of water, ice, wind, and gravity. The rock cycle (Figure 18.3) is a simplified description of the events leading to the formation, alteration, destruction, and reformation of the three major rock groups: igneous, sedimentary, and metamorphic. In the beginning, all rocks on Earth were igneous. Igneous rocks formed as the magma of early Earth cooled and solidified. Today there are many kinds of igneous rocks and their characteristics vary. However, they can be placed in one of two basic groups based on the conditions under which they become solidified. Extrusive rocks arrive on the earth's surface while still in a molten state. Lava, for example, forms new extrusive igneous rocks as it is spewed from volcanoes. Intrusive igneous rocks solidify as they cool slowly beneath the earth while being insulated by nonmagmatic material. Igneous rock such as granite and basalt are created when magma wells up from the earth's mantle and solidifies as it cools above or beneath the surface.

As mountains comprised of igneous rock are weathered by the actions of wind, rain, ice, and acidic solutions created by vegetation, the particles are transported to lower elevations. Sediments transported by rivers and streams to the bottom of ocean floors can accumulate in thick layers over time. The lower deposits are exposed to great pressure due to the sheer weight of the matter above and this causes their particles to adhere and interlock. As silica, calcium carbonate, and iron oxide precipitate from the water into the pore spaces, these cementing agents and the action of pressure consolidate the sediments into sedimentary rock. As rock forms, minerals actually recrystallize as forms that will accommodate a more confined space. The conversion of sediment into solid rock by cementation, compaction, or crystallization is referred to as lithification.

Sedimentary rocks are classified on the basis of mode of formation: mechanical, chemical, or organic. Mechanically accumulated sedimentary rocks are composed of fragments of preexisting rocks in the form of boulders, gravel, sand, silt, and clay. Shale (composed of silt and clay) and sandstone are examples. Sedimentary rocks formed by the deposition or precipitation of soluble materials are said to be chemically accumulated. Calcium carbonate is a common component of such rocks and limestone is a widespread example. Organically accumulated sedimentary rocks are formed from the remains of dead plants or animals. Limestone produced from the skeletal remains of coral and other lime secreting sea animal serves as an example.

Metamorphic rocks were originally something else. Over eons of time, deposits of sedimentary and igneous rocks become increasingly buried until intense heat and pressures transform them into metamorphic rocks. Metamorphism, the transformation of rocks either in texture or mineral composition, can also be influenced by chemically active solutions. Some rocks change in a predictable fashion. Limestone usually becomes marble, sandstone becomes quartzite, and shale often becomes slate. The rock cycle begins again as metamorphic rocks either melt or are uplifted to the earth's surface. Evidence of the rock cycle can be found in various places around the world. For example, new igneous rock material is found at divergent plate boundaries (mid-ocean ridges) and at convergent plate boundaries (volcanic island arcs).

Mineral Deposits

The hydrothermal activity occurring at tectonic plate margins provides some of the richest deposits of economically important metallic ores. Hydrothermal vents occur on the ocean floor along ridges where plates separate. As hot, mineral-laden water rises through fractured rocks and spews out of hydrothermal vents, it deposits metal sulfides around the vents as the water cools. Deposits that are associated with divergent plate boundaries, such as mid-ocean ridges and rift zones, include chemically precipitated copper, lead, and zinc sulfide deposits found within ocean floor sediments.

Where tectonic plates come together, igneous rocks saturated with seawater undergo partial melting with the descent of oceanic lithosphere at a subduction zone. The combination of heat, pressure, and resulting partial melting facilitates the mobilization of metals, which are concentrated and ascend as components of the magma. The metal-rich fluids are eventually released (or escape) from the magma, and the metals are deposited in a host rock. The world's great disseminated copper deposits are found in plutons that have formed above subducting plates. All bodies of intrusive igneous rock, regardless of shape or size, are called plutons. The rocks formed are of intermediate composition, formed by the partial melting of oceanic basalt that was enriched in copper.

Mineral deposits also form where plate movements allow magma to penetrate the earth's surface and form volcanos. As the molten igneous rock inside the volcano cools, minerals with high densities crystallize first and sink to the bottom. Less dense minerals crystallize later and are found near the top of the rock deposit.

Ore deposits are often found along the contact between igneous rocks and the rocks they intrude. This area is characterized by contact metamorphism, which is caused by interactions between heat, pressure, and chemically active fluids of the cooling magma in contact with the surrounding country rock (preexisting solid rock). Contact metamorphic deposits form lead, copper, zinc, and silver deposits when hydrothermal fluids from a pluton, an igneous intrusive mass, rich in these metals replaces the country rock (usually limestone). Hydrothermal fluids carrying metallic ions can also enter fractures or cavities in solid rock, then cool and precipitate these metals as vein deposits. Such deposits can contain gold, silver, lead, zinc, or tin.

In some cases an entire igneous rock mass contains disseminated crystals that may be economically recovered. Perhaps the best-known example of this is the occurrence of diamond crystals found in a coarse-grained igneous rock called kimberlite. It typically is found as a pipe-shaped body of rock that decreases in diameter with depth. Almost the entire kimberlite pipe is the ore deposit, and the diamond crystals are disseminated throughout the rock.

Igneous and metamorphic processes are responsible for producing much of the stone used in the construction industry. Granite, basalt, marble (metamorphosed limestone), slate (metamorphosed shale), and quartzite (metamorphosed sandstone), along with other rocks, are quarried to produce crushed rock and dimension stone in the U.S.

Sedimentary processes also concentrate minerals in deposits. The processes that weather and erode igneous rocks, for example, can deposit them as sediments while removing unwanted materials. Weathering can concentrate some materials to the point that they can be extracted at a profit. Intensive weathering of residual soils (laterite) derived from aluminum-rich igneous rocks can concentrate relatively insoluble hydrated oxides of aluminum and iron. The more soluble elements; such as silica, calcium, and sodium; are selectively removed by soil and biological processes. Weathering and erosion also deposit materials in ocean and streambeds at lower elevations. Sand and gravel, for example, are sedimentary deposits of fine-grained rock. Mineral deposits called "placers" are formed when flowing water separates heavier mineral particles (such as gold) from sediment and drops them on streambeds with little water flow and turbulence. Placer deposits of both gold and diamonds are concentrated by wave action and other coastal processes. Beach sands and near-shore deposits are mined in Africa and other places to acquire these precious minerals.

Rivers and streams usually carry large quantities of material derived from the weathering of rocks. When mineral-laden water pours into shallow marine basins or lakes that become isolated by tectonic activity (uplift), massive amounts of "evaporite" minerals can remain after the water evaporates. In other cases, large inland lakes with no outlets essentially dry up. Examples of minerals found as "evaporite deposits" include potassium and sodium salt, gypsum, sodium, and calcium carbonate, sulfate, borate, nitrate, and limited iodine and strontium compounds. Salt water (brines) derived from wells, thermal springs, inland salt lakes, and the ocean can yield iodine, calcium chloride, and magnesium.

Mineral Distribution and Abundance

The U.S. Geological Survey has adopted specific terminology to provide a uniform description of supplies of actual and potential mineral resources (Figure 18.4). It divides

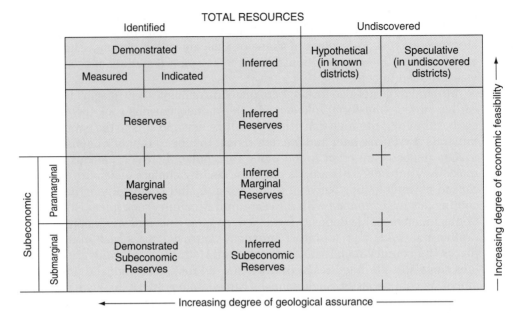

FIGURE 18.4   Mineral resources can be classified as reserves and total resources. Total resources include all the potentially mineable discovered and undiscovered deposits in the world (from U.S. Geologic Survey).

the estimated total resources of a given mineral into identified and undiscovered categories. Identified resources are mineral deposits that have a known location, quality, and quantity. Undiscovered resources are those believed to exist on the basis of geological information, but quality and quantity are unknown.

The amount of a mineral reserve does not remain fixed because economic conditions, exploration effort, and technology change over time. For example, when energy is cheap, marginal ore deposits can become economical and the reserve base expands. The base can shrink when the reverse is true. The system also allows nonprofitable identified resources to be added to the reserve base when ore market values increase.

Because mineral deposits are not evenly deposited in the earth's crust, some nations have rather large supplies of most economic minerals while others must obtain some or most from foreign markets. Figure 18.5 shows the position of the U.S. in the world mineral market by identifying minerals obtained mostly or exclusively by importing. Some minerals are imported into the U.S. because they can be obtained more cheaply elsewhere, while others are simply absent.

Strategic and Critical Minerals

Minerals that can interrupt industrial production when supplies are inadequate are called critical minerals. For the purpose of national defense, some minerals are defined as strategic minerals. Metals needed directly or indirectly to produce arms can be classified as strategic. Examples include some scarce metals needed in large quantity to produce high strength metal or electronic components.

MINING AND MINERAL EXTRACTION _____

What is an Ore?

An ore is any naturally occurring material from which a mineral or minerals can be extracted profitably. A mineral deposit or part of a mineral deposit, consisting of ore, is an ore body.

Ore bodies range in size from a few tons to more than a billion tons. Even to the nongeologist, the word ore has a somewhat straightforward meaning. It indicates that

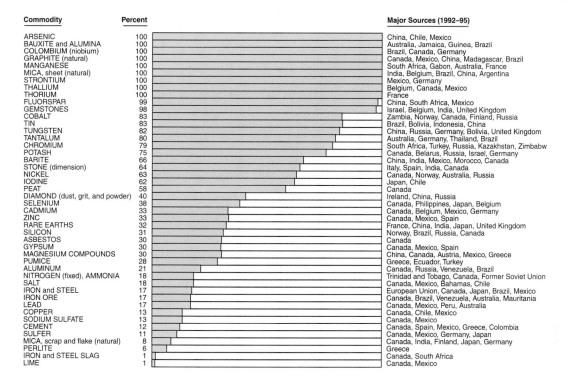

Commodity	Percent	Major Sources (1992–95)
ARSENIC	100	China, Chile, Mexico
BAUXITE and ALUMINA	100	Australia, Jamaica, Guinea, Brazil
COLOMBIUM (niobium)	100	Brazil, Canada, Germany
GRAPHITE (natural)	100	Canada, Mexico, China, Madagascar, Brazil
MANGANESE	100	South Africa, Gabon, Australia, France
MICA, sheet (natural)	100	India, Belgium, Brazil, China, Argentina
STRONTIUM	100	Mexico, Germany
THALLIUM	100	Belgium, Canada, Mexico
THORIUM	100	France
FLUORSPAR	99	China, South Africa, Mexico
GEMSTONES	98	Israel, Belgium, India, United Kingdom
COBALT	83	Zambia, Norway, Canada, Finland, Russia
TIN	83	Brazil, Bolivia, Indonesia, China
TUNGSTEN	82	China, Russia, Germany, Bolivia, United Kingdom
TANTALUM	80	Australia, Germany, Thailand, Brazil
CHROMIUM	79	South Africa, Turkey, Russia, Kazakhstan, Zimbabw
POTASH	75	Canada, Belarus, Russia, Israel, Germany
BARITE	66	China, India, Mexico, Morocco, Canada
STONE (dimension)	64	Italy, Spain, India, Canada
NICKEL	63	Canada, Norway, Australia, Russia
IODINE	62	Japan, Chile
PEAT	58	Canada
DIAMOND (dust, grit, and powder)	40	Ireland, China, Russia
SELENIUM	38	Canada, Philippines, Japan, Belgium
CADMIUM	33	Canada, Belgium, Mexico, Germany
ZINC	33	Canada, Mexico, Spain
RARE EARTHS	32	France, China, India, Japan, United Kingdom
SILICON	31	Norway, Brazil, Russia, Canada
ASBESTOS	30	Canada
GYPSUM	30	Canada, Mexico, Spain
MAGNESIUM COMPOUNDS	30	China, Canada, Austria, Mexico, Greece
PUMICE	28	Greece, Ecuador, Turkey
ALUMINUM	21	Canada, Russia, Venezuela, Brazil
NITROGEN (fixed), AMMONIA	18	Trinidad and Tobago, Canada, Former Soviet Union
SALT	18	Canada, Mexico, Bahamas, Chile
IRON and STEEL	17	European Union, Canada, Japan, Brazil, Mexico
IRON ORE	17	Canada, Brazil, Venezuela, Australia, Mauritania
LEAD	17	Canada, Mexico, Peru, Australia
COPPER	13	Canada, Chile, Mexico
SODIUM SULFATE	13	Canada, Mexico
CEMENT	12	Canada, Spain, Mexico, Greece, Colombia
SULFER	11	Canada, Mexico, Germany, Japan
MICA, scrap and flake (natural)	8	Canada, India, Finland, Japan, Germany
PERLITE	6	Greece
IRON and STEEL SLAG	1	Canada, South Africa
LIME	1	Canada, Mexico

FIGURE 18.5 U.S. net import reliance on selected minerals as a percentage of total consumption in 1996 (in descending order of importance). The shaded area represents net imports (from U.S. Geologic Survey).

TABLE 18.1 Crustal Abundances of Certain Metals and Minimum Contents in Mineable Ores Are Approximated in 1986

Metal	Crustal abundance	Mineral content
	%	
Aluminum	8.2	35
Iron	5.6	16
Magnesium	2.3	3.3
Manganese	0.9	30
Chromium	0.0100	25
Zinc	0.0070	6
Copper	0.0055	0.03
Lead	0.0013	3
Tin	0.00021	1
Beryllium	0.0003	0.05

Source: Cameron 1986.

something of value is concentrated above average crustal amounts (Table 18.1). Metals generally considered scarce include copper, lead, zinc, gold, silver, platinum, uranium, mercury, and molybdenum. In the abundant class we have iron, aluminum, chromium, manganese, titanium, and magnesium. Nonmetallic resources generally are not called ores, but are classified as industrial rocks and minerals.

Surface Mining

Open Pits. In general, deposits within a hundred meters of the surface are extracted from open-pit mines and those at greater depth come from underground mines. Open pits account for 90% of the ore mined in the U.S. but only about 50% of the ore mined in the rest of the world. Iron, copper, and lead ores are commonly obtained from open pits (Figure 18.6).

FIGURE 18.6   An open-pit copper mine near Silver City, New Mexico.

Deposits with intimate mixtures of ore and worthless rock are usually mined by bulk extraction methods whereas rich zones of ore are often mined by more selective methods. Open-pit mining is almost always less selective than underground methods. Because the cost and difficulty of mining increase with depth, extremely deep ore deposits are not mineable at a profit by any method.

Open-pit mines typically remove overburden (the worthless rock that overlies ore) and extract the ore. Rock often must be broken by blasting before it can be removed. The ratio between the volumes of ore and the overburden that must be handled is an important factor in determining the feasibility of mining an ore. For the sake of creating a safe workplace, pits are much wider at the top than at the bottom. One of the largest open-pit mines in the world, the Bingham Mine just outside Salt Lake City, Utah, removes over 225,000 metric ton of rock each day, only 20% of which is ore.

Strip mining is a special open-pit method that is used on shallow, flat-lying ore bodies such as coal. Strip mining is actually one of the most environmentally acceptable forms of mining because it fills the pit and restores the land surface to its original form. Coal extraction will be discussed more fully in chapter 19.

Quarries.   Sand and gravel are removed from thousands of small pits in many parts of the country. Building rocks such as limestone, granite, and marble are taken from larger pits called quarries.

Dredging.   Dredging is a special type of open-pit mining that is used where the inflow of water is too high to be pumped out economically, and the ore is sufficiently unconsolidated to be dug without blasting. Dredging employs chain buckets and draglines to scrape up surface deposits covered with water. For example, it removes sand from streambeds and gravel from placer deposits in streams. Hydraulic mining is a cross between dredging and conventional mining in which a jet of water is used to disaggregate the rock and wash it into a processing facility.

Subsurface Mining

Some mineral deposits lie so deep that surface mining becomes impractical. These deep beds or veins are removed by subsurface mining. For some metallic ores, miners dig a deep vertical shaft, blast tunnels and rooms, and transport the ore to the surface. Certain soluble minerals such as salt are removed from underground deposits by so-

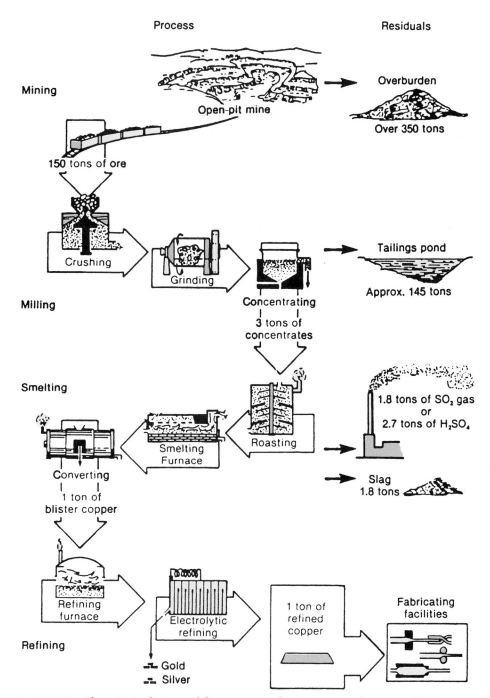

Process

Residuals

Mining

Open-pit mine

Overburden
Over 350 tons

150 tons of ore

Crushing

Grinding

Concentrating
3 tons of
concentrates

Tailings pond
Approx. 145 tons

Milling

Smelting

Roasting

Smelting
Furnace

1.8 tons of SO₂ gas
or
2.7 tons of H₂SO₄

Slag
1.8 tons

Converting
1 ton of
blister copper

Refining
furnace

Electrolytic
refining

1 ton of
refined
copper

Fabricating
facilities

Gold
Silver

Refining

FIGURE 18.7   The principal stages of the copper production process (courtesy of U.S. government).

lution subsurface mining. Compared to surface mining, proportionally more underground mining is done in developing countries where labor costs are relatively lower.

Processing

Extracted minerals usually need to be processed before they can be used, such as copper (Figure 18.7). Rocks taken from quarries may simply require crushing, grinding, and sizing. Metal ore requires considerably more work to remove impurities. In some cases the purified mineral is converted to a different chemical form by smelting or other chemical processes. Usually these processes free a metallic element from the oxygen, sulfur, or other elements with which it is combined in an ore. For example, aluminum is found in ore form as aluminum oxide (Al_2O_3). After the ore is purified and melted, electrical current is passed through the molten oxide to convert it to aluminum metal and oxygen gas. Copper and zinc often are found combined with sulfur, which is removed in smelters by heat or chemical reactions. The physical separation of a mineral from its ore is referred to as beneficiation.

Integrated mills are the traditional method for making steel from raw iron ore. There are three main components to an integrated mill. Coke ovens bake coal to produce coke. A blast furnace makes iron from iron ore, coke, and limestone. A basic oxygen furnace then converts the iron into steel. Minimills (small- to medium-sized steel mills) have lower capital investment and operating cost than integrated mills because they use electric arc furnaces. However, they do not have the facilities necessary to process raw iron ore like integrated mills. Instead they use scrap steel as their primary raw material. While these units have traditionally produced lower grades of steel, they are now competing with integrated units for flat-rolled steel and specialty grades such as stainless steel.

IMPORTANT METALLIC MINERALS

Iron

Numerous metals are known and used, but the most important ones are iron, copper, aluminum, lead, zinc, gold, and silver. Of these, iron must be considered the single most important metal of the modern world.

In fact, about 95% of all metal consumed in the world is iron. More than 70 countries make steel basically from iron. World iron ore and steel production is worth $350 to $400 billion annually, slightly less than world oil production (Kesler 1994). Steel is widely used in constructing buildings, highways, industrial plants, and other structures. Iron and steel products still make up about 75% of the weight of an automobile. However, this figure will continue to fall as lighter materials are substituted to improve fuel consumption.

Ferro-alloy metals are mixed with iron to increase the strength, hardness, and corrosion resistance of steel (Table 18.2). Alloy steel accounts for about 15% of world iron production. Ferro-alloy metals include chromium, manganese, nickel, silicon, cobalt, molybdenum, vanadium, tungsten, columbium, and tellurium. They permit steel to be used in a wide variety of applications, thus expanding its market and displacing use of aluminum and other metals (Schottman 1985). The total value of primary world ferro-alloy metals in these intermediate forms is slightly more than $30 billion, almost 10% that of iron and steel (Kesler 1994). The U.S. is especially vulnerable to supply shortages of alloy metals. For example, the U.S. has virtually no

TABLE 18.2   Alloy Elements Used in Steel

Element	Function in Steel
Aluminum	Deoxidation and grain-size control
Chromium	Hardenability, high-temperature strength, corrosion resistance
Cobalt	High-temperature hardness (HTH)
Columbium	As-rolled strength
Copper	Corrosion resistance and precipitation hardening
Lead	Machinability
Manganese	Deoxidation, sulfur control, hardenability
Molybdenum	Hardenability, HTH, temper brittleness control
Nickel	Hardenability and low-temperature toughness
Rare earths	Inclusion control, ductility, toughness
Silicon	Deoxidation, electric properties
Sulfur	Machinability
Tungsten	HTH and hardenability
Vanadium	Grain-size control, hardenability, HTH

Sources: Kesler 1994 and U. S. Bureau of Mines, *Mineral Facts and Problems.*

manganese reserves and is currently importing all of the manganese it uses. Iron is the fourth most abundant mineral (5%) in the earth's crust and can be formed by sedimentary, hydrothermal, and igneous processes. The chief ores of iron are hematite (Fe_2O_3) and magnetite (Fe_3O_4) (Hoeffer 1995). More than 50 countries produce iron ore (Kuck 1988). Although small iron deposits occur in almost all countries of the world, large deposits have a more restricted distribution.

Aluminum

Aluminum is the second most abundant metallic element in the earth's crust after silicon. However, it is a comparatively new industrial metal that has been produced in commercial quantities for just over 100 years (Plunkert 1993). Aluminum is lightweight, is resistant to atmospheric corrosion, is a good conductor of electricity and is strong when alloyed with other metals. This makes aluminum metal ideal for the manufacture of cars, trucks, aircraft, trains, and ships. It is also used to construct bridges and to produce aluminum siding, mobile homes, wiring, beverage cans, and foil wrap. In the U.S., the container and packaging industry is the largest consumer of aluminum metal, accounting for about 35% of apparent consumption. The main ore from which aluminum is obtained is bauxite (Al_2O_3 ZnH_2O). Australia is a leader in bauxite exports. Aluminum has a sedimentary origin. It is the end product of intense chemical weathering of aluminum-bearing minerals (primarily feldspars).

Copper

Owing to its conducting properties, about 70% of the copper produced is used in the electrical industry (Hoeffer 1995). Most copper ores are sulfides. The principal ones include native copper (Cu), chalcopyrite ($CuFeS_2$), and chalcocite (Cu_2S). Copper can form as an early crystallization mineral of a mafic magma or may be hydrothermal in origin (Hoeffer 1995).

Lead

Refined lead is a soft, heavy metal which was one of the first metals used by man. It has a low melting point, which makes it among the easiest metals to cast, is the most corrosion-resistant common metal, and has unusual electrical properties. About 70% of lead produced finds its way to the transportation industry where it is used to make batteries, fuel tanks, solder, seals, and bearings. Batteries are a main end use. Galena (PbS) is the chief ore of lead. Lead is of hydrothermal origin and is commonly associated with copper and zinc deposits.

Zinc

Zinc is the fourth most widely used metal after iron, aluminum, and copper. About three-fourths is used in metal form and one-fourth in compound form. More than 90% of the metal is used for galvanizing steel and for alloys (Jolly 1993b). The remainder is used to produce dust, oxide, and various chemicals. Most metal products find widespread use in the automotive, construction, electrical, and machinery sectors of the economy. The chief ore of zinc is the mineral sphalerite (ZnS). Zinc is found in association with lead and most lead mines also extract zinc.

Gold and Silver

Gold and silver are among the precious metals. Throughout world history they have been used as the most basic forms of money. Most of the newly refined gold that is fabricated today goes into the manufacture of jewelry. However, because of its superior electrical conductivity and resistance to corrosion, gold is an essential industrial metal. It performs critical functions in computers, communications equipment, spacecraft, and jet aircraft engines. Its use continues in dentistry, but its use as a form of money is now restricted to collectable coins. Gold is usually found unconsolidated with other minerals in the form of nuggets or grains. Hydrothermal veins and placer deposits are main collecting sites. Silver is less expensive and is used in coins, jewelry, tableware, and many other products. Until the 1960s, it was used in U.S. coins but rising prices encouraged speculation and the melting of coins into other products. Silver is found in the native state and in sulfide ores. Most silver is obtained as a by-product of lead and copper mining.

Nonmetallic Resources

For our purposes, nonmetallic resources include industrial rocks and minerals. Notable examples include:

Sand and Gravel. These are used in the preparation of concrete for highway and building construction. Pure quartz sand is used to make window glass.

Building and Crushed Stone. Limestone and granite are typically removed from quarries in blocks to construct buildings. Crushed stone (mostly limestone) is used in forming roadbeds during highway construction.

Rock Salt. Rock salt is a coarse, crystalline halite which forms as an evaporite. It can be obtained from underground mines or extracted from seawater. Rock salt finds its way to the home as table salt.

Gypsum. Gypsum is widely and heavily used to make wallboard and plaster. It is formed as a sedimentary evaporite mineral.

Clay. Clay is used to produce numerous products, including ceramics. It results from chemical weathering of rock-forming minerals.

Agricultural Fertilizers. Phosphate, nitrate, and potassium are used to create crop fertilizers. Phosphate is produced from deposits composed of the remains of certain marine organisms. Nitrate and potassium can form as evaporites.

Recycling

One of the most attractive means of mineral conservation is recycling (reuse) of materials from discarded industrial products. It is doubly attractive, because it can mean partial recycling of the energy used in producing the materials in the first place. Aluminum is the most striking example. As indicated previously, the energy cost of extracting the metal from bauxite is very high. When aluminum is recycled, 90 to 95% of the energy is recovered. Recycling is nothing new. It has been an important industrial activity for decades, but both the energy crunch and environmental considerations have brought efforts to improve and broaden it.

Whether mineral materials can be recycled depends on their patterns of use, which may be either dissipative or conservative. The burning of coal, oil, and natural gas for transport, heating, or power generation, for example, is dissipative. The principal products, carbon dioxide and water, are useless as sources of further energy. Phosphate, potash, and nitrogen compounds spread on fields as fertilizer are largely dissipated and cannot be recycled. Very little of the nonmetallic minerals used can be recycled, although recycling of glass is becoming more important. Concrete and asphalt, for example, are being recycled in road making. The costs of recycling nonmetals are generally far beyond their value in reuse.

Recycling Metals

Opportunities to recycle metals stand in sharp contrast to those presented by nonmetal minerals. Recycled metals become increasingly important as resources are depleted in the U.S. Because metal recycling is cost-effective, progress continues to be made in recycling technology. There is also a growing public awareness concerning energy and material conservation to sustain the planet. Overall, the metal recycling industry has grown to the extent that by the year 2010, almost half of the world's metal will be recycled. We will provide a brief discussion of this industry focusing on some individual metals that are used primarily in metallic form.

Seventy-three of the 90 naturally occurring elements that are found in the earth's crust are metals or metalloids. The quantities of metals used annually in the U.S. range from about 100 million metric tons (Mmt) as in the case of steel, to a few kilograms, as occurs with osmium. In 1990, secondary (recycled) metals produced in the U.S. were worth $37 billion, only $2 billion less than the value of primary, newly mined metals.

The form in which metals are recovered varies widely. The precious metals, for example, usually are separated from other alloy metals and refined as individual metals. The ferro alloys, however, largely remain in steel as it is reprocessed. Twenty-two metals for which secondary recovery is important, in terms of quantity and/or value, are shown in Table 18.3. The list includes the six platinum metals, represented as a group. In both quantity and value iron, with its myriad steel alloys, is more important than all other metals combined. Five metals—iron, aluminum, copper, lead, and zinc—account for well over 99% of the quantity and just over 92% of the value of all secondary metal produced.

The amounts of secondary metal shown in Table 18.3 represent metal obtained from both new and old scrap. The relative contribution of the two types of scrap ranges from nearly all new scrap (e.g., titanium) to nearly all old scrap (e.g., the platinum group).

The contribution of secondary metal to total metal consumed ranges from greater than 70% (lead) down to about 5% (vanadium). Many factors determine the extent of secondary recovery of individual metals, including unit value, the size of the pool-in-use, the cost of collection and transport of scrap, the cost of metallurgical processing, the cost of metal disposal if it is not recycled, and the cost to the environment of nonrecovery.

The platinum metals are also intensively recycled, mainly because of their extremely high unit value. Although the amounts of secondary platinum and other precious metals are scant compared to steel, they account for 4% of the value of all secondary metals.

TABLE 18.3   Recycling of Metals in the United States in 1990

Metal	Secondary (Metal Content Metric Tons)	Percent of Total Secondary Metal	Value of Secondary Metal (Million Dollars)	Percent of Value of Total Secondary Metal
Iron (inc. steel)	55,500,000	91	25,000	68
Aluminum	2,400,000	4	3,900	11
Copper	1,300,000	2	3,600	10
Lead	920,000	2	930	2
Zinc	340,000	1	560	2
Manganese	60,000	—	3	—
Magnesium	54,000	—	170	—
Nickel	25,000	—	220	—
Antimony	20,400	—	35	—
Titanium	15,000	—	140	—
Tin	7,800	—	65	—
Molybdenum	3,000	—	17	—
Tungsten	2,200	—	9	—
Silver	1,700	—	260	
Cobalt	1,600	—	30	—
Cadmium	700		5	
Selenium	100		1	—
Vanadium	100	—	2	—
Chromium	90	—	580	
Mercury	90	—	<1	—
Platinum Group	71	—	660	2
Tantalum	50	—	18	—
Gold	49	—	600	2
Totals (rounded)	60,600,000	100	36,800	100

Source: Bureau of Mines, U.S. Department of Interior (1993).

Iron and Steel

Over 50% of the iron and steel used each year in the U.S. is obtained from scrap. In the 1960s, 35 to 40% of production came from scrap (Houck 1993). About 80% is recycled at steel mills and foundries in the U.S. In addition, steel mills and foundries recycle scrap generated within their own plants.

About 20% of the iron and steel scrap recovered is exported for recycling in other countries. This makes the U.S. the largest exporter of scrap iron and steel.

Iron and steel recycling is a major industry with over 5,000 establishments throughout the country purchasing steel from individuals and businesses and preparing it for reprocessing mills and foundries.

The demolition of steel buildings and industrial plants creates significant amounts of recyclable scrap. Large numbers of old railroad cars are also recycled each year. Junked automobiles are the largest source of obsolete or postconsumer scrap. Except in the largest cities, virtually all automobiles that are junked pass through an automobile dismantling operation before being sold to a shredder operator. The dismantler removes any salable parts from the car, including reusable parts and readily removable nonferrous metal scrap parts. Examples of salable nonferrous metal scrap are the battery, radiator, heavy copper parts such as the alternator, and the catalytic converter, which contains recyclable precious metals. Municipal solid waste contains as much as 10% steel, consisting primarily of used steel or bimetallic (steel body-aluminum lid) beverage cans and steel food cans. The steel is mostly tin plate. Many municipalities separate the steel and bale it for recycling.

Iron and steel scrap is recycled by remelting and casting into semifinished steel forms or castings. Much more scrap is melted in electric furnaces than in any other type of furnace. In 1990, well over one-half the scrap recycled in the U.S. was melted in electric furnaces. When steel is recycled, care is taken to change alloy steels containing recoverable elements into new steel of same or similar alloy composition. Stainless steel contains at least 10.5% and usually 18% or more chromium. Most stainless steel also contains from 6 to 11% nickel and some contains from 1 to 3% molybdenum. More than 70% of the nickel recycled each year comes from stainless steel scrap and alloy steel scrap (Houck 1993).

Aluminum

About 40% of the domestic supply of aluminum metal is metal recovered from both purchased new and old aluminum scrap (Plunkert 1993). Recycling aluminum saves 95% of the energy needed to make aluminum from bauxite ore. In 1991, 2.3 million metric tons of metal valued at an estimated $3.0 billion was recovered from both new and old aluminum scrap.

Since opening the nation's first consumer-based recycling center in 1968, Reynolds Aluminum Company has recycled more than 7.3 billion pounds of aluminum. According to Reynolds, that is enough aluminum to fill 4 million garbage trucks had they not recycled it. They also report that two out of every three aluminum cans are recycled annually. This is higher than the recycling figures for any other beverage container (such as plastic, glass, or steel), and almost double the recycling rate for newspapers. Recycled aluminum cans are almost always made into new cans. It takes 4 pounds of bauxite to make 1 pound of aluminum, and all of it must be imported from other countries. Since 1968, recycling at Reynolds' centers has eliminated the need to mine 30 billion pounds of bauxite.

Copper

In 1991 copper from old and new unalloyed and alloyed scrap was valued at $2.2 billion and accounted for 44% of domestic consumption. Most old scrap must be reprocessed to form pure copper. The rate of old scrap recovery is limited somewhat by copper's use in products that remain in service for a long time, such as home wiring, automobiles, and electrical plants (Jolly 1993a).

Lead

Lead recovered from old scrap accounts for about 67% of the domestic demand in the U.S. (Edelstein 1993). The chief source of old scrap in industrialized countries is lead-acid storage batteries. Scrap lead is also obtained from obsolete manufactured products. Smelters are a source for new lead scrap. A notable feature of the lead industry is the significant reduction in dissipative uses of lead in gasoline, pigments, ammunition, and chemicals. In 1972, dissipative uses consumed 30% of the total domestic lead demand, but only 11% in 1991. This reduction largely resulted from regulated phaseouts of lead in the interest of environmental quality.

ENVIRONMENTAL CONCERNS WITH MINING ACTIVITIES

Mining activities have been the focus of public and media attention for some time. The pressures to curb mining abuses are steadily rising. Criticisms have addressed nearly all aspects of the industry, from exploration activities to the energy needed to create market products. Each of these will be addressed.

Geological Exploration

In their quest for new deposits, geologists explore areas where there is evidence for the occurrence of specific mineral forming processes. Mineral exploration usually has limited impacts on the environment as access roads, survey lines, and deposits are sampled. However, in sensitive areas, such as wetlands or animal breeding areas, there is greater concern.

Mining Extraction

The Problem of Disturbance.   Mining rearranges the landscape and this does not go unnoticed by the public (Figure 18.8). Adverse effects, however, go beyond the cosmetic features of landscapes because mining destroys habitat and, temporarily at least, causes ecosystem function to cease. Soils once permeable to precipitation become water repellant in some cases and soil microbes and fauna important to ecosystem functioning are lost. Disturbance extends beyond the boundary of the mine proper as roads are constructed to carry massive loads of ore. Dust from the entire operation can often be seen for miles. In coastal areas, dredging can damage aquatic and riparian ecosystems. It can effectively change the drainage patterns adversely impacting bottomland forest species through vegetation changes. On the positive side, mining operations involve less land area than one might suspect. Mines occupy about 3,700 km^2 in the U.S., or 0.26% of the land area (Barney 1980).

Corporate proposals to begin mining on federally owned land near Yellowstone National Park brought public complaints serious enough to involve the President of the United States. Although the mining concerns felt the area was fair game because it had been previously mined, public outrage was sufficient to stop the project. Mining conflicts involving federal lands arise as corporations make claims to their right protected under the General Mining Law of 1872. The original intent of the law was to promote mineral exploration and development in the western U.S., to offer an opportunity to obtain a clear title to mines already being worked, and to help settle land disputes. The law granted free access to individuals and corporations to prospect for minerals on public lands, and allowed them, on discovery, to stake a claim on the deposit. Recent public objections to mining operations near Yellowstone National Park characterized the 1872 law as an outmoded and damaging force, serving corporate interests at the expense of ecosystem preservation.

FIGURE 18.8   Surface-mined land left with no protective cover prior to reclamation (courtesy of U.S. government).

Toxicity and Contamination Problems.   The waste material remaining after ore has been pulverized and concentrated for processing is referred to as tailings. Mismanagement of this material decades ago has created difficult problems today. Many mines were abandoned before the existence of waste regulations. These old mines themselves are safety hazards. However, the real concern centers on mine wastes, which contain metals and other potentially toxic chemical compounds that are being dispersed into the surrounding surface, air, and groundwaters. Problems associated with the mining and processing of sulfide ore deposits have been especially troublesome. As the sulfides oxidize and mix with water, sulfuric acid is produced. This further accelerates the weathering process, including the release of numerous toxic metals such as copper, lead, and cadmium. Cleanup of these wastes could cost as much as $70 billion. Some sites will require complete isolation of wastes in new landfills underlain by impermeable clay barriers.

Mining wastes are regulated in the U.S. by the Environmental Protection Agency (EPA) and related state agencies under provisions of the Clean Water Act, Clean Air Act, Comprehensive Environmental Response, Compensation and Liability Act, and Toxic Substance Control Act (see Chapter 2). Regulations now generally require that tailings ponds be lined with plastic or impermeable clay to contain the toxins (Kesler 1994). Abandoned tailings piles are dried out, sealed with impermeable clay to keep them dry, covered with soil, and seeded with plants. Impoundment dams are built to hold the leached water in a pond. The potential for groundwater contamination sometimes requires the installation of monitoring wells. The U.S. Geological Survey, although not a regulatory agency, routinely works with communities to investigate existing or potential problems from mine-related contamination.

Mine Reclamation. For some time it has been necessary to obtain approval of reclamation plans and environmental impact statements before mining begins. Bonds are routinely secured to assure successful reclamation. As a minimum, the land surface must be restored to an acceptable contour to prevent acid mine drainage caused by the weathering of rock and unmined ore (Johnson and Paone 1982, Kesler 1994, Carlson and Swisher 1987). Bonds can also be used to cover the costs of correcting environmental damage such as stream pollution, should it occur. In some cases, sites have been converted to recreational areas or nature parks containing wetlands. Other examples are less impressive and there is uncertainty regarding whether arid land areas can be returned to original vegetation within decades after mining operations cease.

Funding for reclamation of old, abandoned sites has been restricted. Whereas considerable federal funding has gone into the cleanup of chemical wastes, reclamation of mined areas has received less government funding.

Mining the Ocean. Environmental problems have developed as countries have either initiated mining ocean floors or made plans to do so. Objections center on ownership of such deposits and on enforcement of measures to avoid pollution effects. The mining of manganese nodules serves as an example of an economically important ocean mining activity (Baturin 1987, Cronan 1992). Nodules the size of a golf ball or softball consist of thin, concentric layers of manganese and iron oxide minerals. In addition to manganese and iron, the nodules contain minor but economically important amounts of nickel, cobalt, copper, and other metals. Most nodules are found in areas of the deep-sea floor, in water depths of 5 to 7 kilometers (Kesler 1994). Nodules can also occur at the bottoms of some large lakes.

Mineral Processing

Smelters. Smelting is of great environmental concern because it produces gasses and dust. Depending on specific operations, arsenic, mercury, zinc, and other toxics can be released. Smelting releases sulfur dioxide, a corrosive gas, that combines with atmospheric moisture and oxygen to form sulfuric acid. The effects of acid depositions can be rather specific, resulting in the death of plants or aquatic animals, or rather diffuse and difficult to measure in short time frames. Diffuse effects might include soil acidification which adversely affects soil microbes, which contribute to plant nutrient supplies.

Efforts to reduce smelter emissions have centered on controlling the SO_2 content of flue gas. Consequently, smelter emissions have decreased steadily since reaching peak values in the 1970s. Modern smelters can recover 90 to 99% of their sulfur emissions, depending on date of construction. Smelter particulate emissions peaked around 1950 and have decreased steadily since.

Human Health and Injury. A persistent concern is human injury and death caused by mining accidents or the chronic exposure to mine dust in one form or another. Mining regulations now address these hazards and companies take elaborate steps to prevent employee injury. Nevertheless, fatalities do occur when rock, equipment, and operators fail. Mine fires pose a serious threat in some industries, with fatalities usually being caused by carbon monoxide poisoning. In addition, mine dust can cause a host of lung problems. Silicosis is a lung disease caused by insufficient dust control in silica-rich mines. In the energy sector, coal miners suffer black lung disease and uranium mine workers have high rates of lung cancer. Due to improved safety measures, fatalities and injury issues have been in steady retreat in the U.S. for years. Miners working in developing countries have been less fortunate.

Future Mineral Availability

The availability of minerals to future generations is an important concern in natural resource management. Their future availability is difficult to judge owing to the manifold factors that can affect their production. A simplistic approach is to divide the quantity reported to be available by the annual rate of consumption and project when the supply will run out. However, this approach disregards the role of economics, technology, and environmental issues in shaping current availability.

Overall, one can say that material well-being in the U.S. derived from nonfuel minerals has improved over this century. This has occurred despite the fact that we have become increasingly dependent upon other nations. Some experts assume that minerals will remain accessible and that we will continue to prosper. Others believe that several rapidly developing nations, such as China, will need more mineral resources and that we will have less at our disposal. Some believe our society needs to self-impose limitations on our consumption in the interest of ecological sustainability, regardless of external factors.

Optimists look to the past and present as a means of projecting our materials future. They point to the dire predictions concerning mineral shortages that have been sidestepped by advances in technology and expanding free markets. In fact, numerous metals are becoming less expensive to the consumer. Also, it is becoming increasingly possible to engineer substitutes for materials in short supply and to react to increasing mineral prices by substituting cheaper materials. Plastics have been widely substituted for several metals in the last 25 years. Substitutions and technological innovations continue to downsize or "dematerialize" the mass of minerals used in products. For example, lightweight optical fibers with 30 to 40 times the carrying capacity of conventional materials are replacing copper in many segments of the telecommunications infrastructure. Similarly, cars, computers, containers, and other products are becoming lighter, and often smaller, as performance and durability improve. In the steel industry, innovations in powder metallurgy, thin casting, directional solidification, and drop and cold forging have allowed savings up to 50% of material inputs over the past 20 years. Since the early nineteenth century, the ratio of weight to power in industrial boilers has decreased almost 100 times. Overall, the basic issue concerning mineral resources is not how soon they will become exhausted, but whether progress will continue to be made in free trade and technology.

LITERATURE CITED

Barney, G. O. 1980. *The global 2000 report to the President of the United States.* New York, NY: Pergamon Press. 360 pp.

Baturin, G. N. 1987. *The geochemistry of manganese and manganese nodules in the ocean.* Hingham, MA: Kluwer Academic Publishers. 356 pp.

Bureau of Mines, United States Department of Interior. 1993. *Recycled minerals in the United States.* Special Publication, 76 pp.

Carlson, C. L. and J. H. Swisher. 1987. *Innovative approaches to mined land reclamation.* Carbondale, IL: Southern Illinois University Press. 752 pp.

Cronan, D. S. 1992. *Marine minerals in the exclusive economic zones.* London: Chapman and Hall. 292 pp.

Edelstein, D. L. 1993. Lead (pp. 33–35). In *Recycled minerals in the United States.* Bureau of Mines, United States Department of Interior. Special Publication.

Hoeffer, R. L. 1995. *Physical geology.* Springhouse, PA: Springhouse Corporation. 151 pp.

Houck, G. W. 1993. Iron and steel (pp. 27–31). In *Recycled minerals in the United States.* Bureau of Mines, United States Department of Interior. Special Publication.

Johnson, W. and J. Paone. 1982. *Land utilization and reclamation in the mining industry, 1930–1980.* U.S. Bureau of Mines Information Circular, 8862. 22 pp.

Jolly, J. H. 1993a. Copper (pp. 17–22). In *Recycled minerals in the United States.* Bureau of Mines, United States Department of Interior. Special Publication.

Jolly, J. H. 1993b. Zinc (pp. 69–72). In *Recycled minerals in the United States.* Bureau of Mines, United States Department of Interior. Special Publication.

Judson, S., M. E. Kauffman, and L. D. Leet. 1987. *Physical geology.* 7th edition. Englewood Cliffs, NJ: Prentice-Hall. 484 pp.

Kesler, S. E. 1994. *Mineral resources, economics and the environment.* New York, NY: Macmillan. 391 pp.

Kuck, P. H. 1988. Iron ore (pp. 412–423). In *Mineral yearbook.* U.S. Bureau of Mines, United States Department of Interior.

McAlester, A. L. and E. A. Hay. 1975. *Physical geology: Principles and perspectives.* Englewood Cliffs, NJ: Prentice-Hall. 439 pp.

McKnight, T. L. 1990. *Physical geology: A landscape appreciation.* 3rd edition. Englewood Cliffs, NJ: Prentice-Hall. 610 pp.

Plunkert, P. A. 1993. Aluminum (pp. 1–4). In *Recycled minerals in the United States.* Bureau of Mines, United States Department of Interior. Special Publication.

Schottman, F. J. 1985. Iron and steel (pp. 405–424). In *Mineral facts and problems.* U.S. Bureau of Mines, United States Department of Interior.

CHAPTER 19

Nonrenewable Energy Resources

INTRODUCTION

Energy is the capacity to do work. With unlimited energy, the human mind becomes the limit to what can be built, manufactured, or chemically created. With abundant, low-cost energy, for example, desalination plants could transform enough seawater to freshwater to meet the needs of people living in arid lands throughout the globe. Of course, energy is costly and this limits economic development. One could say that we cannot well understand global politics and economics without understanding how nations differ in their energy resources. Similarly, energy is linked directly to major changes in the ways civilizations advance, from primitive people carrying fire with them as they pursue food to nations going to war to protect their energy supplies.

For the present, however, our energy concerns are more than academic and are directed more to the future than the past. This is so because growing dependence on fossil fuels appears to be altering Earth's climate (Karl et al. 1996, Meyers and Schipper 1992, Chapter 6). Our use of energy and the future of the planet have become closely intertwined and solutions are needed to some of the most complex problems faced by humankind. As this problem is discussed and acted upon by local as well as national governments, our lives may be affected. We may be asked to pay more for energy, directly or indirectly, or be rewarded for making energy substitutions. We may be called upon to make informed judgements and to adjust individual priorities. Whatever comes our way, we will be better prepared if we understand how energy is provided and consumed. The aim of this chapter and the next is to provide that basic understanding.

ENERGY USE

Historical Perspective

Primitive humans used fire for warmth, protection, harvesting game, and eventually for meal preparation. The carbon records created by human fire pits across the land are used to date ancient landscapes. As nomadic people settled, fuelwood provided a reli-

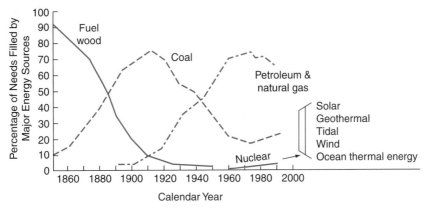

FIGURE 19.1 U.S. energy consumption since 1850 (from Owen and Chiras 1995).

able energy source. As time progressed, wood energy was put to other uses. First there was pottery making and with the dawning of the Bronze Age, primitive metallurgy. In the Mediterranean area, the dependence on fuelwood became so great that vast areas were deforested and became severely eroded. Deforestation led to local fuel crises and forced some of the earliest attempts to recycle metal resources. To reduce the energy needed to produce weapons and primitive tools, bronze was recycled to avoid working directly with ore. In 300 B.C., so much wood was being used for metal smelting that the Roman Senate limited mining (Simon 1996). Numerous examples of Roman architecture also show the need to conserve energy. Glass windows were used to capture solar energy and bathhouses were positioned so that they were passively heated by the sun.

Wood remained the principal source of energy well beyond the Middle Ages. Wood shortages continued to occur, depending on local needs. In England, the shortage of wood used for charcoal in the casting of iron became so acute that in 1599, Parliament passed a law against the cutting of trees for coke used in iron making. The pressure on wood forced an innovation in the smelting process. When it was discovered that blowing machines could be used to eliminate metal impurities caused by smelting with coal, the stage was set for the upcoming Industrial Revolution (Simon 1996). The invention of the steam engine in the late eighteenth century lessened dependence on animals, wind, and water but did not reduce the use of wood energy. Coal mining began in the U.S. around 1860. As recent as 1880, wood gave way to coal as the dominant source for energy in the U.S. (Figure 19.1). To this day, coal remains a key source of electrical power.

Oil made its appearance on the energy scene in 1859 when Edwin Drake drilled the first producing oil well in Titusville, Pennsylvania. More than 10 years later (1870), Standard Oil entered the refining business to produce kerosene for lighting purposes. The future of oil seemed threatened when Thomas Edison invented the lightbulb in 1882. It was just a brief time (1896), however, before Daimler and Benz invented the automobile in Germany and we have been hooked on oil ever since. In the early 1940s, petroleum and natural gas gained the dominant role in energy consumption.

The first self-sustaining chain reaction and consequent controlled release of nuclear energy was achieved in 1942. Within two years, the first atomic bomb was detonated and the Nuclear Age began. In 1954, the Atomic Energy Act was amended to permit the use of nuclear energy for producing electricity. The nuclear industry has yet to play a dominant role in U.S. energy consumption, but maintains a significant and controversial presence in electrical power generation.

Energy Use Today

Energy use can be expressed in numerous ways. The basic unit in the metric system is the joule, which is the force of 1 newton applied over a distance of 1 meter. In

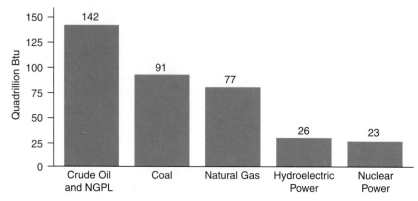

FIGURE 19.2   World primary energy production by source (from EIA 1997).

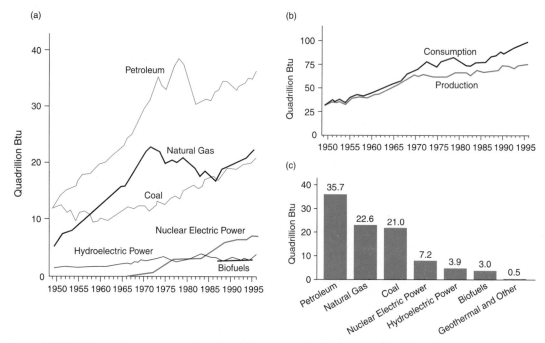

FIGURE 19.3   U.S. energy consumption (a) by source from 1949 to 1996, (b) with energy production from 1949 to 1996, (c) and by source in 1996 (from EIA 1997).

expressing the massive amounts of energy consumed by nations, the unit commonly discussed is the exajoule, or a billion billion joules. For further simplification an exajoule is roughly equal to one quadrillion, or 1,015 British thermal units (Btus). We, therefore, can reference energy production or consumption in numbers of quads of energy. In 1995, crude oil and natural gas plant liquids (NGPL) produced the greatest share of the world's energy supply, 39.6% (Figure 19.2). Coal produced 25.3% and natural gas 21.5%. Hydroelectric power and nuclear power were of much less importance (7.2 and 6.5%, respectively).

In the U.S., coal accounts for 31% of the energy produced followed by natural gas (26.9%), crude oil (18.9%), and others (EIA 1997). With few exceptions, our dependence on petroleum has risen steadily over the past 45 years (Figure 19.3a). Because the growth in nuclear power and other energy sources has not been sufficient to offset our growing energy demands, the U.S. consumes more energy than it produces (Figure 19.3b). This has been the case since the 1950s, but each year for the past decade marked an even greater imbalance.

In 1996, petroleum (38%), natural gas (24.1%), and coal (22.4%) contributed 79% of the energy consumed in the U.S. (Figure 19.3c). Energy derived from nuclear

Fossil fuel power plant

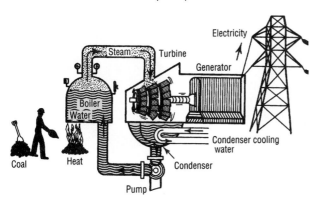

Nuclear power plant
boiling-water reactor (BWR)

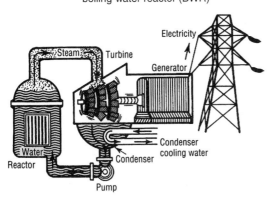

FIGURE 19.4 Electric power plants fueled by coal and nuclear energy work on the same principle. Heat boils water to generate steam pressure, which drives a turbine that generates electricity. To maintain a cycle in the closed system, the steam is condensed by cooling and returned to the boiler (from Owen and Chiras 1995).

power (7.7%) and all forms of renewable power, including hydropower, provided less than 8% (EIA 1997).

Electrical Energy

As individuals we depend on two forms of energy to meet our daily needs. Gasoline fuels our automobiles and electricity meets a host of everyday needs. The electricity reaching us by way of the electrical grid may have been produced by several means, but the basic principle is the same for all. Regardless of the primary energy source, the generation of electricity requires rotation within a generator. Steam-electric generating units burn fossil fuels such as coal, natural gas, and petroleum. The steam turns a turbine that produces electricity by means of an electrical generator. Natural gas and petroleum are also burned in gas turbine generators where the hot gases produced from combustion are used to turn the turbine, which in turn spins the generator to produce electricity. Petroleum is also burned in generating units with internal-combustion engines. As combustion occurs, the engine turns the generator shaft and mechanical energy is thereby transformed into electrical energy. Nuclear power plants seem complex but, like fossil fuels, produce steam that drives turbines (Figure 19.4). Turbines are also driven by hydropower as water falls from one level to the next. Wind generators similarly involve the rotation of blades or propellers.

As electricity flows to where it is needed, more than one primary energy source can feed the commercial grid. We measure the output of electrical energy in kilowatts and megawatts (capacity) or kilowatt-hours and megawatt-hours (demand).

In 1994, fossil fuel accounted for 62% of the world's production of electricity, followed by hydropower (19.22%) and nuclear energy (17.4%). Renewable energy sources other than hydropower, such as wind and photovoltaics, provided less than

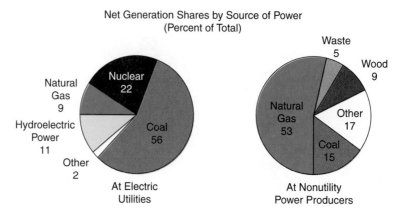

FIGURE 19.5   Electric power generation by energy sources serving utilities and nonutility power producers in 1996 (from EIA 1997).

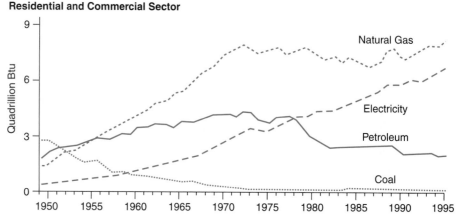

FIGURE 19.6   Energy consumption by residential and commercial use in the U.S., 1949 to 1995 (from EIA 1995).

1%. Compared to the U.S., Western Europe derives a greater share of its total energy from nuclear power (31%). In Central and South America, hydropower generates 76% of the electricity used.

In the U.S., utilities generated 88% of the electricity consumed in 1996. Coal was the fuel used to generate the largest share (56%) of electricity at electric utilities (Figure 19.5). Nonutility power producers used more natural gas than coal and, because they operate on a much smaller scale, are able to use more renewable energy sources. Nuclear energy generated 22% of the electricity produced at utilities.

Energy in the Home, Business, and Industry

The electrical energy consumed by American homes and business places has grown steadily over the last 45 years (Figure 19.6), owing to population growth and the growing numbers of appliances Americans operate. Homes and businesses use other forms of energy and when data are compared in thermal units, more natural gas energy is consumed than is electrical energy. On a per household basis, home energy consumption was actually less in 1993 than it was in 1978. This came about as homeowners, builders, and appliance manufacturers made progress in improving energy efficiency in the face of rising energy costs. In 1993, home space heating consumed the greatest amount of energy (53.2%), followed by appliances (24%), water heating (18.3%), and air conditioning (4.5%) (EIA 1995). On average, each individual consumes three times as much energy now as in 1850. There are now 11-fold more people in the U.S. than in 1850.

For most of the 1980s, homes and businesses used as much energy as the industrial sector (Figure 19.7). From 1950 to 1980 the industrial sector consumed substantially

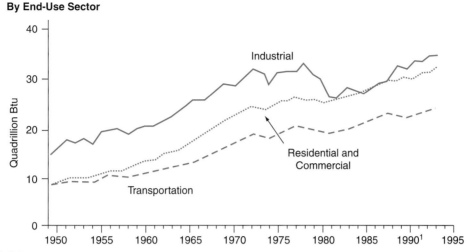

FIGURE 19.7 Energy consumption by end-use sector, 1949 to 1995 (from EIA 1997).

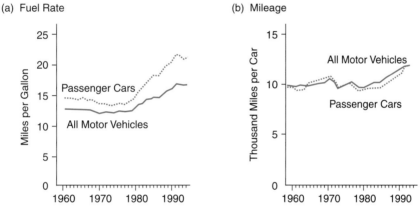

FIGURE 19.8 U.S. motor vehicle statistics, from 1960 through 1994, showing (a) improved auto fuel efficiency achieved from 1980 to 1990 and (b) increasing use of automobiles according to miles traveled.

more. In the face of rising energy prices, industry also took energy conservation measures. Another key point is that our nation is becoming less energy intensive, as measured by the energy needed for each unit of gross domestic product. The greatest reduction in energy intensity followed the period of high oil prices of the 1970s. Since 1985, energy intensity has changed little. The industrial sector's use of electricity continues to grow, but coal consumption is considerably less in 1995 than in 1975. The use of coal in industry has actually been in steady decline since the 1950s. From 1975 to 1995, petroleum and natural gas were consumed at roughly comparable rates by industry. The petroleum and coal products industry use the greatest energy when compared to other industry types. Primary metal and paper production each consume less than one-half the energy used by petroleum or chemical product industries (EIA 1995).

Energy Used for Transportation

The number of automobiles, trucks, buses, and motorcycles registered in the U.S. is about 210 million. One-third this many vehicles were on our roads in 1960. Our consumption of motor oil has likewise increased. Now we need about 10 million barrels per day to keep moving. The volumetric equivalent of one barrel of oil is 42 U.S. gallons. To reduce greenhouse gas emissions and our dependency on foreign oil, the automobile industry has followed government dictates and built more fuel efficient vehicles (Figure 19.8a). However, Americans have chosen to invest their energy savings in driving more miles (Figure 19.8b). Most recently, the rising popularity of less fuel-efficient sport utility vehicles also has contributed to higher consumption. Therefore, our national daily consumption has not been reduced.

ENERGY PRODUCTION

The U.S. draws from diverse energy resources to meet its present energy needs. The finite nature of some energy sources has lead many experts to believe that adjustments will be needed to meet future energy demands. There is also the criticism that our heavy use of fossil fuels is affecting the planet adversely. Before we can grapple with these limitations and concerns, we need to understand how energy is produced. This question will be addressed as we discuss energy resources individually.

Coal

Coal Formation.   Coal is an organic rock, as opposed to most other rocks, such as clays and sandstone, which are inorganic. It contains mostly carbon (C), but also hydrogen (H), oxygen (O), sulfur (S), and nitrogen (N), as well as some inorganic constituents (minerals) and water (H_2O). Coal was formed some tens to hundreds of million years ago as plant material accumulated in swampy environments. Because oxygen is limited in swamps, plant materials do not undergo complete decomposition to carbon dioxide and peat is formed instead. Peat is a yellowish to black, usually fibrous material in which plant fragments are still recognizable. The peat thus formed is subsequently changed into brown coal. As the coalification process continues, cellulose, the dominant material of plants, is chemically converted to CO_2, H_2O, methane (CH_4), and carbon. The first three of these products are volatile and are expelled whereas the carbon remains. Therefore, as coalification proceeds, carbon content increases relative to the other substances just identified.

Coal types are ranked according to the relative amounts of carbon and energy they contain. From the lowest to highest rank more time is required for formation (Table 19.1). The basic process of fuel formation is the same for all the ranks. Owing to their higher energy content, bituminous (soft coal) and anthracite (hard coal) are the stages of coal most often mined, processed, and used as fuel. However, even dried peat is used as a fuel in some countries and lignite can be gasified to produce a clean burning natural gas or liquefied to produce liquid petroleum fractions. The costs associated with gasification and liquefaction processes prohibit their use at current petroleum prices.

Coal contains varying quantities of sulfur which is released to the atmosphere when coal fired power plants produce electricity. In terms of sulfur content by weight, coal is generally classified as low when it contains 1% or less sulfur. Medium sulfur coal contains 1 to 3% sulfur and high sulfur coal contains more than 3%, as measured on a percent by weight basis. Unfortunately, the low-rank coals typically contain much less sulfur than the bituminous coals. Because bituminous coals have relatively high heating values, most electric power plants prefer to use them. Ash is another undesirable component of coal because it reduces energy content and can be a source of environmental pollution. The ash finding its way into coal heaps can be

TABLE 19.1   A Comparison of Coal Carbon and Energy Rank by Type

Coal Type	Carbon Content (%)	Energy Released on Combustion (10^6 Btu per Ton)	Approximate Age (10^6 Years)
Anthracite	85 to 90	22 to 28	60
Bituminous	60 to 85	19 to 30	100
Subbituminous	55 to 60	16 to 24	300
Lignite	50 to 55	9 to 17	350

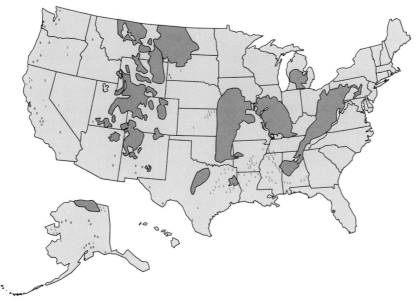

FIGURE 19.9   Coal fields of the U.S. (from EIA 1996).

the result of inadvertent contamination in the mining process or local geology. Moisture content can also vary from one coal source to another. Moisture reduces energy content. Coal rank, sulfur category, and ash and moisture content collectively determine the monetary value of coal.

The Distribution and Abundance of Coal.   Most of the world's coal exists in the Northern Hemisphere. About 85% of the world's coal resources are located in the U.S., the area represented by the former Soviet Union, and China. The U.S. has far more coal than gas and petroleum. Thus, if supplies of gas and petroleum become scarce, coal could be turned to as our major fossil fuel energy source. For the present, coal is the only major U.S. fuel resource that provides a positive trade balance. In 1995, the value of coal exports was $3.3 billion.

A recent estimate of demonstrated reserve base (DRB) of coal resources is 496 billion short tons (EIA 1996) (Figure 19.9). However, almost half of the base is either inaccessible or is likely to be lost in the mining process.

Nearly half (49%) of the DRB is found in the West, with the vast majority in Montana and Wyoming. Eastern Wyoming is best positioned to meet our needs for low sulfur coal. Appalachia contains about 22% of the DRB and the nation's interior region contains about 29% of the DRB.

It is difficult to estimate how long our domestic coal will last. Based on the 1994 production rate, which involves some exports, and the DRB information previously discussed it should last around 250 years. However, as was discussed in the minerals chapter, technology, economics, and environmental concerns are changing at such a rapid pace that estimates remain crude at best.

Coal Mining and Production.   The amount of coal extracted from U.S. mines has increased steadily since the 1960s and in 1994 exceeded 1 billion short tons (Figure 19.10). By a wide margin, bituminous coal has been the coal type most heavily mined. Before the early 1970s, the major fraction of coal was extracted by underground mining. Since then, surface mining has provided a greater and steadily increasing share (Figure 19.10). At present, surface mining accounts for 60% of the coal removed nationwide. In the western states, surface mining accounts for 90% of coal production.

Underground mining, widely used in most eastern coal fields, generally employs one of two methods (Figure 19.11). The "room and pillar method" involves cutting "rooms" into the coal seam leaving large pillars of unmined coal to help support the

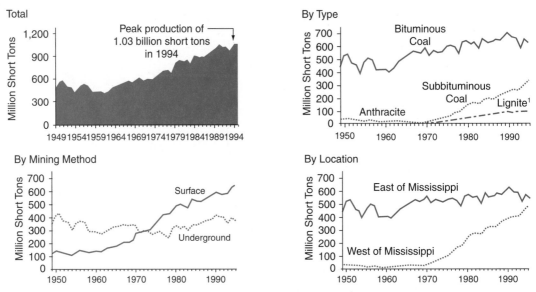

FIGURE 19.10   U.S. coal production, 1949 to 1995, showing production over time (top left), by mining method (lower left), by type of coal (top right), and by location (lower right) (from EIA 1996).

Coal Mining Methods

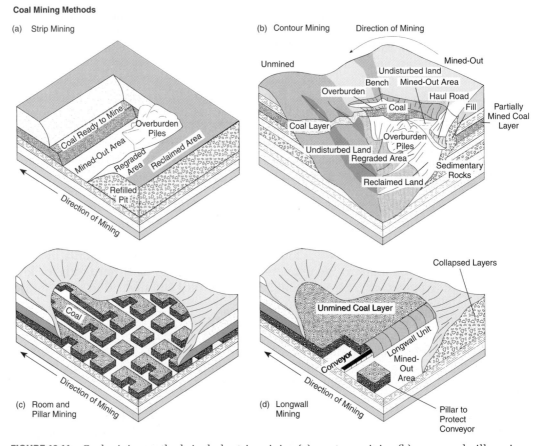

FIGURE 19.11   Coal mining methods include strip mining (a), contour mining (b), room and pillar mining (c), and longwall mining (d) (from Kesler 1994).

ceiling (Radovic and Schobert 1992). Because the pillars are left standing, about 40% of the coal seam is not extracted. The longwall method represents an attempt to overcome this limitation. It employs mechanized mining equipment to carve the coal from the seam so that the ceiling created can collapse as the equipment retreats. Although this method recovers almost all of the coal, physical limitations sometimes will not allow use of all needed equipment.

FIGURE 19.12   Strip-mined rangeland near Colstrip, Montana, which has been revegetated with a mixture of perennial grasses, forbs, and shrubs. Productivity of this land for livestock equals or exceeds that prior to mining.

Surface mining has extraction efficiencies of 80 to 90% (Hall et al. 1992). Compared to underground methods, it uses considerably less labor to extract a given amount of coal. The growth of the western coal industry and its reliance on surface mining explains why both mining efficiency and coal production have increased nationally. While eastern coal production increased only slightly since 1970, western coal has climbed rapidly and, in 1995, nearly equalled eastern production. Almost all of the lignite and subbituminous coal are obtained from strip mines.

In the eastern U.S., contour surface mining is used to extract near-surface coal occurring in hilly country. A wedge of the hill is actually cut away by removing the overburden starting at the outcrop and then proceeding along the contour of the bed in the hillside. Bulldozers and steam shovels are used to remove the overburden covering the coal seams. In the West, flatter terrain permits area strip mining. In many respects it is similar to contour mining, but, as strip mining proceeds, the site is excavated one strip at a time and the overburden from one strip is used to fill the one previously cut.

Although strip mining is highly efficient, its adverse impacts on the environment do not go unchallenged. Strip mining temporarily, at least, disturbs surface and groundwater flow patterns. It causes an abrupt increase in the presence of soil chemical elements in the surface landscape. Without close supervision, some toxic elements can migrate off-site through seepage, runoff, or sediment transport, depending on their mobility and form.

Fertile top soil and overburden must be removed to expose the coal. If the top soil is not properly handled, replaced, and successfully reseeded after mining, large barren areas remain. The water needed to assure successful revegetation success at some western locations is another source of controversy. However, in the driest areas, the more important issue may be whether once productive shrublands will remain reclaimed after operations are abandoned. Mines located near plentiful water sometimes return land to a higher value as viewed by local communities (Figure 19.12) (also see Chapter 2, Figure 2.21).

In 1977, the U.S. Congress passed the Surface Mining Control and Reclamation Act which requires mining companies to reclaim their land. Under this act land mined must be returned to its original contour and revegetated. It also requires control of on-site erosion and chemical contamination of nearby lakes and streams. This was a major step toward reducing the environmental impacts of surface mining. However, the funds provided by a federal tax on coal may not be sufficient to fully restore the disturbed lands (Owen and Chiras 1995).

Issues surrounding coal miner health and safety also cause public concern. Although fatal accidents have steadily decreased, owing to mechanization and improved safety awareness, some risk will always remain. Over 100 miners are killed in mine-related incidents every year in the U.S. One serious threat is the possibility of finely powdered coal dust being ignited to produce a mine explosion. The chronic risk of dust inhalation has been substantially reduced in underground mines but the danger of black lung disease still remains a serious threat to miner health.

Coal Transportation.   Most coal is transported by railway but waterways, highways, or slurry pipelines are also used. Shipments abroad involve ships or barges. Coal transportation is expensive and, depending on rail distance, can double the mine price.

Slurry pipelines have been offered as an alternative method for some situations (Radovic and Schobert 1992). This method involves mixing pulverized coal with roughly an equal weight of water and pumping the resulting slurry through pipelines from the mine to the point of use. One such pipeline is currently in use in the U.S. Slurry pipelines are subject to two criticisms. For one, the use of water in water scarce areas offends the public. For another, with current combustion technology, the coal must be separated from the water before it can be used. More important is the problem of treating the impure water before it can be discharged to the environment (Radovic and Schobert 1992). Finally, the transportation industry serving the coal industry would suffer. Coal is one of the railroad industry's primary clients, accounting for almost 20% of freight revenues (Hall et al. 1992).

Coal Utilization.   Eight percent of the domestic coal produced in 1993 was exported. Less than 1% of the coal used was imported. In the U.S., coal is used in 3 ways: combustion, carbonization, and conversion (gasification or liquefaction).

Coal combustion: Coal combustion is by far the largest and the most important use of coal. More than 85% of the coal consumed domestically in 1995 went to electric utilities. About 8% went to fuel industrial uses (primarily manufacturing), and nearly 4% was used to produce metallurgical coke (EIA 1996). Coal is combined with oxygen to release its energy. The oldest and still most commonly used form of coal combustion is the pulverized coal-fired boiler furnace, used by most utilities in the generation of electricity.

Coal combustion presents the problem of dealing with the ash and the emissions of CO_2, sulfur, and nitrogen oxides created in the burning process. Ash consists of silica, iron, aluminum, and other noncombustible matter that are contained in coal. Ash increases the weight of coal, adds to the cost of handling, and can affect its burning characteristics (Radovic and Schobert 1992). Carbon dioxide is one of the four major greenhouse gases that threaten climate stability. In 1993, coal consumption contributed 35% of the CO_2 released in the U.S. (EIA 1996). CO_2 accounts for 50 to 60% of the greenhouse warming effect. The sulfur released from electric power plants is subsequently converted to SO_2. Both SO_2 and nitrogen oxide cause acid rain and the latter also contributes as much as 5% to the greenhouse effect (see also Chapter 6).

In the U.S., the sulfur content of coals varies from 0.2 to 10% and, unlike carbon and energy content, is not directly related to coal type. Most of the low sulfur coal in the U.S. is found west of the Mississippi River. However, more western coal has to be burned to produce the same energy. Removal of the sulfur is not an easy task. The sulfur attached to coal can be removed by washing and other coal preparation methods. Currently, 40% of the bituminous coal used for power generation is cleaned in some manner. All bituminous coal is crushed to provide uniform size, to increase

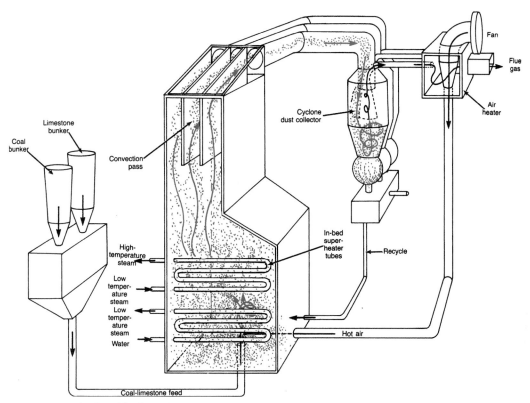

FIGURE 19.13 Fluidized bed combustion. Crushed coal mixed with limestone blown into a furnace. The limestone reacts with sulfur dioxide gases, thus removing most of them from the smoke stack. Air turbulence within the furnace promotes thorough and therefore efficient combustion (from Owen and Chiras 1995).

surface area for combustion, and to remove impurities (Hall et al. 1992). The organic sulfur contained within the coal itself requires post combustion removal, usually by employing scrubbers or liquid filters that put exhaust fumes through a spray of water containing lime. The sulfur reacting with the lime is converted to calcium sulfate, which can be removed as a solid. A newer innovation for sulfur removal involves injecting crushed limestone into a boiler containing powdered coal and combusting the coal at a lower temperature. This reduces the release of both sulfur and nitrogen oxides, another serious pollutant contributing to acid deposition. Fluidized bed combustion (Figure 19.13) can remove 90% of the sulfur, 50% of the nitrogen oxides, and 20% of the CO_2. This method involves blowing pulverized coal into a bed of inert ash or limestone that is fluidized (held in suspension) by the uniform injection of hot air through the bottom of the steam generator. When fluidization occurs, the bed of material expands and exhibits the properties of a liquid. As the coal is burned the sulfur it releases reacts with the limestone and falls to the bottom of the generator as calcium sulfite. Sulfur removal then becomes a straight forward process.

The Clean Air Act of 1990 addressed acid deposition by mandating reductions in both sulfur dioxide and nitrogen oxide levels. It states that by the year 2000, total sulfur emissions must be reduced 10 million tons below 1980 levels. Before 1995, a utility releasing less sulfur than allowed could sell the difference in its allowance to another utility. Allowance trading became a major market activity that continues today, but with the restriction that no new allowances are being given. This puts a permanent cap on the total amount of sulfur released from the coal power industry (see also Chapter 4).

Coal carbonization: The coal carbonization process essentially launched the industrial revolution of the nineteenth century. When coke derived from bituminous coals replaced wood charcoal, the quality of iron and steel products improved greatly. The close proximity of productive bituminous coal mines to the city of Pittsburgh enabled it to become the steel capital of the world. The carbonization process involves heating coal to about 1,000°C in the absence of O_2. If O_2 was present, combustion would transform the coal to ash. Without O_2, only the volatile contents are removed and the remaining product is the fixed carbon, which is called coke (Radovic and Schobert 1992). Coking coals are created from a select group of bituminous coals. Coke (C) is then used in the blast furnace to convert iron ore (FeO) to iron (Fe) as shown by a simple chemical equation:

$$FeO + C = Fe + CO$$

Iron is then converted to steel. Sales of coking coals have declined in the U.S. as the nation's steel industry has downsized. However, this decline in coal consumption has been more than offset by the rising needs for coal in electric power generation (Kesler 1994). About 60% of the coal exported by the U.S. is high-quality metallurgical grade (Hall et al. 1992).

Coal conversion: Coal can be converted to a cleaner and more convenient fuel through either gasification or liquefaction processes. Coal gasification converts coal into synthetic natural gas. Gasification involves the burning of coal at very high temperatures in the presence of hydrogen to produce a gas. This process can yield gasses with varying degrees of heating values, but few can compete with natural gas (Hall et al. 1992). In the direct liquefaction process, pulverized coal is slurried then mixed with hydrogen in the presence of a catalyst under moderate temperature (455°C) and pressure. The vapor and liquid phases produced are then cooled to separate the products and refined to remove any by products in the liquid fuel. Germany liquified coal during World War II to meet its petroleum needs when supplies were interrupted. Another process, indirect liquefaction, first converts coal into a gaseous state using a coal gasifier, then recombines the gaseous molecules into liquid products. Indirect liquefaction has several possible routes into the marketplace. Ultimately, stand-alone plants may be built to produce synthetic liquids as substitutes for crude oil. However, within only a few years, indirect liquefaction may become an integral part of advanced electricity generating power plants. Inherent to the process is the opportunity to store energy from the power plant during off-peak periods when demand for electricity is low. Many of tomorrow's power plants will likely use coal gasifiers to generate combustible gas for high-efficiency gas turbines. When the gas turbines are not using all of the gas from the gasifiers, it could be sent to an indirect liquefaction unit. There it would be converted to methanol and either stored for use later as a turbine fuel or sold commercially.

Overall, gasification and liquefaction technology are commercially available today in several forms and technology is improving. If the price of oil suddenly increases coal conversion will become more attractive. However, when one reviews the environmental impacts of creating synfuels, there are several potential problems. Some argue that the widespread use of synfuels would further accelerate surface mining and that production requires huge amounts of water. Whatever course is taken, coal represents a tremendous energy source because every ton of coal is the equivalent of more than five barrels of crude oil.

Oil and Gas

Oil and gas were formed from the fats and other lipids accumulated as debris of plants, bacteria, phytoplankton, and zooplankton were buried and preserved, along with mud and silt, in ancient marine sediments (Kesler 1994). Microorganisms began converting lipid-rich organic matter to methane as these materials were deposited. However, the richest deposits of oil and natural gas were formed in extensive coastal marine environments with too little oxygen in bottom sediments to support thorough decomposition.

Over long periods of time, kerogen was formed by the leaching action of groundwaters. The organic matter escaping decomposition was more deeply buried beneath sediments. It was converted to kerogen by leaching from groundwaters. As land subsided from the separation of Earth's plates (discussed in Chapter 18), sediment bearing rock strata were exposed to greater temperatures. At temperatures from 50° to 60°C, oil formed. At 100°C most of the oil contained in the kerogen was released and further reactions created natural gas.

As plates collided, immense forces built mountain chains, crumpled rocks into folds, and thrusted rock strata over each other to form complex structures. The sites where rock strata formed arches or domes (i.e., anticlines) became oil-rich pockets. Large, economically recoverable amounts of oil can be trapped in pools or remain in the pores of reservoir rock, porous sandstone, or limestone. Oil not contained in this manner can move freely to the surface where it is further transformed and lost as a recoverable fossil energy resource. Areas such as the La Brea Pits in California mark sites where oil continues to ooze to the surface today. Oil is found in large quantities in the Persian Gulf area because salt deposited by an evaporating sea effectively caps the upward movement of oil. In addition, the region's aridity prevents water from reaching and destroying the resource.

Oil Field Exploration. Oil accumulations, in one form or another, have been found in every continent except Antarctica. Oil does not accumulate as an enormous underground lake. Rather, it has backed up in the pores of reservoir rock, or has accumulated in the space beneath an impermeable structure or stratum. Very often, oil is found in seemingly solid rock, which on close inspection contains minute spaces or pores. Exploration for oil and gas is done mostly by seismic methods. Shock waves from an explosion or vibrating source are sent into the ground from a land-based unit or ship. Further, more conclusive evaluation of potentially productive sites must be done by drilling. Drilling is an extremely expensive process especially costly when no oil and gas are found. Sometimes only gas is found but oil will always be associated with at least some gas.

Several factors determine the value of a crude oil deposit. The location and depth of the oil reservoir are important, as is the hardness of the intervening rock layers that must be drilled. A deposit capable of producing large quantities of gasoline is especially valued. Much of the oil contained in such deposits consists of relatively short carbon chains (i.e., five to nine carbon atoms) and has very low sulfur content. Finally, the current market price of oil is always a key consideration.

Exploration is decided by the prevailing economic conditions created by supply and demand. Exploration boomed in the early 1980s in response to higher oil prices created in part by the oil shortages of the 1970s. Shortages arose as OPEC (Organization of Petroleum Exporting Countries) tried to force higher world oil prices. The percent of successful versus failed exploration wells has actually increased since 1970. Exploration involved more wells as opposed to drilling individual wells to greater depth. Efforts to find oil were successful, despite the fact that our oil resources have been steadily depleted. However, the gushers of yesteryear are no more.

Oil and Gas Production

Drilling: Efforts to pump oil are directed by the ways it is trapped beneath the earth. The U.S. production of crude oil peaked in 1970 (Figure 19.14a) and has been in uninterrupted decline since 1985. The addition of the Alaskan oil field forestalled the decline in total production, but even its production has fallen since the mid 1980s. The overall decrease in total production is due to the falling productivity of individual wells (Figure 19.14b) rather than a decline in the total number of producing wells. Each day the average U.S. well produces about 12 barrels of oil. In Saudi Arabia, each well averages 10,000 barrels per day! Years ago, U.S. oil wells relied to some extent on water and little gas pressure to force oil to the surface. Now, efforts to keep wells productive frequently must involve sophisticated oil recovery techniques, such as steam injection, to

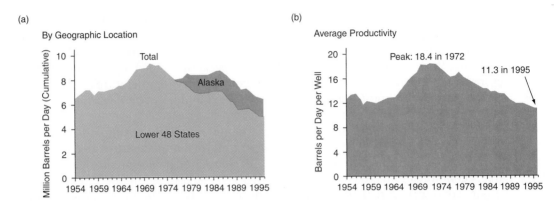

FIGURE 19.14   U.S. crude oil production, 1954–1995 showing (a) production including Alaska's contribution relative to the lower states, and (b) the peak and declining production per well of U.S. oil wells (from EIA 1995).

FIGURE 19.15   Offshore oil production requires expensive platforms.

decrease viscosity and increase flow. Secondary recovery methods extract an additional 10 to 20% of the available oil from a well (Radovic and Schobert 1992).

Today, most U.S. oil production is concentrated in only five states. Roughly 70% of the oil produced in the U.S. in 1989 (about 3 billion barrels) came from Alaska, Texas, California, Louisiana, and Oklahoma. The oil fields of northwestern Pennsylvania, where it all started in 1859, have long been depleted. Currently, major U.S. oil drilling projects are located offshore (Gulf of Mexico, California) and overseas.

Offshore production today provides more than a million barrels of oil each day and has done so since the late 1960s. Production requires the use of expensive platforms (Figure 19.15). Some reflect ingenious engineering and, in scale, may rival the heights of tall city buildings. Some oil occurs in pools underneath the seafloor (in the Point Arguello field, California, for example). Extracting that oil from the earth requires undersea drilling.

Oil and Gas Processing.

Oil: Crude oil is not a chemical compound but rather a mixture of chemical compounds. Some are as simple as methane (CH_4); some are as complex as $C_{85}H_{60}$. Most of the compounds in petroleum contain from five to about 20 carbon atoms (Radovic

U.S. ENERGY SOURCES AND END USES, 1995

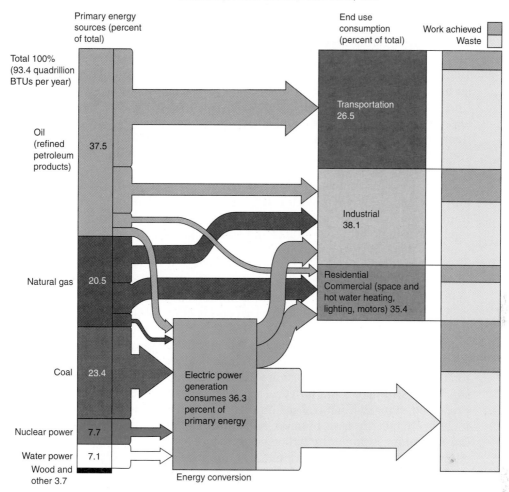

FIGURE 19.16   Refining of crude oil. Major components are removed at various levels, depending on their boiling points, in a giant distillation column (from Nebel and Wright 1996).

and Schobert 1992). Crude oils vary, however, depending on the geological processes that led to their formation. Some are referred to as old-deep oils, meaning they have been acted upon over long time periods and were subjected to high temperatures beneath the earth because of burial depth. By contrast one could speak of young-shallow oils that are highly viscous, high-density materials with a high sulfur content. Consequently, they require more processing to substitute for treatments not provided by nature. Nevertheless, compared to coal the elemental composition of crude oils from different parts of the world shows less variation.

As indicated by the simple chemical formulas just given, crude oils are mostly mixtures of hydrocarbons. In general they contain about 83 to 87% carbon and 11 to 16% hydrogen. They also contain up to 7% oxygen and nitrogen and 4% sulfur and therefore contain dissolved gasses as well as solids and liquids.

Crude petroleum must be processed, or refined, to untap the energy locked up in its molecular structures. The initial phase of refining involves fractional distillation to separate useful products (Figure 19.16). Much of the crude oil present is vaporized by heating. The vapor is passed to a distillation tower where it cools, with lighter molecules, such as gasoline condensing high in the tower and heavier molecules, such as fuel oil, condensing at lower levels. The remaining unvaporized heavy oil is passed to a vacuum distillation tower, where it too is vaporized and condensed into fractions. In the second step, cracking, heavy molecules from distillation are heated under pressure and broken down into smaller molecules that can be used in gasoline and other light hydrocarbon products. The third step, reforming, is a generally similar process in which the actual molecular structure of each product is changed to make it more acceptable to today's markets (Kesler 1994).

Individual refineries can differ as to the types of oil they process and the products they create, but they all usually employ these three basic steps (Radovic and Schobert 1992). Products recovered from these processes include natural gas, gasoline, diesel, jet fuel, light oils (low viscosity), heavy oils (high viscosity), asphalt, fuel oil, gasoline, kerosene, lubricating oils, naphthas, paraffins, petroleum coke, petroleum jelly and wax, as well as feedstock for petrochemical manufacture. The most important refinery product, unleaded gasoline, contains less than 0.05 grams of lead and 0.005 grams of phosphorus per gallon. In addition to transportation fuels, oil is used for home heating and for generating electricity. The fraction of petroleum used for polymeric materials is comparatively small but greatly impacts our lives. Plastics such as polyethylene, polyvinylchloride (PVC), formica, and polystyrene are all derived from petroleum. Numerous synthetic textiles (e.g., dacron, orlon, nylon, and polyesters), synthetic rubber products and some pharmaceuticals, cosmetics, and detergents are derived from petroleum.

Natural gas: Natural gas was produced by decomposing plant and animal remains that were buried in the earth for millions of years by sedimentary deposits. It is therefore a fossil fuel like coal and oil and is often found associated with them. As the gas leaves the well and is transported, it cools and the heavier hydrocarbons liquefy. The gas and crude oil separate as this mixture enters a field separator and is subjected to less pressure. Natural gas can then be drawn off from the top of the separator. Both the condensate and gas are useful products. Natural gas removed from field separators still may contain hydrocarbons heavier than methane and may be further processed to recover natural gas liquids (NGLs). Products recovered from the NGLs include ethane, propane, butanes, and additional natural gas (Leffler 1979). Processing also removes nonhydrocarbon gases, which can be sufficiently abundant to constitute resources in their own right. The removal of essentially all hydrogen sulfide (H_2S) is of particular concern because it is toxic and reacts with moisture in pipelines to create highly corrosive sulfuric acid.

Natural gas currently supplies about 24% of U.S. energy needs. It is used primarily for heating buildings, home cooking, industrial processes, and generating electricity. It presents several advantages over other fossil fuels. It is easier to transport than coal because it can be moved throughout a piped network extending across international boundaries. More importantly, it is a cleaner burning fuel because it contains less carbon relative to hydrogen, is nearly ash free, and has very little sulfur and nitrogen.

Oil shale and tar sands: Because of the declining availability of domestic oil, the industry has been evaluating alternative supplies. Oil shales and tar sands are among the candidate substitutes, but their use involves major compromises. For beginners, both are examples of oil resources that were not fully developed over their geologic past. Oil shale is a sedimentary rock that was formed millions of years ago from the mud at the bottoms of lakes. It contains kerogen, the same solid organic material contained in rocks that produced oil when subjected to higher temperature and pressure. Tar sands contain a thick, oily residue called bitumen. The kerogen and bitumen contained in these resources can be extracted and refined like petroleum. The industrial processes required, however, are energy intensive. Another drawback is that existing deposits will not greatly extend our energy supplies. Tar sands deposits are not extensive and only about one billion barrels could be recovered from U.S. deposits. With current technologies, 80 to 300 billion barrels of oil could be extracted from oil shale thus providing the U.S. with no more than 30 years of oil. Finally, the use of oil shale and tar sands could be one more step toward environmental degradation. For example, the most valuable oil shale deposits in the U.S. occur in Colorado, Wyoming, and Utah where sufficient water for processing is in short supply (Gleick 1994).

The Future of Oil and Natural Gas. Some would characterize the twentieth century as a period of rapid economic development in which the energy needed for industrial expansion was based on oil resources. Tremendous economic development, especially in the U.S., Europe, and Japan, was facilitated by low oil prices. However, it is obvious that our relationship with oil will face forced revisions sometime in the not too distant future. One necessary adjustment will be to find alternate sources of energy as the U.S. becomes increasingly dependent on foreign oil supplies. Granted, the U.S. currently is not facing an oil crisis and this condition will prevail as long as it is able to purchase foreign oil at modest prices. In 1995 the U.S. actually purchased more oil from Venezuela than from Saudi Arabia, whose contribution was matched by Canada. Imports from Mexico, Nigeria, and the United Kingdom were also significant.

A key factor is how long developing countries will be able to supply U.S. needs. Asian countries such as China are becoming oil dependent themselves and may soon begin competing with the U.S. for world supplies. More specifically, it is anticipated that world oil consumption will rise to more than 74 million barrels within the next few years and should reach a range of 89 to 99 million barrels a day by 2015 (EIA 1995). Also, the number of automobiles around the world has doubled since 1973 and is approaching 600 million. By some estimates the world oil supply will last less than 45 years. An optimist would say that developing countries have not intensively explored their natural resources and undiscovered oil could greatly extend oil supplies. A pessimist might say that the world demand for oil will outpace population growth, thus shortening the life of oil in the global economy to less than 50 years.

In 1994, the U.S. obtained 45% of the oil it consumed from foreign sources. By 2005, net imports will account for about 57% (EIA 1995). Once again, the vast stores of oil found in the Middle East will likely be critical in meeting U.S. energy needs in the near future. By the year 2000, OPEC will supply about 35 million barrels per day to U.S. customers. In 2015, barring unforseen developments, OPEC will export about 52 million barrels each day to the U.S. The U.S. has given itself some protection against the oil supply disruptions by creating an oil stockpile, the Strategic Petroleum Reserve. However, it can postpone economic turmoil for only a few months.

The outlook for natural gas is better than for oil, both in the U.S. and abroad. New studies suggest that the total U.S. reserves could be seven times greater than previously estimated and could last about 60 years. Estimates of world natural gas vary, but proven and undiscovered reserves will last about 104 years at the current rate of consumption. Oil shale or tar sand might be substituted for natural gas in some places. However, their distribution is too fragmented for them to become dominant world fossil fuel sources without global cooperation in investment and production.

Fuel Conversion. As discussed, the earth's fossil fuel energy can be separated into vapors (natural gas), liquids (crude oil), and solids (coal, oil shale, and tar sand). The U.S., the former Soviet Union, and China together possess more than 80% of the ultimately recoverable resources. The most important of these fossil fuels are natural gas, petroleum, and coal. However, as domestic and world oil supplies of these resources fade, transitions to other energy resources will be essential. No one knows exactly when these transitions will begin because of the complex of economical, political, and technological factors potentially involved. However, one can look to existing resource uses to identify the most probable possibilities.

To minimize the greenhouse effect, our most immediate need is to move away from fuels that release high amounts of carbon per unit of energy expended. This statement applies to both transportation and electric power generation. Natural gas (methane) is a likely substitute candidate because of its high hydrogen to carbon content. However, methane itself is a stronger greenhouse gas than CO_2. Small leaks in extensive production and delivery systems could be major contributors to atmospheric pollution.

Overall, the U.S. has more coal than oil or natural gas, and more natural gas than oil. Coal conversion to gas or liquids, as previously discussed, therefore remains a strong possibility. Because conversion either adds hydrogen or removes carbon, depending on the process used, it favorably alters the hydrogen to carbon ratio and creates a cleaner fuel. To date, industrial development along these lines in the U.S. has been discouraging. The Great Plains Coal Gasification Plant in Beulah, North Dakota, was completed in the mid 1980s. However, construction and operating costs were higher than anticipated and its product cannot compete financially in today's energy market place. South Africa provides a more encouraging example (Kesler 1994). There, political embargoes on oil imports stemming from apartheid forced the conversion of its abundant coal resource to liquid fuel. Three plants owned by SASOL employ 32,000 people and produce 130 different products, including gasoline, diesel fuel, jet fuel, and motor oil. Its plants are supplied by the largest underground coal mining system in the world.

In the U.S., propane like methane (natural gas) is also a very clean burning fuel and existing automobiles can be retrofitted with the equipment needed to use it. No modifications are needed to burn ethanol, which can be added to conventional gasoline to create a cleaner fuel ("gasohol"). Ethanol and other fuel alcohols can be obtained from sources other than fossil fuels.

Environmental Problems Associated with Exploration and Production.   Oil exploration and production can create several environmental problems. Numerous exploration wells are necessary to estimate the extent of an oil deposit. The initial phases of oil field exploration development therefore can involve considerable road construction and land disturbance. Native wildlife and wilderness can be seriously affected, especially if work is done recklessly. Currently, regulations and company practices are done more carefully, but site disturbance simply cannot be avoided. Improvements in drilling technology have actually reduced the impact of drilling on the environment and large quantities of gas are no longer flared in the night sky. By today's standards this would be wasteful and, in fact, many oil fields have lost most of their associated gas.

The greatest threat posed by oil and gas extraction is the escape of underground fluids and land subsidence. The escape of fluid from offshore drilling operations is actually a rare event. Only one well of the 30,000 drilled between 1964 and 1991 has blown out (Kesler 1994). This period includes the time period in which offshore exploration peaked in the mid-1980s (EIA 1997).

On land, the most troublesome fluid that comes from oil and gas wells is brine—i.e., fluid with concentrated salts. Some brines contain enough sodium, potassium, bromine, or related elements to be of commercial interest. However, most simply create a disposal problem and are injected back into the sedimentary strata from which they came. The early practice of collecting the brine in unlined ponds called oil pits is no longer permitted.

The removal of shallow oil deposits can cause the overlying land surface to subside. Modern reservoir engineering techniques designed to remove oil more efficiently have minimized this problem, but low-lying areas are vulnerable to even small subsidence effects (Kesler 1994).

Oil transportation involves domestic pipelines but also a growing dependence on international shipping. About 200,000 miles of oil pipelines exist in the U.S. The 800-mile Trans-Alaskan pipeline running from Alaska's "north slope" at Prudhoe Bay to the ship port at Valdez is best known to the public. Its construction caused serious protest in the pipeline's planning stages, owing to the delicate ecosystems potentially affected by its construction. Ocean-crossing oil supertankers provide an economical means of oil transport. One tanker can carry as much as 50 million barrels. This is enough to meet U.S. oil needs for three days. However, the rupture of a tanker hull can create an ecological catastrophe costing billions to clean up (Figure 19.17). The much publicized *Exxon Valdez* spill in

FIGURE 19.17 A common murre covered with oil (courtesy of U.S. Fish and Wildlife Service; photo by Jill Parker).

1989 created haunting images of birds, marine mammals, and fish being lost and rescued. Much of the damage caused by spills goes without notice. The larval and egg stages of fish are especially sensitive to the toxic effects of oil, and a surface slick may have its greatest impact on immature fish. Populations of most commercially important species of fish are extremely variable and a spill at just the wrong time can greatly affect their numbers.

The extraction of oil or its transportation in environmentally sensitive areas continues to receive justified public attention. As exploration presses into more remote areas to locate oil supplies, we will increasingly be forced to find a balance between the risk of foreign dependence on energy and potential damage to sensitive ecosystems. Two cases illustrate the dilemmas at hand. One involves oil company desires to obtain oil from Alaska's arctic slope and islands which are located near North America's largest oil resource, the Prudhoe Bay oil field. Protection of ecosystems against the intrusive nature of oil field development continues to block oil field development. Along the coastline of California a similar protest has arisen over development of the Point Arguello field, near Santa Barbara. Such cases justify considerable debate and thorough scientific study.

The facts are not always accurately stated by either overzealous environmentalists or the oil industry. For example, a major leak in an undersea pipe could create severe ecological damage. However, as discussed earlier, such events are rare and the public is not well aware of natural oil emissions from the marine floor.

Transportation Fuel Issues. The U.S. is forced to move ahead with oil development largely because of its excessive dependence on automobiles. The public wants to be free of automobile pollution but does not want to sacrifice the loss of personal freedom the automobile provides. The long lines at gas stations during the very brief oil crisis we experienced in the 1970s demonstrate the extent of our dependence on gasoline. The only people spared this brief period of forced trip planning and high gas prices were those who relied exclusively on mass transportation. It follows that part of the transportation problem can and must be improved by providing more means of public transit. In areas where public transit would be a poor financial investment, solutions must be found elsewhere. From 1974 to 1985, the fuel economy of U.S. cars almost doubled, which represents real but insufficient progress. Fuel modifications and emission retarding devices were also developed. Recent progress in automotive design gives hope that more advances are on the horizon.

Cars are becoming increasingly lighter as the auto industry learns to substitute lightweight, composite materials for more conventional steel. The ability to mold these materials in aerodynamically efficient car bodies is in a state of rapid advancement. Finally, low resistance tires, more aerodynamic designs, reductions in vehicle weight, and advances in electric battery technology are providing realistic opportunities to convert autos to electric power (Schiffer 1994, Sperling 1995). It is likely that a completely new kind of vehicle will begin to appear in automobile showrooms within the next few years. This new class of automobiles will operate more cleanly, and will be quieter and more responsive than current models (Flavin and Lenssen 1994). Both electric and electric-hybrid autos are being designed. However, efforts to redesign the automobile and to spur public transportation will be wasted if people are unwilling to revise their individual driving habits and a recent trend toward preferences for larger and heavier vehicles. Other nations will increasingly encounter the blessings and burdens associated with the automobile. Most projections show rapid increases in the number of automobiles driven in Asia. This will increase demand for fossil fuel and the greenhouse effect simultaneously.

Nuclear Energy

The Nuclear Alternative.   As fossil fuel resources decline, we must look elsewhere for energy resources. Nuclear power is a possible alternative (Figure 19.18). Nuclear power seems capable of eliminating or greatly reducing two of our most pressing problems: fading energy resources and global warming, as related to the combustion of fossil fuels. However, one must understand the basics of the resources and industry involved before drawing conclusions on its viability as an energy solution.

Nuclear power stations and fossil-fueled power stations of similar capacity have many features in common. Both require heat to produce steam to drive turbines and generators. In a nuclear power station, however, the nuclear fission replaces the burning of coal or oil. Nuclear fission reactions involve the basic particles located in the nuclei of heavy elements (i.e., elements having large numbers of protons and neutrons in their nuclei). More specifically, nuclear reactions usually involve ^{235}uranium, which is referred to as an isotope. Atoms of the same element with different atomic weights are called isotopes. Elements are defined by the number of protons and, therefore, all uranium atoms have 92 protons. The superscript number appearing before an element's name gives the atomic mass (the sum of protons and neutrons) of a given isotope. Because a single proton and a single neutron both have an

FIGURE 19.18   A nuclear power plant. Note the conspicuous cooling tower, which is necessary to cool and recondense the steam from the turbines before returning it to the boiler (courtesy of U.S. Dept. of Interior).

atomic weight of about one, the number also indicates the total number of these particles present in the isotope.

In a nuclear reactor, a free neutron collides with ^{235}uranium causing the isotope to split and release neutrons (Figure 19.19). Energy is created as the reaction proceeds because some matter is lost in the fission process. Because more neutrons are released from a fission event than are needed to induce the event, more and more uranium nuclei split in what is described as a "chain reaction."

To provide useful power, the rate of the chain reaction must be controlled. This can be achieved in nuclear reactors by using a "moderator" such as ordinary or "light water," which absorbs the free neutrons emitted as opposed to fissioning. Therefore, it can be said that the chain reaction is controlled by diluting the fissionable uranium atoms with water or other types of nonfissionable atoms. All of these fission fragments created by the fission process have high kinetic energy, which is converted after collisions with other nuclei into the random motion of thermal energy.

In nature uranium exists as ^{235}uranium or ^{238}uranium. The isotope that serves the U.S. nuclear industry, ^{235}uranium, is much rarer in nature than the alternate isotope. The ^{238}uranium accounts for 99.3% of the uranium naturally present in Earth's crust and ^{235}uranium only 0.7%. The ^{238}uranium is not considered a primary nuclear fuel because it nearly always reacts with a free neutron by absorbing it rather than by fissioning.

In U.S. nuclear reactors, uranium fuel is housed in an array of reactor "fuel assemblies." Before the initial start-up of a nuclear power reactor, the core must be loaded with fresh nuclear fuel. The fuel is basically a reservoir from which energy is extracted as long as a chain reaction can be sustained. The reactor itself is housed by a containment building that is sturdy enough to withstand a limited explosion. The aim is to confine any steam or gas explosions that might otherwise release radioactive water to the environment (Mounfield 1991). Assemblies comprise fuel rods created by encasing ^{235}uranium pellets in zirconium alloy. The heat produced by nuclear fission in the reactor core is carried away by the water which, as stated, serves also as a moderator.

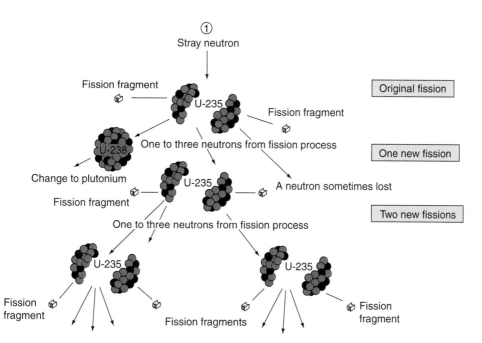

FIGURE 19.19 Nuclear fission involving a chain reaction. A neutron fly off by a ^{235}uranium atom strikes another which undergoes fission, thus releasing fragments or daughter nuclei and one to three neutrons. The process continues as these neutrons cause other ^{235}uranium nuclei to split (from Owen and Chiras 1995).

The process varies depending on the type of reactor. In nuclear plants that involve a "boiling-water reactor," water is converted to steam, which directly drives a steam turbine connected to an electric generator. The steam is then condensed to water and pumped back to the reactor to begin another cycle. In plants employing a "pressurized-water reactor," a device called a "steam generator" prevents radioactive water from directly contacting the turbine. Heat from water that has passed through the reactor core is transferred to water confined in a secondary loop. As heat is added, the steam produced in the secondary loop directly drives a turbine. In the reactor core, cooling water is kept at a very high pressure and is heated to some 600°C.

The boiling-water reactor is potentially more troublesome than the pressurized water reactor because the water serving as coolant and moderator can become radioactive before reaching the turbine. Contamination is possible because slight leaks can occur in the thin cladding of the fuel rods, and from radioactivity induced just outside the cladding. Because the steam that drives the turbine is potentially radioactive, great care must be exerted to avoid steam leaks in the turbine itself. The pressurized water reactor eliminates this problem altogether by confining the water circulating through the reactor core to a separate loop. However, the inclusion of a second loop reduces operational efficiency below that provided by the boiling-water reactor. Other reactor designs have also been developed and used abroad. These include the pressurized heavy-water reactor, the gas-cooled reactor, and the advanced gas-cooled reactor.

Because the coolant water in light-water reactors absorbs free neutrons, the amount of the fissile ^{235}uranium present must be concentrated above its natural level of 0.7%. Were this step omitted, it would be impossible to sustain a chain reaction. The concentration of ^{235}uranium steadily decreases as a reactor operates and energy is produced. At the same time the fertile $_{238}$uranium nuclei present are constantly being converted into fissile ^{239}Pu nuclei, some of which will, in turn, undergo fission and produce energy. While these reactions are taking place, the concentration of neutron-absorbing fission products (also called "poisons") increases within the nuclear fuel assemblies. After years of operation, the concentrations of fissile nuclei and poisons in fuel rods become altered to the extent that the chain reaction will not proceed. In other words, free neutrons are being absorbed or lost faster than they are created by fission events. A reactor reaching this condition must be shut down and refueled.

As one might guess, intervals between refueling are related to the amount of fissile ^{235}uranium loaded into the fuel assemblies at the beginning of the reactor's fuel cycle. Between refueling events, there is a period over which a reactor could operate at full output before a fission chain reaction would cease to be sustained. Its energy reservoir at any given time can therefore be said to contain so many full-power days. The plant's "capacity factor" is defined as the ratio of its actual level of operation to the maximum, full-power level of operation for which it is designed. The rate of fission at any given time is regulated by inserting or removing control rods, which absorb neutrons. Control rods can be inserted or withdrawn from around the fuel, thereby slowing or increasing the rate of fission. A variety of systems are in place to automatically stop the reaction and to provide emergency cooling.

We have just described the fueling factors associated with the converter type reactors presently used in the U.S. However, there is another category, the breeder reactor, which is restricted from use in the U.S. because of security reasons. Breeder reactors get their name from their ability to breed, or create, fuel as they operate. Breeder reactors can be fueled with the more common uranium isotope, ^{238}uranium, and plutonium. The plutonium undergoes fission, much like the ^{235}uranium in a light-water reactor, producing neutrons that can be absorbed by ^{238}uranium to create ^{239}uranium, which eventually becomes plutonium. This reactor requires fast neutrons for absorption by ^{238}uranium, so there is no moderator. The coolant for a breeder reactor is a liquid metal, currently sodium.

The Nuclear Fuel Cycle. The complete nuclear fuel cycle for a typical light-water reactor goes well beyond the boundaries of the nuclear reactor. The cycle is routinely described as having a "front end," which includes the processes required to prepare nuclear fuel for the reactor. The "back end" of the cycle includes the processes required to properly manage the spent nuclear fuel. Because the unused uranium and plutonium in the spent fuel can be reprocessed to recover useful nuclear fuel, we are speaking of a cycle as opposed to a chain of events (see Figure 19.20). Commercial recycling in the U.S. has limited applications, but this is not so in all countries.

The front end: The front end of the cycle involves several phases starting with exploration. Natural bodies of uranium ore must be discovered by drilling and other exploratory techniques. In cases where the quantity is known and is acceptable, and extraction costs are acceptable by current practice, it can be said that a uranium reserve exists. Ore deposits believed to exist through technical reference, but as yet undiscovered and, therefore, unquantified, are called resources. By now, the terms reserves and resources should be familiar to the reader.

The next phase is mining. The methods used to mine uranium ore are similar or identical to those used to extract ores of other metals. However, uranium can also be obtained by solution mining, and as a byproduct phosphate mining. In the U.S., the content of ores typically ranges from 0.05 to 0.3% uranium oxide (^{308}U). In other countries, ore concentrations of ^{308}U can be less, but in general are of a higher grade. Countries with high-grade deposits include Australia, Canada, South Africa, Niger, Namibia, Brazil, France, and Gabon. China and Russia may also have substantial deposits of uranium.

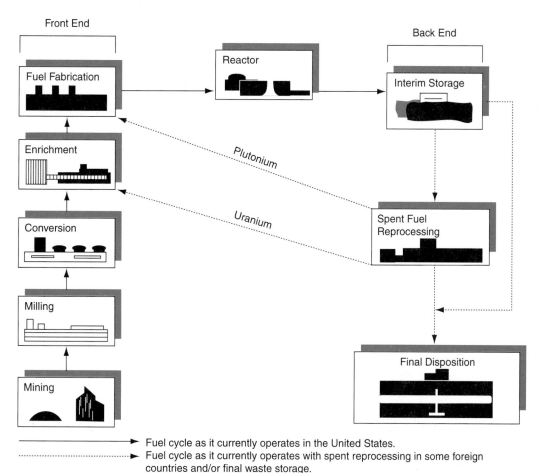

Fuel cycle as it currently operates in the United States.

Fuel cycle as it currently operates with spent reprocessing in some foreign countries and/or final waste storage.

FIGURE 19.20 The nuclear fuel cycle (from EIA 1995).

Milling follows mining. This process involves crushing and grinding the uranium-bearing ore followed by chemical extraction to obtain uranium oxide. This product is also called uranium concentrate or "yellow cake"(^{308}U) is 85% pure uranium in an isotopic mixture of 0.7% fissionable ^{235}U and 99.3% nonfissionable ^{238}U. Milling generally is done near the mining operation. Depending on circumstances, the yellow cake can be marketed like other energy resources.

The next step in the cycle is the chemical conversion of ^{308}U to uranium hexafluoride (^{6}UF), which is transformed from a solid to a gas if it is elevated above room temperature.

Enrichment involves concentrating ^{235}uranium so that enough fuel can be loaded to sustain a nuclear chain reaction in a light-water reactor. One method by which the necessary material can be concentrated, or enriched, involves gaseous diffusion. This consists of passing a "feed stream" of uranium hexaflouride (^{6}UF) gas through a long series of diffusion barriers that pass ^{235}uranium at a faster rate than the heavier ^{238}uranium atoms. This differential treatment progressively increases the percentage of ^{235}uranium in the "product stream." The "waste stream" contains the depleted uranium (that is, uranium having a ^{235}uranium concentration below the natural concentration of 0.7%). The enrichment process requires considerable energy. Alternative means for enrichment include gas centrifuge separation, as done in Europe by commercial operations and a laser separation technique being developed.

The final phase in the front end is fabrication. After the enriched ^{6}UF is changed to an oxide, it is pressed into pellets of ceramic uranium dioxide. The pellets are then encased in tubes of zirconium alloy or stainless steel, which will resist corrosion. After the tubes, also called rods or elements, are mounted into assemblies the processed fuel product is ready for loading into a reactor.

The back end: The back end of the cycle involves fewer steps starting with interim storage. After spent fuel is removed from a reactor, it is stored either at the reactor site or at a facility specifically designed to manage highly radioactive waste. At the reactor site, spent fuel rods are usually stored in water, which serves two purposes. It provides continual cooling for rods still giving off heat created by residual radiation, and shields workers from the radiation otherwise emitted.

Reprocessing follows interim storage. Spent fuel contains significant amounts of fissile (^{235}U, ^{239}Pu), fertile (^{238}U), and other radioactive materials. A "fertile" is capable of producing a fissile isotope. All of these radioactive materials are potentially useful to the nuclear industry and can be chemically recovered from the spent fuel and recycled if the nation in question permits it. At present, spent fuel obtained from nuclear power plants in Europe and Japan is being reprocessed to recover nuclear fuel. One of the issues pertaining to recycling is, of course, national security against atomic terrorism.

The last step is waste disposal. The safe disposal and storage of spent fuel and other high-level nuclear wastes is a highly controversial subject. Nevertheless, these hazardous wastes must be isolated from the environment until the radioactivity they contain decays to a safe level. Under the Nuclear Waste Policy Act of 1982, as amended, the Department of Energy has responsibility for the development of the waste disposal system for spent nuclear fuel and high-level radioactive waste. Current plans call for the ultimate disposal of the wastes in solid form in licensed, deep, stable geologic structures. More will be said about this later.

The Current Status of Nuclear Power.   Nuclear reactors essentially generate electricity by capturing the thermal energy released from their reactor core. Today, nuclear power accounts for over one-fifth of international electricity generation. Since 1972, their contribution to U.S. power generation has grown steadily (Figure 19.21).

At the end of 1995, 437 commercial nuclear units were operating in 31 countries throughout the world. The U.S. leads the world in power generation with its nearest competitor being France (Figure 19.22). In 1995, the 109 nuclear power plants oper-

Nuclear and Total Net Generation of Electricity, 1957–1995

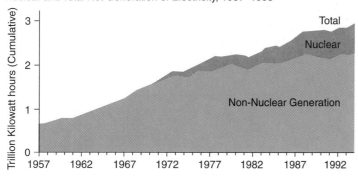

FIGURE 19.21 The relative contribution of nuclear power to the generation of electricity from 1957 to 1992 (from EIA 1995).

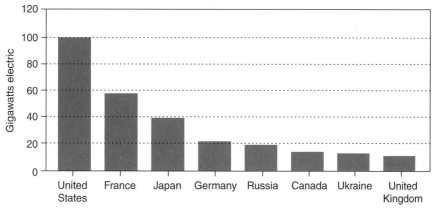

FIGURE 19.22 Nations with the largest nuclear generating capacity in 1995 (from EIA, 1995 with the original source being: International Atomic Energy Agency, "Nuclear Power Reactors in the World." Vienna, Austria, April 1996).

ating in the U.S. met 20% of the nation's demand for electricity. By comparison, France used nuclear reactors to produce 76% of its electricity. In the U.S., nuclear power plants are more numerous in the eastern half of the nation (Figure 19.23).

The Future of Nuclear Power in the United States. When the feasibility of the nuclear fission reaction was confirmed in 1939, it was immediately realized that nuclear energy had the potential to supply tremendous amounts of power. In the 1950s, President Eisenhower believed that nuclear energy would someday be so plentiful that its consumption would not be fully measured. Unfortunately, the nuclear industry suffered some set backs along the way and now its future is altogether uncertain. The U.S. nuclear industry probably will, at best, maintain its contribution to electric power generation in future decades. Several factors have challenged the nuclear industry and collectively weakened its appeal. These factors can be grouped into issues pertaining to economics, public safety, and environmental quality.

Economic Issues. Economic concerns stem from the marginal ability of the nuclear industry to compete with other power industries. The capital-intensive nature of nuclear-power plant construction requires billions in investment dollars. Too often construction costs substantially exceeded initial projections. This problem became worse, not better, over time. The factors responsible for escalating cost include construction delays, midcourse design changes, siting and licensing controversies, and increased labor costs (Holdren 1992). Construction costs in the U.S. increased after 1979, the year of the accident at the Three Mile Island plant in Pennsylvania. Prior to that time, the 63 completed plants required an average construction time of 6.3 years each. From 1979 through 1989, however, the average construction time rose to 11 years for the 47

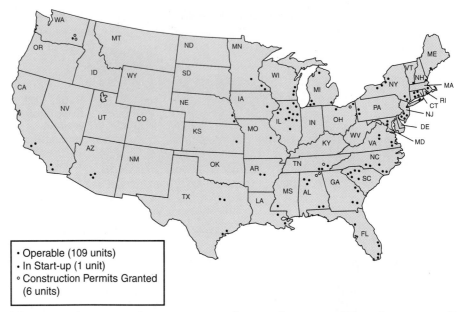

• Operable (109 units)
• In Start-up (1 unit)
∘ Construction Permits Granted
  (6 units)

FIGURE 19.23   Nuclear power plants operating in the United States as of December 31, 1995 (from EIA 1995).

Capacity Factor, 1973–1995

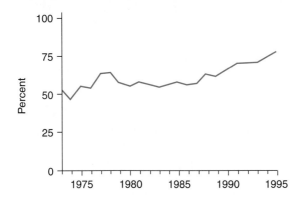

FIGURE 19.24   Nuclear power plant capacity factor from 1973 through 1995 (from EIA 1995).

plants that were built. Whereas coal fired power plants and other utilities also experienced increasing construction costs, they were less severe, making nuclear power comparatively less attractive to investors.

Operating costs have also been a problem. Rather than providing mostly uninterrupted power, plants were too frequently shut down for repairs. One measure of operational efficiency is load factor, which is the ratio between the electricity that a plant actually produces and the amount that it was designed to produce. The average load factor in the U.S. through 1987 was less than 60%, indicating that many of the plants experienced frequent shutdowns. In contrast, Canada's load factor in 1987 was 78.2%. At that time, Belgium led the world with an average lifetime load factor of 81%. Recent trends in the U.S. toward deregulation and privatization of electricity supply systems have increased pressure on nuclear plant operators to be economically competitive with other generating technologies. For this reason and others, load factor and plant capacity have steadily improved since 1987. Now the industry operates near the 75% load-factor level (Figure 19.24).

After a period of substandard performance, some plants have been closed, or decommissioned. The elimination of such plants improves average load factor, but involves, again, tremendous costs to contain radioactive materials. When plants were constructed, it was anticipated that they would eventually return investments after 40 years of service. Plants prematurely decommissioned have failed to reach the mature years of profitability.

NONRENEWABLE ENERGY AND THE ENVIRONMENT

Nuclear Plant Accidents

At present, the public perception of nuclear safety limits potential for growth in the nuclear industry. In 1970, only about 25% of the public opposed building new nuclear plants. From 1975 to 1979 less than 35% of the public opposed the industry. However, after the accident at Three Mile Island in 1979, public support faded even more, and by 1985, 67% of the public opposed nuclear industry growth (Holdren 1992). The accident at Chernobyl, Ukraine, in 1986 further reinforced public distrust.

In the accident at Three Mile Island, 27 tons of fuel melted in the base of the pressurized water containment vessel. Although no significant leaks to the environment were reported, the public remains skeptical about the amount of radioactive material actually reaching the atmosphere (Figure 19.25). Because the adverse effects of radiation exposure appear years after it occurs, surrounding communities remained concerned. The total cost of the cleanup has not been established, but exceeded $1 billion six years after the accident (Hall et al. 1992).

The effects of the Chernobyl accident were more obvious. Twenty people were killed immediately at the site and millions of people were exposed to measurable quantities of radiation. The failed reactor was a light-water graphite reactor with a graphite core and no shielding structure (Edwards 1994). By U.S. and western standards, the design itself was substandard and clearly would be unacceptable in the U.S. A fundamental flaw was the absence of a heavily reinforced containment building to limit radiation exposure if an accident should occur. The same design was used in the construction of many Soviet reactors. Ironically, a safety training exercise triggered the tragedy. The reactor was operating outside its own stated safety regulations when the accident occurred. In the accident, the water coolant pump failed and as the core temperature rose to some 3,000°C, the uranium fuel melted. Several explosions removed the top of the building over the reactor and the graphite ignited. The radioactive particles released to the atmosphere were carried all over northern Europe.

When the U.S. and former Soviet Russia mended their relations, there was a major cutback in atomic weaponry. In the interest of world safety, the United States has extended its nuclear policy to one of improving nuclear safety in Russia and Ukraine. In the initial stages of U.S. assistance, it became clear that Soviet nuclear power plants lacked adequate operational and maintenance procedures, quality assurance programs, and reliable construction designs. A recent report by the World Bank and International Energy Agency identified 25 high-risk reactors in Russia, Ukraine, Armenia, Bulgaria, Slovakia, and Lithuania. These need to be upgraded or replaced at costs estimated to be as high as $24 billion. U.S. interests go beyond the immediate need of protecting the world against radiation exposure. In Ukraine, for example, nuclear energy was consumed regionally while petroleum energy was exported to provide profitable international trade. In the face of a collapsing economy Ukraine and other presently independent states may return to cold war politics and threaten global nuclear destruction.

The Chernobyl accident was the worst disaster in the history of nuclear power (Figure 19.26). The world community is still being challenged by its long-term economic, environmental, biological, and psychological impacts. However, by some assessments, reactors are considered safe because accidents are rare and emissions

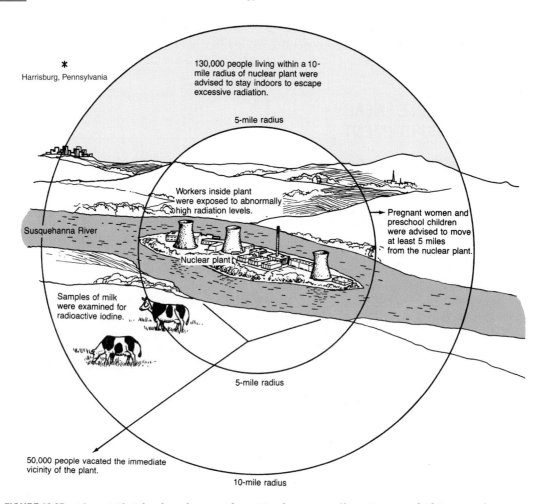

FIGURE 19.25   Three Mile Island nuclear accident, March 28, 1979 (from Owen and Chiras 1995).

of radioactive material is very low under most conditions. In fact, there have only been about 10 accidents where reactors have released radioactive particles into the atmosphere.

It will be hard to repair public distrust in nuclear power plants and to overcome the financial difficulties of the industry. However, this situation may be altered because of factors inside and outside the industry. Inside the industry, plans are to renovate reactor designs to improve safety and reduce operating costs. Newer designs call for smaller plants with passive ("brain dead") controls, which take over in the event of cooling failure. Improvements in computer controls allow reductions in otherwise extensive wiring systems. Moreover, designs allow 50% fewer welds in the structure that encloses the fuel and its cooling water. Structures therefore will be less prone to failure. By simplifying designs and scaling down plant size, the industry is making an overt attempt to reduce capital investment and to reduce construction delays. According to the Electric Power Research Institute (EPRI), a group funded by electric utilities in the U.S. and some foreign countries, reactors should be designed to last for 50 years, to have a load factor of 87%, and be built in less than four and one-half years. The industry hopes to satisfy these criteria as it seeks to restore investor appeal.

Outside the industry, environmental problems created by burning fossil fuels are reaching the point where some measure of risk may, sooner or later, be forced on the public to avoid economic turmoil. The diminishing supplies of U.S. fossil fuel resources will also force major adjustments and will become a jointly considered factor in setting national policy. Abroad, a developing world will likewise struggle to fund energy sources.

FIGURE 19.26 Aerial view of the soviet nuclear reactor building in Chernobyl after the accident (from Owen and Chiras 1995).

About half of the CO_2 released to the atmosphere over the past 150 years was produced by burning fossil fuels, particularly coal. Without nuclear power, the world would have to rely almost entirely on fossil fuels, especially coal, to meet electricity demands. Emissions of carbon dioxide from burning fossil fuels are calculated worldwide to be about 20,000 million tons per year, of which about 45% comes from coal and 40% from oil. Coal-fired electricity generation produces nearly twice as much CO_2 as natural gas per unit of power, but nuclear power contributes no CO_2. If all of the world's nuclear power were replaced by coal-fired power, CO_2 emissions would increase by one third.

Because the U.S. has more extensive supplies of coal than oil or natural gas (EIA 1996), coal provides the logical alternative to nuclear energy. A 1,000 megawatt coal-fired electrical generation plant operating at 75% capacity has a typical fuel requirement of about 2.3 million tons of coal a year. A nuclear power reactor of the same capacity (after its initial fuel loading of uranium) has an annual requirement of about 25 tons of fuel. Compared to U.S. nuclear power, coal-fueled power generation in the United States causes seven times more fatalities per unit of electricity generated. Another factor often omitted in energy discussions is that coal burning often releases more radioactivity than the equivalent nuclear power. Finally, the types of uranium and plutonium used for bombs and electric power generation differ greatly. Bomb-grade uranium is highly enriched (>90% ^{235}U, instead of about 3%) and bomb-grade plutonium must be fairly pure (>90%). The real risk created by building a plant is therefore one of environmental contamination as occurred in Ukraine.

According to the United Nations projections, more than 90% of world population growth in the foreseeable future will be in the less-developed countries. Furthermore, 90% of that growth will be concentrated in areas where nuclear energy would be practical in engineering terms. This explains why China is adding 10 reactors to its three already in use. China's power generation requirement is expected to almost double from 1994 to 2010. Much of the energy needed will be obtained from nuclear power plants. Like many developing countries, China's energy demands are growing even faster than the population (Levine et al. 1992). Energy shortages are aggravated further by the existence of outmoded coal-burning technology, which inefficiently uses resources and contributes avoidable pollution. Although nuclear energy is now marginally competitive in the U.S., this is not so in many places abroad. Some countries lack fossil fuels and the cost of coal transportation rules against its use.

Terrorism

The threat of nuclear terrorism can not be ignored by the world's nuclear community. According to some experts, security standards are too lax (Berkhout and Feiveson 1993, Stansfield 1997). The potential for problems in this area is another mark against the growth of the nuclear industry, here and abroad.

All nuclear power plants use or generate uranium and plutonium that can be employed to manufacture nuclear weapons. In this context, nuclear proponents point to the fact that nuclear power plants have not been a source of nuclear weapons development. In Europe, however, there has been an increase in the number of arrests for nuclear trafficking (Williams and Woessner 1996). In particular, there is the concern that organized crime in Russia will gain access to nuclear materials that, by some descriptions, posed less risk to the world when positioned on missile warheads (Bukharin 1996). The covert nature of smuggling and terrorism may rule out a precise assessment of the threat posed by criminal use of nuclear materials. To reduce access to bomb capable materials, American Presidents have steered the domestic industry away from breeder reactors and have called on the global community to do the same. Nevertheless, enriched uranium and plutonium are part of the fuel reprocessing industry in Europe, at least on a limited scale. As already mentioned, the U.S. is assisting Russia and Ukraine. This cooperation extends to security controls to eliminate theft. An outright ban on the production of weapon-grade plutonium is under mutual study.

Environmental Degradation

Contamination of the environment by nuclear waste remains a hot topic today and is an added disincentive to nuclear industry growth (Holdren 1992). Some view the threats posed by waste to be greater than those posed by nuclear power plants. They cite the fact that nuclear plants do not pose the threat of a nuclear explosion because the fuels are insufficiently concentrated and plant contamination, as occurred in Ukraine, is unusually rare. To some at least, the risk of contamination caused by a transportation accident or waste burial represents a substantial threat to them and their ecosystems.

Such fears are not groundless. There are examples of human death and health impairment associated with management of waste materials. These center on the improper disposal of milling wastes that caused avoidable suffering among the Native Americans living near uranium spoils in the Southwest. They center also on the families exposed to radon gas when spoil materials were used for residential landfills. In Colorado, people actually lived above radioactive material put there to level their homesite. Hopefully, the nuclear industry is now beyond this type of recklessness, but the public rightfully refuses to put safety much beyond their immediate control.

The problems just cited were created by mismanagement in the front end of the nuclear energy cycle and fall under the category of mining and milling. A substantial number of safeguards now regulate spoil materials. Today, the back end of the nuclear energy cycle is the focus of attention.

Nuclear plants generate both low-level and high-level radioactive waste. Low-level waste is material whose radioactivity will decay to safe levels in less than 50 years. It can be buried in shallow landfills. Among the materials included in low-level wastes are isotopes of nickel and cobalt from corroding pipes, and small amounts of radioactive contaminants discarded with worker protective clothing and tools. High-level waste obtained from spent commercial fuel requires centuries to decay to safe levels of radioactivity. It must be stored in sealed containers because of the intensity of its ionizing radiation. The volume of high-level waste produced outside the power industry is 20 times greater, but contains less radioactivity.

Spent fuel rods are viewed as especially hazardous and contaminations caused from their improper disposal raise serious concerns. Such releases would be more threatening than accidents involving extraction and processing because most of the radionuclides produced during these processes occur naturally. Nuclides become more radioactive after undergoing fission in plant operation. On the positive side, the costs of dealing with this high-level waste are built into electricity tariffs. For instance, in the U.S., consumers pay 0.1 cents per kilowatt-hour, which utilities put into a special fund. So far more than $11 billion has been collected.

The public, nuclear industry, and federal and state governments have invested large sums of time and money as the waste issue has gone before Congress, state legislatures, and courts. For the moment, Yucca Mountain in Nevada is scheduled to become a permanent depository site for high-level radioactive waste. Opponents of the site claim that the mountain may not be safe for the thousands of years that the high-level waste must be isolated. Further, the site may not meet the regulations of the Nuclear Regulatory Commission and the Environmental Protection Agency.

The issue of high-level wastes seems all but solved. Spent fuel is currently stored at nuclear plants in pools of water.

The Low-Level Waste Policy Act of 1985 stated that all states either develop their own site or form a compact with other states to select a disposal site of mutual satisfaction. However, this has often been delayed and deadlines have been repeatedly extended.

The Waste Isolation Pilot Plant (or WIPP) located in southeastern New Mexico is designed to dispose of transuranic radioactive waste left from the research and production of nuclear weapons. Transuranic elements have an atomic number higher than that of uranium (atomic number 92) and all such elements are produced artificially and are radioactive. The WIPP site was also hotly debated for years before it became a reality. Nevertheless, project facilities include disposal rooms excavated in an ancient, stable salt formation, 2,150 feet underground. The facility began receiving waste in 1998.

THE IMMEDIATE FUTURE

As one can see, nuclear energy has been surrounded by controversy and disappointments arising not only from public fear but also its inability to compete in the energy sector for private financing. Some would say it has been forced to limp along and suffer failure because government support was lifted before the fledgling industry took root. Public confidence must be restored and this will require continued improvements in plant capacity and safety. A just appraisal also calls on the public to examine the natural sources of radiation present and radiation risks created outside the energy sector from medical exposures, the military, and chemical industry. Even smoke detectors designed to improve home fire safety rely on small quantities of radioactivity.

Fusion energy will vastly eliminate the threat of contamination and will one day draw upon a vast resource found in the sea. However, this technology is decades away because of the challenge of containing reactions that melt metal. Some form of magnetic containment is most promising. In the interim, we must find substitutes for fading sources of fossil fuels, which may be altering the world's climate. Based on present plans, the immediate future for nuclear energy is decommissioning existing plants. We will also have to replace the power lost from nuclear operating licenses that begin expiring in the year 2000. Natural gas provides a short-term solution while helping with the problem of global warming. Over coming decades, other substitutes must be found and renewable energy sources are among the possibilities (see Chapter 20).

LITERATURE CITED

Berkhout, F. and H. Feiveson. 1993. Securing nuclear materials in a changing world. *Annual Review Energy Environment* 18:631–665.

Bukharin, O. 1996. Security of fissle materials in Russia. *Annual Review Energy Environment* 21:467–496.

Edwards, M. 1994. Chernobyl: Living with the monster. *National Geographic* 186(2):100–115.

Energy Information Administration. 1995. *Annual Energy Review 1995.* DOE/EIA–0384(95).

Energy Information Administration. 1996. *U.S. coal reserves: A review and update.* U.S. Department of Energy. DOE/EIA–0529(95), p. 83.

Energy Information Administration. 1997. *Annual Energy Review 1996.* DOE/EIA–0364(96).

Flavin, C. and N. Lenssen. 1994. Reshaping the power industry. In Starke, L., ed. *State of the World 1994.* New York: W. W. Norton.

Gleick, P. H. 1994. Water and energy. *Annual Review Energy Environment* 19:267–299.

Hall, C. A. S., C. J. Cleveland, and R. Kaufmann. 1992. Energy and resource quality. In *The economy and economic process.* University Press of Colorado. 577 pp.

Holdren, J. P. 1992. Radioactive-waste management in the United States: Evolving policy prospects and dilemmas. *Annual Review Energy Environment* 17:235–259.

Karl, T. R., R. W. Knight, D. R. Easterling, and R. G. Quayle. 1996. Indices of climate change for the United States. *Bulletin of the American Meteorological Society.* 77 (2):279–292.

Kesler, S. E. 1994. *Mineral resources, economics, and the environment.* New York, NY: MacMillan. 391 pp.

Leffler, W. L. 1979. *Petroleum refining for the non-technical person.* Tulsa, OK: Petroleum Publishing Co. 159 pp.

Levine, M. D., F. Liu, and J. F. Stinton. 1992. China's energy system. *Annual Review Energy Environment* 17:405–435.

Meyers, S. and L. Schipper. 1992. World energy use in the 1970s and 1980s: Exploring the challenges. *Annual Review Energy Environment* 17:463–505.

Mounfield, P. R. 1991. *World nuclear power.* New York: Routledge.

Nebel, B. J. and R. T. Wright. 1996. *Environmental science.* Upper Saddle River, NJ: Prentice-Hall. 698 pp.

Owen, O. S. and D. D. Chiras. 1995. *Natural resource conservation.* Upper Saddle River, NJ: Prentice-Hall. 586 pp.

Radovic, L. R. and H. H. Schobert. 1992. *Energy and fuels in society.* New York, NY: McGraw-Hill. (College Custom Publishing Series). 422 pp.

Schiffer, M. B. 1994. *Taking charge: The electric automobile in America.* Washington, DC: Smithsonian Institution Press.

Simon, J. 1996. *The ultimate resource.* 2nd edition. Princeton, NJ: Princeton University Press.

Sperling, D. 1995. *Future drive: Electric vehicles and sustainable transportation.* Washington, DC: Island Press.

Stansfield, T. S. 1997. *Caging the nuclear genie: An American challenge for global security.* Boulder, CO: Westview Press.

Williams, P. and P. N. Woessner. 1996. The real threat of nuclear smuggling. *Scientific American* 274(1):40–44.

CHAPTER 20

Renewable Energy: The Sustainable Path to a Secure Energy Future

INTRODUCTION

The U.S. has two basic energy problems. It is becoming less able to meet its needs for oil as its domestic reserves decline. At the same time, the world is bracing itself for climatic change due to carbon dioxide increase from fossil fuel combustion. The U.S. contributes a disproportionate amount of greenhouse gasses to the atmosphere. The one thing that seems certain is that this will force change in the emphasis the U.S. gives to energy issues. In addition, it is highly unlikely that there will be a quick fix because the issues are multinational and often pit technologically advanced countries against those on the cusp of rapid economic development. This was obvious as nations aired their views on greenhouse gas emissions during the United Nations conference (Kyoto Conference) on global warming held in Kyoto, Japan in December, 1997. Before this historic meeting, the U.S. position was one of calling for voluntary emissions reduction programs and monitoring. As a result of the Kyoto Conference, the U.S. has agreed to reduce its emissions to 1990 levels no later than the year 2012. By taking no action, it will dump 25% more carbon dioxide into the atmosphere by 2012 (Figure 20.1). Furthermore, if the U.S. continues along its current path it will become more dependent on foreign oil and, by some estimates, will invite economic calamity. This argument is not without merit because two-thirds of the world's oil is projected to come from the Middle East by the year 2020.

A number of changes will be needed for the U.S. to reduce impacts on climate change and set the stage for energy sufficiency (Thurlow 1990). It seems likely that one response will be to move more aggressively into renewable energy. Today, renewable energy provides only about 8% of the energy consumed in the U.S. (Figure 20.2). Advocates of renewable energy argue that this figure is far too low and that renewable energy has great potential for solving energy dilemma. We will address this issue in some detail in this chapter.

Renewable energy has many advantages. These became more obvious when U.S. energy prices soared in the 1970s and the American public became better informed about environmental issues. One effective response to the energy crisis was simply to conserve home and transportation energy. However, conservation alone will not meet our future needs because the world's population and economies continue to grow.

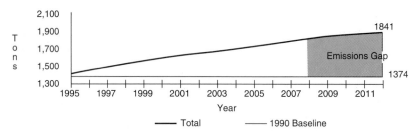

U.S. Carbon Emissions
Million Metric Tons of Carbon

FIGURE 20.1 Graph showing U.S. carbon emissions with and without a federal commitment to return to 1990 levels as discussed during the 1997 Kyoto Conference (from EIA 1997a).

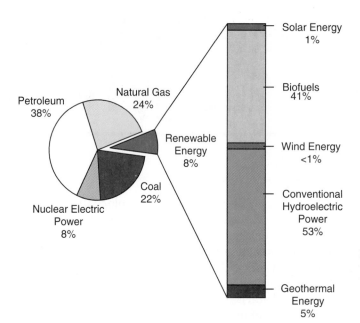

FIGURE 20.2 The contribution of renewable energy to total U.S. energy needs in 1996 (from EIA 1997b).

The single most attractive feature of renewable energy is that it is supplied mostly by the sun and is therefore limitless. The sun drives the wind and part of the hydrologic cycle, which provides hydropower. Also, solar power can be captured as heat, which can be used directly for power generation or converted with no intermediate steps to electricity. The sun also drives photosynthesis, which can be used to create renewable liquid fuels for transportation. Beyond the sun, only gravitational energy remains as a renewable energy source. It is provided through hydropower (falling water) and by the moon's gravitational effect in tidal energy.

Despite the impressive potential of renewable energy sources, some energy analysts regard them as impractical and exotic. History suggests otherwise. In the early nineteenth century, the most common source of energy in the U.S. was firewood. In the same era, stream flow powered the milling of grain and lumber, and wind was used to pump water and propel ship cargo and passengers. Also, greater attention was given to the features of home construction that affected light and temperature comfort. As fossil fuels spurred on modernization, toilsome and time-consuming practices were mostly, but not entirely, abandoned and forgotten. Unfortunately, renewable energy resources were forgotten with the exception of stream power and domestic wood fuel. Although we no longer see stream-powered grist mills, hydroelectric power was harnessed in the early 1900s and remains an important energy source. Also, some 5 million households burn wood as their primary source of heat (Brower 1992), while the pulp and paper industry meets over half of its energy needs by burning process wastes.

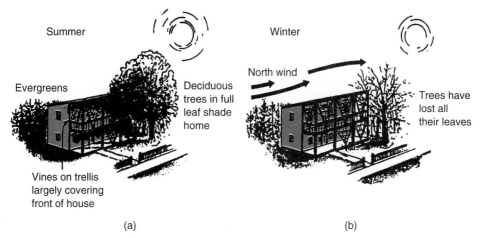

FIGURE 20.3 Landscaping can improve home energy efficiency. In summer (a), deciduous trees or vines can shade a home to reduce interior heat loads. In winter (b), the absence of leaves allows sunlight to provide additional warmth (from Nebel and Wright 1996).

RENEWABLE ENERGY SOURCES

Solar Energy

Uses of solar energy ranges from simple to complex, depending on modes of energy capture, transmission, and utilization. Passive solar heating is on the simple end of the scale and involves designing a building's structure to capture and store solar energy (Bevington and Rosenfeld 1990). Although passive solar designs can greatly reduce energy use in new buildings, little or no extra cost is required. Significant energy reductions in homes and office buildings is no trivial issue. About one-third to one-half of all energy produced in industrialized countries goes toward heating and cooling buildings (Brower 1992). In Boulder, Colorado, all new residential construction must include some passive solar heating.

When solar designs are combined with efficiency measures such as insulation and low-emissivity glass, energy savings are greatly multiplied. Passive solar design is site specific but generally involves orienting a building toward the south to capture the winter sun. South-facing window areas are used for winter heating and natural ventilation. In general, 75% of the heating requirements for a building occur at night. Therefore, designs frequently employ wall or floor materials that can store heat in their mass. Landscaping can be an integral part of the design and can afford seasonal shading and solar exposure as needed to further improve living comfort (Figure 20.3). Overhead shading can reduce a typical home's heating and cooling costs by 15 to 25%. The main constraints on passive solar design are economic and technical in nature. However, the small additional cost of passive design is returned over the life of the home. Materials for passive design are being improved at a steady pace.

Active Versus Passive Heating. Solar collectors are discrete units that collect, store, and distribute solar energy for water heating, space heating, and space cooling (Figure 20.4). Most use pumps or fans to circulate hot water or air through ductwork or plumbing (Figure 20.5). However, some rely entirely on natural convective currents created by temperature gradients. Systems without motorized components are properly called passive systems. Active systems have been used to provide energy to numerous types of buildings, including those serving farms, universities, and hospitals. Often the purpose is to supplement rather than replace more conventional methods of controlling building temperature.

Solar systems are well suited to meeting the hot water demands of home and commercial buildings. Solar water heating units often consist of a series of plastic

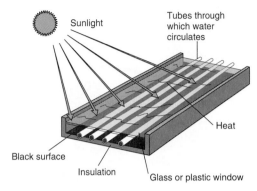

FIGURE 20.4   A flat-plate solar collector. Sunlight is converted to heat as it is absorbed by a black surface. A clear glass or plastic window covering the surface traps heat once sunlight energy enters. Air or water is heated as it passes over and through tubes embedded in the black surface (from Nebel and Wright 1996).

FIGURE 20.5   Solar panels provide the primary source of heat for this modular home (courtesy of U.S. Dept. of Energy).

or metallic pipes that are mounted on building roofs (Figure 20.5). The hot water they produce can substantially reduce energy costs. Some systems can be retrofitted to existing buildings and are becoming increasingly attractive to homeowners as deregulation weakens competition from other energy sources. Solar water heaters are frequently seen in developing countries. The majority of homes in Israel use solar water heaters.

Solar-Thermal Concentrating Systems.   Solar energy can be used to heat and vaporize water or another working fluid to produce electricity. Several approaches have been used for concentrating the solar energy needed. The power tower approach uses an array of sun-tracking mirror assemblies that concentrate sunlight onto a tower-mounted central receiver in which a circulating fluid is heated and used to produce energy (Dracker and De Laquil 1996). The Department of Energy's interest in central receiver systems, such as the power tower approach, has remained high over the years. Experimental systems have been erected in the U.S. and numerous foreign countries.

The solar trough has received both research and commercial attention (Figure 20.6). The curved and highly reflective inner surface of the trough concentrates sunlight onto a line receiver, a vacuum-sealed metal or glass pipe containing fluid. The pipe runs the length of the trough and the fluid contained is heated to temperatures as high as 400°C (750°F) (Brower 1992). A commercial electric generating plant installed in the Mojave Desert of California grouped large numbers of troughs together to generate heat.

FIGURE 20.6 A solar trough. Note how the curved reflector concentrates sunlight on the pipe positioned at the trough's center. The oil heated within the pipe can be used to boil water and generate steam for driving a conventional electrical generator (from Nebel and Wright 1996).

In another approach, parabolic mirrors (heliostats) are used to concentrate solar energy for the purpose of steam generation. The steam is used to generate electricity. Parabolic dishes can also be used for manufacturing metals, glass, cement, and other materials because of the intense heat that can be generated. Such temperatures cannot be obtained from parabolic trough systems. Dishes are also more efficient at generating electricity. Electrical applications employ numerous dishes. The fluid heat gathered by the centrally positioned receiver of each dish is transferred to a central location for steam generation.

Dish/engine systems add an electrical generator to the center of the dish. This technology uses a stretched-membrane parabolic dish to focus solar energy onto a receiver filled with sodium. As intense heat vaporizes the metal, the vapor condenses on the heater head of the attached Sterling-cycle engine. The transferred heat causes the helium gas inside the engine to expand. As the gas expands and contracts in a cyclic manner, the engine's piston is driven to and fro. The piston in turn propels an alternator, which generates electricity. The use of the engine overcomes a serious limitation of solar power—the need to supply power when sunlight is minimal to nonexistent. An uninterrupted power supply is possible because the engine can be powered by conventional fuels when solar energy is absent.

The most novel feature of the Sterling engine is its reliance on external combustion. The engines commonly used involve internal combustion and require fossil fuels in some form. The Sterling engine is not a new idea. Robert Sterling was given a patent for the idea as early as 1816. Dish/engine systems are especially attractive for remote applications and for village communities in Third World countries because of the amounts of electricity they generate. In addition, the engines require little maintenance because they are self-lubricating and, unlike portable generators, run quietly. Sandia Laboratory in Albuquerque, New Mexico is but one of several institutions attempting to further the dish/engine concept. Because the net solar-electric efficiency of dish/engine designs is 29% or better, it offers one of the best approaches among solar thermal technologies. Several companies in the U.S. and Europe are developing dish-engine systems for both remote applications and grid-connected, distributed-power applications (Dracker and De Laquil 1996).

Technologies involving parabolic troughs, central receivers, disks, or disk/engine systems have their greatest appeal in desert areas with intense sun. Because deserts are generally sparsely populated, such technology will have limited value on the national scale. Power could be transmitted to large cities beyond deserts. However, much of the power generated would be lost to internal resistance as it is transmitted over long distances. Regionally, it can substantially decrease reliance on fossil fuels in the southwestern U.S., parts of Australia, the Middle East, and Africa.

Photovoltaics.   The basic unit in a photovoltaic system is a solid-state device called the solar cell (or photovoltaic cell) (Weinberg and Williams 1990). The unique feature of the solar cell is its ability to convert sunlight directly to electricity without any intermediate steps involving heat transfer. Photovoltaic cells are made of semiconductors and light sensitive metals, such as selenium. As electrons in the cell become exited by sunlight, an electrical current is generated. Commercially useful quantities of electricity can be generated by coupling panels of solar cells in vast arrays (Figure 20.7). Solar cells operate best in full sun but can also function on cloudy days. Under the best conditions, photovoltaic cells can convert about 17% of the sun's energy to electricity. Photovoltaics are used across the globe to provide electricity to remote areas. Solar electricity can compete with conventional power in remote areas because the latter would have to be transmitted over large distances at higher cost.

In developing countries, photovoltaics are being strongly encouraged and financially sponsored by the United Nations and individual donor countries. The U.N. estimates that more than two million village communities lack power for basic necessities such as pumping well water. Lighting photovoltaics can bring village communities into national communication networks, thus giving village people a more active voice in democratic governments. Education can also be disseminated to remote areas. In

FIGURE 20.7   Panels used to generate solar electricity (courtesy of U.S. Department of Energy).

industrialized countries, photovoltaics power remote telephones, navigation beacons, electric fences, field data recording devices, and a host of other types of equipment. These applications are cost-effective in the absence of an opportunity to cheaply tap an electric grid. The use of photovoltaic cells in the NASA space program over past decades has been a source for technology development and publicity.

The Future of Solar Energy.   Solar energy must clear some obstacles before making deep penetration into the commercial energy market. One major hurdle is the present inability of solar energy to compete economically with fossil fuels. Photovoltaics are especially affected by high manufacturing costs. As long as solar manufacturing operates at its present volume, the cost of individual units will remain high and so will power plant construction. Under existing technology, considerable work must be done by hand. The preparation of crystalline silicon cells, for example, is a tedious process, which involves growing single crystals and cutting them into wafers. Methods involving materials more suited to mass manufacturing are being developed, but progress has been slow under the present market conditions. As long as other fuel sources remain more economical, photovoltaics will be confined to the fringe of the energy sector in developed countries. There are cases, however, where the addition of solar power to an existing electrical grid can eliminate the need for a new and larger conventional power plant. For example, Pacific Gas and Electric in northern California has found that the higher cost of photovoltaic systems is justified today in areas where they effectively reduce peak loads on transmission and distribution networks that must otherwise be upgraded at even greater cost (Brower 1992). The ability of solar units to provide consistent power, to perform in a predictable manner, and to be relatively free from module failure are also attractive to grid managers.

Existing tax codes also detract from broad market acceptance of solar technologies. Businesses can deduct fuel energy costs but not money invested in capital-intensive solar energy. Immediately following the energy crunch of the 1970s, numerous states provided tax exemptions for solar applications, but many of these were phased out over time. What appears to be missing in tax formulations are the environmental benefits that accrue to the public at large as renewable energy replaces fossil fuels. Solar energy proponents are quick to point out the subsidy paid to the petroleum industry in defending foreign oil fields.

One additional obstacle to application of solar technology is energy storage. This problems exists, because of the intermittent nature of sunlight. Photovoltaic systems only generate power and have no storage capacity. Dish/engine systems can keep operating if provided combustible fuel. Here storage is less critical. A more encompassing view toward energy storage issues might be to consider any energy contributions from nonpolluting means of practical benefit. This seems especially true for systems capable of supplementing peak loads.

The future of solar energy in the U.S. rests mostly in the hands of the federal government. Private enterprise has sufficient capital to fund research on basic components, but power plant demonstration projects are mostly beyond its financial reach. Some financial support from the public at large will, therefore, be needed to promote more immediate progress. Government support is slowly returning to levels comparable to the period immediately following the oil crunch, but overall funding remains low relative to the overall task. Without serious set backs, it appears that solar energy will continue slow penetration into the energy market. By the year 2030, it will provide a substantial contribution to the total energy consumed.

Wind Energy

The Practicality of Wind Energy.   Atmospheric circulation involves massive exchanges of energy provided by the sun. On a global scale, winds are caused by the temperature differences between the earth's latitudes and deflection caused by the rotation of the earth (see Chapter 6). Dry air in the vicinity of 30 degrees north and 30 degrees south sinks and flows toward the equator where it replaces rising hot air. This pattern of air movement is called Hadley circulation. At midlatitudes, between 30 and 70 degrees latitude north and south, air flows toward the poles and is deflected westward, creating a wavelike pattern known as the Rosby circulation (Grubb and Meyer 1993). Many smaller circulation patterns are superimposed on the global systems as a result of regional variations in temperature and pressure. Within regions, mountains and valleys can intensify winds as they are channeled across the earth's topography. Large bodies of water can influence wind direction as well as intensity. Scientists use very sophisticated models to predict winds. For our purposes, we need only to know that wind energy can be captured by wind turbines which convert motion energy to electricity.

As discussed earlier, wind provided a practical source of energy long before the petroleum era. The remains of ancient sailing vessels recovered from the depths of existing lakes or marine shorelines by archaeologists attest to the importance of wind energy to much earlier civilizations. Some date back to 5000 B.C. Archaeologists have also provided evidence of the existence in China of simple windmills dating back at least 2000 years. On American soil, millions of windmills were erected during the late nineteenth century as the West was settled. They provided a simple but critical means of pumping water on farms and ranches. By the turn of the century, wind-driven systems generated electricity in small rural communities. By the 1930s, power grids extended into most of these communities and wind generators were abandoned. As rural electrification programs developed, only the most isolated windmills remained for pumping water.

Until the 1970s, there was little interest in reviving wind power. However, after the 1973 OPEC oil scare, wind, like other alternative energy sources, experienced a rebirth of attention. California became the focal point of wind energy development from 1981 to 1986. More than 15,000 utility-scale wind turbines were installed with a total peak capacity of 1,300 MW. Over the same period, small wind turbines were installed throughout the U.S., mainly in rural homes.

In response to the somewhat abrupt entry of government wind industry incentives, developers went to work quickly to bring wind turbines to the market. In the absence of advanced testing and experience, poorly designed machines were sold and installed. The reputation of the wind power industry suffered as machines broke down and failed to deliver the power predicted. Engineers soon found that sites had to be chosen with more selective criteria and that variable wind created unforseen stresses on equipment. In some cases, units actually flew apart, thus endangering onlookers. In addition to technical problems, there were dishonest salesmen who sought only the generous tax credits offered by a hopeful and trusting public. Only in recent years has the image of the industry been somewhat restored.

The late 1980s presented more challenges to the wind industry as tax credits were eliminated and fossil fuel prices fell. Consequently many companies went out of business. The early years of the auto industry were similarly marked by the gradual elimination of pioneering manufacturers.

The 1990s registered a major reversal in the wind industry's image. This occurred as wind generators became increasingly efficient and durable, and as capital and operating costs were reduced to the extent that wind could compete with other energy sources. By 1996, the cost of wind-powered electricity had dropped to only 5 cents per kilowatt-hour. Projections suggest that a further reduction to about 2 cents by 2010 might be attainable. The U.S. presently leads the world in installed wind capacity. In 1994, California had about 16,000 wind turbines and led all other states in electricity generation by a vast margin (Figure 20.8). Its wind power industry was built when natural gas prices and tax incentives were high. In fact, California contains less than 1% of the total U.S. wind energy potential and at least 16 states have a wind resource base that is equal or greater. Overall, wind is currently among the fastest growing renewable energy sources and this growth is occurring in numerous states.

FIGURE 20.8 Wind turbines on a "wind farm" in the Mojave Desert near Palm Springs, California. California leads other states in electricity generation from wind.

Wind Resources. The amount of wind energy theoretically available for use in the U.S. is enormous. Of course, only a small fraction of this resource can be exploited because of physical constraints on available land and the efficiency of energy extraction. However, the present amount of electricity generated from wind in the U.S. could be easily doubled.

Many regions of the country offer at least some usable wind resources. The Great Plains states have abundant wind resources, followed by other parts of the Midwest, the West, and the Northeast. Compared to other regions, the southern U.S. has less potential for wind energy development. The windiest areas in the U.S. tend to occur along the East and West Coasts, along the Rocky and Appalachian Mountains, and in a comparatively expansive belt embracing the Great Plains and 12 states. Kansas and Texas, followed by North Dakota, have the greatest potential power output for wind generating capability. Montana and Wyoming are also rich in wind resources. Overall, at least 37 states have defined potential for utility-scale wind energy development.

Wind power classes are used to express the wind energy available to specific areas. These range from class 1 (the least amount of energy) to class 7 (the greatest amount of energy). The classification process takes into account average wind speed and its variation over time, and the average density of the air. Wind speed is especially critical because wind energy production is proportional to the cube of the wind speed. This means that a 10% increase in wind speed will yield about a 30% increase in the output of a wind turbine. Because air density decreases with altitude, wind speeds at high altitude sites must exceed those at lower sites to be competitive for development (Rosenberg 1993). In general, areas identified as class 4 and above are regarded as potentially economical for power production with existing technology. Good wind areas cover about 6% of the contiguous U.S. land area. They have the potential to supply more than one and a half times the current electricity consumption of the U.S.

Factors that determine the feasibility of actually harvesting wind power go beyond wind factors. The availability of land for development and the distance to existing transmission lines are especially important. Unfortunately, some of the most promising concentrations of wind energy are found in areas having low population densities and are far removed from major transmission lines.

Wind Technology. Most modern wind turbines resemble traditional farm windmills while others follow a vertical-axis design causing them to resemble an overscaled eggbeater. The latter design was developed by the French inventor Darieus and windmills of this type bear his name. Wind turbines are deceptively simple machines consisting of blades, rotor, transmission, electrical generator, and control system, all mounted on a tower. The tower is needed to prevent topographical irregularities, trees, and buildings from impeding wind speed.

The present generation of wind turbines is characterized by advanced, aerodynamic blade shapes, use of highly advanced materials in blade construction, and sophisticated control equipment (Sorensen 1995). Most turbines are made with three aerodynamic, glass- or carbon-fiber blades 20–30 meters long.

The blades are mounted on a driveshaft, which usually incorporates a gearbox transmission to provide a constant rotor speed. The choice of three blades is a compromise between high power output and the stability and lifetime of the blades (Sorensen 1995).

Wind turbines are available in a variety of sizes that can be matched to power ratings for commercial and private demands. The largest models can have propellers that span more than the length of a football field. Such models can provide power to more than 1,000 homes. The largest wind turbine built in the U.S. today (Zond Z-40, 550 kW), stands over 140 feet tall, uses three blades each over 65 feet long, and in windy locations can produce as much energy each year as is consumed by over 200 homes.

As engineers and scientists gained experience in wind power generation, research attention was redirected to intermediate sizes, which could be more easily financed and maintained in clusters called "wind farms." Intermediate-sized wind turbines capable of producing 100 to 400 kilowatts of electricity serve the needs of the bulk electricity market. Wind farms can employ thousands of machines on towers 30 to 50 meters tall (Figure 20.8). Clustering takes advantage of unusually good wind sites, eases construction, and gives utilities an opportunity to more efficiently manage the operation. Electricity from wind farms is fed into the local utility grid and is distributed to customers just as it is with conventional power plants. In densely populated countries in western Europe, wind farms have been built offshore to overcome public objections to their presence in populated areas.

Small turbines generate 1 to 50 kW of electricity. Units having 1 to 10 kW capacity are ideal for generating electricity in rural or underdeveloped areas without an electricity grid. In developing countries, they compete favorably with photovoltaic systems, diesel generators, and grid extension. In the U.S., small wind turbines can be attractive to rural homeowners and farmers who want to reduce their electricity bills. A small home-sized wind machine has rotors between 8 and 25 feet in diameter and usually stands about 30 feet. It can supply the power needs of an all-electric home or small business. Small wind turbines require less maintenance than larger and more complex machines. In most small wind turbines the rotor drives the generator directly and gears subjected to wear are avoided, thus reducing maintenance.

In the early years of the wind power industry, turbine life span was considerably shorter than predicted and operating and maintenance costs were unacceptably high. By adopting more rugged designs and more advanced materials, systems became highly reliable. New designs improved the efficiency of energy capture while reducing stress on wind turbine components, including drive-train hardware. Vibration fatigue was greatly reduced and the addition of aerodynamic braking protected turbines, which tended to "run away" during high winds. As machine durability and performance improved, the cost of wind power declined greatly. Improvements in the manufacturing process played a major role in improving power productivity. As time progressed, companies became more specialized. Some began concentrating on blades or controls while others became industry leaders in tower and gear mechanisms. Only the more mundane parts of the process, such as foundation and site work, are left to local contractors (Sorensen 1995).

Future of the Wind Power Industry. Wind energy brings numerous benefits to the energy scene. Most obvious perhaps is renewability of the resource and the relative absence of nonpolluting features. Regarding domestic fuel availability, wind power adds diversity to the fuel mix and therefore serves as a hedge against supply interruptions. There are environmental concerns, but these appear trivial against the problems created by fossil fuels.

Concerning the economy, wind energy advocates cite the large numbers of jobs created by the wind industry, both in manufacturing and power plant construction and maintenance. Wind energy actually provides more jobs per dollar invested than any other energy technology (more than five times those from coal or nuclear power).

Two factors that have constrained wind industry growth are cost and the intermittent nature of wind energy supply. Even though the cost of wind power has decreased dramatically in the past 10 years, further reductions will be needed to offset financing difficulties. Wind technology requires a higher initial investment than fossil-fueled generators. About 80% of the cost is the machinery, with the balance being the site preparation and installation. In states where the public has insisted on including external costs, wind power plants have gained a solid footing.

Like solar energy, wind power flows are interrupted and vary in intensity. The value of electricity generated by a wind farm is therefore reduced if winds occur outside periods of peak load. Many experts believe wind power will continue to play an incidental role in the nation's electricity supply until reliable and low-cost energy storage becomes widely available. Some argue, however, that answers to our complex energy problems will not be found in a single energy source of any type and that wind energy can reduce our immediate dependence on fossil fuels. An additional constraint on wind energy growth is the distance of remote energy resources from power grids. Nevertheless, numerous states are able to harvest regionally concentrated wind resources and are in the process of constructing and planning plants.

The overall expectation is that technical and economic issues will be solved in at least some regions and that the contribution of wind power to our total energy demand will continue to grow. The Department of Energy forecasts a 600% increase in wind-energy use in the nation in the next 15 years. By the middle of the next century, wind could be producing 10% of U.S. electricity, or as much as hydroelectric dams produce today. Some fear that growth may expand to the extent that local communities will begin protesting wind farms as they would coal-fired powered plants. Public objections have centered on the loss of aesthetic appeal in natural landscapes, the noise created by wind turbine blades, and the threat posed by turbines to birds. Some have even raised the question of the influence wind farms might impose on global atmospheric circulation, but experts can reliably counter this argument.

On an international scale, countries such as the United Kingdom, Canada, Denmark, and Germany are also pursuing numerous wind projects. Also, the small wind turbine market is showing impressive growth in sales to developing countries.

Hydropower

Present Status. In 1900, 4% of our delivered energy supply (in thermal units) was in the form of hydroelectric power. Eighty years later it is still about 4%, although much larger in absolute terms (Hall et al. 1992). Today hydropower provides more than 50% (50.6% in 1996 of the nation's renewable energy (EIA 1997c). It accounts for slightly more than 10% of the electricity generated in the U.S.

Hydroelectric plants convert the potential energy of stored (or impounded) water into electric energy (Figure 20.9). Numerous hydraulic turbines generate electricity as they are spun by water discharged from a dammed river. In some cases the water used for power generation is returned from a lower to higher reservoir to make energy available during periods of peak electrical demand (Moreira and Poole

FIGURE 20.9   The Fort Randall Dam was built by the Army Corps of Engineers for power production and other uses (courtesy of U.S. Army Corps of Engineers).

1993). In effect, lower cost energy is reinvested to produce higher cost energy. Pumped storage plants, therefore, sometimes intentionally produce less but more valuable energy.

The Near Future of U.S. Hydropower Plants.   Hydropower generation offers several advantages over fossil fuels and provides positive benefits in terms of flood control and recreation development (see Chapter 7). However, it is unlikely that new hydropower plant production will pave the way to growth in the industry. This view is supported by the fact that most highly productive dam sites have already been built. Another constraint is the intensification of public objections to dam construction over the years. One objection is that dam construction reroutes natural flows and that valuable, and sometimes scenic, natural ecosystems are lost in the process. Species can also be lost or threatened. Another objection stems from conflicts over the water resource itself. As water is impounded, some is lost in the form of reservoir surface evaporation and lands once fed by natural flows are cut off from their critical water source. The evaporation of water is directly related to the surface area of the body of water and varies with temperature, wind conditions, and humidity. Evaporative losses are, in fact, a form of consumptive water use, which further aggravates water conflicts among agricultural users, industrial users, commercial users, and nature advocates (Gleick 1994).

Today, public resistance poses a challenge to the relicensing of existing dam sites as well as new dam construction. All considered, growth in hydropower production will most likely occur in the form of modifying existing plants to improve operational efficiency. Many dams built in the past were not equipped to their maximum potential because of insufficient local demand for power. Also, in the past, economic disincentives existed for long distance electric power transmission. Many old facilities were retired during times of less expensive energy. Other dams were built solely for flood control, irrigation, and recreation purposes. With the increased dollar value of electricity and new automatic equipment, it is now economically feasible to equip such sites with power generation equipment and to restore or rebuild old power equipment at retired sites (Hall et al. 1992).

The International Scene.   Numerous developed nations face similar constraints in new dam construction. In Canada, for example, the adverse impacts of hydroelectric development, including displacement of population from flood plains and disruption of fisheries, have led to the suspension of a number of major projects in recent times.

However, public resistance increased after Canada had extensively developed its hydroelectric power resources.

In contrast, undeveloped countries are actively developing hydropower and two-thirds of all energy growth is projected to occur in newly industrializing economies. In Central and South America, hydropower is the dominant source of electricity generation. In 1995, more than 80% of the region's net electricity consumption was attributed to hydropower. In Latin America, 85% of growth in electric capacity from 1990 to 2000 was from hydropower. More dams are being built in Brazil where hydropower already provides 95% of its total energy. Argentina plans to add several small- to medium-scale hydroelectric projects in the near future. Closer to home, Mexico's hydroelectric capacity could double by the year 2010.

The greatest growth is now occurring in Asia and other regions that have seen little dam construction to date. Currently, more than 150 hydroelectric projects are under construction in China. The most prominent is the $30 billion Three Gorges Dam project. Construction began in 1994, and is expected to be completed in 2009. However, the project has been severely criticized in China and abroad and financing has been difficult. Nevertheless, work continues at that site and at several others being constructed on a much smaller scale. India has also initiated dam building at an aggressive pace, but again not without public protest (e.g., the Narmadu dam project). The environmental issues and high costs involved in developing India's hydropower resource make it unlikely that substantial development will occur there. Both China and India depend more on coal for electricity generation than other developing Asian nations. Nepal cites environmental problems as the primary reason for its desire to build a hydroelectric station on the Arun River. The Nepalese government argues that forests, their main energy source, are being depleted and that its extensive hydroelectric potential can be employed to protect its environment while furthering economic development. Numerous other Asian countries are building dams. Hydroelectric projects are planned for the Borneo rain forest (Malaysia), Vietnam, Cambodia, and Thailand. In Africa, hydropower advocates promote the abundant hydroelectric potential existing in Sub-Saharan Africa.

Objections to dam construction in developing countries often center on resettlement issues. Too often the people forced to abandon their homes and land are given insufficient voice in how resettlement is to occur. Consequently, the adverse social and economic effects imposed on the people affected tend to be underestimated (Gutman 1994).

Biomass

Biomass represents a broad category of energy sources and has been the predominant energy source for all but a fraction of recorded history. Biomass includes all forms of nonfossil plant and animal matter that can provide energy. Entire dried plants, plant parts, plant oils, animal fat, animal dung, and other forms of biomass can serve as energy resources (Figure 20.10). The energy obtained from biomass can be used for numerous purposes including heat production, electricity generation, and transportation fuel.

In 1996, biomass supplied 43% of the renewable energy consumed in the U.S. (EIA 1997c). Biomass energy use steadily declined in the U.S. until the 1960s, but since then it has expanded to equal about 4% of the energy demand. On a global scale, biomass ranks fourth as an energy resource, providing about 15% of the world's energy needs (Larson 1993). Biomass energy is especially important to the developing world. For example, cooking with wood accounts for about 60% of all energy consumed in the African Sahel (Aubrecht 1995).

Sources of Biomass Energy.   Biomass energy resources can be divided into two basic categories: (1) forest or agriculture derived plant materials, and (2) animal or plant by-products. These will be discussed.

FIGURE 20.10   Animal dung is an important energy source in some developing countries (courtesy of Agency for International Development).

Energy from forests: When home heating bills began to soar in the 1970s, U.S. households near forested areas began using wood energy more aggressively. Wood-burning stoves regained popularity and were modified to improve home heating efficiency. At the same time, the wood products industry began using more and more of its own waste products to meet both its heating and electrical needs. Today, the paper and pulp industry satisfies more than half of its energy needs using mill wastes. Growth in this type of activity is largely responsible for the gains in biomass energy consumption in the U.S. in recent decades.

Energy can be obtained from forests and plantations in numerous ways (Graham et al. 1993). Logging residues composed of tree tops, larger branches, and dead trees represent one of many forest energy resources. Only about 60% of the production of wood in the U.S. is harvested. The rest, which eventually falls to the ground and de-composes, represents a potential energy resource. Timber-producing forests and plantations are often thinned to improve the growth and quality of remaining trees. The thinnings have limited lumber value but can supply energy. As harvested timber goes through saw mills some material is lost as edgings or uneven slab material. This and the sawdust produced contain useful energy. As lumber reaches milling plants, more of the original material is lost but can be recovered for energy. Mill residues are obtained from a host of industries, including construction lumber plants, furniture makers, and pallet plants. Wood that makes up about 40% of total construction and demolition wastes can provide energy when recovered in waste-to-energy plants. About one-third to one-half of all hardwood lumber consumed in the U.S. is used to construct pallets for shipping. These pallets can provide a concentrated energy source when they wear out. In some cases, trees are grown in intensively managed plantations for the purpose of producing energy. Species suited to short rotation energy production are fast growing and include—but are not limited to—poplars, willows, and eucalyptus.

Energy from agriculture: Corn and other grain crops can supply plant sugars, which can yield alcohol as they ferment. Distillation is used to further concentrate ethanol, which can be used as a biofuel. Euphorbia and sunflowers are among plants that produce oils that can be used in a diesel engine.

Crop- and food-processing wastes: Energy can be obtained from semiprocessed agricultural and food-processing wastes. Bagasse, a crushed plant waste product created in large quantities in sugarcane processing regions, can be an important energy source. By burning bagasse residues, the sugar industry captures energy for plant processing and avoids open-air burning, which fouls the atmosphere. Bagasse supplies about 9% of Hawaii's electricity. Waste can also be used to produce biofuel as is done when cheese wastes are converted to alcohol. The main advantage of these approaches is that the energy resource is already concentrated and energy processing substantially reduces industry pollution. With ample raw material, energy processing costs are mostly limited to construction of an energy plant. Theoretically, corn and other grains could supply around 3 billion gallons of ethanol annually.

Municipal and industrial waste: About half the material contained in municipal and industrial wastes is combustible. The heat content of this material is only about a third that of high quality coal, but this waste is abundant. Because wastes are often more expensive to dispose than to burn, energy extraction can be economically rewarding. Also, any practice which reduces the bulk of waste material otherwise sent to landfills potentially benefits the environment. Almost 140 waste-to-energy facilities were in operation in 1990. Most of these were mass-burn facilities—which consume raw waste—but some used what is called refuse-derived fuel (RDF). This is essentially chopped-up or pelletized waste with most of the metal, glass, and other inorganic matter removed. In contrast to unsorted waste, RDF can be burned in conventional boilers and contains fewer toxic materials that can enter the air. RDF can also be burned with wood chips, coal, or both, in cofired facilities. The pulp and paper industry is the leading user of wastes for energy in the U.S.

Capturing biomass energy: Biomass-based power systems can be applied to a diverse set of energy needs. Biomass can be used for generating power for individual households or village communities, for electricity or steam generation in agricultural applications, or for utility-scale electricity generation. Some biomass systems are rather simple while others are more complex. New types of systems are being developed for industry and transportation as proven methods continue to be improved.

Electricity from biomass: Power generation is perhaps the most readily accessible route to the large-scale expansion of biomass for energy. The technologies for biomass conversion to electricity are direct combustion, gasification, and pyrolysis. Using forced air for more thorough oxidation, direct combustion generates hot flue gasses, which are used to produce steam in the heat exchange sections of boilers. Gasification and pyrolysis have been developed to produce different types of fuels from carbon-rich biomass, which can improve power generation. Gasification was discussed earlier as a means of processing coal energy. The biomass gasification process is similar to coal gasification, described in Chapter 19, but instead involves the conversion of the carbon contained in wood to Syngas. The mixture of carbon monoxide and hydrogen contained in Syngas can be used for either space and industrial process heat or for powering gas turbines, which generate electricity. Biomass gasification/generating systems could be less expensive and more efficient than conventional wood-fired, steam-electric power plants. Pyrolysis is defined as the thermal destruction of organic materials in the absence of oxygen. In pyrolysis processes, indirect heating is also used to convert biomass to a mixture of gasses and organic vapors. Finally, a number of major utilities are evaluating the co-firing of biomass in existing coal-fired power stations because of the environmental benefits that may accrue.

The biomass used to fuel these processes can come from a number of sources, including herbaceous as well as woody plants grown specifically for energy production. For efficiency, the biomass needs to be somewhat uniform to reduce handling, storage, and processing costs. "Energy plantations" of fast growing tree species have received considerable attention in this regard.

Biofuels

Ethanol: Ethanol can be produced by biologically catalyzed reactions. In much the same way that sugars are fermented into beverage ethanol by various organisms, such as yeast and bacteria, sugars can be extracted from sugar crops, such as sugar cane, and fermented into alcohol. For crops such as corn, starch is first broken down to simple glucose sugars by acids or enzymes known as amylases (Wyman et al. 1993).

Brazil became the world leader in ethanol production when it began using sugar from its sugar industry to offset rising trade deficits associated with imported oil (Goldemberg and Monaco 1993). Ethanol consumption equaled gasoline consumption on a volume basis in 1990. The majority of new cars sold in Brazil today are designed to use ethanol (Larson 1993).

Today, two forms of ethanol are produced from biomass: anhydrous (100% ethanol) and hydrous (containing about 5% water). Anhydrous ethanol can be mixed with gasoline to create gasohol, a product sold at filling stations. Gasohol can contain up to a maximum ethanol content of 20%. The addition of ethanol actually boosts fuel octane rating. Hydrous ethanol must be used in engines designed specifically for ethanol. The amount of anhydrous ethanol currently produced in the U.S. corresponds to about 8% of total gasoline used in the U.S. (Larson 1993). Grain surpluses, principally corn, are the main source of feedstock for the ethanol industry.

Cellulosic ethanol is one of the most promising technological options available to reduce transportation-sector greenhouse-gas emissions (Lynd 1996). Without government intervention, it is unlikely that ethanol would have evolved as a transportation fuel in either the U.S. or Brazil. A tax exemption, grain production subsidies, and government-sponsored research sheltered ethanol development in the U.S. In Brazil an abundant supply of sugar cane and high import oil costs drove the government to find a domestic fuel substitute that remains comparatively strong.

Biological gasification (anaerobic digestion):   Biogas (roughly half methane and half CO_2) is produced by the biological process of anaerobic digestion of wet organic matter. Almost any biomass except lignin (a component of wood) can be converted to biogas. Anaerobic digesters are made out of concrete, steel, brick, or plastic. In developing countries, biogas generators, fed with dung and other waste, provide village energy. In industrialized countries, animal wastes, crop residues, carbon-laden industrial processing by-products, and landfill material have been widely used to provide power and eliminate waste. China is served by millions of household and community digesters (Levine et al. 1992). An estimated 5,000 digesters are installed in industrialized countries (Larson 1993). These are typically located at large livestock processing facilities (stockyards) and municipal sewage treatment plants. Generally, anaerobic digestion is economical for on-site power production wherever sufficient wastes are available and there is a ready demand for power (Brower 1992).

The Future of Biomass Energy.   The advantages and disadvantages of biomass energy are as diverse as the energy applications themselves. Many of the approaches discussed benefit the environment by displacing the use of polluting fossil fuels. By extending our energy menu biomass fuels provide hope for finding renewable fuel substitutes. On an industrial scale, the use of plantations and agriculture crops offer several benefits. Plantations can be grown on lands that are marginally productive for agriculture and biofuel grain crops can provide much needed economic benefit to rural communities. Biofuel and plantation energy advocates point to the opportunity to produce energy on some of the lands presently excluded from production by government programs. Today, roughly 33 million hectares of cropland are idle in the U.S. The U.S. Department of Agriculture projects that some 52 million hectares will be idled by 2030. Replacing annual row crop production with longer-rotation woody crop production might help revitalize worn-out agricultural soils. Regarding atmospheric CO_2, energy plantations would not result in a CO_2 net gain because this greenhouse gas would be incorporated into new plant growth as it is combusted.

Critics approach the subject of industrial scale biomass production with less optimism. They cite the potentially polluting effects of applying more fertilizers and herbicides to both grain and tree crops, and problems associated with the expansion of crop monocultures. They also cite the narrow profit margins associated with energy crops when all energy costs are correctly tallied (Larson 1993). Some also question the use of grains for fuel when millions go hungry around the world.

The biofuel industry would actually be better served if plant material other than grain crops could be used. Biofuel would probably be cheaper and would rest on a more stable economic platform because grain prices fluctuate. However, plant parts containing relatively more complex forms of carbons simply do not ferment satisfactorily. Biotechnology may one day provide the ideal mix of microbes for producing ethanol from lignocellulosic biomass, but this presently is not the case. The practice of converting waste to energy has been criticized because of pollution and site location issues. The use of biogas digester technology is, perhaps, least objectionable because it provides the most basic energy needs to impoverished millions. One could say that even dung should be put to a higher purpose. However, it does not lose its crop nutritive value when used for biogas generation and can be withdrawn for fertilizing subsistence crops.

Overall, the future of biomass energy is tied to comparative cost advantages that are likely to occur as petroleum energy costs climb and as environmental issues are resolved on a case-by-case basis. In both developing and industrialized countries, biomass is used much less efficiently than is technically possible and economically feasible. Most likely, gasification practices will continue to replace less efficient burning practices. Further technology improvements will continue to reduce pollution problems which, at present, are much less serious than those associated with fossil fuels. In comparison, wood and other biomass fuels contain less sulfur. Technology will also overcome many critical biofuel conversion limitations. In the long run, the availability of land will be a primary constraint. Reasonable estimates assume that crop residues and energy plantations by the year 2050 can supply 30% of the world's energy needs (Larson 1993).

Geothermal Energy

Geothermal Status. Beneath the earth's surface, temperature increases 1°F for every 75 feet increase in depth. In certain areas, however, temperatures rise much more quickly owing to subsurface irregularities that have brought molten material closer to the crust. Such areas contain enormous amounts of heat which can be used as a reliable energy source. The source of the heat—or geothermal energy—is molten rock in the earth's interior and the radioactive decay of material contained beneath earth's crust. America's first inhabitants, Paleo-Indians, were first to use some forms of geothermal energy. They warmed themselves in hot springs, which occur in areas where groundwater heated by the earth's interior flow to the surface in pools of bubbling water. They, as well as early European explorers, surely took note of the spectacular geyser displays, or jets of water and steam which, like hot springs, occur in areas where the earth's subsurface heat is especially close to the surface. Such areas are called hydrothermal convection zones.

Heated groundwater trapped by impervious rock layers awaited detection by those first able to drill deep enough to release the steam and water superheated by the molten rock beneath. Areas containing trapped geothermal energy are called geopressurized zones. It was eventually determined that energy was also potentially acquired from hot rock zones where magma heats overlying rock.

Most geothermal energy today comes from hydrothermal convection zones because the near surface energy source can be developed without elaborate engineering. At several well known locations throughout the world, steam or hot water from such zones is a primary energy source for an entire community. In Reykjavik, the capital city of Iceland, energy captured from a hydrothermal convection zone is used to heat homes, buildings, public baths, and greenhouses.

Power Production.   The steam or hot water provided by geothermal energy can be used to generate electrical power as was first demonstrated in Larderello, Italy in 1904. Geothermal electricity is now produced in numerous countries throughout the world, including New Zealand, Japan, Mexico, and the Philippines as well as the countries already mentioned. It is no coincidence that these countries have regions of overlying areas known for their volcanic activity. Geothermal energy originates in the earth's interior, where the hottest fluids and rocks at accessible depths are associated with recent volcanic activity. One of the world's largest geothermal projects, "the Geysers," is located in northern California on the slope of an extinct volcano.

Geothermal resources occur as hydrothermal fluids, hot dry rock, geopressured brines, magma, and ambient ground heat. To date, only hydrothermal fluids have been developed commercially for power generation.

Hydrothermal Fluids.   Three technologies are used to convert hydrothermal fluids to electricity. The method applied in each case is determined by the physical state of the heated water available. If the fluid is wholly or mostly steam, conventional steam turbines are used to generate power (Figure 20.11). When the fluid is above 200°C—and is primarily water—flash steam technology is usually employed. This method involves spraying the fluid into a tank held at a much lower pressure than the fluid. This causes some of the fluid to vaporize, or flash, to steam. Once converted to steam, the fluid can be used to power turbines and generate electricity. For water less than 200°C, a secondary or working fluid must be used to convert the energy contained in the water. The hot geothermal fluid vaporizes the working fluid which then drives a turbine and generator.

Steam resources are the easiest to harness for power generation, but are rare. Only one steam field, the Geysers, has been commercially developed in the U.S. It follows that hot water plants—using high- or moderate-temperature geothermal fluids—are more common and are the major source of geothermal power in both the U.S. and the world. At present, hot-water plants operate in California, Hawaii, Nevada, and Utah.

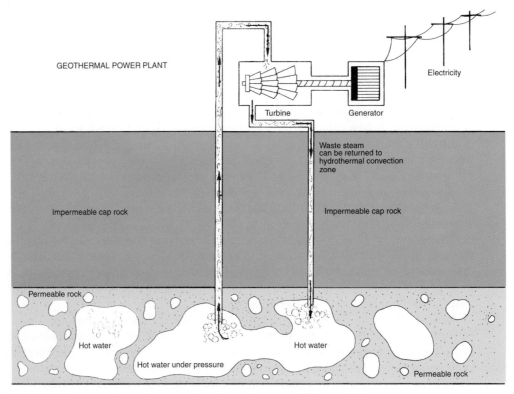

FIGURE 20.11   A geothermal power plant (from Owen and Chiras 1995).

Low- to moderate-temperature (20°C to 150°C) geothermal resources are widespread in the U.S. and are used to provide direct heat for homes and industry. By definition, direct use of geothermal resources does not involve power plants or heat pumps. The heat provided is used directly for warming buildings and greenhouse, for industrial processes and a host of other purposes. Direct-use projects generally rely on resource temperatures between 38°C to 149°C. District heating supplies multiple users with hot water from a utility plant or well field. Heat pumps can also be used to acquire energy from low to moderate temperature geothermal resources. Ground-source heat pumps take advantage of the relatively constant temperature of the earth's interior as a heat source in winter and a heat sink in summer.

Geopressured Brines and Magma. Geopressured brines are hot, pressurized, methane-rich waters found in layers of sandstone 10,000 to 20,000 feet beneath the surface (Carless 1993). Geopressured systems may one day provide a source for power generation. However, the great depth of these sources and pressures involved require engineering technology that is not cost effective at present. About 20% of the geothermal energy found in the U.S. exists in the form of geopressured reservoirs.

Hot Dry Rock. Some areas sit above subsurface geologic formations that are abnormally hot but contain little or no water. To unlock this energy source, water must be pumped into hot rock zone zones then pumped to the surface for heating or electrical power generation. Hot, dry rock resources may supply a significant part of U.S. electric power needs for centuries once technology is developed to make the engineering practices needed commercially acceptable.

The Pros and Cons of Geothermal Energy. Compared to conventional energy sources, geothermal energy offers several advantages. Geothermal electricity is clean, reliable, and—depending on location—cost effective. Geothermal power is cleaner than coal-fired and nuclear power plants. The economics of geothermal power can be attractive because less capital investment is needed. Compared to coal-fired plants, geothermal plants can be 40% less expensive. The capital needed for a nuclear power plant can be three times greater. In addition, geothermal power plants are more reliable than conventional power plants. The new steam plants at The Geysers are operable more than 99% of the time. Finally, the many forms of geothermal energy represent a resource that is both abundant and secure.

Criticisms of geothermal technology center on the chemical contaminants it releases into the environment. Geothermal steam and water frequently contain salts and other contaminants, such as sulfur compounds, which are leached from bedrock minerals. The gaseous pollutants sometimes occurring in the steam—such as hydrogen sulfide, carbon dioxide, ammonia, and methane—require pollution control devices. The corrosive properties of geothermal fluids create both engineering and environmental problems that raise concerns and project expense.

Another limiting feature is the incapacity of steam to retain heat over long transportation distances. This explains why district heating projects are not more geographically extensive. The location of geothermal resources is another issue. Geothermal resources are scattered throughout the more sparsely populated West. Finally, geothermal energy may not always be sustainable. The Geysers in California and a project in New Zealand have both suffered serious losses of power production energy over time. Such cases have negative impacts on the industry's investment image.

The Future of Geothermal Energy. In the U.S., geothermal electric power generation totals about 2200 MW or about the same as four large nuclear power plants. Overall, the current production of geothermal energy from all uses puts this resource third among renewables. Its behind hydropower and biomass but ahead of solar and wind.

Compared to conventional energy, the potential for geothermal energy has been barely tapped and geological exploration continues to locate new resources. Although its development will be confined somewhat within geographic boundaries, geothermal energy represents a significant source for power generation and heat. Also, the cost of generating geothermal energy is anticipated to continue a decreasing trend as technology improves.

Additional Renewable Energy Sources

Tidal Power.   Massive amounts of energy are potentially available from the sea. As we watch large waves pounding a coastline we witness, for example, the action of tremendous forces. The energy potentially available from tidal changes is less striking but, along with wave energy, attracts the attention of energy experts. The technology used to harness tidal energy is rather simple and resembles engineering approaches used to generate electricity from major rivers. A feature that is critical to both is the ability to capture energy from massive amounts of falling water. It follows that both approaches require dams to be constructed. To obtain energy from tidal surges, a dam is built across the mouth of a bay and turbines are mounted in the structure. The in-rushing tide generates power as it passes through turbines. As the tide shifts, the blades can be reversed to capture the energy of the return flow.

The world's first tidal-electric installation was built in 1966 in the La Rance estuary in France. A similar plant was constructed in Russia. The incentives for building such plants are obvious. Tidal power production does not require massive capital, produces no pollutants, and is sustainable. However, the sites capable of producing tides high enough to justify power development are scarce. Only about 24 good sites exist worldwide. To be useful for power production, shoreline topographic features must channel the tide to superior heights. The only location in North America with impressive tides is the Bay of Funday. Here shoreline features channel a tide that approaches 16 meters, the largest in the world. Simply too few sites have such massive tides. Throughout most of the world, the small difference existing between low and high tide (less than two feet) is less than needed for driving turbines. Also, there are environmental and practical concerns. Depending on location, tidal dams can interfere with fish migrations, impede boat traffic, trap sediments, and alter the critical balances of fresh- and saltwater in estuaries. Overall, tidal power is not likely to become a significant contributor to our energy future.

Ocean Thermal Energy Conversion.   The ocean stores vast amounts of energy received from the sun. Ocean thermal-energy conversion (OTEC) technology is directed toward recovery of this energy for power production. The basic aim of OTEC is to exploit the steep temperature gradient found in tropical waters. Project engineering involves floating platforms that draw water from the warmest and coldest depths the location can offer. Pipes extending to a depth of 900 meters (3,000 feet) can be used to exploit cold water. The warm water pumped from immediately beneath the surface platform is used to convert ammonia into a gas, which drives the blades of an electric turbine. The cold water pipe supplies the cooling needed to recondense the gas so that the energy exchange cycle is maintained. Ammonia, the working fluid, is needed because of its low boiling temperature, which is much lower than water. The electric power produced from an OTEC plant can be transferred to shore for various end uses. Another approach, the open or Claude cycle, uses warm seawater to create steam. The warm seawater is sufficient to create steam only because the system employs a vacuum vessel as the evaporator. Because the pressure maintained in the vessel is only 0.03 times atmospheric pressure, the boiling temperature of water is lowered and low pressure steam is produced.

Although OTEC offers the possibility of sustainable energy production, there are serious limitations. Operational efficiency is a key issue because much of the energy obtained must be reinvested in pumping large amounts of water through the system. For example, an experimental plant constructed off the shore of Hawaii used 80% of the energy generated to pump water to the platform. Another limitation is the relative scarcity of locations suitable for positioning OTEC plants. Suitable locations for U.S. development include the Gulf Coast and waters near the shores of Hawaii, Guam, and Puerto Rico.

Again, there are environmental concerns. As cold waters are heated at the OTEC platform, the dissolved carbon dioxide they once contained is released to the atmosphere. Localized effects potentially include disturbance to fisheries and disruptions in the marine food chain stemming from the redistribution of nutrients found in deep waters. The prospects for OTEC therefore seem limited in the broad scope of sustainable energy sources. Extreme care will be needed to avoid ecological disturbances where engineering criteria can be satisfied.

Hydrogen

Status of hydrogen: Hydrogen is believed by many energy experts to be one of the main renewable energy fuels of the future. Its renewability derives from the fact that it can be produced from energy provided by solar, biomass, and other sustainable energy resources. At present, hydrogen is produced primarily by steam reforming of natural gas. Electrolysis is used to obtain extremely pure hydrogen for various pharmaceutical and industrial applications. This process employs electricity to dissociate, or split, water into its hydrogen and oxygen components. Both of these methods are too costly for wide energy applications.

As a renewable fuel, hydrogen may one day be produced on a commercial scale by renewable technologies that can be grouped into three types of processes: photobiological, photoelectrochemical, and thermochemical. Photobiological and photoelectrochemical processes generally would use sunlight to split water into hydrogen and oxygen. Thermochemical processes, including gasification and pyrolysis systems, would use heat to produce hydrogen from sources such as biomass and solid waste.

Some obstacles must be overcome before photobiological technologies can become commercially viable. Most photobiological systems use the natural activity of bacteria and chlorophyll-containing green algae to split water and produce hydrogen (Sandia Laboratories 1995). Current research efforts center on developing strains of microbes, which more efficiently channel the energy obtained from the sun to the reaction responsible for splitting water to produce hydrogen. Photoelectrochemical (PEC) methods employ photovoltaic technology to acquire the energy needed and their energy conversion efficiencies are being improved. Thermal processes such as fast pyrolysis are being tested as a means of using renewable biomass to produce hydrogen. When biomass is heated to moderately high temperature and is vaporized, the vapors can be catalytically reformed to produce hydrogen.

On combustion, hydrogen does not produce particulates, carbon dioxide, and sulfur oxides. Instead, combustion yields energy, water, and a small amount of nitrogen dioxide which is produced when the heat produced causes oxygen and nitrogen in the air to combine. Hydrogen has numerous potential applications, including electrical power generation, space heating, and transportation. Furthermore, the versatility of hydrogen will enable sustainable energy technology to advance on the domestic energy scene more rapidly than would be the case if electricity were the sole means for carrying renewable energy from source to end use.

Hydrogen as a transportation fuel: The ability of hydrogen to serve as a transportation fuel has been demonstrated for many years in the space industry. Liquid hydrogen and liquid oxygen are combined as propulsion fuel for the space shuttle and other rockets. Onboard fuel cells using hydrogen and oxygen also provide most of the shuttle's electric power. Most of the hydrogen used to fuel the nation's space shuttle program is made from natural gas by the steam reforming process. The technology involved is not directly applicable to automobiles. Nevertheless, hydrogen will power pollution-free vehicles of the future. All of the technologies necessary to achieve this goal have been proven in concept. Work is underway to improve the practicality of energy storage and to reduce the costs of the technology for renewably generated hydrogen. In the interim, hydrogen will be increasingly used on a limited scale as an automotive fuel and to demonstrate technological if not economic feasibility. Advanced internal combustion engines will power ultraclean hydrogen transportation vehicles until efficient fuel cells become commercially available. Fuel cells are devices that convert hydrogen gas directly into low-voltage, direct-current electricity. The use of liquid hydrogen to fuel commercial jetliners has already been demonstrated. For cities with heavy jet traffic, liquid hydrogen will contribute substantially to cleaner air.

CONCLUDING REMARKS

As discussed, there are numerous options for producing future energy in a sustainable manner (Figure 20.12). This should allay fears that our lives will be greatly affected as the fossil energy age draws to a close over coming decades. The primary issue before us is how to phase in renewable technologies. Some argue that fossil fuel should be heavily taxed to discourage greenhouse gas emissions. To support their arguments, they cite the external costs of a fossil-fuel economy, which are rarely addressed. Renewable energy advocates have also objected to the way fossil and nuclear fuels concentrate capital and political clout. They prefer a less centralized approach that is more user friendly. The best arguments for a decentralized energy strategy were expressed two decades ago as Lovins (1977) proposed a "soft path" to energy sufficiency. Overall, many of Lovins's arguments remain valid even today. His arguments against nuclear energy appear even stronger now because major accidents have occurred since his remarks were published. Another source of complaint among soft path advocates is the rather low level of financial support provided by our government. In the wake of the 1970s oil crisis, renewable energy has provided much more government financing. In retrospect, these funds were generally well spent. The Department of Energy continues to be at the forefront of renewable energy development, albeit with less resources at hand to foster key demonstration projects.

Proponents of free market control over energy issues see private enterprise and global trade as effective means to energy security. They strenuously object to taxation approaches, which stifle corporate growth and in the long run result in higher costs of consumer goods and services. At the most fundamental level, free market advocates believe that energy scarcity will create opportunities for profit and that solutions will be found and financially rewarded. It might be argued that the recent expansion of British Petroleum into the photovoltaic power sector provides a hint as to what the future holds for us. Major corporations have the resources needed to invest in renewable technology development and to accelerate the transition from fossil fuels. Ultimately, this type of response will not achieve Lovins's (1977) aim of decentralizing economic power. However, the involvement of large corporations raises the hope that development will proceed on a scale that can mean a real difference to environmental quality.

On balance, renewable energy use shows promise for continued growth and provision of numerous future options. However, no single technology is likely to eclipse all others as a viable energy option. We are more likely to see approaches mixed in various ways to take advantage of local energy options (Johansson et al. 1993). The role of hydrogen as an energy carrier could prove an especially important means for meeting our energy needs, including transportation. Overall, progress being made in the renewable energy sector points to realistic opportunities for reducing carbon emissions. According to Ausubel (1996) the world will progressively burn less and less fuel with high carbon to hydrogen ratios. Therefore, decarbonization of the energy systems will probably continue. His analyses imply that " . . . during the next 100 years the human economy will clear most of the carbon from its system and move, via natural gas, to a hydrogen metabolism."

Concerning carbon emissions, recent projections of the Energy Information Administration are less optimistic. By their estimates (EIA 1997d), energy use patterns

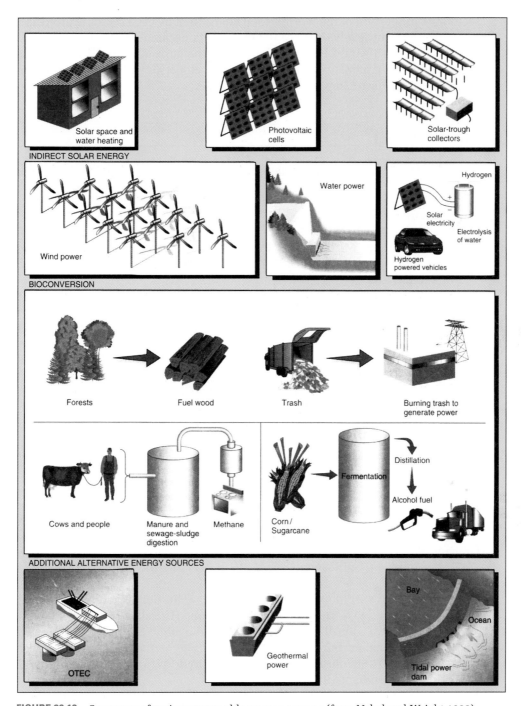

FIGURE 20.12 Summary of various renewable energy sources (from Nebel and Wright 1998).

common to developed countries will spread to newly developing nations (especially those in Asia). Energy consumption will not be restrained by energy costs, which are projected to remain comparatively low. In aggregate, fossil fuel use is expected to grow at the same rate as world energy use over the next 20 years, with natural gas gaining share relative to oil and coal. By the year 2015, oil use will total about 99 million barrels per day, up more than 32 million barrels per day relative to 1993. All forms of energy production except nuclear power are expected to grow over the projection period. However, the use of woody biomass for sequestary carbon may reverse the development of woody biomass for fuels. Even so, renewable energy sources are expected to provide more power in coming decades. Because of growth in most every sector, the proportion of the world's total energy met by each sector is projected to remain near present levels. If these developments became reality, worldwide carbon emissions in 2015 would be 50% higher than the current level. Clearly, the stage is set for some major adjustments in the energy sector if global warming concerns continue to escalate (see Chapter 6).

LITERATURE CITED

Aubrecht, G. J. II. 1995. 2nd edition. *Energy.* Englewood Cliffs, NJ: Prentice-Hall.

Ausubel, J. H. 1996. Can technology spare the earth? *American Scientist* 84:166–178.

Bevington, R. and A. H. Rosenfeld. 1990. Energy for buildings and homes. *Scientific American* 263:77–86.

Brower, M. 1992. rev. ed. *Cool energy: Renewable solutions to environmental problems.* Cambridge, MA: The MIT Press.

Carless, J. 1993. *Renewable energy: A concise guide to green alternatives.* New York: Walker and Co.

Dracker, R. and P. De Laquil III. 1996. Progress commercializing solar-electric power systems. *Annual Review Energy Environment* 21:371–402.

Energy Information Administration. 1997a. Emissions of greenhouse gases in the United States 1996. *Annual energy review 1996.* DOE/EIA-0573(96), 142 pp.

Energy Information Administration. 1997b. *Annual energy review 1996.* DOE/EIA-0364(96).

Energy Information Administration. 1997c. *Renewable energy annual review 1996.* U.S. Department of Energy. DOE/EIA-0603(96), 192 pp.

Energy Information Administration. 1997d. *Early release of the annual energy outlook.* (AEO98).

Gleick, P. H. 1994. Water and energy. *Annual Review Energy Environment* 19:267–299.

Goldemberg, J. and L. C. Monaco. 1993. The Brazilian fuel-alcohol program, (pp. 841–863). In T. B. Johansson, H. Kelly, A. K. N. Reddy, R. H. Williams, and L. Burnham, eds., *Renewable energy: Sources for fuels and electricity.* Washington, DC: Island Press.

Graham, R. L., L. L. Wright, and A. F. Turhollow. 1993. Short rotation woody crops: Opportunities to mitigate atmospheric carbon dioxide buildup. *Climate Change* 22:223–238.

Grubb, M. J. and N. I. Meyer. 1993. Wind energy: Resources, systems, and regional strategies, (pp. 157–212). In T. B. Johansson, H. Kelly, A. K. N. Reddy, R. H. Williams, and L. Burnham, eds., *Renewable energy: Sources for fuels and electricity.* Washington, DC: Island Press.

Gutman, P. S. 1994. Involuntary resettlement in hydropower projects. *Annual Review Energy Environment* 19:189–210.

Hall, C. A. S., C. J. Cleveland, and R. Kaufmann. 1992. Energy and resource quality. *The economy and economic process.* Niwot, CO: University Press of Colorado.

Johansson, T. B., H. Kelly, A. K. N. Reddy, and R. H. Williams. 1993. Renewable fuels and electricity for a growing world economy: Defining and achieving the potential. In T. B. Johansson, H. Kelly, A. K. N. Reddy, R. H. Williams, and L. Burnham, eds., *Renewable energy: Sources for fuels and electricity.* California: Island Press.

Larson. E. 1993. Technology for electricity and fuel production from biomass. *Annual Review Energy Environment* 18:567–630.

Levine, M. D., F. Liu, and J. F. Stinton. 1992. China's energy system. *Annual Review Energy Environment* 17:405–435.

Lovins, A. B. 1977. *Soft energy paths: Towards a durable peace.* Cambridge, MA: Ballinger Publishing.

Lynd, L. R. 1996. Overview and evaluation of fuel ethanol from cellulosic biomass: Technology, economics, the environment and policy. *Annual Review Energy Environment* 21:403–465.

Moreira, J. R. and A. D. Poole. 1993. Hydropower and its constraints, (pp. 73–119). In T. B. Johansson, H. Kelly, A. K. N. Reddy, R. H. Williams, and L. Burnham, eds., *Renewable energy: Sources for fuels and electricity.* Washington, DC: Island Press.

Nebel, B. J. and R. T. Wright. 1996. *Environmental science.* 5th edition. Upper Saddle River , NJ: Prentice-Hall.

Nebel, B. J. and R. T. Wright. 1998. *Environmental science.* 6th edition. Upper Saddle River, NJ: Prentice-Hall.

Owen, O. S. and D. D. Chiras. 1995. *Natural resource conservation.* Upper Saddle River, NJ: Prentice-Hall.

Rosenberg, P. 1993. *The alternative energy handbook.* Lilburn, GA: The Fairmont Press.

Sandia National Laboratories. 1995. Advanced hydrogen technologies. U.S. Department of Energy. DOE/GO–10095–065.

Sorensen, B. 1995. History of, and recent progress in, wind energy utilization. *Annual Review Energy Environment* 20:387–424.

Thurlow, G. 1990. Implications for policy formulation, (pp. 73–76). In G. Thurlow, ed., *Technological responses to the greenhouse effect: Wyatt committee report No. 23.* New York: Elsevier Applied Science.

Weinberg, C. J. and R. H. Williams. 1990. Energy from the sun. *Scientific American* 263:147–151.

Wyman, C. E., R. L. Bain, N. D. Hinman, and D. J. Stevens. 1993. Ethanol and methanol from cellulosic biomass, (pp. 865–923). In T. B. Johansson, H. Kelly, A. K. N. Reddy, and R. H. Williams, and L. Burnham, eds., *Renewable energy: Sources for fuels and electricity.* Washington, DC: Island Press.

Integration of Natural Resources Management

This book has provided a comprehensive introduction to the principles of natural resources management so far as applied primarily in the geographical and political context of the United States. The history of natural resource development and management, summarized here reveals how profoundly natural resource development has shaped human culture and improved material well-being. However, increasing resource demands and scientific awareness also have made it clear that humanity is having a tremendous impact on natural resource sustainability worldwide. Individual chapters have shown that management systems often do not stop at national borders. While ecosystems and their management details vary from those in the U.S., the scientific process and management principles generally apply worldwide. In this section we address some basic social, economic, and ecological issues now facing the globe. We also summarize trends pointing toward greater integration of economies, ecological understanding, and management approaches to a more sustainable natural resource management worldwide. Chapter 21 is an introduction to some of the problems associated with past international resource management, with emphasis on food and fiber production and related trade and distribution. Chapter 22 discusses world economic systems emphasizing the general advantages of market economies over command economies in improving human living conditions. As all of the world seeks the material welfare enjoyed by the present minority in developed nations, the demands on atmospheric and biospheric resources will increase, particularly if the sustainability of resource development does not improve. U.S. national policies applying to international resource management are especially significant because of the disproportional influence they have on international affairs and global resources. Chapter 23 summarizes recent policy pertaining to the goal of sustainable development and growing emphasis on integrated resource management as the means to achieve that goal. Chapter 23 also discusses the importance of technology in the attainment of sustainable development and improved living conditions.

CHAPTER 21

Natural Resources and International Development

Great differences exist between the per capita incomes of developed and developing countries (Schiller 1994). About 3 billion people, or over half of the world population, live in countries where the average income is under $500 per year. In contrast, the average income in the United States was about $25,000 per year in the late 1990s. Unfortunately while living standards continue to improve in the developed countries, they are becoming worse in many of the least developed countries; particularly those in Africa (Figure 21.1). At the same time, certain countries with intermediate levels of development in the Pacific Rim (Korea, Singapore, Taiwan) and portions of Latin America (Chile, Argentina) are experiencing rapid improvement in their living standards.

We will discuss why living standards and economic progress vary widely among different countries. The role of natural resource availability and management in economic development will be considered along with policies that are most likely to promote sustainable development. The main themes of this chapter have been developed mostly from Brown and Wolfe (1985), Miller, (1990), Foster (1992), and Schiller (1994, 2000).

PROBLEMS WITH THIRD WORLD DEVELOPMENT

Most affluent countries have experienced poverty sometime in their past. From the fall of the Roman Empire around 450 A.D. until the mid-1700s, improvements in the human condition were almost nonexistent. Then the Industrial Revolution, starting in Europe (particularly Great Britain), triggered rapid economic growth. The key feature of the Industrial Revolution was the substitution of machines for human labor. The invention of the steam engine in the late 1700s was the most crucial breakthrough in this economic transformation that permitted labor to become more specialized and more efficient. At the same time, it reduced the need for large portions of the human population to work in agriculture (see Chapter 2).

FIGURE 21.1   Although living conditions have continued to improve in developed countries, they have stagnated or declined for many peoples, such as in the photo (courtesy of U.S. Department of Agriculture).

Based on history, we consider seven basic factors to be critical for any country to have major improvement in the human condition (Holechek et al. 1998). These include:

1. National unity
2. Market-oriented economy
3. Democratic form of government
4. Sound education system
5. Protection of private property rights
6. Opportunity for social and economic mobility
7. Level of economic growth exceeds the level of population growth

We will discuss each of these factors.

National Unity

A close look at the countries who are in a economic decline reveals they suffer from internal conflict. This is particularly true of the African countries where tribalism has long been a severe disruptive force. Generally African countries involve a collection of small indigenous tribes that speak different languages, have different religions, and have diverse cultural values. Conflict between these tribal groups has existed for centuries. Leaders from whichever tribe was able to cease power often ruthlessly exploited and persecuted rival groups. Recent holocausts in Ethiopia, Somalia, Rwanda, and Sudan can all be traced to long-standing ethnic and tribal rivalries. This is also true of the Muslim/Christian conflict that has ripped apart Yugoslavia in southeastern Europe and the Catholic/Protestant conflict in Ireland.

In contrast, economically successful countries have been characterized by common languages, customs, values, goals, and traditions that resulted in national unity. Historically ancient Greece, the Roman Empire, Great Britain, the United States, and modern Japan all provide examples of the importance of national unity in economic progress. Recent research by Gerald Scully, an economist with the Dallas-based National Center for Policy Analysis, has corroborated the relationship between cultural unity and economic growth (*Investors Business Daily* 1995).

Market-Oriented Economy

The Industrial Revolution and the rise of market-oriented economies generally occurred together during the 1700s. Adam Smith's classic book *The Wealth of Nations* is considered to be the first formal work on how free markets facilitate economic growth. Smith's work explained how production efficiency could be greatly increased through specialization of labor and how changes in prices of goods give signals to producers to expand or decrease supply in unregulated economies (*Investors Business Daily* 1996). Smith also pointed out how competition and profit incentives improve the availability and quality of goods and lower their cost. Basically Smith delineated the advantage of a capitalist economy in which the factors of production are owned by individuals and basic allocation decisions are made by market forces. Under this competitive system, the more efficient producers prosper while those that are least efficient are allowed to perish.

About 100 years later, Karl Marx wrote *Das Capital,* which challenged capitalism and provided an alternative in the form of communism (Schiller 1994). Marx viewed capitalism as a system in which workers were oppressed by the owners of land and machines (the capitalists). He advocated a classless, stateless economy in which there is no private property. Everyone would share in production and consumption according to their abilities and needs. During the transition from capitalism to communism, Marx believed a strong central government would be needed to give direction to the new society. During this period of socialism, all nonlabor means of production would be owned by the state and the state would make decisions on resource allocation.

Some 34 years after Karl Marx's death (1883), his theories were put to the test after violent revolutions in Russia (1917) and China (1949). The primary deviation from Marx's scenario was that the communist revolutions occurred in underdeveloped rather than developed countries. Although communism was said to be a goal in Russia, China, and other countries in the movement, socialism was applied in actual practice.

The basic goal of the communist movement was an egalitarian society in which all individuals would contribute to productivity, according to their abilities but share equally in the output. However after 72 years, the communist economy of Russia collapsed in 1989. What seemed like a good idea in theory had failed to work in practice. The basic problem with centrally planned (socialist) economies in eastern Europe, China, Cuba, and other parts of the world has centered around inefficiency and lack of individual incentives. Inefficiency comes from the fact that the detailed knowledge of input-output relationships required under central planning is very difficult to obtain. As economies become more complex in terms of variety of goods and services, the degree of miscalculation increases (Schiller 1994). In contrast under market economies, the prices consumers are willing to pay for various goods quickly signals the level of output desired by society.

When prices, production, and incomes are all set by central planners there is little opportunity or incentive for individuals to design new products, improve production efficiency, and better serve the consumer. In other words, there is little reward for risk and innovation. Basically it was severe shortages of food and other consumer goods that caused the downfall of communism in the Soviet Union in the late 1980s (Schiller 1994).

During the 1970s and 1980s, many of the developing countries in Africa, Southeast Asia, and South America had highly socialized economies. The basic flaws of socialism previously discussed retarded economic progress in these countries.

Democratic Form of Government

Democracy involves a form of government in which the citizens have equal opportunity to decide political issues and elect government officials by voting. This process permits orderly change of government when the people it serves believe it necessary. Generally societies lacking democratic processes have ultimately been characterized by failure because nonviolent mechanisms are lacking to replace the inept or corrupt leaders that eventually gain control of government. Totalitarian governments such as Nazi Germany under Adolph Hitler and the Communist Soviet Union under Joseph Stalin have invariably exploited the majority to benefit a small minority. The rapid improvements in the human condition that began in the 1700s in Great Britain and the United States occurred where democratic processes of government and market oriented economies were gradually being established.

Sound Education System

Throughout history knowledge has been the primary source of human progress. Successful civilizations have generally had educational systems that were accessible to high proportions of their citizens and rewarded both knowledge contribution and acquisition. Investment in universal education was the key to the economic successes of Great Britain, the United States, and, most recently, Japan and Germany.

In order for any country to develop a sound educational system accessible to all its citizens, it must have a tax base that will support it. In many of the African countries and parts of South America most of the natural resources, particularly land, are held by the state. Although the land can be used for grazing, farming, mining, and several other purposes, there has generally been no fair and efficient mechanism for the state to tax land users. In contrast in developed countries with market oriented economies, land and other natural resources are largely in private ownership with deeds to these resources on file with local (country) governments. These recorded deeds have long provided a means by which taxes could be assigned to holders to pay for schools, roads, waterworks, and more. These taxes have been primary funds for public education and scientific research.

In contrast, large portions of the populations of developing countries pay no taxes at all, but expect their government to provide them with many of their basic needs, such as education and health care. In such countries, unlimited demands are placed on governments without the means to satisfy them. Without an equitable tax system that makes all citizens accountable for the support of their government, both political and financial stability become unattainable. Flawed systems of taxation explain in part the debt crisis that exists in many developing countries, and why they continue to lack public educational systems that are accessible to everyone.

Protection of Property Rights

Protection of property rights is essential to encourage investment and discourage flight of capital. It is tragic that much of the wealth generated in developing countries is invested in developed countries.

Generally, saving rates are low in developing countries because most of the population lives at the subsistence level and is struggling to survive. Foreign investment is almost essential to increase production opportunities and improve the availability of management, technology, and training of labor. Foreign investment has often been discouraged by developing countries because there has been a belief that foreign investors take away more than they contribute. However, history shows this has not generally been the case. Foreign investment has played a crucial role in the economic development of nearly all successful countries in modern times. This includes development of the United States in the 1800s, much of which was financed by Great Britain, Germany, and other European countries. Without strong assurance of capital protection from expropriation, it is nearly impossible for developing countries to attract internal or external investment. Those developing countries with unstable governments, unpredictable tax and property laws, wildly fluctuating currency values, and totalitarian governments generally have been shunned by foreign investors and have realized the lowest levels of economic growth (Stein 1994).

Many developing countries and some developed countries have resorted to printing money unbacked by additional goods and services to fund their debts. The creation of money by governments to fund their obligations is referred to as debt monetization. This creates inflationary pressures in the economy and devalues the currency. Historically governments that tried to solve their debt problems through monetization have in the end impoverished their people This was true of France prior to the revolution in the late 1700s, Germany after World War I under the Weimar Republic, and in several Latin American (e.g., Argentina, Brazil, Mexico) and African countries after World War II. When foreign investors become uncertain about the future value of a country's currency, they generally avoid it as an investment choice (Schiller 1994). This is because they must convert the foreign country's currency back into their own country's currency to realize their profits.

Opportunity for Social and Economic Mobility

Throughout history, the level of human creativity has been closely associated with the extent that benefits returned to the originator. Much of the innovation from Great Britain and the United States since the late 1700s has been linked to social-political systems that protected the right of individuals to benefit or fail from their actions. A strong point of market economies is that they permit businesses to fail when they are poorly managed or produce obsolete products, and they reward the better managed businesses and those with useful products. This process was referred to as "creative destruction" by the 1920s free-market economist Joseph Schumpter. Generally, the more governments have tried to equalize the rewards for success (sound business decisions), and reduce the penalties for failure (poor business decisions), the less economically successful their societies have been.

Access to education has been one of the primary means by which the impoverished could improve their social and economic position in successful societies. However, laws that prevent discrimination on the basis of race, sex, creed, or religion for jobs, contacts, and other forms of economic opportunity are also of crucial importance.

A critical aspect of economic development is to ensure women have the same educational and economic opportunities as men. Educational discrimination against women leaves half of a nation's human capital illiterate and incapable of passing important skills on to their children (Schiller 1994). Illiterate women tend to have large family sizes, but generally lack the resources and skills needed to provide adequate care for their children. This accentuates scarcity of skilled labor and oversupply of unskilled labor sustaining the cycle of poverty and deprivation.

Level of Economic Growth Exceeds Level of Population Growth

Developed countries are distinguished from developing countries primarily by their ability to increase output at a faster rate than population growth (Schiller 1994). Historically the annual per capita economic growth rate in the United States has averaged about 3½% (Schiller 1994). However, this has slowed down to about 2% since 1980. Table 21.1 provides an overview of recent economic and population growth levels for different parts of the world. The wide disparities in the per capita growth rates of developing countries is explained by differences in social-political policies and rate of population growth.

Generally the higher the rate of population growth, the more difficult it is to increase per capita economic growth on a sustained basis (Brown 1987). Annual human population growth rates below 2% are needed to have rapid improvements in living standards. Otherwise most or all of the increase in economic output goes into immediate consumption and there is little opportunity for savings and investment. Savings and investment are the primary sources of infrastructure, plant, and equipment that will raise the productivity of labor (Schiller 1994).

TABLE 21.1 **Growth Rates in Selected Countries, 1980–1997**

	Average Growth Rate (1980-1997) of		
	GDP	Population	Per Capita GDP
High-income countries			
United States	2.5	1.0	1.5
Japan	1.4	0.3	1.1
Canada	2.1	1.2	0.9
France	1.3	0.5	0.8
Low-income countries			
China	11.9	1.1	10.8
India	5.9	1.8	4.1
Haiti	−3.8	2.1	−5.9
Ethiopia	4.5	2.3	−2.2
Kenya	2.0	2.6	−0.6
Venezuela	1.9	2.2	−0.3
Zimbabwe	2.3	2.0	0.3
Nigeria	2.7	2.9	−0.2

Source: World Bank and Schiller 2000, *World Development Report* 1998/1999.

Various strategies have been used to lower population growth rates in developing countries (Jacobsen 1983, Miller 1990). These have included:

1. Education of people on population problems and benefits of small family size
2. Encouraging couples to delay marriage
3. Provision of cash payments and other benefits to couples who restrict their family sizes
4. Increasing the availability of various birth control methods
5. Provision of cash payment and other benefits to individuals who agree to sterilization
6. Imposition of tax, education, retirement, and other penalties on couples who have more than two children
7. Make abortion legal and readily available
8. Increase education and work opportunities for women
9. Alter social attitudes to favor small instead of large families

China has had the most aggressive population control program of any country in the world (Miller 1990). It is the world's most populous country and has periodically faced the threat of massive famine. After the death of 30 million people from starvation in the 1958–62 famine, China implemented an aggressive policy to reduce population growth and increase food production. Its population control involved:

a. Providing married couples with free birth control alternatives (sterilization, contraceptives, abortion)
b. Provisions of incentives for delay of marriage
c. Increased opportunities for education
d. Rewards for couples who agree to have only one child
e. Penalties for breaking pledges to restrict family size
f. Applying pressure on women pregnant with a third child to have an abortion
g. Training local people to implement the family planning program

The outcome of these programs has been to drop the fertility rate from 5.7 to about 2.3 children per woman by 1994. Although China had hopes of reaching zero population growth by the year 2000, this has not occurred because about one-third of China's population is under 15 years of age.

Although China's program is considered a definite success, its strong coercive elements make it unattractive to most other countries. Experiences from Mexico, Korea, Japan, and Indonesia show the less coercive measures involving education and free contraceptives can be effective if applied well before a country faces mass starvation.

GROWTH STRATEGIES

Developing countries simultaneously encounter numerous barriers that block economic growth. Once a sound socioeconomic-political system is put in place, many of these barriers can often be overcome. A sequence of economic development involving five stages characterized by Walter W. Rostow can occur (Schiller 1994). These include:

a. **Stage 1.** Traditional society with dependence on agriculture and generally low productivity. Here social mobility is low and educational opportunities are limited.

b. **Stage 2.** Agricultural productivity increases because of institutional changes that expand opportunities and permit an entrepreneurial class.

c. **Stage 3.** Savings and investment increase, permitting industrialization. This is usually accompanied by government policies that further enhance growth.

d. **Stage 4.** The industrial growth process becomes more general with the transformation into production of complex goods as well as simple and intermediate goods.

e. **Stage 5.** Most of the population has access to economic opportunity and receives a high level of income. In this stage the longevity by the population is maximized and everyone has access to basic necessities of life (food, shelter, medical care).

Stage 2 is considered critical for the reasons we have previously discussed relating to culture, form of government, and economic system. The repressive institutional rigidity under Stage 1 generally is broken when the majority of the population becomes so impoverished that the risks associated with change exceed the risks associated with continuing the status quo. Throughout history, political upheaval and revolution have been the catalyst for Stage 2.

Development Options

Once the institutional foundation is put in place for economic growth, various choices and tradeoffs must be made regarding whether to emphasize agricultural or industrial production. Increased agricultural productivity generally is thought to be a precursor for industrial expansion (Schiller 1994). This is because food surpluses, which permit farmworkers to switch to urban industrial employment, create a potential for exports, and permit savings in the farm sector that can be used for future investment. Over 70% of the people in most developing countries work in agriculture. Therefore, in contrast with industrialization, agricultural development has the potential to more broadly improve living conditions initially in most countries.

On the other hand, some developing countries in Africa and the Mid-East have low levels of arable land. Countries such as Saudi Arabia, Kuwait, and Libya with high levels of energy and mineral resources that lack suitable farmland have found it effective to develop these resources and invest in human capital rather than to emphasize agricultural expansion. Another choice that confronts developing countries is whether to focus growth in a few sectors or in many sectors simultaneously (Schiller 1994). While it may be ideal for a country to develop all sectors at the same time, this is seldom possible. Experience has shown that the most successful developing countries concentrated their resources in industries where they have the greatest comparative advantage in natural resources, labor, and management (Schiller 1994). In other words, growth in targeted industries should not be greatly limited by shortages of labor, land, technology, management, or some other input. Ideally the targeted industries should be linked to other industries the country could develop. For example, development of the timber resources is compatible with furniture manufacture. Growth in the cattle production industry should accompany development of a meat processing and packing industry.

Another choice that confronts developing countries is whether to emphasize production for domestic consumption or for export. Generally the most successful countries have heavily depended on exports for economic growth (Schiller 1994). This has the advantages of generating foreign currency that can be used to purchase complex goods and technology from other countries and it permits tapping into markets that are much larger than those existing internally. Japan, South Korea, and Taiwan are a few examples of countries that have exported themselves to prosperity since World War II.

The disadvantages of relying on exports include barriers to market access in developed countries, difficulty in producing goods of comparable quality to those in developed countries, and the erratic nature of world demand for exportable products (Schiller 1994). Trade restriction problems have been greatest with textiles, shoes, steel, clothes, sugar, peanuts, and other basic commodities where developed countries such as the United States have established industries that lack a comparative advantage. Here domestic producers apply political pressure for trade barriers to keep out foreign competition.

Quotas in the United States have reduced export earnings of Argentina and Australia from beef production and of the Philippines and several Caribbean countries from sugar production. However in certain cases a developing country may be given a favored nation status that allows them more access to U.S. markets.

Coffee is an agricultural product that cannot be easily grown in the United States and many other developed countries (Schiller 1994). Therefore, it is well suited as an export crop by several African and Latin American countries. Since Brazil is a major coffee producer, its crop can have great impact on other countries. When frost or other factors reduce Brazil's coffee crop, as in 1994, coffee prices go up and other coffee-producing countries benefit. However, when growing conditions are ideal in Brazil, coffee supplies expand, reducing export earnings in countries such as Haiti. Too much dependence on one export leads to sequences of boom and bust that can cause erratic long-term growth and political instability.

Cartels have been formed to stabilize export earnings (Schiller 1994). They basically attempt to limit exports in years of excess supply. The best-known cartel involves the oil-producing countries (OPEC) dominated by Saudi Arabia. Cartels have had only limited success because those countries with faltering economies are strongly tempted to underprice and exceed their voluntary quotas during periods of oversupply.

Rather than take the risks of international trade, many developing countries have concentrated on domestic markets. With the exception of agricultural products, this approach is usually flawed because only a few goods can be produced on a competitive basis with developed countries (Schiller 1994). Domestic demand seldom justifies heavy capital expenditures for production of goods that will be higher priced and of inferior quality to imports.

NATURAL RESOURCES VERSUS ENTREPRENEURSHIP

Without question, countries that are well endowed with natural resources—such as Canada, Australia, New Zealand, Brazil, Argentina, and South Africa—have great advantages in economic development over countries where natural resources are scarce relative to resource demands. However, the successes of Japan, South Korea, and, most recently, China indicate that human entrepreneurial resources are more important than physical resources in economic development.

Hernando De Soto, in his best selling book *The Other Path,* has been one of the most convincing advocates of unlocking the power of markets and people (Schiller 1994). After years of financial success in Europe, he returned to his native Peru and observed that it had a thriving underground economy in spite of repressive government policies. Although Peru had some trade in drugs, most of it centered around essential consumer goods. Government regulation of prices, trade, and other business activities had caused most businesses to go underground. De Soto advocates government adoption of policies that encourage entrepreneurial activities through privatization, reductions in regulation, elimination of price controls, legal protection of property, and infrastructure development. De Soto's book has become the model for market-oriented reforms in Peru, Argentina, Mexico, Russia, and India.

The basic point from De Soto's work is that the greatest natural resource is human ambition, ingenuity, and creativity (Schiller 1994). All things being equal, those countries possessing the highest levels of physical natural resources have the advantage in international development. However countries that adopt democratic, market-oriented development policies can overcome many natural resource limitations. Japan and South Korea are examples of countries that have become world economic powers with limited natural resources.

INTERNATIONAL POLICY

The 1990s have been characterized by a gradual shift from centrally planned to market-oriented economies by most of the world's developing countries. Those countries that have made the most progress with this shift have generally experienced the highest levels of per capita economic growth. At the same time many of these countries have reduced rates of environmental degradation and natural resource depletion because production efficiency is higher under market-oriented systems (Chandler 1987). While markets alone cannot assure environmental and natural resource sustainability, they do impose some self-administered checks on natural resource waste. This is because the resource user has strong incentives for efficiency.

Although in theory centrally planned economies can make resource conservation a high priority, in practice this did not happen (Chandler 1987). Prior to its disintegration in 1991, the former Soviet Union had one of the poorest natural resource conservation records of any country in the world. Their energy consumption rate per unit of gross national product (GNP) was about 50% higher while their sulfur dioxide emissions per unit of GNP were over twice those in the United States. Soil erosion rates on Soviet farmlands far exceeded those in the United States and in west European countries. Prior to market-oriented reforms in the 1970s, China experienced severe forest and cropland degradation and natural resource depletion. Although these problems continue in China, they are now less severe.

Because capital is in short supply in developing countries, loans are important to their development. The World Bank is the primary lending agency to developing countries. It also provides technical assistance to improve the success of the projects it finances. Multinational banks and individual countries also provide similar financial and technical assistance. Developing countries prefer loans with few strings attached. However recent history shows the most successful loans have been those predicated on monetary, market, and environmental reforms.

In the past, developing countries often borrowed beyond their capability to repay their debts (Schiller 1994). Because these debts must be paid back in hard currencies, such as U.S. dollars, export-related projects, such as factories, have taken precedence over projects providing domestic infrastructure, such as schools, roads, and sewage systems.

In the early 1980s, a worldwide recession reduced the capability of Mexico, Brazil, Peru, and other large borrowers to service their debt (Schiller 1994). In response to this crisis, the developed creditor nations adopted a resolution to reduce their exports to developing countries, to increase their imports from developing countries, to place greater emphasis on investment in developing countries, and to increase the level of outright gifts in foreign aid to developing countries.

Naturally, developing countries would like more foreign aid in the form of gifts rather than loans. Foreign aid data in Table 21.2 show that about 57 billion dollars are given to developing countries each year. Because of the low amounts given and the strings that are usually attached, developing countries are now asking for "trade" not aid (Schiller 1994). In other words, they wish to export more of their goods so they can finance internal investment themselves.

The objective of the North American Free Trade Agreement among the United States, Mexico, and Canada is a step toward the trade-over-aid approach. This agreement not only opens U.S. markets to Mexico, but also has important environmental provisions relating to air quality, water quality, and land use.

Latin American countries—particularly Chile, Argentina, Mexico, and Peru—have demonstrated that market reforms can turn around stagnant economic growth, inflation, and capital flight (Stein 1994). However, economic changes in the 1990s involving cutting tax rates, business privatization, lowering of trade barriers, stabilization of currency, and easing restrictions on foreign investment have turned these nations away from the abyss and toward prosperity.

Chile is perhaps the greatest success story of these four countries. During the 1920–73 period, Chile's economy shrank by 10% and verged on hyperinflation under the Marxist policies (Stein 1994). However, after 1973 a string of market reforms were implemented that revived the Chilean economy. Chile has grown an average of 4.5% since 1975 while at the same time slowing its inflation rate. Chile's reforms began with trade barrier reductions in 1973, followed by lower tax rates. During the

TABLE 21.2   Foreign Aid in Relation to GDP, 1990

Country	Official Development Assistance	
	Millions of U.S. Dollars	As a Percentage of GDP
United States	$11,394	0.21
Japan	9,069	0.31
France	9,380	0.79
Germany	6,320	0.42
United Kingdom	2,638	0.27
Canada	2,470	0.44
Netherlands	2,592	0.94
Sweden	2,012	0.90
Australia	955	0.34
Norway	1,205	1.17
Denmark	1,171	0.93
Switzerland	750	0.31
New Zealand	95	0.23
Total foreign aid	$56,632	Average 0.36

Source: World Bank and Schiller 1994; total includes other developed countries.

1980s, Chile replaced its government pension system with a system of individual retirement accounts that use the funds for investment. Chile's pension innovations have served as a model for Mexico and Argentina.

LITERATURE CITED

Brown, L. R. and E. C. Wolf. 1985. *Reversing Africa's decline.* Worldwatch Paper 65. Washington, DC: Worldwatch Institute.

Chandler, W. U. 1987. Designing sustainable economies. In *State of the world.* New York: Worldwatch Institute.

Foster, P. 1992. *The world food problem.* Boulder, CO: Lynne Rienner Publishing.

Jacobsen, J. 1983. *Promoting population stabilization incentives for small families.* Worldwatch Paper 54. New York: Worldwatch Institute.

Holechek, J. L., R. D. Pieper, and C. Herbel. 1998. *Range management principles and practices.* 3rd edition. Upper Saddle River, NJ: Prentice-Hall.

Investor's Business Daily. 1995. Economics and culture. Section B1. September 22, 1995.

Investor's Business Daily. 1996. Guide to the markets. New York: John Wiley & Sons.

Miller, G. T. 1990. *Resource conservation and management.* Belmont, CA: Wadsworth Publishing Co.

Schiller, B. 1994. *The economy today.* 6th edition. New York: McGraw-Hill.

Schiller, B. R. 2000. *The economy today.* 8th edition. New York: McGraw-Hill.

Stein, R. S. 1994. Why Latin America is thriving. *Investor's Business Daily,* January 4, 1994.

CHAPTER 22

Economics and Economic Systems

Because natural resources and other material goods and services are available in limited supplies, but human wants are unlimited, trade-offs or choices must be made regarding what combination of goods and services will be produced by any society. The science of how to allocate scarce resources among competing uses is called economics (Schiller 1994). The science of economics is relatively young, originating when Adam Smith wrote *The Wealth of Nations* in 1776, which described the benefits of market-driven economies.

In this chapter, we will provide a basic overview of economics and economic systems. We explore the issue of natural resources scarcity versus technological scarcity in constraining improvements in human welfare. Finally economic approaches and systems that will meet the needs of the rapidly changing world of the twenty-first century will be explored. While this chapter extends beyond the narrow field of natural resource economics, it illustrates how economic systems enfold natural resource management systems and determine the efficiency of natural resource use and abuse. Our discussion has drawn heavily from Davidson and Rees-Mogg (1987, 1993), Schiller (1994, 2000), and *Investor's Business Daily* (1996).

ECONOMIC TERMINOLOGY

To further develop our discussion of basic economics, it is necessary to define some of the commonly used terminology. The definition of economic terms in the glossary comes from Schiller (2000), which we consider to be an excellent basic textbook on both macroeconomics and microeconomics. Macroeconomics focuses on the economy as a whole while microeconomics deals with the behavior of individual components of the economy. While Chapter 4 addressed benefit-cost microeconomics, this chapter addresses macroeconomics.

BASIC ECONOMIC PRINCIPLES

The two most basic principles of economic science are that resources are scarce and that economic decisions involve opportunity costs (Schiller 1994). Opportunity costs are goods and services foregone when the decision is made to produce more of one item and less of another. The allocation of scarce resources to obtain four basic social goals is the primary purpose of economic science. These four social goals aimed at maximizing human welfare include (Schiller 1994):

1. Economic growth - This involves expanding the supply of goods and services available to the human population.
2. Equitable distribution of income - This involves developing social-political systems that provide everyone with the opportunity to obtain the material goods and services necessary for a quality lifestyle.
3. Full employment - This involves the application of social-political systems that permit everyone capable of working to have a job.
4. Price stability - This involves managing the economy so that periods of rapidly rising or falling prices for goods and services are minimized or avoided.

No socioeconomic system in history has been able to achieve all of the above goals for an extended period of time (20 years or more). However, the United States experienced a period of prosperity from 1953 to 1973 that probably represents the historical pinnacle in terms of near simultaneous achievement of all four economic goals.

Generally market-oriented systems with minimal government intervention maximize efficiency in terms of providing society with the optimal combination goods and services it desires and can afford (see Chapter 4). Both economic growth and price stability tend to be best achieved with market-oriented systems. However equitable distribution of income and full employment are to some extent compromised under market economies. Without some degree of regulation by government severe environmental damage is probable under any economic system. This is because environmental protection is a transaction externality.

An externality is a cost of an economic activity borne by a third party not part of the original transaction. During the "smokestack era" of the late 1800s to the mid-1900s, pollution of rivers and the air by factories in the northeastern U.S. was rampant. Government regulation in the 1970s brought this problem under control. Soil erosion from farming is an externality that is born more by future than present generations. Here again government intervention has been necessary to control the problem. Chapter 4 describes the problem in more detail and solutions based on market incentives.

Market Theories

Adam Smith in 1776 formulated theories that the basic economic questions of what, how, and for whom goods and services are to be produced is best decided by the "invisible hand" (Figure 22.1). This invisible hand was a characterization of how markets operate based on indirect communication between producers and consumers (*Investor's Business Daily* 1996). Under the "invisible hand," increased demand by society for more of something causes prices to rise at the marketplace or shortages will develop. Higher prices send a signal to producers to increase output. Profit potential provides an incentive for producers to respond to this signal. Moreover, failure to sell all the output of some other good generally causes inventories to build up and prices to fall signaling producers to reduce production (see Chapter 4 for additional discussion). Through these "market mechanisms," members of society get the optimum combination of goods and services relative to their incomes. Individuals who have the ability and desire to pay for a good will be able to get it in a market economy. Producers will use the most efficient methods of production to maximize profits. Competition from other producers will tend to keep prices and supply in balance with demand and lead to improvements in the production process.

FIGURE 22.1 Adam Smith, an English economist, formulated basic theories on free markets in 1776. His work titled *An Inquiry Into the Nature and Cause of the Wealth of Nations* is the foundation for the philosophy and workings of market economies (from The Bettmann Archive, IBD 1996).

Adam Smith believed that price signals and consumer demand at the marketplace are far more effective than government in allocating resources. Under market-oriented (capitalistic) economies as envisioned by Smith the factors of production (land, equipment, labor, management) are controlled by individuals rather than the state. Adam Smith's ideas provided the basis for the capitalistic system of government used in the United States after independence from Great Britain was won in 1783. Figure 22.2 shows the increase in inventions and improvement in living standards associated with the development of market economies in western Europe and the United States in the 1700s.

The Marx Alternative to Markets

An economic alternative to market oriented economics was theorized by Karl Marx when he published *Das Kapital* in 1867 (Schiller 1994). Marx developed the rationale for centrally planned or command economies—which are the polar opposite of market economies. Under the system theorized by Marx, the factors of production (land, labor, equipment, management) would be controlled by the government and the government would make all production and distribution decisions for society. This type of system, in which the government owns all nonlabor means of production and decides how resources will be allocated, has been referred to as either communism or socialism.

Communism has many elements of socialism (Schiller 1994). It differs in that theoretically everyone is supposed to share in production and consumption according to their individual abilities and needs. The basic concept of communism is a stateless, classless economy with no private property. Under socialism the state owns all the nonlabor means of production and exercises control over resource allocation. However, resources are only partially distributed according to each person's effort (Schiller 1994). In reality, communism is merely a concept that has never been achieved by any country, including the former Soviet Union and modern China.

Karl Marx thought government should take strong steps to correct the inequities he observed in England and other European countries in the mid-1800s. A few rich people enjoyed extravagant lifestyles while large numbers of people had inadequate

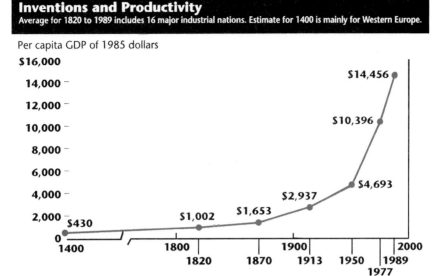

Inventions and Productivity
Average for 1820 to 1989 includes 16 major industrial nations. Estimate for 1400 is mainly for Western Europe.

Source: Data from *Dynamic Forces in Capitalist Development: A Long-Run Comparative View,* Angus Maddison, Oxford University Press, Walton Street, Oxford, 1991.

Year	Invention	Inventor
1440	Printing with movable type	Gutenberg
1642–1671	Calculating machines	Pascal, Leibniz
1764	Spinning Jenny	Hargreaves
1785	Power loom	Cartwright
1825	Steam locomotive	Stephenson
1837	Telegraph	Morse
1856	Bessemer steel process	Bessemer
1865	Antiseptics in surgery	Lister
1876	Telephone	Bell
1877	Internal combustion engine	Otto
1879	Electric light	Edison, Swan
1903	Airplane	Wright
1909	Bakelite (plastic)	Baekeland
1925	Television	Zworykin, Farnsworth
1948	Transistor	Shockley, Brattain, Bardeen
1952	Commercial computer	
1957	Laser	Gould
1958	Integrated circuit	
1965	Minicomputer	
1970	Microprocessor	
1973	Personal computer	
1980's	Genetic engineering	
1990's	World Wide Web	

FIGURE 22.2   The rise of per capita GDP and increase in inventions as market economies began to develop in the 1700s in western Europe and the United States (from IBD 1996).

housing and suffered periodically from starvation under the market-oriented economies that existed at the time. Marx theorized that central planners could produce and distribute a more desirable combination of goods than that achieved under markets. Methods in production could take into account environmental, employment, and other social needs without being determined by cost alone.

Mixed Economies

Various compromises between the extremes of market and command economies have been applied in a wide variety of countries over the last 100 years. So far no economic system has fulfilled all economic needs. However, without question some systems have been more effective than others.

Today no nation relies completely on market or central planning mechanisms in production and allocation decisions (Schiller 1994). Instead, a combination of government regulations and market signals are used to achieve economic and social goals. This integrated type of system is referred to as a mixed economy. The United States is considered to be the most market-oriented economy in the world. Great Britain, Japan, Singapore, Canada, and Australia are considered to be market oriented, but their taxation and regulation levels are higher than in the United States. Russia and China have been attempting to privatize many centrally planned operations such as farming, mining, and marketing, but central planning still strongly dominates in the manufacturing sectors. In Latin America and Europe the state controls many key enterprises, but private markets flourish in agriculture and in the distribution of goods. European economies tend to be the most evenly balanced or mixed.

The degree of taxation is the primary determinant of whether a country has a market, command, or mixed economy. In the United States, people on average pay about 36% of their gross income to government in taxes. In China, the average citizen pays over 70% of his income to the government in taxes. In mixed economies such as Germany, France, Spain, and Sweden, the average citizen pays 40 to 65% of their income in taxes. The government then redistributes the bulk of this income back to the population according to need for healthcare, food, housing, and retirement.

It is now well recognized that both market and centrally planned economies can fail without proper direction from government. The 1930s depression in the United States—when unemployment hit 25% and economic output fell by 10%—is the most impressive example of market failure in the twentieth century (Schiller 1994). The collapse of communism in the former Soviet Union in 1991 provides a primary example of central-planning failure. The challenge for society is to select the appropriate balance of market incentives and government taxation and regulation. It is important to understand economic interventions available to the government and how they work. Historically, the ability of a society to improve its well being has depended far more on its political system and economic management than on its level of natural resources.

The Most Successful Economy in the World

U.S. Economic Status. In many ways, the United States can be called the most successful economy in the history of the world. This generally applies in development of technology, improved living conditions for the poor as well as the rich, improved human health, and improved opportunities for social mobility. The success of the U.S. has been attributed to a political-economic system that emphasizes democracy, fair play, competition, opportunity, individualism, and capitalism. We will now focus on the U.S. in our discussion of economic systems.

People in the U.S. today enjoy a living standard that would have been almost unimaginable 100 years ago (IBD 1996, Figure 22.3). People on average live about 30 years longer than those at the turn of the century, eat better food, and experience much less physical hardship. We have developed more material comforts and conveniences—such as pocket televisions, cellular phones, fax machines, laptop computers, and the Internet—in the last 100 years than in all of previous recorded history. Still, cures for many diseases (AIDS, cancer, Alzheimer's disease, and more) allude us and violent crime and drug use are serious problems throughout the country. Loss of open space, atmospheric and water pollution, and species extinction are new challenges that were of minor concern in the year 1900.

Economic Indicators. Ten basic indicators are commonly used to access the economic health of the economy in the United States at any point in time. These include the following:

1. Gross domestic product
2. Unemployment rate

How Sweet It Is

	1970	1990
Average size of a new home	1,500	2,080
New homes with central a/c	34%	76%
People using computers	<100,000	75.9 mil.
Households with color TV	33.9%	96.1%
Households with cable TV	4 mil.	55 mil.
Households with VCRs	0	67 mil.
Households with two or more vehicles	29.3%	54%
Median household net worth (real)	$24,217	$48,887
Housing units lacking complete plumbing	6.9%	1.1%
Homes lacking a telephone	13%	5.2%
Households owning a microwave oven	<1%	78.8%
Heart transplant procedures	<10	2,125*
Average work week	37.1 hrs.	34.5 hrs.
Average daily time working in the home	3.9 hrs.	3.5 hrs.
Work time to buy gas for 100-mile trip	49 min.	31 min.*
Annual paid vacation and holidays	15.5 days	22.5 days
Number of people retired from work	13.3 mil.	25.3 mil.
Women in the workforce	31.5%	56.6%
Recreational boats owned	8.8 mil.	16 mil.
Manufacturers' shipments of RVs	30,300	226,500
Adult softball teams	29,000	188,000
Recreational golfers	11.2 mil.	27.8 mil.
Attendance at symphonies and orchestras	12.7 mil.	43.6 mil.
Americans finishing high school	51.9%	77.9%
Americans finishing four years of college	13.5%	24.4%
Employee benefits as a share of payroll	13.5%	24.4%
Life expectancy at birth (years)	70.8	75.4
Death rate by natural causes (per 100,000)	714.3	520.2

*Figures are for 1991

Source: Federal Reserve Bank of Dallas

FIGURE 22.3 A comparison of material wealth in the United State in 1970 and 1990 illustrates gains in the living standard of an average U.S. citizen (from IBD 1996).

3. Inflation rate

4. Interest rates

5. Consumer debt level

6. Consumer confidence

7. Stock market

8. Housing starts

9. Retail sales

10. Durable good orders

By any measure the United States is the world's largest economy. In 1996, it produced nearly $7 trillion of output which accounts for nearly one-fourth of the world's entire output (IBD 1996). A recent historical overview of the United States economy is provided in Table 22.1. Gross domestic product (GDP) is used as the primary measure of economic growth in the U.S. and other parts of the world. It is an indicator of how much the average person would get if all output were divided up evenly among the population. Historically the United States economy as measured by GDP has grown at a rate of 3% per year. Because economic growth has exceeded population growth, GDP per capita is three times today what it was in 1900 (Figure 22.4). During the 1990s, the difference between GDP and population growth was 1.5%. Assuming it persists, per capita incomes will double in roughly 40–50 years. In 1996, the per capita GDP in the United States was about $26,000, which is over five times the world average ($4,800).

TABLE 22.1   Overview of the United States Economy Since 1950

	Real GDP growth	Unemployment	Consumer Price Index	Prime Interest Rate	Real Interest Rate	Discount Interest Rate	Return on S&P 500 Stock Index
				%			
1950	8.7	5.3	5.9	2.07	−3.8	1.59	22
1951	8.8	3.3	6.0	2.56	−3.4	1.75	16
1952	4.3	3.0	0.8	3.00	2.2	1.75	12
1953	3.7	2.9	0.7	3.17	2.5	1.99	−7
1954	−0.7	5.5	−0.7	3.05	2.4	1.60	45
1955	5.6	4.4	0.4	3.16	2.8	1.89	26
1956	2.0	4.1	3.0	3.77	0.8	2.77	3
1957	1.8	4.3	2.9	4.20	1.3	3.12	−14
1958	−0.5	6.8	1.8	3.83	2.0	2.15	38
1959	5.5	5.5	1.7	4.48	2.8	3.36	8
1960	2.2	5.5	1.4	4.82	3.4	3.53	−3
1961	2.1	6.7	0.7	4.50	3.8	3.00	23
1962	6.0	5.5	1.3	4.50	3.2	3.00	12
1963	4.3	5.7	1.6	4.50	2.9	3.23	19
1964	5.8	5.2	1.0	4.50	3.5	3.55	13
1965	6.4	4.5	1.9	4.54	2.6	4.04	9
1966	6.4	3.8	3.5	5.63	2.1	4.50	−13
1967	2.6	3.8	3.0	5.61	2.6	4.19	20
1968	4.7	3.6	4.7	6.30	1.6	5.16	8
1969	3.0	3.5	6.2	7.96	1.8	5.87	−11
1970	0	4.9	5.6	7.91	2.3	5.95	0
1971	3.3	5.9	3.3	5.72	2.4	4.88	11
1972	5.4	5.6	3.4	5.25	1.8	4.50	16
1973	5.7	4.9	8.7	8.03	−0.7	6.44	−17
1974	−0.4	5.6	12.3	10.81	−1.5	7.83	−30
1975	−0.6	8.5	6.9	7.86	1.0	6.25	32
1976	5.6	7.7	4.9	6.84	1.9	5.50	19
1977	4.9	7.1	6.7	6.83	0.1	5.46	−12
1978	5.0	6.1	9.0	9.06	0.1	7.46	1
1979	2.9	5.8	13.3	12.67	−0.6	10.28	12
1980	−0.3	7.1	12.5	15.27	2.8	11.77	26
1981	2.5	7.6	8.9	18.87	10.0	13.42	−10
1982	−2.1	9.7	3.8	14.86	11.1	11.02	15
1983	4.0	9.6	3.8	10.79	7.0	8.50	17
1984	7.0	7.5	3.9	12.04	8.1	8.80	1
1985	3.6	7.2	3.8	9.93	6.1	7.69	26
1986	3.1	7.0	1.1	8.83	7.7	6.33	15
1987	2.9	6.2	4.4	8.21	3.8	5.66	2
1988	3.8	5.5	4.6	9.32	4.7	6.20	12
1989	3.4	5.3	4.6	10.87	6.3	6.93	27
1990	1.2	5.5	6.1	10.01	3.9	6.98	−7
1991	−0.9	6.7	3.1	8.46	5.4	5.45	26
1992	2.7	7.4	2.9	6.25	3.4	3.25	4
1993	2.3	6.3	2.7	6.00	3.3	3.00	7
1994	3.5	5.8	2.5	7.20	4.7	3.88	−2
1995	2.3	5.6	2.2	8.50	6.3	5.00	34
1996	3.4	5.3	2.9	8.25	5.3	5.13	20
1997	3.9	4.6	1.7	8.50	6.8	5.00	31
1998	4.2	4.4	1.6	8.12	6.5	4.75	27

Source: Adapted from Schiller 2000.

PROBLEMS WITH MARKET ECONOMIES: THE BUSINESS CYCLE

While the United States is considered to have developed the most successful economy in the history of the world, there have been several periods of severe hardship. Many of today's senior citizens still have vivid memories of the 1930s depression when unemployment rates were over 20% and there was massive financial upheaval (bankruptcies) in the farming, banking, and industrial sectors (Schiller 1994). The Great Depression stands out for its longevity. Generally it is considered to have

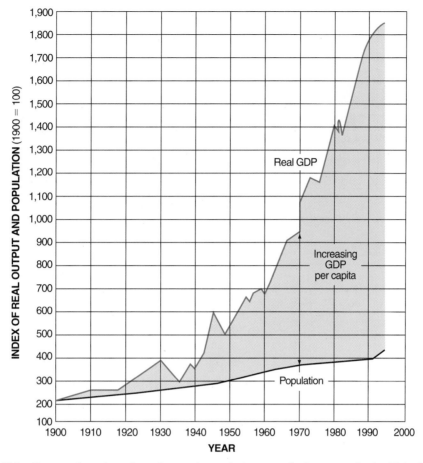

FIGURE 22.4 Output of goods and services and population growth since 1900 in the United States (U.S. Dept. of Labor, adapted from Schiller 1994, 2000).

started with the stock market crash in October of 1929 and ended when the United States became involved in World War II in December 1941.

During the 1920s, the United States seemed to have achieved utopia through its market-oriented system and minimal regulation by government. The economy rapidly expanded for eight years in a row (1921–1929). A variety of new products and technologies such as radios, cars, movies, and various appliances were rapidly improving the human condition and jobs were plentiful. However, the stock market crash and depression in the 1930s caused a major rethinking of economics and government's role in managing the economy.

Stages of the Business Cycle

The rise and fall of gross domestic product that periodically occurs in market economies such as in the 1920s and 1930s is called the business cycle (Figure 22.5). The four basic parts of business cycles include recession, depression, recovery, and boom. When gross domestic product and employment are declining, the economy is said to be in recession. When this condition becomes severe, it is called a depression.

Each of the six basic stages of most business cycles favors different classes of assets (commodities, stocks, bonds, cash, real estate) (Stoken 1984, Pring 1992). During Stage 1 at the bottom of a slump, business becomes leaner and more productive by eliminating unprofitable operations and reducing labors costs (Figure 22.6). In this period, consumer demand is low because of concerns over debt, high unemployment, and high interest rates. Stocks, real estate, and commodities are depressed but high quality bond prices are up.

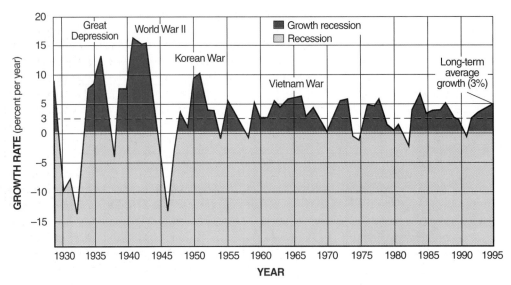

FIGURE 22.5   Business cycles in U.S. history from 1930 to 1995 (from Schiller 1994).

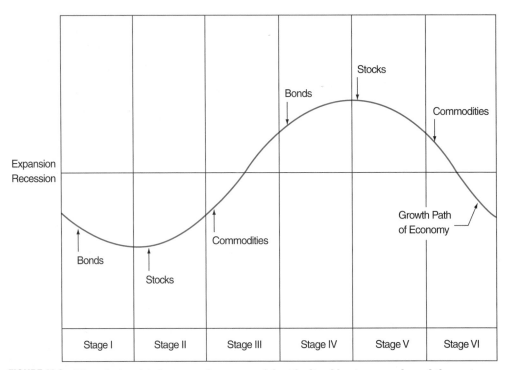

FIGURE 22.6   The relationship between the stages of the idealized business cycle and the various asset classes (from Pring 1992).

Austerity leads to Stage 2, when capital accumulation and lack of credit demand pushes interest rates lower. This causes a mild increase in economic activity. Bankers are cautious because they have just gone through a period of bankruptcies and fore-closures. Therefore, only those businessmen with the highest credit ratings have access to capital. Financial assets, primarily stocks and bonds, do well in Stage 2, but prices of real estate and commodities remain depressed.

Stage 3 marks the beginning of the recovery in commodity and real asset prices. This is due to reduction in inventories and depletion of consumer goods. Bond prices tend to be flat, but stock prices increase because of improvements in corporate earnings. Real estate and commodity prices start to increase in this stage.

Stage 4 brings a high level of confidence about the future of the economy. Consumer and business spending causes interest rates to rise rapidly, which depresses bond prices. Growing inflation and easy credit cause real estate and commodity prices to shoot upwards.

Stage 5 brings the peak in the business cycle. Here optimism about the future has led to recklessness. Credit is too easy to obtain, which causes high inflation and encourages poor business decision making. Commodity prices and real estate peak in this period because of both real and speculative demand. Real demand results from workers experiencing increased wages and access to easy credit. They are in a position to upgrade their standard of living. Speculative demand results from inflation pressures that causes investors to shift into real assets as a hedge against devaluation of the currency. Money flows out of the stock market into short-term money market funds that provide high yields. Long-term bonds are in disfavor due to fears of increasing inflation.

Stage 6 is characterized by a crash in commodity and real estate prices and a general economic downturn due to an oversupply of goods financed by excessive debt. The prosperity of Stages 4 and 5 causes recklessness and overoptimism in bankers, producers, and consumers. The only way the boom can be sustained is with excessively loose credit. If the Federal Reserve maintains the discount rate (cost of money to banks) below the inflation rate, the money supply increases at a more rapid rate than the expansion of the economy. This occurred during the 1970s (Table 22.1). Historically overly loose credit has always caused devaluation of a nation's currency and collapse of its bond market (Davidson and Rees-Mogg 1993). Debtors are always favored over creditors when the government takes the inflationary approach by making the real cost of money negative (prime interest rate minus consumer price index). Investment turns to speculation in real estate, gold, precious metals, chinese ceramics, and other scarce commodities as a hedge against currency devaluation, rather than into creation of real wealth through product development and improved production efficiency.

To contain inflation the Federal Reserve can raise the discount rate well above the inflation rate (typically measured by the consumer price index). This forces bankers to withhold credit, which in turn slows product demand. Commodity and real estate prices fall in response to tighter credit and an oversupply of goods. Falling prices are accentuated by bankruptcies of heavily indebted businesses and consumers that now meet their financial obligations with lower collateral (falling real estate) and less income (lower wages, lower employment levels).

Depressions in the United States

Severe or great depressions involving natural resources have occurred in the United States in the 1780s, 1830s, 1870s, and 1930s. These will be discussed following Batra (1988) and Schultz (1972).

1780s Depression. A severe economic downturn occurred after the American revolutionary army—commanded by George Washington—defeated the British in Yorktown in October 1781. The situation intensified with the signing of the peace treaty in Paris in 1783 that formally ended the Revolutionary War. This depression was caused mainly by a huge deficit in the balance of trade (Batra 1988). After the Revolutionary War, an influx of British manufactured goods into the U.S. was caused by heavy demand for reconstruction of the newly free nation. Prices of American goods (primarily food and fiber resources) were depressed because of protectionist policies by Britain and France and expanded domestic supplies in America due to the ending of war. This problem gradually subsided, starting in 1788, because Americans were given improved access to European markets created by wars (the French Revolution) that reduced Europe's farm production and increased its demand for American exports. It is noteworthy that the 1780s' depression nearly destroyed the newly formed government in the United States. The leadership of President George Washington was critical in holding the country together.

1830s Depression.   During the 1830s, cotton production boomed in the southeastern U.S. following invention of new textile manufacturing technology in England. High cotton prices led to massive land speculation. In the later stages of this boom in 1836, large amounts of land were bought with the sole idea of selling it at a higher price. The banks, which were all private and unregulated at the time, fueled the speculation by printing currency notes that were unbacked by silver or gold. To control the speculation, President Andrew Jackson put into effect policies requiring all loans in currency notes (paper money) be backed by hard money (gold and silver). At the same time, the cotton supply from the U.S. caught up with cotton demand by England. The combination of credit shrinkage and cotton demand cutbacks caused massive insolvency among the land speculators, cotton producers, and several large banks. European financing stopped causing massive bankruptcy throughout the South. Eventually, financial chaos spread to the entire country. The Mexican War of 1845 finally ended this depression.

1870s Depression.   The worst depression of the 1800s began in 1873 and lasted until 1879. It was brought about by unrestrained speculation in railroad construction and a winding down of post–Civil War reconstruction. The root cause was an overheated demand for new resource development and transport to markets. It spurred overexpansion of the money supply and credit by the federal government during and after the Civil War. By 1873, total credit was six times greater than total cash. In September 1873, it became apparent that earnings from all of the malinvestment, particularly in the railroads, could not support the debt levels used to finance it. Stocks on the New York Exchange began a precipitous crash. In 1874, massive bankruptcy resulted in the railroads and connected industries, such as steel rolling mills, machine shops, and foundries. Unemployment was widespread. Hunger and homelessness were beyond the relief capabilities of the charities and churches. Ultimately the farm sector was impacted because of falling food and clothing prices resulting from shrinking demand.

1930s Depression.   The depression of the 1930s was the most severe in the history of the United States in terms of length (1929–41) and depth (18–25% unemployment). The causes of the 1930s depression are more complex than those in the early history of the country. During the 1920s, capitalism in the United States was more restrained than at the turn of the century. President Theodore Roosevelt had put into effect powerful antitrust legislation to break up the monopolies in the railroad, oil, and steel industries. This resulted in more competition and caused wealth to be less concentrated in a few hands. Banking legislation was passed to set up the Federal Reserve Banking System in 1913, which supposedly would restrain speculation. Under this system, the government restricts the amount of money loaned by private banks. They must keep a portion of their deposits in reserve with a central bank controlled by the federal government (Federal Reserve Bank). The nation's money supply and access to credit can supposedly be controlled by the Federal Reserve Bank through changes in the interest rates the central bank charges member banks and through changes in reserve requirements for the member banks (Schiller 1994).

Nevertheless, the root cause of the 1930s depression was excessive credit extended in response to agressive business and consumer demand, which resulted in commodity overproduction and massive bankruptcy during the 1920s. During the 1920s there was a flood of new products, such as automobiles and home appliances. At the same time the housing industry boomed. Purchases of homes, automobiles, and appliances were based on easy assess to credit. This was particularly true during the latter part of the expansion after 1926. Innovations in debt finance such as installment loan purchases and new home mortgages were used to extend the expansion phase of the business cycle beyond its rational limits. Surpluses of raw resources and finished goods—such as oil, tin, rubber, copper, steel, wheat, corn, leather, houses, and autos—began to pile up as consumer demands for new commodities were satisfied. Employers had little choice but to lay off workers, who also were consumers in debt. Basically by 1929, the limits to credit expansion had been reached. Consumers had gone deeply into debt to buy material goods while producers had gone into debt to expand their supply. Without employer and employee income to pay off the debts, divesture and bankruptcy were unavoidable.

The depression progressed rapidly because of the way the stock market and trade were regulated by government. During the last half of the 1920s, a huge boom was underway in financial assets, primarily stocks. This was initially based on the growing profits for companies in the new industries, such as General Motors and Radio Corporation. However, as the expansion continued, the stock market became increasingly speculative. In other words, investors bought stocks based more on the idea that previous price increases would continue rather than on the basis of future earning prospects. At the same time, deterioration of investment conditions in Europe, following World War I, helped concentrate speculation in American financial assets. The problem of stock speculation was furthered by the banks and brokerage houses that required investors only maintain a 10% equity position in stocks they purchased. This practice is referred to as buying on margin.

By 1929, the economic expansion that started in 1922 on sound fundamentals had turned into a speculative stock market binge. The reckoning that caused investors to confront reality came from two events. The first occurred when the New York Federal Reserve Bank raised its discount rate from 5 to 6% to slow down stock market speculation. This caused all interest rates to move upward and thereby tightened access to credit. The second event was the announcement of the Smoot-Hawley Tariff Act.

The day the stock market collapsed—Black Thursday, October 24, 1929—was the day President Herbert Hoover announced he would sign the Smoot-Hawley Tariff Act. This act would increase effective tariff duties on imports by nearly 50%. Naturally, other countries retaliated by imposing barriers to imports from the U.S. In other words, savvy investors understood Smoot-Hawley was bad for business domestically and abroad. It severely reduced foreign buying of U.S. goods; particularly farm products. Some economists believe passage of the Smoot-Hawley Act was the primary reason the worldwide depression of the 1930s became so severe (Davidson and Rees-Mogg 1987).

The stock market crash in October of 1929 started a process credit shrinkage that became self-perpetuating. At the bottom, in 1932, the stock market had lost 95% of its peak value in 1929. As equities fell in value, banks had to convert securities into cash to meet customer withdrawals. This reduced their lending capability. Demand was reduced because of lower credit availability from the financial institutions and because common stocks had been the primary assets of many consumers. The fractional reserve banking established in 1913 turned the leverage of the banking system inside out. Bank failures caused massive withdrawals by consumers. Under deflationary conditions it was rational for consumers to postpone purchases. Because deflation raised real interest rates sharply any investment in productive capacity was a losing proposition. The best returns came from buying government bonds. In the 1930–32 period, the real returns from buying government bonds was 6 to 13%, which is quite high by historical standards.

Natural resource-based industries such as farming and energy (oil) were hit the hardest by the deflation of the 1930s. Part of the problem was caused by severe overexpansion in commodity production capacity prior to the depression and part was due to foreign retaliation in response to the Smoot-Hawley Act, which many farmers had so strongly supported.

Keynesian Economic Approach

The failure of market systems in the 1930s led to a new school of economic thinking based on the views of Sir John Maynard Keynes, a British economist (Schiller 1994).

From Keynes point of view, the depression of the 1930s was caused by a shrinkage of demand. He believed the best way to counteract this problem was for government to reduce taxes and spend large sums of money on public works and "safety net" programs to help the needy. It was thought this approach would stimulate the economy and reduce social unrest. Once business activity resumed and the economy was growing at a healthy pace, the government could keep the economy from "overheating" by reducing its spending and/or raising taxes.

The ideas of John Maynard Keynes were applied by President Roosevelt in the mid-1930s when a variety of federal government public works, safety net, and conservation programs were put into effect. At the same time, Social Security was initiated to provide retirement benefits to the elderly. These programs marked the beginning of the United States' movement away from an unregulated market economy to a more mixed economy.

Although it is generally agreed that Keynesian programs implemented by President Roosevelt in the mid-1930s helped alleviate the hardship of the depression, it was the U.S. entry into World War II that really ended it. World War II set up the United States for a long period of prosperity that lasted into the early 1970s. During World War II the production capacity of most of the world was destroyed or severely impaired but the United States was completely untouched. By the end of the war, the United States had the best factories and technology in the world and there was huge pent-up demand both within and outside of the country. Consumer debt had been nearly eliminated by the inflation and forced savings that resulted from the wartime economy. New efficient technologies had largely replaced those that were old and obsolete. Although wars cause great human suffering and hardship, they also bring economic renewal and revival.

From 1950 until 1973 only two mild recessions occured, in 1954 and 1958 (Table 22.1). During this period, the United States experienced strong economic growth, low unemployment, and price stability. The Keynesian economic policies being used by the government appeared to have finally tamed the business cycle. However, the Korean War (1950–53) and the Vietnam War (1965–72) both provided strong stimulation when the U.S. economy might have otherwise faltered.

Stagflation in the 1970s.   In the early 1970s, a new combination of factors came together that severely dampened the previous prosperity. Rather than depression, the outcome of these factors was a condition called "stagflation." It involved rising prices, rising unemployment, and a low level of economic growth (Schiller 1994). During the 1970s, demand boomed for basic natural resources, such as oil, gold, silver, and land. Farm product prices, particularly for grain, made large increases. This was a decade when scarcity, pollution, and human overpopulation were considered major threats to the future of mankind (Ehrlich 1968, Meadows et al. 1972, Brown 1983). Much of contemporary thought on natural resource management came out of the 1970s. We will make the case that the economic stagflation of the 1970s was caused more by economic failure than by natural resource exhaustion. Davidson and Rees-Moss (1987), Schiller (1994), and IBD (1996) are primary sources of information pertaining to this issue.

The basic cause of the stagflation of the 1970s centered around the difficulties that the U.S. government encountered in the 1960s' funding of the Vietnam War and the "War on Poverty" under Lyndon Johnson's presidency (1963–68). Rather than raise taxes to pay for the Vietnam War and the "great society" social programs, the government under Presidents Johnson and Nixon basically chose to fund them through printing the money. At the same time, a variety of environmental laws were passed that raised the production costs for many basic industries. During this same period, Japan and West Germany emerged as strong competitors with the United States in production of automobiles, electronics, and appliances. By the 1960s, the Japanese and Germans had replaced factories destroyed during World War II with those using the most modern innovations. In contrast, much of the industrial capacity of the United States depended on factories that had minor upgrading since the 1940s.

Therefore the United States increasingly was forced to give up world market share of several manufacturing industries to Japan and West Germany, and became more dependent on agricultural exports.

The expansion of the money supply relative to GDP created pressure for U.S. currency (dollars) to be devaluated against other currencies, particularly the German mark and Japanese yen. Based on an agreement reached in 1944 (Bretton Woods Agreement), the International Monetary Fund (IMF) was established. The IMF fixed all exchange rates in terms of the dollar and the dollar in terms of gold. The dollar became the world's principal reserve currency and the only currency that could be directly redeemed in gold. This allowed the United States government to increase the supply of dollars to cover up to a point its excessive spending during the 1960s. That point was reached in 1971 when the United States could no longer meet the demands for redemption of dollars with gold. In August of 1971, President Richard Nixon took the dollar off the gold standard, which set the stage for floating exchange rates (devaluation of the dollar against other currencies). This set the stage for the series of events that caused the "stagflation" of the 1970s and early 1980s.

The devaluation of the dollar caused U.S. exports—primarily farm products—to become more competitive on world markets while at the same time its primary import—oil—became much more expensive. In order to protect themselves against the declining purchasing power of the dollar, the mideastern oil producing nations formed a cartel (OPEC) and demanded higher oil prices. At the same time climatic adversity in China, India, and Russia in the early 1970s depressed world grain production and greatly elevated U.S. grain exports.

The combination of increased world competition, dollar devaluation, oil supply shocks, world climatic adversity, loose monetary policy by the Federal Reserve, and increased regulation of U.S. industry during the 1970s all explain the high levels of inflation, high unemployment, and reduced level of economic growth compared to the 1940s, 1950s, and 1960s.

In contrast to a depression, under inflationary conditions natural resource-based industries such as farming, mining, energy extraction, and forest products flourish. Gold went from $35 an ounce in 1971 to over $800 an ounce at the high in early 1980. Before the dollar's devaluation in 1971, crude oil fluctuated around $2 to $3 per barrel. By the early 1980s, it shot up to $35 per barrel. Wheat, corn, soybeans, cotton, and cattle are some of the farm commodities that showed large price runups during the inflation of the 1970s.

One factor often overlooked as a cause of the natural resource supply shocks of the 1970s was world communism (Davidson and Rees-Mogg 1987). This was the decade when Marxist centrally planned governments were at their peak on a global basis. In China and the former Soviet Union, both agricultural and oil production were severely depressed because of repressive, inept governments that discouraged human initiative and relied on antiquated production technology and equipment. Without question, the extreme inefficiency in use of natural resources and human labor by the communistic countries in the 1970s contributed to the natural resource supply and demand imbalances of the era.

The Economy of the 1980s and 1990s

In 1980, Ronald Reagan—a Republican—was elected president of the United States. He had many very different ideas on how to run the American economy compared to either his Democratic (Jimmy Carter) or Republican (Richard Nixon, Gerald Ford) predecessors of the 1970s. The Reagan strategy centered around restraining the money supply to control inflation, deregulation, lower taxes, encourage immigration, and expand military spending. This mix of programs has been referred to as "supply side" economics. The most drastic element of his program was using the Federal Reserve Bank to restrain inflation by elevating the discount rate (Table 22.1). Throughout the 1980s, this policy of high real interest rates kept the dollar strong and contained inflation. It caused money to flow into financial assets (stocks and bonds) and out of natural resource assets such as gold, farmland, and oil, and other commodities that are inflation hedges.

The strong dollar caused U.S. farm products to be expensive relative to those of other countries, in world markets such as Canada, Argentina, and Australia. During the 1980s and 1990s world climatic conditions were quite favorable for food production. In this same period new technology and market-oriented reforms in China, India, Argentina, and several other countries resulted in a great boost in world agricultural production. Deregulation of the oil industry in the U.S. in conjunction with improved efficiency in energy use caused precipitous drops in oil prices. Between 1980 and 1983, oil dropped from $35 to $9 a barrel. Overproduction was so severe in the farming sector (see Chapter 7) that farmers were given surplus grain if they agreed not to plant part of their land in 1983. In the Food Security Act of 1985, a major provision was to pay farmers to retire their more erodible lands under the Conservation Reserve Program. Although this program had important conservation benefits, its main objective was to reduce the oversupply of wheat, corn, and other farm products. This program has been continued under the 1990 and 1995 farm bills (see Chapters 8 and 12).

Most of the economic policies of the Reagan administration were retained by President George Bush (1989–1992) and President Bill Clinton (1993–2000). So far these policies (monetary and regulatory restraint, free trade, loose immigration laws) have been effective in prolonging economic growth, minimizing unemployment, and keeping prices stable. A major criticism of both the Reagan and Bush Administrations was the high-level deficit spending. The major departures of the Clinton Administration from policies of Reagan and Bush were fiscal policy restraint (reduced federal budget deficits) and some modest increases in taxes on business and capital gains. However, capital gains taxes were lowered by Congress in 1997. All three presidents have been effective in creating conditions that encourage technological development and application.

Currently the biggest economic concerns are the high debt levels of American consumers, local governments, and the federal government, and a demographic shift toward a higher proportion of old people in the population. Many economists have found great similarities between the 1990s and the 1920s (Batra 1988, Davidson and Rees-Mogg 1987). In both periods, taxes were relatively low on the rich, technological growth was rapid, unskilled workers had little improvement in their wages, agricultural prices were depressed, there was a large runup in the stock market, and consumers went heavily into debt to buy houses and new products.

It is our point of view that macroeconomic policies applied in the United States during the 1980s and 1990s have indeed moderated the excesses associated with business cycles of the past. We believe that the global trend toward accelerated technological advance, market economies, and free trade—at least in the near term—will prevent another severe depression such as in the 1930s. However, we also believe recessions and short-term depressions will always be a part of market economies. Further we feel they serve a useful function that is often overlooked. This will be discussed.

CREATIVE DESTRUCTION AND HUMAN PROGRESS

Any serious study of the origin of the American constitution shows its authors believed that cycles of growth and decay were necessary to maintain a nation's vitality and moral fabric (IBD 1996). A basic premise was that periods of hard times tend to make people strong and responsible while waste and corruption are inevitable after a long run of prosperity. Fundamental parts of early American ideology emphasized freedom to fail, as well as to succeed, and strong belief in the ability of the individual rather than the state.

From its beginning anyone who developed new products in the U.S. could easily market them without interference from the government (IBD 1996). In contrast, innova-

FIGURE 22.7 Joseph Schumpeter, an early twentieth-century U.S. economist, saw capitalism as a dynamic system beset by periodic crises he termed "waves of creative destruction" (from IBD, The Bettmann Archive, 1996).

tion under the totalitarian governments that had existed in Europe since the Dark Ages often resulted in punishment because it was perceived to cause political instability. As an example, in 1579 the German inventor of a new technique for weaving ribbons was ordered strangled because it threatened the economic security of the ruling class.

New technology, when applied, causes old technology to be discarded (IBD 1996). Under democratic market economies this results in wealth constantly being transferred among individuals based on their contribution rather than their heredity. It means at anytime that part of the population contributing the most desired products will experience disproportional prosperity to those whose products have become obsolete. In the U.S., the buggy whip and carriage industries were allowed to disappear at the turn of the century when automobiles replaced the horse as the primary means of transportation. The electric light made life easier for people but it destroyed the oil lamp industry.

In the 1920s, the economist Joseph Schumpeter, used the term "creative destruction" to characterize the wave of product cycles and innovation that occur under democratic market economies (Figure 22.7). Schumpeter argued that innovations tended to appear in clusters facilitated by entrepreneurs (Davidson and Rees-Mogg 1987). The new businesses increase the demand for the means of production and labor. This impacts the old businesses by causing their costs to rise and shifting demand away from their products. As consumer preference shifts toward the new products, the older businesses are forced into consolidation and/or insolvency. The profits from the new businesses encourage competition that leads to improvement in the production process. Soon the lure of easy profits causes an overexpansion of the new businesses. When consumers' need for the new products are satisfied and business has overbuilt production capacity, depression sets in to level the excess and encourage the innovations that lead to the next recovery.

From Schumpeter's perspective, human progress depended heavily on permitting business cycles to work with minimal interference from government. He understood that established business had a vested interest in keeping new products and services off the market. Hence the entrepreneur was an essential factor in overcoming the various roadblocks thrown up by these vested interests. To Schumpeter, stabilized market systems as envisioned by today's economists were a contradiction. The great progress that results from waves of "creative destruction" is only possible under a chaotic, uncontrolled economic system that allows the new to replace the old.

PROBLEMS WITH CENTRALLY PLANNED ECONOMIES

We believe our economic discussion would be incomplete without some analysis of why communism failed in the former Soviet Union. This was true to a lesser extent in China under Mao Tse Tung. After Mao's death in 1976, a gradual process of reform was initiated that has made China's transition to a mixed economy much smoother than in Russia.

Centrally planned economies, such as the former Soviet Union, promised equitable distribution of income and full employment. However, the actual reality of centrally planned economies was that the higher-level bureaucrats enjoyed lavish lifestyles while the common citizen lived in severe poverty when compared to the United States and most other market-oriented countries (Schiller 1994). While nearly everyone in the former Soviet Union held a job, they often held unneeded jobs and/or were highly dissatisfied with their occupation. Although the former Soviet Union could keep prices for basic goods stable and at low levels relative to incomes, most citizens shopped in near empty stores and stood in line for hours to buy bread or shoes, or other necessities. Low-quality goods, scarcity of basic goods such as soap, sugar, and hamburger, and poor systems for distribution of goods have been serious problems for all centrally planned economies (Schiller 1994). Since the 1970s, the poorest people in the United States have had far better health care, food, transportation, and housing than the middle class in the centrally planned economies of the world.

The basic problems with centrally planned economies center around human nature and the difficulties that occur when the government tries to control all aspects of the production/distribution process (Schiller 1994). The hard reality is that what works in theory often does not work in practice.

The idea of everyone contributing what he can based on his abilities but taking only what he needs is at the core of the communist viewpoint (Schiller 1994). However, the problem centers around human motivation. Much more training and mental discipline are required to attain the skills of a rocket scientist or biochemist than to be a janitor or a farm field worker. Without the incentives of higher pay and greater prestige, few people want to go through the rigorous four to six years of training required to be an engineer, doctor, or lawyer. Another communist failure centered around lack of tangible reward for good work habits and willingness to take risk.

Satter (1989) provides an excellent analysis of agricultural failure in the former Soviet Union. Under the Czar in 1900 before the communist revolution, Russia had became the largest grain exporter in the world. By 1989, after 62 years of central planning, Russia was the largest grain importer in the world. It had more land devoted to cereal production than any other country and roughly 25% of its workforce was in agriculture compared to 3% in the United States. Yet Russia's grain output per acre was little more than half the United States' production.

In the former communistic Soviet Union, all that mattered was the production plan. Regional officials told the chairman of each collective farm what to plant and how much. Every aspect of farming was to be directed rigidly like a military operation without the individual farmer making a single decision. Although the Soviet government sank nearly $1 trillion into agriculture between 1971 and 1985, there was practically no increase in production.

Under the Soviet system, farmers were often forced to plant inappropriate crops and comply with rigid deadlines for sowing and harvesting. When the plan said it was time to plow, attempts were made to plow even when the ground was so wet that the tractors became stuck in mud. When the plan said time to harvest, crops were cut—ready or not.

Satter (1989) reported that in every place he visited in Russia, farmers were doing the absolute minimum. They fulfilled the plan for sowing, but ignored it for fertilization because that was harder to monitor. Instead of spreading manure, they burned it so no one would know it was not used. Chemical fertilizer was commonly dumped in the streams.

In characterizing the farmworker, Satter (1989) described a pattern of perpetual idleness and alcoholism. Perhaps most interesting was the work pattern of tractor drivers paid by the acre to plow fields. Near the roads they would cut furrows 9 inches deep, but as soon as they were far enough into the field to avoid detection by the farm chairman they would lift the plow, race the engine, and cut 2-inch furrows with harmful consequences to the crop. Although nearly everyone seemed to make some attempt at working in the morning, by afternoon it was common to observe groups of people standing around smoking and talking.

Satter (1989) found that crops that did get harvested in Russia ran a gauntlet of barriers before reaching consumers. These included a lack of trucks to haul crops to the cities, primitive roads that were mostly unpaved, and a lack of suitable storage facilities. Because of poor storage facilities, nearly a third of the harvested grain, fruit, and vegetable production was lost to spoilage.

Prior to its disintegration in 1991, the former Soviet Union made a token attempt to deal with its worsening food situation by allowing farmers to work small tracts on their own. The problem with this scheme was that it operated within the existing collective-farm system that controlled access to equipment, feed, fertilizer, and seed. Satter (1989) describes how one small farmer was able to prosper under the new system of partial privatization but at the same time suffered because of the disapproval of his neighbors. He had to contend with verbal abuse, intimidation, and even physical assaults from a society that had been brainwashed for years to equate personal ambition with greed and prosperity with profiteering.

Environmental Problems in the Former Soviet Union

Environmental pollution has been a major criticism of market-oriented economies. However, major reductions in air and water pollution levels have been achieved in the United States over the past 25 years as a result of government regulations, improved production technologies, and tax incentives (see Chapter 4). In contrast, pollution problems in most communistic countries—particularly Russia and China—increased in the same period (Chandler 1987, French 1991). One part of the problem centers around failure of the centrally planned economies to replace old inefficient production processes and equipment with new and improved ones. Another problem is the lack of accountability of any organization or individual for environmental damage.

By the 1980s throughout Russia and China, it was nearly impossible to find a major river that was unpolluted (French 1991). Water use in steel mills was 50 to 150% higher than in the United States. Sulfur dioxide and nitrogen oxide emissions approached lethal levels in some areas. Toxic emissions of lead and fluoride were permitted in quantities hazardous to human health.

One outcome of environmental decline in Russia has been a shortened average human life span compared to market-oriented economies such as the United States, Great Britain, or Canada (Table 22.2). Reductions in infant mortality in the former Soviet Union have been much lower than those in the United States, Canada, Japan, and the western European countries over the past 25 years (World Resources Institute 1994).

The Collapse of Communism

Because of the various problems previously discussed, the economic collapse of communism was certain by the late 1980s. Just prior to the disintegration of the Soviet Union in 1991, the average Soviet citizen had no telephone, no car, few appliances, and crowded housing (Schiller 1994). Basic necessities—such as soap, sugar, salt, and meat—were rationed and often unavailable. Even the Soviet government

TABLE 22.2   Human Life Expectancy at Birth for Selected Countries

| | Life Expectancy At Birth | |
Noncommunist Countries	1970–1975	1990–1995
United States	71.3	75.9
Great Britain	72.0	76.2
France	72.4	76.9
Spain	72.9	77.6
Canada	73.1	77.4
Mexico	62.9	70.3
Brazil	59.8	66.2
Japan	73.3	78.7
Sweden	74.7	77.9
Communist Countries		
Russia	X	70.0
China	63.2	70.9
Cuba	70.9	75.7
North Korea	61.5	70.8
Vietnam	50.3	63.8

Source: United States Government Population Division.

categorized 40% of its human population as poor. This forced the realization that communism had been a failure and resulted in the new commonwealth of independent states replacing the former Soviet Union.

Today all the various countries in eastern Europe that comprised the former Soviet Union are restructuring and applying various degrees of market reforms (Schiller 1994). The transition from communistic rule to mixed or market economies has not been smooth. Political, judicial, and financial changes are accompanying the economic changes. Converting homes, industries, and land into private ownership has been a difficult task. A major problem in Russia has been the lack of an institutional structure to facilitate market transactions and private property rights (Schiller 1994). The conversion of inefficient state-owned enterprises into productive, profitable operations has caused massive unemployment and factory shutdowns. Implementation of a fair and workable tax system has been another daunting challenge. This in part contributed to the near bankruptcy of the government in 1998. Failure to privatize many large, unprofitable state-owned enterprises continues to drain government resources. The lack of personnel with management skills to run potentially profitable enterprises is also a serious problem.

In spite of these setbacks in Russia, market reforms are gaining ground. Russia has retained its new democratic form of government, and the 1996 election showed the majority want democratic market reforms to continue and do not want to return to the communistic system. If the process of "creative destruction" can go on for another 5 or 10 years without political and social upheaval, many economists believe Russia and most of its former states will begin to experience economic growth and prosperity. Because of its vast oil, mineral, forest, and farmland resources, whatever happens in Russia will be critical to the economic welfare of the rest of the world in the twenty-first century.

PROBLEMS WITH MIXED ECONOMIES

Sweden, Austria, Denmark, the Netherlands, France, and Germany are examples of the mixed economies in Europe that provide a stronger safety net (unemployment, health care, retirement, education, benefits) than the U.S. In order to pay for these benefits, the average worker is taxed at a rate between 45% and 65%, depending on country, compared to 36% in the U.S. (Stein 1993). The United Kingdom (Great Britain) which is the most market oriented of the European economies has an average tax rate of 39%.

The primary problems with the mixed economies in Europe have been a low level of economic growth, low growth in jobs, high levels of unemployment, and more rapidly rising government debt levels than in the U.S. (Stein 1993). GDP growth has averaged about 1.5% for European mixed economies over the past 10 years compared to 2.6% for the U.S. Unemployment across the mixed economies of Europe has averaged 7–12% compared to 6% in the U.S. The United Kingdom, with the lowest tax rate of the European economies, has had the best economic growth rate (about 2.0%) but the 10-year unemployment rate is near 8%.

The generous safety nets and higher taxation levels in the European mixed economies have reduced the incentives to work. Unemployed workers in western Europe can collect 40 to 65% of their salary in government compensation (Stein 1993).

Perhaps most alarming is the trend toward increasing levels of unemployment in the European mixed economies. Unemployment data from the European Union show an upward drift from 7–8% in the 1980s to 10–12% in the 1990s (Stein 1993). Apparently the generous unemployment benefits reduce incentives to find work once workers are laid off. Long-term unemployment causes them to gradually lose their skills and work ethic. This makes them less employable. Studies indicate employers prefer to hire workers with the least amount of time unemployed.

The conservative viewpoint is that too much economic security will destroy American character (Davidson and Rees-Mogg 1993, Oliver 1994a, b). Individualism has been a key element in the economic success of the U.S. Basically, individuals take responsibilities for their actions with little dependency on the state. They expect to fully enjoy the rewards from their thrift and enterprise, but at the same time they accept personal accountability for reckless risk-taking and hard luck.

In contrast the egalitarian, or liberal, viewpoint centers around the idea that the state should be responsible for the welfare of its citizens and should equalize to some extent good fortunes of some with the misfortunes of others (Oliver 1994a, b). In other words, the state is expected to protect people from the consequences of their own decisions.

Another concern with egalitarian or socialistic type of economic system is that taxes needed to pay for public security reduce the pool of capital (savings) available for technological development and enterprise expansion (risk-taking) in private enterprises (Davidson and Rees-Mogg 1993). The high debt levels and low economic growth rates in the mixed European economies to some extent support this concern.

Defenders of the present system in the U.S. argue that provision of opportunity to the disadvantaged is important in maintaining social stability and vitality in any society. Human productivity is likely to be higher if they can take limited risks knowing there is some degree of government support if they fail.

The degree to which success should be taxed and misfortune should be subsidized remains a controversial question. Current thinking by some leading market economists in America, such as Milton Friedman and Gary Becker, is that both taxes and the safety net should be reduced. People should have the option of determining their own degree of economic security. Those wanting more security would live a more austere life, have higher savings, and more catastrophic insurance. Under this approach taxation levels reduced to somewhere between 15 and 25% of average income would provide essential government services such as law enforcement, defense, education, and infrastructure with some help to the poor in terms of food and shelter. Charities and churches would assume a bigger role in helping the needy as was the case during the 1800s and early 1900s. Under this approach, the safety net would be similar to that during the early 1960s before President Lyndon Johnson's "war on poverty."

Presently Congress appears to be moving in this direction with welfare reform. However, failure to make major changes in government social security and health-care programs rule out tax reductions much below the present 36% level.

IMPORTANCE OF
INTERNATIONAL TRADE AND COMPETITION

International Trade

We consider removal of trade barriers to be one of the most critical factors in improving global human welfare. This is because world natural resources are unevenly distributed. Countries such as the United States and Canada that are well suited for wheat, cattle, and timber production have little capacity to produce coffee, bananas, cocoa, or rubber. Russia and other countries in the former Soviet Union have vast reserves of minerals, timber, and fossil fuels, but lack the technologies to extract and transport these resources to factories for conversion into usable products. Densely populated countries—such as Japan, Germany, and Great Britain—have low levels of mineral, timber, and energy resources, but have high levels of technological and capital resources. Through free-trade capital, management, and labor resources can move to where they will be most efficiently deployed. Different countries gain access to products and services they would not have if each country tried to be self sufficient. At the same time, the competition that results from free trade creates strong incentives for industries to use the best production processes and locate where production costs will be lowest. Coffee, aluminum, chrome, tin, tea, and rubber are just a few of the natural resource based commodities that we do not have in the U.S., but are a critical part of our welfare (Schiller 1994).

Another important concept in free trade centers around the benefits of specialization (Schiller 1994). It is much more efficient for a country or an individual to produce a few things and let the marketplace supply the rest. A key concept of Adam Smith's classic work *The Wealth of Nations* is that the organization of labor to produce different things in mass quantities is the key to rapid human progress. Specialization allows different companies and countries to produce different kinds of computers, automobiles, telephones, televisions, and more.

However, the question that comes up is why should a country import things that it also exports. The answer centers around product differentiation and efficiency. Consumers have access to more options and variety if they can buy either Japanese or U.S. cars. This also applies to television sets, computers, cameras, food, and other commodities. It is also true that each country is capable of producing all these goods, but the quantity that can be most efficiently produced will vary considerably between countries.

Schiller (1994) provides an example using bread and wine, which are both produced in the U.S. and France. He demonstrates that France—which is best at producing wine—and the U.S.—which is best at producing bread—both benefit by engaging in trade. Although each country will continue to produce both bread and wine, under international trade their outputs of these products will be adjusted so each country benefits. Schiller (1994) makes the point that, by engaging in trade of goods both countries can produce, consumers in each country benefit by getting more of both products at lower cost.

Comparative advantage refers to the ability of a country to produce specific goods at a lower cost than another country (Schiller 1994). The U.S. has the most land highly suitable for wheat farming, but France has the most land well suited for growing grapes. By growing crops that are most suited to land type and engaging in trade, each country does better than if each used poorly suited lands to grow certain crops because they do not engage in trade (e.g., land well suited for grapes is not necessarily a good place to grow wheat). French wine and bread do not taste exactly the same as American wine and bread. Consumers in both countries undoubtedly enjoy the variety that comes from international trade.

International Competition

By engaging in international trade, countries force businesses in their countries to become more efficient and responsive to consumer needs. The automobile industries in the U.S. and Japan provide an example of how consumers can benefit from international trade and competition. Until the late 1960s, three U.S. companies (General Motors, Ford, and Chrysler) built most of the cars sold in North America. However, during the 1970s, the type of car demanded by U.S. citizens changed from large, heavy, and fuel inefficient to the opposite. This was due to increased gas prices, a flattening of consumer income growth, and the desire for cars that were more environmentally friendly. The small, compact Japanese cars met this requirement. During the 1970s, the American automobile companies lost over a third of their market share to the Japanese. However, during the 1980s and 1990s, the American companies made major improvements in their cars partly in response to Japanese competition. The outcome of international trade in the automobile industry is that American consumers have access to cheaper, better made, more fuel efficient cars that come in a wider variety of sizes and styles and last longer than they did 30 years ago.

International Trade Policies. Without question, free trade results in increased consumption and lower prices. However, individual companies and corporations that have inferior products apply constant political pressure for trade restrictions in the forms of quotas, tariffs, embargos, and other barriers. As we have previously pointed out, the Smoot-Hawley Tariff Act of 1930 contributed much to the Great Depression. The Smoot-Hawley Act raised tariffs to an average 60% and eliminated most imports (Schiller 1994). This act was designed to raise domestic employment and demand for goods. However, it transferred the unemployment problem to other countries and made them less able to purchase U.S. goods. This hardship caused other countries to retaliate with trade restrictions of their own.

U.S. imports from other countries fell 29% while exports fell 33% in the 1930 to 1931 period (Schiller 1994). Worldwide trade declined about 30%, causing massive unemployment and economic retrenchment. Since 1934, the U.S. and most other countries have pursued policies geared towards reducing trade barriers.

Since World War II, broad trade agreements that apply across industries and countries have been the objective of world trade negotiations (Schiller 1994). The General Agreement on Tariffs and Trade (GATT) was signed by 23 of the world's largest trading countries in 1947. Its objective is to pursue free trade and market access for all of its members (now 107 nations). Initially GATT focused on manufactured goods, but more recently agricultural products, music, computer software, movies, and copyrighted books have been included. Although many trade barriers still exist, GATT has resulted in definite progress toward free trade.

The North American Free Trade Agreement (NAFTA), signed in 1992, involves trade agreements among the U.S., Canada, and Mexico. Its goal is the elimination of all trade barriers among these countries within 15 years.

NAFTA is expected to increase employment and lower the cost of goods in all three countries (Schiller 1994). However, in the U.S., jobs in textile and electronic industries are being lost—while in Mexico, job loss has occurred in farming and banking.

The transition into free trade is painful for those who lose their jobs. In order to smooth the transition, the U.S. government has used a variety of programs that include retraining assistance, job search aid, subsidies for relocation, and a period of cash income payments.

ECONOMIES IN THE TWENTY-FIRST CENTURY

The main criticism of a nearly pure market economy—such as what existed in the United States prior to the depression—centers around instability manifest in the business cycle. Throughout its history the United States has been through a series of boom and bust periods caused by development of new technologies, excessive availability of credit, overexpansion by business, and overindebtedness by producers and consumers. These periods were invariably followed by periods of economic retrenchment when the oversupply of goods and excessive debt were leveled by bankruptcy, payback, and consolidation. In other words periods of excess were followed by periods of hardship. A major goal of economic policy since the depression of the 1930s has been to dampen the extremes of the business cycle.

The opposite extreme from a pure market economy is a centrally planned or command economy. Experiences from the former Soviet Union and China have shown centrally planned (communistic) economies to be a failure. Although they achieved the goal of full employment, they have been inferior to market systems in terms providing an improving quality of life to their citizens. Scarcity of goods and services has been a major problem under communistic systems. Technology development and equitable distribution of goods have been poorly achieved. Environmental degradation has been much more severe under centrally planned economies in Russia and China compared to the United States, Canada, and Great Britain.

During the 1980s, the world has moved away from centrally planned economies to mixed or market economies involving a combination of market incentives and government directives. It is recognized that some degree of intervention by government is needed to prevent environmental degradation, provide a safety net for the poor, provide opportunity for upward mobility, prevent excessive concentration of wealth, and to moderate the business cycles. Since the early 1980s, the United States has been fairly successful in achieving these goals while at the same time maintaining incentives for technological development, efficiency, and industry.

A variety of new approaches that are directed by government, but have market incentives, are being tried to deal with natural resource problems (see Chapter 4). Reducing trade barriers and improving cooperation among countries has tremendous potential to reduce environmental degradation and improve human welfare. These have been discussed in other chapters on soil, farmland, rangeland, forest, wildlife, fishery, air, water, energy, and mineral resources.

LITERATURE CITED

Batra, R. 1988. *Surviving the Great Depression of 1990.* New York: Bantam Doubleday Dell Publishing Group.

Brown, L. R. 1983. *Population policies for a new economic era.* World Watch Paper No. 53. Washington, DC: Worldwatch Institute.

Chandler, W. U. 1987. Designing sustainable economies. In *State of the world: 1988.* New York: Worldwatch Institute, W. W. Norton & Co.

Davidson, J. D. and W. Rees-Mogg. 1987. *Blood in the streets.* New York: Simon and Schuster.

Davidson, J. D. and W. Rees-Mogg. 1993. *The great reckoning.* New York: Simon and Schuster.

Ehrlich, P. R. 1968. *The population bomb.* New York: Sierra Club/Ballantine Books.

French, H. F. 1991. Restoring the East European and Soviet environments. In *State of the World: 1991.* New York: Worldwatch Institute, W. W. Norton & Co.

Investor's Business Daily. 1996. Guide to the markets. New York: John Wiley & Sons.

Meadows, D. H., D. L. Meadows, J. Randers, and W. W. Behrens III. 1972. *The limits to growth.* New York: The New American Library.

Oliver, C. 1994a. Socialism vs. capitalism debate. *Investor's Business Daily* 10(248):1–2 (March 3, 1994).

Oliver, C. 1994b. The morality of free markets. *Investor's Business Daily* 11(79):1–2 (August 1, 1994).

Pring, M. J. 1992. *The all-season investor.* New York: John Wiley & Sons.

Satter, D. 1989. Why Russia can't feed itself. *Reader's Digest* 68 (October): 61–66.

Schiller, B. R. 1994. *The economy today.* 6th edition. New York: McGraw-Hill.

Schiller, B. R. 2000. *The economy today.* 8th edition. New York: McGraw-Hill.

Schultz, H. D. 1972. *Panics and crashes.* New Rochelle, NY: Arlington House.

Stein, R. S. 1993. Europe's wilting welfare state. *Investor's Business Daily* 10(161):1–2. (Nov. 12, 1993).

Stoken, D. A. 1984. *Strategic investment timing.* New York: Macmillan.

World Resources Institute. 1994. *World resources: 1994–95.* New York: Oxford University Press.

CHAPTER 23

Sustainable Development, Technology, and the Future

OVERVIEW

We demonstrated in preceding chapters that the natural resource heritage of the United States is rich, diverse, and dynamic. Natural resource management is an information-dependent adaptive process based on scientific principles that extend back to management foundations established over a century ago. We have emphasized that ecosystem services associated with biotic productivity, materials recycling, maintenance of atmospheric composition, environmental stability, and maintenance of genetic diversity all depend on maintaining the functional integrity of ecosystems. Understanding of the functional whole is critical in the sustainable management of natural resources.

In preceding chapters, we have shown effective management depends on understanding ecological and economic principles and their integration into comprehensive management processes. Many advances have been made in the development of natural resources management and conservation principles over the past century. However, we continue to ask the same question now that was asked at the beginning of the twentieth century. Is the present development of natural resources sustainable and, if not, what is needed to make it so?

What has changed most over the last century is knowledge of natural process, development of technology, and the rapidity with which humanity can change the world for better or for worse. Only in the most recent decades have we come to realize how interconnected geophysical, ecological, and anthropogenic processes really are, and how much sustained production of goods and services (Table 23.1) depends on the functioning of properly managed ecosystems.

The functional integrity of Earth processes determines the quantity and quality of natural resource goods and services provided in support of human welfare. The shallow layer of air, land, and water in which we live has serviced humanity for many thousands of years in ways that ecologists and economists are just beginning

TABLE 23.1 Examples of Ecosystem Processes, Services, and Goods

Processes
Production of Organic Matter
Decomposition of Organic Matter
Nutrient Cycling
Grazing Regime
Fire Regime
Hydrologic Regime
Infiltration
Runoff
Evapotranspiration
Soil Erosion Regime
Adaptations and the Evolutionary Process

Services
Maintenance of Atmospheric Quality
Control and Amelioration of Climate
Regulation of Freshwater Supplies
Origin and Maintenance of Soils (and their buffering capacity)
Detoxification and Degradation of Wastes (Pollution dilution)
Natural Control of Pathogenic and Parasitic Organisms (Pest control)
Pollination of Cultivated and Wild Plants
Purification of Air and Water
Renewal of Soil and Water Fertility
Retention and Delivery of Nutrients to Plants by Soils and Water
Genetic Resources (Improve existing and developing new domestic plants and animals)
Aesthetic, Cultural, Spiritual Renewal
Recreational Services

Goods
Foods (mammals, birds, fish, shellfish and other invertebrates, plants, fruits, nuts, spices)
Fibers (cotton, flax, hemp, wool, cashmere, silk)
Fuels (botanochemicals)
Pharmaceuticals and Medicines (psychoactive drugs, Codeine, diuretics, pain killers, antibiotics)
Building Materials (lumber and other woody materials, resins, glues, shellac)
Industrial Products (waxes, rubber, dyes, vegetable fats and oils)
Cooking Oils (plant and animal fats and oils)

Sources: Daily (1997) and West (1995).

to appreciate. Technology has yet to cost-effectively replace many basic ecosystem services, such as clean air, productive soil, clean water, uncontaminated living space, environmental stability, and genetic information. Because of incomplete understanding of human impacts on natural ecosystem functions, future improvement in human welfare is being increasingly jeopardized by environmental degradation. The world's natural heritage teeters at the edge of rapid and irretrievable loss.

Because of unprecedented human population numbers coupled with awesome technology, humankind is impacting the world on a scale far surpassing all previous centuries combined. Nowhere is this more clear at a local level than when rural land suitable for forestry, range, outdoor recreation, or wilderness biodiversity is converted to urban use. In the global view, climate change, biodiversity loss, and resource sustainability have become the major environmental concerns.

While technology advances have done much to diffuse concerns about nonrenewable resources, the greatest concerns are now more broadly based in sustaining renewable resources for the services they provide. Trends in technological advance promise a brighter future for human welfare (e.g., Naisbitt and Aburdene 1990), but the future of humanity will also depend on understanding of ecological, economic, and political systems. The natural resource managers of the twenty-first century must learn from past error and build on past success to meet future challenges.

The world is now a much smaller place than it was just a century ago. What once took months and even years to accomplish in communication and travel now takes nanoseconds to hours. No significant national economy can stand completely independent. Economic depression in Southeast Asia, turmoil in Russia, instability in

Brazil, and consolidation in European currencies make headline news in the U.S. because of their potential implications for the world economy.

Through collective global output, carbon dioxide concentrations have increased over the last century to the highest levels in 160,000 years, based on studies of glacial ice cores. The extent to which the U.S. may be pressured through international politics to cut back emissions of greenhouse gasses is closely related to rain forest management in South America and industrialization in Asia and other parts of the world. The more forest, peat bogs and other natural carbon sinks are converted to carbon dioxide, the more fossil fuel use will need to be slowed, if stabilization of greenhouse emissions becomes a dedicated international policy objective.

The rate of biodiversity loss is unprecedented in the history of human kind, especially in underdeveloped nations. Poorer countries often can be coerced through trade sanctions by richer countries to protect endangered species or to reduce global carbon emissions. However, the morality of such policies is questionable. When ecosystems are protected through international law, it is often unclear who benefits, who bares the costs, and how to proceed fairly. Sound decision making will depend increasingly on the quality of information provided by natural resource ecologists, economists, policy specialists, planners, and managers working with each other and with other disciplines.

New strategies involving a more integrated ecosystems approach to natural resource management, referred to as sustainable development, will be discussed in this chapter. Then we will explore the role of technology in natural resource management. In closing, we will consider what might happen over the next 100 years.

SUSTAINABLE DEVELOPMENT

Defining Sustainable Development

The global concept of sustainable development has rapidly gained the attention of natural resource policy analysts. It is likely to be one of the important concepts in the twenty-first century. Literature referring to "sustainability" has proliferated during the past decade with many variations in its meaning. Thought-provoking discussions on sustainable development are presented by Muschett (1997) and Daly (1997). Sustainable development involves integrating the management of ecological sustainability and economic development to obtain continuous improvement in human welfare on a worldwide basis.

In 1996, the President's Council on Sustainable Development (1996) published a report with goals and recommended actions oriented toward national development. The report encapsulates much of contemporary thinking about sustainable development. The council included a cross section of U.S. leaders representing diverse governmental, business, environmental, civil rights, labor, and cultural interests. Their report identifies many of the important natural resource management issues in the decades to come. The council accepted the definition of sustainable development used by the World Commission on Environment and Development (1987): ". . . to meet the needs of the present without compromising the ability of future generations to meet their own needs" (page iv, President's Council on Sustainable Development 1996).

National Sustainable Development Goals

The President's Council set the desired future condition of the U.S. in its vision statement: "A sustainable United States will have a growing economy that provides equitable opportunities for satisfying livelihoods and a safe, healthy, high quality of life for current and future generations. Our nation will protect its environment, its natural resource base, and the functions and viability of natural systems on which all life depends" (page iv, President's Council on Sustainable Development 1996). Toward accomplishing that vision, the council established 10 goals among which several pertain directly to natural resources management.

Leading the list of goals is achieving a healthy environment for all and an economy of sustained growth that provides a high quality of life for all. Natural resources are to be used, conserved, protected, and restored with assurance of long-term benefits. In addition, the U.S. needs to "move toward stabilization of U.S. population" and lead implementation of global sustainable development policies in the international arena. The global intent of sustainable development is to maintain a rate of resource development into beneficial products that will ultimately provide a high quality of life worldwide. With specific attention to ecosystem management, The President's Council on Sustainable Development (1996, page 119) recommends that the U.S. "Enhance, restore, and sustain the health, productivity, and biodiversity of terrestrial and aquatic ecosystems through cooperative efforts to use the best ecological, social and economic information to manage natural resources." Formal education should "emphasize systems thinking and interdisciplinary approaches . . . and pursue experimental, hands-on learning at all levels" (The President's Council on Sustainable Development 1996, page 74).

Underlying the concept of a sustainable economy, as espoused in the council's report, is the assumption that resources are plentiful enough for continued growth of personal wealth in the U.S. and in other nations of the world. Because the world's national economies are now so closely linked, the concept of sustainable development is unworkable in anything but a global perspective. It is doubtful that any nation, no matter how economically strong, can sustain development indefinitely in a world that otherwise fails to do so. Although an ultimate aim of international sustainable development is improved distribution of the world's natural resource-based wealth, the rate at which that can happen is unclear—as are all of the trade-offs that might be required. Natural resources must be developed and used much more efficiently everywhere if the developing nations are to economically catch up.

Worldwide economic sustainability to a large extent depends on the assumption that the science and technology of resource supply will continue to advance faster than the worldwide demand for resources. The dimensions and uncertainty of this achievement has lead some to conclude that a sustainable world economy will require a lower level of material demand (e.g., Daly 1997). Natural resources analysts will better clarify what is possible and what is not in the coming decades as advances in science and technology close the gap in uncertain knowledge. Sustainability will require greater integration of economists and ecologists in conceptualizing the systems that maintain and improve human welfare.

National Strategies for Sustainable Development

The strategies presented by the President's Council on Sustainable Development (1996) emphasize cooperative partnerships among governments at all levels. Government should promote collaborative multistakeholder approaches to natural resources management. Formal education should "emphasize systems thinking and interdisciplinary approaches . . . and pursue experimental, hands-on learning at all levels . . ." (page 74). The council emphasizes integration of environmental, economic, and policy needs through public involvement and objective science. It also encourages more use of market incentives for attaining environmental goals and encouraging greater flexibility for private business in demonstrating accountability.

The President's Council promotes the concept of extended product responsibility for environmental effects, which is a shared responsibility among manufacturers, suppliers, users, and disposers of the products. It promotes industrial ecology, which studies how to integrate solutions of production inefficiency and waste generation. In the view of the President's Council, government should facilitate a system of collaborative regional planning "accounts" that transcends jurisdictional boundaries and estimates the benefits and costs of alternative land uses. Government also should work with business to improve the design of sustainable human communities and associated landscapes with emphasis on providing alternatives to urban sprawl, new highway construction, and equitableness of public works delivered to communities of different economic status. The council's report emphasizes the importance of personal stewardship in use of the nation's natural resources. They promote voluntary collaborative approaches to protection, restoration, and monitoring of natural resources and resolution of conflicts.

The President's Council on Sustainable Development (1996, page 119) recommends that the U.S. "Enhance, restore, and sustain the health, productivity, and biodiversity of terrestrial and aquatic ecosystems through cooperative efforts to use the best ecological, social, and economic information to manage natural resources." It also recommended that agencies at all levels of government should facilitate local use of ecosystem approaches by eliminating administrative obstacles and by providing information and technical assistance. They should also encourage development of shared goals and responsibilities for restoring damaged ecosystems and by developing collaborative partnerships and compatible databases. Within the ecosystem context, government should encourage sustainable development of agriculture, forests, fisheries, and biodiversity and more efficient use of depletable resources through recycling and other strategies.

Sustainable Development and Conservation

The concept of sustainable development is an outgrowth of the early twentieth-century concept of resource conservation in the U.S. and a large body of subsequent conservation law. An assortment of laws pertaining to public land management in the 1960s and 1970s were precursors to the larger, global view that is now emerging. While the conservation philosophies of the past focused on public lands management and agricultural practices, sustainable development also emphasizes more responsible use of all privately held resources. Sustainable development also extends conservation principles from the national to the global arena. It places greater emphasis on a systems approach and on public policy that works through market incentives.

Although the concepts of sustainable development have roots extending back a century or more, ecological and economic sciences are now more advanced. However, many controversial points in scientific understanding and human idealism remain to be resolved. One particularly troublesome difficulty is the inability to compare ecosystem services that can be priced in economic terms with those that cannot, such as biodiversity. Another problem is determining how the costs of managing for sustainability can be fairly distributed among all public interests. A particular problem in the U.S. is the private land "takings" issue associated with lost economic value of land when its use becomes restricted through law.

Information Deficiencies and Sustainable Development

The rate of sustainable development that can be achieved is contingent on resource status. A complete inventory of resource supply and use rate is yet to be accomplished, even in the developed world. This deficiency of information is especially great for environmental resources, particularly biodiversity. However, some progress has been made and the results are unsettling. Costanza et al. (1997), in an attempt to draw attention to the value of natural services, developed an "order-of-magnitude" estimate of ecosystem service values that rivals the total economic activity of the world (the Gross World Product). Regardless of unavoidable inaccuracies, they made the point that present measures of economic activity do not fully account for the loss of existing environmental services that occur in natural resource development. Much more ecological and economic research needs completion before an accurate accounting is possible and the risks of unsound decisions can be assessed with assurance.

Among the most debated environmental costs are those associated with the loss of global biodiversity (Wilson 1988, Noss and Cooperider 1994). Two-thirds of the world's genetic information could be lost sometime in the twenty-first century based on present extinction rates. With that loss goes the raw material of many potential new products, such as medications, fibers, foods, lubricants, and natural agents of pest control. The potential for genetic engineering depends on the natural occurrence of desirable genes. An international conference in Rio de Janeiro demonstrates the difficulties inherent in global biodiversity protection. Underdeveloped nations, mostly in the Tropics, hold claim to the greatest reservoirs of biodiversity. They want some form of compensation for beneficial discoveries based on unique chemical, physical, or biological attributes of native species in their countries. Nations with low diversity and more potential to develop new products from nonnative sources, have been less enthusiastic about sharing the benefits of genetic discoveries.

Biodiversity is important in reducing environmental disturbances, such as flooding, fires, and storms. Although the services provided by diverse communities are well documented—such as stabilization of eroding soils and storage of nutrients—their economic values are difficult to fully quantify. We have barely developed a preliminary list of the ecosystem services provided (e.g., Daily 1997). We know little about the cost-effectiveness of natural services compared to engineered alternatives. Much has been said, for example, about the importance of floodplain and wetlands in maintaining water supply and trapping nutrients and contaminants. However, there is as yet no complete assessment of how natural services compare to reservoirs, sewage treatment plants, and other humanly modified ecosystems.

Restoration of natural service is emerging as an important resource management strategy. However, restoration science is relatively recent and much remains to be learned. Some types of resource restoration are easier to accomplish than others. Rebuilding vegetative biomass for soil formation, erosion control, and sequestering greenhouse carbon will take a much shorter time than rebuilding biodiversity once it is lost. However, complete recovery of ecosystem biomass, soils, and water quality is contingent on biodiversity in ways that are now incompletely understood.

Sustainable Development in U.S. River Corridors

As indicated in the President's Council on Sustainable Development (1996), sustainable development is relevant in all aspects of natural resources management in the U.S. Nowhere, however, is sustainable development more evident than in the river corridors of the U.S. Here urban, agricultural, forest, recreational, wildlife, fisheries, and water management activities converge where a disproportionate number of endangered species reside. River corridors provide ecosystem connections that support many aquatic and terrestrial species at risk of extinction. River management provides some of the most insightful trends because many different types of land and water resource use interact and the economic stakes are especially high. Improved management of river corridors, including restoration of desirable natural services, is the object of recent federal legislation involving several federal resource agencies.

Anthropogenic changes in the river corridors exacted insidious costs from environmental services, which have yet to be totaled. Degradation of certain river and floodplain services has been associated with agricultural and urban run off, urban sprawl, forest management, and water resources management. Past environmental legislation—the Clean Water Act in particular—has improved human welfare in river corridors. While dams and levees have improved navigation, reduced flood damage and increased water supplies, they also have contributed to the fragmentation of aquatic systems and the gradual decline of numerous aquatic species (Noss and Cooperider 1994). Until recently, the biodiversity losses were overlooked or viewed as unfortunate, but justifiable costs for the greater benefit derived from dams and levees.

Most recently, restoration has been initiated in some of the ecosystems where environmental degradation has been particularly costly. Although most existing restoration projects are local and small, some in progress are substantial, such as proposed restoration of the Everglades (see Davis and Ogden 1994 for background). The Kissimmee River in Florida, located in the upper part of the Everglades watershed, is being returned to its natural configuration, having been previously straightened by dredging and confined by levees (Toth 1995).

Additional engineering for flood control has reached a point where the costs exceed the benefits in many river systems. While removal of large dams or levee systems was at one time unthinkable, it is now being practiced, planned, or contemplated in North Carolina, Maine, Florida, Idaho, and California. Because of concern over losing salmonid diversity in the Columbia River watershed, river managers and stakeholder publics are considering operational modification or bypassing four hydroelectric reservoirs in the Snake River (a tributary to the Columbia); perhaps ultimately removing them.

It was not until the Endangered Species Act of 1973 that the costs associated with possible species extinction were seriously considered in the mix of developmental costs and benefits. We have also learned much more about other natural service values of floodplains. In the past, small towns sometimes were moved so reservoirs could be constructed to prevent downstream areas from flooding. It has only been within the last decade that homes and small villages have been moved out of floodplains to maintain ecosystem integrity and to avoid future flooding costs. Future management is likely to increasingly include breaching and set back of levees and removal of old dams as we gain improved understanding of the total costs and benefits of river corridor services. These efforts will require comprehensive management planning and integration at scales rarely attempted in the past.

Land Control and Sustainable Development

One of the practical problems associated with sustainable development is the issue of property rights. The amendment to the Clean Water Act of 1977, called the Comprehensive Wetlands Conservation and Management Act of 1995, makes it clear that wetland protection should not inhibit economic development and should protect private property rights (Hammer 1996). This includes compensation to the owner if the property value is diminished by 20% or more. Most wetlands involve private property. The intent of wetland regulatory law is no net loss of wetland function and value. Therefore, wetland filling for development requires mitigation by creating or restoring wetlands elsewhere. Similar issues continue to emerge with respect to declaring critical habitat for threatened and endangered species on private lands. Even more difficulty is encountered when threats originate in remote areas, as is often the case with watersheds.

How much value do wetlands, an endangered species, or other environmental services contribute to sustaining national and global wealth? Is it enough to justify payment to private property owners at rates determined by real estate values? Protection and restoration of ecosystem services may require conversion of privately owned lands in part or entirely to public ownership. The present inability to accurately value many environmental resources is of concern because of the need to provide fair monetary compensation to private owners for deprivation. Restoration of valued natural services is not likely to progress extensively until such values can be assessed accurately and private landowners are compensated fairly for the benefits they must forego in the national interest.

Costanza et al. (1997) have written a highly controversial and provocative article describing the values of ecosystem services. This article has drawn widespread criticism from many economists who believe the methods used were technically flawed. Costanza et al. (1997) estimated that the sum total value of all of the world's ecosystem services was conservatively worth about $33 trillion, more than the world's Gross Domestic Product. The article gained support from those who believe that approximate and flawed estimates are better than none at all. The article drew attention to the prevalent inclination to overlook and undervalue natural services in the pursuit of economic development. Thus, the value of the paper is in calling attention to policy needs and methodological deficiencies pertinent to the economics of sustainable development. A future challenge in sustainable development will be to provide a much better understanding of ecosystem services than now exists (Cairns 1996).

ECOSYSTEM MANAGEMENT

Defining Ecosystem Management

The primary means for pursuing sustainable development of natural resources is through a process now commonly known either as ecosystem management or as integrated resource management. Ecosystem management is the process of identifying management problems and implementing solutions in the context of communities and environmental variables in a specified area. The management space is defined by ecosystem boundaries instead of political and jurisdictional boundaries. Because ecosystems are influenced by material and energy flows originating outside their boundaries, much of ecosystem management must address interactions across boundaries, including the actions of living organisms and, most obviously, the actions of humanity. Franklin (1997) pointed out that a common theme of ecosystem management is managing toward the goal of sustainable ecosystems. Haney and Boyce (1997) and Vogt et al. (1997) emphasized that social goals are usually incorporated, including resource output goals linked closely to the concept of sustainable development.

Ecosystem management is a concept that goes back at least 30 years (Van Dyne 1969) but has only in the last decade been adopted as a philosophical approach to federal management of resources. In 1995, a Memorandum of Understanding agreeing to use of an ecosystems approach to management was signed by the federal agencies that are involved with natural resources management. One intent of ecosystem management is development of a vision of desired managed condition that is shared by all stakeholders. Because of its complexity, ecosystem management can be nothing other than an interdisciplinary and interagency team approach to public resource management. In addition to ecological attributes, Grumbine (1994) found that ecosystem management implied scientific research and monitoring, interagency cooperation, organizational change, and management goals determined by human values. Because ecosystem boundaries cross administrative boundaries, one of the most challenging aspects of ecosystem management is the need for interagency and interdisciplinary team integration. Typically a diverse array of stakeholder needs and agency functions need to be accommodated. Because detailed understanding of ecosystem process at a local level is rare, research and monitoring need to be integral parts of effective ecosystem management.

It has only been in recent decades that understanding of ecosystem function and linkages to other ecosystems have become scientifically well enough understood that ecosystem management could be considered a reasonable approach to management. An important stimulus for ecosystem management has been associated with the shortcomings of the Endangered Species Act of 1973. Because the Act provides no authority for preventing losses before species become threatened or endangered, it is a last resort for biodiversity protection. Over 20 years of listing 727 species, only 16 species were delisted, 7 because of extinction (Nester 1997), and over 3,000 are being considered for listing. Ecosystem management is viewed as a promising way to identify stresses before they result in threatened or endangered (T&E) status and, through comprehensive planning, prevent species from becoming endangered. However, much confusion remains about ecosystem management, because ecosystem functions and structure are not completely understood.

One major problem derives from the open-system nature of most ecosystems, meaning that boundaries can be placed on ecosystems in a variety of justifiable ways depending on management objectives and perspectives. The diverse stakeholders in ecosystem management typically identify and emphasize different problems. Thus flexibility is important and Vogt et al. (1997) believe that precise definition of ecosystem management is likely to introduce too much specificity and rigidity in a concept that derives meaning from comprehensiveness and inclusiveness. On the other hand, when ecosystem management is designed too generally, as is the tendency, it can lose utility as a guiding concept. Franklin (1997) emphasized the importance of choosing the right scale proportional to the problem. He cited as an example the problems associated with managing the large mammals of Yellowstone Park, which range through the "greater Yellowstone ecosystem" surrounding and including the park.

Integrating Resources Management into Ecosystems Management

The concept of integrated resource management is an approach that often parallels and links closely with the concept of ecosystem management. In some applications they are virtually indistinguishable. Integrated resource management is the synthesis of resource management system fragments into a functional whole. Thus, the objective of integrated resource management is to bring all of the management fragments into a planning and implementation whole that more effectively sustains resource outputs than the traditional fragmented approach to management. By its very nature, ecosystem management is a form of integrated resource management that develops a systems synthesis across the full range of resources yielded from the ecosystem.

The integrated approach to natural resource management has been advanced most clearly in watershed, coastal-zone, agricultural and forest resource management (see, for example, Heathcoate 1998; Kidd and Pimentel 1992). The concept of more integrated public management of resources extends back at least to the surge of conservation laws passed in the 1930s. Many of these laws involved coordination of agencies in attaining water, soil, and wildlife resource ideals. Managing certain federal lands for multiple use goes back to 1960 legislation that requires coordinated integration of stakeholder demands and assessments of resource supply. In 1965, this concept also was incorporated into federal water resource planning legislation in river basins with the intent of improving water supply for agricultural, municipal, industrial, navigational, and recreational uses while protecting environment quality. The scope of that approach has increased through the years as social awareness has grown about the values of other watershed resources and their interaction with water supply. For agricultural and forestry resources, Kidd and Pimentel (1992, page 22) define integrated resource management in terms of "the design and operation of agricultural systems so that all of the parts of each system operate together to enhance productivity and sustainability."

Compared to integrated resource management, ecosystem management is a more inclusive concept, which emphasizes optimization of the whole array of resource outputs as objects of management instead of some specific selection. Increasingly the outputs are described as ecosystem services. In Franklin's (1997) view, sustainability pertains to the potential of ecosystems for producing all goods and services for indefinitely long periods (see also Daily 1997). A key challenge for ecosystem management is to integrate all of the different resource systems associated with range, forest, farm, outdoor recreation, urban, environmental, fish, wildlife and biodiversity resources into a holistic framework. That holistic level of interaction among management agencies and stakeholder groups remains more an aspiration than a reality because of the different resource perspectives.

Agencies come to ecosystem management with different expectations based on their different concepts of mission. It is common for those agencies and disciplines concerned with water-based resources to accept a watershed management approach as the ecosystem approach. However, for those agencies concerned with the movements of large mammals or birds, the watershed approach has little meaning. Terrestrial migrants, such as waterfowl, pay little attention to watershed boundaries. Similarly, such boundaries have little meaning to those concerned about atmospheric and oceanic resources. Thus, ecosystem management must somehow layer different ecosystem perspectives over one another and integrate the management of those layers if the optimal emphases on diverse resources and human services is to result. A computer tool with potential for facilitating this approach is geographic information systems (GIS).

Some confusion exists because some philosophers equate ecosystem management with restoration and maintenance of natural ecosystems. Ecosystems, however, may be managed to emphasize goods and service provision without necessarily providing them in the wilderness state. These services include both market services, such as timber and food fishes, and non-market services, such as clean water supply, recreation, and biodiversity. Whereas many services can be priced either directly through the economic market, or indirectly through benefits analysis, other service values, such as biodiversity provision, have as yet to be satisfactorily estimated in monetary terms. For agencies and stakeholders organized according to special interest advocacy this integrative concept is difficult to assimilate. However, welfare economists have wrestled for decades with developing a benefits-cost paradigm for facilitating such integration (see, for example, Loomis 1994) and progress is being made.

Because the ecosystem concept is so comprehensive, practitioners of ecosystem management need to have more than an introductory knowledge with ecological science and more than an intuitive sense of how to facilitate positive stakeholder input into planning and management process. It requires an ability to work in partnerships to identify all stakeholder problems and their interactions. It also requires framing solutions in a shared vision of desired ecosystem function while sustaining options for future generations. The concept requires managers who can interact effectively in an interdisciplinary and often contentious environment. While natural resource managers need special expertise in a particular area, they also need to be broadly enough informed so they can contribute effectively on interdisciplinary teams.

Ecosystem Health and Adaptive Management

Two concepts closely associated with ecosystem management are ecosystem health and adaptive management. All strategies of ecosystem management are increasingly being judged in terms of the ecosystem health that results. Costanza et al. (1992) indicate that self-maintenance signified ecosystem health. A degraded state is one in which functions could not be sustained indefinitely without costly management intervention. Self-sustaining states have real cost advantages over ones that must be maintained at human expense.

Whereas scientifically developed cause-and-effect understanding should underlie ecosystem-based natural resource management (see, for example, Boyce and Haney 1997; Vogt et al. 1996), controlled experimentation at ecosystems scale often is technically infeasible and prohibitively expensive. An alternative to controlled experimentation is adaptive management (Holling 1978, Walters 1986). In adaptive management, managed ecosystems are scientifically monitored to assess the effectiveness of their management and the results analyzed and published to improve management. It might be thought of as learning by doing and redoing from what was learned. Adaptive management is a long-established concept, if otherwise labeled such as "experience is the best teacher." The more formal process espoused by Holling (1978) and Walters (1986) promotes an organized approach that includes formal concept development, often in the form of a mathematical model of how management is believed to work in the ecosystem context, measurement of system outputs indicative of management effectiveness, analysis of results, sharing of results through a scientific peer review process, and modification of the model for future management action.

The institutionalization of adaptive management has been slow to occur, however. One of the primary fears of agencies with a management authority is that it will divert scarce management funds into research with relatively little management application. There is concern that society and policy makers have insufficient understanding of science and scientific process to tolerate the uncertainty implied by a significant diversion of funds into investigation of results. There is much to be said about that position, which identifies one of the greatest impediments to future resource management progress in democratic societies—the growing gap between scientific management of resources and the ability of the voting public to understand the needs, risks and benefits associated with scientific management of the human environment. Thus ecosystems management education cannot stop at the level of resource management professionals.

TECHNOLOGY

In addition to more effective global strategies, the future effectiveness of natural resource management in sustaining ecosystem services will depend on human mastery of its technology. The history of technology reveals it has been an environmental double-edged sword, producing both problems and solutions. Many technological advances are underway that could facilitate more effective resource mangement, once they are carefully integrated into a comprehensive management planning process.

The Importance of Computers

Natural resource managers usually must gather, organize, and analyze large amounts of information to predict the outcome of different management practices. In the early years, data were entered and stored on paper forms and analyzed by hand. Invention of adding machines greatly facilitated development of modern data management and analytical methods. Since the first commercial computers were built following World War II computer technology has advanced at a phenomenal rate. This has allowed information processing at speeds several orders of magnitude greater than before World War II.

Computers "process" information input through software programs that direct computer operations. Computers are unsurpassed as a means for organizing and managing data, including storage and retrieval when needed. Portable computers are now routinely used in various natural resource field assessments. Data are either entered by hand or by way of a wide variety of electronic meters, including digital cameras, and then carried or relayed to a larger processor computer.

For much natural resources management, data are increasingly being organized through technology GIS. GIS data in the form of maps may be analyzed for variables such as vegetation types, wildlife species, land use, soil types, geological formations, mineral deposits, watershed processes, land ownerships and jurisdictions, and proposed development. The advantages of this type of data organization are numerous considering the potential for computer software to evaluate systematic relationships. One of the more important uses at this time is the identification of threatening land-use patterns with respect to the distribution of rare and endangered species and ecosystems.

Systems Analysis

Successful natural resource management increasingly depends on current knowledge of biological and financial outcomes of various management practices. In tomorrow's world, any successful natural resource program will depend on highly skilled personnel who can use computers to assess the risk/reward ratios of various management options. We consider a strong background in computer science, business management, and economics as important to today's natural resource manager as knowledge in soil science, range science, ecology, agronomy, forestry, wildlife management, and land use planning.

Natural resource managers usually must consider large amounts of information to predict the outcome of different management practices. These managers can and do benefit from computer data compilation and synthesis systems geared toward practical problem solving. Since the 1970s, computers have been used increasingly by the natural resource managers in both data storage and data analysis. George Van Dyne is considered the pioneer in development of computer applications to natural resource problems. During the 1960s and 1970s, he developed computerized models that integrate knowledge on various aspects of rangeland ecosystems into a framework for management decisions (Van Dyne 1966a). The approach adapted by Van Dyne and others to facilitate analysis of ecological process is referred to as *systems analysis* and is discussed in detail, as it relates to natural resources, by Van Dyne (1966b), Van Dyne (1969), and Shugart and O'Neill (1979).

Analysis of management systems is the process of defining goals and objectives for ecosystem management and identifying procedures for their efficient accomplishment. The most important attributes of systems analysis are that it provides a unifying structure for interrelating facts and observations, and that it serves as a dynamic force as new facts and observations become available.

Models

An ecosystem model is a plan or method of organizing information about an ecosystem (Figure 23.1). It is an abstraction of a real system. It does not represent the system in its entirety, but it does simulate the ecosystem in aspects regarded as essential by the modeler. For example, no conceivable model could possibly represent all the plant and animal species present in a ecosystem, but it must be reliable in representing the behavior of that portion of the system being simulated. The medium of the abstraction is the various symbols of mathematics, and the real system variables have analogous mathematical variable counterparts. Thus a model is a mathematical representation of the portion of the ecosystem under study.

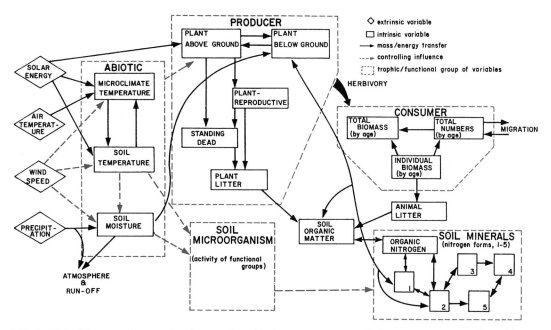

FIGURE 23.1   Diagram of a grassland ecosystem. Each arrow corresponds to one or more equations (from Bledsoe and Jameson 1969).

Models can be classified as predictive or theoretical, based on their intended use. Predictive models predict the future behavior of variables, while theoretical models provide insight into how the system functions. Predictive models are validated by establishing, to a reasonable degree, the degree of accuracy to which the model simulates the system's behavior and the conditions over which the model can be used. The more complex the ecosystem, the more difficult it is to predict the results of management programs. Models provide a tool for examining the effects of many manipulations on the ecosystem and determining the results without doing field trials. However, enough field validation must be done to ensure that the model represents the behavior of the system. Validation of theoretical models consists of attempting to disprove the model (i.e., to find a single instance in which the model is not similar to data from the real world) so that confidence in theory can be established. Most ecosystem models combine predictive and theoretical functions. A computer is needed to handle the huge volume of data acquired for each ecosystem unit. Models are useful for analysis of numerous natural resources phenomena both in and outside an ecosystem context.

Artificial Intelligence

The use of artificial intelligence in management of natural resources began with the development of expert systems for problem solving and decision making. An *expert system* is a type of artificial intelligence program that follows a few general procedures for solving problems. It uses facts (often obtained from written reports), experience, and models stored in the memory of a computer by human experts. There are two ways in which the computer can arrive at conclusions. It can reason forward, going from facts to solution, or it can work from a hypothetical solution to seek supporting evidence. Using the two reasoning approaches to solve a particular problem, the computer suggests a set of hypotheses based on data input by the user. The system then considers each hypothesis in turn, attempting to find a specific solution. The artificial intelligence programs analyze decisions, can interpret the meanings, or can ask appropriate questions. Thus the computer aids managers who often reach conclusions from partial or uncertain evidence by following possible lines of reasoning. Also, the system can add to its database as it gains experience. An expert system was used to develop a rangeland simulation model (Ritchie 1989). Simulation allows the manager to examine entire ecosystems and to alter various parameters and determine their biological and financial outcomes. Major management adjustments may be recommended as a result of even minor adjustments to variables. Maintaining a high degree of accuracy in decision-support functions requires a commitment to collect, develop, and verify the experts' knowledge.

The use of expert systems led to the development of other artificial intelligence procedures pertinent to the management of natural resources. Some of these are: (1) integrated expert systems, which link management models with natural resource models; (2) intelligent GIS files; and (3) artificial intelligence modeling of ecosystem components and their interaction with the environment (Coulson et al. 1987). Expert systems have led to knowledge systems, intelligent decision systems, creativity systems, and literature research systems (Truett 1989). Although many advances have been made in artificial intelligence, their promise remains far from fully realized.

Virtual Reality

Virtual reality (VR) is a computer simulation usually experienced through headgear, goggles, and sensory gloves that lets users feel they are in another place. VR devices let you see, hear, feel, and interact with real or abstract data from computer-generated models. Perspective views may be generated by combining digital terrain data with satellite imagery, scanned aerial photographs or maps, airborne video, ground-penetrating radar, and GIS files (Truman et al. 1988, Everitt et al. 1990, Smith et al. 1991). The three-dimensional views are used to visualize and verify the results of the natural resource models. Database visualization in 3-D enables the decision makers and the general public to more readily understand complex data.

Economic Analysis

Natural resource economics is used to determine efficiency and equity. Efficiency involves allocation of scarce resources among competing uses for defined periods to produce the greatest quantity of net output or products from a given amount of input. Often, maximum profits occur at some productivity level below maximum productivity. The manager must determine the level of production that yields the greatest net returns. Equity concerns the distribution of products among competing consumers and owners of resources (Schiller 1994). Resource allocation affects both efficiency and equity (Gardner 1984). In some situations, efficient production is a more desirable goal than maximizing gross production.

Many practices necessary to sustain natural resources—such as farmland, rangeland, forest, and wildlife—are difficult to evaluate economically. Benefits such as the esthetic quality of a pristine alpine lake, an uncluttered grassland vista, or saving an endangered species are difficult to quantify monetarily. Often it takes many years to realize the benefits of farming practices that reduce soil erosion, or grazing practices that will improve rangeland. In many cases, what is most profitable in the short run may be unprofitable in the long run and will adversely affect several future generations. Soil on farmlands, rangelands, and forests may be considered a nonrenewable resource because it takes many years to form through geological and biological processes. However, soil losses due to a few years of improper grazing, logging, or farming practices can erase what took nature thousands of years to build. This represents a cost to future societies as well as to landowners.

Under some conditions, such as war or climatic adversity, short-term necessity may take precedence over long-term sustainability. Computers provide an essential tool in determining the cost/return ratios of various management alternatives to find the ideal combination of short- and long-term benefits.

Models of ecosystem processes have been linked to economic and management process to aid management policy analysis (Figure 23.2). Cole et al. (1990), for example, describe a model for analyzing the sport fishery of New Mexico, the benefits of which depend greatly on integrating water management with biological management and management of an array of amenities including boat ramps, campgrounds, and sanitary facilities. Once developed, models of this type can be analyzed for the most benifical combination of management provisions. They are, however, data demanding, and require an investment in adaptive management that has yet to be widely accepted in natural resource management.

Farming is so sophisticated now that many farmers have computerized maps of soil characteristics of their individual fields. These maps are used to precisely determine what level of fertilization and irrigation will maximize yield. Modern irrigation equipment uses these computerized maps to allocate nutrients and water to each part of the field so financial returns will be maximized.

THE FUTURE

When looking to the future there is always a danger of omitting an event of great importance or including something that does not happen. However, we feel that it is very important to look at the future and discuss some possible events that may be important.

Over the past 50 years, agriculture and forestry in developed countries has been transformed from a resource-based industry to a science-based industry. Future improvements in human welfare will depend very heavily on maintaining or increasing the role of technological advance. Any natural resource management strategy must be developed with some idea of what the future (twenty-first century) will be like socially,

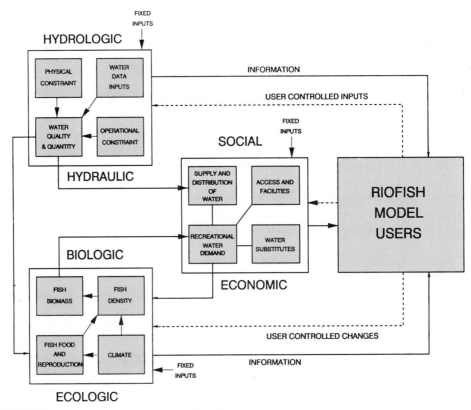

FIGURE 23.2   The integration of ecological and economic process into managment systems models can facilitate policy analysis concerned with providing efficient and sustained benefits. This diagram illustrates the main interaction among geophysical, ecological, economic, and management components in a state sport-fishery managment system (from Cole et al. 1990).

politically, culturally, and economically. It is important to recognize that social demand determines the value of any natural resource product or commodity. Therefore natural resource scientists, managers, and producers must have at least a basic understanding of consumer needs and demand in the future. This is particularly important when the long production cycles of many natural resource products are taken into account, such as for oil, water, timber, fish, and various types of wildlife.

Because of massive technological breakthroughs, an economic boom was predicted for the 1990s that will last far into the twenty-first century. Naisbitt and Aburdene (1990), in their book on future megatrends, make the following statement:

"The global boom of the 1990s will be free of the limits on growth we have known in the past. There will be an abundance of natural resources throughout the 1990s from agricultural products and raw materials to oil. Everything that comes out of the ground will be in oversupply for the balance of this century and probably much longer. We will need fewer raw materials, as we have been moving away from material-intensive products for decades. A prototypical example of the shift away from the material intensive products is fiber-optic cable. Just seventy pounds of fiber-optic cable can transmit as many messages as 1 ton of copper wire. Equally important, those 70 pounds of fiber-optic cable require less than 5% of the energy needed to produce one ton of copper wire. There will be no energy crisis to impede the 1990s global boom. Each year since 1979, the United States has used less energy than the year before."

So far, that prediction has been mostly realized, although there are signs that sustainablility of economic development may be undergoing strain. The recent depression in growth of Asian economies is one such sign, as is the recent, upswing in automotive fuel use in the U.S. An additional concern is accelerating loss of biodiversity providing resource development options for the future. A spin-off from the new economic boom is a renewed concern and appreciation for the natural environment. On this subject, Naisbitt and Aburdene (1990) state:

"The world's preoccupation with defense and the cold war, which is receding, is being replaced by concerns about the destruction of our natural environment, now our most important common problem."

In his vision of the future beyond the year 2000, Drexler et al. (1991) stated that nanotechnology (extreme miniaturization) will permit the human race to feed itself with ordinary, naturally grown, pesticide-free foods while returning more than 90% of today's agricultural lands back to nature. If present trends continue, natural resource managers will lose credibility if they use the gloom and doom predictions since the time of Malthus to justify future research and management. Some economists consider the only food problems of today's world to be those of distribution and political incompetence. However, many ecologists and environmentalists question whether the technological breakthroughs in the future can keep up with those of the past.

Nature—the environment in a natural state—is increasingly attaining monetary value as a source of wealth itself (Rolston 1989). The trend is toward cleaner and greener wealth being equated with greater wealth (Drexler et al. 1991). We believe peoples throughout the world will continue to increase the importance they place on environmental quality relative to material goods in their concept of personal and economic well-being. This does not mean that natural resource managers can ignore the importance of basic commodity production. We believe short periods of food, energy, and timber scarcity will occur in the twenty-first century, as they have in other centuries, because of climactic, political, and other types of adversity. However commodity production will increasingly be considered as only one aspect of natural resource management, particularly in the more developed countries.

A new electronic heartland was predicted for the United States by Naisbitt and Aburdene (1990). They stated: "Linked by telephones, fax machines, Federal Express, and computers, a new breed of information worker is reorganizing the landscape of America. Free to live almost anywhere, more and more individuals are deciding to live in small cities and towns and rural areas. A new electronic heartland is spreading throughout developed countries around the globe, especially in the United States. Quality-of-life rural areas are as technologically linked to urban centers as are other cities. This megatrend of the next millennium is laying the ground work for the decline of cities." More recently Swasy (1994) made a similar statement: "Today's white-collar boom towns were yesterday's cow pastures." In the future, this will only mean greater demands on natural ecosystems, with the preservation of open space turning into perhaps the biggest natural resource challenge.

Another prediction we will make is that the 1990s will be considered the beginning of the age of biological engineering. We may have the ability to create organisms by genetic engineering to produce whatever product society demands (Naisbitt and Aburdene 1990). Many scientists involved in genetic engineering are making this prediction. Boyd and Samid (1993) state: "The amazing pace of discovery within the field of molecular biology has expanded the understanding of as well as the potential for genetic engineering in laboratory and domestic animals. The future of the planet earth and the species that inhabit it can be significantly affected by these technologies. The use of transgenic animals is limited only in the human imagination."

Nanotechnology. Breakthroughs in molecular biology could lead to a prosperity never before experienced by humankind (Drexler 1986, Drexler et al. 1991). They could completely change production processes as we know them today. These new mind-bending technological possibilities are explored by Drexler (1986) and Drexler et al. (1991).

Molecular biology is presently causing a revolution in agriculture. The secrets of DNA are rapidly being unlocked with the aid of computers and biotechnology. The new industry of genetic engineering has developed pest-resistant strains of crops and hormones that dramatically increase meat and milk yields from livestock. These innovations and the use of computerized machinery are reducing the use of fertilizers, pesticides, and water for crop production. As the genes of food and fiber crops are redesigned, food processing requirements could shrink. Meats, cereal grains, fruits, and various vegetables may have flavorings programmed in.

However in 50 years, these developments may seem crude and simplistic compared to what lies ahead. Drexler (1986) points out that scientists in the United States and Japan are involved in molecular engineering efforts that could result in using the techniques of genetic coding to form microscopic computers. The potential outcome is computers that are so tiny they would fit into a human cell.

Today production processes are organized around processes that rearrange masses of atoms (Davidson and Rees-Mogg 1987). These manipulations involve heating, cooling, forging, casting, pressing, pounding, and chemical mixing. People involved in these acts work with trillions of atoms. In contrast, the new nanotechnologies (microscopic computers) will build products based on the manipulation of individual atoms. While genetic engineering is the first step in this process, the techniques could be used to make nonliving products. Keep in mind that genetic codes appear to be merely "machine tool systems" for putting together living molecules. There is no known reason once these processes are understood they cannot be used for nonliving things. Drexler (1986) makes the following statement regarding the outcome of nanotechnology:

"AI (artificial intelligence) systems will bring still swifter automated engineering, evolving technological ideas at a pace set by systems a million times faster than a human brain. The rate of technological advance will then quicken to a great forward leap: in a brief time many areas of technology will advance to the limits set by natural law . . . This information is a dizzying prospect. Beyond it, if we survive, lies a world with replicating assemblers, able to make whatever they are told to make, without need for human labor."

Practically any product desired by humankind could be built by molecular assemblers programmed with artificial intelligence. Imagine products such as watches, washing machines, and stoves with perfectly fitting parts assembled with almost no human participation. If this scenario should become reality, scarcity of material goods would virtually be a thing of the past.

Human life spans might be more than doubled from a new medicine practiced at the level of individual human cells. Molecular computers and machines in individual cells could monitor their performance and preserve them indefinitely. We do recognize, however, that, although nanotechnology has great possibilities, it may never come to pass as it's presently envisioned.

Optimism versus Pessimism.   The darker side of the future centers around economic cycles and demographic changes (human population increase). Generally those who have predicted gloom and doom for the future, such as Malthus and Ehrlich (1968), have focused on the human population overrunning the supply of natural resources. However, throughout the history of the world, we have made the case that economic chicanery and mismanagement have had far more to do with human suffering and environmental deterioration than natural resource scarcity. The declines of ancient societies in Egypt, Greece, and Italy; the extended Dark Ages of Europe; the French Revolution; the rise and fall of Adolph Hitler in Nazi Germany; the 1930s' depression in the United States; and the rise and fall of Communism in the Soviet Union and China all provide examples of great human hardship that resulted from a failure of government policies, economic systems, and political leaders.

Ever since the late 1960s, more pessimistic analysts have predicted that consumer and public debt expansion in the United States and several other developed countries would lead to a severe worldwide economic depression (Davidson and Rees-Mogg 1993). However, the day of reckoning has been delayed by productivity increases, inflation, a wide variety of innovative ways to expand credit, and better systems of banking. Based on history, all debt is leveled sooner or later by payback, default, and/or inflation. Many economists believe the greatest problem that confronts the United States over the next 30 years will be how to deal with its debt problem (Casey 1993, Davidson and Rees-Mogg 1993).

In closing, we believe that the interaction of technological development, information transfer, and application of improved economic and management systems will have far more to do with world human condition in the near and distant future than a scarcity of natural resources or indebtedness. We are optimistic about the future of humanity because of major changes in most developed and developing countries that reflect a trend toward democracy, market economies, protection of individual rights, racial equality, sexual equality, free trade, increased global competition, improved financial management and zero population growth. At the same time, there is growing appreciation for natural environments and the products they contribute. We believe major advances in human thinking will continue to lead to a sustainable and improving future. This will mean finding the appropriate balance among the human population size, its material needs, and its environmental needs for the existing level of technology. While the future, as always, is uncertain, we look forward to it with cautious optimism, and have no desire to return to the past when life was shorter, involved more physical hardship, and its opportunities were far more limited.

LITERATURE CITED

Bledsoe, L. and D. A. Jameson. 1969. Model structure for a grassland ecosystem. In R. L. Dix and R. G. Beidleman, eds., *The grassland ecosystem: A preliminary synthesis.* Range Science Series No. 2. Colorado State University, Fort Collins. 417 pp.

Boyce, M. S and A. Haney, eds. 1997. *Ecosystem management: Applications for sustainable forest and wildlife resources.* New Haven, CN: Yale University Press.

Boyd, A. L. and D. Samid. 1993. Review: Molecular biology of transgenic animals. *Journal Animal Science Supplement* 3(71):1–9

Cairns, J. Jr. 1996. Determining the balance between technological and ecosystem services, (pp. 13–30). In P. C. Schulze, ed., *Engineering within ecological constraints.* Washington, DC: National Academy of Engineering, National Academy Press.

Casey, D. 1993. *Crisis investing for the rest of the 90's.* New York: Carol Publishing Group.

Cole, R. A., F. A. Ward, T. J. Ward, and R. M. Wilson. 1990. *Development of an interdisciplinary planning model for water and fishery management.* Water Resources Bulletin 26:597–609.

Costanza, R. B., G. Norton, and B. D. Haskell, eds. 1992. *Ecosystem health.* Washington, DC: Island Press.

Costanza, R., R. D'Arge, R. De Groot, S. Farber, M. Grasso, B. Hannon, K. Limburg, S. Haeem, R. V. O'Neill, J. Paruelo, R. G. Raskin, P. Sutton, and M. van den Belt. 1997. The value of the world's ecosystem services and natural capita. *Nature* 387:253–259.

Coulson, R. N., J. L. Folse, and D. K. Loh. 1987. Artificial intelligence and natural resource management. *Science* 237:262–267.

Daily, G. C., ed. 1997. *Nature's services. Societal dependence on natural ecosystems.* Washington, DC: Island Press.

Daly, H. 1997. *Beyond growth: The economics of sustainable development.* Boston: Beacon Press.

Davidson, J. D. and W. Rees-Mogg. 1987. *Blood in the streets.* New York: Summit Books—Simon & Schuster.

Davidson, J. D. and W. Rees-Mogg. 1993. *The great reckoning.* New York: Simon & Schuster.

Davis, S. M. and J. C. Ogden, eds. 1994. *Everglades: The ecosystem and its restoration.* Delray Beach, FL: St. Lucie Press.

Drexler, K. E. 1986. *Engines of creation.* New York: Anchor Press/Doubleday Publishing Co.

Drexler, K. E., C. Peterson, and G. Pergait. 1991. *Unbounding the future: The nanotechnology revolution.* New York: William Morrow and Company.

Ehrlich, P. R. 1968. *The population bomb.* New York: Ballantine Books.

Everitt, J. H., K. Lulla, D. E. Escobar, and A. J. Richardson. 1990. Aerospace video imagining systems for rangeland management. *Photo Engineering and Rem. Gens.* 56:343–349.

Franklin, J. F. 1997. Ecosystem management: An overview (pp. 21–53). In M. S. Boyce and A. Haney eds., *Ecosystem management: Applications for sustainable forest and wildlife resources.* New Haven, CN: Yale University Press.

Gardner, B. D. 1984. *The role of economic analysis in public range management. Developing strategies for rangeland management.* National Research Council/National Academy of Sciences, eds. Boulder, CO: Westview Press.

Grumbine, R. E. 1994. What is ecosystem management? *Conservation Biology* 8:27–38.

Hammer, D. A. 1996. *Creating freshwater wetlands.* 2nd edition. New York: CRC Press.

Haney, A. and M. S. Boyce 1997. Chapter 1 Introduction (pp. 1–17). In M. S. Boyce and A. Haney eds., *Ecosystem management: Applications for sustainable forest and wildlife resources.* New Haven, CN: Yale University Press.

Heathcoate, I. W. 1998. *Integrated watershed management: Principles and practice.* New York, NY; John Wiley & Sons.

Holling, C. S., ed. 1978. *Adaptive environmental assessment and management.* New York, NY: John Wiley and Sons.

Kidd, C. V. and D. Pimentel. 1992. *Integrated resource management: Agroforestry for development.* New York, NY: Academic Press.

Loomis, J. B. 1994. *Integrated public lands management: Principles and applications to national forests, wildlife refuges, and BLM lands.* New York, NY: Columbia University Press.

Muschett, F. D. 1997. *Principles of sustainable development.* Delray Beach, FL: St. Lucy Press

Naisbitt, J. and P. Aburdene. 1990. *Megatrends 2000: Ten new directions for the 1990s.* New York: William Morrow and Company.

Nester, W. R. 1997. *The war for America's natural resources.* New York, NY: St. Martin's Press.

Noss, R. F. and A. Y. Cooperider. 1994. *Saving nature's legacy: Protecting and restoring biodiversity.* Washington, DC: Island Press.

The President's Council on Sustainable Development. 1996. *Sustainable America: A new consensus for prosperity, opportunity, and a healthy environment for the future.* Washington, DC: U.S. Government Printing Office, Superintendent of Documents.

Ritchie, J. R. 1989. An expert system for a rangeland simulation model. *Ecological Modeling* 46:91–105.

Rolston, H. III. 1989. *Philosophy gone wild.* Buffalo, NY: Prometheus Books.

Schiller, B. S. 1994. *The economy today.* 6th edition. New York: McGraw-Hill.

Shugart, H. H. and R. V. O'Neill, eds. 1979. *Systems ecology.* Stroudsburg, PA: Dowden, Hutchinson & Ross.

Smith, S. M., H. E. Schreier, and S. Brown. 1991. Spatial analysis of forage parameters use geographic information system and image-analysis techniques. *Grass and Forage Science* 46:183–189.

Swasy, A. 1994. America's 20 white-collar addresses. *The Wall Street Journal* 129: B1, B3.

Toth, L. 1995. Principles and guidelines for restoration of river/floodplain ecosystems—Kissimmee River, Florida (pp. 49–73). In John Cairns Jr., ed., *Rehabilitating damaged ecosystems,* 2nd ed. Ann Arbor, MI: CRC Press, Lewis Publishers.

Truett, W. L. 1989. Artificial intelligence and applications across the 1990s: An applied approach. *American Laboratory* 21(4):40–47.

Truman, C. C., H. G. Perkins, L. F. Asmeissen, and H. D. Allison. 1988. Using ground-penetrating radar to investigate variability in soil properties. *Journal of Soil and Water Conservation* 43:341–345.

Van Dyne, G. M. 1966a. Application and integration of multiple linear regression and linear programming in renewable resource analyses. *Journal of Range Management* 19:356–362.

Van Dyne, G. M. 1966b. *Ecosystems, systems ecology, and systems ecologists.* ORNL–3957. Oak Ridge, TN: Oak Ridge National Laboratory.

Van Dyne, G. M., ed. 1969. *The ecosystem concept in natural resource management.* New York: Academic Press.

Vogt, K. A., J. C. Gordon, J. P Wargo, D. J. Vogt, H. Asbjornsen, P. A. Palmiotto, H. J. Clark, J. L. O'Hara, W. S. Keeton, T. Petel-Weynand, and E. Witten. 1997. *Ecosystems: Balancing science with management.* New York, NY: Springer.

Walters, C. 1986. *Adaptive management of renewable resources.* New York, NY: Macmillan Publishing Company.

West, N. E., ed. 1995. *Biodiversity on rangelands.* Logan, UT: College of Natural Resources, Utah State University.

Wilson, E. O., ed. 1988. *Biodiversity.* Washington, DC: National Academy Press.

World Commission on Environment and Development. 1987. *Our common future.* Oxford, England: Oxford University Press.

Glossary

abiotic Without life.

abyssal zone The deep, offshore ocean occupied by dark, cold waters and ocean bottom beyond the shallower depths surrounding continents and islands.

acid A chemical compound that disassociates into hydrogen ions and ions of other elements. Acid strength is in proportion to the concentration of hydrogen ions free for chemical reaction and measured by pH.

acid rain Precipitation made more acidic than environmental background levels by natural and artificial emissions that form sulfuric, nitric, and other strong acids. Fossil fuel power plants and automotive emissions are major sources.

adaptation A life-promoting adjustment to environment. Evolutionary adaptation derives from genetic change and retention of genes in a population, resulting from greater reproduction by the most environmentally fit members.

administration The process of executing policy through organizational communication and coordination into an effective management force. Also, the executive body that leads in goal development, comprehensive planning, implementation of policy, and performance evaluation.

aerobic Relying on oxygen for normal function, such as aerobic organisms. *See also* **respiration.**

age structure The distribution of population membership among age categories. Stable populations form a pyramid structure with young organisms at the base.

agricultural revolution The first known transition from hunter-gatherer subsistence to crop culture and animal husbandry about 10,000 years ago in the Middle East.

air Low density material composing the atmosphere made up mostly of nitrogen, oxygen, water vapor, carbon dioxide, inert gasses, and a large number of trace gaseous and particulate (e.g., dust, pollen) materials of natural and artificial origin.

alfisols Deep, organic soils with moderately developed soil horizons associated with deciduous forests.

algae Important primary producer organisms without complex vascular tissue to conduct nutrients and metabolic gasses as in the higher mosses, ferns, and flowering plants. They vary in size from microscopic plankton to marine kelp many meters long.

alkalinity Chemical compounds containing the potential to form a hydroxide base in water with the capacity to reduce acidity and raise pH. Carbonates and bicarbonates are common forms of alkalinity in natural waters.

alluvium   Unconsolidated earth of mostly gravel, sand, and clay sizes that accumulates as a consequence of erosion, transport and deposition by running water at the foot of mountains and in flood plains, river valleys, and deltas.

alpha particle   Proton emitted as radiation from the nucleus of a radioisotope.

anadromous   Pertaining to migration between freshwater, where reproduction occurs, and marine environments, where maturation occurs. Many salmon and sturgeon species are anadromous.

anaerobic   Metabolism that relies on chemistry without oxygen to function. Anaerobic bacteria, for example, are able to decompose organic matter without oxygen present, albeit much less efficiently than aerobic decomposition.

angiosperms   Those flowering plants that develop seeds in an ovary.

animal unit   In range management, a standard unit of animal mass for comparing the grazing intensity of different animal sizes based on a cow and calf combination of about 1,000 pounds. Animal-unit months are used to correct for time differences in grazing use.

aquaculture   The husbandry of aquatic species for production of food, sport, and other goods and services.

aquifer   A geological formation that contains groundwater. It may be unconfined above an impermeable layer, appear at the surface as groundwater springs, or be confined between two impermeable layers. Water wells penetrate aquifers.

arable   Can be cultivated by ploughing or tilling.

aridisols   Basic, desert soils with poorly developed horizons. Some form a subsurface layer of impermeable calcium carbonates (caliche).

artesian   Confined aquifers and their natural springs and artificial wells in which water under pressure boils up through openings in the upper impermeable layer containing the aquifer.

aspect   In climatology and ecology, refers to directional orientation of slopes, especially with respect to the sun.

assimilation   In physiology and ecology, the incorporation of nutrient into living biomass where it provides for respiration and production. Also, in environmental science, the ability of an ecosystem to take up and neutralize contaminant effects.

atmosphere   The circulating air enveloping the earth and made up mostly of nitrogen (78%) and oxygen (21%). The atmosphere includes the lower troposphere, where most weather occurs, and the upper stratosphere.

atomic mass   The sum of neutrons and protons in the atom's nucleus.

atoms   The basic components of all matter, themselves composed of a nucleus of neutrons and protons orbited by electrons. Each chemical element is composed of atoms with a unique number of protons and a unique atomic weight.

autotrophic   Adjective for primary production.

baby boom   A spurt of human reproduction and population growth; in the U.S., the most dramatic baby boom occurred during the two decades following World War II.

bacteria   A diverse group of single-celled organisms, of which some cause disease, others are primary producers, and most decompose organic matter into inorganic components.

barrier island   Elongated coastal deposits, mostly of sand, eroded from continental and offshore locations. Barrier islands often protect estuarine wetlands and other coastal lowlands from the full force of storms.

base   A chemical compound that disassociates into hydroxide ions and ions of other elements. Bases react with acids to form neutral salts.

bathyal Pertaining to the cold, dimly lit ocean zone too dark for photosynthesis.

benefit-cost analysis (BCA) An economic comparison of alternative actions to determine which one provides society with the most beneficial use of its resources.

benthic Pertaining to the bottom of water bodies, such as a benthic habitat.

benthos Bottom organisms and associated organic matter of aquatic ecosystems.

beta particle Electrons emitted from the nucleus of a radioisotope.

bioamplification Biomagnification.

biodegradable Anything that can be reduced to smaller and simpler form by living processes including, most importantly, biological decomposition.

biodiversity The variety of life forms. Biodiversity occurs at genetic, population, species, community, and ecosystem levels. Among many possible expressions, the most frequent is the number of species (the species richness).

biofuel Any combustible, nonfossil organic matter used as fuel (e.g., wood), especially gas and liquid fuels produced from agricultural crops.

biogas A mixture of methane and CO_2 produced by anaerobic decomposition of wet, nonwoody organic matter.

biogeochemical Pertaining to processes of material flow, transformation, and cycling through living and nonliving chemical forms in the biosphere. *See also* **cycle.**

biological oxygen demand (BOD) Under controlled conditions, the respiratory uptake of oxygen during aerobic consumption and decomposition of organic matter in water.

biomagnification Increased concentration of chemical contaminants in organisms as chemicals pass through the food web to higher trophic levels. Certain contaminants, such as DDT and mercury, accumulate in living tissue because they are excreted at a slower rate than they are assimilated.

biomanipulation Management of ecosystem biomass, production, and resource partitioning through the food web usually by adding to or deleting from upper trophic levels, such as by stocking fish or fishing for them.

biomass The volume or weight of living organic matter, usually expressed as dry weight or carbon weight per unit area or volume, such as kilograms (kg)/hectare (ha) or grams (g)/meter cubed (m^3).

biome Terrestrial ecosystem defined by the dominant vegetation form occupying the landscape, such as forest, woodland, grassland, desert shrub, and tundra.

biosphere Those interactive parts of the earth's lithosphere, atmosphere, and hydrosphere where life is supported and where life influences Earth forms and processes.

biotic potential The capacity of populations to expand in abundance and to recover abundance when more favorable environments are provided.

birth rate Population regeneration during some specified time interval (e.g., number of eggs hatched/year and number of live births/month).

bitumen High-sulfur, heavy oil extracted from tar sand.

bluegreen algae The Cyanobacteria. A bacterial group of primary producers, many species of which are capable of fixing nitrogen gas directly into organic nitrogen.

bog A wetland type typically characterized by sphagnum mosses and other acid-tolerant plants, low calcium, deep peat accumulation, and acidic water-stained tea brown by dissolved organic matter.

breeder reactor A nuclear power reactor that, as it operates, produces fissionable plutonium-239 by irradiating nonfissionable uranium-238.

brownfields   Abandoned industrial sites characterized by persistent physical and chemical alterations of the host ecosystem.

browser   In wildlife and range management, those animals that consume the growing tips of woody plants (e.g., deer).

buffer   In water chemistry, a chemical compound that reacts with strong acids or bases to form compounds that reduce the rate of pH change.

business cycle   Fluctuation in the economy between recession or depression and recovery and expansion caused by changing supply and demand relationships.

caecum   A large digestive organ in the alimentary canal of certain herbivores, which increases digestion efficiency of fibrous plants. Horses and ruffed grouse have caecum digestion.

calorie   The energy that raises the temperature of one gram of pure water 1°C.

calorimeter   A device for measuring calories released upon material combustion.

cambium   In woody plants, a thin layer of cells actively growing between the inner wood and outer bark.

canopy   The uppermost layer of branches and leaves in a forest—the overstory.

capital   The properties in natural resources, buildings, equipment, and other materials needed to develop and distribute goods and services.

capitalism   An approach to economic development based on a market economy and the freedom to invest in private capital for production of goods and services.

carbon   The element contributing most to organic matter including living biomass, detritus, and fossil fuel. Carbon also occurs in carbonate rock, carbon dioxide, and other inorganic matter. Carbon is cycled through ecosystems by primary production, consumption, and decomposition.

carbon sinks   A natural, long-term storage location for carbon, such as soil, forest, bog peat, and deep ocean.

carcinogen   Cancer-causing agent.

carnivore   Consumers of animal matter situated in the third trophic level and higher.

carrying capacity   The ability of a specified ecosystem condition to support a population, usually expressed as the maximum population number or weight that can be sustained indefinitely.

catadromous   Aquatic migration in which the species, such as the American eel, develops to maturity in fresh inland waters, descends to the oceans to reproduce and returns to points of origin in juvenile form.

chain reaction   A sequence of nuclear fissions created as the products of each fission cause subsequent nuclear fission in the critical mass of a fissionable material.

chemosynthesis   Primary production based in the use of chemically bound energy, often in compounds of iron or sulfur, to drive the uptake of nutrient elements. Chemosynthesis can occur in darkness and is much rarer than photosynthesis.

chlorofluorocarbons (CFCs)   Widely used synthetic compounds of carbon, chlorine, and fluorine that destroy atmospheric ozone through a chemical reaction.

chlorophyll   Photosensitive pigments, at least one form of which appears to be common to all organisms that undergo photosynthesis. The chlorophyll in cells is the site of active photosynthesis and an indicator of primary production and biomass.

clear-cutting.   Cutting all trees in an area and removing them.

climate   The long-term pattern of weather determining mean and extreme seasonal conditions of atmospheric radiation, temperature, precipitation, wind, humidity, and other weather phenomena.

climax community The last stage of natural community succession toward a dynamic equilibrium with environmental conditions in terrestrial ecosystems. The climax stage usually is more structurally diverse, more productive, and more efficient in cycling nutrients than earlier stages of community succession.

climographs A graphic representation of population distribution with respect to environmental temperature and precipitation.

coal The most used fossil fuel, coal is a solid composed of combustible fossil plant matter with high carbon content. A variety of forms differ in hardness, chemical composition, and energy content. The carbon, sulfur, and nitrogen content are sources of air pollution.

coal liquefaction An artificial process of coal conversion to synthetic oil, alcohol, and other liquid fuels.

coke The highly carbonized energy-concentrated product of heating coal to about 1000°C in the absence of O_2.

coliform bacteria Usually harmless bacteria used as an indicator of fecal waste contamination in water, soils, and food.

command economy An economic system in which an administrative authority determines the production rate and kinds of goods and services and their prices. *See also* **communism; socialism.**

commerce (commercial) Flow of market goods and services. Commercial fishing and hunting are for market trade and commercial extinction occurs when plants or animals are too scarce to harvest profitably.

commodity Something of use and an object of trade.

common use *See* **commons.**

commons Property open to common use without restriction or assigned responsibility.

communism In political ideology, a classless social system in which all members of society are the state and share equally in resources and production capacity according to need.

community In ecology, an assemblage of plant, animal, and microbial populations interacting through nutritional webs and other means to form a generally identifiable whole wherever similar environments and ecological connections occur.

compensation point That point at which light intensity becomes too little for photosynthesis to sustain primary production.

competition Striving to gain use of the same scarce resources. In ecology, competition for food, space, and other resources occurs among individuals within species and between species. In economics, competition for greater purchasing power occurs among individuals offering up the same quality of goods and services for a lower price.

compost Organic matter stored and treated to enhance partial decomposition for use as a soil conditioner and fertilizer.

comprehensive planning An organization-wide planning process that addresses the relative contribution of programs to meeting organizational mission, appropriate emphasis on future programs, development of new programs, and the retirement of program elements that no longer justify their costs.

concentration Amount of one material within a specified volume or weight of another material. Concentrations in liquids and gasses typically are expressed volumetrically, such as milligrams (mg)/liter (l). Concentrations in solids often are expressed by weight, such as kilograms (kg) /metric ton (t).

coniferous Pertaining to trees that produce seed in cones, have narrow, needlelike leaves, and relatively soft wood. Most are evergreen, gradually losing old leaves and adding new leaves.

conservation   A philosophy and approach to resource management that seeks the greatest human benefit from resource use over the long run through efficient use and protection of renewable resource regeneration capacity.

conservation biology   The study and management of phenomena associated with sustaining biodiversity in all of its many expressions from genetic diversity to ecosystem diversity.

conservation movement   A moral movement first involving a loose coalition of nineteenth-century scientists, educators, publishers, editors, and politicians who espoused efficient natural resource development and use for the greatest long-term benefit to society.

conservation of energy   From the laws of thermodynamics. Energy is neither created nor destroyed but is simply transformed. Energy transformation efficiency cannot exceed 100% and the input energy in any transformation is degraded to a less intense form able to do less work.

conservation of matter   A natural law stating that matter is neither created nor destroyed in any physical or chemical transformation.

conservation tillage   *See* **tillage.**

consumer price index   An index to the annual amount that prices increase because of inflation.

consumers   In economics, all users of all kinds of resources. In ecology, those users of food resources that ingest organic matter and assimilate it internally.

consumptive use   Any resource use that results in lost resource utility.

continental shelf   Those parts of continents submerged by shallow ocean.

continuous grazing   In range management, the grazing of a particular pasture throughout the grazing season, year after year.

contour farming   Ploughing and planting that follow the land contour and create mini-terraces that slow water runoff and soil erosion.

conventional-tillage farming   Land cultivation by plowing, disking it to break up the soil clods, then smoothing it.

cooling tower   An engineered structure designed to cool thermal effluents from steam-generation power plants and other industries through evaporation or atmospheric heat exchange.

coppice method   Tree regeneration by sprouting from stumps and roots following cutting.

Coriolis force   The force from the rotation of the earth that causes a deflection of air and water movements to the right and counterclockwise in the Northern Hemisphere and to the left and clockwise in the Southern Hemisphere. Its effect shows in large atmospheric, lake, and ocean circulation patterns.

cover   In wildlife and fishery science, those attributes of habitat that provide refuge from the physical environment and predation.

critical mass   The fissionable material needed for nuclear chain reaction.

crop rotation   Alternating field planting between crops with the intent of sustaining soil structure and fertility. Nitrogen-fixing beans and alfalfa often are planted alternately with other crops, for example.

crown fire   Fire in the forest overstory.

cycle   In ecology, material cycles involve the movement—mostly by air and water transport—of elements through different forms and locations in ecosystems and their environments. Gaseous cycles include atmospheric phases such as for oxygen, carbon, nitrogen, and hydrogen. In contrast, phosphorus and other sedimentary-cycle nutrients have no gaseous phase and full cycling is very slow, depending on continental erosion and uplift.

Cynanobacteria Bluegreen algae.

DDT A synthetic pesticide, Dichlorodiphenyltrichloroethane, now outlawed in the U.S. because of environmental damage, but widely used elsewhere in the world.

death rate Mortality. The number or fraction of deaths in a population, occurring in some specified period of time; frequently one year.

deciduous Describes woody plants that shed all of their leaves at one time just before extreme cold or dry seasons. Examples include maples, oaks, poplars, alders, and many other broad-leaved trees, and a few conifers, such as larches.

decomposers Organisms that externally secrete enzymes on to adjacent organic matter where it is digested to molecule sizes that can be assimilated through cell membranes. The vast majority are bacteria and fungi.

decreaser In range management, those highly palatable plants that decrease as grazing increases in intensity.

deferred-rotation A grazing management system in which livestock are rotated between two or among more range areas to increase the long-term efficiency of range conversion into livestock production.

deforestation Forest loss or removal.

degradation Reduction in form, capacity, potential, or value, such as the breakdown of organic matter by decomposition, the erosion of continents, the lowering of energy available to do work, or the lowering of potential for resource sustainability.

demand In economics, the desire shown for goods and services indicated by the willingness to pay for them. *See* **supply and demand.**

demographics Statistics of human populations including age, sex, ethnicity, income, education, health, and various other descriptive attributes.

dendrology The study of trees leading to their taxonomic classification.

denitrification In the nitrogen cycle, a decomposition process that reduces solid forms of nitrogen to nitrogen gas, which can return to the atmosphere.

density Number or weight of items per standard areal unit (e.g., hectare, acre) or volume (e.g., cubic meter).

density dependent Ecological processes that depend on population density, such as prey mortality due to predation.

density independent Ecological processes that are independent of population density, such as weather effects where cover is not a limiting factor.

depletion rate (time) Reduction in material amount within a specified time; most usually applied to resource supply and concentrations.

depression In economics, a condition characterized by overproduction of goods and services for demand, sluggish trade, lowered employment, and lower average income and buying power.

desalination Salt removal by distillation or other techniques.

desert A biome characterized by low vegetation density, a mix of woody and herbaceous species, and succulent or waxy-leaved species. The natural community is adapted to spotty precipitation typically totaling to less than an annual mean of 300 mm.

desertification Transformation of nondesert ecosystems to desertlike conditions with less vegetation, less organic matter in the soil, more rapid drying of the soil, and more wind and water erosion. Desertification typically is caused by overgrazing and improper crop culture.

detritivore   A detritus-eating animal that derives part of its nutrition from decomposers.

detritus   Dead organic matter usually in some state of decomposition by associated bacteria or fungi.

deuterium   An isotope of hydrogen.

dimictic   Twice mixing; pertaining to lakes that mix each spring and fall and remain stratified in summer and under winter ice.

disclimax   Conditions of community succession different from climatic climax and maintained by chronic soil conditions or by fire, flood, storm, tidal flux, and other periodic environmental disturbance.

discounted interest rate   *See* **interest rate.**

dish engine   An external combustion engine formed from a parabolic dish, which concentrates solar energy, and a receiver that captures and transfers the heat to a chamber where heated and cooled helium gas expands and contracts, driving a piston. The dish engine can be linked to an alternator to generate electricity.

diversity   The variety of form and function in ecosystems, societies and any other manifestation. Opposite of monotony.

drainage basin   Watershed.

dredging   A process of shoveling or pumping sediment out of aquatic environments for purposes of mining, navigation improvement, or beach replacement and augmentation.

drift net   A net usually made of fine monofilament synthetic material that is set adrift to entangle fish and invertebrates for commercial harvest.

drip irrigation   Precise and efficient delivery of water to plant roots through a system of slowly leaking hoses and outlet fixtures.

drought   A prolonged period of dry weather sometimes defined by a precipitation less than 75% of average.

dust bowl   The result of extreme drought of the early 1930s centered in the south-central U.S. plains. Named after huge dust storms caused in part by unwise farming of shortgrass prairie.

dynamic equilibrium   Fluctuation without consistent trend around some stable condition.

ecology   The study of interactions among living organisms and their physical, chemical, and biological environment.

economic growth   Increased value of goods and services indicated by such measures as the Gross National Product.

economic system   An organized complex of processes and institutions based in human preferences, motivation, and choices made in developing natural resources, capital goods (physical facilities), and labor for production, distribution, and consumption of goods and services. Often referred to simply as "the economy."

economics   The study of demand for, supply of, and trade in goods and services. Economics facilitates more beneficial decisions where resources are scarce.

ecosphere   Biosphere.

ecosystem   A cohesive, self-regulating, functional, and adaptive organization formed from the interactions of natural communities and their habitats (physical environments). Ecosystems are hierarchical (i.e., communities and habitats interact to form larger systems, such as the different communities and habitats in a lake or in a terrestrial biome). The ecosphere is the largest ecosystem.

ecosystem management   A holistic approach to sustainable natural resource management, which seeks resolution of resource use conflicts based on sound understanding of systems process. It identifies and works with the natural integration of ecological and social processes and within and across boundaries of natural systems, such as watersheds and biomes. *See* **integrated resource management.**

ecotone A zone of gradual transition between two ecosystems showing attributes of both as well as unique attributes.

edaphic Pertaining to soils, such as the nutrient and structural environment of plant roots.

edge In ecology, a sharp transition and boundary between ecosystems.

efficiency Output divided by input. Efficiency can be physical (e.g., energy transformation from one form to another), ecological (e.g., ingested food conversion to production), and economic (e.g., benefits for the costs).

electrofishing A technique used to sample freshwater fishes by passing battery of gasoline generated electricity through metal curtains, seines, or, most commonly, loops at the ends of insulated handheld poles.

electromagnetic Forces that include solar radiation of different wavelengths and energy level. Radiation prominently includes, in increasing wavelength, ultraviolet, visible light, infrared, and radio waves.

electron A negatively charged particle of nearly no mass that orbits the nucleus of an atom. Electron sharing by atoms is the means by which the molecules of complex structures are formed.

electrostatic precipitator A device installed in smokestacks for removing materials from emissions.

elements In chemistry, the 92 naturally occurring atomic structures that form the basic building material of all natural abiotic and biotic structure. Some of the most important for life processes include carbon (C), hydrogen (H), oxygen (O), nitrogen (N), and phosphorus (P). Elements combine to form chemical compounds.

emigration Population departure from an area.

emissions trading A market approach to pollution control in which credits are earned whenever a polluting source reduces emissions by more than the legally required amount. The credits might be saved for later emission use or sold to others.

endangered species A designation applied when wild populations could soon become extinct in their natural range. The designation usually includes unique subspecies. In the U.S., listing for protection and recovery is authorized under the federal Endangered Species Act and individual state laws.

energetics Processes involving energy transformation.

energy Force that can move or change matter and cause heat transfer between objects of different temperature; i.e., the capacity to do work. Fundamental forces in nature include electromagnetic energy (e.g., visible light), gravity, and the strong and weak forces in atoms.

entisols Young volcanic soils of the Rocky Mountains little more than organically enriched regolith without horizon development.

environment The sum of material and energy conditions surrounding and permeating any object or assemblage of objects, such as organisms and ecosystems.

environmental resistance The sum of all environmental factors limiting populations and biotic communities.

environmental scanning A wide-ranging review of trends and influential events in a management environment with emphasis on opportunities for and threats to the organization's mission, goals, and objectives.

environmentalists   A political grouping of people who typically favor policies that protect natural environment and partial or full restoration of altered environments to their natural condition.

epilimnion   In lakes, a warm-water layer of lower density that floats on a colder layer of higher density (hypolimnion), creating stratification in waters too deep for wind action to effectively mix all the way to the bottom.

epipelagic   Pertaining to the uppermost layer of the open ocean.

erosion   The physical or chemical displacement of rock, soil, and organic matter. Physical erosion results from flowing water, wind, and particles in suspension. Chemical erosion results from aqueous solutions that dissolve minerals. Soil erosion occurs uniformly as sheet erosion, concentrates in the small channels of rill erosion, and gouges out deep channels in gully erosion.

estuary   A transitional ecosystem between freshwater flows and ocean characterized by tidal flux and a moving wedge of relatively dense saline water under less dense freshwater. Aquatic organisms in esturaries typically are broadly tolerant of salinity and depth changes.

euphotic zone   That part of aquatic ecosystems illuminated enough to sustain primary production.

eutrophication   Increased primary production, decomposition, and associated physical, chemical, and biological changes in aquatic ecosystems resulting from nutrient enrichment. Eutrophication may be natural or culturally caused by fertilizer, organic waste, and treated sewage, among other sources.

eutrophy   A state of high nutrient enrichment and high aquatic productivity.

evaporation   A physical process in which enough heat is gained in water or other material to vaporize it, in the process transferring heat from the material to the gas.

evapotranspiration   The joint processes of evaporation and transpiration.

even-aged   About the same age. Term used in forestry for trees.

evergreens   Trees that do not shed their leaves all at once or seasonally, including most coniferous trees and many broadleaf semitropical and tropical trees.

evolution   Organic evolution is the process by which life adapts to environment through change in genetic structure and associated anatomical, physiological, and behavioral attributes. The most adaptive traits are retained by differential reproductive success and gene transfer to successive generations.

exotic species   Organisms of nonnative and usually foreign origin.

external benefit   *See* **externality.**

external cost   *See* **externality.**

externality (ies)   In economics, either the exclusion of benefits or costs from consideration in the price of an economic transaction. For example, pollution cleanup costs incurred by society are excluded from the price of goods charged a company and people other than those benefitting from the goods subsidize the low price. In turn, the buyers of the goods are externally benefitted by society's subsidy.

extinction   Total loss of a genetically unique group of organisms at subspecies and species levels of genetic distinction. Extinction may be local or global and is assumed global when not stated otherwise.

fecundity   The number of embryos per mature adult in a population. Fecundity is a measure of population reproductive potential and can be used to estimate birthrate if embryonic mortality is known.

feedlot   An enclosure where livestock, sometimes in the thousands, are intensively husbanded.

fingerling A postlarval immature game fish about as long as any one of the fingers on a man's hand, or typically between about 2 and 5 inches.

fish ladder A structure designed to allow fish passage around or over a dam or other obstruction. Typically a series of steplike pools is constructed up an incline.

fisheries A location where, or a group of organisms from which, aquatic species (not necessarily true fishes) are actually or potentially captured by humans. Fisheries exist for commercial sale, sport, and subsistence.

fission In atomic energy, the splitting of a fissionable isotope by a neutron or other atomic particle.

fixed costs Capital investment, rent, and other costs that are independent of variation in goods production.

flashflood A rapidly rising flood following a storm event typically in steep and impermeable watersheds, such as in mountainous desert.

floodplain The relatively flat land adjacent to streams and rivers that is formed and maintained by periodic flooding.

fluidized-bed combustion (FBC) A process for removing sulfur dioxide from emissions of burning coal. Powdered coal and limestone are mixed in a suspension maintained by forced air as the mixture is burned. The limestone reacts with the sulfur dioxide to form a calcium sulfate precipitate.

flyways Routes repeatedly flown by birds in migration from one area to another. Four distinct flyways are recognized for waterfowl of North America: Pacific, Central, Mississippi, and Atlantic.

food chain The predator-prey feeding sequence from plant to herbivore, herbivore to carnivore, and carnivore to top carnivore. Recognizable food chains are rarely longer than five feeding levels.

food web The interconnection of two or more food chains. *See* **food chain.**

forage Plants fed on by herbivores; most often applied to the foods of large grazing and browsing livestock and wildlife.

forb A term used in range science that lumps together all of the broadleaf, branching, herbaceous vegetation of grasslands to distinguish them from the grasses, rushes, sedges, and other grasslike herbaceous plants.

forecasting Trend and other analysis used in planning to project future conditions shaping the organizational environment and shaped by the organization.

forest A biome classification identified by the dominance of medium to tall woody plants, forests occur in areas of moderate to high precipitation.

forestry The practice of forest resource management.

fossil fuel Incompletely decomposed organic matter altered in structure and chemistry by long exposure to an anoxic, pressurized environment under the weight of sedimentary rock. They include especially petroleum, natural gas, and coal.

fragmentation The loss of wholeness in ecosystems, habitat, law, and management that threatens ecosystem integrity and dependent functions and services; most especially biodiversity.

fry In fisheries, the larval stage of fish following hatching and before all embryonic characteristics disappear.

function In ecology, the rates and directions of material change that sustain generally predictable systems interactions and outputs. Energy drives function.

fungi One of the life kingdoms, most fungi live by decomposing organic matter while some are parasitic. Fungi penetrate detritus and hosts with threadlike hyphae. Examples are mushrooms, molds, rusts, and yeasts.

game   When applied to wildlife and fish, animals that are harvested for both sport and food.

gamma rays   High-energy electromagnetic radiation from radioisotopes. Gamma rays readily penetrate and damage living tissue.

gasohol   A fuel mixture of about one part methyl alcohol and five to ten parts gasoline.

GDP   *See* **gross domestic product.**

generalists   Broadly adapted species that can take advantage of a wide range of ecological resources, although less efficiently in each resource category than narrowly adapted specialists.

genes   The sites on deoxyribonucleic acid (DNA) molecules that are inherited through reproduction and direct the structural development and functioning of all life.

genetic diversity   The variety of genetic expression in the deoxyribonucleic acid (DNA) of chromosomes within species and all other life classifications.

genetic engineering   Alteration of genetic material with the intent of changing organism traits to improve attributes such as food production and disease prevention.

geographic information system (GIS)   A computer-mediated approach to organizing information through maps, plans, and other means of spatial referencing.

geothermal   Earth-generated heat. Useful geothermal energy typically transferred from molten rock, extruded from deep mantle sources, into solid rock, liquid water, and steam.

goal   A desired achievement typically expressed in general and holistic terms. Goals identify broad resource categories (e.g., range forage, ecosystems), direction of change (e.g., more, fewer), and approaches (e.g., stock, restore, protect, educate) to achievement by objective and leave specific details and schedules to the objectives.

goods   In economics, material wealth that can be traded and sold; merchandise. Goods can be natural resources used as they are or in manufacture.

gram   A unit of weight measure equivalent to 0.001 kilograms, 1,000 milligrams, 0.0022 pounds, and 0.035 ounces.

grassland   A biome (terrestrial ecosystem) dominated by grasses and forbs, and found in areas of moderate precipitation.

grazer   In wildlife and range management, those animals that consume forbs and grasses, such as cattle. In ecology, any animal that consumes a primary producer in the *grazing* food chain, such as herbivorous zooplankton.

green revolution   The relatively dramatic increase in grain and other crop yields following World War II. The green revolution was based on more effective integration of improved soil preparation, fertilizers, irrigation, pesticides, plant breeding, and selection for desirable crop attributes. Often associated was costly soil erosion, pesticide contamination, fertilizer pollution of runoff waters, and high fossil fuel energy consumption.

greenhouse effect   Atmospheric warming as a consequence of differential transmission of electromagnetic radiation by atmospheric gasses known as greenhouse gasses.

greenhouse gasses   Atmospheric gasses that cause the greenhouse effect; examples are carbon dioxide, water vapor, chlorofluorocarbons, methane, and nitrous oxide.

greenway   Linear corridors of lawn, garden, and natural area often paralleling roads, walkways, bicycle paths, drainages, utility rights-of-way, and other natural space compatible with urban development.

gross domestic product (GDP)   The total monetary value of all goods and services priced inside a nation's borders over the course of one year. It approximates the sum of income from all sources. Average *per capita* GDP is a measure of average individual welfare. Real GDP is adjusted for the estimated rate of inflation.

gross national product   The gross domestic product of nations.

gross production   *See* **production.**

groundwater   Water that flows at various velocities through and accumulates in the ground. While some groundwater flows relatively rapidly to surface springs—depending on soil, slope, and geology—other water accumulates in deep aquifers where it moves very slowly.

gymnosperms   Plants with seeds exposed; usually in a cone and not in an ovary.

habitat   The environment either actually or potentially lived in (inhabited) by a population or by a biotic community.

habitat conservation plans   A process through which land containing critical habitat of a federally threatened or endangered species can be used for other purposes while assuring protection of the species despite some incidental loss of population members.

hardwoods   The wood or whole plants of any of a wide variety of broad-leaved angiosperms with dense arrangements of vessels and tracheids.

hazardous materials   Used, stored, or disposed materials that are toxic or otherwise potentially harmful to people, property, and supporting ecosystems. Legally, hazardous materials in the U.S. are any on a long list declared hazardous and regulated by the federal EPA and state agencies.

heartwood   Inner wood, typically darker in color, its vessels and tracheids have filled and hardened to strengthen the trunk.

heat   The kinetic energy held in matter, which flows spontaneously between objects holding different concentrations, i.e., temperatures, toward a condition of uniform concentration.

hectare   A unit of areal measure equal to 10,000 m^2 and about 2.5 acres.

herbivore   Consumers of living primary producers; examples are deer, cattle, grasshoppers, and certain zooplankton.

heterotroph   Secondary producers.

histosols   Peaty soils of very high organic matter content found primarily in lowland areas with swamps and bogs.

holistic grazing   An approach to grazing that usually involves short and intense grazing in rotation among numerous areas and is compatible with forage growing cycles and nutritional value.

homeostasis   Self-regulation.

horizon   In soil science, a distinct soil stratum with unique soil attributes that set it apart.

host   The living nutritional source for a parasite.

humidity   The concentration of water vapor in the air.

humus   A mix of decaying organic matter of generally unrecognizable origin, inorganic matter, small consumers, and decomposers that occurs in topsoil just below the plant litter. Humus typically has an open, well-aerated structure that holds water and binds nutrients well.

hydrocarbons   A general class of chemical compounds with a high proportion of carbon, hydrogen, and oxygen.

hydroelectric   Electrical power produced from the force of gravity operating on water falling through turbines connected by drive shafts to generators.

hydrologic cycle   Movement of water from earth to atmosphere through evapotranspiration and back again through precipitation.

hydrology   Water science and study.

hydropower   Any mechanical or electrical power derived from falling water.

hydrosphere   The water world, including all of the ice, liquid water, and water vapor in, on, and above the earth as well as the water included in living things.

hypolimnion   The colder and more dense layer of water found at lake bottoms during summer. Because the hypolimnion does not mix with overlying water it often develops marked differences in water chemistry and life processes.

igneous rock   Magma of diverse composition extruded from the earth's mantle and cooled into surface and subsurface formations. Granite, basalt, schist, and gneiss are common forms.

immigration   Movement of new population members, including humans, into an established population or area.

impoundment   Reservoir.

inceptisols   Slightly more developed variations of entisols found on volcanic ash primarily in the Pacific Northwest.

increaser   In range management, plants that are slightly to highly unpalatable to grazing animals and increase in abundance as grazing intensity increases.

incremental benefits   Marginal benefits.

incremental cost   *See* **marginal benefits.**

industrial revolution   The rapid transition from agricultural economies to economies based on manufacture of material goods. The harnessing of water and steam power in the seventeenth and eighteenth centuries initiated and accelerated the industrial revolution.

infiltration   In watershed science, the penetration of soil surface by water.

inflation   Rate of increase in the average price charged for all goods and services.

information age   A time, now continuing, in which information processes dominate work and lifestyle. The information age rose to prominence during the middle twentieth century, when the economy became most influenced by information services facilitated by major advances in computing and communications technology.

inorganic   Composed without carbon.

institutions   In economics, practices or organizations for governing human behavior. Markets and government regulations are two primary examples.

integrated management   A holistic approach to planning and implementation across organizational authorities with the intent on capturing all important interactive elements of the management focus, such as integrated coastal zone management.

integrated pest management   Combining chemical, crop cultivation, predator, and other biological methods of pest control to promote the most cost-effective sustainable agriculture.

integrated resource management   A systematic approach to natural resource management, which seeks greater efficiency and sustainability through comprehensive understanding of interactions among discrete areas of resource management. *See* **ecosystem management.**

integrity   In environmental law and ecological science, the completeness of ecological structure and function resulting in an adaptive, self-regulating ecosystem.

intercropping   Cultivating a mix of two or more crops in the same field, usually for improved pest and soil management.

interest rate   The price charged to borrow money for use. The real interest rate is corrected for inflation. The prime interest rate is charged by the largest banks to the best customers. It establishes a basis for comparison of all other interest rates and is monitored and managed by the Federal Reserve, which is the government lending institution of last resort. The Federal Reserve lends at a discounted interest rate a fraction lower than the prime.

internal combustion engine Form of power conversion to the piston drive of a crankshaft by repeated explosion of combustible fuel inside a confined space.

intertidal Pertaining to the space and ecological processes that occur between low and high tides.

intrinsic rate of increase In population ecology, the maximum rate a population can increase under the most favorable conditions of environment, population density, and other population attributes.

invasive species Nonnative aggressively colonizing species that displace native species. Examples include zebra mussels, purple loosestrife, Kentucky bluegrass, and starlings.

inventory Taking stock; in management planning, the gathering and organizing of information about existing natural and human resources, facilities, finances, trends, issues, opportunities, threats, organizational strengths and weakness, and other knowledge pertaining to management effectiveness and efficiency.

inversion *See* **thermal inversion.**

ion An electrically charged atom, either positive or negative.

island biogeography The study of the effects of island size and distance from the mainland on species colonization rates, extinction rates, and diversity. The principles extend to any habitat fragment isolated in a very different landscape setting (matrix), such as lakes in a forest or mountaintops in a desert.

isotope A variant form of a chemical element with the same number of protons but a different number of neutrons in the atom nucleus.

issue A disruption in planning continuity that begs attention and response. Issues may be either threats to or opportunities for goal accomplishment.

keystone species Like the keystone that holds up a stone arch, a keystone species has disproportionate effect on sustaining the integrity of an ecosystem. Beaver, for example, engineer the pond ecosystems that many species depend on. Humans are the most influential keystone species on Earth.

kilo- A word prefix that signifies 1,000, such as kilocalorie (1,000 calories), kilowatt (1,000 watts), and kilogram (1,000 grams).

kinetic energy Energy contained in matter above a temperature of absolute zero.

labor All forms of physical and mental work applied in an economic system.

lake Any closed, water-filled basin with variable flow direction differing substantially from the ocean in chemical composition. Lakes are of artificial and natural origin, temporary and permanent, fresh and salty, including small ponds and the very large "great" lakes.

lampreys A family of fishes, many of which are external parasites of other fish species and are often lethal in effect. The marine lamprey is especially large and lethal.

land-use planning A process by which alternative land-use proposals are evaluated and a plan is selected for the most cost-efficient provision of all services desired by the community. Land-use planning typically centers on coordination of private development with public infrastructure and environment.

landfill A common means for disposing of solid wastes by burial in and under soil and unconsolidated regolith.

landscape ecology   Ecological study of arrangements and interactions among like and different vegetation, topographic, anthropogenic, and other geographic features, typically at a regional scale.

leaching   The dissolving and removal of materials by a solvent, such as water.

levee   An artificial or natural wall of earth bordering a shoreline. Artificial levees are constructed to restrict flooding.

life expectancy   The length of life expected at any particular age based on the average length of life under the expected environmental conditions.

limiting factor   A principle in population ecology, related to tolerance, which states that too little or too much of any one factor can prevent further growth and expansion of a population regardless of how near to optimum all other factors may be.

limnetic zone   The offshore surface waters of a lake illuminated enough to sustain photosynthesis, or to about 1% of the illumination above the lake surface.

liquefied natural gas (LNG)   Liquefaction of natural gas by intense cooling.

liter   A unit measure of volume equivalent to 0.001 m^3, 1,000 milliliters, and 1.10 quarts.

lithosphere   The world of inorganic solid and molten rock and its breakdown products, including the inorganic parts of soil.

littoral zone   Waters illuminated enough to sustain photosynthesis on the bottom, or to about 1% of illumination above the water surface.

loam   A soil texture with near equal mix of sand and clay.

long-range planning   Far future assessment of trends and events that could profoundly influence organizational management effectiveness in the long run.

magma   Molten matter in the earth's mantle, and in fissures leading upward toward the Earth's surface, including volcanic vents.

management   For natural resources, the planning and implementation of change with the intent of directing resource outputs toward beneficial outcomes. Usually preceded by a resource focus, such as wildlife, forest, range, urban land, recreation, fishery, or, more holistically, ecosystem management.

marginal benefits   In economics, the benefits gained from a specified input, the marginal cost. Wise economic decisions are based on maximizing the marginal net benefit, which is the marginal gross benefit in excess of marginal cost.

marginal cost   *See* **marginal benefits.**

market economy   An economic system in which individual buyers and sellers price, buy, and trade goods and services freely based on individual knowledge of competitive offers. Unlike farmer's markets, most competitive markets are not physically consolidated and rely on advertising media to compare prices and qualities.

market hunting   Hunting with intent to sell wildlife products.

marketable pollution permits   A tradeable allocation of air, water, soil, or other space for energy or material disposal once government decides what can be safely assimilated. The allocation may be sold to others to the advantage of both parties.

marsh   A wetland dominated by herbaceous plants such as cattails, sedges, and grasses tolerant of wet soils.

mass   The material, or matter, in a defined volume or area.

mass transit   Buses, trains, trolleys, and other forms of transportation that carry a large mass of people.

maturation   In population biology, becoming reproductively capable. In community ecology, the process of succession to the climax stage.

maximum sustained yield   *See* **sustained yield.**

mega-   A prefix meaning very large or more specifically, 1,000,000. For example, power plant output is reported in megawatts (millions of watts).

megapolis (also megalopolis)   An outsized city typically created from suburban sprawl among once separate cities.

mensuration   The measurement of timbered area, whole forests, single trees, logs, and other pieces and units of forest products.

meristem   Actively growing plant tissue of shoots and roots.

mesic   Moist; with precipitation typically over 500 mm.

mesopelagic   A dimly lit oceanic zone too dark for photosynthesis where many species find refuge during the day before rising to the epipelagic zone to feed at night.

metamorphic rock   Rock transformed heat and pressure from original igneous or sedimentary form typically to forms with greater density. Examples are marble from granite and slate from shale.

meter   A unit of linear measure equal to 100 centimeters, 1,000 millimeters, 39.4 inches, and 1.09 yards.

micro-   A word prefix for very small or more specifically, 1/1,000,000 (0.000001). A micrometer is 1 millionth of a meter, equal to 1 micron.

migration   Repeated population movements between two or more locations. Migrations typically occur in response to seasonal temperature and wetness cues or physiological cues. They may follow the same track coming and going or follow a circle without backtracking.

milli-   A prefix meaning one thousandth (0.001), such as milligram (one thousandth of a gram).

mineral   Any naturally-occurring element or compound usually crystalline and solid in structure and inorganic in composition. Although organic, coal is sometimes included.

mission   The organizational purpose. For management organizations, the organization's "reason for being" often is captured in a mission statement, which ideally guides goal and objective development.

model   A simplified abstraction of reality that intends to capture the primary attributes of real-world form and function for research and management analysis. Models may be physical, such as water-resource engineering models, or mathematical, such as many geophysical, ecological, and economic process models.

molecule   Atomic union of the same or different chemical elements. By sharing electrons, two atoms of hydrogen and one atom of oxygen form a molecule of water.

mollisols   Natural grassland soils, which typically are deep, rich in organic matter, alkaline, and show moderate horizon development in a soil profile.

monoculture   One-crop cultivation, most evident in areas where a single crop dominates the landscape, such as corn and wheat in parts of the midwestern U.S.

monopoly   In economics, control over the supply of specific goods and services in the absence of competition.

mortality   Population death rate.

multiple use   A land management approach, promoted in certain federal land-management legislation, which advocates an overlay of compatible land uses resulting in the greatest total profit on private lands or public benefit on public lands.

municipal   Pertaining to village, town, and city service organizations and regional coordinative institutions.

mutagen   Anything that increases the rate of mutation from background rates.

mutation   Material change, most often applied to genetic and chromosomal change in living cells leading to structural and functional change in living organisms.

nano-   A word prefix for a unit of measure equal to 0.000000001 (1 billionth) and for very small things measured in billionths of a meter (e.g., nanoplankton, nanotechnology).

natality   Population birthrate.

natural gas   Underground deposits of gas consisting of 50% to 90% methane and small amounts of heavier gaseous hydrocarbons such as propane and butane. Natural gas is a fossil fuel like coal and petroleum and is often found associated with them.

natural law   A 100% predictable natural process under specified conditions.

natural radioactivity   Nuclear change in which unstable nuclei of atoms spontaneously emit mass, energy, or both at a fixed rate.

natural resources   Actually or potentially useful materials and energy untransformed by manufacturing for the most part. The undeveloped land, water, air, and life, in all their diverse forms, are the most basic of natural resources.

natural selection   The process by which some genes and gene combinations in a population are passed on to the next generation more often than others because the parents having the most adaptive gene-linked traits survive and reproduce more successfully.

natural system   Complex natural organization with structures and functions that generate predictable material and energy outputs when provided historic environmental inputs of material and energy.

nekton   The swimming inhabitants of water.

neritic zone   The ocean zone closest to the intertidal zone near shore and including the relatively shallow waters of the continental shelves and similar depths around offshore islands.

net production   *See* **production.**

networking   Informal and formal interaction within and across administrative lines of authority, often by way of advanced communication technologies. Worldwide networks are rapidly developing through personal computers and the Internet.

neutron   An uncharged particle with a relative mass of 1.0 found in the nuclei of atoms.

niche   In ecology, how a species fits into an ecosystem both physically and functionally. The physical (or environmental) niche defines the tolerable environment for all life stages and successful reproduction. The functional niche is what the species does in an ecosystem, including how it competes for resources and influences its environment.

nitrogen   A required and often limiting life nutrient found sparsely in inorganic form and commonly in gaseous form. Nitrogen undergoes a complex cycle through relatively small quantities of inorganic solids, organic form, and gas. Bacteria play an essential role in the cycling rate.

nitrogen fixation   The transformation of gaseous nitrogen to inorganic form by lighting-caused oxidation and to organic form by bacterial (including bluegreen algae) assimilation.

nonconsumptive use   Resource use that does not affect resource abundance, such as certain recreational use of wilderness and water use where evaporation loss is nil.

nonpoint source   A diffuse source of materials, often pollutants, such as entire watersheds or the windswept landscapes contributing to atmospheric dust. Many of the remaining pollutants in water are from nonpoint sources.

nonrenewable resources   A nonregenerative resource, which is irreversibly diminished with consumptive use resulting in a useless form.

nuclear energy   Nuclear fission, fusion, or spontaneous release of radiation.

nuclear fission   The nuclei of certain isotopes with large mass numbers split apart into two lighter nuclei when struck by a neutron. This process releases neutrons and a large amount of energy.

nuclear fusion   The forcing of nuclei of two elements at very high temperature into a fusion forming a nucleus heavier than either of the initial nuclei. Much energy is released in the process.

nucleus   In physics, the concentration of mass at the center of an atom, composed of positively charged protons and (except for hydrogen) uncharged neutrons. Each proton and neutron has a relative mass of 1.0. Atomic mass of elements varies as the number of protons and neutrons (atomic number) in the nucleus varies.

nutrient   An element or compound needed by all living things for material growth and other life functions.

objective   A measurable, highly specified intended result of planning and action, including schedules for completion (e.g., increase black duck population number by 50% within a decade). Objective specification seeks general direction from goals.

oil   *See* **petroleum.**

oil shale   A sedimentary rock containing kerogen, a waxy organic residue that can be vaporized upon heating and condensed to an oil.

old-growth forest   Forest in a mature or climax successional stage in which some trees are approaching maximum age. In the U.S., it typically has not been previously cut or intentionally burned, at least by Europeans.

oligopoly   Determination of the supply of a good or service by a few suppliers; a cartel.

oligotrophic   Waters of low nutrient concentration and low productivity.

omnivore   Animals that can feed from both primary and secondary production sources (e.g., vegetable and meat). Examples are pigs and people.

operations planning   The process of organizing specific actions and facilities to accomplish management objectives, including the assignment of budgets to specific projects and tasks.

opportunity   Open access. Opportunities taken are offset by lost opportunities, or tradeoffs. Taking the opportunity to benefit from building on land, for example, foregoes the opportunity to benefit from farming it.

opportunity cost   The value lost in any economic transaction resulting in foregone opportunities. For example, the cost of developing a mine in a valuable wilderness area should include the opportunity cost, i.e., the value lost with the wilderness.

optimum sustained yield   *See* **sustained yield.**

ore   Minerals with concentrations of extractable metal.

organic   Matter produced by life processes that always contains carbon and hydrogen, typically oxygen (hydrocarbons), and other elements less consistently.

orographic effect   Topography influences precipitation because air masses cool as they move upward over mountains.

overburden   Layer of soil and rock overlying a mineral deposit that is removed during surface mining.

overfishing   Reducing fish supply by fishing to the point where population sustainability is threatened or is at too low a level to sustain beneficial resource use.

overgrazing Reducing forage supply by grazing to the point where sustainable forage is threatened or is at too low a level to sustain beneficial use.

overpopulated For organisms, including humans, too many individuals for supporting resources to sustain at some prescribed quality level.

oxisols Highly developed soils of the wet Tropics; the surface horizon is leached by rain of most nutrients forming a soil of low fertility and farming suitability.

oxygen A basic element in life structure, a by-product of photosynthesis, and the final receptor for carbon elimination (forming carbon dioxide) in aerobic metabolic cycles. Because photosynthesis breaks down carbon dioxide and aerobic respiration reforms it, the oxygen and carbon cycles are closely linked as oxygen and carbon move between organic states in the biosphere and gaseous states in the atmosphere.

ozone A compound of three oxygen atoms (O_3), which form a layer in the stratosphere that filters out life-harming ultraviolet radiation from the sun. Ozone also is generated from by-products of internal combustion in the lower atmosphere, where it becomes a harmful pollutant in high enough concentrations.

ozone hole A depletion of ozone in the outer atmosphere linked to emission of artificially synthesized chlorofluorocarbon gases, such as Freon.

P/B ratio The production/biomass ratio. A measure of population or community turnover rate of biomass. Low P/B indicates low recovery rate and relatively high vulnerability to resource overuse.

P/E ratio Climatic classification for vegetation development potential based on the ratio of precipitation to evaporation. Desert P/E is low; rain forest P/E is high.

parasite Organisms that live on (ectoparasite) or in (endoparasite) a living host and consume from it without killing it directly—at least immediately. Parasites vary in the harm they cause hosts, some feeding only on dead tissue, with little other effect, and others feeding slowly on vital tissue until the host ultimately dies.

passive solar heating The controlled trapping and conversion to heat of solar radiation by greenhouse or other material absorption, usually for heating building interiors.

pathogen Disease-causing organisms.

pelagic Pertaining to open lake and ocean waters offshore and off bottom.

per capita For each individual, as in mean *per capita* income. Latin, meaning "by head."

percolation In watershed science, the downward movement of water through soil.

permafrost Permanent layer of ice beneath ground surface in arctic and alpine regions. Permafrost is an environmental determinant of tundra ecosystem form and function.

petrochemicals Petroleum-derived chemicals used to make many products including gasoline, kerosene, lubricant oils, plastics, fertilizers, pesticides, synthetic fibers, paints, and various medications.

petroleum Literally (from the French), rock oil. A slippery, flammable, liquid fossil residue varying in thickness and color and containing hydrocarbon compounds with traces of sulfur, nitrogen, and other biochemical elements. Petroleum is a source of petrochemicals.

pH A measure of acid and base strength expressed as a log-base 10 transformation of the hydrogen concentration. A solution of hydrogen ion concentration equal 0.0001 molecular weight of water has a pH of 4.0. Pure water equally dissociates into hydrogen ions (acid) and hydroxide ions (base) resulting in a neutral pH of 7.0.

phosphorus An element in solid form that frequently limits primary production because it occurs relatively sparsely in soil and water. It is a major component of commercial fertilizers and the first nutrient to suspect as a cause of natural or cultural eutrophication of freshwater.

photosynthesis   A process of organic synthesis in which solar radiation provides energy for combining the carbon from carbon dioxide (CO_2) and the hydrogen and oxygen of water (H_2O) to produce glucose ($C_6H_{12}O_6$) and other simple compounds with oxygen released back to the environment.

photovoltaic   Pertaining to a solid-state device, the solar cell (or photovoltaic cell), which converts sunlight directly to electricity via electron excitement in selenium or other sensitive metal.

phyto-   A word prefix meaning plant or plantlike (e.g., phytoplankton, phytobenthos).

pioneer community   First interactive association of primary producers, consumers, and decomposers and first stage of primary succession in an area previously devoid of life.

plankton   Plant, animal, bacteria, and other organisms largely adrift in lake, large river, and ocean currents. Although most members (plankters) are microscopic, many species are large enough to see, and some marine jellyfish are very large. Many plankters can control their position in the water column.

plate tectonics   *See* **tectonics.**

point source   In environmental law and technology, any consolidated and readily identifiable origin of pollutants, such as drain pipes, smokestacks, and landfill or feedlot drainage.

policy   A specific or collective course of action defined for individuals and organizations. Policy consists of the rules, regulations, codes, and other officially sanctioned guidance that directs behavior within organizations, both public and private.

politics   The art and science of guiding, enforcing, shaping, or circumventing policy through analysis, persuasion, and, sometimes, deception.

pollution   An alteration of matter or energy level that diminishes resource value, including threats to present or future human health, economic welfare, and other quality of life.

pollution tax   A tax levied on the discharge of harmful waste that is set equal to the environmental cost incurred. Such taxes stimulate pollution control by the taxed parties.

polyclimax   An alternative to the succession concept of a single, climate-controlled climax, which proposes that many variations of community composition and form may appear in equilibrium with climate, depending on circumstances.

polyculture   Mixed-crop interspersion often used in tropical areas to sustain production.

population   In ecology and genetics, a reproductively interactive group of organisms.

population dynamics   Changes in population numbers and related measures as a consequence of changed birthrate, death rate, emigration, and immigration.

porosity   The volume and other dimensions of open-space networks (interstices), typically filled with gasses or liquids, in otherwise solid material.

potential energy   Stored energy materially constrained from kinetic expression such as the constraint of a dam on water exposed to gravity or the forces contained in the structure of atoms and molecules.

precipitation   In a climatic reference, the fall of liquid or solid form (condensing fog, rain, sleet, hale, and snow) from the atmosphere. Dry precipitation is dust or other nonaqueous fallout. Chemical precipitation follows reaction in a solute that forms nonsoluble matter.

predation   In the usual sense, the killing of an animal by another animal and subsequent consumption, either whole or in parts.

prescribed burning   Intentional burning under controlled conditions to prevent wildfire and, usually, to encourage the effect of natural burning in terrestrial ecosystems where fire plays an important role in sustaining fire-adapted vegetation.

preservation   Reservation and maintenance of resources by forestalling use and destruction.

prey   In the usual sense, an animal taken alive and killed by another animal for food. Also, the act of predation, i.e., to prey on.

primary production   The conversion of inorganic nutrient elements into organic matter by solar energy via photosynthesis or chemical energy via chemosynthesis (some bacteria).

prime interest rate   *See* **interest rate.**

prior appropriation   In water law, the granting of water use priority, especially in time of water shortage, in accord with the time of first documented claim for use. The senior users get their allotment before the junior users.

private property   Land, resources, manufactured goods, and ideas owned by a fraction of the public, either an individual or group, authorized exclusive control over use and disposal within bounds imposed by public law.

production   Output. In economics, production is the output of goods and services, usually specified for some area and time period. In ecology, production is the output of living matter. Gross production includes that matter which is respired away in metabolism. Net production is gross production minus respiration, including all biomass consumed and decomposed by the next trophic level.

productivity   Ecological productivity is the rate at which output of living matter is generated. In economics, it is the output per unit input. *See also* **production.**

profile   Vertical cross-section revealing strata and other structure in soil, water, forest, and other systems.

profit   Income in excess of costs including salaries, materials, rent, and all other costs.

profit incentive   Motivation by promise of greater profit.

profundal zone   Offshore zone of a lake too dark for photosynthesis and including both benthic environments and associated waters.

program planning   Planning organized around provision of specific customer services and associated benefits. Program planning typically is linked to specific funding sources, management goals, and projects.

progressive movement   A political movement, which, at the turn of the twentieth century, promoted resource conservation and fairer distribution of opportunity and wealth among human resources.

project planning   The process of evaluating alternative approaches to goal achievement through design that accomplishes a specific set of objectives. Projects result in a demonstrable change such as physical creation, restoration, or protection; new technologies, procedures, or basic knowledge; or other outputs serving organizational goals.

public   All citizens in a declared jurisdiction, such as a city, state, or nation.

public domain   Land and other environment ultimately owned and controlled by the public, but not assigned to a specific management authority.

public service   Aid offered with the intent of improving the general public welfare.

public trust   The authority placed in government agencies to manage properties owned collectively by the public for the public, both present and future.

purse seine A commercial fishing net set at the surface and towed in a large circle around fish then drawn to closure under the fish and winched in. Purse seines are used to capture schooling fish such as herring, mackerel, and tuna.

pyrolysis In biofuel technology, the thermal destruction of organic materials in the absence of oxygen to convert biomass to a mixture of gases and organic vapors.

radiation Movement of particles or energy waves outward in all directions from an emitting source such as the sun (electromagnetic radiation) and atomic fission (both particle and energy radiation).

radiation balance An indicator of energy level and dynamics in objects and areas determined from comparison of solar or other radiation input to radiation output.

radioactive isotope *See* **radioactivity; isotope.**

radioactivity The spontaneous radiation of particles and energy from unstable nuclei of a radioisotope. The major forms of radiation include energy in the form of gamma rays and alpha and beta particles.

rain shadow The effect of topography on air movements resulting in low precipitation on the downwind side of mountains.

range A place that may be roamed over at large; more specifically, any land area producing forage available for on-site feeding by grazing or browsing animals.

rational policy analysis A policy approach that considers all of the relevant alternatives, identifies and evaluates all the consequences that would follow from the adoption of each alternative, and selects that alternative with the most preferable consequences.

real interest rate *See* **interest rate.**

recharge To restore potential use; especially in groundwater hydrology or aquifer replenishment. Areas of high-infiltration soil, regolith, and fissured rock form "recharge zones" of exceptionally high groundwater contribution.

reclamation In environmental and natural resource management, a goal and action taken to improve the services of an artificially degraded ecosystem by creating an ecosystem different from the original in form and function.

recycling In environmental management, the collection, reprocessing, and reuse of materials in the same or modified form.

redundancy In community ecology, the backing up of one species' functions by other species in a community.

regional planning Geographical planning organized within natural or political boundaries and usually focused on modifying the distribution, type, and intensity of land uses for the general improvement of public welfare.

regolith Unconsolidated inorganic material overlaying rock formations and often topped with stratified soils (with identifiable horizons). In young landscapes, such as those recently exposed by glaciers, regolith often grades upward into unstratified soils of very low organic content.

renewable energy Biochemical energy that can be renewed through life process or abiotic sources of energy of continuous supply, such as solar radiation and gravity.

renewable resource A resource that regenerates through biological reproduction or through biogeophysical cycles, such as forest, wildlife, soil, and water resources.

rent seeking In economics, the action of special interests who use their private resources to lobby government to pass laws that economically benefit the group at a net cost to society.

reserve Identified resources or resource areas set aside for some future use.

reservoir   A storage place. In water resources management, a reservoir is an artificial lake, usually a stream or river backed up behind a dam in an impoundment.

resilience   Ability to bounce back to original shape or condition, including the self-restoration of a population, community, or ecosystem following temporary exposure to environmental stress.

resource partitioning   In ecology, the allocation of spatial, nutritional, and other ecological resources among the populations of a natural community.

respiration   A process of energy release from and material change in chemical compounds as they are reorganized in metabolic pathways. In aerobic respiration, atmospheric oxygen helps drive metabolic cycles by removing toxic buildup of waste carbon and hydrogen by-products in the form of carbon dioxide and water vapor.

rest-rotation grazing   In range management, a grazing system that excludes pasture use for 12 months while other pastures absorb the grazing load.

restoration   In resource management, a goal and actions taken to reestablish the original population, environment, or ecosystem following anthropogenic change.

riparian   Pertaining to stream and riverbanks or, less commonly, to lake and estuary shores. Riparian vegetation often is distinct from adjacent uplands because of groundwater and flooding influences.

riparian right   Water laws based on water use by riparian landowners and return in a reuseable state. Riparian law is common in humid regions where little water is lost in use. *See* **prior appropriation** *for comparison.*

river   Any water-filled, open channel in which the direction of flow is determined mostly by gravity in the absence of tidal effects. Rivers and streams are loosely differentiated according to relative size, but inconsistently so among regions.

river continuum   The generally predictable continuous transport and change of materials, habitats, and associated aquatic communities from the smallest riparian-dominated headwater streams to the largest rivers.

rock   Natural solids composed of one or more minerals in the lithosphere.

rock cycle   *See* **tectonic.**

roundwood   Raw timber product; logs.

ruminants   Certain herbivorous hoofed mammals with a multichambered stomach, which allows more thorough plant digestion through swallowing, partial digestion, regurgitation, thorough chewing (of cud), swallowing again, microbial fermentation, and final digestion. Sheep and cattle are ruminants.

runoff   Intermittent and perennial surface flow of water on watershed surfaces and through stream channels to surface bodies of water, soil percolation, and groundwater aquifers.

salinity   Saltiness; usually of water or soil. In seawater and estuaries, salinity is expressed as parts per thousand based on weight. The oceans average near 35 parts per thousand (3.5%) salinity.

salinization   Salt accumulation, as in soil.

saltwater intrusion   Movement of saltwater into fresh groundwater usually by over pumping the groundwater, which essentially sucks saltwater into the freshwater aquifer.

sanitary landfill   A place of solid waste disposal in which waste is spread in thin layers between layers of soil or fine regolith and care is taken to prevent erosion and contamination of ground- and surface waters.

sapwood   Outer wood, typically lighter in color, that conducts water and dissolved materials.

satellite city   A city that grows at the edge of a well-established city, usually at a major intersection with a superhighway encircling the original city.

savannah   Biome condition in which grassland and open woodland mix.

seagrass   Several plant genera (e.g., *Zostera, Thalassia, Phyllospadix*) that have adapted to shallow estuaries and coastal ocean.

secondary consumers   Carnivorous consumers at the third or higher trophic level.

secondary forest   Immature forest believed to have started from secondary succession on a site disturbed by cutting, fire, storm, or other natural or artificial cause.

secondary producers   Heterotrophs, i.e., organisms that consume or decompose the organic matter that originated through primary production (autotrophs).

secondary succession   *See* **succession.**

sediment   In earth processes, settled particulate material at the bottom of a water body or left behind by receding waters, such as the sediment left after a flood.

sedimentary   Pertaining to sediment and especially to limestones, dolamites, sandstones, and shales that form following sedimentation under the pressure of accumulated sediment.

seed-tree method   Trees are left during cutting to seed the next tree generation.

selective method   Harvest only of trees specified for reasons of management or product quality, such as trees of a certain species, size, maturity, or condition.

semiarid   Between xeric (arid desert) and mesic (moist to wet) conditions, with between 300 and 500 mm of precipitation.

sere (seral)   The sequence of community changes that link successional stages through time from pioneer to climax.

sewage treatment   Partial to complete removal and neutralization of sewage wastes from water by a variety of means including primary (filtration and sedimentation), secondary (bacterial decomposition promoted by turbulent mixing), and advanced treatment (chemical and biological removal of inorganic nutrients).

sheet erosion   *See* **erosion.**

shelterbelt   Windbreak.

shelterwood method   A rotational selective cutting designed to leave a seed source and protective cover for forest regeneration.

short-duration grazing   In range management, a grazing system in which each pasture is grazed briefly and intensively, followed by a long period of nonuse.

shrublands   Land where the predominate vegetation is less than 3–4 meters tall and woody.

silviculture   Forest culture, which manages forest growth, health, and composition, and draws heavily on ecology, soils, and dendrology.

skidding   A method of moving logs by dragging.

sludge   In sewage treatment, the sediment from treated wastewater, usually rich in organic matter, bacteria, viruses, and various natural and synthetic chemicals, many of which are toxic. Sludge often presents a recycling or disposal dilemma.

smog   Atmospheric contamination with a mixture of irritating gasses, colloids, suspended solids, and acidic droplets usually appearing as a haze of various tints. Smog from industrial sources is predominantly sulfur dioxide and suspended solids. Photochemical smog is the result of solar radiation causing chemical reaction of hydrocarbons and nitrogen oxides.

socialism   A social system in which the government owns all or most of the natural resources and economic production capacity.

softwoods   In forest management, trees with low density and easy-to-work wood; in contrast with hardwoods. So many softwoods of the U.S. are conifers that the two designations are nearly the same.

soil   Complex mixture of fine inorganic particles (mostly clay, silt, and sand), decaying organic matter, water, air, roots, bacteria, fungal filaments, and other living organisms.

solar cell   *See* **photovoltaic cell.**

solar collector   Device for collecting radiant energy from the sun and converting it into heat.

solar energy   Electromagnetic energy directly from the sun. Other forms of energy derive indirectly from solar energy such as wind, ocean and pond thermal gradients, and biomass.

solar heating   Heat in solar energy captured by solar collectors and stored in water, rocks, or other materials for later distribution. Solar heating is done by active means, which rely on additional energy to move the heat, and by passive means without auxiliary energy sources.

solar furnace   Creation of high temperature by artificial focus of solar radiation for electric generation or other purpose.

solar pond   Shallow waterbody managed to create strong thermal stratification and transform the energy extracted from the thermal gradient to electricity generation.

solid waste   Any unwanted or discarded material that is not a liquid or a gas.

specialists   In ecology, narrowly adapted species that take advantage of a small selection of ecological resources, although much more efficiently than broadly adapted generalists.

speciation   Evolution of new species through genetic recombination, mutation, and natural selection of adaptive characteristics by environmental conditions.

species   The most basic classification of living things, which ideally includes all organisms with potential for successful reproductive interaction under natural environmental conditions, and excluding all others.

species diversity   The variety of different species, which is determined by the richness of species numbers and the relative evenness of species abundances. Greatest diversity occurs where richness and evenness of abundance are both high.

spectrophotometry   Measurement technology based on the wavelengths of visible light.

spillover benefits   Benefits accrued to others as a by-product of economic trade, such as the bird watching farmer who invests more in careful pesticide use, and thereby benefits other bird watchers and hunters at no expense to them. *See also* **externality.**

spillover costs   Costs passed off to others as a by-product of economic trade, such as the costs a power company is not charged for the fish killed by thermal pollution and acid rain, or the respiratory disease and climate warming caused by air pollution. *See also* **externality.**

spodosols   Soils with moderate horizon development in areas supporting coniferous forests in the northeastern U.S. They have a leached mineral horizon with a distinct organic horizon and relatively low clay accumulation.

stability   In ecology, maintenance of function despite environmental stress.

stakeholder   Any individual or group significantly affected by management decisions either positively or negatively. Stakeholders include recipients of management services, competitors, partners, and anyone suffering or benefitting from anticipated side effects.

stand   In plant ecology and forest and range management, a homogeneous and close grouping of plants of the same species.

standing crop *See* **biomass.**

stocking rate Number of a particular kind of animal grazing on a given area of rangeland.

stomata Small openings, especially those in plant leaves, which allow uptake of carbon dioxide for photosynthesis.

strategic planning Planning that identifies appropriate approaches to goal accomplishment on a time horizon typically 5 to 10 years in the future.

strategy An approach to goal accomplishment.

stratification Layering, as in the thermal layers formed in lakes and oceans, the story levels in forests, and the horizons of soils.

stratosphere Rarefied outer part of the atmosphere, extending from about 12 to 30 miles above the earth's surface.

strip cropping Planting different crops in alternating rows to help reduce depletion of soil nutrients and need for pesticide applications.

strip mining Removal (stripping) of surface material, the overburden, and digging out the underlying coal, iron, copper, stone, and other materials. Resources are mined from the surface by digging deep trenches, large open pits, or strips along mountain contours.

structure The spatial organization of matter, including form and arrangement.

sublimation Direct vaporization of ice and snow, bypassing the liquid form of water.

subsidence The sinking of part of the earth's crust due to natural processes or to underground excavation, such as a coal mine, or removal of groundwater.

subsistence Sustaining food and other basic needs for oneself and family using agricultural, wildlife, fisheries, forest, range, water, and other resources.

subspecies Genetically discrete populations within a species typically associated with one region in the species' range.

suburban sprawl The often environmentally destructive and socially undesirable spread of domestic housing and retail services without benefit of careful regional planning.

succession In ecology, the more-or-less orderly and gradual replacement of ecosystem form and function through species colonization and population aging, as each successional stage alters the site in favor of the next. Primary succession starts from bare rock, regolith or open water. Secondary succession starts from some stage of less mature community development following a disturbance, such as fire.

superfund The informal abbreviation given to a fund created for hazardous waste cleanup under federal legislation enacted in 1980.

supply and demand A basic theory ("law") of economics which states that where price is the only variable controlling consumer demand for and supplier provision of a good or service, and the price increases, the demand decreases and supply increases. Also, as price decreases demand increases and supply decreases. A market equilibrium results where supply equals demand.

surface fire Forest fire limited to near the ground surface in litter and understory.

surface mining Removal of soil, subsoil, and other strata and then extracting a mineral deposit found fairly close to the earth's surface. *See* **strip mining.**

sustainable agriculture movement An approach to agriculture that substitutes polyculture over monoculture and minimizes use of fossil fuels, inorganic fertilizer, pesticide use, and irrigation water.

sustainable development An evolving philosophy of global resource development that promotes provision for the present generation while protecting opportunities for future generations. Among the main tenets are developing more efficient resource use, maintaining a healthy environment, controlling population growth, and conserving resource options.

sustained yield   Maintenance of renewable resource harvest or other take with no indicated reduction of future yield potential. Maximum sustained yield is the greatest possible quantity of resource that may be taken over the long run. Optimum sustained yield is the amount of resource that can be harvested for maximum long-term profit or public benefit. Optimum yield is usually less than, and no more than, maximum sustained yield.

swamp   Wetland characterized by emergent woody vegetation.

SWOT analysis   A method for inventorying important influences on organizational management effectiveness in four categories: organizational strengths (S) and weaknesses (W), and the opportunities (O) and threats (T) originating outside the organization.

symbionts   Species that functionally complement and mutually benefit one another such as the reproductive gain of flowers pollinated by the consumers of their nectar.

synfuels   Synthetic fuels developed by gasification and liquefaction technology of coal, tars, and other organic matter.

systems analysis   A comprehensive and systematic approach to examining complex systems interactions either directly or through use of mathematical models. For natural resource management, analysis usually involves controlling or monitoring system inputs (usually some material, energy or information flow), monitoring system outputs, and modifying the system to evaluate the effectiveness and efficiency of alternative management approaches.

tailings   Mineral remains following the processing of ore, usually piled or washed into collecting basins. While many are no more harmful than eyesores, other abandoned tailings in numerous locations have been found to leach or radiate environmental contaminants.

tar sand   A natural mixture of fine sand, clays, and bitumin that can be heat-processed to extract and refine oil products.

tectonic(s).   In Earth science, pertaining to Earth crustal structure and movement. Plate tectonics involve movement of the large solid plates that form the crust, float on the molten mantel, shift against one another and ride over, or slide under each other. Plate tectonics result from volcanic process and cause earthquakes, lift mountains, rift valleys, and move continents (continental drift).

temperature   Heat concentration. At 15°C, one calorie of heat absorbed by 1 ml (1 cm^3) of water raises the temperature 1°C.

terracing   The slope is converted into an alternating series of broad, level benches and steep slopes.

tertiary sewage treatment   *See* **sewage treatment.**

thermal inversion   The layering of cooler dense air below warmer and lighter air during atmospheric calm. Inversions usually occur where cool air can drain off mountains and under the warmer air at lower elevations.

thermal pollution   Wastewater return flow or other addition of heat that reduces resource values in receiving ecosystems.

thermocline   The transition zone between warm surface water and colder deep water in lakes and oceans. The density change in the thermocline results in a stable stratification and isolation of the layers.

threatened species One step removed from endangered species status. A legal definition declared by responsible agencies (for U.S. federal law, the Fish and Wildlife Service) when a species is vulnerable to changes that could endanger its continued existence. *See also* **endangered species.**

tide The periodic fluctuation of oceanic water elevation along coasts created by interaction of Earth revolution, shoreline shape, and lunar gravity.

till In agriculture, to plow and otherwise work the soil. In geology, the mixed clay, sand, gravel, and boulders resulting from past glacial deposition.

tillage Mechanized soil disturbance for crop cultivation by ploughing, disking, and other means. Conservation tillage reduces fuel energy needs and erosion loss by minimizing or eliminating mechanized soil disturbance.

tolerance In ecology, the range of environmental variation over which an organism or population can function. Usually tolerance is judged with respect to the limits set by a single parameter, such as temperature, and is revealed by distribution, relative abundance, and experiment. *See also* **limiting factor.**

tonne A metric ton, which is equal to 1,000 kg and 2,200 pounds (long ton).

tracheids Tubelike structures that conduct water and waterborne material through the wood of gymnosperms and angiosperms. In cross-sections, they appear as tiny pores of different dimensions.

tradeoffs The losses in benefit incurred by actions taken for any gain in benefit.

tragedy of the commons Coined by Garrett Hardin, who explained why the sustainability of natural resources held in common by all (the commons), but free to anyone, is rapidly degraded (the tragedy). Examples include the present status of international oceanic fisheries and past game and range resource use on U.S. public domain.

transcendentalism Frequently associated with Emerson, a nineteenth-century philosophical movement that was a precursor of the preservation movement led by Muir and others.

transpiration A physiological process in which water vapor is pumped out leaf openings and water flow up through the plant is maintained.

trawl A form of fishing net that is dragged behind a boat on or above bottom then winched aboard to remove the catch.

trend analysis Examination of chronological changes and their interactions for insight into future conditions.

tritium (T- hydrogen-3) A heavy isotope of hydrogen with an atomic mass number of 3.

trophic Pertaining to food; used alone or as a suffix in other words, such as eutrophic.

trophic dynamics Food web processes characterized by energy and material flow through trophic levels and populations in natural communities.

trophic levels Nutritional levels in a natural community, including primary production at the first level, herbivore production at the second level, and carnivore production at higher levels. Because most energy and carbon is lost at each energy transfer, the depiction of community trophic levels in a diagram forms a pyramid typically with fewer than five levels.

troposphere The lower atmosphere, which contains most atmospheric mass and weather phenomena. It lies below the stratosphere.

turnover rate The number of times a population or group of populations is totally replaced through death and birth during some specified period.

ultisols   Forest and savannah soils of the southeastern U.S. with a leached acidic upper horizon and an often impermeable lower horizon rich in clay.

underpriced resources   Resources with prices that do not reflect all costs associated with their development. The costs incurred by acid rain, for example, are not included in the price of coal. Underpricing favors resource exploitation over conservation.

understory   In ecology and forestry, the lower strata of forest vegetation beneath the highest stratum, known as the forest canopy or overstory.

uneven-aged management   Method of forest management in which trees in a given stand are maintained at many ages and sizes to permit continuous natural regeneration.

upwelling   Locations along continental shelves where ocean currents force cold, nutrient-rich water to the surface, usually on the continent's west side.

urban areas   Concentrations of human populations in towns and cities and their suburban peripheries.

urbanization   The net movement of people into town and city concentrations. In the U.S., urbanization rate is estimated by the change in percentage of the population living in officially defined urban settlements of 2,500 people or more.

vegetation   The collective structure of plant communities.

vertisols   Soils, mostly of the southcentral U.S., with a high clay content that shrinks and cracks when wet.

vessels   Elongated cells important in conducting water and waterborne materials in angiosperm wood. They contribute to the appearance of pores in cross-sections.

viable population   A population that is self-sustaining and not at risk of extinction.

vision   In the planning process, vision is a concept of the future desired condition for the management system of an organization, including most importantly the condition of the managed resources. Shared vision planning involves many partners and stakeholders who, in a crucial first step, develop one shared vision, held in common, for some desired future condition.

watershed   The region delivering water and its transported material to any selected point along a topographic gradient, usually a stream or river channel. Watersheds have depth as well as topographic surface, and groundwater as well as surface water is shed from both the surface and underlying rock formations.

watt   Measure of power used for electricity, equivalent to one joule-second.

weather   Day to day dynamics of the atmosphere including air circulation patterns, temperature variation, precipitation, and other measures. *See* **climate.**

weathering   Physical and chemical disintegration of solid materials by natural processes of abrasion, impact, freezing and thawing, and erosion.

wetland   Land area temporarily or permanently flooded by marine or inland waters to depths shallow enough for rooted wetland plants to occur under otherwise favorable conditions. Wetlands include marshes, bogs, swamps, and barrens, such as mudflats.

wilderness   Immense wild area with no vehicular access and only temporary human visitation. Under the Wilderness Act of 1964, wilderness is a public land-use classification for lands sufficiently remote and unmodified by use.

wildlife   In the broadest sense, all free-ranging wild organisms, including plants. More usually, the term is applied to attractive or otherwise valued wild animals in terrestrial ecosystems, especially birds and mammals.

wind farm   Wind turbines grouped in suitably windy locations to "harvest" reliable wind resources for electricity generation.

wind turbines   Wind-driven generators of electricity, consisting of blades, rotor, transmission, electrical generator, and control system, all mounted on a tower.

windbreak   Any impediment to wind, but commonly rows of dense trees or tall hedges planted to reduce wind erosion on cultivated land or to protect homes in open country.

woodland   A terrestrial ecosystem classification for any of a number of plant communities dominated by short trees and shrubs, often in a mix with grasslands and savannahs.

work   In physics, the outcome of force exerted on mass expressed in motion. The power held in energy is defined in terms of the work it can do.

xeric   Dry; annual precipitation less than 300 mm.

xeriscaping   Urban landscaping sustainable under desert climates.

yarding   In forestry, the process of moving logs from where they fell to a transport point or landing.

yellowcake   Uranium oxide. Also called uranium concentrate, yellowcake is 85% pure uranium in an isotopic mixture of 0.7 percent fissionable uranium-235 and 99.3% nonfissionable uranium-238.

zoning   Land-use regulation, planning, and management by zone. Although the term is most often used in municipal and county regulation of private land use, any designated land use of public lands also is a form of zoned use.

zoo-   A word prefix meaning animal (e.g., zooplankton, zoobenthos).

Index

abiotic factors, and forest
protection, 324
abiotic resources, 2
acid mine drainage, 200–201
acid precipitation, 173–174
adaptive management, 677–678
adders, environmental, 112
administration
fishery management, 529
organizational, 145–147
advocacy groups, 132–133
aesthetics, in urban ecosystem
management, 445,
446–447
age class distribution, 311–312
agriculture, 368
biomass energy from, 618
in developing countries,
377–379
farmland policy, 382–389
green revolution, 370–372
history, 369
irrigation, 193, 205–206, 377
in New Zealand, 389–390
pesticide controversies,
379–382
sustainable, 382
types of, 369–370
in United States, 372–377
on urban interface, 406
See also soil
agroforestry, 378
air, 10
and conservation economics,
100–101
and urban ecosystem
management, 454–458
alfisols, 225, 226–227
allelopathic plants, 381
allotments, custodial, 363–364
allotment vacancy, 363
alpine tundra, 271–272
altitude, 154
aluminum, 563, 566
anaerobic digestion, 620
andisols, 225
animal damage control, 474–476
animals, rangeland, ecology of,
342–346

aquaculture, 513–514
aquatic ecosystems
estuarine, 282–284
marine, 284–288
stream and river, 279–282
wetland, 272–279
aquatic primary succession,
88–89
aquatic resources, and urban
ecosystem
management, 454. *See
also* fish resources
aridisols, 225, 228–229
artificial intelligence, 680
aspect, 154–155
atmospheric composition and
pressure, 152
atmospheric contaminants, and
urban ecosystem
management, 455–458
atmospheric pollution, and
agriculture, 377
atmospheric resources. *See*
climate
autecology, 60
authority, 145–147
automobile, and urban decline,
435–436
average set-asides, 384–386

Benefit-Cost Analysis (BCA),
115–120
benefit-cost ratio, 118
benefits, spillover, 108
benthic zone, 284
benthos, 284
bioassessment, in fisheries,
526–527
biochemical oxygen demand
(BOD), 200
biodiversity, 69–71
defined, 534–535
and endangered species
management, 533–534,
545–546
in fisheries, 517–519,
522–523
international cooperation, 535

loss of, 535
services and value, 536–538
biofuels, 620
biological diversity, 10–11
biological gasification, 620
biological pest control, 381
biomass, 65, 66–67
biomass energy, 617–621
biomes, 69, 253
biotic communities, 76–77
biotic resources, 2
bogs, 273
boreal forest, 298–299
brines, geopressured, 623
broad-leaved evergreen forests,
temperate, 300
browsers, 343
brush control, 364
budget policy, 142
business
energy use, nonrenewable,
576–577
and private resource
management systems,
131–132
business cycle, 650–653

California annual grassland,
258–259
carbon, 83
carbonization, coal, 583–584
carnivores, 75
carrying capacity, 347
choices, in conservation
economics, 99–108
clear-cutting, 315–316, 326–327
climate, 151
change and human activities,
169–176
elements, 151–152, 156–162
factors, 151–156
grassland, 247, 250
instability and natural
resource management,
166–169
types of, 162–163
United States, 163–166
climatic history, global, 167–168

climax, 87
climax theory, 90–92
coal
distribution and abundance of, 579
formation of, 578–579
mining and production, 579–582
transportation of, 582
utilization of, 582–584
cold desert, 263–265
colonization (up to 1776), 19–26
combustion, coal, 582–583
commercial forest land, 307
commercialization, and wildlife management, 497–500
commercial water use, 206–207
communism, collapse of, 661–662
communities, 68–69
biodiversity, 69–71
biotic, 76–77
competitive interactions, 75
habitat, 69
riparian, 280–281
successional, wildlife associations with, 489–491
trophic functions, 71–75
comparative digestive systems, and range animal ecology, 342–343
competition
within communities, 75
international, 665
comprehensive organizational planning, 139–140
computers, 678
conflict resolution, for public land management, 410
coniferous forests, temperate, 300
conservation
and depression, 43–44
and endangered species management, 544–545
and sustainable development, 672
of water, 213
See also wildlife conservation
conservation biology, 77
conservation economics, 99
Benefit-Cost Analysis, 115–120
decision making
hard choices, 99–102
opportunities displaced, 102–108
future, 108
and governments, 114–117
market, harnessing power of, 109–112
regulations, incentive-based, 112–113
conservation stocking rate, 363
conservation tillage, 234–240
consumption, of natural resources, 2, 4

consumptive use, of water, 205, 207
contaminants, atmospheric, and urban ecosystem management, 455–458
contamination
mining, 568
nuclear, 602
urban water wastes, 451–452
continental climate, 163
continents, distribution of, 153–154
continuous grazing, 351
contour farming, 236
convective storms, 162
Convention on International Trade in Endangered Species (CITES), 545
convergent storms, 162
conversion
coal, 584
fuel, 590
Cooperative Wildlife Research Units, 471
copper, 563, 566
coppice forest methods, 317
coral, 287–288
costs
Benefit-Cost Analysis, 115–120
for marketable pollution permit systems, 110, 111–112
spillover, 108
cost subsidies, agricultural, 386–387
cover, habitat as, 487–488
creative destruction, 658–660
crop insurance, 387
crop-processing wastes, 619
crops, 213
cultivation methods, 381
cultural eutrophication, 86
custodial allotments, active management of, 363–364
cutting methods, 315–317
cuttings, emergency, 319–320
cycling, 82–86

deciduous forests
temperate, 299
tropical, 302
decision making, in conservation economics, 99–108
decomposers, 73–75
deferred-rotation grazing, 351–352
degree of slope, 155
demand, 117
democracy, 25–26, 635–636
density, 61, 312–313
depressions, 43–44, 653–655
desalinization, 213
desert climate, 163
desertification, 175–176

deserts
cold, 263–265
hot, 260–263
desert shrublands, 251–252, 338–339
developing countries
agriculture in, 377–379
natural resources management
economic growth strategies, 639–641
entrepreneurship versus, 641
international policy, 641–643
problems with, 633–639
digestive systems, and range animal ecology, 342–343, 344–345
disease, and forest protection, 324
disease-causing organisms, 198
disengaging, 389
dispersal, 61–62
distribution, 61–63
domestic water use, 206–207
downtown renewal, 437–438
dredging, 196, 560
drilling, 585–586
drought, 161, 357
drought-resistant crops, 213

Earth, latitude, shape, and rotation, 153
eastern climate (U.S.), 166
eastern deciduous forest, 270–271
ecology, 60
in colonization period (up to 1776), 19–22
communities, 68–69
biodiversity, 69–71
biotic, 76–77
competitive interactions, 75
habitat, 69
trophic functions, 71–75
ecosystems, 77–78
energy flow, 79–82
human component, 94–96
material flow, storage, and cycling, 82–86
succession, 87–94
in gilded age (1861–1899), 32–33
in neoprogressive period (1945–present), 48–50
populations, 61–63
energetics, biomass, and production, 65–67
strategies, 67–68
structure and dynamics, 63–65
in progressive period (1900–1945), 40–41
rangeland, 340–341
rangeland animal, 342–346
in westward expansion era (1776–1860), 26–27

economic analysis, 115, 681
economic growth, in developing
        countries, 637–639
economic indicators, 648–649
economic mobility, in developing
        countries, 637
economics
        in colonization period (up to
                1776), 23–25
        creative destruction and
                human progress,
                658–660
        in gilded age (1861–1899),
                33–37
        Keynesian, 655–657
        in neoprogressive period
                (1945–present), 50–51
        and nuclear energy,
                597–599
        principles, 644–649
        in progressive period
                (1900–1945), 41–44
        recreational, 411–412
        terminology, 644
        in westward expansion era
                (1776–1860), 28–29
        See also conservation
                economics
economies
        centrally planned, 660–662
        future, 666
        market, 23–24, 635,
                650–658
        mixed, 647–648, 662–664
        United States, 648–649
ecosystem health, 677–678
ecosystem management
        defined, 675–676
        ecosystem health and
                adaptive management,
                677–678
        and resources management,
                676–677
ecosystems, 2, 77–78, 246
        aquatic
                estuarine, 282–284
                marine, 284–288
                stream and river, 279–282
                wetland, 272–279
        and conservation economics,
                100, 102
        endangered, 546–547
        energy flow, 79–82
        fisheries, 524–525
        human component, 94–96
        and in-stream water uses, 207
        material flow, storage, and
                cycling, 82–86
        riparian, 102
        succession, 87–94
        terrestrial, 246–253
                deserts, 260–265
                forests, 324–330
                grasslands, 253–260
                tundra, 271–272
                woodlands, 265–271
        urban, 421–423 (see also
                urban ecosystem
                management)

ecotones, 76
ecotourism, 12
edge, concept of, 488–489
education
        in developing countries, 636
        fishery, 529
        wildlife, 484–485
electricity/electric power
        from biomass energy, 619
        and conservation
                economics, 101
        environmental adders for, 112
        nonrenewable, 575–576
El Nino, 168
emissions trading, 113
empire building, 114–115
endangered ecosystems,
        546–547
endangered species, 480–481
        defined, 538–539
        and extinction, 539–542
Endangered Species Act, 472,
        543–544
endangered species
        management, 533–534,
        543–546
energetics, 65–66
energy flow, 79–82
energy flux, in urban
        ecosystem, 421
energy production,
        nonrenewable
coal, 578–584
nuclear energy, 592–599
oil and gas, 584–592
energy, 11
        and conservation
                economics, 100
        nonrenewable, 572
                coal, 578–584
                and environment,
                        599–603
                future of, 603
                nuclear energy, 592–599
                oil and gas, 584–592
                use of, 572–579
        renewable, 605–606,
                626–628
                biomass, 617–621
                geothermal, 621–624
                hydrogen, 625–626
                hydropower, 615–617
                ocean thermal energy
                        conversion, 624–625
                solar, 607–611
                tidal power, 624
                wind, 611–615
        and urban ecosystem
                management, 455
engineering, water management,
        194–197
entisols, 225
entrepreneurship, and natural
        resources, 641
environment, and nonrenewable
        energy, 599–603
environmental adders, 112
environmental degradation,
        nuclear, 602–603

environmentalists, in gilded age
        (1861–1899), 37
environmental movement, in
        westward expansion
        era (1776–1860), 31–32
environmental policy. See
        policy
environmental scanning, 134
equable climate, 162
erodible lands, retirement of,
        239–240
erosion, 186–187
estuarine ecosystems, 282–284
ethanol, 620
European feudalism, 24
European market economy, 24
eutrophic lakes, 85–86
evenness of abundance, 69
evergreen forests
        temperate broad-leaved, 300
        tropical, 300–301
excess capacity, 377
exotic pests, and forest
        protection, 324
exotics, nonpredatory, 542
exploration
        geological, 567
        oil and gas, 585, 590–591
exports, agricultural (U.S.),
        376–377
extinction, causes of, 539–542
extraction. See mining

Farm Bill (1996), 244
farming, 187
farmland, 11
farmland policy, 382–389
Federal Aid in Wildlife
        Restoration Act, 471
federal government, policy role
        of, 142
federal lands, and livestock
        grazing, 360–361
federal rangeland management,
        and public opinion, 408
federal recreational
        management, 397,
        399–404
federal water resources
        agencies, 402
fertility, of soil, 224, 240–241
feudalism, European, 24
fire, and succession, 92 (see also
        forest fire)
fish
        endangered, and
                conservation
                economics, 101
        in neoprogressive period
                (1945–present), 56
        in progressive period
                (1900–1945), 47–48
Fish and Wildlife Coordination
        Act, 471
fisheries, 12
        biodiversity issues, 517–519
        future issues, 530
        limits, 511–513

management, and science, 519–525
North American, 508
technological revolution, 508–510
fishery allocation, 507–508
fishery management
administration, 529
bioassessment, 526–527
habitat assessment, 527–528
resource demand assessment, 528
fishery professionals, 525–529
fishery resource, 504–505
fish farming. See aquaculture
fishing
for food and other goods, 506–514
recreational, 514–516
unregulated, 541
flood damage, 193, 453–454
Florida climate (U.S.), 166
food
and conservation economics, 101
fishing for, 506–514
and wildlife conservation, 469–472
food-processing wastes, 619
food production. See agriculture
Food Securities Act, 472
food sources, and wildlife habitat management, 486
forage selection, by range ungulates, 343
forbs, 247, 344
forecasting, 135
forest distribution, 297–302
forest ecosystem management, 324–330
forest fire, 321, 327–328
forest land, 11
forest management, 54, 56, 309–314
forest preservation, in gilded age (1861–1899), 40
forest products, 296
forest protection, 321–324
forest resources, in progressive period (1900–1945), 46
forests, 252, 293
biomass energy from, 618
and conservation economics, 101
eastern deciduous, 270–271
emergency cuttings and thinnings, 319–320
global problems, 330–331
intermediate treatments, 319
as rangeland, 339
and recycling wastepaper, 331–332
southern pine, 269–270
stand management, 315–317
timber harvesting, 320
tree planting, 320–321
tree structure and function, 294–296
United States, 302–308

western coniferous, 268–269
forest succession, 330
fossil fuels, and global warming, 171–172
frontal storm systems, 161–162
frost-free period, 157–158
fuel
nuclear, 595–596
transportation, 591–592, 625–626
fuel conversion, 589–590

gap dynamics, 328–329
gas, natural, 584–592
General Agreement on Tariffs and Trade (GATT), 388
geographic planning, 140–141
geological exploration, 567
geological foundations, for mineral resources, 551–558
geopressured brines, 623
geothermal energy, 621–624
gilded age (1861–1899), 32–40
glacial ice, 181
global warming
and fossil fuels, 171–172
implications of, 172–173
goals, 136
gold, 563–564
government controls, 103
governments
and conservation, 114–117
democratic, 635–636
disengaging from agriculture, 389–390
policy roles of, 142
and rangeland management, 359
and urban land-use planning, 442–444
grasses, 247, 343–344
grasslands, 247
California annual grassland, 258–259
climate and soils, 247, 250
northern mixed prairie, 256–257
palouse prairie, 259–260
as rangeland, 337–338
shortgrass prairie, 257–258
southern mixed prairie, 255–256
tallgrass prairie, 253, 255
grazers, 343
grazing
purchasing privileges for, 362–363
and range plants, 340–341
and succession, 92–93
grazing capacity, 347
grazing lands, private, and recreation, 405–406
grazing systems, 349–356
Great Basin climate (U.S.), 164–165
Great Depression, 43–44, 654–655

greenhouse effect, and radiation balance, 169–170
green revolution, 370–372
groundwater, 182
development of, 213
and urban ecosystem management, 450–451
growth, in developing countries, 637–641
guilds, 68
gully reclamation, 238

habitat
community, 69
fisheries, 521
population, 63
wildlife, and water allocation, 208–210
habitat assessment and management, fisheries, 527–528
habitat conservation plans, for endangered species, 544–545
habitat fragmentation, 491–492
habitat management and assessment, wildlife, 483, 486–494
habitat modification, and extinction, 539–541
harvest management
fisheries, 523–524
wildlife, 495–497
harvesting
icebergs, 214
timber, 320
water, 190, 192
hazards, and recreational management, 414–415
heat, and urban ecosystem management, 454–455
heating, active vs. passive, 607–608
herbivores, 75
high intensity–low frequency grazing, 353
histosols, 225
homes, nonrenewable energy use, 576–577
home water treatment, 203
Hopkin's bioclimatic law, 154
hot desert, 260–263
hot dry rock, 623
housing subsidies, in urban infrastructure, 432
human population increase, 7–8
human progress, and creative destruction, 658–660
humans
and climatic change, 169–176
and ecosystem, 94–96
and mining concerns, 569
humidity, 162, 454–455
hunting, unregulated, 541
hydro-dam licensing, 102
hydrogen, as energy source, 625–626
hydrologic cycle, 183

hydropower, 615–617
  and conservation economics, 101, 102
  and in-stream water use, 207
hydrothermal fluids, 622–623

icebergs, harvesting, 214
imports, agricultural, 386
inceptisols, 225
incremental benefits and costs, 118
independence, 23
Industrial Revolution, 24
industrial wastes, 619
industry
  energy use, nonrenewable, 576–577
  and water use, 206
infiltration, 185–186
information deficiencies, and sustainable development, 672–673
information flow, 145–147
information management, 147, 415
insects
  hormones, 381
  and forest protection, 321–323
  sterilization, 381
institutions, design of, 102–103
in-stream use, 203, 207
integrated pest management, 381–382
integrated resource management planning, 140–141
integrated water resource management, 192–194
integrative recreational management planning, 412
intercropping, 378
intermediate feeders, 343
intermediate treatments, 319
international competition, 665
international trade, 664–665
intertidal ecosystems, 285–287
intrinsic rate of increase, 64
introduction
  fisheries, 520–521
  of predators, 541
inventory planning environments, 133
iron, 562–563
iron, 84–86, 562–563, 566
irrigation, 193, 205–206, 377
island biogeography, 77
island species, vulnerability of, 77

kelp forests, 286–287
Keynesian economic approach, 655–657
keystone species, 68

Lacey Act, 471
lakes, 182, 277–279

land area, 8–12
land control, and sustainable development, 674
land ownership, 25
landscape
  colonization period (up to 1776), 19–22
  gilded age (1861–1899), 32–33
  neoprogressive period (1945–present), 48–50
  progressive period (1900–1945), 40–41
  westward expansion era (1776–1860), 26–27
landscape ecology, 77
land-use planning. See urban land-use planning
land-use practices, managing, 187–189
land uses, 8–12
land wastes, 448–449
La Nina–El Nino Cycle, 168
latitude, 153
law, common elements of, 144
leaching, 224
lead, 563, 567
lead banking, 113
legislation, 143–144
levees, construction of, 196
livestock, 101, 102
livestock distribution, 348–349
livestock grazing, 188–189
livestock production, 356–358
local governments
  policy roles of, 142
  water resources agencies, 403–404
logging, 187

magma, 623
management systems
  boundaries, 126–128
  organizations as, 124–126
managers, recreational resource, 412–413
mandates, 134
marine currents, 155
marine ecosystems, 284–288
Marine Mammal Protection Act, 472
marketable pollution permit systems, 110–113
market economies, 23–24, 635, 650–658
marketing orders, 386
market-oriented economy, 635
markets, 99, 103
  and conservation, 107
  influences on, 106
  limits of, 108
  Marx alternative to, 646–647
  mechanics of, 104
  and pollution control, 109–112
  and production, 104–106
  and Smith, Adam, 106

See also conservation economics; economics
market theories, 645–646
marshes, 274–275
Marx, Karl, 646–647
material flow, 82–83
material flux, in urban ecosystem, 421
matrix, 77
maximum net present value (MNPV), 118
Mediterranean climate, 163
Merrill three-herd/four-pasture system, 353
metallic minerals, 562–567
Migratory Bird Act, 471
Migratory Bird Hunting Stamp, 471
Migratory Bird Treaty Act, 471
mineral deposits, 556–557
mineral extraction. See mining
mineral processing, 561–562, 569
mineral resources, 551
  geological foundations, 551–558
  in gilded age (1861–1899), 40
  metallic, 562–567
  mining and extraction, 558–562, 567–570
  in neoprogressive period (1945–present), 54
  nonmetallic, 564
  in progressive period (1900–1945), 45–46
minerals, 11
  distribution and abundance, 557–558
  future availability of, 569–570
  realm of, 551–553
  strategic and critical, 558
mine reclamation, 569
mining
  coal, 579–582
  concerns with, 567–570
  and conservation economics, 102
  ore, 558–559
  processing, 561–562
  subsurface, 560–561
  surface, 559–560
  and water use, 206
mission, 134, 135
mixed economies, 647–648, 662–664
models, 679–680
mollisols, 225–226
mountain browse, 266–267
municipal wastes, 619

nanotechnology, 683–684
National Environmental Policy Act, 472
National Forest Service, 399–401
national parks
  in gilded age (1861–1899), 40
  in neoprogressive period (1945–present), 56

in progressive period
(1900–1945), 46
National Park Service, 401–402
National Resources
Conservation Service
(NRCS), 241–242
national unity, 634–635
natural aesthetics, and urban
ecosystem
management, 446–447
natural gas, 584–592
naturally bounded planning, 140
natural parks, 12
natural resource management,
5–7
advocacy systems, 132–133
and climatic instability,
166–169
and developing countries
economic growth
strategies, 639–641
entrepreneurship
versus, 641
international policy,
641–643
problems with, 633–639
and ecosystem management,
676–677
historical perspective
colonization (up to 1776),
19–26
gilded age (1861–1899),
32–40
neoprogressive period
(1945–present), 48–57
progressive period
(1900–1945), 40–48
westward expansion
(1776–1860), 26–32
and human population
increase, 7–8
information sources, 13
intent, 135–138
internal environment,
128–131
organizational
administration,
145–147
organizational function,
122–124
organizational planning
boundaries, 138–140
planning process, 133–135
policy, 141–144
private business systems,
131–132
public systems, 131
and recreation, 414–415
regional planning, 140–141
and succession, 93–94
systems, 124–128
natural resources
defined, 1–4
and entrepreneurship, 641
future outlook, 12
and land uses, 8–12
and profit incentives,
108–109
regulation of, 25–26

underpriced, 108
and urban services
expectations,
420–421
See also specific resources
navigation, 193, 207, 453–454
nekton, 284
neoprogressive period
(1945–present), 48–57
neritic zone, 284
networking, 147
New Zealand, agriculture in,
389
nitrogen, 83–84
nongovernment stakeholders, in
urban land-use
planning, 442
nonmetallic mineral
resources, 564
nonpredatory exotics, 542
nonrenewable energy. See
energy, nonrenewable
nonvegetated wetlands, 277
North American Free
Trade Agreement
(NAFTA), 388
North American Waterfowl
Management Plan,
472, 479
northern mixed prairie, 256–257
nuclear energy
economic issues, 597–599
as energy alternative,
592–594
and environment, 599–603
fuel cycle, 595–596
future of, 603
status of, 596–597
in United States, 597
nuclear fuel cycle, 595–596
nuclear plant accidents,
599–601
nuclear terrorism, 602
nutrients, 199
nutrient spiraling, 83
nutrition, and rangeland
management, 345–346

oak woodland, 267–268
objectives, setting, 137
oceanic zone, 284
oceans, 182, 284–285, 569
ocean thermal energy
conversion, 624–625
off-stream use, 203, 205–207
oil, 584–592
oil field exploration, 585
oil shale, 588–589
omnivores, 75
open pits, 559–560
operations planning, 137
opportunities
in conservation economics,
102–108
and natural resource use, 129
optimism, 684–685
ore, defined, 558–559
organic matter, soil, 224

organizational administration,
145–147
organizational function,
elements of, 122–124
organizational integration,
forces in, 122–124
organizational performance
evaluation, 138
organizational planning
boundaries, 138–140
organizations, as management
systems, 124–126
orographic storms, 162
outdoor recreation. See
recreation
overpumping, 208
ownership, of forests, 308
oxisols, 225, 228
oxygen, 83
ozone depletion, 175

pacific climate (U.S.), 163–164
palouse prairie, 259–260
paper use, reducing, 332
parks, 12, 40, 46, 56
partitioning, 82
patchiness, 77
P/B ratio, 67
pelagic zone, 284
percolation, 185–186
performance evaluation, 138
periodicity, 88
permit systems, marketable, for
pollution, 110–113
pessimism, 684–685
pesticides, 379–382
pest management, 381–382
pests, exotic, and forest
protection, 324
pH, soil, 224
phosphorus, 84–86
photosynthesis, 71
photovoltaics, 610–611
physical niche, 63
piñon-juniper woodland,
265–266
plankton, 284
planning
comprehensive
organizational,
139–140
endangered species
management, 544–545
long-range, 137
operations, 137
program, 139
project, 138–139
recreational, 410–411, 412
regional, 140–141
strategic, 136–137
urban land-use
contemporary, 441–444
historic, 438–441
regional challenges, 458
planning process, 133–135
planning professionals, urban
land-use, 441–442
planning vision, 136

plants
  allelopathic, 381
  range, 340–341, 358
  See also vegetation
plant stratification, 88
plate boundaries, 554
plate tectonics, 553–554
poisonous plants, 358
polar climate, 163
policy, 141–143
  colonization period (up to 1776), 25–26
  endangered species, 543–546
  farmland, 382–389
  in gilded age (1861–1899), 37–40
  international resource management, 641–643
  international trade, 665
  in neoprogressive period (1945–present), 52–57
  in progressive period (1900–1945), 44–48
  for rangeland management, 359–364
  soil, 242–245
  in westward expansion era (1776–1860), 29–32
politically bounded planning, 140
pollution
  atmospheric, 377
  thermal, 452
  water, 201
pollution control, incentive-based, 109–113
pollution taxes, 109–110, 112
polyclimax viewpoint, 92
polyculture, 378–379
population assessment and management, by wildlife managers, 482–483
population growth, in developing countries, 637–639
population increase, human, 7–8
populations
  distribution and dispersal, 61–63
  dynamics, 63–65
  energetics, biomass, and production, 65–67
  habitat and physical niche, 63
  identity, 61
  strategies, 67–68
  structure, 63
  wildlife, management of, 494–497
prairie. *See* grasslands
precipitation, 158, 161
  acid, 173–174
  systems, 161–162
predators, introduced, 541
Predatory Mammal Control Program, 471
primary producers, 71
primary productivity, 71
private business management systems, 131–132

private grazing lands, and recreation, 405–406
private interdependence, in urban infrastructure, 428–430
private policy, 142–143
private recreation opportunities, and tourism, 404
private resource management, 5–6
problem identification, 122
problem solving, 122
processing
  extracted minerals, 561–562, 569
  natural gas, 588
  oil, 586–588
producers, 71–73
production
  and markets, 104–106
  organic, 67
  rangeland livestock, 356–358
  See also energy production
production costs, agricultural, 375–376
productivity, 67
profit incentives, and natural resources, 108–109
program planning, 139
progressive period (1900–1945), 40–48
project planning, 138–139
property rights, in developing countries, 636–637
public interdependence, in urban infrastructure, 428–430
public land disposal, in gilded age (1861–1899), 37
public lands
  fisheries, 522
  mining vs. recreation, 102
  recreational challenges on, 404–410
public management systems, 131
public opinion, and federal rangeland management, 408
public policy, 142–143
public resource management, 5–6

quarries, 560

radiation, solar, 152–153
radiation balance, 169–170
rainmaking, 213–214
ranching, and recreation, 408
range animal ecology, 342–346
rangeland, 11
  condition and trend, 341–342
  defined, 335–336
  types of, 337–339
rangeland ecology, 340–341
rangeland livestock production, 356–358
rangeland management
  defined, 336

future of, 364–366
  government policy, 359–364
  grazing systems, 349–356
  historical perspective, 339–340
  livestock distribution, 348–349
  and public opinion, 408
  and scenic beauty, 406–407
  stocking rate, 346–348
rangeland vegetation, controlling, 358–359
range management, in neoprogressive period (1945–present), 54, 56
range plants
  grazing effects on, 340–341
  poisonous, 358
range resources, in progressive period (1900–1945), 46–47
reclamation, and urban ecosystem management, 453–454
recreation, 12
  attributes of, 395–396
  and conservation economics, 101, 102
  defined, 392–395
  future demands for, 415–417
  in gilded age (1861–1899), 40
  importance of, 395
  and in-stream water use, 207
  and integrated water resource management, 193
  and land use management, 189
  in neoprogressive period (1945–present), 56
  on public lands, 404–410
  and ranching, 408
  on rangelands, 364
  resource conflicts and resolution, 396
  urban, and natural aesthetics, 446–447
  and wildlife conservation, 472–474
recreational economics, 411–412
recreational management, 410–415
  federal, 399–404
  historical perspectives, 397–399
  and wildlife conservation, 473–474
recreational planning, 410–411, 412
recreational resource managers, 412–413
recycling
  metals, 564–567
  mineral materials, 564
  wastepaper, 331–332
regional planning, 140–141, 458
regolith, 219
regulation(s)
  agricultural (U.S.), 374
  incentive-based, 109–113

reintroduction, in wildlife population management, 497
relative humidity, 162
renewable energy. *See* energy, renewable
renewal, of natural resources, 2, 4
rent seeking, 114
reproduction methods, 315–317
research
    fishery, 529
    wildlife, 484–485
reservoirs, construction of, 194–195
resource demand assessment, and fisheries, 528
resource managers, recreational, 412–413
resource partitioning, 82
resources. *See* natural resources
resource user satisfaction
    fisheries, 528–529
    wildlife, 484
rest-rotation grazing, 352
riparian communities/zones, 280–281
    and conservation economics, 102
    grazing systems for, 355–356
    and water resource management, 190
river ecosystems, 279–282
rivers, 182, 673–674
rock cycle, 555–556
rocks, 551–553, 623
runoff, 186–187, 452

sagebrush shrub steppe, 263–264
salinity, 200
salinization, 210
salt desert shrubland, 264–265
salt-resistant crops, 213
savannah woodlands, 252, 339
scarcity, 117
scenic beauty, and range management, 406–407
science, fishery, 519–525
seasonal-suitability grazing, 353
secondary succession, 89–90
sedimentary cycle, 84
seed-tree method, 316
selection method, 317
seral stages, 87
set-asides, 384–386
sewage water, reclamation of, 213
shade tolerance, 314
shelterwood method, 316
short-duration grazing, 353–355
shortgrass prairie, 257–258
shrubs, 247, 344
silts, 199
silver, 563–564
silviculture, 310, 315
site quality, classification based on, 313–314
size class distribution, 311–312
slope, degree of, 155
smelters, 569

social mobility, in developing countries, 637
soil, 10
    characteristics, 222–224
    classification, 224–229
    defined, 219
    grassland, 247, 250
    and National Resources Conservation Service (NRCS), 241–242
soil depth, 223–224
soil erosion, 229–231
    controlling, 234–241
    in United States, 231–234
soil fertility, 224, 240–241
soil formation, 220–222
soil organic matter, 224
soil pH, 224
soil policy, 242–245
soil profile, 219–220
soil resources, in progressive period (1900–1945), 46–47
soil structure, 223
soil texture, 222–223
solar energy, 607–611
solar radiation, 152–153
solar-thermal concentrating systems, 608–610
solids
    dissolved, 200
    suspended, 199
solid waste
    taxes on, 112
    and urban ecosystem management, 448–449
southern mixed prairie, 255–256
southern pine forest, 269–270
southwestern climate (U.S.), 165–166
Soviet Union (former), environmental problems in, 661
species composition, 312
species richness, 69
spillover costs and benefits, 108
spodosols, 225, 228
sportfishing, 514–516
spotted owls, and conservation economics, 100
sprawl, urban, 436–437
stagflation, 656–657
stakeholders, 135, 442
standing crop. *See* biomass
stand management, 315–317
stands, classification of, 310–311
state governments
    policy roles of, 142
    water resources agencies, 403–404
steel, 100–101, 566
stocking
    fisheries, 520–521
    in wildlife population management, 497
stocking rate, 346–348
    conservation, 363
    and grazing systems, 356

storage, 82–86
storms, types of, 161–162
strategic planning, 136–137
stratification, 88
stream drainage channelization, 196–197
stream ecosystems, 279–282
strip-cropping, 236
subsidies
    agricultural, 386–387
    housing, 432
subsurface mining, 560–561
suburban sprawl, 436–437
succession, 87–94
    forest, 330
    wildlife associations with, 489–491
sulfur, 83–84
sulfur dioxide allowance trading, 113
sunlight, and urban ecosystem management, 454–455
superhighways, 433–434
surface mining, 559–560
surface water impairment, 197–198
surface waters, and urban ecosystem management, 451
suspended solids, 199
sustainable agriculture, 382
sustainable development, 668–675
swamps, 275–276
SWOT analysis, 135
synecology, 60
systems analysis, 678–679

tallgrass prairie, 253, 255
tar sands, 588–589
taxpayer ignorance, 114
technology, 668–670
    artificial intelligence, 680
    computers, 678
    economic analysis, 681
    models, 679–680
    nanotechnology, 683–684
    in progressive period (1900–1945), 41–43
    systems analysis, 678–679
    virtual reality, 680–681
    wind, 614
tectonics, 553–554
temperate broad-leaved evergreen forests, 300
temperate coniferous forests, 300
temperate deciduous forests, 299
temperate mixed forests, 300
temperature, 157–158
terracing, 237–238
terrestrial ecosystems
    deserts, 260–265
    grasslands, 253–260
    overview, 246–253
    tundra, 271–272
    woodlands, 265–271
terrestrial primary succession, 87–88

terrorism, nuclear, 602
thermal discharges, 201
thermal pollution, 452
thermoelectricity, 205
thinnings, emergency, 319–320
third world countries. *See*
        developing countries
threatened species, 480–481
tidal power, 624
tillage, conservation, 234–240
timber, and conservation
        economics, 100
timber harvesting, 320
timber removals, 308
timber resources, in United
        States, 306–308
topography, 154–155, 349
tourism, 12
    and private recreation
        opportunities, 404
    in gilded age (1861–1899), 40
toxicity, from mining, 568
toxic materials, 200
trade, international, 664–665
trade policy, agricultural,
        388–389
transportation
    energy use, nonrenewable,
        577–579
    and urban ecosystem
        management, 445–446
    in urban infrastructure,
        430–431
    of urban water wastes,
        451–452
transportation fuel, 591–592,
        625–626
treaties, endangered species,
        543–544
tree planting, 320–321
tree structure and function,
        294–296
trophic-dynamics, 79–80
trophic efficiency, 80–82
trophic functions, producers,
        71–73
trophic levels, 75
trophic producers, decomposers,
        73–75
tropical deciduous forests, 302
tropical dry climate, 163
tropical evergreen forests,
        300–301
tropical wet climate, 163
tundra, 252–253, 271–272, 339

ultisols, 225, 228
United States
    agriculture problems,
        372–377
    biomes of, 253
    climatic types in, 163–166
    economic depressions in,
        653–655
    economy, 648–649
    forest land area in, 306
    forests of, 302–306
    hydropower, 616

natural resource management
        history
    colonization (up to 1776),
        19–26
    gilded age (1861–1899),
        32–40
    neoprogressive period
        (1945–present), 48–57
    progressive period
        (1900–1945), 40–48
    westward expansion
        (1776–1860), 26–32
    nuclear power in, 597
    soil erosion, 231–234
    sportfishing, 514–516
    sustainable development,
        670–672, 673–674
    timber resources, 306–308
    water use
        future demand, 213–215
        problems and conflicts,
            207–212
        types of, 203, 205–207
U.S. Constitution, and policy
        development, 143
U.S. Department of Agriculture
        (USDA), National
        Resources
        Conservation Service,
        241–242
urban decline, 435–436
urban design, and ecological
        services, 423
urban ecosystem, 421–423
urban ecosystem management
    air, 454–458
    integrating services, 444–445
    land, 445–449
    water, 449–454
urban form and function,
        424–428
urban growth, 434–435
urban impacts, extent of, 422
urban infrastructure,
        development of,
        428–434
urban interface, agriculture
        on, 406
urban landscape, 420–428
urban land use, 11, 189
urban land-use planning
    contemporary, 441–444
    historic, 438–441
    regional challenges, 458
urban recreation, 446–447
urban renewal, 437–438
urban runoff, 452
urban sprawl, 374–375, 436–437
urban wildlife management,
        447–448
utilities, in urban infrastructure,
        431–432

Vale Rangeland Rehabilitation
        Program, 361–362
vegetation
    as climatic factor, 156
    rangeland, 358–359

and recreational
        management, 414
vegetation-type conversion, 190
vertisols, 225
virtual reality, 680–681
visitor management, 413–414

waste heat, and urban ecosystem
        management, 452
wastepaper, recycling, 331–332
wastes
    for biomass energy, 619
    and urban ecosystem
        management,
        448–449, 451
water, 10, 178–180
    forms and distribution,
        181–182
    ownership of, 211–212
    properties of, 180
    and urban ecosystem
        management, 449–454
    and wildlife habitat
        management, 487
water allocation, and wildlife
        habitat, 208–210
water conservation, 213
water distribution, and
        rangeland
        management, 348–349
water harvesting, 190, 192
water pollution
    sources, 201
    taxes on, 112
water quality management,
        197–203
water resources
    in gilded age (1861–1899), 40
    in neoprogressive period
        (1945–present), 54
    in progressive period
        (1900–1945), 45–46
water resources management
    agencies, 192
    agencies, 402–404
    engineering, 194–197
    hydrologic cycle, 183
    land use practices, 187–189
    objectives, 192–194
    special techniques, 190, 192
    watershed, 183–187
    wetland and riparian areas,
        189–190
watershed management,
        184–185, 190, 192
watershed processes, 185–187
watersheds, 183–184, 201
water supply
    and irrigation, 193, 377
    and urban ecosystem
        management, 449–450
water transport, long-distance,
        214–215
water treatment, 201, 203
water use
    future demand, 213–215
    problems and conflicts,
        207–212

types of, 203, 205–207
wealth, 23
western coniferous forest,
    268–269
westward expansion
    (1776–1860), 26–32
wetlands, 189, 272–279
wildlands, 12
wildlife, 12
    categorization of, 486
    conflicts, 463–465
    and conservation
        economics, 101
    in neoprogressive period
        (1945–present), 56
    professional concept of, 465
    in progressive period
        (1900–1945), 47–48
    public concept of, 465–466
    values, 463–465, 467–468
wildlife authority, 464–465
wildlife conservation
    historic and legislative
        perspectives
        animal damage control,
            474–476
        foods and other goods,
            469–472

management in twentieth
    century, 476–480
recreation, 472–474
threatened and
    endangered species,
    480–481
as wildlife management
    philosophy, 466–468
wildlife habitat, and water
    allocation, 208–210
wildlife management
    categorization of wildlife, 486
    challenges and trends in,
        500–502
    and commercialization,
        497–500
    conservation philosophy
        of, 466
    habitat, 486–494
    of populations, 494–497
    public input, 485–486
    supply and demand, 485
    twentieth-century
        perspectives, 476–480
    urban, 447–448
wildlife manager,
    responsibilities as,
    481–485

wildlife population
    management, 494–497
wildlife resource depletion,
    469–471
wind, 156–157, 454–455
windbreaks, 239
wind energy, 611–615
wind resources, 613
wind technology, 614
wolves, endangered, and
    conservation
    economics, 101
woodlands
    eastern deciduous forest,
        270–271
    mountain browse, 266–267
    oak, 267–268
    piñon-juniper, 265–266
    savannah, 252, 339
    southern pine forest,
        269–270
    western coniferous forest,
        268–269
World War II, and natural
    resource
    management, 48
zinc, 563